FUNDAMENTALS OF
RADIO AND ELECTRONICS

Contributing authors

RAYMOND F. GUY
Senior Staff Engineer
National Broadcasting Company, Inc.

EDWARD C. JORDAN
Professor and Head
  of Electrical Engineering
University of Illinois

PAUL H. NELSON
Guest Professor, Department of Electronics
  and Electrical Communication Engineering
Indian Institute of Technology
Kharagpur, India

FRED H. PUMPHREY
Dean of Engineering
Alabama Polytechnic Institute

JOHN D. RYDER
Dean of Engineering and
  Professor of Electrical Engineering
Michigan State University

FUNDAMENTALS OF RADIO AND ELECTRONICS

Second Edition

W. L. EVERITT, Editor

Dean, College of Engineering
University of Illinois

Modern Asia Edition

PRENTICE-HALL INC., Englewood Cliffs, N. J.
CHARLES E. TUTTLE COMPANY, Tokyo, Japan

This *Modern Asia Edition* is a complete and unabridged photo-offset reproduction of the latest American edition, specially published for sale in the Far East. It is the only authorized edition so published and is offered in the public interest at a great reduction in price.

Inquiries or suggestions concerning *Modern Asia Editions* will be welcomed by the Charles E. Tuttle Company, 15 Edogawa-cho, Bunkyo-ku, Tokyo, Japan.

© 1942, 1958
BY
PRENTICE-HALL, INC.
Englewood Cliffs, N.J.

ALL RIGHTS RESERVED

No part of this book may be reproduced in any form, by mimeograph or any other means, without permission in writing from the publishers.

First printing (Modern Asia Edition), 1959
Fourth printing, 1961

PRINTED IN JAPAN

PREFACE

At the centennial of engineering in Chicago in 1952, I presented a paper on "Radio, the Founder of a Dynasty." Radio has nurtured and developed the broad concepts of the science and technology now commonly referred to as electronics. It is the father of the vacuum tube and has made possible measurement and control by means of signals, as well as transmittance of intelligence. Therefore, in revising the text originally titled *Fundamentals of Radio*, we broadened its coverage to include the general field of electronics, which, while it has its roots in radio, has expanded explosively since 1942. It now has an impact on many fields of human endeavor.

The wide acceptance of the first edition showed that there was a need for an intermediate-level book which would be accurate in its presentation of concepts but which could also be used by people at all technical levels, both for self-study and for classroom instruction. In order to make the text as self-contained as possible, we have retained two features of the first edition: a short résumé of the most important topics in mathematics that are needed for an adequate discussion of electronics, and a brief discussion of the electric-circuit fundamentals required.

The second edition has been greatly enlarged to cover the many important developments which have taken place since 1942. Outstanding among these developments have been the invention and application of transistors as a supplement to or replacement of vacuum tubes, the expansion of television (including color), radar and navigational aids, further development of electronic means for controlling and measuring almost everything that can be controlled or measured, and improvement in older methods of communication.

W. L. EVERITT

Urbana, Illinois
1958

CONTENTS

1. MATHEMATICS OF RADIO AND ELECTRONICS................... 1

Digits and their uses, 1. Place notation and counting, 2. Scientific notation, 2. Accuracy and significant figures, 5. The four fundamental operations, 6. Logarithms, 9. Antilogarithms, 11. Binary number system, 20. Symbols of algebra, 24. Some laws of algebra, 25. Solving equations, 29. Quadratic equations, 31. Trigonometry, 33. Vector addition, 39. Rates of change, 40. Integrals, 43. The exponential e^t, 44. Sines, cosines, and their derivatives, 46.

2. DIRECT-CURRENT CIRCUITS... 48

Introduction, 48. Electrical quantities, 49. Ohm's law, 50. Series circuits, 52. Parallel circuits, 53. Series-parallel circuits, 54. Determination of resistance, 56. Power and energy, 58. Kirchhoff's laws, 61. Method of superposition, 62. Symbols and abbreviations, 63. Batteries, 64. Internal resistance and polarization, 69. Electromagnetism, 69. Iron as a conductor of magnetism, 71. Magnetic characteristics of iron, 72. Dynamo and transformer steel, 74. Permanent magnet steels, 74. Permalloy, 74. D-c generators, 74. D-c motor, 77. Dynamotors, 77. Electric meters, 78. Voltmeter resistance requirements, 80. Volt-ohm-milliampere meters, 81.

3. CIRCUITS WITH TIME-VARYING VOLTAGES....................... 83

Time-varying voltages and currents in radio, 83. Types of time-varying voltages, 83. Electromagnetic characteristics of a coil, 85. The time constant of a coil, 88. The energy stored in the magnetic field, 89. Magnitude of inductance, 89. Electrostatic characteristics of a capacitor, 91. Dielectric constant, 91. Materials and dimensions determining capacitance, 92. Capacitive circuit response, 95. Alternating-current circuits, 97. Adding alternating currents, 98. A-c circuit with resistance, 99. Peak, average, and rms values of a-c waves, 99. Rate of change of current in a sine wave, 101. Coil response to a-c currents and voltages, 102. Reactance voltage, 103. Resistance and inductance in series with a-c applied voltage, 104. Response of capacitors to sinusoidal a-c, 105. Resistance and capacitance in series with a-c applied voltage, 107. Resistance, inductance, and capacitance in series with a-c applied voltage, 107. Introduction to phasor circuit techniques, 108. Phasor solution of resistance and capacitive circuits, 109. Phasor solution of general series circuits, 110. Series resonant circuits, 112. Parallel circuits, 115. Parallel resonant circuits, 116. Impedance matching, 117. Wave meters, 118. Mutual inductance, 118. Use and characteristics of the transformer, 120. Coefficient of coupling, 122. Selectivity and coupling, 123. Alternating-current meters, 124. Inductance and capacitance measurement, 127. Circuit response to complex waves, 128.

4. VACUUM TUBE AND TRANSISTOR PRINCIPLES.................... 131

Introduction, 131. Electron emission, 131. Thermionic emitters, 133. Physical construction of cathodes, 133. Diodes, 134. Space charge, 135. Plate resistance, 136. Rectifying action, 137. Triodes, 137. Triode characteristic curves, 137. Tube parameters or coefficients, 139. The load line, 141. The equivalent circuit of a vacuum tube, 142. Tetrodes, 143. Plate characteristics of a tetrode, 144. Pentodes, 145. Variable-mu or remote-cutoff pentodes, 148. Beam power tubes, 149. Dual-purpose tubes, 150. Limitations in tube operating conditions, 151. Cathode-ray tubes, kinescopes, 152. Iconoscopes, 155. The orthicon, 156. The klystron, 157. The magnetron, 157. Gas tubes, 158. Semiconductors, 159. Semiconductor diodes, 163. The transistor, 165.

5. RECTIFIED POWER SUPPLIES.. 169

Half-wave rectifier, 169. Full-wave rectifier, 170. Filter circuits, 170. Rectifier tubes, 173. Voltage-doubler rectifier, 174. Voltage quadruplers and multipliers, 175. Bridge rectifier, 176. Regulated power supplies, 177. Series-tube regulators, 178.

6. TRANSMISSION AND RECORDING OF SOUND..................... 180

Nature of sound, 180. Distortion, 181. High-fidelity requirements, 183. Limited frequency range equipment, 183. Microphones, 183. Phonograph pickups, 185. Tape recording, 187. Reproducers, 187. Telephone receivers, 187. Loud-speakers, 188. Telephone circuits, 189. Common-battery circuits, 189. Telephone lines, 190. The decibel, 191. Losses in telephone lines, 191. Repeaters, 192.

7. AUDIO AND VIDEO AMPLIFIERS.................................. 194

Fundamentals of amplifiers, 194. Vacuum-tube amplifiers, 194. Classification of amplifiers, 195. Resistance-capacitance coupled amplifier, 196. Frequency response of a resistance-capacitance coupled amplifier, 199. Gain of an audio amplifier, 200. The pentode amplifier, 201. Multistage audio amplifiers, 202. Hum and tube noise, 204. Over-all gain and frequency response, 205. Nonlinear distortion in audio amplifiers, 206. Dynamic characteristics of a vacuum-tube circuit, 207. Selection of the operating point, 209. Avoiding nonlinear distortion, 210. Transformer-coupled amplifier, 210. Push-pull circuits, 214.

Power amplifiers, 215. Inverse feedback, 216. Cathode-follower amplifier 219. Grounded-grid circuit, 222. Phase inverter, 223. Video amplifiers, 225. Compensated video amplifiers, 227. Direct-current amplifiers, 230. Public-address systems, 232. Transistor audio amplifiers, 240. Basic transistor circuits, 241. Equivalent for the common-base transistor circuit, 241. Graphic characteristics, 247. The common-emitter circuit, 248. Common-collector circuit, 249. Bias circuits for transistors, 251. Multistage transistor amplifiers, 253. Multistage R-C coupled transistor amplifiers, 255. Transformer-coupled transistor amplifiers, 256. Push-pull and complementary symmetry transistor circuits, 257.

8. PULSE AND SWITCHING CIRCUITS 263

Simple pulse-forming circuits, 263. Clamping circuits or the d-c restorer, 266. Clipping, 268. Oscillograph sweep voltages, 269. Synchronization of the sweep, 271. Further sweep circuits, 271. The blocking oscillator, 273. Eccles-Jordan trigger circuit, 274. Scaling circuits, 275. Multivibrators, 276. Synchronization of the multivibrator, 279. Transistor triggers, 279.

9. ELECTROMAGNETIC WAVES 282

Nature of waves in any medium, 282. Transverse and longitudinal waves, 284. Phase in wave motion, 285. Reflected waves and standing waves, 286. Electromagnetic waves on wires, 288. Sound waves and electromagnetic waves, 289. Standing waves, 292. Characteristic impedance Z_0, 293. Phase of voltage and current, 294. Standing wave ratio, 295. Coaxial transmission lines, 297. Waveguides, 298. Waves in three dimensions, 301. Dimensions of an antenna, 303. Radiation resistance, 304. Mechanism of radiation, 305. Direction of the electric and magnetic fields, 306. The receiving antenna, 307.

10. TRANSMISSION AND RECEPTION OF SIGNALS BY RADIO 309

Electromagnetic waves as message carriers, 309. Modulation, 309. Amplitude modulation, 309. Frequency modulation, 311. Phase modulation, 311. Pulse modulation, 311. The components of a radio communications system, 312. Types of service, 314. Forms of transmission, 314. Scope of radio communications, 317. Sideband frequencies, 318. Single and vestigial sideband transmission, 319. Suppressed carrier transmission, 320. Advantages of single-sideband suppressed carrier transmission, 320. Concept of the radio channel, 321. Practical considerations of channel width, 321. Radio frequency spectrum, 322. Shared use of the radio frequency spectrum, 323. Regulation of radio, 323. Radio transmitters, 324. Simple radio receivers, 325. Superheterodyne receivers, 326. Problems of receiver selectivity, 328. Example of a modern overseas radio communications system, 330. Example of a police radio system, 334.

Contents

11. AMPLITUDE-MODULATION DETECTORS, RADIO-FREQUENCY AMPLIFIERS AND MODULATORS.................................... 338

Detection or demodulation, 338. Requirements for rectifiers, 340. Crystal detectors, 340. Diode tube detectors, 341. Triode detectors, 343. Heterodyne detection of continuous waves, 347. Detection of interrupted continuous-wave signals, 349. Radio frequency amplifiers, 349. Difference between radio and audio amplifiers, 350. Resonant coupling for impedance matching, 351. Harmonic suppression, 353. Radio amplifier circuits, 354. Amplifier input circuits, 355. Single-ended output circuits, 356. Grid-bias voltage supplies, 357. Radio amplifier classifications, 359. Application notes, 360. Class A radio amplifiers, 360. Amplitude distortion in radio frequency amplifiers, 363. Improper operation, 363. Heat dissipation limits in power tubes, 366. Neutralization, 366. Class A pentode radio amplifier, 368. Shielding to prevent self-oscillation, 370. Shielding to prevent radiation, 370. Tetrode amplifiers, 371. Filtering d-c supply circuits to radio amplifiers, 371. Amplifier tube complements, 372. Parallel operation of power tubes, 373. Push-pull operation of power tubes, 374. High-level and low-level modulation, 376. Class B radio amplifiers, 376. Class B modulators, 379. Class C radio amplifiers, 381. Balanced modulators, 385. Pentode radio amplifiers, Class B and Class C, 386. Wide-band television intermediate frequency amplifiers, 386. Doherty high-efficiency radio amplifier, 389. Chiriex high-efficiency r-f modulating system, 391. Problems of designing radio-frequency amplifiers at the higher frequencies, 393. Grounded grid radio amplifiers, 395. Cathode coupled amplifiers, 397. Grid-modulated amplifiers, 399. Plate modulation of screen-grid tubes, 401. Parasitic oscillations, 401.

12. AMPLITUDE MODULATION RADIO TRANSMITTERS................. 405

Influence of transmitter function on design, 405. TYPICAL TRANSFORMER PERFORMANCE REQUIREMENTS: Audio amplification, 405. Input impedance, 406. Audio-frequency response, 406. Audio distortion, 406. Transmitter noise, 407. Modulation system, 407. Carrier frequency, 407. Carrier frequency stability, 407. Type of output circuit, 408. Output impedance, 408. Carrier amplitude shift, 408. Power rating, 409. Maximum ambient temperature, 409. Power supply, 409. Power consumption, 409. Other specifications, 410. Crystal carrier frequency oscillators, 410. The simple vacuum tube oscillator, 411. Quartz crystals, 413. The piezoelectric effect, 413. Buffer amplifiers, 417. Driver amplifiers, 419. Amplitude-modulated radio-frequency amplifiers, 421. Neutralizing of r-f amplifiers, 424. Power supply regulation, 425. Inverse feedback in transmitters, 426. Schematic diagram of a 10-kw transmitter, 428. First audio stage, 428. Second audio amplifier, 428. Third audio stage, 429. Power supplies, 429. Safety interlocks, 435. Shielding, 435. Telegraph transmitters, 436. Keying methods, 436. Frequency multipliers, 437. Transmitter powers, 438. Single-sideband suppressed carrier system, 439. Ex-

Contents

ample of a high-power international broadcasting installation, 443. R-f transmission lines, 446. Open wire lines, 446. Coaxial transmission lines, 449. Antenna tuning and matching, 449. Tower lighting, 449.

13. AMPLITUDE MODULATION RADIO RECEIVERS 451

Sensitivity, 451. Noise, 452. Static, 452. Thermal agitation noise, 452. Tube noise, 453. Receiver noise factor, 454. Measuring receiver noise factor, 455. Receiver hum noise, 456. Automatic volume control, 457. Automatic volume control figure of merit, 458. Tuned r-f receivers, 459. The superheterodyne receiver, 459. Receiver oscillator radiation, 460. Spurious response, 461. Image frequency response, 462. Example of superheterodyne, 462. Multiple functions in radio receivers — selectivity, 465. Multi-band receivers, 467. Receivers utilizing transistors, 467. A transistor superheterodyne receiver, 470. Muting system, 472. Diversity receiving system, 473. Selective fading, 474. Transoceanic receivers, 475.

14. FREQUENCY MODULATION 478

General principles, 478. Other factors in frequency-modulation systems, 483. Frequency-modulation systems for communication services, 484. Frequency-modulation transmitters, 485. Frequency-modulation receivers, 488. Typical frequency-modulation receiver, 491.

15. MONOCHROME TELEVISION 493

Introduction, 493. Television sound channel, 494. Sound pre-emphasis, 495. Scanning, 495. Human vision and television scanning, 495. Horizontal scanning rate, 498. Horizontal synchronizing pulse, 498. Vertical blanking and synchronizing, 500. Simultaneous vertical and horizontal synchronizing, 501. Equalizing pulses, 501. Interlaced horizontal scanning, 502. Scanning rates, 504. Bandwidth requirements of a television system, 504. Vestigial-sideband transmission, 506. Vestigial-sideband filter, 506. Diplexing, 506. Television transmitting antenna, 507. Television transmitting station, 508. Television camera, 508. Television camera tube, 508. Monitoring and mixing equipment, 510. Negative transmission, 511. Composite video signal, 511. Television receivers, 513. Picture tube and controls, 515. Sync-separating circuits, 515. Flyback high-voltage supply, 515. Damping tube, 517. Reinsertion of the d-c component, 517. Intercarrier-sound receiver, 519. Typical television receiver circuit, 520. Large-screen and theater television, 521. Industrial and closed-circuit television, 522. Televising motion-picture film, 522.

16. COLOR TELEVISION 525

Introduction, 525. Some properties of light, 525. Primitive color-television system, 526. Color in television, 527. Compatibility in color television, 530. The frequency-interleaving principle, 531. Resolution of color by the eye, 533. The color video signal, 534. Color camera, 534. The matrix equations, 535. Transmitting the color signal, 537. Color-matrix circuit, 539. Modulation of the chrominance subcarrier, 541. Synchronizing color burst, 542. Color receiver, 544. Compatibility in receivers, 544. Picture detector and composite video amplifier, 545. Chrominance demodulators, 547. Receiver color matrix, 547. Chrominance sync channel, 549. Three-color picture tube, 549. Other systems of color television, 552.

17. VACUUM-TUBE INSTRUMENTS 553

Oscillators, 553. Inductance-capacitance feedback oscillators, 555. Radio-frequency oscillators or signal generators, 556. Frequency-modulated oscillators, 558. Resistance-capacitance tuned oscillators, 560. Grid-dip oscillator, 561. Frequency-measuring equipment, 561. Cathode-ray oscilloscope, 563. Vacuum-tube voltmeters, 567. Tunable vacuum-tube voltmeters, 570. The Q-meter, 570. Tube-testing equipment, 572.

18. ULTRAHIGH-FREQUENCY AND MICROWAVE CIRCUITS 575

Limitations of ordinary circuit elements, 575. Transmission lines and waveguides as circuit elements, 576. Phase relation of voltage and current on a transmission line, 577. Transmission line sections as reactance elements, 580. Direction of energy flow, 584. The resonant quarter-wave line, 585. The quarter-wave as a matching section, 589. Examples of matching with quarter-wave lines, 590. The half-wavelength section, 592. Stub-line matching and double-stub tuners, 594. Waveguide microwave circuits, 595. Printed or strip-line microwave circuits, 597. Uhf and microwave generators, 598. Other uhf and microwave devices, 609.

19. RADIO WAVE PROPAGATION 615

General nature of propagation, 615. Polarization, 616. Power density and Poynting vector, 616. The surface wave, 617. The sky wave, 618. The ionosphere, 619. Effect of the ionosphere on the sky wave, 620. Critical frequencies, 623. Absorption in the ionosphere, 625. Regular variations in the ionosphere, 625. Fading, 627. Reduction of fading, 629. Static and man-made noise, 629. Noise-reducing systems, 631. Vhf, uhf and shf propagation, 632. Direct and ground-reflected waves, 632. Diffraction and refraction, 634. Tropospheric waves, 635. Forward-scatter propagation, 636. Summary of radio wave propagation, 637.

20. RADIO ANTENNAS 640

Functions of an antenna system, 640. The elevated half-wave antenna, 640. Radiation characteristics of a half-wave antenna, 642. The grounded antenna, 643. Antennas of other heights, 645. Dipole and monopole antennas, 645. Losses and efficiency, 645. Ground systems, 646. Coupling networks, 647. Directional antenna systems, 653. Directivity of a single linear antenna, 653. Vertical antennas spaced one-half wavelength, 655. Other phases and spacings, 657. Patterns in the plane of the antenna axes, 658. Colinear array, 659. Rectangular or mattress array, 660. Effect of the ground on vertical radiation patterns, 661. Antenna gain and effective area. Microwave antennas, 664. Parabolic reflectors and paraboloids, 665. Horns, 666. Microwave lenses, 667. Slot antennas, 668. Wide-band and special-purpose antennas, 670. Folded dipole antenna, 671. Television transmitting antennas, 673. Television receiving antennas, 675. Navigational antennas, 680. Direction finders and radio compasses, 680.

21. RADAR, RADIO RELAYS, RADIO AIDS TO NAVIGATION, PULSE COMMUNICATION.. 684

Radar, 684. Essential components of a simple radar system, 685. Video amplifier, 687. Pulse repetition rate, 687. Time base, 688. Pulse duration — separation of objects, 688. Radio frequencies used for carrier, 690. Beam antennas, 690. Antenna power gain, 691. Antenna beam width, 693. Reflection from objects, 696. Radar distance range, 698. The radar equation, 699. Minimum usable received signal power, 700. Distance ranging and direction, 702. Plan-position radar, 702. Speed of rotation of beam, 705. Image-persistence time, 705. A shipboard radar system, 705. Weather radar, 707. Isoecho contour, 708. Precision approach radar, 711. Other radar applications, 711. Radar beacons, 712. Tubes used in radar, 713. Radio relay systems, 713. Radio relay antennas, 714. Example of a simple television radio relay system, 715. Multiplexing, 716. Grouping, 717. Subdivisions of message frequency channels, 718. Relay frequency staggering, 718. Relay design, 719. Bell System radio relay system, 720. Radio relay noise, 724. Pulse communication, 725. Pulse code modulation, 727. Code grouping, 727. Pulse power, 728. Time division multiplexing, 728. Performance, 730. Loran navigational aid, 730. Identification of stations and lines of position, 732. Function of the receiver indicator, 732. Vhf omnidirectional range system (VOR), 734. DME: distance-measuring equipment, 735. Other navigational aids: ILS, 735.

22. INDUSTRIAL APPLICATIONS.. 737

Introduction, 737. High-frequency induction heaters, 738. Diathermy, 743. Electron timers, 744. Cycle timers, 749. Electronic methods of industrial measurement, 749. Temperature measurement using electronic methods, 749. Millivoltmeter with electronic follower, 750. Potentiometer voltage measurements, 753. Capacitor potentiometer methods, 754. Types of voltage-sensitive elements, 755. Impedance-varying transducers, 755. Bridge measurement of impedance, 756.

Strain gage measurements, 757. The tools of control, 757. Gas triodes or thyratrons, 757. A phase-shifting circuit, 760. The ignitron tube, 761. The magnetic amplifier, 762. Photoelectric measurement and control, 766. Electronic control of motors, 768. Servomechanisms, 769. Electronic research instruments, 771.

APPENDIX. SAFETY AND SPECIAL RADIO SERVICES.................. 777

Aviation Services, 777. Marine radio services, 779. Public safety radio services, 782. Disaster Communications Service, 784. Amateur Radio Service, 785. Radio Amateur Civil Emergency Service, 785. Land Transportation Radio Services, 786. Industrial radio services, 787. Citizens' Radio Service, 790. Regulation of these stations, 791. How to apply for a license, 792.

INDEX.. 793

CHAPTER 1

MATHEMATICS OF RADIO AND ELECTRONICS

Certain concepts and rules of mathematics are needed to understand radio and electronics. These rules reduce facts about circuits, components, and their behavior into a consistent and well-organized body of theories and laws to explain how a radio or an electronic system works. The discussion in this chapter is necessarily condensed, but the ideas presented here are fundamental to any serious work in the field of electronics. Our mathematical starting point is the concept of a set of numbers.

DIGITS AND THEIR USES

The decimal system utilizes ten *digits* which are written 0, 1, 2, 3, 4, 5, 6, 7, 8, 9. The word digit means any number less than ten, and is also used as the name of a finger (a digit) on our hands. The fact that we have ten fingers is probably the reason that ten is used as the *base* for our *decimal number system*.

Let us consider first the tallying of objects. This does not involve counting them; man used a tallying system for keeping track of his possessions long before he could count. In tallying we make a mark or tally for each object in a collection until all these objects have been accounted for. There is a one-to-one correspondence between the marks and the objects; this one-to-one relationship is a definition of equality.

A one-to-one correspondence may be set up between numbers and segments (parts) of a line which extends without limit in both directions. Figure 1-1 shows how this is done. Distances along a line are marked with numbers which are a measure of the length of the line from a reference point or zero. Each number is represented by a certain specified length or segment of the line. The *unit of length* is the distance from the reference point (0) to 1. The diagram also illustrates the use of numbers in measuring distances or other quantities as contrasted with their use in counting.

Distances measured to the right of a reference point (0) are called *positive* and marked with the sign + ; distances measured to the left are negative and are marked with the sign − . Thus, +1, +2, +3, +4, +5, +6, and so on, are positive numbers and −1, −2, −3, −4, −5, −6, etc., are negative numbers. If no sign is written in front of a number it is under-

1

stood to be positive. The *absolute value* of a positive or a negative number is the magnitude or value of the number without regard to its direction; that is, without the positive or negative sign.

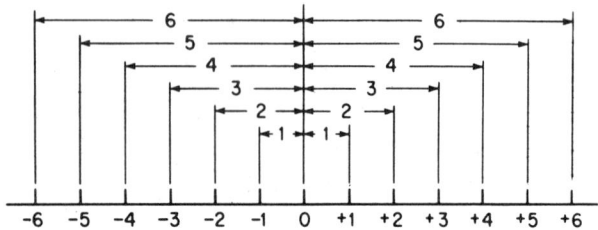

FIG. 1-1. Graph showing positive and negative numbers.

PLACE NOTATION AND COUNTING

We count up to nine objects by using the digits 1, 2, 3, 4, 5, 6, 7, 8, 9 in ordered succession. If more than nine objects are to be counted the *place* (or location) of the symbol (digit) has a special meaning. For example, the number 1,492 means one thousand (1,000) combined with four hundreds (400), nine tens (90), and two ones (2). When added as a sum we have

   1,000 → one thousand
     400 → four hundreds
      90 → nine tens
       2 → two ones
   1,492

```
↑↑↑↑— ones place      ⎫
 ││└── tens place     ⎬  PLACE VALUES
 │└─── hundreds place ⎪
 └──── thousands place⎭
```

The position or location of the digit in the number denotes the *place value*. Each place has one tenth of the value of its neighbor to the left and ten times the value of its neighbor to the right in the decimal system. There is no limit to the size — large or small — of numbers which can be written in this place system.

The digit zero (0) is a number like the rest of the digits, and in the place system means that no tens, hundreds, thousands, etc., are to be used where it occurs. For example, in the number five hundred no tens or units are required; this fact is indicated by writing a zero in the tens place and a zero in the units place, as in 500.

A dot called a *decimal point* is written to the right of the ones place; the places farther to the right are one tenth (0.1), one hundredth (0.01), and so on without limit.

SCIENTIFIC NOTATION

Two methods are used to avoid the difficulty and risk of error resulting from writing out numbers which have a great many zeros before or after the decimal point.

Chapter 1   **Mathematics of Radio and Electronics**   3

The first method, called *scientific notation*, is to write a number less than 10 and multiply it by 10 as many times as necessary to locate the decimal point correctly. That is, 250 is written as $2.5 \times 10 \times 10 = 250$. Instead of writing out this multiplication, a small number (*superscript* or *exponent*) is written above and to the right of 10 to show how many times 10 is to be multiplied together as a factor in this product. For example, 10^2 means 10 taken twice or $10^2 = 10 \times 10 = 100$; $10^3 = 10 \times 10 \times 10 = 1,000$; and so forth. The (^{2}) and the (^{3}) are positive exponents. If the number is less than 1 the exponent is a negative number. For example, 0.05 means 5×10^{-2}; $0.003 = 3 \times 10^{-3}$; and so on. The following table shows how this multiplication is worked out in *powers of ten*.

TABLE 1-1

Large Numbers		Small Numbers
$10^1 =$ 10	$10^0 = 1$	$10^{-1} = 0.1$
$10^2 =$ 100		$10^{-2} = 0.01$
$10^3 =$ 1,000		$10^{-3} = 0.001$
$10^4 =$ 10,000		$10^{-4} = 0.0001$
$10^5 =$ 100,000		$10^{-5} = 0.00001$
$10^6 =$ 1,000,000		$10^{-6} = 0.000001$

When 1 is added to the exponent of a number to the left of an equality sign in this table, the number to the right of the sign is multiplied by 10; when 1 is subtracted from the exponent of a number to the left of the equal sign, the number to the right is divided by 10. Any number on the right of an equal sign may be represented by the corresponding number on the left; that is, for 100 write 10^2, for 1,000 write 10^3, and so forth. When used this way 10 is called the *base*.

A rule for writing numbers in scientific notation is:

1. (a) For *large numbers* (10 or larger) move the decimal point to the left enough places to form a new number which lies between 1 and 10. (b) For *small numbers* (less than 1) move the decimal point to the right enough places to form a new number which lies between 1 and 10.

2. Multiply this new number by 10 with an exponent numerically equal to the number of places the decimal point was moved.

3. Make the exponent positive if the decimal point was moved to the left and make it negative if the decimal point was moved to the right.

To write 2,500 in scientific notation, move the decimal point three places to the left to get 2.5 (part 1 of rule); since the decimal point is moved three places to the left, multiply by 10^3 making the exponent positive to get $2.5 \times 10^3 = 2,500$ (by parts 2 and 3 of the rule). Similarly, $2,500,000 = 2.5 \times 1,000,000$. From the table, $1,000,000 = 10^6$. Thus, $2,500,000$ is written 2.5×10^6.

To write 0.0025 in a similar way, remember that 0.0025 is 2.5 times 0.001. From the table 0.001 is 10^{-3}. Therefore 0.0025 is written 2.5×10^{-3}

in scientific notation. For 0.0000025 the system works even better to reduce the number of zeros. This number is 2.5 times 0.000001; from the table 0.000001 is 10^{-6}; therefore, $0.0000025 = 2.5 \times 10^{-6}$.

The rule is reversed to change from scientific notation to ordinary place notation. Move the decimal point (to the right if the exponent is positive, to the left if it is negative) as many places numerically as indicated by the exponent of 10, supplying zeros as necessary. For example, $2.56 \times 10^2 = 256$; $1.86 \times 10^5 = 186,000$; $7.854 \times 10^{-3} = 0.007854$; $2.4 \times 10^{-5} = 0.000024$.

Numbers written in scientific notation are advantageous when multiplying and dividing. Suppose that 2.2×10^3 is to be multiplied by 3×10^2. First multiply 2.2 by 3, which gives 6.6. For the 10's, consider that $10^3 = 1,000$ and $10^2 = 100$. The product $1,000 \times 100 = 100,000 = 10^5$ (from the table). But $10^5 = 10^{2+3}$. Rewriting the solution gives

$$(2.2 \times 10^3)(3. \times 10^2) = (2.2 \times 3)(10^3 \times 10^2)$$
$$= 6.6 \times 10^{2+3} = 6.6 \times 10^5.$$

As another example, let 2×10^{-2} be multiplied by 5×10^3. First multiply 2×5 to get 10. Then $10^{-2} \times 10^3 = 10^{-2+3} = 10^1$. The answer is thus $10 \times 10 = 100$ or 1.0×10^2.

Extending the same idea, for any given base, such as 10, the following rules of operation with exponents apply:

To *multiply* powers of a base, add exponents, taking into account their signs: $10^4 \times 10^2 = 10^{4+2} = 10^6$; $10^4 \times 10^{-2} = 10^{4-2} = 10^2$; $10^{-4} \times 10^2 = 10^{-4+2} = 10^{-2}$.

To *divide* powers of a base, subtract exponents, taking into account their signs: $10^4/10^2 = 10^{4-2} = 10^2$; $10^4/10^{-2} = 10^{4+2} = 10^6$; $10^{-4}/10^2 = 10^{-4-2} = 10^{-6}$.

The second method of writing very large or very small quantities is done by adding prefixes to their names. A table of such prefixes follows.

TABLE 1-2

Multiply the known number of →	mega- M	kilo- k	unit	centi- c	milli- m	micro- μ	milli- micro- mμ	micro- micro- $\mu\mu$
to obtain the unknown number of	by ↓	by ↓	by ↓	by ↓	by ↓	by ↓	by ↓	by ↓
mega-........M	1	10^{-3}	10^{-6}	10^{-8}	10^{-9}	10^{-12}	10^{-15}	10^{-18}
kilo-..........k	10^3	1	10^{-3}	10^{-5}	10^{-6}	10^{-9}	10^{-12}	10^{-15}
unit...........	10^6	10^3	1	10^{-2}	10^{-3}	10^{-6}	10^{-9}	10^{-12}
centi-........c	10^8	10^5	10^2	1	10^{-1}	10^{-4}	10^{-7}	10^{-10}
milli-........m	10^9	10^6	10^3	10	1	10^{-3}	10^{-6}	10^{-9}
micro-.......μ	10^{12}	10^9	10^6	10^4	10^3	1	10^{-3}	10^{-6}
millimicro-....mμ	10^{15}	10^{12}	10^9	10^7	10^6	10^3	1	10^{-3}
micromicro-...$\mu\mu$	10^{18}	10^{15}	10^{12}	10^{10}	10^9	10^6	10^3	1

Use the table in this way. Suppose that one million cycles is to be converted to kilocycles. As stated in the table, "Multiply the known number of" cycles by 10^{-3} (third column, second entry) "to obtain the unknown number of" kilocycles. That is, 10^6 cycles = $10^6 \times 10^{-3}$ kilocycles = 10^3 kilocycles = one thousand kilocycles. Again, to convert 1,000 kilocycles to megacycles, multiply the known number of kilocycles by 10^{-3} (second column, first entry) to obtain the unknown number of megacycles. This gives 10^3 kilocycles = $10^3 \times 10^{-3}$ megacycles = 10^0 megacycle = 1 megacycle. One kilocycle is one thousand cycles; one microfarad is one millionth of a farad; one millihenry is one thousandth of a henry, and so on for other quantities. The meaning of the terms cycles, farads, and henrys is explained later.

ACCURACY AND SIGNIFICANT FIGURES

Most numbers used in electronic work are obtained by measurement. It is necessary, therefore, to look at the amount of accuracy which may be expected from these measurements, and especially how many decimal places should be used in the calculations based upon these measurements.

Sometimes meters have an accuracy stated as "Maximum error 3% of full-scale reading." This means that with a full-scale reading of 100 v (volts), for example, the actual voltage when the meter reads full scale is between 97 v and 103 v, 3 v being 3% of the full-scale reading. Unfortunately, the same 3-v error may also happen at lower scale readings, and if this instrument reads 10 v, the actual voltage may be between 7 v and 13 v.

To see what else this may lead to, *relative error* and *percentage of error* must be considered. Suppose that a meter reads 16.0 v and the reading should be 16.2 v. The *error* is 16.2 − 16.0 = 0.2 v, and the *relative error* is

$$\frac{\text{error}}{\text{true value}} = \frac{0.2}{16.2} = \frac{1}{81} = 0.0123.$$

Multiplying the last figure by 100 to convert it to percentage gives a *percentage of error* which is about $1\frac{1}{4}\%$. If another meter, used on a power line, read 16,000 v when the correct reading was 16,200 v, the error would be 16,200 − 16,000 = 200 v. The relative error would be

$$\frac{200}{16,200} = \frac{1}{81} = 0.0123,$$

or about $1\frac{1}{4}\%$ again. The position of the decimal point or the size of the quantities involved have nothing to do with the *relative error*.

The diameter of the earth has been carefully measured by accurate instruments. However, its average diameter is often stated as 8,000 miles, which is accurate only to the nearest thousand miles. A more accurate figure would be 7,900 miles, accurate to the nearest hundred miles; or 7,930 miles, accurate to the nearest ten miles; or even, 7,927, accurate to the nearest mile. Yet 8,000 miles is a useful expression if it is remembered that only the 8 in the thousands place means very much, that the zeros are there just to keep the 8 in its place. Another way of saying this is that 8 is the significant figure, the zeros not being significant. It is agreed that the significant figures shall be only those digits determined by measurement. In the expression 7,930 miles for the earth's diameter, only the figures 7,93 are significant; in 7,927, all figures are significant. When we are paying attention to significant figures, zeros following the last of the other digits do not count unless it is so stated. For example, the following numbers have two significant figures: 17,000; 0.00057; 95; 23,000,000. With certain instruments measurements may be made which are accurate to five significant figures, such as 60,103. If the instrument had read 60,100, the last two zeros would not count as significant figures unless the fact was expressly so stated. Measurements and calculations in electronics seldom are carried to more than three significant figures.

Now if two numbers which represent quantities obtained by measurement, each with three significant figures, are multiplied together, the product may consist of as many as six digits, such as 32.4 × 41.4 = 1,341.36. There is nothing in the process of multiplying which will increase the accuracy of the original measurements; therefore, the product is accurate only to three significant figures and should be written 1,340.

In general, the result of a calculation should be *rounded off* to the number of significant figures in the least accurate of the measured values used in the calculation. In rounding off numbers it is usual to take the next larger number if the last digit is greater than 5; to take the next smaller number if the last digit is less than 5; and if the last digit is 5 the preceding digit is increased by 1 if it is odd and left unaltered if it is even. Thus, 124 would be rounded off to 120 with two significant figures; 127 would be rounded off to 130; and 125 to 120; but 135 would be rounded off to 140.

THE FOUR FUNDAMENTAL OPERATIONS

As long as whole objects only are to be counted, the natural numbers or *integers* are enough; that is, 1, 2, 3, 4, 5, 6, 7, 8, 9, 10, 11, and so on. These numbers and others are used in the four fundamental operations of addition, subtraction, multiplication, and division.

Addition is needed when two groups of things are to be combined. Of course, each group might be counted and then the combined group counted again, but a trial will show that instead of the objects themselves being counted, the numbers representing them may be added in a manner that never varies. That is, 20 apples added to 15 apples, 20 radios and 15 radios, or 20 objects of any kind combined with 15 objects of the same kind always amount to 35 objects in the whole group. Two conclusions follow from this process: (1) operations with numbers themselves, rather than with the objects they represent, may be performed; (2) if the operations are correctly carried out, the results are always right. Taking advantage of the properties of the system of numbers, it is necessary only to memorize sums of digits like 1 plus 1 equals 2, 1 plus 2 equals 3, and other basic combinations in order to add any numbers whatsoever.

Subtraction is the opposite or inverse of addition. If a television set has 35 tubes and a deluxe radio receiver has 24, the difference between the numbers of tubes is 11, which is obtained by subtracting 24 from 35. To check the correctness of the work, add the difference, 11, to the smaller number, 24, and the larger number, 35, is obtained.

Multiplication may be thought of as continued addition. If troops march past four abreast they may be counted as $4 + 4 + 4 + 4 = 16$ if there happen to be four ranks. This is the same as counting the four ranks only and multiplying by the number in each rank; that is, 4 times 4 equals 16.

Division is the inverse of multiplication, just as subtraction is the inverse of addition. Thus, if 60 is divided by 12 the result is 5; the inverse operation, 12 multiplied by 5, gives 60. Division by zero (0) is not possible because there is no number which, when multiplied by zero, will give any number except zero itself.

All of the ideas discussed are well known and are easy to understand; but they must be stated here as a foundation for other ideas, no more difficult to learn, ideas needed for an understanding of radio and electronics.

To save time in writing mathematics, certain *symbols of operation* are used. Among these are:

For addition: the plus sign (+). $21 + 12 = 33$.
For subtraction: the minus sign (−). $18 - 8 = 10$.
For multiplication: the multiplication sign (×), sometimes the dot (·); sometimes the quantities to be multiplied are simply written side by side. $5 \times 6 = 30$ or $5 \cdot 6 = 30$; $4 \times a = 4a$.
For division: the division sign ÷, the bar —, or the mark /. Thus $28 \div 7 = 4$, $\frac{28}{7} = 4$, $28/7 = 4$.

Sometimes, to avoid confusion, a complicated expression must be enclosed in parentheses ( ), brackets [ ], or braces { }. For example, consider $15 \times 5 - 2$. Does it mean $75 - 2$ or 15×3? By using parentheses confusion is avoided: $(15 \times 5) - 2 = 75 - 2 = 73$, $15 \times (5 - 2) = 15 \times 3 = 45$. When we use parentheses in a multiplication we can omit the × sign: $15(5 - 2)$ means the same as $15 \times (5 - 2)$, $(5)(6)$ means the same as 5×6 or $5 \cdot 6$.

If one integer, called the *numerator*, is divided by another integer, called the *denominator*, the indicated division is called a *fraction*. If the numerator is smaller than the denominator, the fraction is said to be a *proper fraction;* if the numerator is the larger, the fraction is said to be an *improper fraction*. Thus, $\frac{1}{2}, \frac{2}{3}, \frac{5}{6}, \frac{81}{82}$ are proper fractions; $\frac{3}{2}, \frac{5}{3}, \frac{11}{6}, \frac{201}{82}$ are improper fractions. The latter are frequently reduced to the sum of an integer and a proper fraction: $\frac{3}{2} = \frac{2}{2} + \frac{1}{2} = 1\frac{1}{2}$; $\frac{11}{6} = \frac{6}{6} + \frac{5}{6} = 1\frac{5}{6}$; $\frac{201}{82} = \frac{164}{82} + \frac{37}{82} = 2\frac{37}{82}$.

FRACTIONS

		Example
Common fractions	Proper common fraction:	$\frac{3}{7}$ Numerator Denominator
	Improper common fraction:	$\frac{7}{3}$ Numerator Denominator
Decimal fraction		0.0625

The four fundamental operations may be applied to fractions as well as to integers, but some care is necessary. For example, to add $\frac{3}{7}$ and $\frac{2}{5}$ they must be reduced to a *common denominator*, thus:

$$\frac{3}{7} + \frac{2}{5} = \frac{3 \times 5}{7 \times 5} + \frac{2 \times 7}{7 \times 5} = \frac{15 + 14}{35} = \frac{29}{35}.$$

The same process is needed in subtracting fractions; to subtract $\frac{4}{9}$ from $\frac{7}{8}$, proceed as follows:

$$\frac{7}{8} - \frac{4}{9} = \frac{7 \times 9}{8 \times 9} - \frac{4 \times 8}{8 \times 9} = \frac{63 - 32}{72} = \frac{31}{72}.$$

Fractions may be written in several forms which mean the same thing; thus $\frac{1}{2}$ is the same as $\frac{2}{4}$, since upon dividing the numerator and denominator of the latter by 2 the original $\frac{1}{2}$ is obtained.

Multiplication of fractions may be done without reducing to a common denominator; for example, $\frac{3}{4}$ times $\frac{5}{8}$ becomes

$$\frac{3}{4} \times \frac{5}{8} = \frac{3 \times 5}{4 \times 8} = \frac{15}{32}.$$

Division of fractions is easy if one rule is used: invert the *divisor* (fraction divided into another) and then multiply. Thus, dividing $\frac{3}{4}$ by $\frac{5}{8}$,

$$\frac{3}{4} \div \frac{5}{8} = \frac{3}{4} \times \frac{8}{5} = \frac{3 \times 8}{4 \times 5} = \frac{24}{20} = \frac{6}{5} = 1\frac{1}{5}.$$

The above are called *common fractions*. Another type of fraction is the *decimal fraction*, which utilizes the place values of the number system to better advantage.

If the numerator and denominator of $\frac{1}{4}$ be multiplied by 25, the fraction becomes $\frac{25}{100}$, which also equals $\frac{2}{10} + \frac{5}{100}$. This may be written 0.25, by extending the idea of place values discussed on p. 2. Any common fraction may be converted to a decimal fraction by dividing the numerator by the denominator. For example, $\frac{7}{8} = 0.875$; $\frac{11}{64} = 0.171875$; $\frac{1}{3} = 0.333 \cdots$,

Chapter 1   **Mathematics of Radio and Electronics**   9

the dots indicating that no matter how long the division is continued, there will be still more digits.

As soon as common fractions have been converted to decimal fractions, they may be added, subtracted, multiplied, and divided just like integers. The following examples show the process:

$$0.875 + 0.125 = 1.000 \qquad 0.21 \times 0.3 = 0.063$$
$$0.625 - 0.0625 = 0.5625 \qquad 1.5 \div 0.5 = 3.0$$

It will be seen that in adding or subtracting, the periods or *decimal points* are always lined up with one another; in multiplying, the number of *decimal places* (digits to the right of the decimal point) in the result or product is the sum of the number of decimal places in the numbers multiplied together; in dividing, the decimal point may be located by setting the decimal point to the right in both the divisor and the *dividend* (number divided into) until the divisor is no longer a fraction, and locating the decimal point in the result at this place. An example will show how this is done.

To divide 1.728 by 0.12, write the figures either

$$0.12 \overline{)1.728(} \qquad \text{or} \qquad 0.12\overline{)1.728}$$

Move the decimal point to the right in both numbers until the divisor (0.12) is no longer a fraction, thus:

```
                                        14.4
    12)172.8(14.4        or        12)172.8
       12                              12
       --                              --
       52                              52
       48                              48
       --                              --
        48                              48
        48                              48
```

LOGARITHMS

Table 1-1 shows how certain numbers may be written in scientific notation. A similar system may be used to simplify multiplying, dividing, and other operations with numbers.

It may be seen from Table 1-1A, which repeats part of Table 1-1, that $10^0 = 1$ and $10^1 = 10$. A number between 1 and 10 must be represented

TABLE 1-1A

$10^{0.3010}$	2	10^3	1000
$10^{0.4771}$	3	10^2	100
$10^{0.6021}$	4	10^1	10
$10^{0.6990}$	5	10^0	1
$10^{0.7782}$	6	10^{-1}	0.1
$10^{0.8451}$	7	10^{-2}	0.01
$10^{0.9031}$	8	10^{-3}	0.001
$10^{0.9542}$	9	etc.	

by 10 raised to some power between 0 and 1 as indicated by an exponent. This is shown in more detail in the left-hand column of Table 1-1A, where the exponents for the integers between 1 and 10 are given.

For example, 2 lies between 1 and 10; it may be represented by $10^{0.3010}$, in which the exponent lies between 0 and 1. Likewise, 3 can be written as $10^{0.4771}$ and similarly for the other numbers in the left-hand column of Table 1-1A. Since 10 (the *base*) occurs in every instance where this system is used, it is convenient to call the exponent a *logarithm* to the base 10 and not to write the base at all. The logarithms of the integers from 1 to 10 are the following (logarithm to the base 10 is abbreviated "log"):

$$\log 1 = 0.0000 \quad \log 4 = 0.6021 \quad \log 7 = 0.8451$$
$$\log 2 = 0.3010 \quad \log 5 = 0.6990 \quad \log 8 = 0.9031$$
$$\log 3 = 0.4771 \quad \log 6 = 0.7782 \quad \log 9 = 0.9542$$

Now from the table of powers of 10 the exponent (or logarithm) of numbers between 10 and 100 must lie between 1 and 2. But 20 is the same fraction of the distance from 10 to 100 that 2 is from 1 to 10. Therefore the logarithm of 20 is 1.3010 and similarly

$$\log 200 = 2.3010; \quad \log 2{,}000 = 3.3010$$

The part of the logarithm to the right of the decimal point remains the same as long as the digits in the original number remain the same; the part to the left of the decimal point changes by 1 whenever the number is multiplied or divided by 10.

The decimal fraction or right-hand part of the logarithm is given in the table to four decimal places and is called the *mantissa*. The integral part to the left of the decimal point is called the *characteristic* and is found in the following way:

(a) Move the decimal point in the number until the number remaining is between 1 and 10.

(b) Count the number of places the decimal point has been moved and call this number the characteristic.

(c) Make the characteristic positive if the decimal point was moved to the left and negative if it was moved to the right.

(d) A negative characteristic is written as a positive one, with 10 subtracted from the entire log.

Example: Find log 4,570 in the table of logarithms (Table 1-3, pp. 12-13).

(a) Find 45 in the left-hand column of Table 1-3 and move across this line to the column headed by 7 to find 6599. Write this as 0.6599.

(b) Move the decimal point to the left three places to give 4.570, a number between 1 and 10. The characteristic is 3 and positive.

(c) The complete logarithm is $3 + 0.6599 = 3.6599$.

Example: Find log 0.00121.

(a) Find 12 in the left-hand column, go across this line to the column headed 1 and find 0.0828.

(b) Move the decimal point to the right three places, to give 1.21. The characteristic is 3 and negative, but is written 7 (mantissa) − 10.

(c) log 0.00121 = − 3 + .0828 = 7.0828 − 10.

If the exact value cannot be found in the table, take the value nearest it; or *interpolate* as follows:

Example: Find log 0.7854.

From the table of logarithms we find log 0.786, which is too large, and log 0.785, which is too small. We take the difference of these two values

$$\begin{aligned} \log 0.786 &= 9.8954 - 10 \\ \log 0.785 &= 9.8949 - 10 \\ \text{Difference} &= \overline{0.0005} \end{aligned}$$

Since 0.7854 is 0.4 of the interval between 0.785 and 0.786, 0.4 × 0.0005 = 0.0002 must be added to the logarithm of the smaller number. The required log 0.7854 is thus 9.8951 − 10, to the accuracy of the table used.

ANTILOGARITHMS

The number corresponding to a given logarithm is called the *antilogarithm* and is found as follows:

(a) Find the mantissa of the logarithm in the body of the table. Move across to the left-hand column for the first two figures of the antilog and note the column heading which is the third figure; consider the result as a number between 1 and 10.

(b) Move the decimal point to the right as many places as the characteristic when the latter is positive and to the left when it is negative.

Example: Find antilog 2.5877.

(a) Find 0.5877 in line 38 and column 7; consider this as 3.87.

(b) Move the decimal point 2 places to the right, which gives 387, the required antilog.

Example: Find antilog 7.3243 − 10.

(a) Find 0.3243 in line 21, column 1; consider this as 2.11.

(b) The characteristic is 7 − 10 = −3. Therefore, move the decimal point 3 places to the left to get 0.00211 as the required antilog.

If the exact value cannot be found in the table, take the value nearest to it; or interpolate as follows:

Table 1-3
LOGARITHMS

N	0	1	2	3	4	5	6	7	8	9
10	0000	0043	0086	0128	0170	0212	0253	0294	0334	0374
11	0414	0453	0492	0531	0569	0607	0645	0682	0719	0755
12	0792	0828	0864	0899	0934	0969	1004	1038	1072	1106
13	1139	1173	1206	1239	1271	1303	1335	1367	1399	1430
14	1461	1492	1523	1553	1584	1614	1644	1673	1703	1732
15	1761	1790	1818	1847	1875	1903	1931	1959	1987	2014
16	2041	2068	2095	2122	2148	2175	2201	2227	2253	2279
17	2304	2330	2355	2380	2405	2430	2455	2480	2504	2529
18	2553	2577	2601	2625	2648	2672	2695	2718	2742	2765
19	2788	2810	2833	2856	2878	2900	2923	2945	2967	2989
20	3010	3032	3054	3075	3096	3118	3139	3160	3181	3201
21	3222	3243	3263	3284	3304	3324	3345	3365	3385	3404
22	3424	3444	3464	3483	3502	3522	3541	3560	3579	3598
23	3617	3636	3655	3674	3692	3711	3729	3747	3766	3784
24	3802	3820	3838	3856	3874	3892	3909	3927	3945	3962
25	3979	3997	4014	4031	4048	4065	4082	4099	4116	4133
26	4150	4166	4183	4200	4216	4232	4249	4265	4281	4298
27	4314	4330	4346	4362	4378	4393	4409	4425	4440	4456
28	4472	4487	4502	4518	4533	4548	4564	4579	4594	4609
29	4624	4639	4654	4669	4683	4698	4713	4728	4742	4757
30	4771	4786	4800	4814	4829	4843	4857	4871	4886	4900
31	4914	4928	4942	4955	4969	4983	4997	5011	5024	5038
32	5051	5065	5079	5092	5105	5119	5132	5145	5159	5172
33	5185	5198	5211	5224	5237	5250	5263	5276	5289	5302
34	5315	5328	5340	5353	5366	5378	5391	5403	5416	5428
35	5441	5453	5465	5478	5490	5502	5514	5527	5539	5551
36	5563	5575	5587	5599	5611	5623	5635	5647	5658	5670
37	5682	5694	5705	5717	5729	5740	5752	5763	5775	5786
38	5798	5809	5821	5832	5843	5855	5866	5877	5888	5899
39	5911	5922	5933	5944	5955	5966	5977	5988	5999	6010
40	6021	6031	6042	6053	6064	6075	6085	6096	6107	6117
41	6128	6138	6149	6160	6170	6180	6191	6201	6212	6222
42	6232	6243	6253	6263	6274	6284	6294	6304	6314	6325
43	6335	6345	6355	6365	6375	6385	6395	6405	6415	6425
44	6435	6444	6454	6464	6474	6484	6493	6503	6513	6522
45	6532	6542	6551	6561	6571	6580	6590	6599	6609	6618
46	6628	6637	6646	6656	6665	6675	6684	6693	6702	6712
47	6721	6730	6739	6749	6758	6767	6776	6785	6794	6803
48	6812	6821	6830	6839	6848	6857	6866	6875	6884	6893
49	6902	6911	6920	6928	6937	6946	6955	6964	6972	6981
50	6990	6998	7007	7016	7024	7033	7042	7050	7059	7067
51	7076	7084	7093	7101	7110	7118	7126	7135	7143	7152
52	7160	7168	7177	7185	7193	7202	7210	7218	7226	7235
53	7243	7251	7259	7267	7275	7284	7292	7300	7308	7316
54	7324	7332	7340	7348	7356	7364	7372	7380	7388	7396
N	0	1	2	3	4	5	6	7	8	9

TABLE 1-3 (*Continued*)
LOGARITHMS

N	0	1	2	3	4	5	6	7	8	9
55	7404	7412	7419	7427	7435	7443	7451	7459	7466	7474
56	7482	7490	7497	7505	7513	7520	7528	7536	7543	7551
57	7559	7566	7574	7582	7589	7597	7604	7612	7619	7627
58	7634	7642	7649	7657	7664	7672	7679	7686	7694	7701
59	7709	7716	7723	7731	7738	7745	7752	7760	7767	7774
60	7782	7789	7796	7803	7810	7818	7825	7832	7839	7846
61	7853	7860	7868	7875	7882	7889	7896	7903	7910	7917
62	7924	7931	7938	7945	7952	7959	7966	7973	7980	7987
63	7993	8000	8007	8014	8021	8028	8035	8041	8048	8055
64	8062	8069	8075	8082	8089	8096	8102	8109	8116	8122
65	8129	8136	8142	8149	8156	8162	8169	8176	8182	8189
66	8195	8202	8209	8215	8222	8228	8235	8241	8248	8254
67	8261	8267	8274	8280	8287	8293	8299	8306	8312	8319
68	8325	8331	8338	8344	8351	8357	8363	8370	8376	8382
69	8388	8395	8401	8407	8414	8420	8426	8432	8439	8445
70	8451	8457	8463	8470	8476	8482	8488	8494	8500	8506
71	8513	8519	8525	8531	8537	8543	8549	8555	8561	8567
72	8573	8579	8585	8591	8597	8603	8609	8615	8621	8627
73	8633	8639	8645	8651	8657	8663	8669	8675	8681	8686
74	8692	8698	8704	8710	8716	8722	8727	8733	7398	8745
75	8751	8756	8762	8768	8774	8779	8785	8791	8797	8802
76	8808	8814	8820	8825	8831	8837	8842	8848	8854	8859
77	8865	8871	8876	8882	8887	8893	8899	8904	8910	8915
78	8921	8927	8932	8938	8943	8949	8954	8960	8965	8971
79	8976	8982	8987	8993	8998	9004	9009	9015	9020	9025
80	9031	9036	9042	9047	9053	9058	9063	9069	9074	9079
81	9085	9090	9096	9101	9106	9112	9117	9122	9128	9133
82	9138	9143	9149	9154	9159	9165	9170	9175	9180	9186
83	9191	9196	9201	9206	9212	9217	9222	9227	9232	9238
84	9243	9248	9253	9258	9263	9269	9274	9279	9284	9289
85	9294	9299	9304	9309	9315	9320	9325	9330	9335	9340
86	9345	9350	9355	9360	9365	9370	9375	9380	9385	9390
87	9395	9400	9405	9410	9415	9420	9425	9430	9435	9440
88	9445	9450	9455	9460	9465	9469	9474	9479	9484	9489
89	9494	9499	9504	9509	9513	9518	9523	9528	9533	9538
90	9542	9547	9552	9557	9562	9566	9571	9576	9581	9586
91	9590	9595	9600	9605	9609	9614	9619	9624	9628	9633
92	9638	9643	9647	9652	9657	9661	9666	9671	9675	9680
93	9685	9689	9694	9699	9703	9708	9713	9717	9722	9727
94	9731	9736	9741	9745	9750	9754	9759	9763	9768	9773
95	9777	9782	9786	9791	9795	9800	9805	9809	9814	9818
96	9823	9827	9832	9836	9841	9845	9850	9854	9859	9863
97	9868	9872	9877	9881	9886	9890	9894	9899	9903	9908
98	9912	9917	9921	9926	9930	9934	9939	9943	9948	9952
99	9956	9961	9965	9969	9974	9978	9983	9987	9991	9996
N	0	1	2	3	4	5	6	7	8	9

Example: Find antilog 0.4208.

$$\begin{array}{ll} \log 2.64 = 0.4216 & \log 2.64 = 0.4216 \\ \log 2.63 = 0.4200 & \text{given } \log = 0.4208 \\ \text{Difference} = \overline{0.0016} & \text{Difference} = \overline{0.0008} \end{array}$$

Since $0.0008 \div 0.0016 = 0.5$, the required antilog must lie 0.5 of the way from 2.63 to 2.64, or 2.635. The characteristic is zero so the decimal point does not have to be moved.

Logarithms may be used to make multiplication and division easier. The rules are the same as those stated for exponents. That is, *to multiply, add the logarithms of the numbers* and look up the antilogarithm of the result. The work of multiplying 479 by 89 may be arranged as follows:

$$\begin{array}{l} \log 479 = 2.6803 \\ \log 89 = 1.9494 \\ \log \text{ of product} = \overline{4.6297} \\ \text{antilog } 4.6297 = 42{,}630. \end{array}$$

To divide, subtract logs; to divide 479 by 890, arrange thus:

$$\begin{array}{l} \log 479 = 12.6803 - 10 \\ \log 890 = 2.9494 \\ \log \text{ of quotient} = \overline{9.7309 - 10} \\ \text{antilog } 9.7309 - 10 = 0.538. \end{array}$$

Since the difference of the logs is a negative number here, 10 is added at the left and subtracted to the right of log 479 (this does not alter its value) and the log of the quotient shown is obtained.

Logs are very useful in finding higher powers and roots of numbers. Suppose that 24^5 is needed. This could be found exactly by multiplying $24 \times 24 \times 24 \times 24 \times 24$, but logs make the operation much easier. Thus:

$$\begin{array}{l} \log 24 = 1.3802 \\ \text{multiply by} 5 \\ \log (24^5) = 5 \log 24 = \overline{6.9010} \\ \text{antilog } 6.9010 = 7{,}962{,}000. \end{array}$$

Other powers are also easy: for example, $2^{0.5} = 2^{\frac{1}{2}} = \sqrt{2}$.

$$\begin{array}{l} \log 2 = 0.3010 \\ \text{multiply by} 0.5 \\ \log (2^{0.5}) = \overline{0.1505} \\ \text{antilog } 0.1505 = 1.414 \end{array}$$

This is actually the process of extracting a square root, because

$$2^{0.5} \times 2^{0.5} = 2^{0.5+0.5} = 2^1 = 2.$$

Logarithms may be plotted graphically, and the resulting curve may be used as a table or otherwise. Instead of using equal divisions on the paper in plotting this graph, it is handier to use paper which is divided according to the logarithms of the numbers, as shown in Fig. 1-2. Entering the chart at 2, on the bottom, proceed upward to the line and then to the left to find 0.3, which is log 2. To find antilog 0.8, enter the chart at 0.8 on

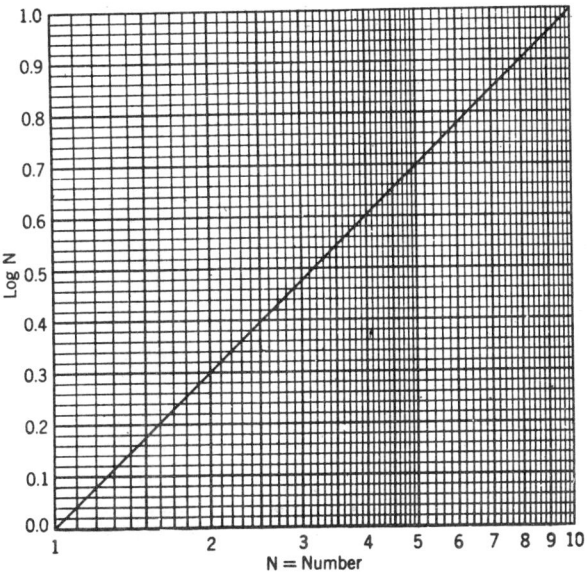

Fig. 1-2. Curve showing logarithms of numbers.

the left, proceed horizontally to the line, and then down to 6.4 as required.

The power level in a radio communication system often has a wide range even though the total energy involved may be small. It is convenient to use a logarithmic unit which depends upon the power ratios rather than upon power magnitudes. Such a unit is the *decibel* (abbreviated db) and it is defined by the equation

$$db = 10 \log_{10} P_1/P_2,$$

where db is the number of decibels, P_1 is the larger amount of power, and P_2 is the smaller amount of power being compared. If one signal has twice as much power as another the ratio $P_1/P_2 = 2$; $\log_{10} 2 = 0.3010$, and therefore the number of decibels is $10 \times 0.3010 = 3$. This may be seen from the chart showing the relation between power ratio and decibels, Fig. 1-3. If the output power in a given system is greater than the input power, the system is said to have a gain and the number of decibels is positive; if the output is less than the input, the system is said to have a loss, and the gain is expressed in negative decibels ($-$db).

As has been shown, charts may be made with divisions which are logarithmic. For some purposes it is more convenient to use simply a scale which is divided into parts proportional to the logarithms of numbers, such

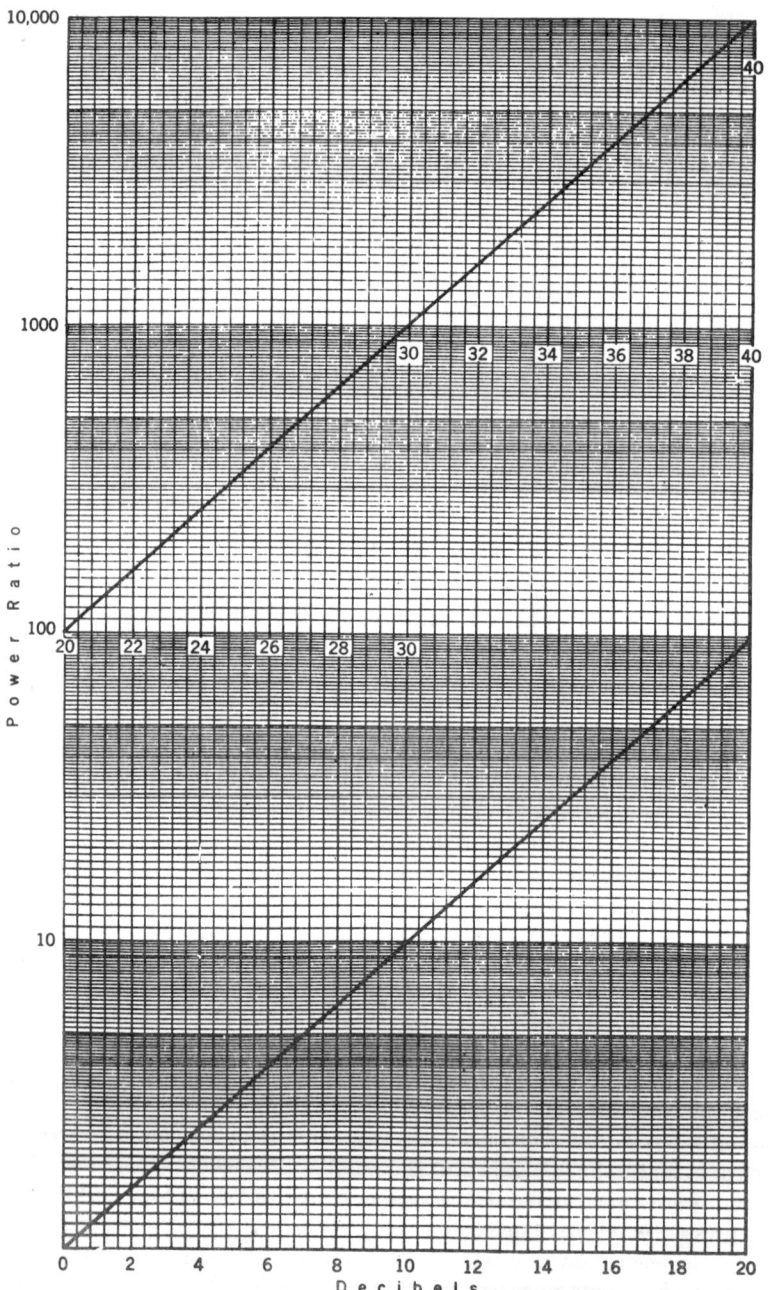

Fig. 1-3. Relation between power ratio and decibels.

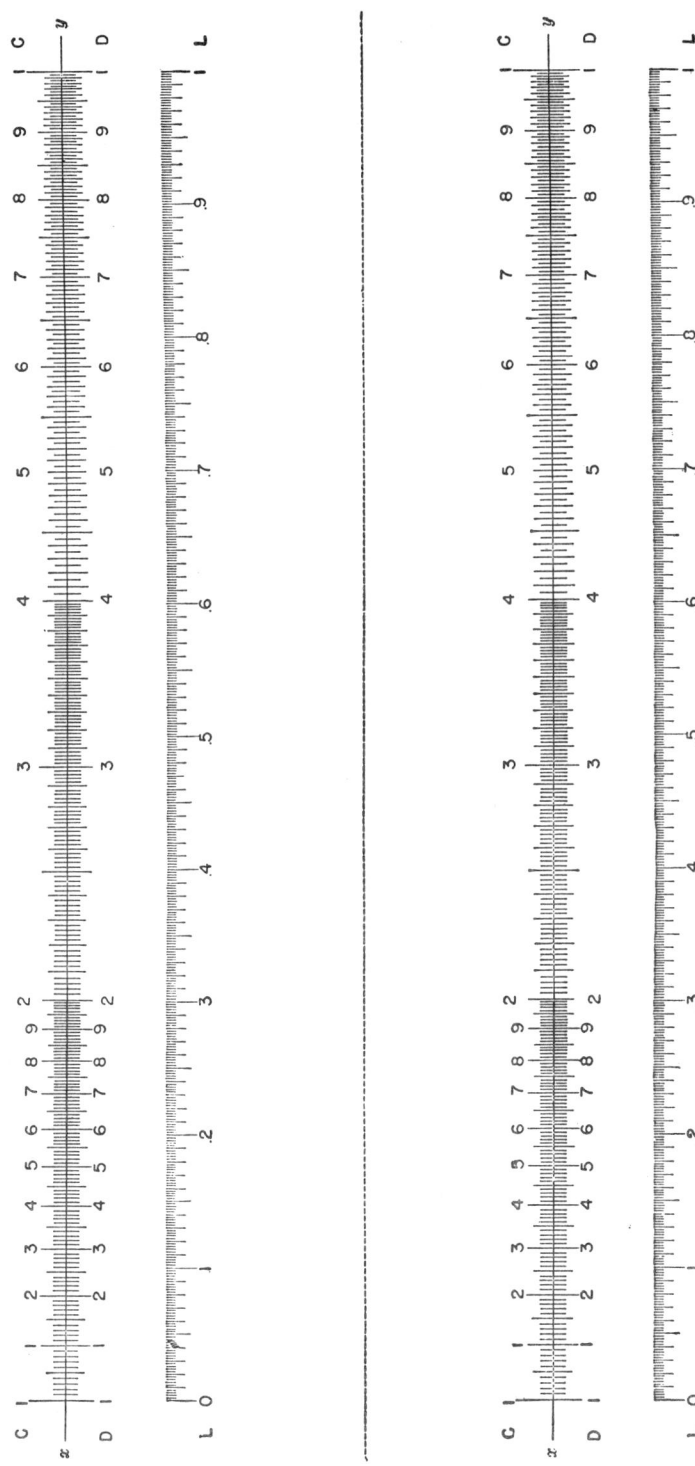

FIG. 1-4. Typical slide rule scales. (The second one may be cut out and pasted on cardboard or wood to make a model slide rule. Also cut along line x-y.)

Chapter 1   **Mathematics of Radio and Electronics**   19

as Fig. 1-4. It is seen that the point marked 2 is about 0.3 of the length of the scale from the left-hand end; 4 is located 0.6 of the length from the left, and so on. Other divisions are marked according to the following system. Considering the 1 at the left-hand end of the scale (called the *index*) to represent 100, the main divisions will be 1, 2, 3, · · · , to represent 100, 200, 300, · · · . Between 1 and 2 are other divisions marked from 1 to 9, which represent 110, 120, 130, · · · . These are further divided with a third set of marks which do not have numbers but represent 101, 102, 103 · · · , to 199. Between the 2 and 4 the subdivisions with the longest lines represent 210, 220, 230, · · · , which are again divided to represent 202, 204, 206, · · · , etc. The smallest divisions here represent numbers twice as large as those in the portion of the scale marked 1 to 2. From 4 to the right-hand end of the scale the major divisions are also by 10's (410, 420, 430 · · ·) but the smallest division represents only 5; that is, 405, 415,

FIG. 1-5. Logarithmic scale with certain points marked.

425, · · · with 10's divisions between. In Fig. 1-5 is a scale with the following points marked: A 365, B 327, C 263, D 1,745, E 1,347, F 305, G 207, H 1,078, I 435, J 427.

The numbers at either end of the scale (called the *left-hand* and *right-hand index*, respectively) may be multiplied or divided by any power of 10; that is, the left-hand index may represent 1.0, 100, 10^6, 0.001, and all other points on the scale will have the same multiplier. This property and other properties of logarithms permit two such scales as have been described to be used for multiplication, division, and other operations with numbers except addition and subtraction.

If the scale marked C (Fig. 1-4) be cut out on line xy, the C scale may be matched with the D scale or moved to any position along the latter. For example, if the 1 on the C scale (called C for brevity) be placed opposite 2 on D, 2 on C will be opposite 4 on D, 3 opposite 6, and so on. But 4 is 2 × 2, and actually log 2 on C has been added mechanically to log 2 on D and the result, quite properly, is log 4. This process (adding logarithms mechanically) is the basis of the *slide rule*, which makes tedious calculations much easier.

Division may be performed with equal ease. For example, to divide 9 by 3, set 3 on C over 9 on D and read 3, the answer, under the left-hand index (1) on C.

This same setting also gives the result of dividing 6 by 2, 7.5 by 2.5,

36 by 12, and many other combinations. This property may be used to work problems in proportion, such as

$$3:9 = x:42,$$

finding the answer 14 above 42 on the D scale.

The location of the decimal point may be found by rules given in slide-rule instruction books, but is often located by inspection. Thus, in multiplying 195 by 24, an approximate calculation (made mentally) with 200 × 20 shows that the answer should be somewhere near 4,000, and actually is 4,680.

Common fractions are easily converted to decimal fractions on these scales, by dividing the numerator by the denominator. That is, to convert $\frac{1}{16}$ to a decimal fraction, set 16 on C over 1 on D and find 0.0625 on D under the right-hand index of C, after locating the decimal point.

The relation between the divisions on the D scale and the logarithms of numbers may be seen by finding the logarithms on the scale of equal parts, marked L. For example, opposite 2 on D find 0.301 on L, opposite 3 on D find 0.477 on L, and so forth.

BINARY NUMBER SYSTEM

In recent years machines called *digital computers* have been designed and built to do many types of computations with superhuman speed and accuracy. These machines utilize switches, neon lamps, electron switching circuitry, relays, and other devices which operate in a two-position or "on-or-off" fashion. While any number can be used as the base of a number system, the binary (two-digit) system is particularly suited to these computers and functions better for this purpose than the ordinary decimal (base 10) system.

To "program" a digital computer (that is, to feed information or a problem into such a machine) it is necessary to convert numbers from the decimal system to the binary system. This is done in the following way.

Suppose that five switches, marked A, B, C, D, E, are available and that each has an "on" position and an "off" position. These five switches can be used to "count" all numbers in the binary system up to 32 in the decimal system. When all five switches are open, the binary number zero is indicated corresponding to zero in the decimal system (Table 1-4, line *a*). For the decimal number 1, the first switch on the right (E) is closed, as indicated (line *b*). The next step must be to open switch E (since this is the only way to change its condition, and at the same time to close switch D (line *c*) corresponding to the decimal number 2. We have now stepped over in the binary place system from $2^0 = 1$ to $2^1 = 2$ (line *c*). Next, switch E is closed again to indicate the decimal number 3 so that both switches are closed as shown in line *d* of Table 1-4. Both switches must now be opened and the next switch to the left (C) closed, to indicate the decimal number 4.

Chapter 1   Mathematics of Radio and Electronics   21

The place value of switch C is $2^2 = 4$, analogous to $10^2 = 100$ in the decimal system.

TABLE 1-4

Decimal Number (base 10)		Binary Number (base 2) A B C D E	
0	Note: 0 means switch open;	0 0 0 0 0	(a)
1	1 means switch closed.	0 0 0 0 1	(b)
2		0 0 0 1 0	(c)
3		0 0 0 1 1	(d)
4		0 0 1 0 0	(e)
5		0 0 1 0 1	
6		0 0 1 1 0	
7		0 0 1 1 1	
8		0 1 0 0 0	(f)
9		0 1 0 0 1	(g)
10		0 1 0 1 0	(h)
11		0 1 0 1 1	
12		0 1 1 0 0	
13		0 1 1 0 1	
14		0 1 1 1 0	
15		0 1 1 1 1	
16		1 0 0 0 0	
..		...	
25		1 1 0 0 1	
..		...	
31		1 1 1 1 1	

Place values: 10^0, 10^1 ; 2^0, 2^1, 2^2, 2^3, 2^4

Proceeding in a similar way, the decimal numbers 5, 6, and 7 can be formed as indicated in Table 1-4. For the latter (7) three switches C, D, E are closed and it is necessary to step over one place, corresponding to $2^3 = 8$ (switch B), to get 8 in the decimal system. The rest of the numbers up to 16 follow; for 16 in the decimal system we go to 2^4 and close switch A.

All the decimal numbers including 31 can be formed in the binary system with these five switches; for 32, which is equivalent to 2^5, we must step over to the left one more place in the binary system, and this requires another switch. The system illustrated with five switches counts only to 31.

Any number can be expressed in the binary system by selection and position of the two digits 0 and 1; by repeating them as multiples of 2 raised to a power which is indicated by the place value of the digit in the number. An exponent also can be used to indicate the place value, as in examples. Analogously, any number is expressed in the decimal system by selecting the correct digit from 0 through 9 and multiplying it by 10 raised to a power indicated by an exponent (for illustration) or by its place value in the number. For example, 1,492 in decimal notation is

$$
\begin{array}{rl}
1 \times 10^3 =& 1{,}000 \\
4 \times 10^2 =& 400 \\
9 \times 10^1 =& 90 \\
2 \times 10^0 =& 2 \\
\hline
1{,}492 \qquad =& 1{,}492
\end{array}
$$

In the same way the binary number 011001 means

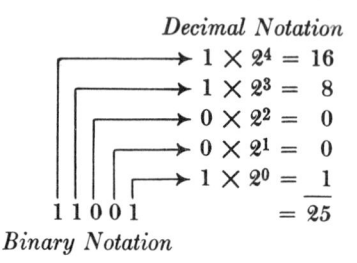

Decimal Notation

$$
\begin{array}{rl}
1 \times 2^4 =& 16 \\
1 \times 2^3 =& 8 \\
0 \times 2^2 =& 0 \\
0 \times 2^1 =& 0 \\
1 \times 2^0 =& 1 \\
\hline
1\,1\,0\,0\,1 \qquad =& 25
\end{array}
$$

Binary Notation

A table of powers of 2 up to 2^{12} is given so that additional examples can be worked out.

TABLE 1-5
POWERS OF 2

$2^0 = 1$

$2^1 = 2$	$2^{-1} = 0.5$
$2^2 = 4$	$2^{-2} = 0.25$
$2^3 = 8$	$2^{-3} = 0.125$
$2^4 = 16$	$2^{-4} = 0.0625$
$2^5 = 32$	$2^{-5} = 0.03125$
$2^6 = 64$	$2^{-6} = 0.015625$
$2^7 = 128$	$2^{-7} = 0.0078125$
$2^8 = 256$	$2^{-8} = 0.00390625$
$2^9 = 512$	$2^{-9} = 0.001953125$
$2^{10} = 1024$	$2^{-10} = 0.0009765625$
$2^{11} = 2048$	$2^{-11} = 0.00048828125$
$2^{12} = 4096$	$2^{-12} = 0.000244140625$

Places to the right of the decimal point represent decimal fractions; that is, these places correspond to negative powers of 10, such as $10^{-1} = 0.1$, $10^{-2} = 0.01$, etc. In the same way a binary point is used to set off values less than unity such as $2^{-1} = \frac{1}{2} = 0.5$ (in decimal the system), $2^{-2} = \frac{1}{2}^2 = \frac{1}{4} = 0.25$, and so on. As an example, 3.625 in the decimal system is written as binary 11.101, as shown at the top of the next page.

Chapter 1 **Mathematics of Radio and Electronics** 23

$$\begin{array}{r}
\text{Decimal Notation}\\
1 \times 2^1 = 2.000\\
1 \times 2^0 = 1.000\\
1 \times 2^{-1} = 0.500\\
0 \times 2^{-2} = 0.000\\
1 \times 2^{-3} = 0.125\\
\hline
1\ 1\ .\ 1\ 0\ 1 \qquad = 3.625
\end{array}$$

Binary Notation

This may be worked out by starting with the number in the decimal system as 3.625. From Table 1-4 we find that decimal 3 is written as 11.000 in the binary system. Now it is only necessary to find the binary equivalent of 0.625. Since 2^{-1} is $\frac{1}{2}$ or 0.5 in the decimal system, we may subtract this from 0.625, or 0.625 − 0.500 = 0.125. We now have information to write 3.5 as 11.1 in the binary system. Continuing with the example, if we try to subtract $2^{-2} = 0.25$ from 0.125 we see that it is too large to give a positive remainder; so we go on to $2^{-3} = 0.125$. Subtracting this from 0.125 gives no remainder and we can now write decimal 3.625 as binary 11.101. Other examples may be worked out by the same process with the aid of Table 1-5.

Addition in the binary system requires a binary-addition table which is shown in Table 1-6. The last entry in this table means that every time the

TABLE 1-6
ADDITION TABLE FOR BINARY NUMBERS

$0 + 0 = 0$

$0 + 1 = 1$

$1 + 0 = 1$

$1 + 1 = 10$

	0	1
0	0	1
1	1	10

sum $1 + 1$ occurs, we write 0 in the column where it happens and carry forward 1 to the next column. This is analogous to carrying 10's in decimal addition. For example:

$$\begin{array}{r}
\text{Binary Notation} \quad \text{Decimal Notation}\\
11.01 = 3.25\\
100.11 = 4.75\\
\hline
\textit{The sum:}\quad 1000.00 = 8.00
\end{array}$$

The 1's in the right-hand column are added first, to give 0 and carry 1 to the next column to the left. When this carried 1 is added to the 1 in the next column we get 0 and carry 1 to the next column (third from the left). Again the sum is 0 and carry a 1, and so on. Other sums can be worked out in the same way.

TABLE 1-7
MULTIPLICATION TABLE FOR BINARY NUMBERS

$0 \times 0 = 0$

$0 \times 1 = 0$

$1 \times 0 = 0$

$1 \times 1 = 1$

	0	1
0	0	0
1	0	1

To perform binary multiplication the very short binary-multiplication table is used (Table 1-7).

For example, to multiply binary 0111 = 7 decimal by binary 0010 = 2 decimal, set the numbers down just as in ordinary multiplication. Perform the partial products by the multiplication-table results, and then add the partial products. This is worked out below.

```
Binary Notation     Decimal Notation
    00111                  7
    00010                  2
    00000                 14
    00111
   00000
  00000
  00001110 = 14   (See Table 1-4)
```

The open switches to the left (represented by 0's) are not needed for the binary number 01110 = 14 decimal as given in Table 1-4

It seems unlikely that the binary system will soon be used for everyday problems in place of the familiar decimal system. If we had been born with only fists instead of fingers, we might have learned to count by twos instead of tens, and then if India's outstanding mathematical contribution of the positional notation had still been forthcoming, we and our children would have had to learn only the simple multiplication of Table 1-7. Hence it seems logical that if we are going to build new brains ourselves in the form of electronic computers, we should design them for the simplest arithmetic, which is the binary system.

Digital computers and similar electronic computing devices use binary numbers because this permits simpler switching elements inside the machine than is possible with the decimal system. It also permits simpler storage systems for the numbers that result from the computations. Every computer must be programed — that is, it must be told what to do — and part of the programing often involves changing from decimal to binary numbers and back, so that the involved arithmetic operations can be made in the binary system and the answer delivered in the usual decimal notation.

SYMBOLS OF ALGEBRA

Algebra is a special form of language, which uses *symbols* in place of words to express ideas.

Symbols are used in mathematics so that ideas may be put down on paper and talked about easily; using words for the same purpose would often cause confusion. What method shall be used to choose these symbols? First, symbols must be as brief as possible; second, their meaning must be generally agreed upon by those who use them. One choice is to use the initial letter of a word, like R for resistance, C for capacitance, W for watts, P for power. Sometimes symbols for certain things have been used so long they are accepted almost universally; for example in electrical terminology, I for current, E for voltage, L for inductance, and so on. (The meaning of these terms will be explained later.) Other *symbols* generally indicate *numbers;* such as x, y, z (last few letters of the alphabet), which usually mean quantities *unknown* or *variable;* a, b, c (first letters of alphabet) often stand for *known* or *unvarying* quantities. Mathematics packs much meaning into one symbol to make the symbol useful where words or numbers would not serve.

The symbols used in mathematics are precise in meaning; words also are used with meanings more precise and more limited than they may have in everyday conversation, and sometimes these meanings are different from the common ones. Some definitions of such words follow.

An *algebraic expression* is any combination of numbers, letters used for numbers, and signs of operation written according to the rules of algebra; like $10x$, $t + t$, $mx + b$. Note that $10x$ means 10 times x; this is a product and the 10 and the x are the two factors. *Factors* of a product are the numbers which, when multiplied together, form the product. In algebra numbers and letters or combinations of letters written together without signs of operation are to be multiplied; xy means x times y, $10a$ means 10 times a. In the product $10a$, 10 is the *numerical coefficient* or factor and a is the *literal* (letter) *coefficient* or factor. When x is written alone, the coefficient 1 is understood; x means 1 times x. A *term* is any expression like $10x$, t, mx, b, $5r$, and so forth. *Similar terms* are those with the same letter or letters, such as $10x$ and mx. If terms do not have the same letters (or literal coefficients) they are *dissimilar terms*. Similar terms may be combined by *adding* the *numerical coefficients;* remember that subtracting is the same as adding a term with a minus sign ahead of it. For example:

$$3x + 4x = (3 + 4)x = 7x; \qquad 3a - 2a = (3 - 2)a = 1a = a.$$

SOME LAWS OF ALGEBRA

In *addition* terms may be combined or grouped in any order. That is, $a + b = b + a$ and $a + b + c = (a + b) + c = a + (b + c)$. If dissimilar terms occur, they may be combined in groups of similar terms; for example, $50a + 98b + 5a + 3b + a = (50 + 5 + 1)a + (98 + 3)b = 56a + 101b$.

Multiplication and *division* are carried out with the coefficients of similar terms, like

$$7 \times 2a = 14a; \qquad \tfrac{1}{2} \times 10w = 5w; \qquad \frac{10x}{2} = 5x;$$

$$\frac{2(6x - 3x)}{3} = 2x; \qquad \frac{18x}{6x} = 3$$

Just as $10 \times 10 = 10^2$, $a \times a = a^2$, and $a \times a = a^3$. Similar terms with the same exponents may be combined; that is, $3a^2 + 4a^2 = 7a^2$.

Expressions containing several terms are called *polynomials*, for example $3x^3 + 4x^2 - 5x + 2$, which is a polynomial in *descending powers* of x. Polynomials may be added by combining similar terms as follows:

$$\begin{array}{r} 3x^3 + 4x^2 - 5x + 2 \\ -2x^3 - x^2 + 8x - 1 \\ \hline x^3 + 3x^2 + 3x + 1 \end{array}$$

To make the step from arithmetic to algebra easier, here is another example worked out. Suppose 2,565 is to be added to 5,331. In the positional and scientific notation these numbers can now be written

$$\begin{array}{r} 2 \times 10^3 + 5 \times 10^2 + 6 \times 10 + 5 \\ 5 \times 10^3 + 3 \times 10^2 + 3 \times 10 + 1 \\ \hline 7 \times 10^3 + 8 \times 10^2 + 9 \times 10 + 6 \end{array}$$

Incidentally, numbers are usually read in this form: seven thousand, eight hundred, and ninety-six. The additions of the digits are done from memory and the 10's with exponents are carried through without change. Now if x be substituted for 10, the resulting *algebraic expression* may be added in the same way.

$$\begin{array}{r} 2x^3 + 5x^2 + 6x + 5 \\ 5x^3 + 3x^2 + 3x + 1 \\ \hline 7x^3 + 8x^2 + 9x + 6 \end{array}$$

Now any other number, such as 2 or 3, may be put in place of x and the result will still be correct. The x stands for any number.

The same rules for exponents apply as stated earlier for the base 10; that is, $a^2 \times a^3 = a^{2+3} = a^5$; $\dfrac{a^2}{a^3} = a^{-1}$; $(a^2)^3 = a^2 \times a^2 \times a^2 = a^6$; $a^{\frac{1}{2}} \times a^{\frac{1}{2}} = a = \sqrt{a} \times \sqrt{a}$; $a^{\frac{1}{2}} = \sqrt{a}$.

RULE OF SIGNS. In multiplying or dividing, if the terms have *like signs* the result is *positive* (has plus sign); if the terms have *unlike signs* the result is *negative* (minus sign).

$(+a) \times (+b) = +ab$ $(+a) \div (+b) = +\dfrac{a}{b}$

$(-a) \times (-b) = +ab$ $(-a) \div (-b) = +\dfrac{a}{b}$

$(+a) \times (-b) = -ab$ $(+a) \div (-b) = -\dfrac{a}{b}$

$(-a) \times (+b) = -ab$ $(-a) \div (+b) = -\dfrac{a}{b}$

Example:
$(+3a^2b) \times (-2a^3b^2c) = (+3)(-2)(a^2)(a^3)(b)(b^2)(c) = -6a^5b^3c$.

Removal of parentheses (and of other signs of grouping) from expressions is easy if certain rules are followed. These signs of grouping are often used to indicate multiplication of terms, and in the course of solving a problem it may be necessary that they be removed systematically. Therefore:

(1) Apply the rule of signs.

Example: $\quad 2(a + b - c) = 2a + 2b - 2c.$

Also, $\quad -3a(a^2 - 3a + 5) = -3a^3 + 9a^2 - 15a.$

(2) If no coefficient is indicated for the group, consider the coefficient to be 1.

Example: $\quad (a + b) = 1(a + b) = a + b.$

Also, $\quad -(x - 2y) = -1(x - 2y) = -x + 2y.$

(3) Perform indicated multiplications or divisions first; then addition or subtraction.

Example: $\quad 5 - 3(a + 2b) = 5 - 3a - 6b.$

(4) Remove signs of grouping one set at a time, starting with the innermost set.

Example:

$-3c[a + 2b(b - c)] = -3c[a + 2b^2 - 2bc] = -3ac - 6b^2c + 6bc^2.$

In the multiplication of algebraic expressions certain combinations occur so often that they need special study, so that they may be known and the answers obtained quickly. These combinations are often called *special products*. If the special product is recognized, its factors may often be found by inspection. *Factoring* is defined as the process of finding two or more expressions whose product is a given expression.

The simplest sort of factoring is dividing out a *common factor*. For example, to factor

$$3x^4 - 9x^3 + 12x^2$$

it is possible to divide out a 3, an x^2, or a $3x^2$. The latter, being the largest, is usually the factor desired; thus,

$$3x^4 - 9x^3 + 12x^2 = 3(x^4 - 3x^3 + 4x^2)$$
$$= x^2(3x^2 - 9x + 12)$$
$$= 3x^2(x^2 - 3x + 4).$$

Another product which occurs frequently is the product of two *binomials* (expressions having two terms).

$$\begin{array}{r} 2x + 6 \\ x - 5 \\ \hline 2x^2 + 6x \\ -10x - 30 \\ \hline 2x^2 - 4x - 30 \end{array}$$

or
$$(2x + 6)(x - 5) = 2x^2 - 4x - 80.$$
$$\underset{A}{\uparrow}\quad\underset{B}{\uparrow}\underset{C}{\uparrow}\quad\underset{D}{\uparrow}$$

From this example it may be seen that (1) the product usually will have three terms; (2) the first term is the product of the first terms in the binomials; (3) the second term is the algebraic sum of the products of the two outer terms and the two inner terms, that is, (A)(D) + (B)(C); (4) the last term is the product of the last terms of the binomials. The rule of signs must be applied at all times.

This method may be used in reverse; that is, given the product the factors may be found. For example, to factor $9x^2 + 18x + 5$, two binomials are needed having their first terms factors of $9x^2$, having their second terms factors of 5, and having $18x$ as the sum of the product of the two outer terms plus the product of the two inner terms of the trial factors (A)(D) + (B)(C). The correct factors are found by trying out different combinations until the right ones are obtained. Several possible combinations may be set down, but only one meets the third requirement.

	Trial factors	Inner term	Conclusion
First attempt............	$(x + 5)(9x + 1)$	$x + 45x = 46x$	No good.
Second attempt.........	$(x + 1)(9x + 5)$	$5x + 9x = 14x$	No good.
Third attempt..........	$(3x + 1)(3x + 5)$	$15x + 3x = 18x$	Correct.

Certain binomials occur often as equal factors of squares. These are of the type $(a + b)$ or $(a - b)$. The square of either of these consists of the sum of three terms: (1) the square of the first term in the binomial; (2) twice the product of the two terms of the binomial; (3) the square of the second term of the binomial. That is,

$$(a + b)^2 = (a + b)(a + b) = a^2 + 2ab + b^2;$$
$$(a - b)^2 = (a - b)(a - b) = a^2 - 2ab + b^2.$$

This process is sometimes used with numbers as a short-cut method to find their squares. For instance:

$$(23)^2 = (20 + 3)^2 = (20)^2 + 2(20)(3) + 3^2$$
$$= 400 + 120 + 9 = 529;$$
$$(29)^2 = (30 - 1)^2 = (30)^2 - 2(30)(1) + 1^2$$
$$= 900 - 60 + 1 = 841.$$

Using the formulas for these squares is easier than multiplying out the usual way. Diagrams illustrating the binomial squares are shown in Fig. 1-6.

It is easy to factor expressions of this sort; simply take the square root of the first and the last terms and write the sign of the middle term between them. For example:

Chapter 1 — Mathematics of Radio and Electronics

$$a^2 + 6a + 9 = (a + 3)^2,$$
$$4x^2 - 12x + 9 = (2x - 3)^2.$$

Note: Be sure that the middle term is twice the product of the square roots of the first and last terms.

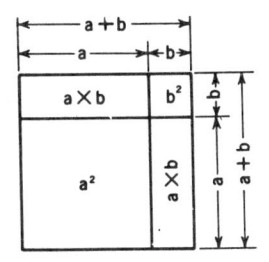

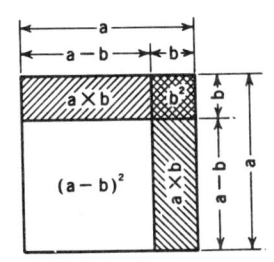

Fig. 1-6. Illustrating binomial squares by areas.

Another useful combination is the result of multiplying the sum and difference of the same two terms. Thus,

$$(a + b)(a - b) = a^2 - b^2.$$

Factoring the difference of two squares is done by taking the square root of each square and making the factors the sum and difference of these roots.

$$4n^2 - 36 = (2n + 6)(2n - 6)$$

This special product may also be used in arithmetical problems. If the product of 31 and 29 be desired, it may be considered as the product $(30 + 1) \times (30 - 1) = 900 - 1 = 899$.

SOLVING EQUATIONS

The two parts of the *equation* separated by the equality sign are usually called the *members;* either *left-hand* and *right-hand* member or *first* and *second* member, respectively. Any operation of algebra may be performed on one member of an equation as long as the same operation is performed on the other member. However, division by 0 is not possible. That is, the *laws of operation* state that if *equals* are added to, or subtracted from, or multiplied by, or divided by *equals*, the results are *equal*.

Example: Suppose an equation

$$\lambda = \frac{300,000}{f}$$

is given and λ (read lambda) is given as 600; how may f be found?

Solution: First multiply by f on both sides of the equal sign,

$$f\lambda = \frac{300,000f}{f},$$

and then divide both members of the equation by λ:

$$f\frac{\lambda}{\lambda} = (300,000)\frac{f}{f\lambda};$$

but $\frac{\lambda}{\lambda} = 1$ and $\frac{f}{f} = 1$, so that

$$f = \frac{300,000}{\lambda}.$$

Substitute the value of $\lambda = 600$ in this equation and obtain

$$f = \frac{300,000}{600} = 500.$$

An important formula in radio is Ohm's law, usually written in symbols as $E = IR$. The meaning of the symbols will be explained later, but this equation will serve as another example of the operations which may be performed on an equation.

Following the laws of operation, let both sides be divided by I. That is,

$$\frac{E}{I} = \frac{IR}{I} = R, \quad \text{or} \quad R = \frac{E}{I}.$$

Again, let both sides be divided by R, and obtain

$$\frac{E}{R} = \frac{IR}{R} = I, \quad \text{or} \quad I = \frac{E}{R}.$$

Still another change may be made; suppose both sides of the equation are multiplied by I; then

$$EI = (I)(IR) = I^2R.$$

Just above, it is shown that $I = \frac{E}{R}$; substituting this value for I in the last equation makes it

$$E\frac{E}{R} = \frac{E^2}{R} = EI = I^2R.$$

The rule for the last operation is that *things equal to the same thing are equal to each other*.

Sometimes it is not possible to evaluate an equation exactly because some factors in it are not exact; for example, a formula containing $\pi = 3.1416$ (approximately), such as

$$X_L = 2\pi f L \quad \text{or} \quad X_C = \frac{1}{2\pi f C}.$$

In this case, the number used for π should have the same number of significant figures as the rest of the data; 6.28 is often used for 2π.

Clearing fractions is often necessary in solving equations. Consider the equation

$$R = \frac{1}{\frac{1}{R_1} + \frac{1}{R_2}}$$

The first step is to place the denominator over its own common denominator, $R_1 R_2$:

$$R = \frac{1}{\frac{R_2 + R_1}{R_1 R_2}}.$$

Finally, the denominator is inverted and multiplied by the numerator, which in this case is 1:

$$R = \frac{R_1 R_2}{R_2 + R_1}(1) = \frac{R_1 R_2}{R_1 + R_2}.$$

QUADRATIC EQUATIONS

An equation of the type

$$ax^2 + bx + c = 0$$

is called a *quadratic equation*. The coefficients a, b, c are any numbers, positive or negative, and x is a variable whose value is to be found. Special equations of this type have already been considered on p. 28, but sometimes it is difficult to find the necessary factors. *To solve the quadratic equation*, that is, to find a formula which will always give the correct value of x, first divide the entire equation by a, the coefficient of x^2, to give

$$x^2 + \frac{b}{a}x + \frac{c}{a} = 0.$$

Now subtract the term $\frac{c}{a}$ from both sides of the equation:

$$x^2 + \frac{b}{a}x = -\frac{c}{a}.$$

The third term in a perfect square is the square of half the coefficient of x. Performing this operation, and adding this term to both sides of the equation, we get

$$x^2 + \frac{b}{a}x + \frac{b^2}{4a^2} = \frac{b^2}{4a^2} - \frac{c}{a}.$$

The square root of both sides may now be taken. The left-hand side has now been made into a perfect square, so it may be factored; on the right-hand side the operation can only be indicated. Since both $+$ and $-$

quantities have the same sign when squared (rule of signs), both signs must be used before the square-root sign or *radical* ($\sqrt{\ }$) on the right. So

$$x + \frac{b}{2a} = \pm \sqrt{\frac{b^2 - 4ac}{4a^2}},$$

or $\quad x = \dfrac{-b + \sqrt{b^2 - 4ac}}{2a} \quad$ and $\quad x = \dfrac{-b - \sqrt{b^2 - 4ac}}{2a}.$

The process used to work out this formula is known as "completing the square." Now an example of its use will be given.

Consider the equation

$$4x^2 + 12x + 9 = 0.$$

This is a quadratic form, with $a = 4$, $b = 12$, and $c = 9$. Putting these figures in the formula

$$x = \frac{-12 \pm \sqrt{(12)^2 - 4(4)(9)}}{2(4)}$$

$$= \frac{-12 \pm \sqrt{144 - 144}}{8} = -\frac{12}{8} = -\frac{3}{2}.$$

Since equations do not always yield a solution as readily as the one above, consider another example. Given the equation

$$2x^2 + 5x - 12 = 0,$$

it is seen that $a = 2$, $b = 5$, $c = -12$. Substituting these values in the formula

$$x = \frac{-5 \pm \sqrt{5^2 - 2(4)(-12)}}{2(2)} = \frac{-5 \pm \sqrt{25 + 96}}{4}.$$

With the plus sign,

$$x = \frac{-5 + 11}{4} = \frac{6}{4} = \frac{3}{2}.$$

With the minus sign,

$$x = \frac{-5 - 11}{4} = \frac{-16}{4} = -4.$$

As a check on the result, substitute each of the values of x in the given equation:

$x = -4$	$x = 1\frac{1}{2} = \frac{3}{2}$
$2(-4)^2 + 5(-4) - 12 = 0$	$2(\frac{3}{2})^2 + 5(\frac{3}{2}) - 12 = 0$
$2(16) - 20 - 12 = 0$	$2(\frac{9}{4}) + \frac{15}{2} - 12 = 0$
$32 - 32 = 0$	$12 - 12 = 0$
$0 = 0$	$0 = 0$

Chapter 1   **Mathematics of Radio and Electronics**   33

TRIGONOMETRY

Trigonometry is that branch of mathematics which deals with the properties of triangles. Actually, the ideas of trigonometry have been extended to solve many problems in radio, as will be seen later.

A *right triangle* is a triangle in which one of the angles is a *right angle* or 90°, like the triangle in Fig. 1-7. For convenience, the angles are marked with the capital letters A, B, C (with C at the right angle) and the side opposite each angle is marked with the corresponding small letter a, b, c.

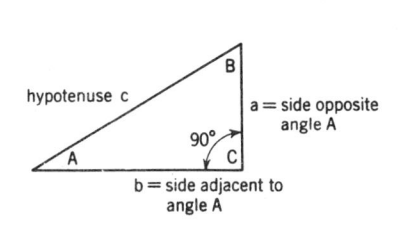

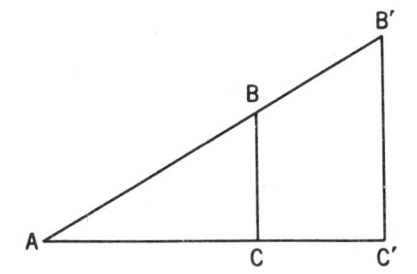

FIG. 1-7. Right triangle with parts labeled.

FIG. 1-8. Triangles ABC and $AB'C'$ are similar.

In geometry it is shown that the sum of the angles in any triangle is 180°; that is, $A + B + C = 180°$. But since $C = 90°$, $A + B = 180° - C = 180° - 90° = 90°$. Therefore A and B each must be less than 90° (*acute angles*) and the following relations must be true:

$$A + B = 90°; \quad A = 90° - B; \quad B = 90° - A.$$

Now consider two triangles like ABC and $AB'C'$ in Fig. 1-8, one of which is larger than the other. In geometry it is shown that these two triangles, ABC and $AB'C'$, are *similar*, that is, their angles are equal and their sides are proportional to one another. Therefore, it may be said that

$$\frac{BC}{AC} = \frac{B'C'}{AC'} = \text{a constant.}$$

No matter what the size of triangle $AB'C'$, as long as the shape is the same as that of the triangle ABC this ratio will be true. If the angles were changed, the ratio would change also. There is a definite relation between an acute angle of a right triangle and the ratios of the sides.

To talk about these ratios (also called *trigonometric functions*), it is handy to name them according to the following definitions, which refer to Fig. 1-7.

If A is an acute angle in any right triangle, then by definition:

$$\text{sine of } A = \frac{\text{length of side opposite angle } A}{\text{length of hypotenuse}} = \sin A = \frac{a}{c};$$

$$\text{cosine of } A = \frac{\text{length of side adjacent to angle } A}{\text{length of hypotenuse}} = \cos A = \frac{b}{c};$$

$$\text{tangent of } A = \frac{\text{length of side opposite angle } A}{\text{length of side adjacent to angle } A} = \tan A = \frac{a}{b};$$

$$\text{sine of } B = \frac{b}{c};$$

$$\text{cosine of } B = \frac{a}{c};$$

$$\text{tangent of } B = \frac{b}{a}.$$

The abbreviations sin, cos, and tan are commonly used in place of the corresponding complete names sine, cosine and tangent. Other ratios may be formed but are not needed here. It is important to remember that these ratios depend only upon the angle in the right triangle and not upon the size of the triangle; the ratios are numbers.

It may be shown that in any *right triangle* the square on the *hypotenuse* is equal to the *sum of the squares* on the *two sides;* this generalization is called the *hypotenuse rule* or the *Theorem of Pythagoras.* That is,

$$a^2 + b^2 = c^2,$$

from which

$$a = \sqrt{c^2 - b^2} \quad \text{and} \quad b = \sqrt{c^2 - a^2}.$$

Values for the ratios or *trigonometric functions* may be worked out for certain angles quite easily. Consider a triangle with three equal sides, each one unit in length. Since the sides are equal the angles opposite them must be equal. The sum of the three equal angles is 180° and so each angle

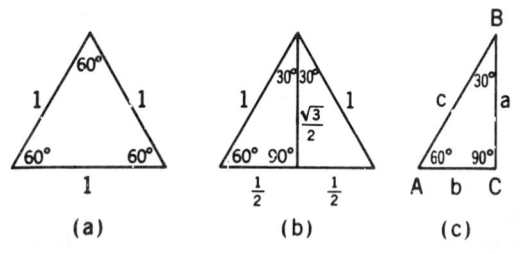

FIG. 1-9. Equilateral triangle and 30°-60° triangle.

must be 60°, as shown in Fig. 1-9. A line drawn from the top of the triangle to the center of the lower side will divide the lower side (or base) into two portions, each $\frac{1}{2}$ unit long, and will divide the triangle into two portions, each of which is a right triangle. Taking one portion of the triangle (Fig. 1-9c) and lettering the angles and sides as stated at the beginning of this section, the ratios can be evaluated for the angles in this triangle. From the hypotenuse rule the side a is

Chapter 1   **Mathematics of Radio and Electronics**   35

$$a = \sqrt{c^2 - b^2} = \sqrt{1^2 - (\tfrac{1}{2})^2} = \sqrt{1 - \tfrac{1}{4}} = \sqrt{\tfrac{3}{4}} = \frac{\sqrt{3}}{2} = 0.87.$$

From the definition,

$$\sin A = \frac{a}{c} = \frac{0.87}{1} = 0.87 = \frac{\sqrt{3}}{2} = \sin 60°.$$

From the definition,

$$\cos A = \frac{b}{c} = \frac{1}{2} \div 1 = 0.50 = \cos 60°.$$

Likewise,

$$\tan A = \frac{a}{b} = \frac{\sqrt{3}}{2} \div \frac{1}{2} = \sqrt{3} = 1.73 = \tan 60°.$$

From the fact that angle $A = 60°$ it is found that $\sin 60° = 0.87$, $\cos 60° = 0.50$, and $\tan 60° = 1.73$.

The same triangle may be used to get the values for the ratios or trigonometric functions for 30°. The side b is opposite angle B and the side a is adjacent to angle B; angle B is 30°. From the definition,

$$\sin B = \sin 30° = \frac{b}{c} = 0.50;$$

also,

$$\cos B = \cos 30° = \frac{a}{c} = 0.87;$$

and

$$\tan B = \tan 30° = \frac{b}{a} = 0.58.$$

Tables like Table 1-8 are available to solve problems. To use the table, proceed as follows:

(1) To find the functions of 37°: go down the left-hand column to 37° and on this line find $\sin 37° = 0.6018$, $\cos 37° = 0.7986$, and $\tan 37° = 0.7536$.

(2) To find the angle whose cosine is 0.8988: go down the column of cosines until 0.8988 is found and then proceed to the left on this line to find 26°; thus $\cos 26° = 0.8988$.

(3) To find $\cos 16.5°$ (16°30′): This cosine is not listed, but since 16.5° lies 0.5 or one half the interval between 16° and 17°, the cosine of 16.5° may be assumed to lie 0.5 of the interval between $\cos 16°$ and $\cos 17°$. To calculate $\cos 16.5°$ it is necessary to subtract the proportional part of the difference from the value of $\cos 16°$ because the cosines decrease as the angle increases. This operation is called *interpolation*, and is performed as follows:

(a) $\cos 16° = 0.9613$ (d) $\cos 16° = 0.9613$
(b) $\cos 17° = 0.9563$ (e) $0.5 \times 0.0050 = \overline{0.0025}$
(c) Difference $= \overline{0.0050}$ (f) $\cos 16.5° = \overline{0.9588}$

TABLE 1-8
SINES, COSINES, AND TANGENTS

Degrees	Sine $\left(\dfrac{opp.}{hyp.}\right)$	Cosine $\left(\dfrac{adj.}{hyp.}\right)$	Tangent $\left(\dfrac{opp.}{adj.}\right)$	Degrees	Sine $\left(\dfrac{opp.}{hyp.}\right)$	Cosine $\left(\dfrac{adj.}{hyp.}\right)$	Tangent $\left(\dfrac{opp.}{adj.}\right)$
0	.0000	1.0000	.0000	45	.7071	7071	1.0000
1	.0175	.9998	.0175	46	.7193	.6947	1.0355
2	.0349	.9994	.0349	47	.7314	.6820	1.0724
3	.0523	.9986	.0524	48	.7431	.6691	1.1106
4	.0698	.9976	.0699	49	.7547	.6561	1.1504
5	.0872	.9962	.0875	50	.7660	.6428	1.1918
6	.1045	.9945	.1051	51	.7771	.6293	1.2349
7	.1219	.9925	.1228	52	.7880	.6157	1.2799
8	.1392	.9903	.1405	53	.7986	.6018	1.3270
9	.1564	.9877	.1584	54	.8090	.5878	1.3764
10	.1736	.9848	.1763	55	.8192	.5736	1.4281
11	.1908	.9816	.1944	56	.8290	.5592	1.4826
12	.2079	.9781	.2126	57	.8387	.5446	1.5399
13	.2250	.9744	.2309	58	.8480	.5299	1.6003
14	.2419	.9703	.2493	59	.8572	.5150	1.6643
15	.2588	.9659	.2679	60	.8660	.5000	1.7321
16	.2756	.9613	.2867	61	.8746	.4848	1.8040
17	.2924	.9563	.3057	62	.8829	.4695	1.8807
18	.3090	.9511	.3249	63	.8910	.4540	1.9626
19	.3256	.9455	.3443	64	.8988	.4384	2.0503
20	.3420	.9397	.3640	65	.9063	.4226	2.1445
21	.3584	.9336	.3839	66	.9135	.4067	2.2460
22	.3746	.9272	.4040	67	.9205	.3907	2.3559
23	.3907	.9205	.4245	68	.9272	.3746	2.4751
24	.4067	.9135	.4452	69	.9336	.3584	2.6051
25	.4226	.9063	.4663	70	.9397	.3420	2.7475
26	.4384	.8988	.4877	71	.9455	.3256	2.9042
27	.4540	.8910	.5095	72	.9511	.3090	3.0777
28	.4695	.8829	.5317	73	.9563	.2924	3.2709
29	.4848	.8746	.5543	74	.9613	.2756	3.4874
30	.5000	.8660	.5774	75	.9659	.2588	3.7321
31	.5150	.8572	.6009	76	.9703	.2419	4.0108
32	.5299	.8480	.6249	77	.9744	.2250	4.3315
33	.5446	.8387	.6494	78	.9781	.2079	4.7046
34	.5592	.8290	.6745	79	.9816	.1908	5.1446
35	.5736	.8192	.7002	80	.9848	.1736	5.6713
36	.5878	.8090	.7265	81	.9877	.1564	6.3138
37	.6018	.7986	.7536	82	.9903	.1392	7.1154
38	.6157	.7880	.7813	83	.9925	.1219	8.1443
39	.6293	.7771	.8098	84	.9945	.1045	9.5144
40	.6428	.7660	.8391	85	.9962	.0872	11.4301
41	.6561	.7547	.8693	86	.9976	.0698	14.3007
42	.6691	.7431	.9004	87	.9986	.0523	19.0811
43	.6820	.7314	.9325	88	.9994	.0349	28.6363
44	.6947	.7193	.9657	89	.9998	.0175	57.2900
45	.7071	.7071	1.0000	90	1.0000	.0000	.....

Mathematics of Radio and Electronics

(4) To find tan 35.66° (35°40'): Since the tangent of an angle increases as the angle increases, the proportional part of the difference is added to the value of tan 35° to give tan 35.66°.

    (a)   tan 36° = 0.7265   (d)   tan 35° = 0.7002
    (b)   tan 35° = 0.7002   (e)   0.66 × 0.0263 = 0.0175
    (c)   Difference = 0.0263   (f)   tan 35.66 = 0.7177

(5) Find angle A if sin A = 0.6626. From the table it is seen that A lies between 41° and 42°.

    (a)   sin 42° = 0.6691   (d)   sin A = 0.6626
    (b)   sin 41° = 0.6561   (e)   sin 41° = 0.6561
    (c)   Difference = 0.0130   (f)   Difference = 0.0065

    (g) Since A is between 41° and 42°, $A = 41° + \dfrac{0.0065}{0.0130} \times 1°$; The 1° is the difference between 41° and 42°.

    (h) $A = 41° + 0.5° = 41.5°$ or $41° + 30' = 41°30'$.

Another system for finding the values of the trigonometric functions is to draw a curve of sines. Suppose a circle is drawn whose radius is 1 unit long (Fig. 1-10), with center at A. Then draw a diameter AD and continue

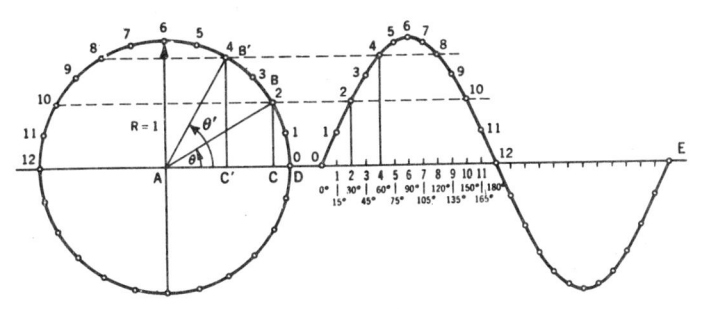

FIG. 1-10. Construction for sine curve.

this line to E; draw another line at a right angle to the diameter and to the end of the radius (CB). The triangle ACB is a right triangle and sin θ = CB/AB. The symbol θ is the Greek letter theta and is used as the symbol for any angle here. But $AB = 1$ (unity in length) and so sin θ = CB. If the radius were drawn in another position (AB'), the line from B' to the diameter at C' would again be sin $\theta' = C'B'$, and so on, no matter what the angle θ happens to be. If the entire circle be divided into parts (say 24) and the radius considered as if moving in a direction opposite to that of the clock hands it will occupy each of these positions in succession. Divisions are marked on the line DE with the numbers of degrees corresponding to the positions of the radius. By drawing lines horizontally from the end of the radius in the various positions to the vertical lines as shown, a curve of sines, or a *sine curve*, will be formed when the points are joined by a smooth curve.

Other information may be obtained from such a diagram. Following the lines used in making the diagram, it may be seen that sin 120° = sin 60°, sin 150° = sin 30°, and so on. This suggests a general statement that sin (180° − A) = sin A. Many similar rules have been worked out.

In Fig. 1-10, cos θ = AC. If the radius were again rotated, and the lengths AC, AC', and so forth, plotted vertically above the degree marks on DE, a *cosine curve* would be obtained. The two curves may be compared; the shape is exactly the same, but when the sine curve is zero the cosine curve has the value 1, and so on. That is, the cosine curve would be the same as the sine curve but shifted forward by 90°.

For some purposes it is convenient to use a special unit to measure angles instead of the usual degrees. If the radius of a circle is bent around the circumference of the circle like a flexible rule, it will fit 2π times without overlapping (π = 3.1416, approximately). That is, 2π radii will be needed to make a curved line as long as the circumference. The relation is true no matter what the size of the circle. The radius, wrapped around the circle, will be the curved edge of a piece of pie, with an angle at the center (Fig. 1-11). This angle is called a *radian*. One radian is about 57.3°. There will

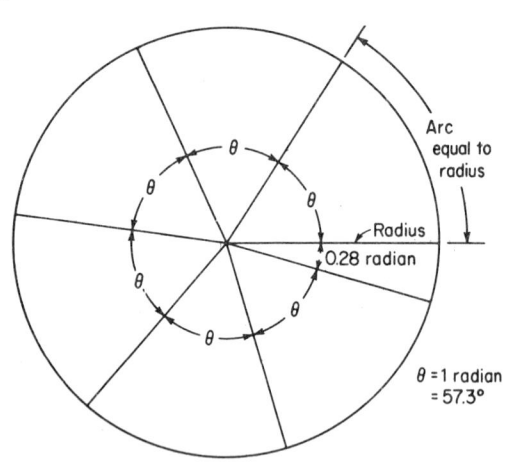

Fig. 1-11. Measure of angle in radians.

be 2π radians in the whole circumference or 360°. To *convert degrees* to *radians*, divide the number of degrees by 57.3; to *convert radians* to *degrees*, multiply the number of radians by 57.3. When we said that a cosine curve is similar to a sine curve except for a shift of 90°, we could have referred instead to a shift of $\pi/2$ radians; this is expressed in mathematical symbols as cos θ = sin ($\theta + \pi/2$).

Chapter 1   *Mathematics of Radio and Electronics*   39

VECTOR ADDITION

Some relationships to be explained later are indicated in the triangle in Fig. 1-12.

$$\frac{R}{Z} = \cos \theta; \quad R = Z \cos \theta; \quad Z = \frac{R}{\cos \theta};$$

$$\frac{X}{Z} = \sin \theta; \quad X = Z \sin \theta; \quad Z = \frac{X}{\sin \theta};$$

$$Z = \sqrt{R^2 + X^2}; \quad \tan \theta = \frac{X}{R}.$$

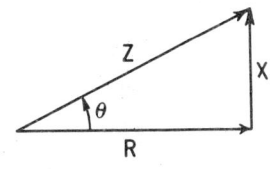

FIG. 1-12. Impedance triangle with parts marked.

The symbol θ is the Greek letter theta.

Suppose it is stated that $R = 8$ and $X = 6$, which are to be combined to give $Z = 10$. Certainly these cannot be added by arithmetic because this would make the sum $8 + 6 = 14$. Quantities which have a definite direction as well as size, such as **R**, **X**, **Z**, are called *vectors* or, in certain electrical quantities, *phasors*, and are added by *vector*, or *phasor, addition*. To add **R** to **X** in this fashion, place them as shown in the triangle (the arrowhead indicating the direction of each) and draw a line from the beginning of **R** to the end of **X**. This will be **Z**, the *vector sum* of **R** and **X**, and its length is $Z = \sqrt{R^2 + X^2} = \sqrt{8^2 + 6^2} = 10$, as required.

It is customary to set vectors in boldface type. In writing, a short bar, as $\overline{X}$, over the vector distinguishes it from nonvector expressions, as X.

GRAPHS AND CURVES

The relationship between two quantities which vary in some interdependent fashion is often displayed by graphs or curves. This display may be as simple as the line used at the beginning of this chapter to show the one-to-one correspondence between numbers and line segments. Often a set of *rectangular coordinates* is used, and the directions are labelled x and y. Again, velocity may be related to time and distance by a graph. For example, if an object travels at a speed of 3 meters per second, then a distance of 3 meters will be covered in one second, a distance of 6 meters in 2 seconds, a distance of 9 meters after a lapse of 3 seconds, and so on.

Rather than write this *functional relationship* in words it is much more convenient to show it by means of a graph, as in Fig. 1-13. This curve is plotted from the statement that distance is the product of speed and time. Time units are marked out as *abscissas* (horizontally) and there is a one-to-one correspondence between time intervals and distance intervals. The latter are indicated as *ordinates* (vertically) on the graph. A line is drawn across the graph to indicate the relation between time intervals and distance intervals. For example, at the point a (2.5 seconds) the dashed line may be followed up to the line or curve OG; another dashed line now leads to the left to mark the point 7.5 meters. This is the distance that the object travels in 2.5 seconds. A relationship which can be represented by a straight line, as in this instance, is said to be *linear*. Other relationships exist which may be identified by the special sort of curvature which the line

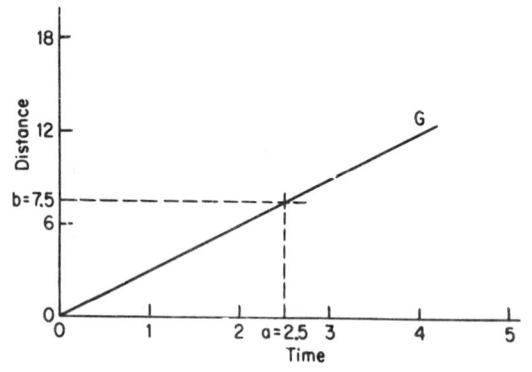

FIG. 1-13. Distance and time of travel.

describing them has; others may be quite irregular, depending upon the data being plotted. Often the data are obtained from observation or experiments in the physical world.

RATES OF CHANGE

The concept of *rate of change* (which is also called a *derivative*) is very important. The rate of change of distance with respect to time (such as speed in miles per hour) may be found from the curve in Fig. 1-13. For example, the distance an object travels, say 12 meters, in a given time (4 seconds — from Fig. 1-13) may be represented as a ratio.

Distance/time = 12/4 = 3 meters per second, which is the speed, or rate of change of distance with respect to time. Taking another point, say

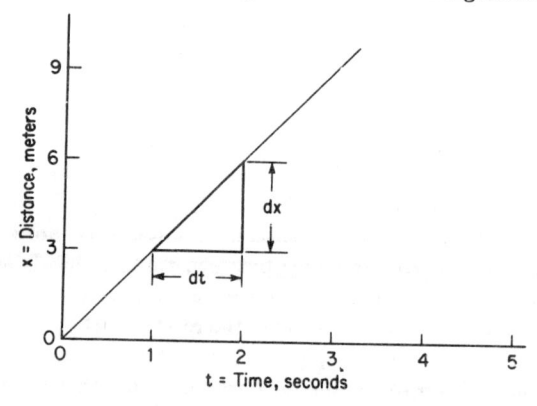

FIG. 1-14. Constant rate of change.

18 meters, and dividing by the corresponding time (6 seconds) gives
$$\frac{\text{distance}}{\text{time}} = \frac{18}{6} = 3 \text{ meters per second.}$$
In this special case the rate of change of distance with respect to time is a constant and may be plotted as a graph like that shown in Fig. 1-14.

The same ideas may be expressed by symbols and made more general in application. Suppose that the distance in the example is represented by the symbol x and time by the symbol t. Next select a particular distance dx in Fig. 1-14, which is taken to be $6 - 3 = 3$ meters. The corresponding time interval is called dt, and is found from Fig. 1-14 as $2 - 1 = 1$ second. The quotient dx/dt represents the rate of change of distance with respect to time, and in this instance,
$$\frac{dx \text{ (meters)}}{dt \text{ (seconds)}} = \frac{3 \text{ meters}}{1 \text{ second}} = 3 \text{ meters per second.}$$
This example is one in which dx/dt is a constant, since it has the same value for any point along the curve relating distance to time. The ratio dx/dt is called the *slope* of the curve, a constant in this case.

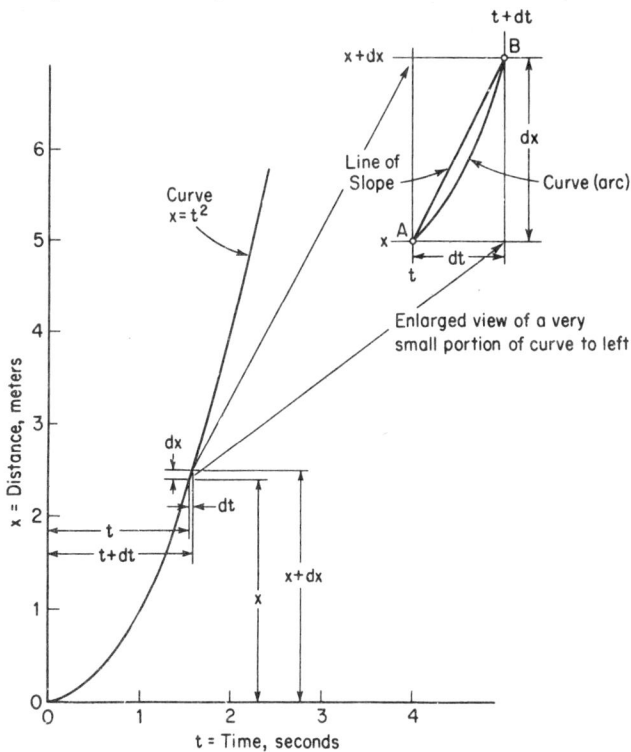

Fig. 1-15. Curve for defining derivative.

42 **Mathematics of Radio and Electronics** Chapter 1

The relationship between two varying quantities is not necessarily the same at all times. Figure 1-15 shows a curve of this sort, where doubling the time causes a fourfold increase in distance, multiplying the time by 3 makes the distance increase ninefold, and so on. This may be stated symbolically by the relation

$$x = t^2. \tag{1.1}$$

It is no longer possible to relate the distance intervals to the time intervals simply by dividing difference of distance by the corresponding difference of time; the slope of the curve is constantly changing. It is still possible to find the slope at any point by a process called finding the derivative of a function.

Figure 1-15 shows in the upper right-hand corner a small portion of the curve to an enlarged scale. It can be seen that along a very short length of the curve (or arc) the curvature of the arc is so small that it nearly coincides with the straight line drawn between points A and B. The distance dx now is a small change or *increment* in the increasing-x direction and dt is a corresponding small change or increment in the increasing-t direction, both being measured from the point A. The point A is described by the values x and t on the curve. The point B is described by the values $x + dx$ and $t + dt$. Putting these new values into equation 1.1 gives

$$x + dx = (t + dt)^2 = (t + dt)(t + dt) = t^2 + 2t\,dt + (dt)^2. \tag{1.2}$$

Since dt is only a very small fraction of t, the quantity $(dt)^2$ is a still smaller fraction of t, and since dt can be made as small as desired, the term $(dt)^2$ can be dropped from this equation without causing any appreciable error. This gives

$$x + dx = t^2 + 2t\,dt.$$

Now to leave dx by itself on the left-hand side of this equation, subtract x from the left-hand side and its equivalent t^2 from the right-hand side to get

$$dx = 2t\,dt.$$

This operation is equivalent to getting the distance dx and the time dt in Fig. 1-15, from which the slope is found to be dx/dt or, in this case, by dividing both sides by dt,

$$dx/dt = 2t.$$

The process just outlined is called finding the *derivative* of a function.

It should be noted that both dx and therefore dt may be made smaller and smaller until finally (in the limit) each approaches zero in value. However, the quotient (ratio) dx/dt still has meaning and a value, and this value is precisely the derivative or slope of the curve at a point along the curve.

Chapter 1 **Mathematics of Radio and Electronics** 43

While the example given uses distance (x) and time (t) as the quantities which vary, the same principles apply no matter what the variables in the problem happen to be. Time is perhaps the commonest *independent variable* in radio and electronic work. These ideas (functions, variation, rate of change) are extended over a vast field of quantities which vary in certain ways in physics, engineering, and radio.

Figure 1-16 shows a curve of $dx/dt = 2t$ which is the derivative (slope) of the curve $x = t^2$ (shown as a dashed line here). It is possible to repeat this operation; that is, to take a second derivative of a function. The

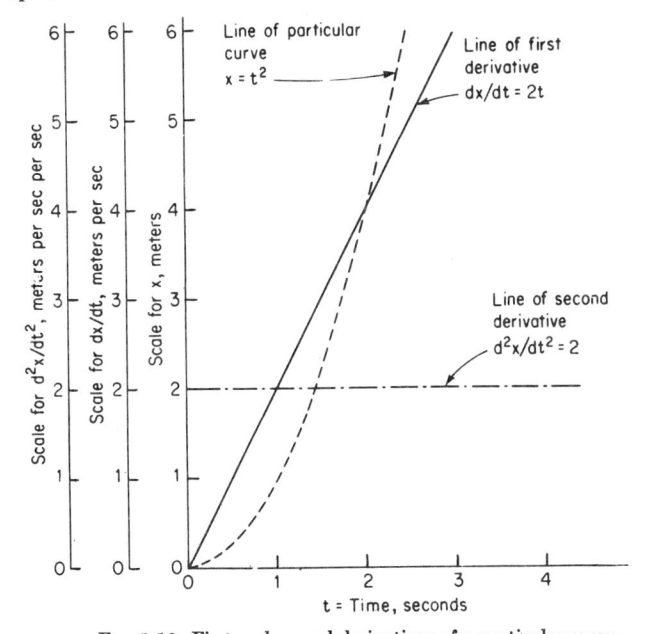

FIG. 1-16. First and second derivatives of a particular curve.

second derivative is written as d^2x/dt^2 which is read as "the second derivative of x with respect to t." It is actually the slope of the slope curve. Since the slope in this particular case is constant, the graph for the slope is simply a constant value of 2 (meters per second per second).

The derivative of a constant is zero since anything which is constant does not change and therefore can have no rate of change. We can take derivatives of any function which has a derivative until its rate of change is a constant. Beyond this we cannot go, because all further derivatives have zero as a value.

INTEGRALS

The process of differentiation (finding the rate of change) can be done in reverse; that is, the problem "given a rate of change, what is the function of which this is the rate of change?" can often (but not always) be solved. The function which represents the answer to this question is called an *integral* or *anti-derivative* and finding the answer is called *integration*.

Suppose that the function $dx/dt = t^2$ is given as a derivative (rate of change) of an unknown function; what is this unknown function? From our experimentation with derivatives we might assume that it is t^3. However, taking the derivative of this function results in $3t^2$. This is not what was assumed as the derivative but is different from it only in having the factor 3. If we try again with the assumption that $t^3/3$ is the integral and take the integral of $t^3/3$ we get $3t^2/3$, which is t^2, precisely the function taken at the beginning of this paragraph as the function $dx/dt = t^2$. From this experiment we conclude that the integral of the function $dx/dt = t^2$ is, in fact, $t^3/3$.

However, if we start with the function $x = t^3/3 + 5$ and take its derivative, the result again is just t^2 because the derivative of a constant is zero. The answer to the question "given a rate of change (derivative), what is the function of which this is the rate of change?" is always uncertain to the extent of a constant which must be added. The constant itself must be determined by other conditions in the original problem.

Certain symbols are universally used to indicate these operations. In the preceding example, the problem is stated by the notation

$$x = \int t^2 \, dt$$

and the answer is $x = t^3/3$.

The sign $\int$ is read "the integral of" and has the effect of canceling the operation of taking a derivative.

Mathematics which is concerned mostly with finding integrals and with their applications is called *integral calculus*.

THE EXPONENTIAL e^t

It has already been shown that 10 is used as the base for the decimal number system and as a base for logarithms ($\log_{10}$), and 2 as a base for a binary system of numbers. The symbol e, or sometimes ϵ, is used to represent a very special number which is used as a base for another system of logarithms (ln) and the function e^t (sometimes called the "growth function") has important properties which are useful in electronic work. The number is $e = 2.71828\ldots$ to five-place accuracy.

One way to define e^t is by means of a sum of a series of an infinite number of terms formed in a particular fashion. Such a series is

$$e^t = 1 + t + \frac{t^2}{2!} + \frac{t^3}{3!} + \frac{t^4}{4!} + \cdots + \frac{t^n}{n!}$$
$$\text{a}\text{b}\text{c}\text{d}\text{e}\text{f}\text{g}$$

Chapter 1   **Mathematics of Radio and Electronics**   45

where (a) t represents any number;
(b) is the constant term in the series;
(c) $2! = 1 \times 2 = 2$ and is called "2 factorial";
(d) $3! = 1 \times 2 \times 3 = 6$ and is called "3 factorial";
(e) $4! = 1 \times 2 \times 3 \times 4 = 24$ and is called "4 factorial";
(f) "$\cdots$" means that as many terms of similar form may be written in place of this symbol as desired;
(g) the last term is the "general term" and shows how any term following the constant term is formed.

As an illustration this series may be used to calculate the value of e^t to any desired accuracy. For example, suppose that we wish to know the value of e^t, a number, when $t = 1$. Replacing t by this value (1) everywhere in the series gives the following result. The computation is arranged in columns for convenience in carrying out the work.

Term in Series	Replace t by 1	Column for Addition
1	1	1.00000
t	1	1.00000
$t^2/2!$	1/2	0.50000
$t^3/3!$	1/6	0.16667
$t^4/4!$	1/24	0.04167
$t^5/5!$	1/120	0.00833
$t^6/6!$	1/720	0.00139
$t^7/7!$	1/5040	0.00019
$t^8/8!$	1/40320	0.00002
$\cdots$		2.71827

This number, $e = 2.71828 \cdots$ to five decimal places, is the base of the "natural" or Naperian logarithms. The computation carried out here appears to be in error only in the fifth decimal place. This error can be reduced by using more decimal places in each of the terms, and more terms.

For later work we need to know the value of the derivative of e^t. Writing the series again

$$e^t = 1 + t + \frac{t^2}{2!} + \frac{t^3}{3!} + \frac{t^4}{4!} + \cdots.$$

It is permissible in this case to take the derivative of each term and use the results to form a new series. Doing this gives

$$\underset{a}{\frac{d(e^t)}{dt}} = \underset{b}{0} + \underset{c}{1} + \underset{d}{\frac{2t}{2!}} + \underset{e}{\frac{3t^2}{3!}} + \underset{f}{\frac{4t^3}{4!}} + \cdots.$$

where (a) $d(e^t)/dt$ may be read "the derivative of e to the t power with respect to t";
(b) the derivative of a constant (that is, 1) is zero;
(c) the derivative of t with respect to t is 1;

46 *Mathematics of Radio and Electronics* Chapter 1

(d) the derivative of t^2 is $2t$ and the 2 cancels the 2 in the denominator to give t as the value of this term;

(e) the derivative of t^3 is $3t$; the 3 cancels the 3 in the denominator which is $3! = 1 \times 2 \times 3$ and gives $t^2/2$ as the value of this term;

(f) the derivative of t^4 is $4t^3$, the 4 cancels the factor 4 in the denominator which is $4! = 1 \times 2 \times 3 \times 4$, and gives $t^3/3!$ as the value of this term.

Collecting these terms and rewriting the result gives

$$\frac{d(e^t)}{dt} = 1 + t + \frac{t^2}{2!} + \frac{t^3}{3!} + \cdots,$$

which is precisely the series for e^t itself. This property — the derivative of a function is equal to the function itself — is one of the special attributes of e^t.

SINES, COSINES, AND THEIR DERIVATIVES

When an angle θ is expressed in radian measure it is possible to define cosine θ by a series with an infinite number of terms and use as many of these terms as we desire to calculate a value for the cosine. This series is

$$\cos \theta = 1 - \frac{\theta^2}{2!} + \frac{\theta^4}{4!} + \frac{\theta^6}{6!} + \cdots.$$

As an example, assume that the angle $\theta = 1$ radian, which is about $57.3°$. Substituting this value (1) in the series and arranging in columns for convenience in adding we have:

Term in Series	Replace θ by 1	Columns for Addition	
		(+)	(−)
1	1	1.00000	
$\theta^2/2!$	1/2		0.50000
$\theta^4/4!$	1/24	0.04167	
$\theta^6/6!$	1/720		0.00139
$\theta^8/8!$	1/40320	0.00002	
No further terms come within the five decimal places used:		1.04169	0.50139 *Sums of columns*
		0.50139	
Difference between sums of columns:		0.54030	

From a table we find that the value of cos 1 radian = 0.54030 to five decimal places; this is verified exactly by the calculation above.

The defining series for sine with θ in radian measure is

$$\sin \theta = \theta - \frac{\theta^3}{3!} + \frac{\theta^5}{5!} - \frac{\theta^7}{7!} + \cdots.$$

It is permissible to take the derivative of each term in this series and combine the results to form a new series for the derivative of sin θ. Thus

$$\underset{a}{\frac{d(\sin \theta)}{d\theta}} = \underset{b}{1} - \underset{c}{\frac{3\theta^2}{3!}} + \underset{d}{\frac{5\theta^4}{5!}} - \underset{e}{\frac{7\theta^6}{7!}} + \cdots,$$

where (a) $d(\sin \theta)/d\theta$ may be read "the derivative of sine θ with respect to θ";
(b) the derivative of a constant (1) is zero;
(c) the derivative of θ^3 is $3\theta^2$, the 3 cancels a 3 in the denominator which is $3! = 1 \times 2 \times 3$ to give $\theta^2/2!$ as the value of this term;
(d) the derivative of θ^5 is $5\theta^4$, the 5 cancels the 5 in the denominator which is $5! = 1 \times 2 \times 3 \times 4 \times 5$ to give $\theta^4/4!$ as the value of this term;
(e) the derivative of θ^7 is $7\theta^6$, the 7 cancels the 7 in the denominator which is $7! = 1 \times 2 \times 3 \times 4 \times 5 \times 6 \times 7$ to give $\theta^6/6!$ as the value of this term.

It may be seen when these terms are collected that this process gives just the series for the cosine or

$$\frac{d(\sin \theta)}{d\theta} = \cos \theta = 1 - \frac{\theta^2}{2!} + \frac{\theta^4}{4!} - \frac{\theta^6}{6!} + \cdots = \sin\left(\theta + \frac{\pi}{2}\right).$$

If the same process is used on the cosine series, the result will be

$$\frac{d(\cos \theta)}{d\theta} = -\sin \theta = \cos\left(\theta - \frac{\pi}{2}\right).$$

The fact that the derivative or rate of change of a sine function is a cosine function and vice versa (except for a negative sign) is of tremendous significance in electrical engineering. It will be shown later that the current through a capacitor is proportional to the time rate of change of the voltage across it. Hence, if an alternating voltage which varies in proportion to the sine of the time elapsing is impressed across a capacitor, the current which flows will vary in proportion to the cosine of the time elapsing.

CHAPTER 2

DIRECT-CURRENT CIRCUITS

INTRODUCTION

Radio and Electronics is the science of communication and control including the use of electromagnetic vibrations in space. It is the problem of the designers and operators of radio equipment to set up vibrations at the proper frequencies and of sufficient power so that they can be detected by electric circuits tuned to respond to these particular vibrations. The proper care and manipulation of the complicated radio and industrial electronic apparatus of the present time requires an extensive knowledge of electric circuits, vacuum tubes, and other equipment. The basis of all work in electronics is an understanding of the character of electric phenomena, particularly the way in which electric circuits respond to electric impulses.

One hundred years ago electricity was thought to be some peculiar kind of fluid which flowed in wires very much as water or oil flows in pipes. This idea permitted scientists to explain many of their experiments with electricity. As more and more new experiments were tried, the fluid theory did not continue to explain all the results that were obtained, and the new theories that evolved to explain the results became more and more complicated. At the present time so many different experiments have been tried, and the theories used to explain them have become so complicated, that much study involving higher mathematics is required to understand these advanced theories. Fortunately it is necessary for the electrical technician to understand only the more elementary of them in order to work effectively.

The old belief that electric current is a fluid flow caused by an electric pressure and opposed by a resistance in the wire is still very effective in solving many of the problems of the electrician and the radio operator. It is now known, however, that electric current is not a true fluid but that it consists of the drift of millions of negatively charged particles along a wire. These negatively charged particles are called *electrons* and are so extremely small that they flow through the spaces between the atoms of the conductor.

In most metals one or two electrons in the outer portion of the atom are very loosely bound to the nucleus of the atom. As a result large numbers of these electrons are free to drift about in the interatomic space. Materials of this type are called *conductors*. If the electrons are attracted by connecting the conductor to a battery they will accelerate because of the force of attraction, but the final speed attained by any individual electron will be small because it will bump into one of the many atoms before it has gone very far and bounce off in another direction. The energy which the electron had absorbed by reason of its acceleration is given up to the atom and appears in the form of heat.

Some substances have a molecular structure in which the electrons are so closely bound to the molecule that a very high electric pressure is required to tear them away. Substances of this type are called *insulators* and are used in electric circuits to keep the electron flow restricted to paths desired by the designer. Other substances, intermediate between conductors and insulators, are called semiconductors and are discussed in Chap. 4.

Although it is interesting to know that electrons float about inside metal conductors, a complete understanding is not essential to a workable knowledge of electric circuits. Occasional reference may be made to the above statements to give a clearer explanation of physical occurrences. The simple fluid theory, however, will be the basis for most of the following development.

ELECTRICAL QUANTITIES

In order to discuss electrical phenomena intelligently, it is necessary to define some electrical quantities and specify the units in which these quantities are measured. Four of these quantities and their units of measurement are given below. Other definitions will be added as needed.

The coulomb. The unit of electrical charge or the quantity of excess (or deficiency of) electrons is called the *coulomb*. It is the charge which would be obtained by collecting on or removing from a single charged body approximately 6.3 quintillion (6.3×10^{18}) free electrons. This is a large unit and is seldom used in elementary radio calculation. It is important, however, as a basis for other units.

The ampere. The unit of electric flow is called the *ampere* (amp). If one coulomb of charge passes a given point on a wire in one second, then one ampere is said to flow. In other words, the ampere is a special name given to a *coulomb per second*. It is specified by international agreement as the constant current which will deposit silver at the rate of 0.001118 gram per second, as this definition gives a means of obtaining a standard anywhere, while the counting of the electrons would be difficult. The conventional positive direction of current is opposite to the direction of electron flow, i.e., the electron is a *negative* charge.

The ohm. The unit of resistance to electric flow is called the *ohm* and it is specified by international agreement as the resistance which, at a temperature of 0°C, is offered to the flow of current by a column of mercury of uniform cross section, of a length of 106.3 cm, and having a mass of 14.45 g. The magnitude of the cross section so specified is essentially 1 sq mm.

The volt The unit of electric pressure is the *volt* (v) and is the pressure which will force one ampere to flow through a resistance of one ohm.

Some idea of the size of the coulomb may be gained from the fact that if a charge of 6.8 millionths of a coulomb were placed one foot from a similar charge there would be a repelling force of one pound acting between them. The ampere is most easily visualized by referring to a 100-watt electric light which takes about one ampere of current. The favorite illustration for the volt is that an ordinary dry cell has 1.5 v. The normal pressure for domestic electric service is 120 v. The resistance of an electric toaster or flat iron is about 25 ohms.

OHM'S LAW

In order to obtain a better understanding of electrical circuits, use will be made of the fluid analogy. In Fig. 2-1 a pump is shown driven by a motor. This pump is used to circulate oil* through a cooling coil

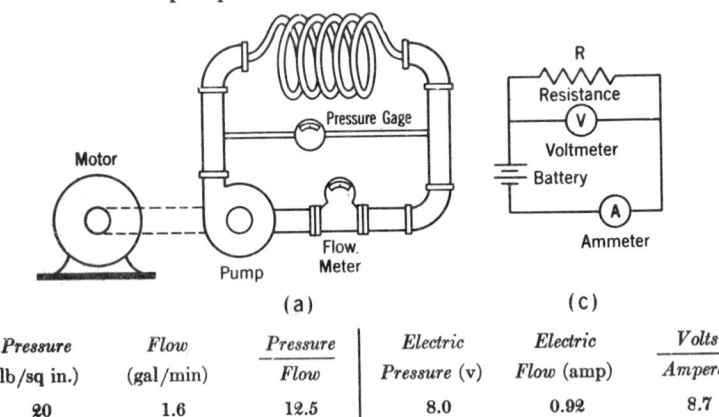

(a)

Pressure (lb/sq in.)	Flow (gal/min)	Pressure / Flow	Electric Pressure (v)	Electric Flow (amp)	Volts / Amperes
20	1.6	12.5	8.0	0.92	8.7
35	2.8	12.5	12.0	1.38	8.7
75	6.0	12.5	16.8	1.93	8.7
100	8.0	12.5	22.4	2.57	8.7
150	12.0	12.5	47.0	5.40	8.7

(b)   Fig. 2-1. Analogy between hydraulic and electric circuits.   (d)

of small copper tubing. A gauge is connected to the ends of the copper tubing to measure the difference in pressure across the coil, and a flow meter is inserted in the pipe to measure the rate at which the oil flows through the tubing.

If the speed of the pump is changed and if readings are taken of the pressure gauge and flow meter at each pump speed, a set of data will be obtained as shown in the table in Fig. 2-1(b). It is important to observe that at each pump speed the pressure divided by the flow gives the same

* Oil is used instead of water because it is a liquid of high viscosity and will obey Ohm's law of the hydraulic circuit.

result. In this case that value is 12.5. The flow at any pressure can, of course, be found by dividing the pressure by 12.5. If the flow is desired at some pressure other than those tested, it might also be obtained by dividing the pressure by 12.5.

Example: What is the flow at 50 lb per square inch pressure? According to the relation stated above,

$$\text{flow} = \frac{\text{pressure}}{12.5} = \frac{50}{12.5} = 4 \text{ gal per minute,}$$

pressure being in pounds per square inch. The constant 12.5* is characteristic of this particular size and length of tubing and so can be called the resistance of the coil of tubing.

To the right of this simple hydraulic circuit is shown a similar electric circuit. A battery supplies the electric pressure that causes an electric current to flow through a coil of copper wire, indicated as R. The meter used to measure the electric pressure in volts is called a *voltmeter*. The meter used to measure the current in amperes is called an *ammeter*. If taps are arranged on the battery so that different voltages may be applied to the coil of wire, then a set of readings of volts and corresponding amperes can be made. These readings would be comparable to the pressure and flow readings of the hydraulic circuit. In the electric circuit the voltmeter reading divided by the ammeter reading is always 8.7 and this constant is called the resistance. It is seen by this analogy that in the electric circuit also it is possible to predict the current flow with any given voltage. For instance, if the current corresponding to 65 v were desired, then

$$I = \frac{E}{8.7} = \frac{65}{8.7} = 7.5 \text{ amp,}$$

E being in volts. The resistance is measured in ohms. The formal statement of the relationship observed above is as follows:

The current in amperes is equal to the pressure in volts divided by the resistance in ohms.

This statement is known as *Ohm's law* and is the basis for much of electric-circuit theory. It may be expressed mathematically in three forms:

$$I = \frac{E}{R}, \qquad R = \frac{E}{I}, \qquad E = RI.$$

Although this is the general rule of behavior of electric circuits, there are many exceptions, and often the operation of important commercial equipment will depend upon these cases of unusual behavior.

---

* This constant depends also upon the viscosity of the fluid. In electricity the variable corresponding to viscosity does not occur.

SERIES CIRCUITS

Electric conductors may be connected following one another so that any current flowing through one must flow through the other. This is shown in Fig. 2-2. When circuits are connected in this manner the resistors are said to be connected in series. The combined or equivalent resistance of R_1 and R_2 connected as in Fig. 2-2 is

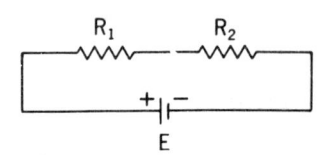

$$R_{total} = R_1 + R_2.$$

Fig 2-2. Electric circuit with resistors connected in series.

The electrical pressure across R_1, when added to the electrical pressure across R_2 will equal the total battery pressure.

According to Ohm's law the pressure or voltage across R_1 is $R_1 I$ and the voltage across R_2 is $R_2 I$. Since the current is the same in both resistors, the voltage across the individual resistors will be proportional to their resistances. Also, the proportion of the total voltage across R_1 will be $R_1/(R_1 + R_2)$. This relationship is used many times in radio, and such a combination of resistors to give a reduced voltage is known as a *potentiometer* or *voltage divider*.

Example: If a resistor of 100,000 ohms is connected in series with a resistor of 5,000 ohms across a 90-v battery (a) what current will flow? (b) what would the voltage across the 5,000-ohm resistor be if used to control a vacuum tube?

Solution: The equivalent resistance of the two resistors in series is

$$R_{total} = 100,000 + 5,000 = 105,000 \text{ ohms.}$$

By Ohm's law, the current will be

$$I = \frac{E}{R} = \frac{90}{105,000} = 0.00086 \text{ amp.}$$

The voltage across the 5,000-ohm resistor is

$$E = E_{battery} \frac{R_2}{R_1 + R_2} = 90 \frac{5,000}{105,000} = 4.28 \text{ v.}$$

Exercise 2-1. A vacuum-tube filament takes 0.9 amp at 6.3 v. (a) What is the resistance? (b) What additional resistance would be required if the filament were to be supplied from a 12-v battery?

Exercise 2-2. The heaters of five vacuum tubes are to be supplied in series. They require a current of 0.3 amp and each has a resistance of 21 ohms. What is the total resistance and what voltage must be used to supply the necessary current?

Exercise 2-3. How much resistance would be placed in series with a 50,000-ohm resistor to obtain a voltage of 8.4 across the 50,000-ohm resistor when a 45-v battery is the source of pressure?

Chapter 2   *Direct-Current Circuits*   53

PARALLEL CIRCUITS

Resistors in electric circuits may be connected in parallel as shown in Fig. 2-3. When resistors are connected in this manner the battery voltage is impressed across each resistor just as if the other resistor were not there. The current in each resistor is determined by Ohm's law as,

$$I_1 = \frac{E}{R_1}, \quad I_2 = \frac{E}{R_2}, \quad \text{and} \quad I_3 = \frac{E}{R_3}.$$

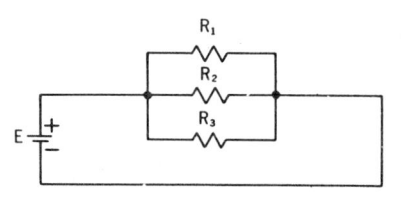

Fig. 2-3. Electric circuit with resistors connected in parallel.

The total current in the circuit is the sum of the currents in the individual resistors, so

$$I_{\text{total}} = I_1 + I_2 + I_3 = \frac{E}{R_1} + \frac{E}{R_2} + \frac{E}{R_3} = E\left(\frac{1}{R_1} + \frac{1}{R_2} + \frac{1}{R_3}\right)$$

The quantity $1/R$ is a constant called the *conductance* and is indicated by the symbol G. It is that characteristic of a resistor which when multiplied by the voltage gives the current. The unit of conductance is called the *mho*. This is recognized as the ohm spelled backwards and the name was chosen to be a reminder that the *mho* was the reciprocal of the ohm. Since the sum of the currents is equal to the total current, the sum of the conductances in a parallel circuit is equal to the total conductance.

$$G_{\text{total}} = G_1 + G_2 + G_3 \cdots.$$

The proportion of the total current which is flowing in any one resistor is the conductance of that circuit element divided by the total conductance of the circuits which are in parallel. Expressed mathematically, this is

$$\frac{I_1}{I_{\text{total}}} = \frac{EG_1}{E(G_1 + G_2 + G_3 \cdots)} = \frac{G_1}{G_1 + G_2 + G_3 \cdots}.$$

Example: Determine the equivalent resistance of the following four resistors connected in parallel. What proportion of the current will flow through the 8-ohm resistor?

$$R_1 = 20 \text{ ohms} \quad\quad R_2 = 25 \text{ ohms}$$
$$R_3 = 12 \text{ ohms} \quad\quad R_4 = 8 \text{ ohms}$$

Solution:

$$G_1 = \frac{1}{20} = 0.050 \quad\quad G_2 = \frac{1}{25} = 0.040$$

$$G_3 = \frac{1}{12} = 0.083 \quad\quad G_4 = \frac{1}{8} = 0.125$$

Total conductance is

$$G_{\text{total}} = G_1 + G_2 + G_3 + G_4 = 0.298.$$

Equivalent resistance is

$$R_{eq} = \frac{1}{0.298} = 3.35 \text{ ohms.}$$

Proportion of current flowing through the 8-ohm resistor is

$$\frac{I_{8\text{-ohm}}}{I_{total}} = \frac{G_4}{G_1 + G_2 + G_3 + G_4} = \frac{0.125}{0.298} = 0.418 = 41.8\%.$$

The special case where only two resistors are connected in parallel is so very common that it probably constitutes the majority of the problems in parallel circuits. In this special case

$$I_{total} = I_1 + I_2 = \frac{E}{R_1} + \frac{E}{R_2}$$

$$= E\left(\frac{1}{R_1} + \frac{1}{R_2}\right) = E\left(\frac{R_2 + R_1}{R_1 R_2}\right) = \frac{E}{R_1 R_2/(R_1 + R_2)}.$$

Therefore, $$R_{eq} = \frac{R_1 R_2}{R_1 + R_2}.$$

This equivalent resistance will act in every way just as the two resistors in parallel, and so it is very common for engineers to refer to the equivalent resistance of two resistors in parallel as the product of the resistances divided by the sum of the resistances. It should be remembered, however, that this formula applies only to the case of two resistors in parallel.

Exercise 2-4. Two resistors, one of 50 ohms and one of 20 ohms, are connected in parallel across a 100-v line. What is the total current and the equivalent resistance?

Exercise 2-5. Two circuit elements are connected in parallel across a 240-v line. One has a conductance of 0.0063 mho and the other a conductance of 0.0172 mho. What current will be taken from the line? What proportion of this current will flow through the element having a conductance of 0.0172 mho?

Exercise 2-6. A 90-v battery supplies a total current of 0.134 amp to two parallel resistance elements. If one of the resistance elements takes 0.039 amp, what is the resistance of each element?

Exercise 2-7. Five resistors are connected in parallel. What is the total current and the proportion of this current going through R_3 when a potential of 24 v is applied? The resistors are: $R_1 = 4$ ohms, $R_2 = 7$ ohms, $R_3 = 22$ ohms, $R_4 = 10$ ohms, $R_5 = 65$ ohms.

SERIES-PARALLEL CIRCUITS

Many times it is desirable to use combinations of series and parallel arrangements of resistors in radio equipment. The procedure used to solve circuits of this type is to combine the parallel resistors into an equivalent resistor and add this value of resistance to that of the series resistors. The

Chapter 2   *Direct-Current Circuits*   55

total current produced by the impressed voltage is then determined. This total current will divide in a parallel circuit in proportion to the conductances and so the current in any one of the resistors may be found.

Example: In the circuit shown in Fig. 2-4, determine the current in the 7-ohm resistor.

Solution:

$$G_2 = \frac{1}{10} = 0.100, \quad G_3 = \frac{1}{7} = 0.142, \quad G_4 = \frac{1}{12} = 0.083.$$

Equivalent conductance is

$$G_{eq} = 0.100 + 0.142 + 0.083 = 0.325.$$

Equivalent resistance is

$$R_{eq} = \frac{1}{0.325} = 3.08.$$

Total resistance is

$$R_t = 3.08 + 3 = 6.08.$$

Fig. 2-4. Electric circuit with resistors connected in series-parallel.

Total current is

$$I_t = \frac{E}{R_t} = \frac{45}{6.08} = 7.40.$$

Current in 7-ohm resistor is

$$I_{7\text{-ohm}} = \frac{G_3}{G_{eq}} I_t = \frac{0.142}{0.325} \times 7.40 = 3.23 \text{ amp.}$$

An alternate method of obtaining the current in any branch of a parallel circuit is to determine the voltage across the circuit from the equivalent IR drop and then divide this voltage by the resistance to obtain the current in that circuit element.

Example: In the circuit shown in Fig. 2-5 determine the current in the 10-ohm resistor.

Solution: Equivalent resistance of parallel circuit is

$$R_{eq} = \frac{10 \times 15}{10 + 15} = \frac{150}{25} = 6 \text{ ohms.}$$

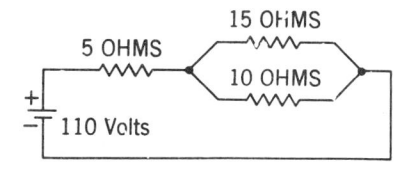

Fig. 2-5. Electric circuit with resistors connected in series-parallel.

Total resistance of circuit is

$$R_t = 5 + 6 = 11 \text{ ohms.}$$

Total current is

$$I_t = \frac{E}{R_t} = \frac{110}{11} = 10 \text{ amp.}$$

Volts across parallel resistors are

$$E_p = IR_{eq} = 10 \times 6 = 60.$$

Current through 10-ohm resistor is

$$I_{10\text{-ohm}} = \frac{E_p}{R_{10\text{-ohm}}} = \frac{60}{10} = 6 \text{ amp}.$$

Exercise 2-8. Determine the total current and the current in the 5-ohm resistor of the circuit of Fig. 2-6.

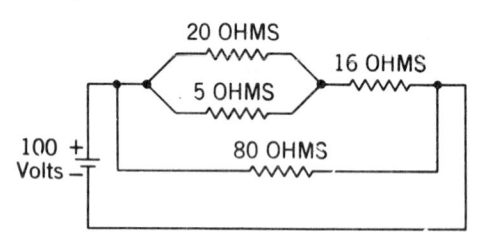

Fig. 2-6. The electric circuit for Exercise 2-8.

Exercise 2-9. Determine the total current and the current in the 25-ohm resistor of the circuit shown in Fig. 2-7.

Exercise 2-10. What is the difference of potential across ab and across bc of the circuit shown in Fig. 2-8?

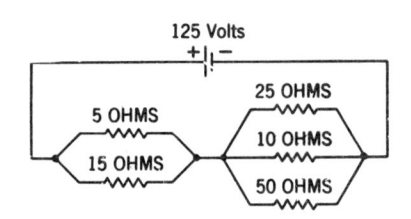

Fig. 2-7. The electric circuit for Exercise 2-9.

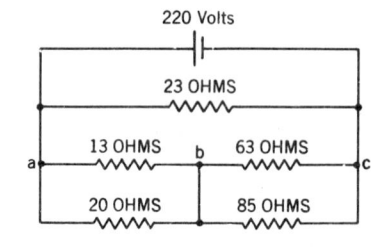

Fig. 2-8. The electric circuit for Exercise 2-10.

DETERMINATION OF RESISTANCE

Since copper wire is used so extensively in radio circuits it is important to know the resistance of various sizes of wires. This is given in Table 2-1 with several other items of information for various sizes of copper magnet wire.

Occasionally materials other than copper are used, and if the resistance relative to that of copper is known, the resistance of wires of these other materials may be found by multiplying the resistance of copper wire of the same size by the relative resistance. A list of relative resistances is given in Table 2-2.

Direct-Current Circuits

TABLE 2-1
PROPERTIES OF COPPER MAGNET WIRE

Size of wire (AWG)	Diameter (mils*)	Ohms per 1,000 ft at 20°C	Pounds per 1,000 ft	Diameter C† (mils)	Diameter E‡ (mils)	Diameter EC§ (mils)
6	162.0	0.3951	79.46	170.0		
8	128.5	0.6282	49.97	134.0	131.0	136.0
10	101.9	0.9989	31.43	107.0	104.0	109.0
12	80.81	1.588	19.77	85.8	83.0	88.0
14	64.08	2.525	12.43	69.1	66.1	71.1
16	50.82	4.016	7.818	55.8	52.6	57.6
18	40.30	6.385	4.917	45.3	42.0	47.0
20	31.96	10.15	3.092	37.0	33.5	38.0
22	25.35	16.14	1.542	29.4	26.8	31.3
24	20.10	25.67	1.223	24.1	21.3	25.8
26	15.94	40.81	0.7692	19.9	17.0	21.5
28	12.64	64.90	0.4837	16.6	13.6	17.6
30	10.03	103.2	0.3042	14.0	10.8	14.8
32	7.950	164.1	0.1913	12.0	8.75	12.8
34	6.305	260.9	0.1203	10.3	7.01	11.0
36	5.000	414.8	0.0757	9.00	5.60	9.60
38	3.965	659.6	0.0476	7.97	4.47	8.47
40	3.145	1049.0	0.0299	7.15	3.55	7.55

* 1 mil = 0.001 in.
† C means single cotton covered.
‡ E means enameled.
§ EC means enameled with single cotton covering.

TABLE 2-2
RESISTANCE DATA ON SOME OF THE MORE COMMON METALS AND ALLOYS

Pure metals	Resistance relative to copper	Temperature coefficient of resistance (20°C)	Alloys	Resistance relative to copper	Temperature coefficient of resistance (20°C)
Iron	5.80	0.0050	Radiohm	77	0.0007
Zinc	3.43	0.0035	Nichrome	65	0.00017
Tungsten	3.20	0.0045	Advance	28	0.00002
Aluminum	1.55	0.0040	High brass	4.8	0.0016
Gold	1.40	0.0034	Low brass	3.8	0.0017
Silver	0.943	0.0038	Commercial bronze	2.4	0.0020

Note: Small variations in purity or composition may change these values appreciably.

The resistances given here are at normal or room temperature and are satisfactory for most use. In most cases the resistance of metals increases with temperature and if the temperature is very high or if very accurate results are required, temperature effects must be considered.

POWER AND ENERGY

Electron flow through conductors was described in the introduction of this chapter. The agitation or the increase in the random movement of the molecules because they are being continuously hit or bumped by the drifting electrons was noted, and also that this results in an increase in temperature. The passage of current through a conductor having resistance is always associated with such a generation of heat. The relation between the current, voltage, and resistance of the circuit and the conversion of electric energy into heat is an important element in the study of electric circuits.

Referring again to the hydraulic circuit of Fig. 2-1, it is known that for constant flow the energy put into the circuit by the pump will be doubled if the pressure is doubled. It will also be doubled if the pressure remains constant and the flow is doubled. A similar variation in power exists in the case of the electric circuit. The power converted into heat in a resistor may be said to be directly proportional to the product of the current and the voltage. Expressed mathematically, this is

$$P = E \times I,$$

P being expressed in watts (w), E in volts, and I in amperes. This statement may then be used as a definition of a watt (w). *The watt is the rate at which electric energy is being supplied when a current of one ampere is flowing at a pressure of one volt.*

Several additional equations for power may be derived from the above statement by the use of Ohm's law. These are very useful when the information supplied is not in volts and amperes. These equations are:

$$P = E \times I = IR \times I = I^2R. \quad (\text{since } E = IR)$$

$$P = E \times I = E \times \frac{E}{R} = \frac{E^2}{R}. \quad \left(\text{since } I = \frac{E}{R}\right)$$

Since power is the rate at which energy is being transferred, the total energy is the product of the power and the time. Thus, a small unit of energy is the *watt-second* or *joule*. The more common unit, however, is a much larger one known as the *kilowatt-hour*. This unit specifies an energy equivalent to 1,000 w, or one kilowatt continued over a period of one hour. It is this unit which is the basis of most of the bills for electric energy issued by the power companies.

Example: The cathode of a type 850 vacuum tube requires 3.25 amp at 10 v. What is the power requirement for the cathode heater?

Solution: $\quad P = EI = 10 \times 3.25 = 32.5$ w.

SIZE AND RATING OF RESISTORS

Resistors play such an important part in electronic circuits that it is important to know something of their limitations. As has been explained, electrical energy is converted into heat when current flows in a resistor, and so a resistor will normally operate above the temperature of its surroundings. Each resistor will have a maximum temperature to which it may be raised without damage to the resistance wire or insulation.

Many of the resistors consist of nichrome or other resistance wire wound around a ceramic tube. The wire and tube are then covered with a low-temperature glaze, as shown on the three resistors at the left of Fig. 2-9.

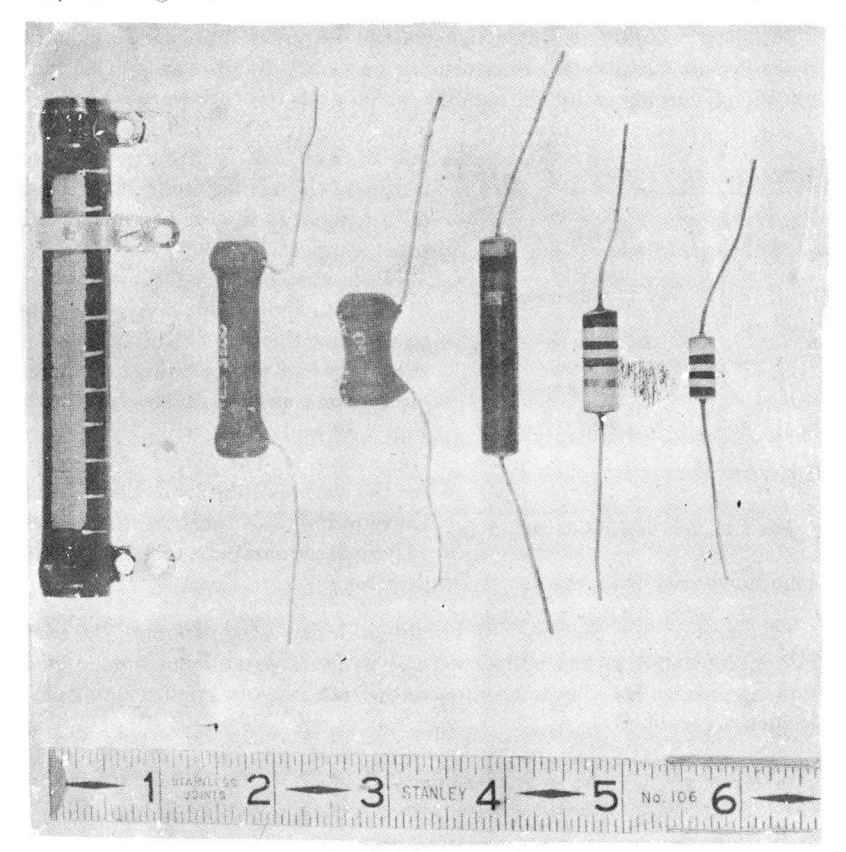

FIG. 2-9. Typical resistors used in radio and electronic equipment.

Resistors of this type will normally have a rating of from 5 to 25 w. This means they will radiate 5 to 25 w to the surrounding air before they will get so hot that they will be damaged. Thus, if the second resistor from the left is rated 4,000 ohms and 10 w, it would carry current computed thus:

$$I^2R = P, \quad I^2 \times 4{,}000 = 10, \quad I^2 = \frac{10}{4{,}000}.$$

$$I = \sqrt{\frac{1}{400}} = \frac{1}{20} = 0.05 \text{ amp.}$$

The voltage drop across the resistor when carrying this maximum or limiting current is

$$E = IR = 0.05 \times 4{,}000 = 200 \text{ v.}$$

Such a resistor should not be placed, therefore, across a voltage that is greater than this limiting value.

In most radio sets, a single high-voltage d-c power supply is provided, as described in Chapter 5. It is usually necessary to provide one or more additional and lower d-c voltages for screen grids or other purposes. These voltages are normally provided by a voltage-divider. The resistor at the left of Fig. 2-9 is designed for this purpose, the center connection being adjustable. It is also possible to add other connections.

It is customary for these dividers to carry an appreciable current in order to stabilize the supply voltage, and so they have high power ratings. A typical circuit for a voltage divider is shown in Fig. 2-10 and analyzed in the illustrative example. The power requirements for the various tubes are specified and the required resistances are determined. (These determinations are usually made experimentally with the use of a voltmeter.)

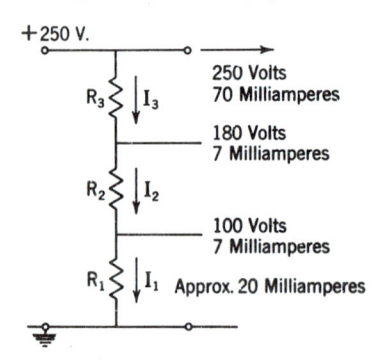

FIG. 2-10. A voltage divider circuit.

Example: For purposes of circuit stability a continuous "bleeder" current of about 20 milliamperes (ma) is desirable and is assumed as I_1 flowing through R_1. The magnitude of the resistance and power rating of R_1 is then determined.

$$R_1 = \frac{E_1}{I_1} = \frac{100}{0.020} = 5{,}000 \text{ ohms}, \quad P = EI = 100 \times 0.02 = 2 \text{ w.}$$

It is then required to determine R_2. Since the current in R_2 consists not only of the bleeder current but also of the 7 ma being supplied at 100 v, the current must be

$$I_2 = 0.020 + 0.007 = 0.027 \text{ amp}, \quad R_2 = \frac{E_2}{I_2} = \frac{80}{0.027} = 3{,}000 \text{ ohms},$$

$$E_2 = 180 - 100 = 80 \text{ v}, \quad P = EI = 80 \times 0.027 = 2.2 \text{ w.}$$

Finally, R_3 is determined in a similar manner.

$I_3 = 0.027 + 0.007 = 0.034$ amp, $\quad R_3 = \dfrac{E_3}{I_3} = \dfrac{70}{0.034} = 2{,}000$ ohms,

$E_3 = 250 - 180 = 70$ v, $\quad\quad\quad P = 70 \times 0.034 = 2.5$ w.

Total power $= 2.0 + 2.2 + 2.5 = 3.7$ w.

Calculations are approximate only, since standard resistors must be used.

The three resistors at the right of Fig. 2-9 are of a different type. They are composed of carbonized strips of material that will normally have very high resistance and correspondingly lower current-carrying capacity. Since the carbonized material has a tendency to change its resistance with temperature, it is usual to limit the operating temperature to a lower value than is permitted with the ceramic tube, wire-wound type. The resistors shown would probably range from 5-w to 0.5-w ratings. Thus, if the center resistor had a resistance of 1 megohm and a power rating of 1 w, the current and potential ratings, as computed above, would be 1 ma and 1,000 v. The color bands shown on the figure are a code to specify the magnitude of the resistance.

FUSES

One important use of the heating effect of the electric current is in the insertion of a resistance unit with small current-carrying capacity into the circuit, so that when the current goes beyond a certain predetermined amount the resistor is melted and opens the circuit. This resistance unit is called a *fuse* and is used to protect the more expensive equipment from harmful effects of excessive current. Fuses are of many different types, and range in size from a few milliamperes up to hundreds of amperes. Since they are placed in a circuit to protect the equipment, they should not be replaced by larger fuses or by heavy conductors because they were specifically designed to open the circuit under overload conditions. Oversize fuses or solid jumpers defeat the purpose of the fuses and permit operation at overload with consequent damage to equipment.

KIRCHHOFF'S LAWS

Two rules or laws known as Kirchhoff's laws are important in solving complicated electric circuits. These laws were implied in the solutions of series and parallel circuits but are stated explicitly as follows:

(1) *The current flowing into any junction of an electric circuit is equal to the current flowing out of that junction.*

(2) *The sum of the battery or generator voltages around any closed circuit is equal to the sum of the voltage drops in resistances around the same circuit.*

Example: The use of these laws is illustrated by determining the current flow in the 4-ohm resistance of Fig. 2-11. The currents are shown as I_{ab}, meaning that the current specified is the current flowing from a to b, I_{cb} the current flowing from c to b, and I_{bd} the current from b to d.

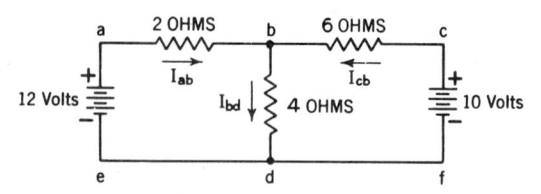

FIG. 2-11. A circuit illustrating the use of Kirchhoff's laws.

Using Kirchhoff's first law at the point b,

$$I_{ab} + I_{cb} = I_{bd}.$$

Using Kirchhoff's second law around circuit $eabd$,

$$2I_{ab} + 4I_{bd} = 12.$$

Using Kirchhoff's second law around the circuit $fcbd$,

$$6I_{cb} + 4I_{bd} = 10.$$

The above three equations have three unknown currents, I_{ab}, I_{cb}, and I_{bd}, and their value may be found by solving the three simultaneous equations. Since $I_{ab} = I_{bd} - I_{cb}$, the second equation may be written as

$$2I_{bd} - 2I_{cb} + 4I_{bd} = 12,$$

or

$$6I_{bd} - 2I_{cb} = 12.$$

Multiplying both sides by 3 and adding to the third equation of the set of three above,

$$\begin{aligned} 18I_{bd} - 6I_{cb} &= 36 \\ 4I_{bd} + 6I_{cb} &= 10 \\ \hline 22I_{bd} \phantom{+6I_{cb}} &= 46 \end{aligned}$$

$$I_{bd} = \tfrac{46}{22} = \tfrac{23}{11} = 2\tfrac{1}{11}.$$

METHOD OF SUPERPOSITION

Another method of obtaining the solution for a circuit having several voltages is based on a principle often used in electrical theory. It states that *the current in a wire is the sum of the currents produced by each voltage acting by itself and with the other voltages shorted out.*

Example: In Fig. 2-11, the portion of I_{bd} caused by the 12-v battery can be obtained by shorting out the 10-v battery.

The equivalent resistance of the 6-ohm and 4-ohm resistances in parallel is $(6 \times 4)/10 = 2.4$. This is in series with the 2-ohm resistance, so that the total current is

$$I = \frac{12}{4.4} = \frac{30}{11}.$$

Since six tenths of this current goes through the 4-ohm resistance, the current contribution by the 12-v battery is

$$\tfrac{30}{11} \times \tfrac{6}{10} = \tfrac{18}{11} = 1\tfrac{7}{11} \text{ amp.}$$

The portion of I_{bd} caused by the 10-v battery can be obtained by shorting out the 12-v battery. The equivalent parallel resistance is

$$\frac{2 \times 4}{6} = \frac{8}{6} = 1\frac{1}{3}.$$

This is in series with the 6-ohm resistance, so that the total current is

$$I = \frac{10}{7\tfrac{1}{3}} = \frac{30}{22} = \frac{15}{11}.$$

Only one third of this current will go through the 4-ohm resistance. Hence, the contribution of the 10-v battery is

$$\tfrac{15}{11} \times \tfrac{1}{3} = \tfrac{5}{11}.$$

The total current is, then,

$$1\tfrac{7}{11} + \tfrac{5}{11} = \tfrac{23}{11} = 2\tfrac{1}{11} \text{ amp.}$$

The answer by this method is, of course, the same as that obtained by the use of Kirchhoff's laws.

Exercise 2-11. Determine the voltage across ab of Fig. 2-12 by using both Kirchhoff's laws and by the method of superposition.

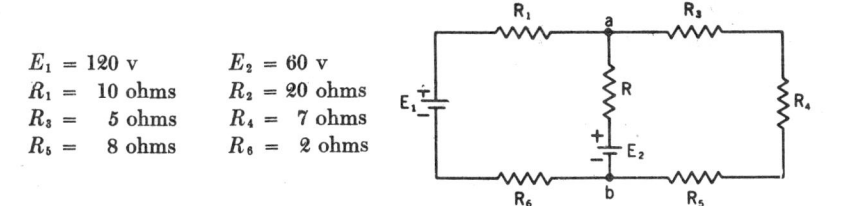

$E_1 = 120$ v $\qquad E_2 = 60$ v
$R_1 = 10$ ohms $\qquad R_2 = 20$ ohms
$R_3 = 5$ ohms $\qquad R_4 = 7$ ohms
$R_5 = 8$ ohms $\qquad R_6 = 2$ ohms

FIG. 2-12. A circuit for Exercise 2-11.

SYMBOLS AND ABBREVIATIONS

Many kinds of equipment are used in radio work and the circuits are often quite complicated. In order to simplify diagrams, a standard set of symbols is used to indicate the circuit elements. These are assembled in Figs. 2-13(a) and (b). Below the symbol is given the name of the element and the name and abbreviations of the units where possible.

The very wide range of magnitudes of quantities used in radio has led to the adoption of many units which are decimal parts of the basic unit. Thus *milli-* placed before the name of a unit such as millivolt (mv) means that one millivolt is one thousandth of a volt. A list of prefixes and the size of the new unit in terms of the original unit is listed in Table 1-2. The use of such units saves many troublesome decimals. *It is necessary, however, to remember that the circuit laws are based on ohms, amperes, and volts, and other units must be converted to these before circuit problems can be solved.*

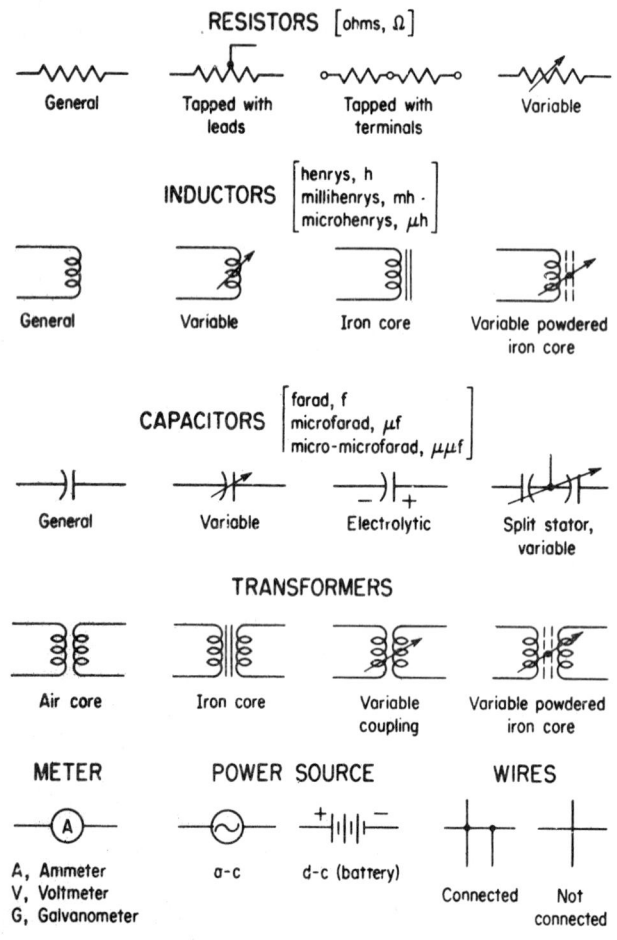

FIG. 2-13. Electrical symbols and abbreviations.

BATTERIES

An important source of electric pressure for radio purposes is the electric cell or battery of cells. Batteries are of two types. One of these, known as the *primary battery*, uses up the original materials and is thrown away after its useful life is accomplished. The common commercial form for this type is the *dry battery* or *dry cell*, which is used in portable sets and elsewhere when small amounts of power are required over a short period. The other

Direct-Current Circuits

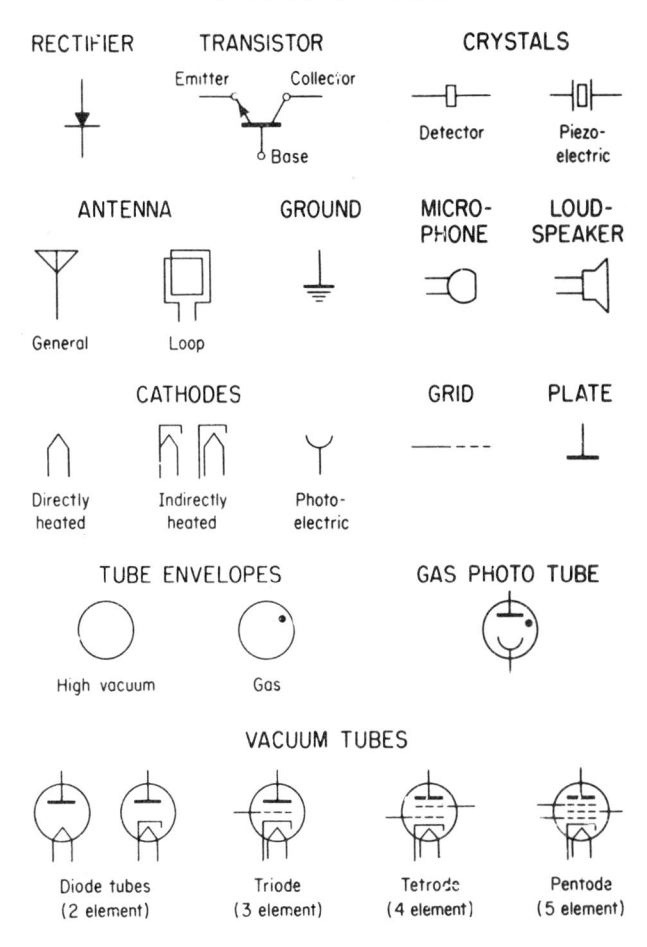

RECTIFIER

TRANSISTOR — Emitter, Collector, Base

CRYSTALS — Detector, Piezo-electric

ANTENNA — General, Loop

GROUND

MICROPHONE

LOUDSPEAKER

CATHODES — Directly heated, Indirectly heated, Photo-electric

GRID

PLATE

TUBE ENVELOPES — High vacuum, Gas

GAS PHOTO TUBE

VACUUM TUBES — Diode tubes (2 element), Triode (3 element), Tetrode (4 element), Pentode (5 element)

FIG. 2-13 (contd.)

type, known as the *secondary* or *storage battery*, may be recharged by forcing current through it in the reverse direction after it has supplied its normal amount of electric energy. Usually storage batteries are of larger current capacity than dry batteries and are used as main sources of power on portable transmitters and receivers where it is possible to recharge the batteries conveniently, as in automobiles.

Dry cell. A diagram of the construction of a dry cell is shown in Fig. 2-14. The positive electrode of the dry cell is a large carbon rod located in the center of a zinc container that forms the negative electrode. The electrolyte, which is a dilute solution of ammonium chloride (sal ammoniac) is mixed with some porous inert material to form a paste which is placed on the inside of the zinc container. Between this paste and the carbon rod is a porous mass of manganese dioxide and carbon granules. This material is called a depolarizer and its function is to absorb the ammonia and hydrogen gas that is given off at the positive electrode and tends to insulate the electrode. The top of the cell is sealed by an insulating compound so that

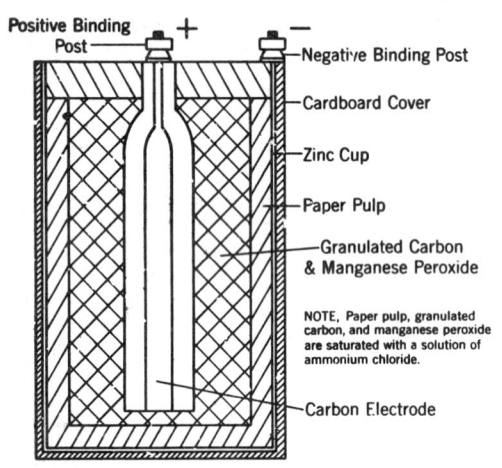

FIG. 2-14. The construction of a dry cell.

only the terminals connected to the carbon rod and to the zinc container are visible. The zinc container is then placed in a cardboard carton, which acts as a protection and as an insulator. The larger dry cells, approximately $2\frac{1}{2}$ inches in diameter and 6 inches tall, are supplied as individual cells. For larger voltages where very small currents are sufficient, a number of small cells are connected in series and mounted in a common container. Several terminals will probably be brought out in order to supply different voltages. This unit is the well-known B battery used to supply voltage to the plates of vacuum tubes, and to transistors in portable equipment.

Storage batteries. The *lead storage battery*, which is the most common type, consists of a positive plate of lead peroxide and a negative plate of porous or spongy lead in a dilute solution of sulphuric acid. The usual construction of batteries of the portable type is known as the pasted-plate construction. In this a lead and antimony alloy is used to make a grid of the type shown in Fig. 2-15. This grid is filled in with a paste which, after an electrochemical forming process, gives lead peroxide on the positive plate. This lead peroxide is the active chemical material, while the grid acts as physical support for the chemical and as electrical conductor for the current that is developed by chemical action. A similar grid is used for the negative plate but a different paste is used, which after forming consists mainly of pure lead in very porous condition.

Chapter 2 　　　　　　　　　 *Direct-Current Circuits* 　　　　　　　　　 67

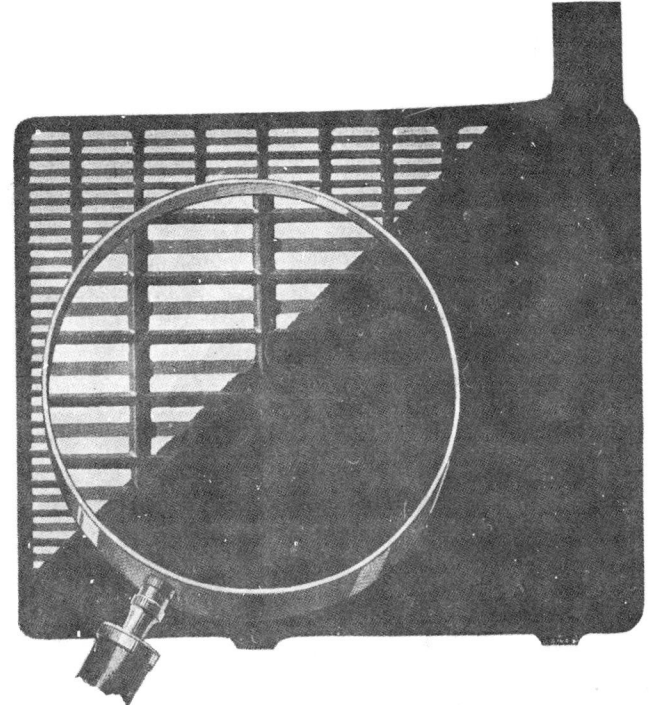

Fig. 2-15. Grid of a storage battery partly filled with active material. (Courtesy Electric Storage Battery Co.)

In Fig. 2-16 is shown a completed positive and negative plate, while in Fig. 2-17 several of these positive and negative plates are assembled into groups. In Fig. 2-18, porous wood and fiberglas separators are shown. The plates of the positive group are nested between the negative plates, being separated from them by the thin and porous separators made of either wood, glass, rubber, or combinations of these. The whole unit is then set in a rubber jar with a cover to support the terminals and is sealed into the main jar as shown in Fig. 2-19. The sulphuric acid electrolyte is introduced through the filling hole in the cover.

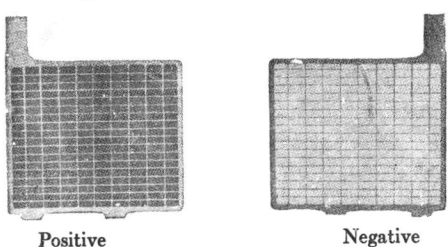

Positive 　　　　　 Negative

Fig. 2-16. Completed storage battery plates. (Courtesy Electric Storage Battery Co.)

68　　　　　　　　　　*Direct-Current Circuits*　　　　　　　　Chapter 2

Negative　　　　　　　　　　Positive

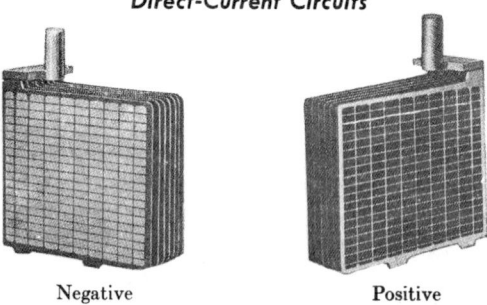

Fig. 2-17. Storage battery plates assembled in groups. (Courtesy Electric Storage Battery Co.)

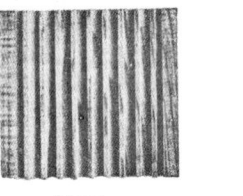

Wood　　　　　　　　　　Fiberglas

Fig. 2-18. Separators for insulating positive and negative plates in storage batteries. (Courtesy Electric Storage Battery Co.)

This construction permits maximum surface area to be exposed to the electrolyte with a minimum of distance between the positive and negative plates, this distance being only the thickness of the separators. The separators are sufficiently porous that they offer little hindrance to the flow of ions which are the carriers of electric charge, but they do maintain mechanical and electrical isolation of the electrodes.

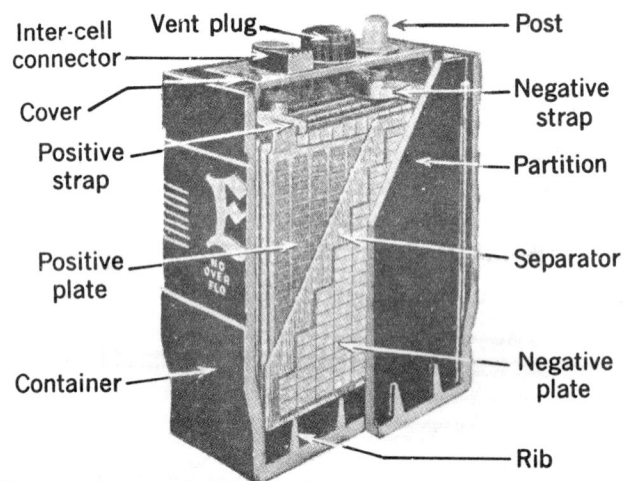

Fig. 2-19. An assembled lead storage cell (cut away to show the construction). (Courtesy Electric Storage Battery Co.)

The chemical reactions that take place are rather complicated but the final result is that both plates change their active material to lead sulphate as the battery is discharged. This lead sulphate has a tendency to harden, crystallize, and expand if permitted to stand. For this reason it is important to keep lead batteries in a charged condition. A battery which remains in a discharged condition soon will have buckled plates and reduced capacity, and will have to be discarded long before its normal life span.

INTERNAL RESISTANCE AND POLARIZATION

In both the dry cell and the storage battery the source of energy is chemical, and the products of chemical reaction may tend to insulate the electrodes or may tend to cause chemical or ion concentrations that reduce the terminal voltage when current is drawn from the battery. This action is known as polarization and in some cases may continue for some time after the current drain has been stopped. Usually the construction of the battery will permit these effects to be neutralized after a time. This entire process produces an effect similar to, and is sometimes called, the "internal resistance" of the battery. Polarization is much more pronounced in the dry cell than in the storage battery.

ELECTROMAGNETISM

A phenomenon which is closely associated with electricity is *magnetism*. The horseshoe magnet and its ability to pick up iron and steel objects is universally familiar. Apparently this attraction is caused by some form of disturbance in the space surrounding the magnet. This disturbance is described as a magnetic field and is indicated by lines connecting the north and south poles, as shown in Fig. 2-20. The density of the lines indicates

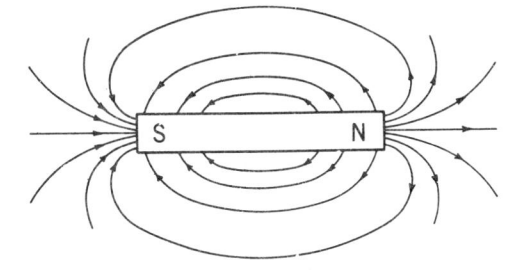

FIG. 2-20. The magnetic field of an isolated bar magnet.

the strength of the field, i.e., the relative force on another small magnet which might be introduced into the field. The direction of the lines indicates the direction of the field at any point, i.e., the direction in which a small magnet, free to turn, such as a compass, would point. The magnet shown in Fig. 2-20 is not bent as is the horseshoe magnet, so is called a bar magnet.

In Fig. 2-21 two magnets are placed with the north pole of one near the south pole of the other. The lines of magnetic flux flow from the north to the south pole, and since the poles attract one another we say that the lines of magnetic flux tend to shorten themselves, much as stretched rubber bands might do. In Fig. 2-22 another arrangement of two bar magnets is shown, wherein the north pole of each magnet is near the north pole of the other. The lines of magnetic flux must leave both north poles and find their way back to the south poles. In this case they are squeezed sidewise against each other, and since the poles repel one another we say here that the lines of magnetic flux tend to exert a sidewise push or lateral force upon one another.

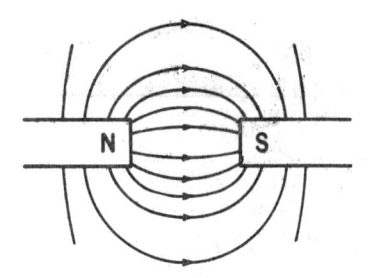

Fig. 2-21. The magnetic field about two unlike poles.

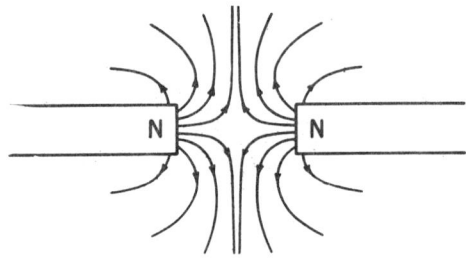

Fig. 2-22. The magnetic field about two like poles.

In order that the density of these lines of magnetic flux be meaningful as a concept of field strength, the lines must be sufficiently numerous to adequately show both magnitude and direction of the field. The maxwell is a very small unit of magnetic flux, so when one line is used to represent one maxwell the concept is very effective. The weber is a much larger unit of magnetic flux (10^8 maxwells) but is convenient in calculating electromagnetic problems. For purposes of computation the weber is therefore used as the standard unit of magnetic flux.

Experiments indicate that wires carrying electric current have magnetic effects. In Fig. 2-23 a wire is shown carrying electric current from the right to the left. The north pole of a pocket compass located above the wire will point away from the observer and indicates that the arrows on the flux are

Fig. 2-23. The direction of magnetic flux caused by electric current is indicated by the right-hand rule.

in the same direction. This effect gives rise to a rule that is commonly used to remember the magnetic effect of an electric current. *If the right hand is placed around the wire with the thumb pointing in the direction of the current, then the fingers will indicate the direction of magnetic flux.* If the wire which is carrying electric current is wound around a cylindrical tube, the magnetic effects of the various turns of wire aid each other and produce a field very similar to that of a bar magnet, as is shown in Fig. 2-24. The strength of the

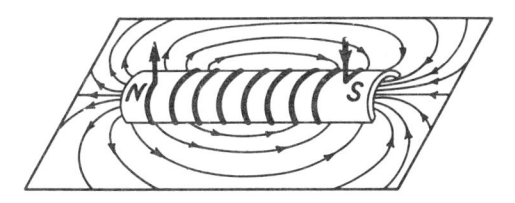

FIG. 2-24. The magnetic field about a coil of wire carrying current.

magnetic field depends upon the magnitude of the current and the number of turns in the coil. This relation has caused the *ampere-turn* to be adopted as a common unit of magnetic pressure or magnetomotive force. The ampere-turn may be defined as the magnetic pressure produced by a current of one ampere flowing in a loop of one turn. Thus a current of 3 amperes flowing in a coil of 5 turns would produce a magnetic pressure of 3 times 5 or 15 ampere-turns.

IRON AS A CONDUCTOR OF MAGNETISM

If an iron rod, not previously magnetized, is placed in a coil that is carrying current, it is observed that the strength of the magnetic field is greatly increased. This increase is due to the unusual ability of iron to act as a "conductor" of magnetic flux. The ratio of the flux produced in iron to that produced in air under similar magnetic pressure varies, but will range between several hundred and six or seven thousand for most commercial steels.

A simple illustration of the manner in which iron and steel are used to control magnetic flux in electric equipment is the electric contactor switch shown in Fig. 2-25. Here an electric coil is placed on a U-shaped iron assembly, and an iron armature is hinged so that it normally will fall open owing to a spring. When the coil is energized with electric current, the flux set up will cause the armature to be attracted and the contactor will be closed.

FIG. 2-25. A magnetically operated relay. (Courtesy Struthers-Dunn, Inc.)

MAGNETIC CHARACTERISTICS OF IRON

Since iron and steel are of such importance in the construction and operation of electrical equipment it is of value to study the properties of iron and of various other types of magnetic materials having iron as their base.

Scientific investigation has shown that iron has the characteristic of increasing the magnetic effect of an electromagnet because the iron molecule is itself a tiny electromagnet. In soft iron, these molecules are free to assume a random arrangement and so no external effect is noticed. When the iron is placed in a coil of wire carrying electric current, groups of molecules called *domains* tend to arrange themselves in line with the field, and thus their

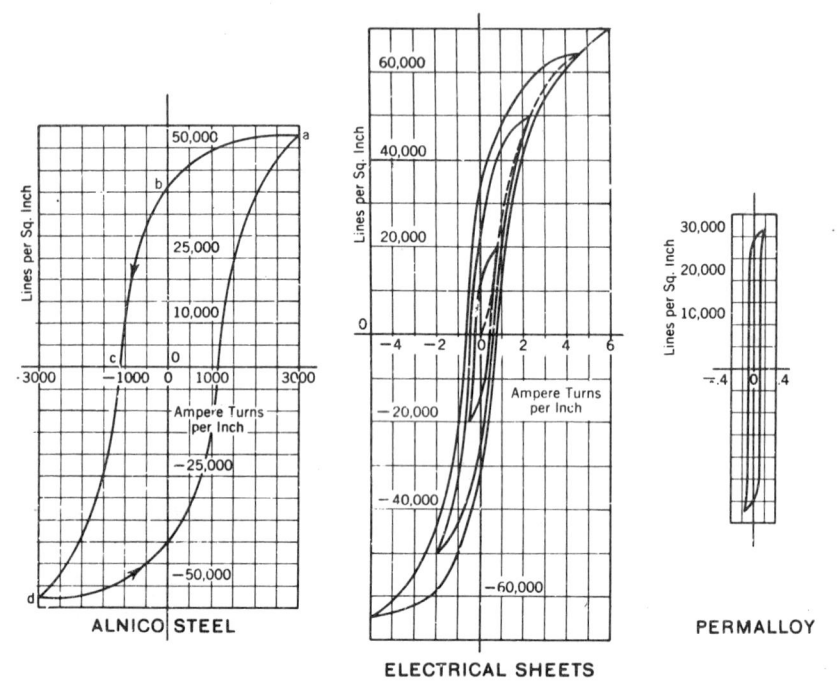

FIG. 2-26. Hysteresis loops of three typical magnetic materials.

magnetic pressure or magnetomotive force is added to that of the coil. When the current is stopped or when the iron bar is removed from the coil, the molecules in the iron again tend to take on a random arrangement. This process is not complete, however, since friction causes many of the domains to retain their alignment and some amount of magnetic effect. This remainder is known as residual magnetism. If the soft iron is replaced by hardened carbon steel, then the alignment of domains will not be as complete as in the case of the soft iron because of the vastly greater internal friction. When the hardened steel is removed from the coil it will, however, retain most of the magnetism and may be called a *permanent* magnet. The internal friction or resistance to magnetic alignment is called *hysteresis*, and a curve showing the variation of magnetism or magnetic flux density with magnetizing pressure or force is known as a hysteresis loop.

Figure 2-26 shows several hysteresis loops that are typical of three distinctly different kinds of magnetic materials. The one in the center represents the kind of steel used for motors, generators, and transformers. The one on the left is representative of permanent magnets, while the one on the right is used for coils and transformers in telephone circuits where the currents are very small. The hysteresis loop is determined by increasing the magnetizing force and measuring the magnetic flux until the maximum is reached at a in the left loop of Fig. 2-26. The magnetizing force is then reduced and the flux drops to b, after which the magnetizing force is reversed

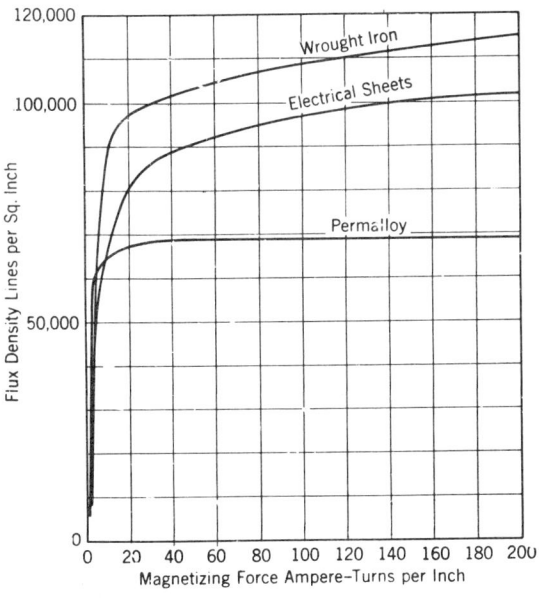

FIG. 2-27. Magnetization curves of three typical magnetic materials.

and increased to c and then on to d, which is the maximum in the opposite direction. The force is then reduced, reversed, and increased again to a. It will be seen that more magnetizing force is required for the same magnetic effect when the flux is increasing than when it is decreasing. The horizontal distance between the sides of the loop is then a measure of the friction or hysteresis. The actual energy loss for each magnetic reversal is proportional to the area of the loop. The residual magnetism is ob. The reversed magnetizing force necessary to bring the flux back to zero is oc and is called the *coercive force.* The average amount of flux produced by a magnetizing force that is reversing will be a curve drawn through the points of the hysteresis loops as shown in the center of Fig. 2-26. Such a curve is called a *magnetization curve* and is important in determining the action of an iron core in a transformer. Magnetization curves for several different magnetic materials are given in Fig. 2-27.

DYNAMO AND TRANSFORMER STEEL

Since the energy loss is proportional to the area of the loop, it is desirable that the steels used for a-c machines and transformers have narrow hysteresis loops. It has been found that the electrical and magnetic characteristics are improved by adding some silicon to the steel. Most steel for dynamos and transformers is used in the form of sheet punchings which are stacked up to give an adequate magnetic path. Such construction is called a laminated type of magnetic structure, and the material for it is supplied in large sheets known commercially as electrical sheets.

PERMANENT MAGNET STEELS

How hardened carbon steel may form permanent magnets has already been mentioned and explains the phenomenon of magnetizing the blades of pocket knives, hardened screwdrivers, and other tools. It has been found that an alloy of iron with aluminum, nickel, and cobalt, when properly heat-treated, gives very high-strength permanent magnets. These are known as alnico magnets and are widely used. Other alloys including combinations of tungsten, chrome, and cobalt are also extensively used. They are characterized by high retentivity and high coercive force.

PERMALLOY

Still another type of magnetic material is known as permalloy and is an alloy of iron with nickel which is carefully heat-treated to give high magnetic flux with very small magnetizing force and to have a very narrow hysteresis loop. This material is used in coils and transformers on telephone circuits where the currents are quite weak.

Very thin sheets of this material may be used in alternating-current coils up to a frequency of 10,000 cycles. For use at higher frequencies the material is finely ground, covered with insulation, and mixed with a binder. This mixture is then pressed into the desired shape and baked to give mechanical strength. Cores of this type may be used at frequencies up to 100 megacycles.

D-C GENERATORS

When a wire is moved across a magnetic field as shown in Fig. 2-28, an electric pressure or electromotive force is set up within the wire. This action is the basis for the generation of nearly all electric power and so is very important. The amount of voltage or pressure set up is dependent upon the strength of the field, the length of the wire that is in the field, and upon the velocity of the wire across the field. Since the strength of the magnetic field is represented by the density of the lines of flux, the voltage can then be said to be proportional to the rate of cutting flux.

Associated with this generating action is another action, based upon the fact that a force is produced on a wire located in a magnetic field and carrying electric current. If the two ends of the wire are connected to an external electrical element to complete a circuit, the current produced will flow in the same direction as the voltage that is being generated. The force exerted on the wire will be opposite to the direction of movement. In other words, if the conductor is moving up through the magnetic field of Fig. 2-28, a voltage will be generated tending to send current outward from the page. If a current is permitted to flow, this current will produce a force that opposes the motion. This condition is required by the law of conservation of energy and explains the transformation of mechanical energy to electrical energy in an electric generator. It should be observed that the voltage being generated is not influenced by the current flowing but is dependent only on the strength of the magnetic field and the speed of the generator. Likewise, the force or torque is independent of the speed of the generator, but does depend upon the magnitude of the field and of the current.

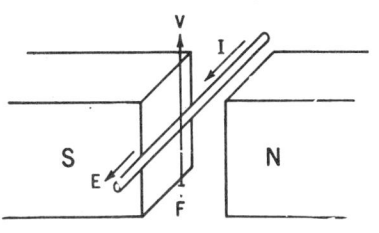

FIG. 2-28. The relation between (a) the magnetic field, motion, and voltage and (b) the magnetic field, current, and force in a generator.

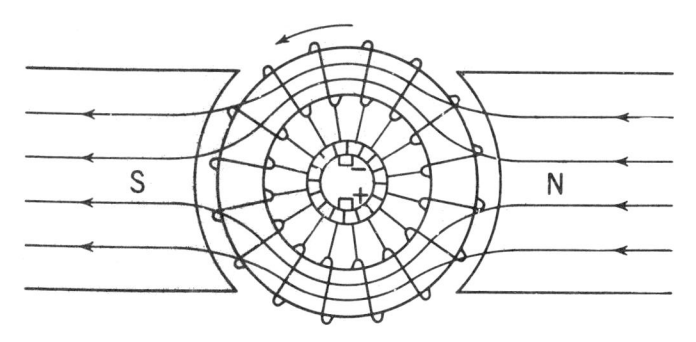

FIG. 2-29. An elementary type of d-c generator.

In Fig. 2-29 is shown an elementary form of direct-current generator. A cylindrical iron ring is mounted on bearings so that it can rotate between the two magnetic poles. The flux enters the ring on the side of the north pole and leaves at the south pole. A wire is wound around the ring as indicated in the diagram, and each turn is connected to an insulated copper bar in an assembly called the *commutator*. The commutator is composed of a number of tapered copper bars separated from each other by mica insulation, clamped together, and machined to form a smooth cylindrical surface. Carbon blocks known as brushes are held in stationary supports and make contact with the copper bars on the rotating commutator.

When the armature is rotated, the conductors on the outside surface cut the flux under the poles and voltage or electric pressure is produced. The conductors under the north pole produce a voltage outward from the page. Each of these is added to the other, so that the difference of pressure between the brushes is the sum of the voltage produced by all the conductors on that side of the armature. The voltages generated by the conductors under the south pole are into the page, and these will also be additive, so the difference in pressure is the same as it was for the side under the north pole. The direction of current flow will thus be out of the bottom brush and into the top brush for both windings.

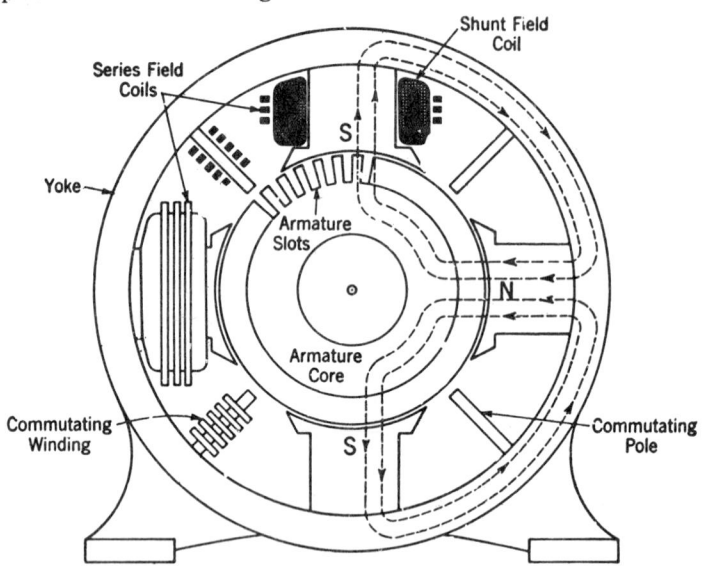

Fig. 2-30. Typical construction of a commercial d-c generator, showing the various parts of the generator except the armature coils and commutator.

A careful study of the diagram shows that the rotation will change the relative position of the individual conductors on the surface of the armature but will not alter the voltage generated with respect to the brushes. This, then, is a direct-current generator.

When the brushes are connected to an external circuit, a current will flow through the external circuit from the positive to the negative brush,

Chapter 2　　　　　　　*Direct-Current Circuits*　　　　　77

and this same current will flow from negative to positive within the machine, dividing equally between the two sides of the machine. The current flows in the same direction in the windings as the voltage generated, so the force produced by the conductors is opposed to the motion of the conductors. This set of forces must be overcome by some form of motor or engine, usually called the prime mover. The prime mover supplies the mechanical energy that is converted to electrical energy in the generator. A diagram showing the typical construction of a commercial generator is shown in Fig. 2-30.

D-C MOTOR

If the generator discussed above is connected to a direct-current power line, and if the engine driving it is then disconnected, the generator will start to slow down. Since the voltage is directly proportional to the speed of the generator, as soon as the generator slows down the voltage produced in the windings is reduced, and current will be forced through the windings by the larger external voltage, in a direction opposite to the direction of voltage previously generated. This change causes a force to be exerted on the conductors in the same direction as the motion (see Fig. 2-31), and so the machine continues to run. If a load of some type, such as a pump, is connected to the electric machine it slows down still more so that a larger current can flow and produce more force tending to continue the rotation. When an electric machine or dynamo is operating in this manner it is called a *motor*.

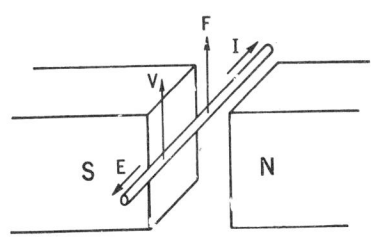

FIG. 2-31. The relation between (a) the magnetic field, motion, and voltage and (b) the magnetic field, current, and force in a motor.

DYNAMOTORS

In radio transmitters, which use vacuum tubes, quite high voltages are required, with a power demand beyond the capacity of small B batteries. In order to meet this need in mobile equipment, a high-voltage generator may be driven by a low-voltage motor, which is supplied from the storage battery. Since it is desirable to reduce weight as much as possible in portable equipment, it is customary to combine the two machines on one magnetic frame.

The motor winding is composed of a few turns of large wire connected to a commutator on one end. The generator winding consists of a large number of turns of small wire mounted on the same iron core but connected to a commutator on the other end of the machine. A motor-generator set of this type is called a dynamotor.

In commercial machines a drum type of armature core is used and the conductors are placed in slots. Figure 2-32 shows dynamotors used for mobile radio power supply.

FIG. 2-32. Two commercial type dynamotors. (Courtesy Pioneer Gen-E-Motor Corp.)

ELECTRIC METERS

The practical use of the circuit theories that have been discussed depends to a great extent upon an ability to measure the magnitudes of the currents, the voltages, and the resistances of equipment in actual operation. The ability to analyze and interpret correctly the variations of meter readings is the ultimate justification for circuit analysis. A brief discussion of the theory and construction of direct-current meters is therefore of interest.

Nearly all of today's direct-current meters are based on a design that was developed by Arsene d'Arsonval in 1881 and so are called *D'Arsonval-type* meters. The diagrammatic construction of this type of instrument is shown in Fig. 2-33. The main parts of the instrument are the permanent magnet M to which the soft-iron pole pieces P are fastened. The soft-iron core C is held by nonmagnetic supports that are bolted to the pole pieces. The magnetic assembly is an integral unit with a uniform air gap between the core and

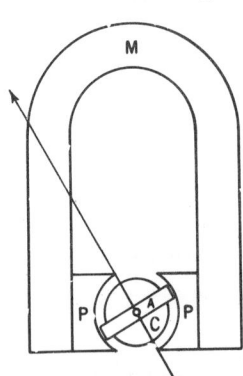

FIG. 2-33. Diagram of a permanent-magnet moving-coil type of meter.

Chapter 2 *Direct-Current Circuits* 79

the pole pieces. This design produces a magnetic field in the air gap that is constant in magnitude and radial in direction.

A coil assembly such as that shown in Fig. 2-34 is supported in jeweled bearings so that it is free to rotate back and forth in the air gap. The coil itself is composed of many turns of fine wire. The hairsprings located at the top and bottom hold the coil in the zero position when no current is flowing in the coil. They also act as insulated connections to the two ends of the coil. The counter-balanced pointer indicates the movement of the coil on a calibrated scale.

When a current flows in the coil, forces are exerted that cause the coil to rotate in the air gap. This rotation is limited by the tension in the hairsprings. The movement of the pointer is proportional to the force exerted on the coil and thus is proportional to the current. The normal direct-current meter is, therefore, a current-measuring instrument. By means of auxiliary circuits, however, it may be made to act as either a voltmeter or ohmmeter. A phantom view of a commercial meter is shown in Fig. 2-35.

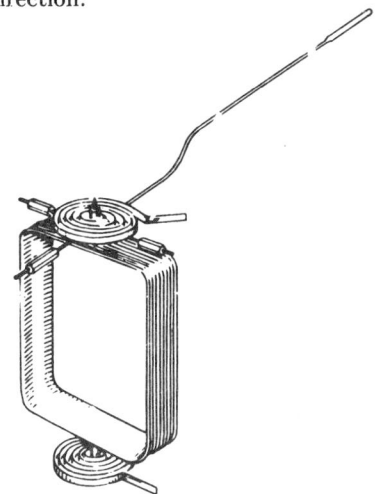

FIG. 2-34. The moving coil assembly for a D'Arsonval type of meter.

Ammeter. Since only very small magnitudes of currents are necessary to cause the meter to read, it is customary to use very low resistances to carry the main portion of the current. The meter, being connected across this low resistance or shunt, takes a small but fixed proportion of the main current. By using shunts of varying resistances the same meter can be used for a number of different current ranges.

Voltmeter. If a resistor is placed in series with the meter, the voltage of a circuit can be determined by measuring the current flowing through the resistor. Additional voltage ranges may be obtained by simply adding resistors in series.

Ohmmeter. If a small battery is located in the meter case and a resistor is connected in series of sufficient magnitude to limit the meter current to full scale when the terminals are shorted, then when additional resistance is inserted between the ohmmeter terminals the current will be reduced and the pointer on the meter will move down the scale. These re-

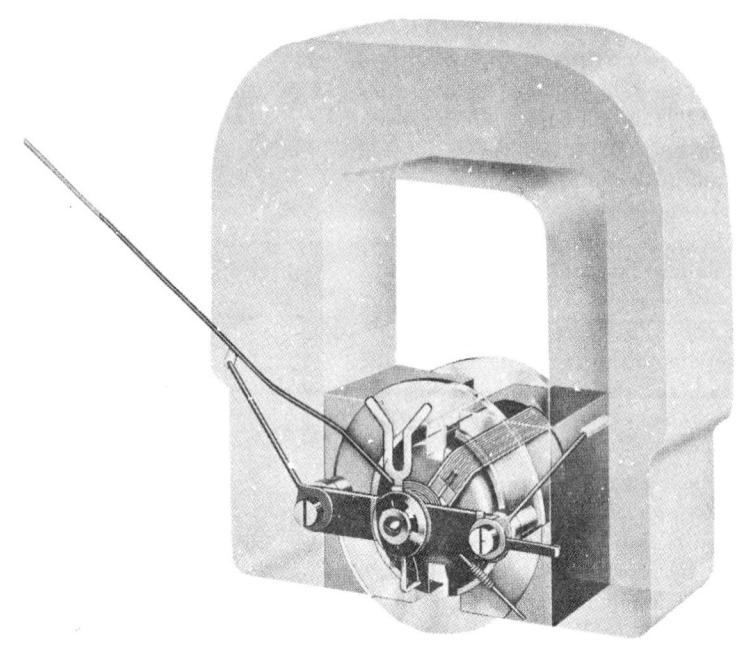

FIG. 2-35. A phantom view of a permanent-magnet moving-coil meter. (Courtesy Weston Electrical Instrument Corp.)

FIG. 2-36. A typical radio service multimeter. (Courtesy Simpson Electric Corp.)

duced readings may be calibrated in terms of external resistance; such a meter is called an ohmmeter.

VOLTMETER RESISTANCE REQUIREMENTS

Since many of the resistors used in electronic equipment have very large resistance values, it must be realized that placing a voltmeter in parallel with them may appreciably change the circuit conditions. For instance, many voltmeters designed for power use have a sensitivity of only 200 ohms per volt. A voltmeter with a full scale of 250 v, therefore, would have a total resistance of only 50,000 ohms. If this meter were used to measure the voltage drop in a resistor of the same magnitude, the voltage might be reduced to

Chapter 2   *Direct-Current Circuits*   81

half value when the meter was connected to the circuit.

High-resistance meters have been developed that have sensitivities of 20,000 ohms per volt. Such a meter would have a total resistance of 5 megohms for the 250-v scale and so would not cause excessive error when connected to most radio circuits. Such meters are more delicate than meters which have a lower voltage sensitivity and so must be handled with reasonable care in order to give continuing satisfactory service.

VOLT-OHM-MILLIAMPERE METERS

In radio testing so many different measurements must be made in analyzing trouble that a formidable array of instruments would be required if a separate instrument were used for each function and range. As a result, numerous manufacturers supply a single instrument with a large number of circuit combinations so that it may be used as a voltmeter, an ammeter, and an ohmmeter.* The basic instrument is a high-sensitivity permanent-magnet moving-coil meter that will have a resistance of about 2,000 ohms and will give full-scale deflections with 50 microamperes (μa).

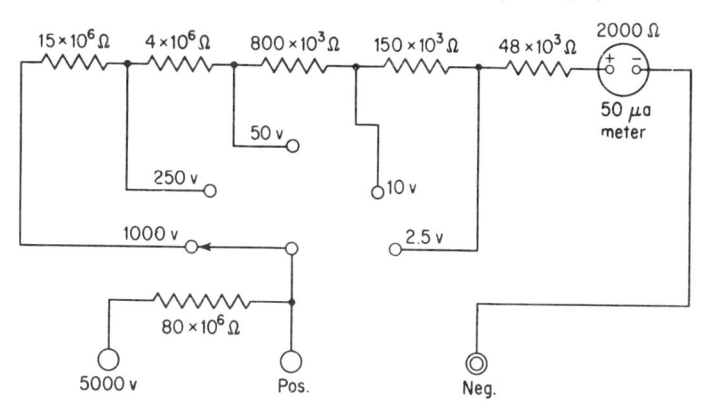

Fig. 2-37. A simplified multirange d-c voltmeter circuit.

The simplified voltmeter circuit for such a meter is given in Fig. 2-37. A rotary switch connects the meter terminals to the proper series resistor for the range indicated.

The simplified diagram of the meter when used for current measurement is given in Fig. 2-38. Here it is seen that as the range for the current is increased, the shunted resistance is decreased so that full-scale deflection is obtained at about 250-mv drop. It is often desirable to have special connections for the high current setting.

* Most of these meters are provided with rectifiers so that a-c volts and output can also be measured. See Chapter 3.

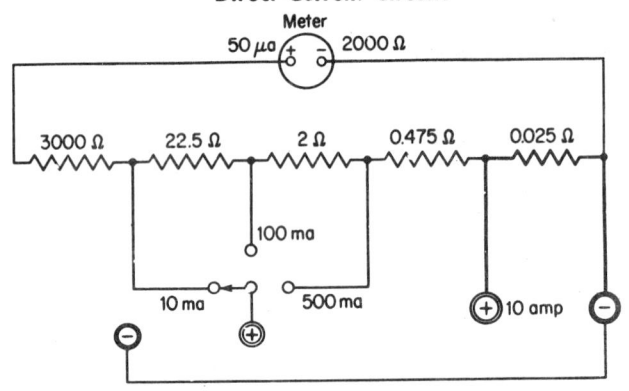

Fig. 2-38. A simplified multirange d-c ammeter and milliammeter circuit.

The ohmmeter connections are shown in Fig. 2-39. This is a more complicated circuit than was referred to above. The zero adjustment makes possible correction for voltage variation of the internal batteries.

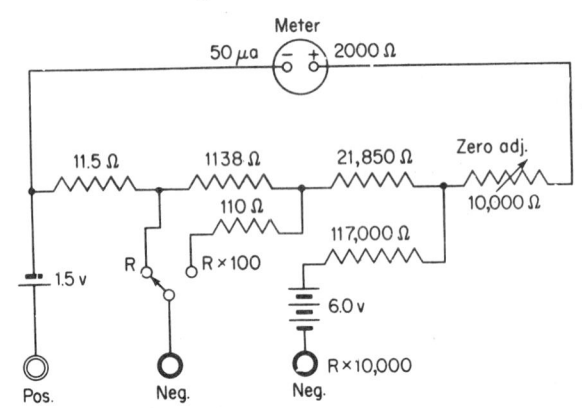

Fig. 2-39. A simplified multirange ohmmeter circuit.

ANSWERS TO EXERCISES

2-1. (a) 7 ohms; (b) 6.33 ohms.
2-2. 105 ohms; 31.5 v.
2-3. 218,000 ohms.
2-4. 7 amp; 14.28 ohms.
2-5. 5.64 amp; 73.2%.
2-6. 2,308 ohms; 947 ohms.
2-7. 13.29 amp; 8.2%.
2-8. 6.25 amp; 4 amp.
2-9. 12.5 amp; 3.125 amp.
2-10. $E_{ab} = 39.3$ v; $E_{bc} = 180.7$ v.
2-11. $E_{ab} = 70.9$ v.

CHAPTER 3

CIRCUITS WITH TIME-VARYING VOLTAGES

TIME-VARYING VOLTAGES AND CURRENTS IN RADIO

The direct or continuous flow of electric current discussed in Chapter 2 is the basis of most of the electric power supply in radio and electronic equipment. It is the controlled variation of current in electric circuits, however, that results in the transmission of intelligence by radio and other communication systems. The study of time-varying currents and voltages, their effects on different types of circuits, and the response of different types of circuit elements to time-varying voltages are all of great importance to the student of radio and electronics.

TYPES OF TIME-VARYING VOLTAGES

Time-varying voltages are composed of three fundamental types of variation, or combinations of these fundamental types. The first type is a suddenly impressed d-c voltage. This is the type of voltage variation that

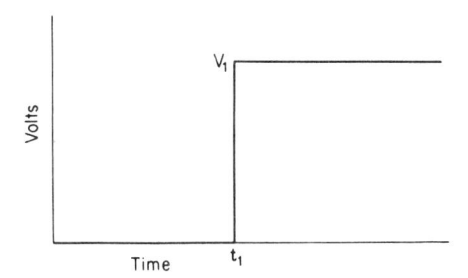

FIG. 3-1. The time-voltage curve of a suddenly impressed d-c voltage.

occurs when a switch on a d-c circuit is closed. It is represented graphically in Fig. 3-1. When the switch is closed at time t_1, the voltage suddenly rises to a constant voltage V_1.

A second type of time-varying voltage is one that varies sinusoidally with time. This means that the voltage is alternating and that the magnitude varies in accord with the sine curve, as explained in Chapter 1. The characteristic of sinusoidal waves that is particularly significant is that the rate of change of a sinusoidally varying quantity is also a sine wave, as will be explained later. The graphical representation of a sinusoidally

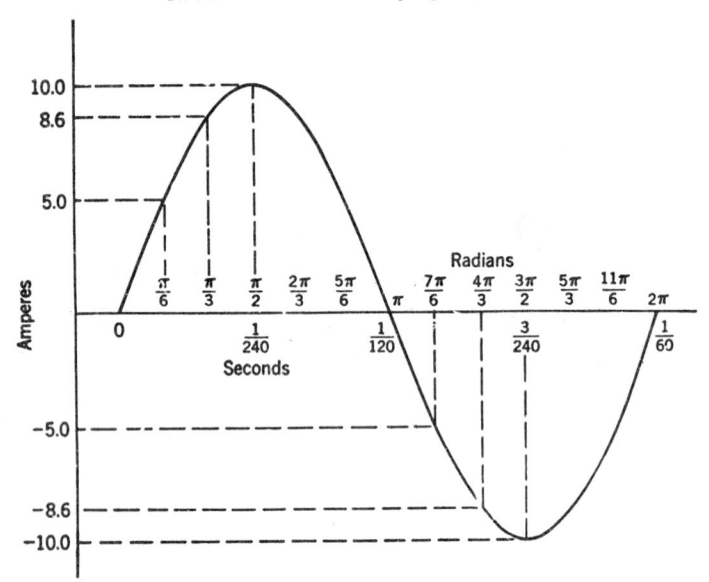

84   *Circuits with Time-Varying Voltages*   Chapter 3

Fig. 3-2. Sinusoidal variation with time.

varying quantity (current in this case) is given in Fig. 3-2. The frequency of variation is 60 cycles.* The positive maximum of 10 amp is reached one quarter of a cycle (or 1/240 second) after the zero current. This curve is similar to that of Fig. 1-10.

The third fundamental type of time-varying voltage is that which is composed of a number of sine waves of multiple frequencies superposed on

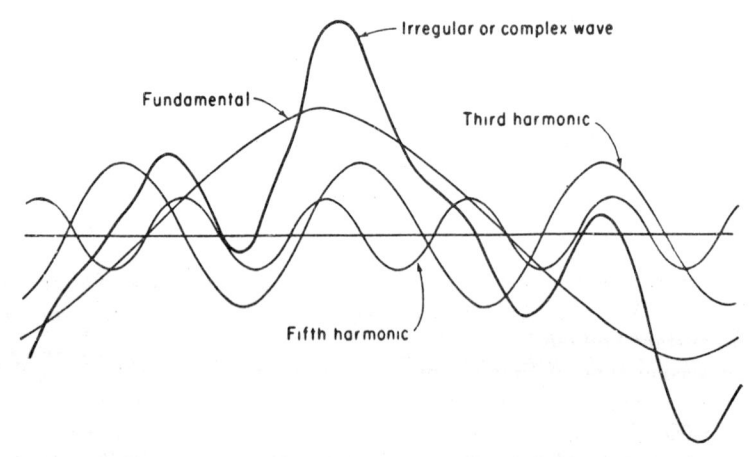

Fig. 3-3. A complex wave formed from the addition of harmonic sine waves.

*While the technically correct expressions for frequency are "cycles per second," "kilocycles per second," and "megacycles per second," the shorter terms of "cycles," "kilocycles (kc)," and "megacycles" are acceptable by common usage, and are employed in this book.

Chapter 3 **Circuits with Time-Varying Voltages** 85

each other. These waves of multiple frequencies are called *harmonics*. Thus, a wave having three times the frequency of the basic or fundamental wave is called a third harmonic, and a wave having seven times the fundamental frequency is a seventh harmonic. Such a combination of waves is shown in Fig. 3-3. The irregular resultant wave is the instantaneous sum of a fundamental, a third, and a fifth harmonic.

Any recurring irregular wave may be represented by a series of fundamental and harmonic waves. By breaking up irregular waves into such a harmonic series, it is possible to predict the response of circuits to them.

Since coils and capacitors as well as resistors are the building blocks of radio circuits, it will be necessary to study the nature of these circuit elements and their response to the different types of time-varying voltages.

ELECTROMAGNETIC CHARACTERISTICS OF A COIL

Suppose a wire is wound on a cardboard tube as shown in Fig. 3-4. The ends of the wire are connected to a galvanometer or sensitive d-c meter. A permanent magnet is also assumed as part of the experimental equipment.

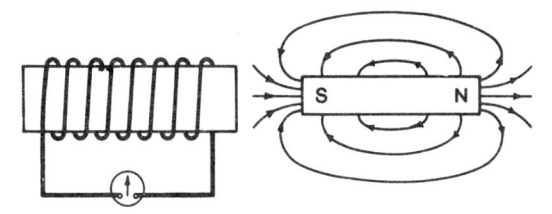

FIG. 3-4. A coil and bar magnet.

When the bar magnet is suddenly moved into the coil, as shown in Fig. 3-5, the galvanometer pointer is observed to swing to the right, which indicates that a voltage has been induced in the coil and a current forced around the circuit. When the magnet is permitted to remain at rest in the coil, as in Fig. 3-6, no current or voltage is observed in the galvanometer.

FIG. 3-5. The bar magnet being inserted in the coil.

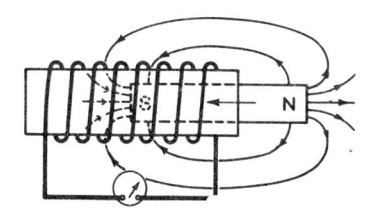

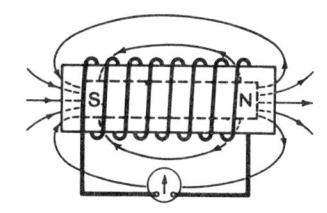

FIG. 3-6. The bar magnet at rest in the coil.

When the bar magnet is suddenly withdrawn from the coil, as shown in Fig. 3-7, the galvanometer pointer swings to the left, which indicates that a voltage has been induced in a direction opposite to that produced when the magnet was being inserted into the coil.

86   Circuits with Time-Varying Voltages   Chapter 3

From these experiments several conclusions may be drawn. First, relative motion between a coil and magnetic field produces a voltage in the coil. Secondly, the current is reversed when the relative motion is reversed. If the direction of the current flow had been determined, it would have been found to be in such a direction as to produce a force that opposed the motion of the magnet. This latter characteristic is of such extensive application that it is given the status of a physical law, called *Lenz's law*. It may be stated in a more general form as follows: *When a voltage is induced in a coil as a result of any variation of the magnetic field with respect to the coil, the voltage is in such a direction that the resulting current tends to oppose the change in the magnetic field.* In Chapter 2 the force on the generator was found to be in a direction that opposed the motion and is one of the many illustrations of this general law.

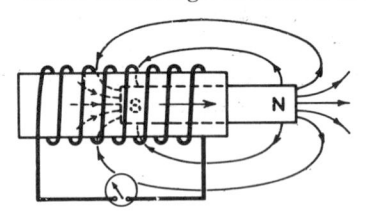

FIG. 3-7. The bar magnet being withdrawn from the coil.

It will be noted, when inserting and withdrawing the magnet from the coil, that the swing of the galvanometer is proportional to the rate at which the magnet is moved with respect to the coil. In other words, the voltage induced in the coil is proportional to the *rate of change of flux*.

If two coils had been used, one having half the number of turns of the other, the galvanometer pointer would have been moved only half as far with the coil having half the turns. (Assume the same rate of moving the magnet.) The voltage induced in a coil is therefore proportional to the rate of change of flux and to the number of turns in the coil. Stated mathematically, this is

$$e = N \frac{d\phi}{dt}$$

where e is the voltage induced in the coil, N is the number of turns, ϕ is the flux in webers, and t is the time in seconds. *This then defines the weber as that flux which when changing at unit rate per second will induce one volt in a single turn of a coil.*

The response of a coil to a suddenly impressed d-c voltage may now be studied. Let the galvanometer in series with the coil be replaced by a battery, an ammeter, and a switch, as indicated in Fig. 3-8. When the switch is closed, the voltage across the coil jumps suddenly to the battery voltage, as indicated in Fig. 3-1. A current starts to flow and a magnetic field is set up, as shown in Figs. 2-24 and 3-8. This field is identical with that of Fig. 3-6; so it is reasonable to expect that in setting

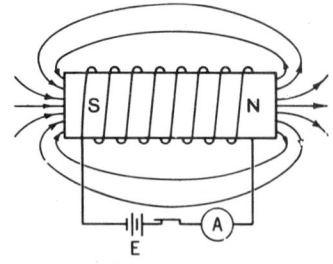

FIG. 3-8. The magnetic field of a coil carrying current.

up this field, opposing reactions in accord with Lenz's law will occur. That they do occur may be verified by inserting an ammeter (with very low inertia so that it will respond almost instantaneously to the current flow) in the circuit and recording the variation of current with time. The ammeter (or oscillograph) shows that the current does not suddenly jump to the value that is determined by the resistance, but starts to rise at a very definite rate and then gradually approaches the final value of current as determined by the resistance. The time variation of this current rise is shown in Fig. 3-9(a). In coils having few turns the time taken to reach the final value of current is very small, possibly about one hundred-thousandth of a second.

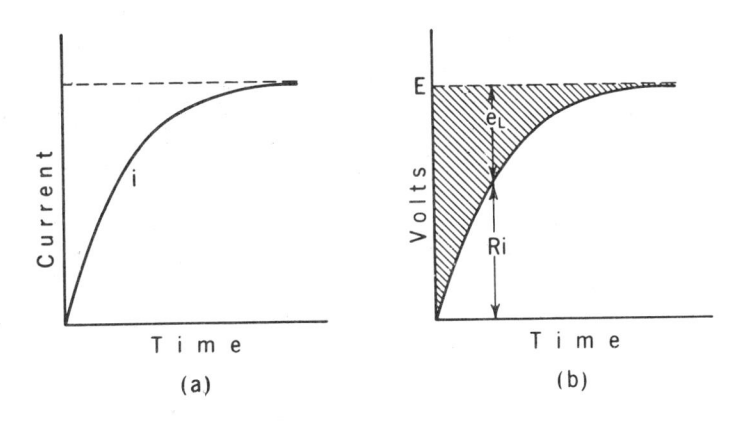

FIG. 3-9. Transient conditions in an inductive circuit. (a) Current rise and (b) voltage distribution.

The mechanism of this reaction is indicated in Fig. 3-9(b) where the Ri drop is plotted against time. The difference between this Ri drop and the constant impressed voltage is called the voltage of self-induction. In Fig. 3-9(b) it is labeled as e_L because e is the usual symbol for instantaneous voltage and L is the symbol for inductance. This voltage of self-induction, or counter-voltage as it is sometimes called, is proportional to the rate at which current (and therefore flux in the coil) is increasing. This leads to the definition of unit inductance as follows:

A coil of wire has an inductance of one henry when a current change of one ampere per second causes one volt to be induced in the coil. Stated mathematically

$$e = L\frac{di}{dt}$$

where L is the inductance in henrys.

THE TIME CONSTANT OF A COIL

The equation for the circuit of Fig. 3-8 may be solved by calculus and is

$$i = \frac{E}{R} - \frac{E}{R}\epsilon^{-Rt/L} = \frac{E}{R}(1 - \epsilon^{-Rt/L}).$$

It is observed from this equation that, if the inductance is large in proportion to the resistance, more time in seconds will be required to reduce the *transient* portion of the current to a negligible value. Thus, the ratio of L to R is known as the time constant of the circuit and indicates the value of time required to bring the exponent of e to -1, at which time the current has reached $1 - (1/\epsilon)$, or 63% of its final value.

This is very convenient since the form of the current rise in an inductor with series resistance is always the same. A curve can be plotted as in Fig. 3-10 in which the vertical coordinate is the percentage of the final value of current and the horizontal coordinate is in time units. When these units are multiplied by the time constant the actual time to reach any percentage of the final value of current may be found.

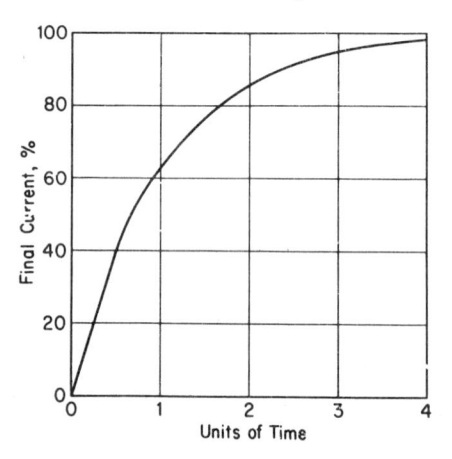

Fig. 3-10. Rate of rise of current in a coil. Time in seconds is equal to units of time multiplied by the time constant L/R.

The manner in which this curve is used may be illustrated as follows. A coil having an inductance of 0.006 henry and a resistance of 10 ohms will have an L to R ratio or time constant of 0.0006 sec. Unit time in Fig. 3-10 then represents 0.0006 sec. It is desired to determine the current flowing one thousandth of a second after the closing of a switch that connects the coil to a 20-v circuit.

The final current is

$$I = \frac{E}{R} = \frac{20}{10} = 2 \text{ amp.}$$

Chapter 3   **Circuits with Time-Varying Voltages**

The time in units along the time scale is

$$t = \frac{0.001}{0.0006} = 1.66 \text{ units.}$$

The current from Fig. 3-10 for 1.66 units is 81% of the final value. Therefore, the desired current value is

$$i = 0.81 \times 2 = 1.62 \text{ amp.}$$

THE ENERGY STORED IN THE MAGNETIC FIELD

The rate at which energy is put into the coil in Fig. 3-9 is

$$P = Ei$$

which is the product of the constant voltage and the instantaneous current. The voltage is divided into two parts, one of which is Ri. The product of this voltage and the instantaneous current gives the power which goes into heat. The product of e_L and instantaneous current gives the energy which is stored in the magnetic field and which can be returned to the electric circuit when the current drops to zero. The magnitude of this stored energy can be determined mathematically, and is

$$W_L = \frac{LI^2}{2}.$$

If the inductance is quite high and current is also fairly large, a considerable amount of energy is stored. If the circuit is suddenly opened, the flux caused by it will drop and a high voltage will result. The energy for this high voltage comes from the magnetic field. The spark coil on the automobile is one example in which this characteristic of energy storage in a coil is used.

Energy storage in a coil in an electric circuit corresponds to the kinetic energy of a moving mass, and so it is sometimes called the inertia element in an electric circuit.

MAGNITUDE OF INDUCTANCE

The magnitude of the inductance of a coil is a physical characteristic of the coil and of the magnetic circuit, just as the inertia of a flywheel is a physical characteristic of the size, shape, and material used in its construction. If additional turns of wire are placed on the coil without appreciably changing any of the other dimensions, the magnetic field will be increased in proportion to the increased number of turns. The magnetic field also reacts on the increased number of turns. The induced voltage will, therefore, be due to an increased magnetic field acting on an increased number of turns, so the magnitude of the voltage will be proportional to the square of the number of turns on the coil. When it is desired to change the inductance of a coil having a known number of turns, a close approximation may be made by using the relationship that, when the turns are close together,

$$L = KN^2$$

where K is a constant and N is the number of turns. If the number of turns is doubled, the inductance is increased to four times its original value, and if the number of turns is made three times as large, the inductance is increased to nine times.

Where very large values of inductance are desired, and where the frequency is not too great, it is customary to place the coil on an iron core. In Chapter 2 it was learned that iron would pass magnetic flux of several thousand times the amount that would flow in air. Therefore, inserting an iron core may increase the inductance a thousand times. Since the amount of flux in iron is not directly proportional to the current in the coil, as shown in the magnetizing curve of Fig. 2-27, it is common practice to insert an air gap in the magnetic circuit so that the flux may be maintained more nearly proportional to the current. This arrangement will permit the inductance to be quite constant over a reasonable variation of load current. Since there is a loss in the iron for every cycle of alternating current, this loss becomes excessive at high frequencies, and ordinary iron cannot be used. It is, therefore, unusual for ordinary iron cores to be used above 10,000 cycles, but iron-dust or permalloy dust cores are often used at high frequencies, and magnetic ceramic material called ferrites at still higher frequencies.

The commercial construction of coils for radio, therefore, depends greatly on the frequency range for which they are designed. Fig. 3-11 illustrates two of the many different types of inductors available for radio use.

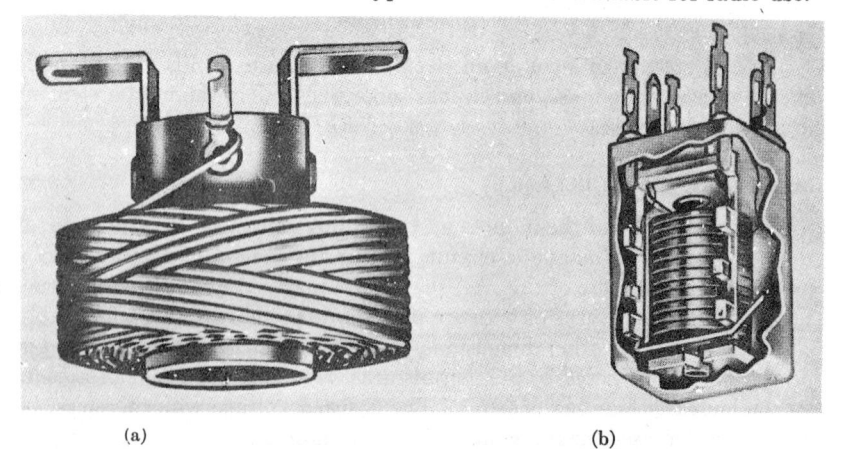

(a)   (b)

FIG. 3-11. Radio coils. (a) A transistor intermediate frequency transformer for 455 kc; (b) a radio-frequency filter choke.   (Courtesy J. W. Miller Co.)

Chapter 3 **Circuits with Time-Varying Voltages** 91

ELECTROSTATIC CHARACTERISTICS OF A CAPACITOR

When two conducting plates are placed close to but insulated from each other, they form what is known as a *capacitor*. An electric charge can be stored on these plates and will be retained as long as the plates of the capacitor are insulated from each other. One of the most fundamental of all electrical concepts is the repelling action of charges of the same polarity and the attractive forces between charges of opposite polarity. If, then, the charged plate having an excess of electrons (or negative charge) is connected to the charged plate having a deficiency of electrons (or positive charge), the repelling and attracting action of the charges will cause large numbers of electrons to flow through the conductor. This flow of electrons constitutes an electric current which will be forced through the conductor against the resistance drop and will, thus, cause heat to be developed. This heat energy was stored in the capacitor as potential energy because of the repelling force of the charge, and this pressure or potential is proportional to the charge. The actual magnitude is dependent upon a proportionality constant that is determined by the size of the capacitor plates, the distance between them, and the insulating material. This constant is called the capacitance and is indicated by the symbol C. Expressed mathematically, the relation is

$$Q = EC,$$

Q being measured in coulombs, E in volts, and C in farads (f).

The unit of capacitance is called the *farad* and may be defined as *that capacitance which will permit the storage of one coulomb of charge at a potential of one volt.*

DIELECTRIC CONSTANT

The above analysis was based entirely upon the effect of the charges within the conductors and so is valid for conductors in a high vacuum; an illustration is the interelectrode capacitance of vacuum tubes. Experimental studies show that the same analysis is also valid for conductors in air. When some liquid or solid insulators are placed between the plates of the capacitor, however, it is found that the capacitance is considerably increased; this increase permits additional energy storage.

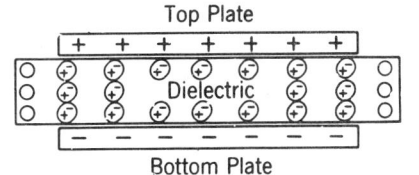

FIG. 3-12. An elementary concept of the molecular tension in the dielectric of a capacitor.

The mechanism of this storage is indicated in Fig. 3-12. The positive charge on the upper plate attracts the negative electrons in the molecules of the dielectric. These electrons are bound so tightly to the molecule that they cannot flow as do the electrons in metals; but the force of attraction exercised by the upper plate and the repulsion exercised by the lower plate combine to produce a strain in the molecules of the dielectric similar to the strain in a spring which has been stretched. The extent to which the effect of the repelling action of the charges in the capacitor plates can be neutralized by the strained condition of the dielectric is dependent upon the physical characteristics of the dielectric. An index of this ability of a dielectric to change the capacitance of a capacitor is known as the *dielectric constant*. It may be defined as *the ratio of the capacitance of a capacitor with the dielectric being considered to the capacitance of the same capacitor if air or a vacuum were used as the insulating medium*. A list of the dielectric constants of several common insulators is given in Table 3-1. In many cases varying methods of manufacture or varying quality cause a range of values.

TABLE 3-1
DIELECTRIC CONSTANTS OF INSULATION MATERIALS

Material	Dielectric Constant
Air	1
Glass	6-9
Porcelain	5-7
Steatite	5-6
Mica	6-7
Polystyrene	2.6
Bakelite	5-15

MATERIALS AND DIMENSIONS DETERMINING CAPACITANCE

Many of the capacitors used in radio are flatplate capacitors, and the capacitance of these may be determined from their dimensions by means of

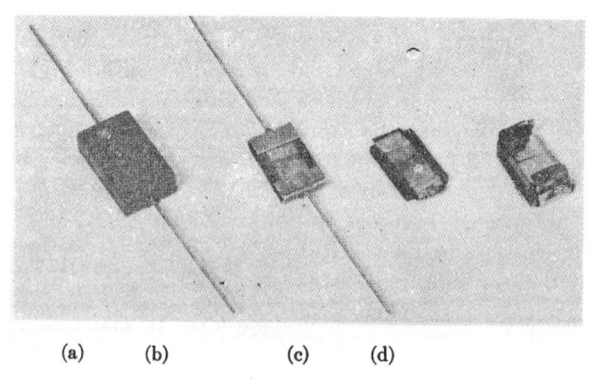

(a) (b)   (c)   (d)

FIG. 3-13. The construction of a mica capacitor. (a) The foil and mica plates are stacked; (b) the ends of the foil of the same polarity are welded together (foil of opposite polarity is brought out to opposite ends); (c) leads are attached to the foil ends; (d) the entire assembly is hermetically sealed. (Courtesy Sprague Electric Co.)

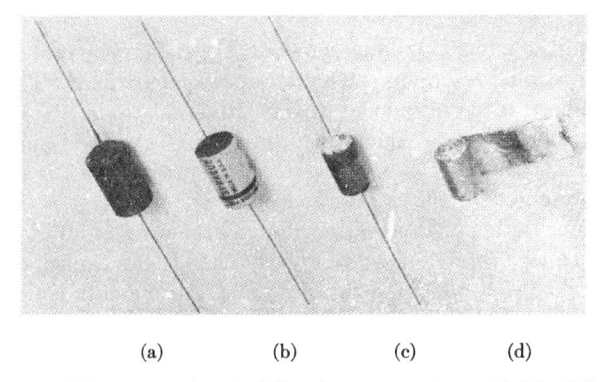

Chapter 3 *Circuits with Time-Varying Voltages* 93

(a) (b) (c) (d)

FIG. 3-14. The construction of a foil-and-paper capacitor. (a) The foil and paper are rolled so that the foil of opposite polarity extends on opposite ends; (b) the leads are attached; (c) the roll is placed in a cardboard sleeve; (d) dipped in wax to seal against moisture. (Courtèsy Sprague Electric Co.)

the formula

$$C = 2{,}248 \frac{AK}{t} 10^{-16} \text{ f}.$$

In this equation, A is the area in square inches of the dielectric under stress, t is the thickness of the dielectric in inches, and K is the dielectric constant determined from Table 3-1. The commercial construction of capacitors varies but often takes the form of sheets of metal foil separated by sheets of mica or by impregnated paper. The typical mica construction is shown in Fig. 3-13. For large values of capacitance the mica construction is usually

FIG. 3-15. A split stator tuning capacitor. (Courtesy Hammarlund Mfg. Co.)

too costly and so paper is used as shown in Fig. 3-14. For very high values of capacitance and low cost, electrolytic capacitors are used. These capacitors use a sheet of aluminum as the positive electrode. The molecular layer of aluminum oxide which is formed by electrolytic action acts as the dielectric. (Since it is so very thin the capacitance is very large.) The electrolyte acts as the negative electrode.

Variable or tuning capacitors usually take the form of interleaved metal plates that can be rotated so as to greatly change the capacitance. Often several are mounted on the same shaft and the magnitudes of both are shifted simultaneously. Two such capacitors mounted on the same shaft are shown in Fig. 3-15.

Certain ceramic materials, particularly those containing titanium compounds, have been found to have dielectric constants of several thousand. It is possible, therefore, to produce midget capacitors of high capacitance and the construction is illustrated in Fig. 3-16.

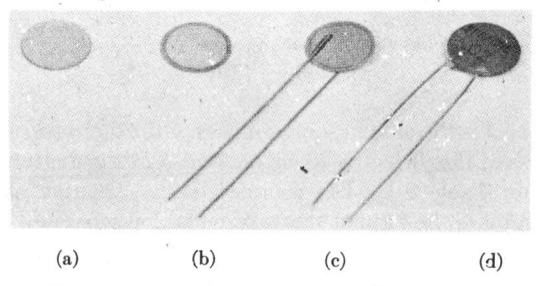

(a)   (b)   (c)   (d)

Fig. 3-16. The construction of a ceramic capacitor. (a) The bare ceramic disc; (b) a silvered electrode is deposited on each side of the disc; (c) leads are attached; (d) capacitor is sealed with resin dip. (Courtesy Sprague Electric Co.)

The farad is such a large unit of capacitance that it is almost never used in radio practice. The microfarad (μf) and the micromicrofarad ($\mu\mu$f) are the units used most extensively. These units, however, must be converted to farads when substituting in equations unless the equations are converted so that the smaller units may be used directly. The value of $1\mu\text{f} = 10^{-6}$ f, and of $1\ \mu\mu\text{f} = 10^{-12}$ f.

Capacitors, when connected in parallel, have a capacitance equal to the sum of the capacitances of each individual condenser. When the capacitors are connected in series, however, the equivalent capacitance must be obtained by the use of the equation

$$\frac{1}{C_{\text{eq}}} = \frac{1}{C_1} + \frac{1}{C_2} + \frac{1}{C_3} \cdots$$

This equation is analogous to the equation for the equivalent conductance of several resistors connected in series as developed in Chapter 2.

CAPACITIVE CIRCUIT RESPONSE

All capacitors are connected to circuits by wires that have some resistance, so that these circuits may be considered to be capacitors in series with resistors. In most electronic circuits additional resistive elements will be added to the circuit to control the rate of current flow to and from the capacitor.

Since the voltage across the capacitor is proportional to the charge, any change of voltage requires a change of charge, and this constitutes an electric current. The magnitude of the current flow is dependent upon the series resistance and the difference of potential between the circuit and the

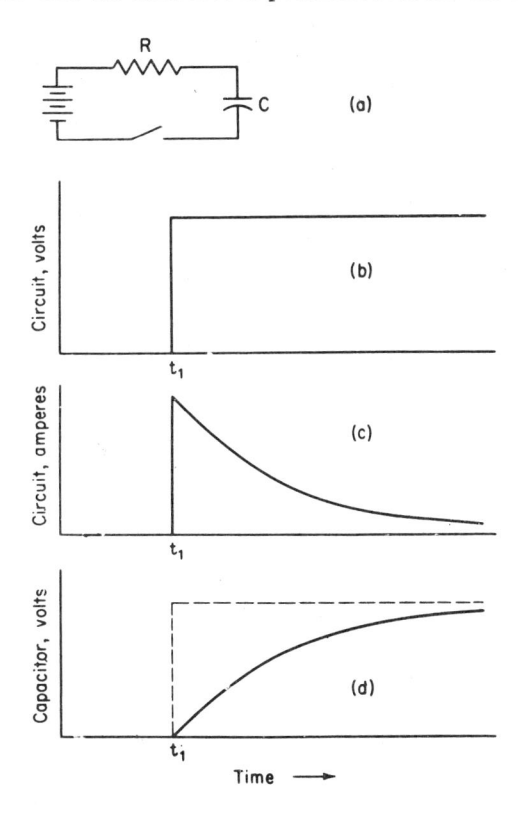

Fig. 3-17. Response of a capacitor-resistor circuit to a suddenly impressed d-c voltage.

capacitor. In the case of the simple resistive-capacitive circuit of Fig. 3-17(a), the constant circuit voltage is suddenly applied when the switch is closed. The time-variation of applied voltage is shown in part (b).

The current may be obtained mathematically, and will be:

$$i = \frac{E}{R}\epsilon^{-(t-t_1)/RC}$$

This is shown in Fig. 3-17(c). The time constant for this circuit is RC just as the time constant for the inductive circuit is L/R.

The voltage across the capacitor will be

$$E_c = E\left(1 - \epsilon^{-(t-t_1)/RC}\right)$$

and is shown in Fig. 3-17(d).

A qualitative analysis of the phenomena is as follows:

(a) When the switch is closed there is no charge on the capacitor, so the current jumps to the value $i = E/R$ as indicated in Fig. 3-17(c).

(b) This current flow causes a charge to collect on the capacitor and thus builds up a counter voltage equal to q/C.

(c) This counter voltage progressively reduces the current and therefore the rate of voltage rise on the capacitor.

(d) Thus the voltage across the capacitor increases rapidly at first. Then the rate of increase slows down as the capacitor voltage approaches the input voltage.

(e) The current after jumping to the value E/R decreases rapidly at first and then gradually approaches zero.

Since the voltage rise across a capacitor is extensively used in electronic circuits as a timing device, it is desirable to be able to determine the time for a certain percentage of voltage rise. This can be done with the aid of

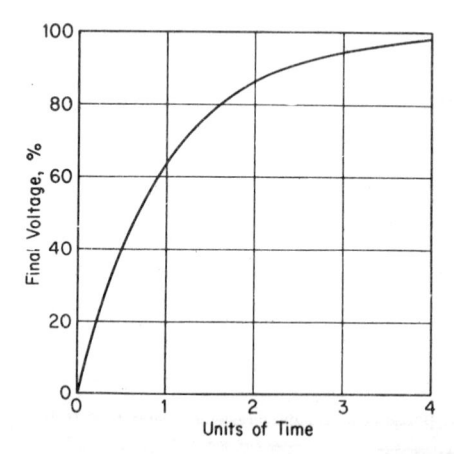

Fig. 3-18. Rate of rise of voltage on a capacitor. Time in seconds is equal to units of time multiplied by the time constant RC.

Fig. 3-18, which is similar to Fig. 3-10. The time for any percentage of voltage rise will give a certain value in units of time. When this value is multiplied by the time constant RC, the actual time to reach this voltage is obtained.

Chapter 3   Circuits with Time-Varying Voltages

Example: A switch closes, placing 50 v across a series circuit composed of a 100,000-ohm resistor and a 0.5-µf capacitor. How long will it take for the voltage across the capacitor to rise to 20 v?

Solution: Twenty volts is 40% of the final voltage, and the curve of Fig. 3-18 shows that it will rise to 40% of the final value in 0.5 unit of time. The actual time is then

$$0.5 \times RC = 0.5 \times 100,000 \times 0.5 \times 10^{-6} = 0.025 \text{ sec.}$$

It is often important to know what circuit constants should be chosen for a timing circuit. This is illustrated in the following example.

Example: A trigger circuit requires 60 v to operate. A constant voltage of 100 v is to be impressed on a series RC circuit. Select circuit constants for a time delay of 0.008 sec.

Solution: The trigger voltage is 60% of the final voltage, so the time unit required is 0.90.

$$0.90 \times RC = 0.008 \quad \text{and} \quad RC = 0.0089.$$

An infinite number of solutions is possible, so that it is necessary to select a suitable capacitor and then determine the resistor to go with it.

Let $C = 0.01$ µf, then

$$R = \frac{0.0089}{C} = \frac{0.0089}{10^{-8}} = 0.0089 \times 10^{8} = 890,000 \text{ ohms.}$$

Since such timing circuits are extensively used in electronic equipment, the above method of calculation, which does not require the use of calculus, is very convenient.

ALTERNATING-CURRENT CIRCUITS

The second type of time-varying circuits, and the most important for analyzing electronic circuits, is the sinusoidal variation. This type of variation occurs extensively in nature. When a pebble is dropped in still water, the ripples are sine waves. A tuning fork produces pressure waves that vary sinusoidally with time. The balance wheel in a watch oscillates sinusoidally and energy is alternately stored in the hair spring and then in the movement of the wheel. This oscillation of energy is characteristic of sine waves and constitutes an important phase of study of electric circuits.

The representation of a sine wave was developed in Chapter 1 using a rotating radius. It was also shown that the rate of change of a sine wave was another sine wave that was displaced 90° from the first. These concepts will be used as the basis for a system of analysis of the behavior of electrical circuits when sinusoidal voltages are impressed.

The rotating radius used in the construction of the sine wave is called a phasor in electric-circuit terminology and is used to represent the magnitude and phase position of an electric current or voltage. A specific

Circuits with Time-Varying Voltages

example involves a current that varies from 10 amp maximum in one direction to 10 amp maximum in the opposite direction at a frequency of 60 cycles, which is the standard power frequency in the United States. The length of the radius will represent the maximum value of the current as

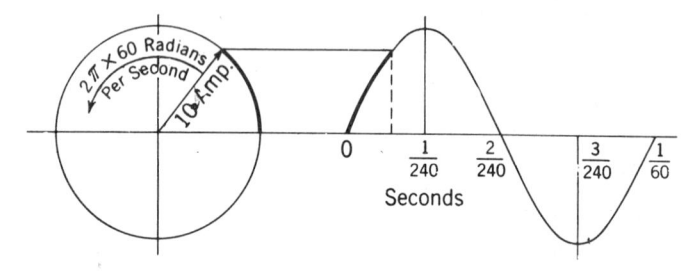

FIG. 3-19. A rotating phasor and sine wave show the variation of current in a conductor that is carrying alternating current.

shown in Fig. 3-19. It will rotate at $2\pi 60$ radians per second so that it will complete one revolution in one sixtieth of a second and sixty revolutions in one second.

ADDING ALTERNATING CURRENTS

It is not possible to add alternating currents in the same way that direct currents are added because they do not always reach their maximum values at the same time in different branches of the circuit. To illustrate this difference, let it be assumed that an a-c circuit exists such that a current of 5 amp maximum is added to the previous current of 10 amp maximum. The 5 amp current is also of 60 cycles but comes to its maximum 60° or $\pi/3$

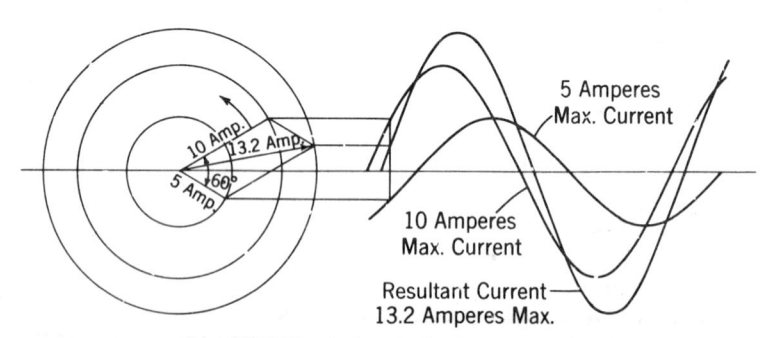

FIG. 3-20. The addition of alternating currents.

radians after the 10 amp current. In Fig. 3-20 the phasor for each current is shown, with the 5 amp phasor lagging 60° behind the 10-amp phasor. The current at each instant and the instantaneous summation of the two currents are also shown by means of the sine waves. It is seen that this addition of instantaneous currents gives another sine wave. This additive behavior is one of the many interesting and convenient characteristics of sine waves. Another of these sine-wave characteristics is that if a parallelogram is completed using the original phasors as two sides, the diagonal will be a new phasor which will generate the wave of the instantaneous sum of the currents.

It is seen from the foregoing that either the sine wave or the rotating phasor may be used to represent an alternating-current quantity. Sometimes one and sometimes the other is more convenient or more effective in showing the characteristics that are to be studied. Hereafter the method will be used which seems to be most effective, and only occasionally will both representations be used on the same problem. Both the phasor and wave type of diagrams may be used for a-c voltages as well as for currents. Both types are particularly useful in showing the magnitude and phase relationships of currents and voltages when both are included in the same diagram. It is common practice in such a diagram to use one scale for voltages and a different scale for currents.

A-C CIRCUIT WITH RESISTANCE

When a sinusoidal a-c voltage is impressed on a circuit of resistance only, such as a lamp or a heater, the current will be sinusoidal and the current maximum will occur at the instant of maximum voltage. Thus it is observed that Ohm's law holds true instantaneously in a-c as well as in d-c

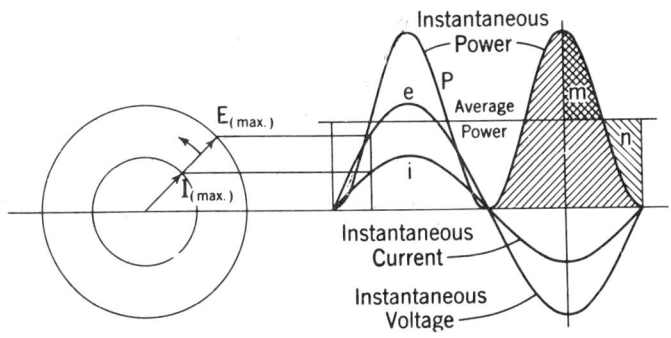

FIG. 3-21. Current, voltage, and power in a resistive circuit.

circuits. The diagram in Fig. 3-21 shows this situation. In addition, a curve showing the instantaneous product of current and voltage is given. This product is the instantaneous power and varies from zero to maximum and back again twice in every cycle.

PEAK, AVERAGE, AND RMS VALUES OF A-C WAVES

In the discussion of alternating-current waves, so far, the magnitude has been specified in terms of the maximum or peak value of the wave. This is one way of stating a-c magnitudes. This method may be rather confusing or misleading, however, in dealing with average power output since, in a resistance circuit such as is shown above, only the peak power is equal to the product of the maximum current and maximum voltage:

$$P_{peak} = E_{max} I_{max}.$$

The average power, or heating effect, is the main consideration in many studies of a-c circuits. In Fig. 3-21 the instantaneous power is represented by a sine wave of double frequency. A horizontal line drawn through the center of this sine wave will show the average power. Since this axis of the wave is equally distant from the peaks it follows that the axis is one half of the peak power. This ratio is also verified graphically by the observation that the double cross-hatched section labeled m will just fit into the section labeled n. The area under the power wave is a measure of the energy, so the average power is a straight line which has the same area beneath it as the power wave. This is the horizontal line drawn at one half the peak power. Average power for a resistance circuit may, therefore, be specified as

$$P_{average} = \frac{E_{max} I_{max}}{2} = \frac{E_{max}}{\sqrt{2}} \times \frac{I_{max}}{\sqrt{2}},$$

since $2 = \sqrt{2} \times \sqrt{2}$. The values represented by $E_{max}/\sqrt{2}$ and $I_{max}/\sqrt{2}$ are values of current and voltage that will give average power when multiplied together and are called, therefore, *effective* values of current or voltage. Then the current having 10-amp maximum would have an effective value of

$$I_{effective} = \frac{I_{max}}{\sqrt{2}} = \frac{10}{\sqrt{2}} = 7.07 \text{ amp.}$$

Effective and rms are terms used interchangeably. Rms means *root mean square;* more specifically, the square root of the average of the squares. This concept gives rise to an alternate development for effective values. Since it is desirable to have $P = I^2 R$ in a-c as well as d-c, the I for a-c will have to be such as to give the same average power as the d-c. The instantaneous power is $i^2 R$, where i is the instantaneous current, so if the values of i^2 are averaged over a half period, the resulting magnitude must be equal to I^2. This average is given in Table 3-1 for the 10-amp maximum current referred to previously. It is found that the average of the i^2 is 50 and the square root is 7.07 amp. This conclusion confirms the previous analysis and indicates the reason for the use of the abbreviation rms. The effective values are used so extensively in engineering work that unless otherwise specified it is customary to give the value of current or voltage in effective rms amperes or volts. Electric meters are likewise calibrated in rms values unless otherwise marked.

In a few applications, in power supplies and rectifiers for example, the average electron drift or flow is the real measure of effectiveness of the

Chapter 3   *Circuits with Time-Varying Voltages*   101

TABLE 3-1

DETERMINATION OF AVERAGE AND RMS VALUES
OF A SINE CURRENT HAVING 10 AMP MAXIMUM
VALUE

Degrees	i (amperes)	i^2
10	1.74	3.03
20	3.42	11.79
30	5.00	25.00
40	6.43	41.35
50	7.66	58.67
60	8.66	75.00
70	9.40	88.36
80	9.86	97.22
90	10.00	100.00
100	9.86	97.22
110	9.40	88.36
120	8.66	75.00
130	7.66	58.67
140	6.43	41.35
150	5.00	25.00
160	3.42	11.79
170	1.74	3.03
180	0.00	0.00
Sum........	114.34	900.8
Average.....	6.36	50.0

Equivalent direct current = $\sqrt{50.0}$ = 7.07 amp.

current and for these the true average of the half wave is used. As is shown in Table 3-1, 6.36 amp d-c will give a cumulative electron drift equivalent to a rectified alternating current having a maximum value of 10 amp. The average alternating current is, therefore, specified as 0.636 of the maximum value of a sine wave.

RATE OF CHANGE OF CURRENT IN A SINE WAVE

An important characteristic of a-c circuits depends upon the rate at which the current is increasing or decreasing. This rate of change was discussed in Chapter 1; expressed mathematically it is

$$\frac{d(I_m \sin 2\pi ft)}{dt} = 2\pi f I_m \cos 2\pi ft.$$

In order to emphasize this relation it will also be developed from a simple sine curve of current as shown in Fig. 3-22. Since the vertical change in the sine wave represents the change in current and the horizontal change represents the change in time, the current change divided by the corresponding time change is by definition the rate of change of the current. This ratio is also the slope or steepness of the sine curve.

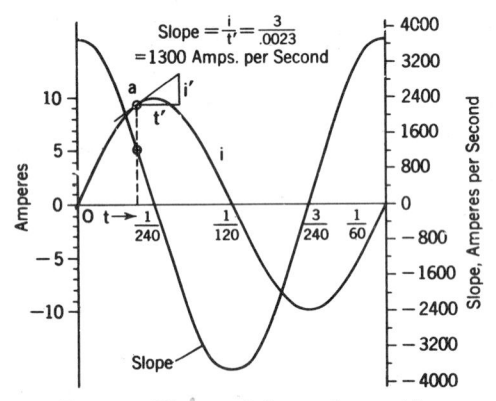

Fig. 3-22. The rate of change of current in a-c.

At the point a a tangent is drawn and the slope can be determined by dividing the distance i' in amperes by the distance t' in seconds. If this same procedure is followed for a number of points on the current curve and the results plotted against the corresponding time, a curve will be obtained for the slope. This curve will be a sine wave displaced by 90°. This displacement is another of the significant characteristics of the sine curve. It will be observed that the maximum value is at the time the current is passing through zero, which means that it is a sine wave that leads or precedes the current wave by a time-phase angle of 90° or $\pi/2$ radians. The maximum value of the curve of the rate of change of current is $2\pi f$ times the maximum of the current curve where f is the frequency. This maximum magnitude may be explained by the fact that the rate of change when the current is going through zero is represented by the terminus of the phasor moving at full velocity. Since its angular velocity is $2\pi f$ radians or f complete revolutions a second, its velocity or rate of change will be $2\pi f$ times its length, and the length is equal to the maximum value of the current. This relationship is very important, as it is the basis for the determination of the reactance of a-c circuits.

COIL RESPONSE TO A-C CURRENTS AND VOLTAGES

The response of a coil of inductance L, and carrying a current i, will be studied with the aid of Fig. 3-23. The instantaneous current is represented by the sine wave marked i. Since the voltage induced in a coil is equal to

Chapter 3   *Circuits with Time-Varying Voltages*   103

the inductance times the rate of change of current (as specified in the definition of the henry), the magnitude of the induced voltage is

$$E_{L(\text{max})} = 2\pi f\, I_{\text{max}} L = (2\pi f L) I_{\text{max}}.$$

The quantity $2\pi f L$ is called the *inductive reactance*, or often just the *reactance* of the coil. The usual symbol for this term is X or X_L, the subscript L being

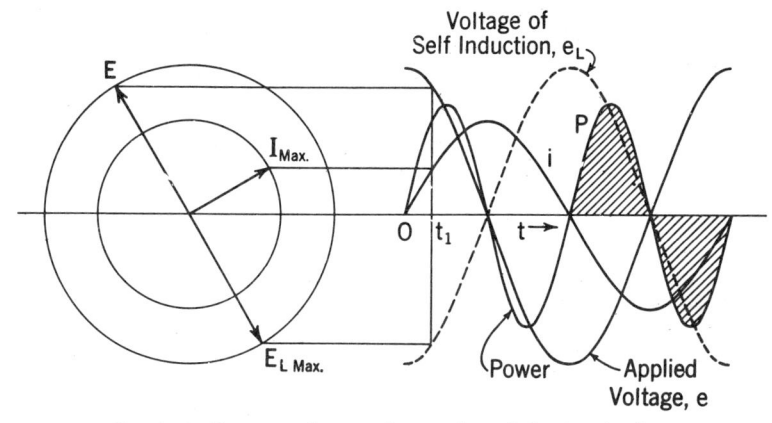

Fig. 3-23. Current, voltage, and power in an inductive circuit.

used to indicate that it is a reactance caused by an inductance coil. The reactance is expressed in ohms, just as was resistance, because the product of the current and reactance gives the magnitude of the reactance voltage.

Lenz's law states that this voltage is in such a direction as to cause a current that would oppose the change of flux. At time t_1 of Fig. 3-23, the current is increasing, so a current that would oppose the change of flux would be a negative current and the voltage of self-induction e_L is also negative. This instantaneous voltage wave is shown on the diagram as a dashed line and is the negative of the rate of change of current, so

$$e_L = -L\frac{di}{dt}.$$

REACTANCE VOLTAGE

Although the voltage of self-induction is extremely important in understanding the mechanism of the voltage-current relationships in an inductance coil, it is itself seldom used in circuit analysis. Usually the designer is interested in how much current will flow when a certain voltage is impressed on the circuit, and the emphasis is therefore upon the impressed voltage and not upon the internal opposing voltage. If the resistance of the coil is negligibly small, and if an a-c voltage is impressed upon the coil, the alternating current flowing will increase until the impressed voltage is neutralized by the voltage of self-induction. Under these circumstances the voltage of self-induction is equal at every instant to the impressed voltage, so that the impressed voltage wave e may be drawn on the diagram of Fig. 3-23 as equal and opposite to the e_L wave. (The assumption that the resistance is negligibly small may seem idealistic to the beginning student, but many of the most effective methods of measurement at high frequency are based on this assumption.)

The instantaneous power input to the coil is obtained from the instantaneous product of the current and voltage waves. This is shown in Fig. 3-23 and is a sine wave of double frequency whose average value is zero. In other words, the positive loop indicates that energy is being stored in the magnetic field, and the negative loop shows that an equal amount of energy is returned to the circuit or source. The energy is stored during the time the current (and magnetic field) is increasing and is returned when the current is decreasing in magnitude. This ability of the magnetic field to absorb and return energy gives a coil electrical characteristics that are similar to mechanical inertia.

To summarize, it has been shown that the current in a coil across which a sinusoidal voltage is impressed lags 90° behind the applied voltage, that the net power input is zero, and that the reactance is directly proportional to frequency. (Because of the increase of reactance with frequency, a coil is often used to reduce the high-frequency current in a circuit or branch of a circuit. Such coils are called *chokes* because they choke off high-frequency current.) A radio-frequency choke was shown in Fig. 3-11(b). When coils are used as chokes at power frequency they always use an iron core to increase the inductance.

RESISTANCE AND INDUCTANCE IN SERIES WITH A-C APPLIED VOLTAGE

When an a-c current flows through a resistor and inductor in series, as shown in Fig. 3-24, the voltage across the resistor will be in phase with the current, as shown by the dashed curve e_r. The voltage across the coil will be 90° ahead of the current, as is shown by the dashed curve e_L, where e_L is now used to represent the impressed voltage necessary to force the current through the inductor. The total voltage will be the sum of these two voltages, which is also a sine wave leading the current by an angle that is dependent upon the relative magnitudes of drops across the coil and the resistor. This angle is shown on the diagram as θ*, and it may be demonstrated that

$$\tan \theta = \frac{X}{R}.$$

It has been assumed in the study of coils so far that the resistance of the coil was so small that it might be neglected. When this is not true, the resistance is usually considered as being in series with the reactance, and the coil itself acts as a reactance and resistance in series. In this case, the ratio

---

* θ is the Greek letter theta and is commonly used to indicate an angle.

of the voltage across the coil to the current flow in the coil is neither a resistance nor a reactance, but is given the name of *impedance*. It is measured in ohms just as are resistance and reactance.

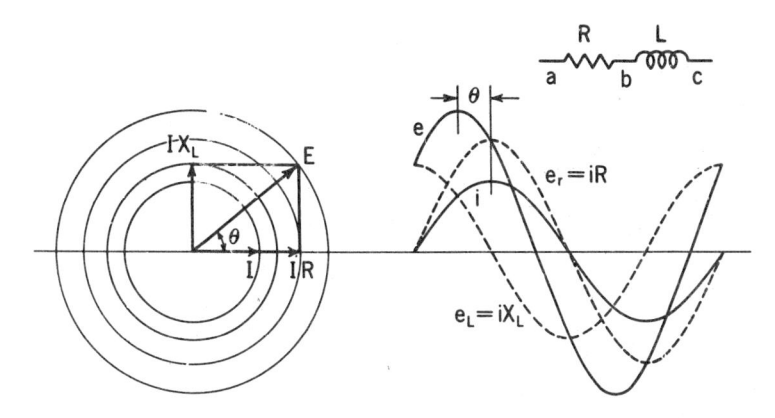

Fig. 3-24. Alternating current and voltage with resistance and inductance in series.

Reference to Fig. 3-24 will show that the maximum value of the voltage is not the sum of the maximum values of the resistance drop and the reactance drop. If one is large with respect to the other, the smaller one will have very little effect on the magnitude of the voltage drop (or impedance) but will affect the phase angle. The magnitude of the impedance which is referred to as Z may be found as follows:

$$Z = \sqrt{R^2 + X_L^2}.$$

In most radio equipment it is desirable to have a large ratio of X to R for coils. This characteristic of coils is so important that it has been given a special designation. The ratio of reactance to resistance is a figure of merit of the coil design and is called the Q^* (quotient) of the coil.

RESPONSE OF CAPACITORS TO SINUSOIDAL A-C

Since the instantaneous charge on a given capacitor is proportional to the voltage at any instant, the charge will increase and decrease as the voltage increases and decreases. In Fig. 3-25 the voltage has been plotted as a sine wave and the charge as another sine wave of different magnitude but in phase with the voltage. If the charge on the capacitor is continually changing, the conductor connecting the capacitor to the remainder of the circuit must have a flow of electrons to and from the capacitor. The current in the circuit is, therefore, the rate at which the charge on the capacitor is being increased or decreased. It was learned earlier that the rate of change

* This Q has no relation to the symbol Q for charge.

of a sine wave is another sine wave 90° ahead of the original wave in time phase, so the current in the capacitor circuit may be shown in Fig. 3-25 as a sine wave *leading* the wave of charge and voltage by 90° and labeled i. (It may be noted that this current is opposite in time phase to the current in an inductance, which lags 90° behind the voltage.)

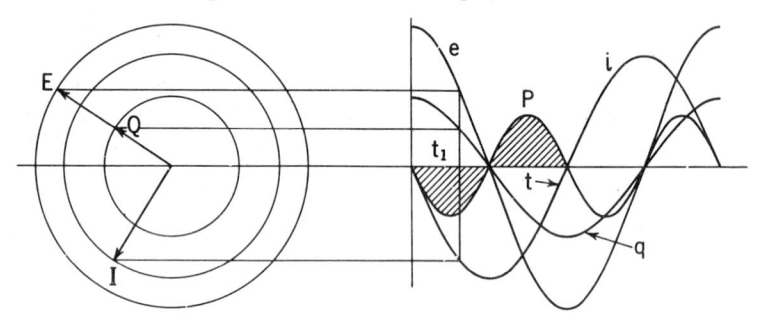

Fig. 3-25. Current, voltage, and power in a capacitive circuit.

The instantaneous power is also plotted in Fig. 3-25, and it is found to be a sine wave of double the voltage frequency, having zero average power as in the case of the inductance coil shown in Fig. 3-23. A careful study, however, will show that for the quarter cycle when the voltage is going from positive maximum to zero, the power is negative in the capacitor circuit while it is positive in the circuit containing the inductance coil. In the following quarter cycle it is positive in the capacitor circuit and negative in the inductive circuit. This is an important characteristic of these circuit elements, because it permits periodic power transfer from one element to the other and is the basis for the oscillations of tuned circuits used so extensively in radio equipment.

In a preceding paragraph it was determined qualitatively that the current was equal to the *rate of change* of the charge. It remains to determine the numerical relationships which involve the frequency. In the discussion on the rate of change of a sine wave it was found that the maximum value for the rate of change is $2\pi f$ times the maximum value of the original sine wave. The maximum rate of change of the charge Q is, therefore, $2\pi f$ times $Q_{\max}$ and this is likewise the maximum value of the current. Stated mathematically, this relation is

$$I_{\max} = 2\pi f Q_{\max}.$$

It is known that the charge is equal to the product of the voltage and capacitance; thus

$$Q_{\max} = CE_{\max}.$$

Substituting this value in the above equation,

$$I_{\max} = (2\pi f C) E_{\max},$$

where $2\pi fC$ is a constant giving the relation between the current and the voltage. Since the reactance of a capacitive circuit is that value by which the current must be multiplied in order to obtain the voltage, the reactance of a capacitor may be specified as $1/(2\pi fC)$. In mathematical form this statement is

$$E = IX_C = \frac{I}{2\pi fC},$$

so that
$$X_C = \frac{I}{2\pi fC}.$$

Since both the capacitance and frequency are in the denominator, it follows that the impedance is decreased with increase of frequency and is decreased also with an increase in capacitance. This is the opposite of the reactance of an inductance, which increases both with an increase of frequency and of inductance.

RESISTANCE AND CAPACITANCE IN SERIES WITH A-C APPLIED VOLTAGE

When resistance is connected in series with a capacitor, a situation is obtained similar in many ways to that existing in the case of a resistance in series with an inductance. The voltage across the resistance is in phase with the current, *while the voltage across the capacitor lags 90° behind the current.* The total voltage, being the sum of the two component voltages, also lags behind the current. This angle of lag, or the angle by which the current leads the voltage, has a tangent, the value of which is X_C/R, or

$$\tan \theta = \frac{X_C}{R}.$$

In the design and manufacture of capacitors it is easy to keep the resistance low, so that it is usual to have the capacitive reactance very large with respect to effective resistance. In other words, the Q (ratio of X_C to R) of capacitors is normally very large. Circuits composed of resistors and capacitors in series are very common in radio circuits. The equivalent impedance of such circuits is

$$Z = \sqrt{R^2 + X_C^2}.$$

RESISTANCE, INDUCTANCE, AND CAPACITANCE IN SERIES WITH A-C APPLIED VOLTAGE

When a number of circuit elements are connected in series, the current in all of the elements must be the same. The voltage across each element is determined by the current and the impedance and reactance characteristics of the element. The voltage of the coil reactance is of opposite phase to that of the capacitor-reactance, so that these voltages will be directly subtractive. Thus

$$Z = \sqrt{R^2 + (X_L - X_C)^2}.$$

The total voltage will be the instantaneous sum of all the voltages across the individual elements. Since the inductive reactance drop is often neutralized by capacitive reactance drop, it is very common for the voltage across the individual circuit elements to be greater than the impressed voltage where high-Q coils are used.

INTRODUCTION TO PHASOR CIRCUIT TECHNIQUES

The preceding discussion of simple electrical circuits when a-c voltages are applied is adequate for certain types of work with radio and electronic circuits. When numerical values of circuit relationships are required the above equations tend to be somewhat clumsy and inadequate, so special methods have been developed using phasors. We have already seen how a rotating phasor may be used to represent an alternating current or voltage. Usually the rotational characteristics are assumed but not used. Ordinarily either a current or voltage phasor is selected as reference and the other phasors of voltage and current are drawn with respect to this reference. Such a diagram may be likened to a snapshot of the group of rotating phasors, and it permits the relations of magnitude and phase to be represented easily and effectively.

The method of adding alternating currents that are not in phase has already been discussed. It was found that the phasors may be added by completing the parallelogram and drawing the diagonal. It is possible to define the impedance of a circuit as a special kind of phasor so that the same circuit procedures that were used in d-c can be applied to a-c circuits.

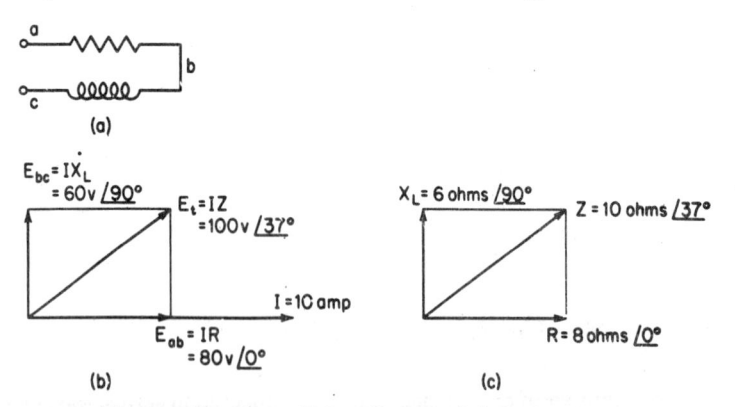

Fig. 3-26. The phasor diagram for a simple inductive circuit.

The solution of a simple a-c circuit by use of phasors will illustrate the method. Let it be assumed that a current of 10 amp (rms value) is flowing through a resistor of 8 ohms and a coil having a reactance (X_L equal to $2\pi f L$) of 6 ohms, as shown in the circuit of Fig. 3-26(a). The current is selected as the reference phasor and drawn horizontally to the right using a length of 10 units equal to the rms amperes. The voltage required to force this current through the resistor is equal to IR and this is in phase with the current as explained previously. The voltage required to force the current through the coil is equal to IX and must reach a maximum time 90 degrees before the current. Since the phasors are always assumed to rotate in a counter-clockwise direction, the phasor for the coil voltage E_{bc} is drawn vertically upward and is said to lead the current phasor by 90°. The total voltage required to produce the 10-amp current is then the sum of the two voltage phasors and is found by completing the parallelogram. This resultant phasor is the impedance drop as explained previously.

Since the current is a common element in all of these voltages, the phasor diagram in part (c) of Fig. 3-26, which is identical with the voltage diagram except for the scale, may be said to be an impedance diagram. It will be noted that the resistance does not change the phase of the voltage with respect to the current. The inductive reactance phasor, however, rotates the voltage 90° in a counter-clockwise direction with respect to the current. The phasor sum of the resistance and reactance is a phasor that rotates the resultant impressed voltage by an angle θ, whose tangent is X_L divided by R. The impedance also has a magnitude element so that the magnitude of the voltage impressed is equal to the product of the current in amperes and the impedance in ohms.

The impedance may therefore be defined as a device (mathematically known as an operator) that is applied to the current to get both the magnitude and phase of the voltage. When the current is multiplied by the impedance the magnitude is the product of the magnitudes of the two phasors, and the phase of the voltage is the phase of the current rotated through the phase angle of the impedance.

Usually the voltage is the known phasor and the current is desired. Since

$$E = IZ, \quad \text{then} \quad I = \frac{E}{Z}.$$

The division of the voltage by Z is the inverse of multiplication, so the magnitude of the current phasor is obtained by dividing the magnitude of the voltage by the magnitude of the impedance, and the phase of the current is found by rotating the current phasor in the negative or clockwise direction by the angle of the impedance.

PHASOR SOLUTION OF RESISTANCE AND CAPACITIVE CIRCUITS

A similar procedure is followed in solving circuits involving a resistor and capacitor in series. Let the resistor in this case have a resistance of 3 ohms and the capacitive reactance X_C (equal to $1/2\pi fC$) be 4 ohms. Let the current of 10 amp be used as the reference phasor in Fig. 3-27(b). The voltage across the resistor is 30 v in phase with the current, and the voltage necessary to force the current through the capacitor is 40 v. This capacitor voltage will reach its maximum later than the current maximum, so the phasor representing the voltage will lag behind the current phasor by 90° and so is drawn vertically downward. The sum found by completing the parallelogram is 50 v, and this voltage lags 53° behind the current; or the current leads the voltage by 53°

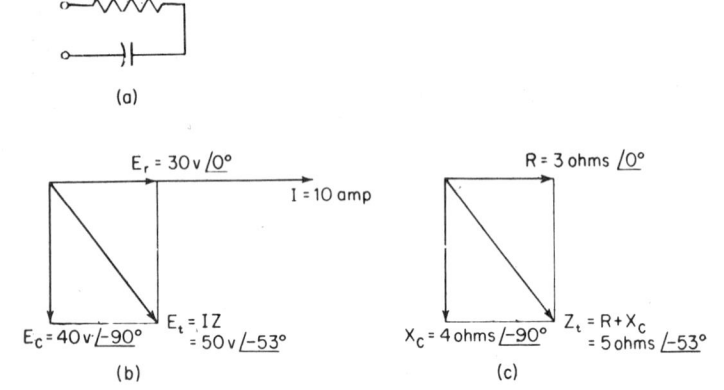

Fig. 3-27. The phasor diagram for a simple capacitive circuit.

When the current phasor is multiplied by the capacitive reactance, the magnitude is changed by the product of the magnitudes and the phase is rotated 90° in the clockwise direction. When the current phasor is multiplied by the impedance Z, the magnitude of the voltage is the product of the current and impedance magnitudes, and the phase is rotated by the angle θ whose tangent is X divided by R.

When the voltage is known, the current is found by dividing the voltage by the impedance just as in the inductive circuit. In this case, the impedance has a negative angle, so the current phasor is rotated counterclockwise so that it leads the voltage.

PHASOR SOLUTION OF GENERAL SERIES CIRCUITS

In any series circuit the current is common to all circuit elements and the total voltage is the sum of the phasor voltages across each of the circuit components. In most coils the resistance of the coil is not negligible and so both resistance and inductance must be considered. In series circuits the coil is represented by a series resistance and inductive reactance that will give the true voltage drop across the coil.

The addition of impedance phasors have been obtained by completing the parallelogram, but a more convenient method is to draw these phasors in sequence with their correct magnitude and phase direction. The resultant impedance is then the phasor from the origin to the arrowhead of the final phasor.

Chapter 3 · Circuits with Time-Varying Voltages 111

These methods will be demonstrated by an example with several circuit elements in series.

Example: A coil having a resistance of 20 ohms and an inductance of 8 millihenrys is connected in series with a resistor of 30 ohms and a capacitor of 0.30 μf. The circuit is connected to a power source of 250 v at a frequency of 2,000 cycles. Determine the magnitude of the current in amperes and the phase of the current with respect to the voltage.

Solution: First, calculate the inductive reactance.

$$X_L = 2\pi f L = 2\pi\, 2{,}000 \times 0.008 = 100.5 \text{ ohms.}$$

Second, calculate the capacitive reactance.

$$X_C = \frac{1}{2\pi f C} = \frac{1}{2\pi\, 2{,}000 \times 3 \times 10^{-7}} = 265 \text{ ohms.}$$

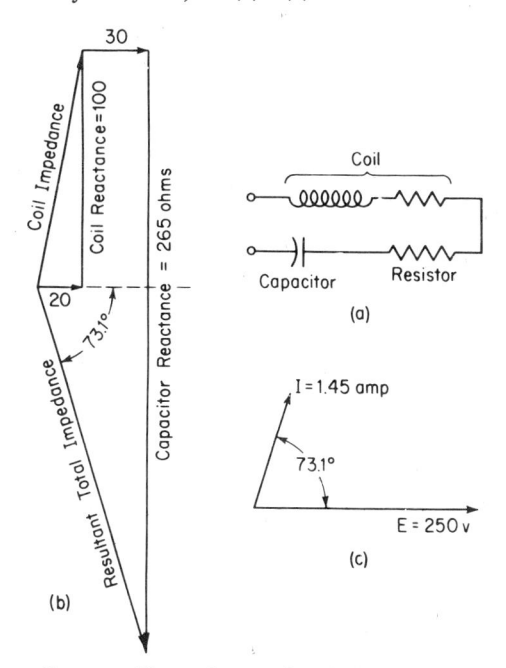

Fig. 3-28. Phasor diagrams for a series circuit.

Third, draw the phasor diagram of the impedance. This may be done graphically as shown in Fig. 3-28(a) or may be calculated as

$$Z = \sqrt{(R_{\text{coil}} + R_R)^2 + (X_L - X_C)^2} = \sqrt{(20 + 30)^2 + (100 - 265)^2}$$
$$= \sqrt{50^2 + 165^2} = 172$$

$$\theta = \tan^{-1}\frac{165}{50} = 73.1° \text{ (counter-clockwise).}$$

This same result is obtained from measured values on the diagram if it is accurately drawn.

Fourth, determine the current.

$$I = \frac{E}{Z} = \frac{250 \text{ v}}{172 \text{ ohms} \:/\:-73.1°} = 1.45 \text{ amp} \:/73.1°.$$

Thus the current will have a magnitude of 1.45 amp and will lead the impressed voltage by an angle of 73.1° as shown by the phasor diagram of Fig. 3-28(b).

An interesting peculiarity of the series circuit in the above example is that the voltage across the capacitor is much greater than the impressed voltage. The drop across the capacitor is

$$E_C = 1X_C = 1.45 \times 265 = 384 \text{ v}.$$

This multiplication of voltage can be much greater if the capacitor reactance is made equal to the coil reactance and if the circuit resistance is made as low as possible.

SERIES RESONANT CIRCUITS

The term *resonance* comes from the characteristic of some objects which respond to sound. Usually these objects respond to sounds of certain pitches only, and are said to be "tuned" to those pitches. The responses to sounds are of such high frequency and small amplitude that it is difficult to

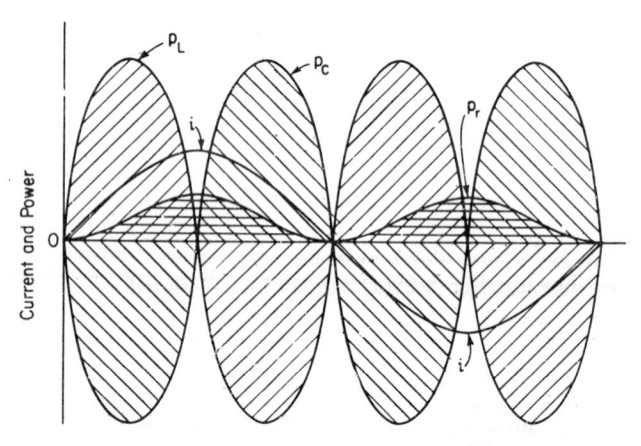

FIG. 3-29. Instantaneous power in a series-resonant circuit of low resistance.

observe them visually. A similar phenomenon is the pendulum clock. Here the pendulum swings back and forth, shifting kinetic energy (motion of the pendulum in the center of the swing) to stored or potential energy (raised position of the pendulum at the end of the swing). The slight impulse from an escapement wheel is all that is needed to keep this pendulum swinging back and forth indefinitely. The amplitude of the vibration or movement is much greater than could be given by a single impulse from the escapement wheel, but the regular impulse from the escapement is sufficient to replace the losses of energy in the friction and air resistance of the pendulum.

A similar phenomenon may be produced in electric circuits which have both inductance and capacitance. Both of these circuit elements store energy for one half cycle and return it on the next half cycle. Since the capacitor is storing energy when the inductance is returning energy, it is possible for the capacitor and inductance to pass large amounts of energy from one to the other with the outside circuit supplying only the losses.

In Fig. 3-29 a series resonant circuit is shown with resistance which is small with respect to the inductive and capacitive reactances. The instantaneous power for each circuit element is plotted, and it is noted that the energy being absorbed by the coil is just equal to that being supplied by the capacitor. A quarter cycle later the capacitor will absorb energy which is supplied by the coil. True resonance occurs only when the inductive

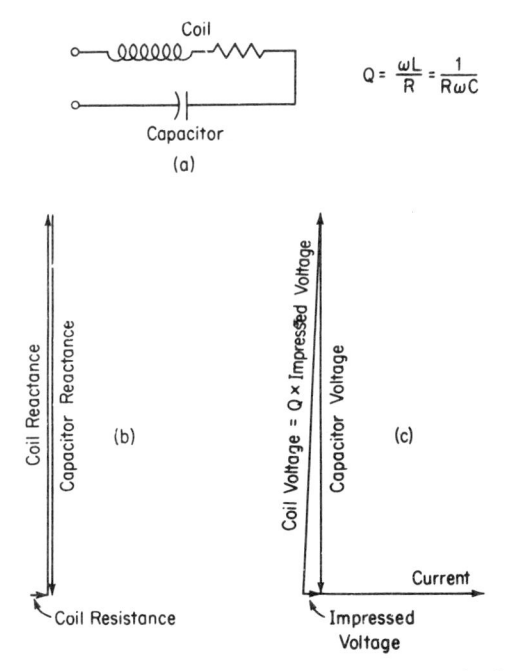

$$Q = \frac{\omega L}{R} = \frac{1}{R\omega C}$$

FIG. 3-30. The phasor diagram for a series-resonant circuit.

reactance is exactly neutralized by the capacitive reactance and the effective impedance of the circuit is equal to the resistance. Under these conditions, large amounts of energy are passed from coil to capacitor and back again, as shown in Fig. 3-29. In most resonant circuits the resistance of the capacitor is negligible and so the Q of the circuit equals the Q of the coil. The voltage across each element at resonance will be Q times the impressed voltage, as shown by the phasor diagram of Fig. 3-30.

The adjustment to the resonant condition may be made by varying the inductance, the capacitance, or the frequency. When it is desired to obtain resonance at a specific frequency, it is usual to adjust the inductance or capacitance. This is called *tuning the circuit*. When the frequency is varied (but constant voltage maintained), the current will vary in accord with the curves shown in Fig. 3-31. The magnitude of the peak current is dependent

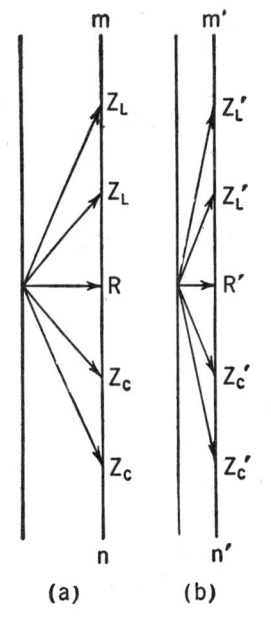

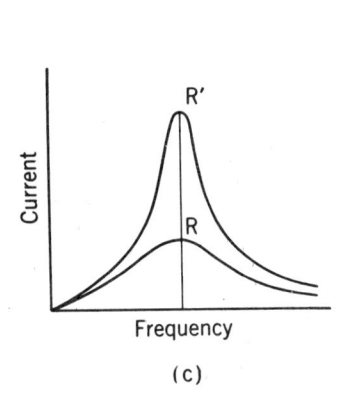

Fig. 3-31. The variation of current with frequency in a series-resonant circuit with constant voltage impressed.

upon the effective resistance, so that the sharpness of the peak of the curve increases as the resistance is reduced. The curve that is labeled R has a coil Q of about 3 while that of R' has a Q of 10. The Q of good coils at radio frequency will range from 100 to 300. With such high values of Q the resonant peak becomes very sharp, and the circuit is said to be highly selective with respect to frequency.

At radio frequencies (10 kc to 10 megacycles) the losses in the coil insulation and the rapid increase in the resistance owing to skin effect cause the effective resistance to increase about as fast as the reactance, so that the Q of the coil remains relatively constant for large changes in frequency.

Chapter 3　　*Circuits with Time-Varying Voltages*　　115

With high values of Q it is possible to use the series resonant circuit as a voltage multiplier. Since the voltage across the coil or capacitor is many times the input voltage, this larger voltage may be used to control the grid of a vacuum tube. Although such a circuit is a good "voltage transformer," it is not usually desirable where a power conversion is required.

The above analysis has assumed that a constant potential source of power was available. In radio circuits this is seldom true, as the source of the voltage may have considerable impedance. In series resonant circuits, the response depends upon the resistance of the entire circuit, so it is just as important to have low resistance in the source as in the load circuit. A vacuum tube with a high plate resistance is not, therefore, a suitable source* for a series resonant circuit if sharp frequency response is desired.

PARALLEL CIRCUITS

When a-c circuits are connected in parallel, the same general rule applies as for resistances in parallel; that is, the current flowing through each part of the parallel circuit is determined independently. The total

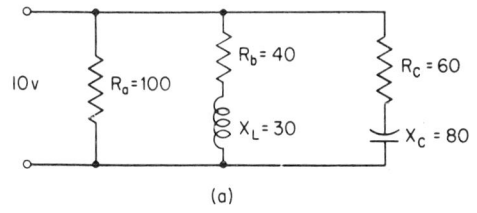

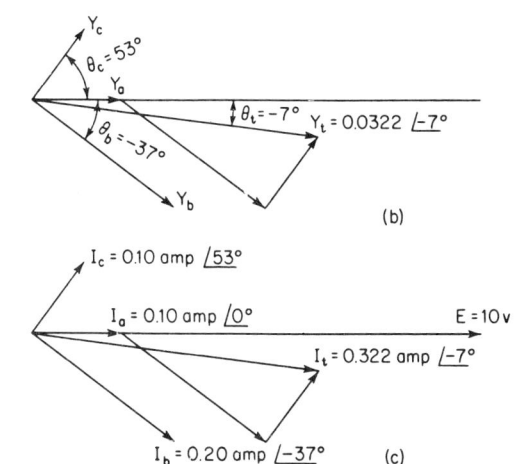

FIG. 3-32. The phasor solution of a parallel circuit problem.

* See discussion of "impedance matching," p. 117.

instantaneous current is, then, the sum of the instantaneous currents in all of the parallel branches.

These currents are not usually in phase, so the effective values cannot be added directly. When the phasor representation of the currents are added vectorially, however, correct results are obtained. Figure 3-32 shows a parallel circuit having three branches. The total current is

$$I_{total} = \frac{E}{Z_a} + \frac{E}{Z_b} + \frac{E}{Z_c} = E\left(\frac{1}{Z_a} + \frac{1}{Z_b} + \frac{1}{Z_c}\right)$$

The quantity $1/Z$, which is a phasor, is called the admittance and is represented by the symbol Y. The magnitude of Y is $1/Z$ and the angle is the same as the angle of Z, but it is in the reverse direction. The total admittance of a circuit is the sum of the individual admittances and may be expressed mathematically as

$$I_{total} = E(Y_a + Y_b + Y_c) = EY_{total}.$$

Example: Determine the magnitude and phase of the current in the circuit of Fig. 3-32(a).

Solution: Determine the admittances of the individual branches.

$$Y_a = \frac{1}{Z_a} = \frac{1}{100\ /0°} = 0.01\ /0°$$

$$Y_b = \frac{1}{Z_b} = \frac{1}{50\ /37°} = 0.02\ /-37°$$

$$Y_c = \frac{1}{Z_c} = \frac{1}{100\ /-53°} = 0.01\ /53°$$

These phasors are shown in (b) of the diagram and the vector sum of these phasors is also shown as $Y_t = 0.0322\ /-7°$. This addition may be accomplished by determining the horizontal and vertical components by means of trigonometric tables or slide rule. The addition can be made graphically also, and this is probably the most simple method. To make the graphical solution, it is necessary to use a sharp pencil, an accurate ruler, and a good protractor for measuring the angles if results are to be accurate.

In Fig. 3-32(c) the phasor diagram of voltage and currents is given. The voltage is used as the reference phasor because it is known. The currents in the individual branches are shown and the phasor addition of the currents gives the total current. This current diagram is identical to the admittance diagram except for a change of scale.

PARALLEL RESONANT CIRCUITS

As indicated previously, resonant circuits are extensively used in radio equipment, and when a coil and capacitor of the same reactance are connected in parallel across a circuit, it acts as an extremely high impedance at

Chapter 3   Circuits with Time-Varying Voltages   117

the resonant frequency. This will be illustrated by connecting the coil and capacitor of Fig. 3-28 in parallel across a circuit of 100 v as shown in Fig. 3-33. The admittance of the coil is

$$Y_L = \frac{1}{Z_L} = \frac{1}{104 \,/87.2°} = 0.0096 \,/-87.2°.$$

The admittance of the capacitor is

$$Y_c = \frac{1}{Z_c} = \frac{1}{104 \,/-90°} = 0.0096 \,/90°.$$

The total admittance is the vector sum of these two.

$$Y_t = 0.00046 \,/0°$$

The equivalent impedance is the reciprocal of the admittance, so that

$$Z_t = \frac{1}{Y_t} = \frac{1}{0.00046} = 2{,}170 \text{ ohms.}$$

With coils having a Q of 100 or more the equivalent impedance becomes very high, and this is a useful characteristic of the parallel resonant circuit. It is interesting that, in a parallel resonant circuit, the smaller the resistance is, the higher is the equivalent impedance. A circuit of this type has very high frequency selectivity when used with a source of high resistance. A parallel resonant circuit is therefore particularly suited for use as the load circuit on a vacuum tube and is so used in tuned radio-frequency amplifiers, as will be described in Chapter 11.

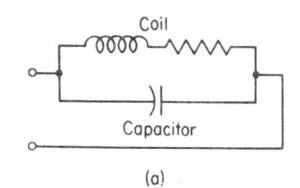

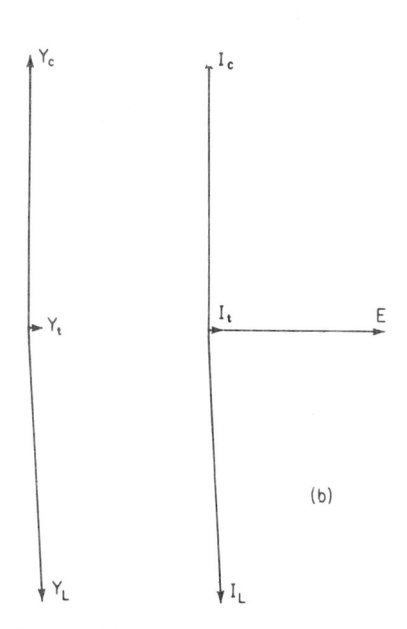

FIG. 3-33. The phasor diagram for a parallel resonant circuit.

IMPEDANCE MATCHING

In most electronic circuits the power involved is small, and so power efficiency is, by itself, of little importance. It is, however, important to pass as much power through the circuit as is possible in order to obtain the maximum signal. It can be proved that this maximum power transfer occurs when the impedance of the load is equal in magnitude to the impedance of the power source and has a capacitive reactance equal to the inductive reactance of the power source, or an inductive reactance equal to the capacitive reactance of the power source. This equalization of the impedances of the load and generator is known as *impedance matching*. In order to match impedances, the series resonant circuit should be used for power sources of low impedance and the parallel (or antiresonant) circuit should be used for high-impedance sources.

The use of parallel resonant circuits for impedance-matching purposes is quite common. Such a circuit is shown in Fig. 3-34. The high-impedance circuit is the standard parallel resonant circuit while the low-impedance circuit is tapped off the inductance. If this circuit is viewed from the low-

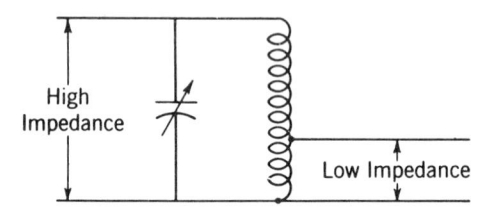

Fig. 3-34. Impedance matching using a parallel resonant circuit.

impedance connection, it is difficult to determine whether it is a series or a parallel resonant circuit, although it is unquestionably in resonance. This circuit can be used either to step up the impedance or to step it down.

WAVE METERS

The resonant circuit has extensive use as a wave or frequency meter for approximate measurements. The meter consists of a coil and a variable capacitor which is calibrated in terms of the frequency at which it resonates with the coil. Such a meter is placed close to a source of radio-frequency power and the capacitor adjusted until the indicator (a lamp or meter) has a maximum reading. This maximum indicates that the circuit is in resonance.

MUTUAL INDUCTANCE

If an inductance coil such as has been studied in Fig. 3-8 has a second coil wound around it, as shown in Fig. 3-35, and if an alternating current flows in the coil, a voltage will be induced in the secondary coil. This secondary voltage is caused by the flux of the primary coil, which is increasing and decreasing sinusoidally with time just as the current is varying. This varying flux produces a voltage in the primary winding which, by Lenz's law, tends to send a current through the coil in such a direction as to oppose the change of flux. Since the secondary coil is wound very closely around the primary, most of the flux which threads through the primary also threads through or links the secondary. The rate of change of this flux produces a voltage in the secondary. Since the flux is directly proportional to the current in the primary, the secondary voltage is directly proportional to the rate of change of primary current.

When two coils are in such a position that a change of current in one will produce a change of flux linking the second, the two coils are said to have *mutual inductance*. This mutual inductance is measured in the same units as self-inductance. Thus, *when a rate of change of one ampere per second in the primary coil will produce one volt in the secondary coil, the two are said to have one henry of mutual inductance.*

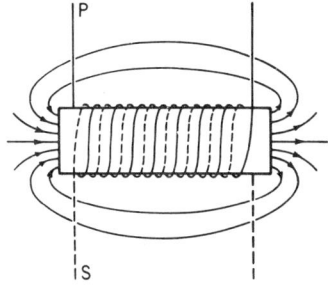

FIG. 3-35. A mutual inductance (air core transformer)

When two coils are wound very closely together, as in the above illustration, they are said to be closely coupled. However, when the two coils are located so that only a small part of the flux produced by the primary coil links the secondary, the coils are said to be loosely coupled. Loose coupling is shown in Fig. 3-36, where the two coils are placed some distance apart although still on the same axis.

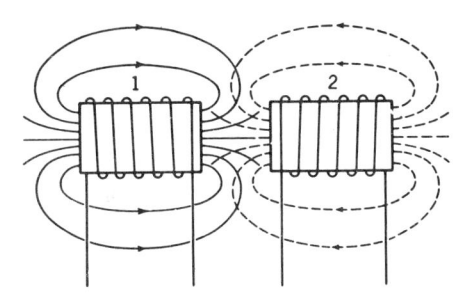

FIG. 3-36. A mutual inductance (loose coupling)

When dealing with the single coil, it was found that, neglecting resistance, the current in the coil increased until the voltage of self-inductance neutralized the impressed voltage. When a second coil is added, as in Fig. 3-35, it produces no effect on the primary as long as it is open-circuited. If, however, resistance or some other form of load is connected across the secondary terminals, a current will flow. This current, by Lenz's law, is in such a direction as to oppose the change of flux which is causing it. This tends to reduce the flux in the primary, which in turn tends to reduce the counter-voltage. As soon as the counter-voltage drops even a very small amount, an additional primary current is drawn from the line to neutralize the magnetomotive force of the secondary current and to bring the primary flux back to its original value. This increases the induced electromotive force until it is again equal to the impressed voltage. If the secondary is supplying power to a resistive load the secondary current will be in phase with the voltage, and the primary current which neutralizes it will also be in phase with the voltage; power will thus be fed into the primary and transferred to the secondary through the medium of the common flux. Such an arrangement of closely coupled coils is called a *transformer*. If the flux is in air, as in the diagram of Fig. 3-35, it is known as an *air-core transformer*. If an iron magnetic path is provided for the flux, the transformer is known as an *iron-core transformer*.

USE AND CHARACTERISTICS OF THE TRANSFORMER

Since in most transformers, particularly those having iron cores, the coupling is very close, for the preliminary discussion it will be assumed that all of the flux threads through or links both coils. If this condition exists, the rate of change of flux for each turn is the same, regardless of whether the turn is the primary or secondary. The induced voltage across the coil will, therefore, be proportional to the number of turns. This characteristic of a transformer is used to increase or decrease the a-c voltage. If it is desired to increase the voltage fed into the grid of a tube, a transformer may be used with many more turns on the secondary than on the primary. Or, if a high-impedance vacuum tube is being used to supply a loudspeaker where considerable current is required at low voltage, the transformer coil having the large number of turns will be connected to the vacuum tube and the coil having only a few turns will be connected to the loudspeaker. This is another instance of impedance matching.

A more careful study of the effects of the current in the transformer will now be made. It was learned earlier that the magnetic effects were proportional both to the current and to the number of turns. This means that the magnetic effect is proportional to the product of current and turns, usually called ampere-turns. Since the ampere-turns must be equal for neutralization, the currents in the primary and secondary will be in an inverse ratio to the number of turns. Thus, if the primary has three times as many turns as the secondary, it will require only one third as many amperes to obtain the same number of ampere-turns. It should be remembered that this neutralization is not complete. A certain amount of current will flow in the primary even when the secondary is open because flux must be produced to set up the counter-voltage of self-induction with which to neutralize the impressed voltage. This current is known as the *exciting current*. A very convenient rule in transformer analysis is that, *neglecting exciting current, the ampere-turns on the primary are just equal and opposite to the ampere-turns on the secondary*. Since the voltage is proportional to the number of turns and the current is inversely proportional to the number of turns, the product, or the volt-amperes, of primary and secondary are equal.

Iron-core transformers — power. In low-frequency transformers, and especially in power transformers, iron cores are used to reduce the amount of exciting current required and thus to improve the efficiency. In most cases, these transformers are designed for a certain definite maximum voltage on the winding. The voltage must be limited because the ratio of the exciting current to flux increases very rapidly when the flux density goes beyond certain limiting values, as is indicated in the magnetization curve of Fig. 2-27. Here it may be seen that the flux density rises very rapidly with increase of ampere-turns until the density has reached about 80,000 lines per square inch. Thereafter, it rises much less rapidly, and if a high flux density is required to neutralize the impressed voltage, extremely large exciting currents will result. If, for instance, a 110-v transformer were connected to a 220-v power line, the flux density in the core would have to be doubled in order to produce the necessary voltage of self-induction. If the transformer were designed to operate at a maximum flux density of 65,000 lines per square inch, the required density at 220 v would be 130,000, which is beyond the limits of the diagram. In circumstances of this kind the exciting current becomes so great that the I^2R losses will cause the transformer to overheat and be damaged. It is very important, therefore, that power transformers should be operated at their correct voltage and frequency rating.

Iron-core transformers — audio. The preceding analysis assumed that power in unlimited quantities was available at a constant voltage, which is a true condition for most power lines. In audio amplifiers, however, transformers are also used for matching impedances, and in this case the power source is usually a vacuum tube with relatively high plate resistance. The current in the primary is therefore dependent upon the strength of the signal. If the strength of the signal is limited, then there is low flux density in the audio transformer and the secondary signal faithfully reproduces the primary signal. If, however, the signal strength becomes excessive, the flux density reaches the saturation point and the flux is no longer proportional to the primary current, so the secondary ceases to reproduce the signal of the primary. This effect is known as *distortion* and is one of the many limitations to accurate reproduction of signals.

Air-core transformers. In most air-core transformers, the assumption that all of the flux links both windings is so far from true that results based on this analysis are not valid. It continues to be true, however, that the signal impressed on the primary causes a voltage and current to be set up in the secondary. The current in the secondary also causes a voltage to be produced in the primary which affects the primary current. The mathematical analysis of this type of circuit is beyond the scope of this text, but the meaning of certain terms, and a qualitative discussion of results, are given in the following pages.

COEFFICIENT OF COUPLING

In Fig. 3-36 it is seen that only a small part of the flux from coil 1 threads through, or links coil 2. If the total flux of coil 1 is designated as ϕ_1,* and that part of the flux which links coil 2 as ϕ_{12}, then the ratio of ϕ_{12} to ϕ_1 is a measure of the magnetic mutuality of the two coils. This ratio is also equal to the ratio of ϕ_{21} to ϕ_2, when ϕ_{21} is that part of the flux (caused by current in coil 2) which links coil 1 and ϕ_2 is the total flux in coil 2. This ratio is called the *coefficient of coupling of the two coils* and may be expressed mathematically as follows:

$$k = \frac{\phi_{12}}{\phi_1} = \frac{\phi_{21}}{\phi_2}.$$

It may be shown that the square of the mutual inductance is equal to the square of the coefficient of coupling multiplied by the product of the self-inductance of the two coils. Thus

$$M^2 = k^2 L_1 L_2 \quad \text{and} \quad k = \frac{M}{\sqrt{L_1 L_2}}.$$

If coils 1 and 2 are connected in series so that the fluxes add, the currents in both are the same, and the equivalent inductance is the sum of L_1 and L_2

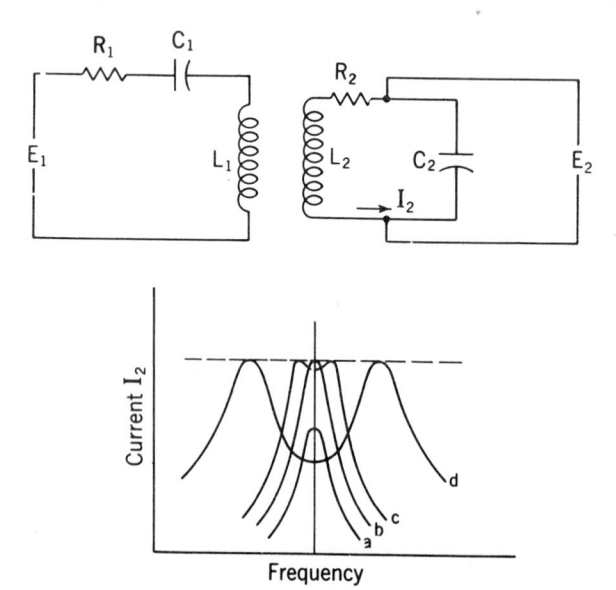

FIG. 3-37. Resonance curves for tuned circuits with different amounts of mutual coupling.

* ϕ is the Greek letter phi and is the symbol generally used for flux.

plus the mutual inductance of coil 1 with respect to 2 and the mutual inductance of coil 2 with respect to 1. This becomes

$$L_t = L_1 + L_2 + 2M.$$

If the two coils are connected so that the self- and mutual-inductance fluxes are in opposition, then

$$L_t' = L_1 + L_2 - 2M.$$

If the second equation is subtracted from the first, then

$$4M = L_t - L_t', \qquad M = \frac{L_t - L_t'}{4}.$$

This relationship is the basis for an experimental determination of the magnitude of mutual inductance.

SELECTIVITY AND COUPLING

Although it is desirable to have a very high value of coupling coefficient for power transformers, close coupling is not always desirable. In order to obtain a high degree of selectivity with respect to certain frequencies, the resonant circuits must be free to oscillate at their normal frequencies without too much interference from the currents in the mutual-inductance circuit element. Thus, if in Fig. 3-37 both the primary and secondary are tuned to resonance at a desired frequency, and if a signal of this frequency is connected to the primary, then a slight current and voltage will appear in the secondary even though it is some distance away. If the secondary is moved closer, this voltage E_2 will increase because more excitation is received from the primary coil to take care of the losses in the secondary oscillatory circuit. This increase does not continue indefinitely, however, for after reaching a certain critical value of mutual inductance the current and voltage in the secondary decline with increase of mutual inductance. The reactions of the secondary current on the primary circuit are such that the primary circuit is no longer in resonance at that frequency, and the primary current is reduced so much that the increased coupling reduces the secondary current and voltage.

If the frequency characteristics for these several conditions of mutual-inductance coupling are obtained from test, it is found that with a very low coefficient of coupling a curve is obtained similar to *a* in Fig. 3-37. It will be noted that it is very sensitive to changes in frequency, but the coupling is *insufficient* to give the maximum signal. If the coupling is increased until the *critical* value is reached, as in curve *b* of the figure, the maximum signal strength comes through at the tuned frequency. This condition is also very sensitive to frequency changes and is desirable where only the one definite response is required. If the coupling is increased just slightly beyond critical then the frequency response is fairly constant over a narrow band, dropping steeply on either side, as shown in Fig. 3-37, curve *c*. This circuit is extensively used in radio circuits, as indicated by the tuning coils and capacitors shown in Fig. 3-38. The mutual coupling is varied by adjusting the position of a powdered iron core by the screw at the top of the coil assembly. If the coupling is increased still more, a frequency characteristic similar to *d* in Fig. 3-37 is obtained, which has a definite double resonance response. This last type of response is an exaggerated form of the previous one and is useful only in showing the true character of the previous response.

ALTERNATING-CURRENT METERS

Most a-c meters are current-measuring devices. To this extent they are similar to d-c meters, but the similarity is limited, because the rapid oscillations of the current do not permit the use of the D'Arsonval or permanent magnet-moving coil type of meter. It is necessary to arrange some mechanism which will give torque in the same direction in spite of the reversal of current. Four main types of indicating meters are used for this purpose: the *dynamometer*, the *iron-vane*, the *rectifier*, and the *thermocouple* types of a-c meters.

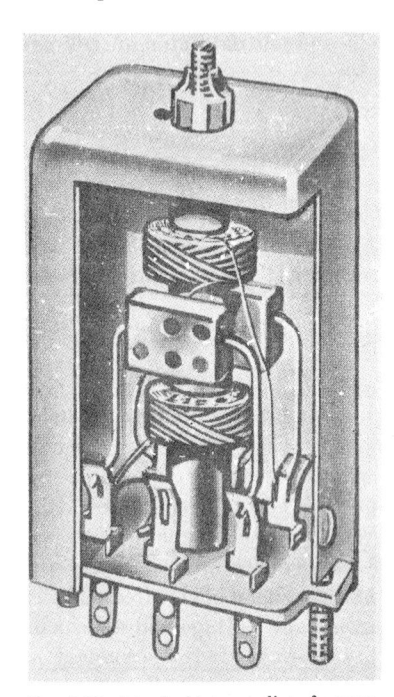

FIG. 3-38. A typical intermediate-frequency transformer with capacitor tuning.

The dynamometer type. In the dynamometer type of instrument a pair of coils carrying the alternating current to be measured produces a magnetic field. These coils are the heavy coils shown in Fig. 3-39. The same current or a portion of it is carried down to the movable coil through the springs. In this type of instrument the reversals of current in both coils come at the same instant so that the torque is always in the same direction and roughly proportional to the square of the current. This meter will operate satisfactorily on direct current, but it is not as sensitive as the usual d-c instrument. It may be used either as an ammeter or as a voltmeter, but its commonest use is as a wattmeter.

When it is used as a wattmeter, the load current is fed through the main field coils and the moving coil is connected in series with a resistance across the line. Since power is $EI \cos \theta$, the torque and hence the reading will be

Chapter 3   *Circuits with Time-Varying Voltages*   125

directly proportional to the power.

The chief disadvantage of this type of instrument is the cost of construction. Its use is limited, therefore, to wattmeters and high-precision laboratory meters.

The iron-vane type. The iron-vane type of instrument takes many different forms. All have a soft-iron vane attached to a shaft mounted in jeweled bearings. A spring provides the restoring torque and a pointer indicates the deflection. The torque to operate the meter depends upon the magnetic response of the soft-iron vane to the magnetic field set up by a fixed coil. A meter of this type is shown in Fig. 3-40. The altrenating current is flowing in the coil and magnetizes both the stationary and movable soft-iron plates. Since the polarity of both plates is the same, they will repel each other even though that polarity is reversing rapidly. No electrical connection is necessary to the moving element. When it is desired to use such a meter as an ammeter, a few turns of large wire are used for the coil. When used for a voltmeter, many turns of fine wire are used in the coil construction.

Rectifier-type meters. It is possible to rectify alternating current and use the standard type of d-c meter. When this is to be done it is common practice to include a small rectifier in the case of the instrument. A diagrammatic sketch of the circuit for such a meter is given in Fig. 3-41. At the left of the diagram is an illustration of the copper-oxide rectifier, approximately to full scale. Meters of this type are usually

FIG. 3-39. Dynamometer type of meter. (Courtesy Weston Electrical Instrument Corp.)

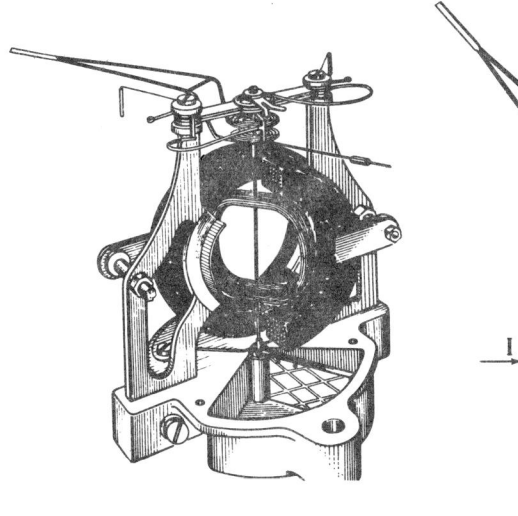

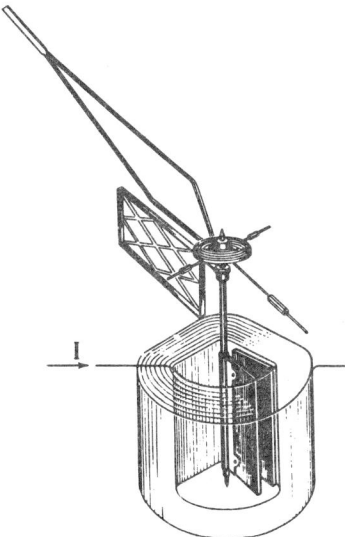

FIG. 3-40. Iron-vane type of meter. (Courtesy Weston Electrical Instrument Corp.)

limited to a few milliamperes, and so are used for low values of current and as voltmeters. Crystal rectifiers are sometimes substituted for the copper-oxide type. Meters of this type may be used at extremely high radio frequencies.

Rectifier-type meters are used as the a-c measuring element for the universal meters so extensively used in radio testing. In this meter the high-sensitivity d-c instrument is paralleled by a resistor in order to minimize the effect of the non-linearity of rectifier resistance. A typical circuit is shown in Fig. 3-42. In this case, R_1 is adjusted along with the rectifier and R_2 to give full-scale deflection at 2.5 v and to draw one milliampere in the meter circuit. The meter assembly then has a sensitivity of 1,000 ohms per volt, and the external resistors, as shown, give the proper voltage ranges. Such a meter measures the average value of the rectified wave and so is subject to wave-form errors.

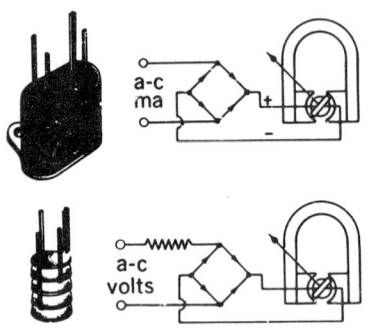

Fig. 3-41. Rectifier type of meter. (Courtesy Weston Electrical Instrument Corp.)

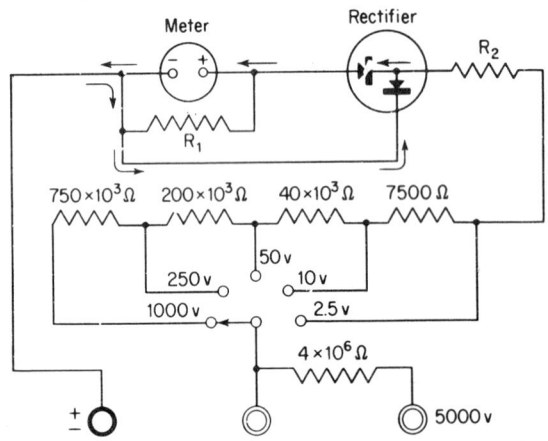

Fig. 3-42. A simplified diagram of a multirange a-c voltmeter with rectifier as used in a universal type of meter.

Thermocouple-type meters. This type of meter uses the thermocouple principle to supply a very sensitive d-c meter. A diagrammatic sketch is shown in Fig. 3-43. The current to be measured flows from A to B, heating up the resistance wire. The thermocouple has its hot junction at E and the cold junctions at C and D. Since this meter depends only on the heating effect, it is particularly adapted to current measurements at high frequencies. Meters operating satisfactorily up to 100 megacycles may be obtained.

Chapter 3 Circuits with Time-Varying Voltages   127

INDUCTANCE AND CAPACITANCE MEASUREMENT

Although it is quite possible to obtain satisfactory measurements of resistances, currents, and voltages with the above meters, they do not lend themselves easily to the measurement of the inductance of coils or the capacitance of capacitors. These measurements are usually more easily obtained either by the use of an a-c type of Wheatstone bridge, or by a method which uses resonance as the measurement device.

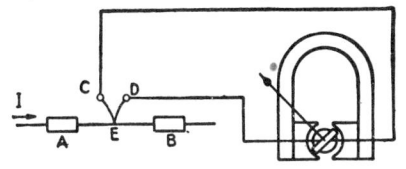

FIG. 3-43. Thermocouple type of meter. (Courtesy Weston Electrical Instrument Corp.)

When a bridge is used, as shown in Fig. 3-44, an a-c voltage, usually from an oscillator, is applied to the mn terminals of the bridge. The standard inductance S is then adjusted so that no signal is obtained in the headphones across op. The impedance drop across on is then the same as

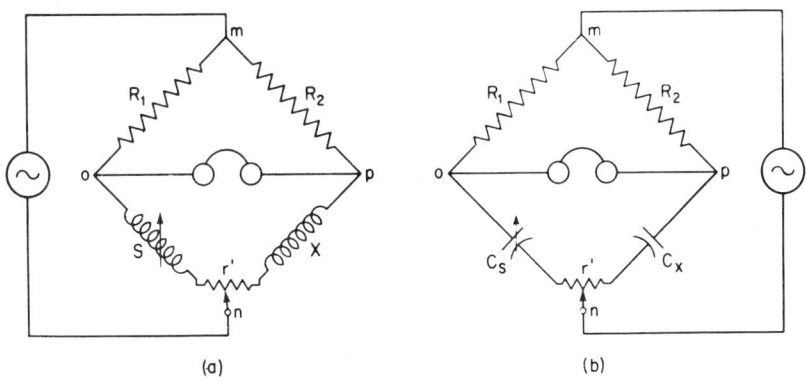

FIG. 3-44. Simple comparative inductance and capacitance bridges using variable standard inductors and capacitors.

that across pn. (The small resistor r' is arranged so that the power factor of the standard and unknown inductor may be made the same and thus obtain a perfect balance.) When this balance has been obtained

$$I_1 R_1 = I_2 R_2 \quad \text{and} \quad I_1 (2\pi f) S = I_2 (2\pi f) X$$

or

$$I_1 \frac{1}{(2\pi f) C_S} = I_2 \frac{1}{(2\pi f) C_X}.$$

Dividing the second equations by the first and solving for X and C_X,

$$\frac{I_1 (2\pi f) S}{I_1 R_1} = \frac{I_2 (2\pi f) X}{I_2 R_2} \quad \text{and} \quad \frac{I_1}{I_1 R_1 (2\pi f) C_S} = \frac{I_2}{I_2 R_2 (2\pi f) C_X}$$

from which

$$X = S \frac{R_2}{R_1} \quad \text{and} \quad C_X = C_S \frac{R_1}{R_2}.$$

At radio frequencies the capacitance between turns on coils tends to greatly reduce the effective inductive reactance of these coils. The effective inductance varies greatly with frequency, therefore, and so it is necessary to make measurements at the frequency to which the coils will be connected. Such measurements are often made most satisfactorily by measuring the capacitance required to produce resonance at the frequency involved. The resonant inductance (and the Q of the coil) may then be determined by meters. If the capacitance and the frequency are known, then the effective inductance is

$$L = \frac{1}{(2\pi f)^2 C}.$$

It should be remembered that the inductance in the equation above is in henrys and the capacitance is in farads.

CIRCUIT RESPONSE TO COMPLEX WAVES

Voice and music waves are reproduced by the microphone as waves of electric current and voltage. These waves are extremely complicated and their numerical calculation is beyond the tools of circuit analysis developed in this text. The manner of handling the analysis may be indicated, however, and some conclusions with regard to response drawn.

It was shown at the beginning of the chapter that complex waves, such as music and voice waves, are composed of combinations of sine waves of different frequencies. In sound reproduction, for instance, these frequencies normally range from about 30 cycles to 10- or 12,000 cycles. In studying the response of circuits to these complex waves, it is usual, therefore, to divide the impressed signal into its component frequencies, and to determine the response of the circuit to each component frequency.

In Chapter 2 the Principle of Superposition was stated and a numerical example worked out on d-c. This same principle holds for a-c of different frequencies. The resultant current is the instantaneous sum of all the currents produced by the component voltages impressed on the circuit. This is rather difficult to visualize and almost impossible to use with constantly varying signals, such as are obtained in music and voice transmission. It is usual, therefore, to analyze the response of the circuit for magnitude (and sometimes for phase) over the frequency range in which one is interested. Such frequency-response curves for a transformer-coupled amplifier are shown in Chapter 7.

Another instance of circuit response to a complex wave is the response of a tuned radio-frequency amplifier to the antenna voltage. The antenna will pick up all the waves being sent out by radio transmitters in the area and apply these voltages to the input circuit of the radio. When the circuit is tuned it will respond most effectively to the resonant frequency, and so the signal at that frequency will be heard. Also, the others will be so poorly received that they will not be heard.

Chapter 3 *Circuits with Time-Varying Voltages* 129

Still another illustration of this type of complex voltage is the case of the output or load circuits of vacuum tubes. In this case, the output current is a combination of d-c and a-c or signal-frequency current. The circuit response to these components is normally such that the d-c component has no effect on the following stage of amplification or other circuit arrangement.

The principle of superposition may normally be used in the analysis of radio circuits, but there are times when it does not hold. If the d-c current flowing in a transformer or iron-core choke produces saturation in the iron, then it affects the inductance and in this way changes the a-c circuit response. This qualification is important to those who design audio transformers, but does not usually concern those who operate or maintain the equipment.

PROBLEMS

1. Why is the rise of the current in an inductor delayed when a d-c voltage is suddenly applied?

2. A 50-millihenry coil has a resistance of 2 ohms. If an emf of 10 v d-c is applied, how long will it take for the current to reach 4 amps? *Answer:* 0.04 sec.

3. Why is the rise of voltage on a capacitor delayed when a d-c voltage is suddenly applied?

4. A resistor of 100,000 ohms is in series with a one-microfarad capacitor. If an emf of 100 v d-c is applied to this circuit, how long will it take for the voltage to reach 50 v? *Answer:* 0.07 sec.

5. What is the maximum rate of change of current in a coil if the maximum current is 40 ma and the frequency is 6,000 cycles? If the coil has an inductance of 20 millihenrys, what is the maximum voltage of self-induction? *Answer:* (a) 1,510 amp per sec, and (b) 30.2 v.

6. What voltage having a frequency of 650 kc would be required to force 5 ma of current through an inductance of 0.9 millihenry? *Answer:* 18.4 v.

7. Twelve volts with a frequency of 450 kc are applied across a radio-frequency choke coil of 2.0 millihenrys. How much radio-frequency (450 kc) current will pass the choke? *Answer:* 2.12 ma.

8. A 10-millihenry coil has a resistance of 50 ohms. What 800-cycle voltage will be required to produce a current of 100 ma? *Answer:* 7.1 v.

9. How much current will be drawn from a 550-kc circuit of 20 v by a 0.014-μf capacitor? *Answer:* 0.97 amp.

10. How much current will be drawn from a 550-kc circuit of 20 v by a series circuit composed of a 0.014-μf capacitor and a 5.8-microhenry inductor having a resistance of 1.0 ohm? *Answer:* 17.5 amp.

11. What is the Q of a coil of 10 microhenrys and an effective resistance of 0.13 ohm at a frequency of 500 kc? *Answer:* 175.

12. Why is it desirable to have a high Q on coils that are used in radio circuits? What is the effect in parallel resonant circuits?

13. What size capacitor will give series resonance with the coil and frequency of Problem 11? What is the circuit impedance under this resonant condition? *Answer:* 0.0101 µf, 0.13 ohm.

14. What size capacitor will be required to give parallel resonance with the coil and frequency of Problem 11? What is the circuit impedance under this resonant condition? *Answer:* 0.0101 µf, 5,500 ohms.

CHAPTER 4

VACUUM TUBE AND TRANSISTOR PRINCIPLES

INTRODUCTION

De Forest's invention of the triode, wherein an electron stream in vacuum is controlled by a voltage on a grid, has led to the modern family of electron tubes, the basic elements of our modern systems of electronics and communication.

Electron tubes are made in a great variety of sizes, from the tiny subminiatures of the hearing aid and proximity fuse, to the large air- and water-cooled tubes employed in high-power radio transmitters. They function as oscillators, detectors, amplifiers, modulators, rectifiers, as components of electronic instruments, radio receivers, radar, computers, and in many other systems. It is desirable to have a good knowledge of their operations and limitations before attempting to apply them.

Electron tubes operate by emission of free electrons into an evacuated or gas-filled envelope. These electrons then travel toward more positive electrodes in the tube, and in their progress are acted upon and controlled by voltages on electrodes placed in the electron path.

Recent years have seen the development of semiconductor devices, particularly the *transistor*, and their application to many systems formerly served by vacuum tubes. The transistor operates by reason of controlled conduction in a material which is nominally a poor electrical conductor usually termed a "semiconductor." While still under active development, the transistor has already achieved considerable importance.

ELECTRON EMISSION

Metallic conductors are composed of atoms bound into regular location patterns called crystals, with a great number of free electrons (usually one or more per atom) in continual random motion and not closely bound to any particular atom. There are, however, surface forces which at normal temperature prevent escape of electrons from the metal surface. If the temperature of the metal is raised sufficiently, some of the heat energy is given to the free electrons as additional kinetic energy, and some of these electrons may then succeed in overcoming the surface forces and escape from the conductor surface. This process is known as *thermionic emission*, and various metals differ widely in the ease with which electrons can be emitted from their surfaces.

The emission property is a function of the metal temperature and of the surface conditions and cleanliness. Richardson's law states the relation between temperature and current in amperes per square centimeter emitted from a metal at temperature T, in degrees Kelvin (degrees Centigrade plus 273°) as

$$I = AT^2 \epsilon^{-b_0/T} \quad \text{amp/cm}^2 \qquad (4\text{-}1)$$

where ϵ is the base of natural logarithms with a value 2.718. Values of A and b_o for usual emitting materials are:

	A	b_o
Tungsten	60.2	52,400
Thoriated tungsten	3.0	30,500
Barium oxide	0.01	20,000

and these emission constants vary considerably for other materials.

The rapid variation of the exponential term as temperature changes indicates that small changes in temperature or heating-power input can

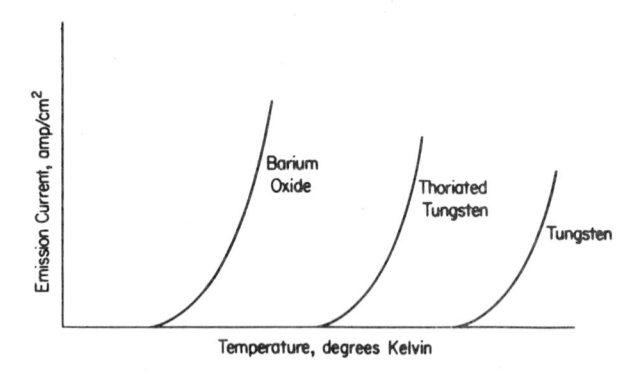

Fig. 4-1. Emission characteristics of three principal filament materials.

radically alter the emission, as indicated in Fig. 4-1. Careful control of input heating power is necessary.

Energy of emission can also be supplied to the electrons in the metal in several other ways, and although the emitted electrons behave the same in all cases, these emissions are identified by names indicating the source of the energy. Examples are: *photoelectric emission* in which light falls on the emitting surface to provide the energy; *secondary emission* in which electrons are bounced out of a surface after bombardment or collision with a *primary* or incident beam of high-energy electrons or gas ions; and *high-*

field emission, in which the electrons are almost literally pulled out by a very strong electric field.

THERMIONIC EMITTERS

The materials of major importance to thermionic emission in present-day tubes are tungsten, thoriated tungsten, and oxide-coated metals.

Tungsten. Emitters or cathodes made of tungsten are operated at temperatures of about 2,500°K, and thus they require rather large amounts of heating power. Emission currents of 4 to 10 ma per watt of heating power can be obtained. Tungsten is used where tube voltages are high and where a possibility of bombardment by gas ions exists. It is used largely in X-ray tubes, high-voltage rectifiers, and a few types of high-power transmitting tubes.

Thoriated tungsten. This material is tungsten containing 1 to 3% of thorium oxide. After suitable heat treatment in tube processing, the thorium oxide changes to thorium, and the thorium atoms migrate to the filament surface where they form a thin activated layer giving electron emission at temperatures of about 1,900°K. Emission currents of 50 to 100 ma per watt of heating power are obtainable, making the thoriated-tungsten cathode a more efficient emitter than pure tungsten. The thorium layer may be stripped off or damaged by bombardment with stray gas ions; thus a thoriated tungsten emitter can be used only in tubes having an excellent vacuum. Most tubes intended for transmitting service at powers from 50 watts to many kilowatts employ such a source.

Oxide-coated emitters. Cathodes of this type employ an inert metal base, usually a nickel alloy, covered with a layer of barium or strontium oxides. Oxide coatings operate at temperatures of only 1,150°K, and emissions of 100 to 1,000 ma per watt of heating power are expected. Almost all receiving tubes, and small transmitting tubes, employ this cathode because of its high efficiency with respect to heating power.

PHYSICAL CONSTRUCTION OF CATHODES

Two types of cathodes are employed in modern tubes, the filamentary and the indirectly heated. The filamentary cathode consists of a wire filament, usually bent in the form of an inverted V or a W, heated by passage of an electric current. The emission occurs directly from the filament, and thus both heating and emission currents flow together in the wire.

The current flow is controlled partly by the potentials of other tube elements with respect to the cathode, and if the filament is heated with alternating current, the potentials of the other elements with respect to the terminals of the filament are continually varying, these fluctuations causing *hum* in the tube output. To reduce these fluctuations it is common practice to connect the other element circuits to a center tap on the transformer winding which supplies the filament-heating current. This point is always at the *average* potential of the entire filament. Since the alternating current goes from maximum to zero twice per cycle, the temperature of the filament will also have cyclical fluctuations. The current emitted will vary cyclically and produce hum in the tube output. It is usual to employ a-c heated filaments only at radio frequencies where power-frequency hum is readily filtered out, or in special circuits which balance out or cancel the hum. In hum-critical applications the filaments are heated with direct current or the indirectly heated cathode is used.

The indirectly heated cathode consists of a thin nickel or alloy sleeve coated with the emitting oxides, and is heated by ceramic-insulated tungsten heater wires folded or coiled inside the sleeve. Since no heating current flows through the sleeve the cathode is at a uniform potential, or is said to be an *equipotential cathode*. The thermal inertia introduced by the metal sleeve eliminates variations of temperature during a cycle of the heater current and thus a major cause of hum is eliminated. Almost all types of tubes employed in receivers or low-frequency amplifiers utilize this construction. A disadvantage is the slight time delay of 15 to 30 seconds which occurs before the cathode is hot enough for electron emission.

DIODES

Thermionic tubes are usually classified according to the number of active elements employed. The simplest is the diode with two active elements, a heated cathode and an anode or plate. If the anode is made

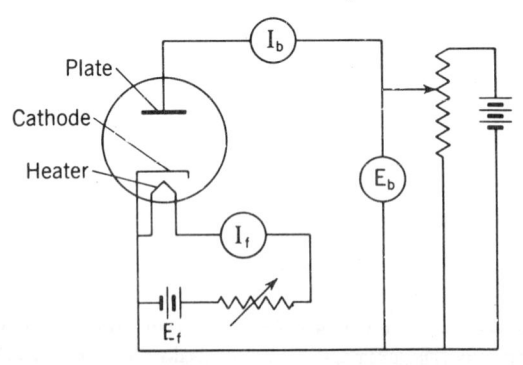

FIG. 4-2. Circuit for obtaining diode characteristics.

positive with respect to the cathode by connection of a battery or other source of potential between them, then electrons emitted by the heated cathode will be attracted to the positive anode. This movement of electrons between cathode and anode constitutes an electric current flow. The magnitude of the current depends on the number of electrons passing a given point per second, and this in turn is a function of cathode temperature and anode-cathode potential.

The relation may be investigated by use of a diode in the circuit of Fig. 4-2. If the heater current is held constant at a value I_{f_1}, thus maintaining constant cathode temperature, and the plate voltage E_b is increased from zero, the plate current I_b will increase in the manner shown by curve a of Fig. 4-3. If the heater current is increased to a new value I_{f_2}, thereby raising the cathode temperature, curve b of the figure results. The two curves coincide at smaller anode voltages, and each becomes horizontal in the upper region, but at different anode current values. In these horizontal regions the plate current is said to be *temperature limited;* the effect is due to the fact that all the electrons which the cathode is able to emit at this temperature are reaching the anode, and consequently the current can increase no further.

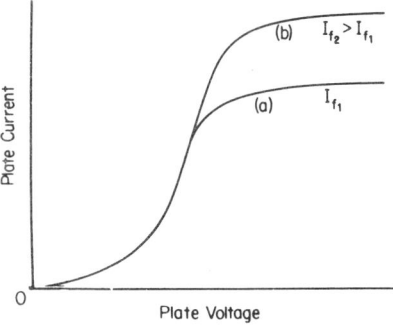

FIG. 4-3. Effect of temperature limitation on diode current.

If, however, the cathode temperature is raised by increasing the heater current, more electrons are emitted, as predicted by Eq. 4.1, and the temperature-limited value of current is increased.

SPACE CHARGE

An explanation for the shape of the lower portion of the curve in Fig. 4-3 is still required, i.e., why are the lower values of anode voltage unable to attract to the anode all the electrons which are known to be available by emission? It is found that the effect is due to the presence of a cloud of electrons in the region near the cathode surface, and this cloud is called the *space charge*.

As the electrons are emitted they have only very small velocities. Upon leaving the cathode surface they are acted upon by two forces, one tending to pull out the electrons, the other tending to force them to return to the cathode. The first is due to the attraction of the positive potential existing on the anode. The second force exists because of the repulsion from the negative electrons already in the space. The plate current flowing at any anode voltage is then due to a balance between these forces. An increase of anode voltage reduces the number of electrons in the space charge which reduces the repelling force, thus permitting more electrons to leave the cathode to supply the increased demand for plate current. A reduction of anode potential decreases the attractive force, leaving more electrons in the space and increasing the repelling space charge; thus, fewer electrons are emitted and the anode current is reduced.

It can be seen that no more electrons are permitted to leave the cathode than are needed at that instant to supply the plate current, even though the temperature is high enough to cause emission of a much higher current. This effect of space-charge limitation is extremely important, as the usual working range of most vacuum tubes is in the region in which space charge is effective. It is also this effect which causes most vacuum tube characteristics to be curves. The applications in which the emission is limited by temperature saturation are rare.

PLATE RESISTANCE

Although the relationship between plate current and plate voltage is not one of simple proportion, and will not appear on a curve as a straight line, it is nevertheless possible to speak of the resistance between the plate and filament of the tube, usually referred to as the "plate resistance." The ratio of anode potential to anode current is known as the d-c or steady-current plate resistance, and is shown in Fig. 4-4 as OA/AP. The concept of d-c

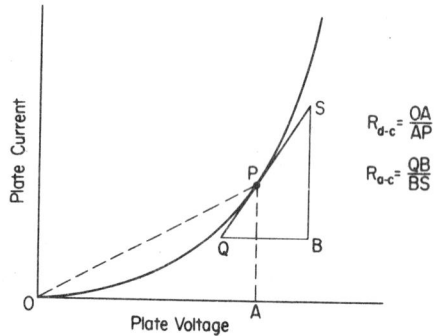

FIG. 4-4. Graphical construction to show d-c and a-c plate resistance of a diode.

plate resistance is useful in some cases where the current through the tube is steady or constant, but a more important quantity is the a-c plate resistance, or the ratio between a small *change* in plate voltage and the corresponding *change* in the plate current. In the neighborhood of point P in Fig. 4-4, the a-c plate resistance is given as QB/BS. For most diodes, the d-c plate resistance is approximately twice as great as the a-c plate resistance. If an a-c voltage, superimposed upon a larger d-c voltage, is impressed on a tube, the corresponding a-c and d-c components of current would be determined by the respective resistances.

Because the relation between i_b and e_b is generally somewhat curved, it is clear that neither of these resistances is constant for all applied voltages, and both tend toward lower values as plate voltage and plate current are increased.

RECTIFYING ACTION

If the plate of the diode is made negative with respect to the cathode, the electrons will be driven back to the cathode and no plate current will flow. Hence, if an alternating voltage is applied to the plate, current will flow only in the positive half cycles, and it will consist of a succession of pulses, always in the same direction. The tube therefore functions as a *rectifier*, in that an alternating voltage applied to the tube produces a unidirectional current. Wide use is made of this characteristic to obtain direct voltages and currents from an a-c source for supplying radio equipment and amplifiers.

TRIODES

A triode is a tube which contains a third element, the grid, located between cathode and plate. The grid usually takes the form of a helix or spiral of fine wire, so that electrons may pass freely through it. Since the grid is nearer to the cathode, the potential of the grid has a greater effect in controlling electron flow than does the plate potential. If the grid is made negative to the cathode, its repelling effect will partly nullify the attractive effect of the positive anode potential on the electrons at the cathode, and the anode current will be less than if the grid were not present. The grid may be made sufficiently negative to repel all the electrons and stop or *cut off* the anode current. Variation of grid potential will cause a similar variation in the value of anode current, and thus the grid is able to *control* the anode current.

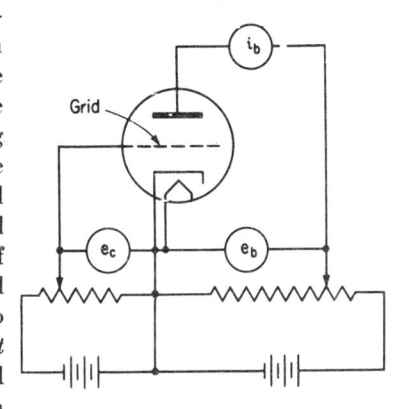

FIG. 4-5. Circuit for obtaining triode characteristics.

If the grid potential is negative with respect to the cathode, as is the case in many electronic circuits, electrons will not be attracted to the grid itself, and there will be practically no grid current.

TRIODE CHARACTERISTIC CURVES

The effects of grid and plate potentials on the plate current can be studied by means of the circuit of Fig. 4-5. In this setup, if the plate potential e_b is held constant and the grid potential varied, one of the curves of Fig. 4-6 is obtained. The other curves are found in the same way, by using different values of e_b. Such a set of curves, showing the relationship between the plate current and the grid potential for constant values of plate voltage, is known as the grid-plate family or the *transfer characteristic* of the tube. The latter name is derived from the fact that they show the transfer relations between a voltage in the grid circuit and current in the plate circuit. *Mutual* characteristic is another term occasionally used. By use of this family, and by interpolating between curves where necessary, it is possible to find the plate current if the grid and plate voltages are known.

It may be noted that the curves tend to be nearly straight lines in the upper portions, and that the curves for various plate potentials are nearly

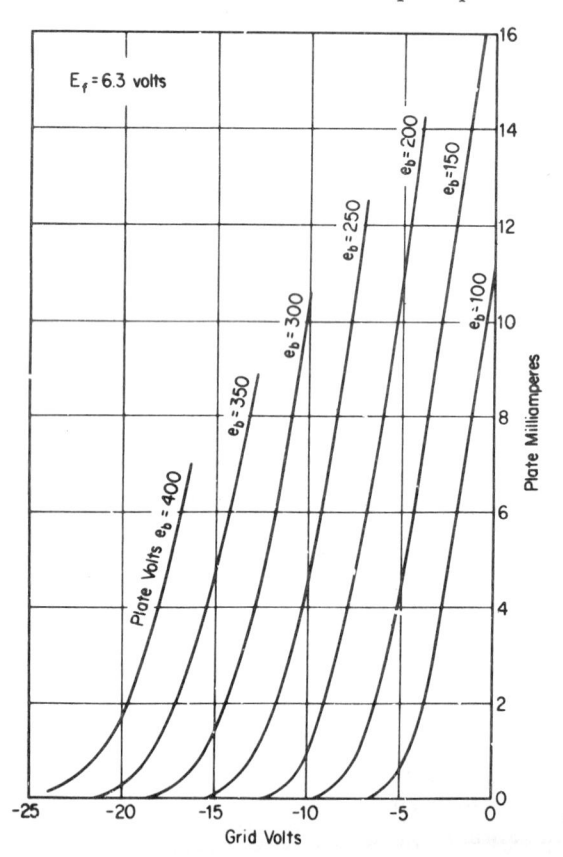

FIG. 4-6. Grid-plate characteristics of a typical triode.

parallel. It will be shown later that these properties are important where the tube is to be used as an amplifier, and where it is desirable to keep the distortion of the amplified signal **as** small as possible.

Another way of presenting the same information is shown in Fig. 4-7. Here the grid potential is held constant for each curve, and the plate potential is permitted to vary. This set of curves is known as the family of *plate characteristics*, and for many purposes it is more useful than the mutual characteristics. The curves of this family resemble in shape those of

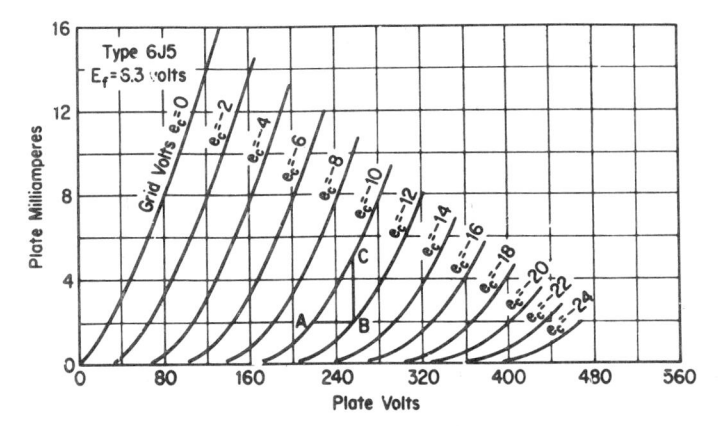

FIG. 4-7. Plate characteristics of triode of Fig. 4-6.

the grid-plate family. The effect of increasing the negative grid potential is chiefly to shift the curves toward the right on the diagram, without causing much change in form.

TUBE PARAMETERS OR COEFFICIENTS

Three important ratios, obtainable from either curve family, are performance coefficients helpful in analyzing and predicting tube operation. These ratios or parameters, are the *amplification factor*, the *transconductance* or mutual conductance, and the internal a-c *plate resistance*.

The amplification factor, symbolized by the Greek letter μ (pronounced *mew*), is defined as the ratio of plate-voltage change to grid-voltage change, when plate current is maintained constant. It is a measure of the relative effectiveness of the grid as compared with the plate in controlling flow of plate current. In Fig. 4-7 the plate currents at A and B are the same, and by the above definition

$$\mu = \frac{AB}{\Delta e_c}$$

where Δe_c represents the difference between the grid potentials of the two curves through A and B. In triodes, μ ranges in value from 2 to 100, with most tubes included in the range 10 to 40. For most triodes, the amplification factor is almost constant for all operating conditions, except at very low plate currents.

Grid-plate transconductance, or mutual conductance, g_m, is the ratio of plate-current change to grid-voltage change producing it, when plate voltage is held constant. It is a measure of the effectiveness of the grid in controlling the anode current. In Fig. 4-7,

$$g_m = \frac{BC}{\Delta e_c}$$

where BC is measured in correct units. When BC is stated in microamperes the value of g_m is in microamperes per volt change, and its units are *micromhos*. For tubes in current use, its value ranges from a few hundred to above 40,000. The value of g_m varies considerably, depending largely on the value of plate current.

Internal a-c plate resistance, r_p, is given by the ratio of plate-voltage change to plate-current change producing it, grid voltage being held constant. The plate resistance is stated in ohms, and in Fig. 4-7

$$r_p = \frac{AB}{BC}$$

where AB is measured in volts and BC in amperes. In most triodes, r_p ranges in value between 300 and 100,000 ohms.

By manipulation of the above three definitions, it can be seen that the following relationship exists between the three tube coefficients:

$$\mu = g_m r_p.$$

It is thus sufficient to specify any pair of these quantities, since the third may be computed from them.

In most tube applications the grid voltage is varied by an input signal, and the plate current and plate voltage are not constant but vary with time between maximum and minimum values, as shown in Fig. 4-8 (reproduced

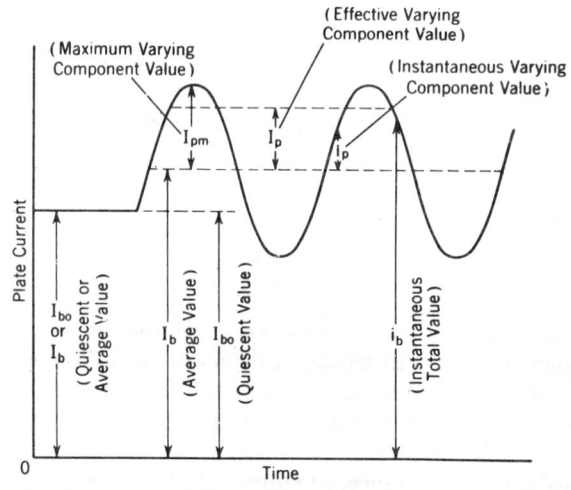

FIG. 4-8. Components of plate current and their standard abbreviations.

from the IRE Electronic Standards). It is helpful to consider that such a current or voltage is composed of a d-c average component I_b, together with an a-c or varying component of peak value I_{pm} or effective value I_p. In all but a few cases the a-c component is the one of interest, but it should be noted that the d-c component is necessary, since this determines the portion of the tube characteristics in which the operation takes place. It should also be noticed that the average value I_b may change when a signal is applied.

A typical instance is given in Fig. 4-9, which shows a triode tube with external resistance R in its plate circuit. E_s is an alternating voltage, the signal, which causes the instantaneous grid potential to fluctuate up and down about its average value E_c. The plate current i_b will vary correspondingly, and it may be regarded as having the components I_b (d-c) and I_p (a-c), as shown in Fig. 4-8. This current in flowing through the load resistance R produces varying amounts of IR drop, and corresponding variations in plate potential will take place. It should be noted that the average or d-c component of plate potential E_b is less than the B-supply voltage E_{bb} by the d-c component of drop in the load resistor R. The a-c components of plate current and plate potential are the useful output of the tube, owing to the input signal E_s.

THE LOAD LINE

A graphical construction may be employed to find the output current and voltage values for any operating condition. This method is illustrated in Fig. 4-10, which consists of the plate family of curves of the tube in the circuit of Fig. 4-9, and a superimposed *load line*. The load line is a graph of the equation

$$e_b = E_{bb} - Ri_b,$$

representing the voltages appearing in the plate circuit of Fig. 4-9. Any point on the line represents a possible combination of plate voltage and plate current, and no other combinations are possible.

It will be apparent from the equation, or by inspection of the circuit, that when $i_b = 0$, $e_b = E_{bb}$. The load line therefore intercepts the horizontal axis at this scale value. Similarly, the vertical intercept occurs at

$$i_b = \frac{E_{bb}}{R}.$$

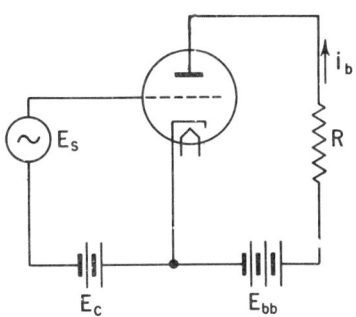

FIG. 4-9. Triode circuit with resistance load.

The load line is most readily drawn by use of these two points, and is straight because R is a constant resistance.

If the grid-bias voltage is E_c, the zero-signal or *quiescent* condition is represented by the point P, which is the intersection of the load line with the plate characteristic corresponding to E_c. This point will be referred to as the *operating* point, or the quiescent (Q) point, and when the signal voltage E_s is applied, the instantaneous relations between plate current and

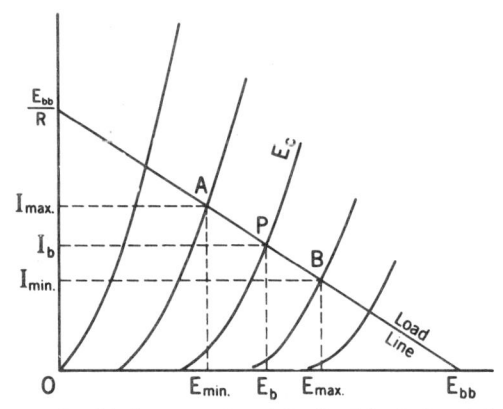

FIG. 4-10. Graphical construction of the "load line" of a triode circuit.

plate potential will correspond to a point oscillating about P along the load line. If the range of grid voltage variation is limited by the grid voltage lines intersecting the load line at A and B, then these points fix the limits of oscillation of plate current and voltage, the plate potential fluctuating between E_{min} and E_{max}. The peak value of a-c plate voltage will be

$$E_{pm} = \frac{E_{max} - E_{min}}{2}.$$

The average or d-c component of plate potential is E_b, and for a given load line this depends on the choice of E_c.

Steady and alternating components of plate current can be read from the diagram in the same manner.

The operation of a tube with resistance load can also be represented by a line drawn on the plot of mutual characteristics, but in this case the line depends not only on the B-supply voltage and the load resistance, but also on the tube characteristics. Since these are not usually straight, such a line is not straight and must be laboriously plotted point by point. This line is known as a *dynamic transfer* characteristic of a tube and its method of construction is explained in Chapter 7.

THE EQUIVALENT CIRCUIT OF A VACUUM TUBE

A second method of determining the a-c output of a vacuum tube is by means of the *equivalent circuit*. This is shown in Fig. 4-11, corresponding to the actual circuit of Fig. 4-9. With certain reasonable assumptions it may be shown that the equivalent circuit leads to the same values of current and voltage in the load resistor R as the a-c components of these quantities in the actual circuit.

The assumptions required are that μ, g_m, and r_p remain constant throughout the range of operation, or (stated in another way) that the curves of the tube characteristics be straight, parallel, and equally spaced in the operating region. These statements can be considered as true for very small signals, but when large signal swings are to be handled the validity of the assumptions should be examined in each case.

The *voltage gain* of the circuit of Fig. 4-9 is defined as the ratio of the output a-c voltage E_o between cathode and anode to the applied signal voltage on the grid. This voltage amplification will now be computed by use of the equivalent circuit.

Since r_p and R form a simple series circuit, the a-c current is given by

$$I_p = \frac{\mu E_s}{r_p + R}.$$

This current in R produces a voltage drop from K to P, which correspond to the cathode and plate of the tube, respectively. The a-c potential of P with respect to K is

$$E_o = -I_p R = \frac{-\mu E_s R}{r_p + R},$$

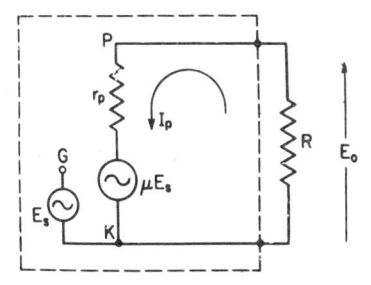

Fig. 4-11. Equivalent circuit corresponding to Fig. 4-9.

the minus sign indicating that the potential of point P is below that of K, as would be expected with the indicated direction of current. The voltage gain, or ratio of E_o to E_s is

$$A = \frac{E_o}{E_s} = \frac{-\mu R}{r_p + R}.$$

The negative sign indicates that E_o and E_s are opposite in phase, as is found experimentally. This fact can also be noted in Fig. 4-10, and it is of importance in amplifiers employing feedback, as will be shown in Chapter 7.

TETRODES

Since the a-c grid and plate potentials are opposite in phase, the a-c potential difference between these two elements inside the tube is considerable, being equal in magnitude to the sum of the a-c input and output voltages. On account of the electrostatic capacitance between grid and plate, this potential difference produces a current flow from the output circuit to the grid through that capacitance, and under certain load conditions such reaction or feedback causes the amplifier to become unstable.

The instability may be avoided by inserting an electrostatic shield in the form of a *screen grid* between the plate and the original grid, which in this arrangement is termed the *control grid*. A tube of this type is known as a *tetrode*, or *screen-grid tube*.

To permit the flow of reasonable amounts of electron current, the screen must be operated at a positive potential. For adequate shielding of the control grid, however, the screen potential should have no a-c or signal component, or must be at cathode potential for a-c, and this calls for special consideration in circuit design.

The fact that the screen grid acts to shield the control grid from variations of plate potential implies also that the cathode is shielded from the plate. As a consequence, electron emission is affected only very slightly by the plate potential; it is determined almost altogether by the control-grid and screen-grid potentials. A portion of the electron stream is deflected to the screen on account of its positive potential, and forms the screen current; the remainder, usually a much larger part, flows to the plate.

In terms of tube characteristics, since a relatively large change of plate potential causes very little change in plate current, the value of r_p, the internal plate resistance, is much larger than for a triode. For the same reason, the amplification factor μ is correspondingly greater for a tetrode than for a triode. The value of g_m being equal to the ratio μ/r_p is thus little changed from the values to be expected for triodes.

PLATE CHARACTERISTICS OF A TETRODE

A typical plate voltage–plate current characteristic for a screen-grid tetrode is drawn in Fig. 4-12 for a fixed value of the control- and screen-grid voltages. For anode voltages above D, anode current is nearly independent of anode voltage, as explained above, and this region constitutes the useful working range for amplifier service.

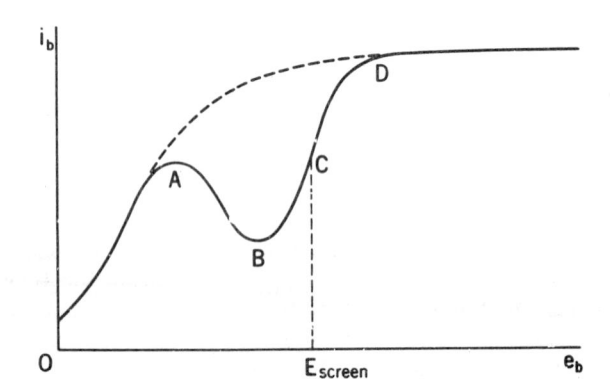

Fig. 4-12. Typical plate voltage-plate current characteristic of a tetrode for fixed control- and screen-grid voltages.

Below D, the action is complicated by the presence of secondary emission of electrons from the anode. That is, if an electron strikes the anode with sufficient velocity, it will dislodge one or more electrons from the anode surface. If at any instant the plate potential is below that of the screen, the secondary electrons will be attracted to the screen, will add to the screen current, and cause a decrease in anode current since the electrons left the anode. This effect is evidenced by the dip in plate-current values shown between A and C, Fig. 4-12. Below A, the primary electrons arrive with velocity too low to dislodge any secondary electrons. Above D, secondary electrons are emitted, but since the anode potential is above that of the screen, all secondary electrons are returned to the anode, and plate and screen currents remain unaffected.

This dip in the plate-current curve is undesirable because it limits the region in which the tube characteristics are straight lines, or in which μ, r_p, and g_m are constants, thus limiting the area of tube operation. Because of this the tetrode has been largely replaced by the pentode, or five-element tube.

PENTODES

As mentioned, one effect of the dip in the plate characteristic owing to secondary emission is to limit the operating region and the available a-c output, since true amplifier action does not extend into the warped region of the curves. The range of operation would be extended materially if the secondary electrons could be sent back to the plate instead of going over to the screen, in which case the plate characteristic would have the form of the dotted curve in Fig. 4-12.

This desirable result is accomplished by the insertion of a third grid, the *suppressor*, between plate and screen grid. The suppressor is usually made with wider spacing than the other two grids, and it is connected directly to the cathode, so that its potential is zero. It has very little effect on the emission of secondary electrons from the plate, but it exerts a powerful control upon them as soon as they are produced. Even at low plate potentials the electric field between plate and suppressor is in the proper direction to send electrons toward the plate, and for this reason the secondary electrons return to the plate instead of going to other tube elements.

Tubes of this type are known as *pentodes*. Figure 4-13 shows the plate characteristics of the 6AU6, which is a typical voltage-amplifier pentode. By comparing Figs. 4-13 and 4-12 it can be seen that the dip has disappeared and that the pentode can furnish a larger a-c output voltage than can the tetrode.

Values of μ, and r_p are high, just as for the tetrode, and values of μ may reach 2,000. The a-c plate resistance of many pentodes may exceed 1 megohm.

In any circuit involving high gain it is important to isolate the input leads from the output connections, either by shielding or physical separation. If this is not done, the advantage of internal shielding by the screen is lost, and instability occurs. Special shielding for this same purpose is provided internally between the leads in the tube.

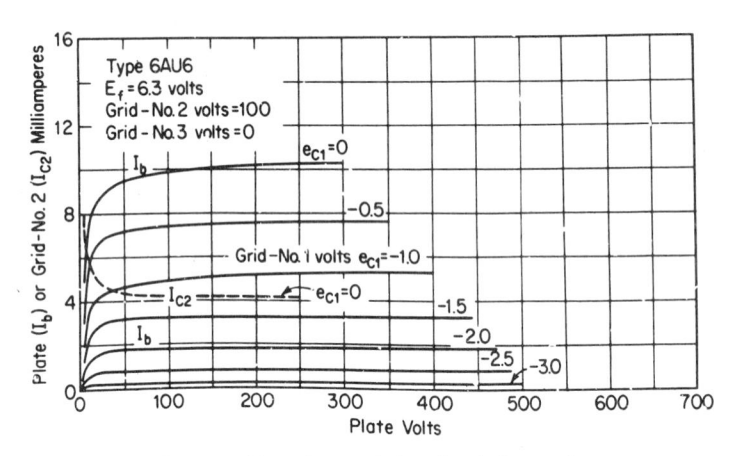

Fig. 4-13. Plate characteristics of typical pentode.

The performance of a pentode amplifier may be studied in the same way as was that of a triode, either by use of the load line or by the equivalent circuit. In the latter case, a simplification is possible on account of the large value of a-c plate resistance of the pentode. In the expression for voltage amplification,

$$A = \frac{-\mu R_L}{r_p + R_L},$$

it is frequently true that r_p is so large in comparison with R_L that the latter may be neglected. By reason of the relationship between the tube constants, it is then possible to write:

$$A = \frac{-\mu R_L}{r_p} = -g_m R_L.$$

It is frequently desirable to obtain the positive voltage for the screen grid from the B-supply source, and a convenient way of doing this is shown in Fig. 4-14. The required screen voltage is usually less than the B voltage available, and the magnitude of R_s is selected so as to produce the necessary amount of drop, which is due to the flow of screen current. Thus, if E_{bb} is 300 v and the desired screen potential is 100 v, the required drop is 200 v. If the screen current at the working voltages is 0.5 ma, this drop will require a dropping resistor of 400,000 ohms.

Without some provision for keeping the screen potential constant, however, the presence of this screen dropping-resistor would cause a loss of output. This loss can be understood if it is remembered that increase of control-grid potential is accompanied by increase in screen current, and this increase in turn causes a fall in screen potential owing to the greater drop in R_s. This fall in turn has the effect of lowering the plate current, and thus counteracting in part the effect of the original signal on the plate current.

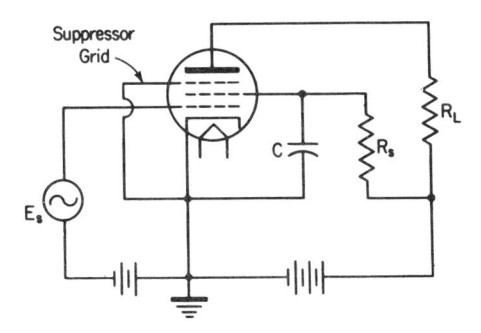

FIG. 4-14. Typical screen supply and circuit for a pentode.

It is possible to avoid this difficulty by the use of the capacitor C, shown in Fig. 4-14. If this capacitor is large enough, it will hold the potential of the screen almost constant, by by-passing the fluctuations in screen current directly to the cathode, instead of permitting them to flow through R_s. To determine the proper size of C, a satisfactory rule is that its reactance should not be more than one-fifth the resistance in shunt with it, at the lowest frequency to be handled. In computing this reactance it is necessary to remember that there is an effective a-c resistance within the tube between screen and cathode, exactly analogous to the a-c plate resistance. Suppose this resistance in the present case to be 100,000 ohms. Inspection of the circuit shows that the resistance shunting capacitor C consists of the internal screen resistance in parallel with R_s. (The source E_{bb} is not included, since there is no a-c potential across it.) The effective resistance across the capacitor is therefore

$$\frac{100{,}000 \times 400{,}000}{100{,}000 + 400{,}000} = 80{,}000 \text{ ohms.}$$

By the above rule, the reactance of C must not exceed 16,000 ohms (one fifth of 80,000 ohms) at the lowest working frequency. If the tube is part of an audio-frequency amplifier passing frequencies down to 50 cycles, the required capacitance is

$$C = \frac{10^6}{2\pi \times 50 \times 16{,}000} = 0.2 \ \mu\text{f.}$$

If a by-pass capacitor of smaller size is used, the gain of the amplifier stage will be satisfactory at the higher frequencies, but will fall off at the low end.

VARIABLE-MU OR REMOTE-CUTOFF PENTODES

In certain pentodes, such as the 6BA6, the grid wire helix has a variable pitch, so that some of the turns are closer together than others. The closely spaced turns have greater control over electron flow from the cathode than those with wider separation, and as a result the grid-plate transfer characteristics are considerably modified.

As the control grid is made more negative, cutoff of the electron stream is approached much more gradually than if the grid spacing were uniformly close. Tubes of this type are known as *variable-mu*, or *remote-cutoff*,

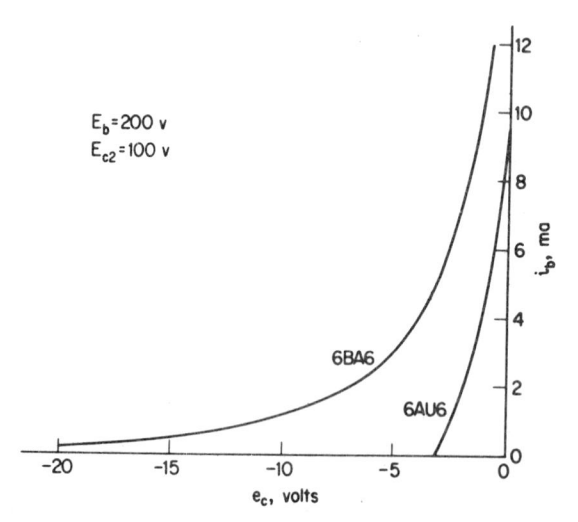

Fig. 4-15. Grid-plate transfer characteristics of 6AU6 (sharp cut-off) and 6BA6 (remote cut-off).

pentodes. A typical example is the 6BA6, and a transfer characteristic for this tube is compared with a similar curve for the sharp cutoff type 6AU6, in Fig. 4-15. The more gradual approach to cutoff inherent in the variable-mu type is evident.

This property can be utilized to control the gain of the tube. For a pentode, the approximate expression for voltage amplification applies:

$$A = -g_m R_L.$$

The value of g_m, however, varies along the curve, being maximum at zero grid-bias voltage, and decreasing steadily to a much smaller value as the grid is made more negative. This can readily be seen from the curve if it is recalled that g_m, the transconductance, is equal numerically to the slope of the transfer characteristic. It then becomes possible to control the gain of such an amplifier by simple adjustment of the d-c grid-bias voltage. In radio receivers, which are required to respond to signals covering a wide range of intensity, the grid bias of certain amplifier tubes is made to depend upon the amplifier output, and as a result the sensitivity is automatically reduced when a strong signal is being received. This type of control, known as *automatic volume control* or *avc*, is incorporated in practically all modern receivers.

BEAM POWER TUBES

Instead of using a suppressor grid to control the secondary emission from the anode, it is possible to obtain the same effect by shaping the tube elements in such a way as to control the space charge near the plate. Figure 4-16 shows the internal structure of the 6L6, a typical beam power tube, and also the distribution of electrons within the tube. The beam-forming plate shown in this figure is connected to the cathode, and its potential is therefore zero. The field produced by this combination of elements is such as to cause a concentration of electrons to occur near the plate, as indicated in the diagram, and thereby to produce a region of minimum potential there. As long as the plate potential is higher than the potential minimum resulting from electron concentration, secondary

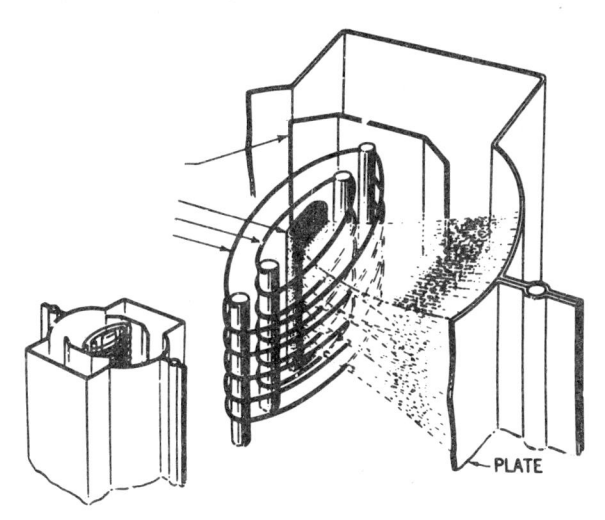

Fig. 4-16. Structure of a 6L6 beam-power tube.

electrons will return to the plate, just as if a suppressor grid were present. The characteristic curves of this tube are shown in Fig. 4-17, and the sharp break on each of the curves is the point at which screen voltage and plate potential minimum are equal.

Figure 4-16 also shows the accurate focusing of the electron beam, causing it to pass between the turns of the screen grid. When this is done,

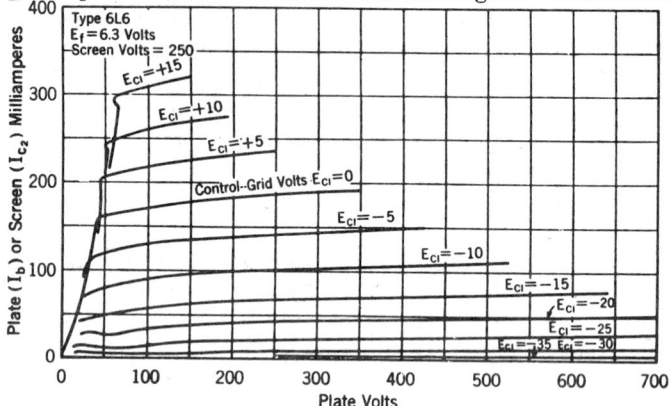

Fig. 4-17. Plate characteristic of 6L6 beam-power tube.

the screen current is much reduced as compared with earlier tubes, and more output power is available for a given amount of peak cathode emission.

DUAL-PURPOSE TUBES

For reasons of economy or space-saving, several functions that would otherwise be accomplished by two or more tubes may be handled by a single multipurpose tube. Such a tube may consist of the elements of two or more tubes all mounted within a single envelope, each unit acting independently of the others, or it may be a combination that depends for its operation on interaction of some sort between the several elements.

An example of the first class is the twin triode, such as the 6SN7 or 12AU7, which contains all the elements of two entirely distinct triodes, except for common heater connections. Such tubes may be used for any application calling for two similar triodes, with resultant space saving and shorter leads. Another example is the 6AT6 or the 12SQ7, which contain two diodes and a triode, but with a common cathode.

The second class of multipurpose tubes can be illustrated by the 6BE6, called a pentagrid converter. As seen in Fig. 4-18, it contains five grids, which are referred to by number counting from cathode towards anode. In a typical frequency converter application, grid No. 1 (G_1) and grid No. 2 (G_2) serve as control grid and anode respectively of a triode oscillator. Grids G_3 and G_5 are tied together and serve as a screen, shielding grids G_2 and G_4 from each other and from the plate. Grid G_4 receives the incoming signal, and its a-c voltage modulates or varies the stream of electrons passing from the oscillator section to the anode. The anode current then varies as a result of the combination of the effects of the oscillator voltage and the signal voltage, but the only type of coupling between the two signal sources is by the electron flow; electrically the two circuits are isolated by the shielding action of the two screen grids. Among other tubes depending on similar electron coupling are the 6L7, 12SA7, and the 6K8, which may be studied in the tube manuals.

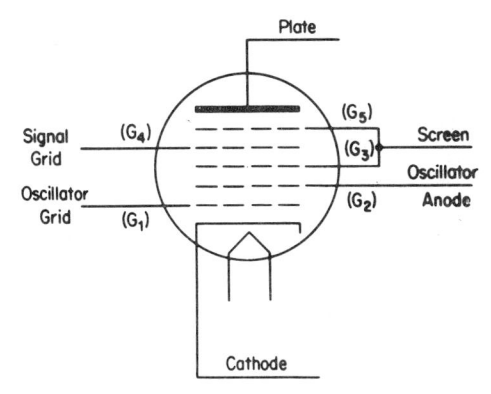

FIG. 4-18. Pentagrid converter tube.

LIMITATIONS IN TUBE OPERATING CONDITIONS

Certain limits must be imposed on the operating conditions of all tubes if destruction of the tube is to be avoided and reasonable life obtained.

One limit is the peak or maximum emission current available from the cathode. If an attempt is made to exceed the rated current value by applying excessive voltages to tubes having oxide coatings or thoriated filaments, the emitting materials may be stripped from the surface of the cathode.

Another limitation is the maximum voltage (called the *peak inverse voltage* in rectifier tubes) that can be impressed across the tube elements without causing an electrical breakdown or arc.

A limit which combines both current and voltage is *plate dissipation*. When an electron arrives at the anode, its kinetic energy of motion is transformed to heat and this raises the temperature of the anode structure. If the temperature goes too high the plate material may emit gases, destroying the vacuum, or the plate may even melt. To remove this heat from the tube is a major problem in the design of large transmitting tubes. Smaller tubes are cooled by radiation of their heat from the plate; larger tubes are cooled by forced air or water circulation around the anode (see Figs. 4-19 and 4-20).

The maximum amount of heat that can be safely dissipated by the plate of any tube is stated in terms of the corresponding electrical power, in watts, and is referred to as the *maximum plate dissipation*. With tubes containing grid structures, the heating of each grid must also be taken into account. The heat input to the anode or any electrode is equal in watts to the product of the current in amperes reaching the electrode and the voltage between the electrode and the cathode or electron source.

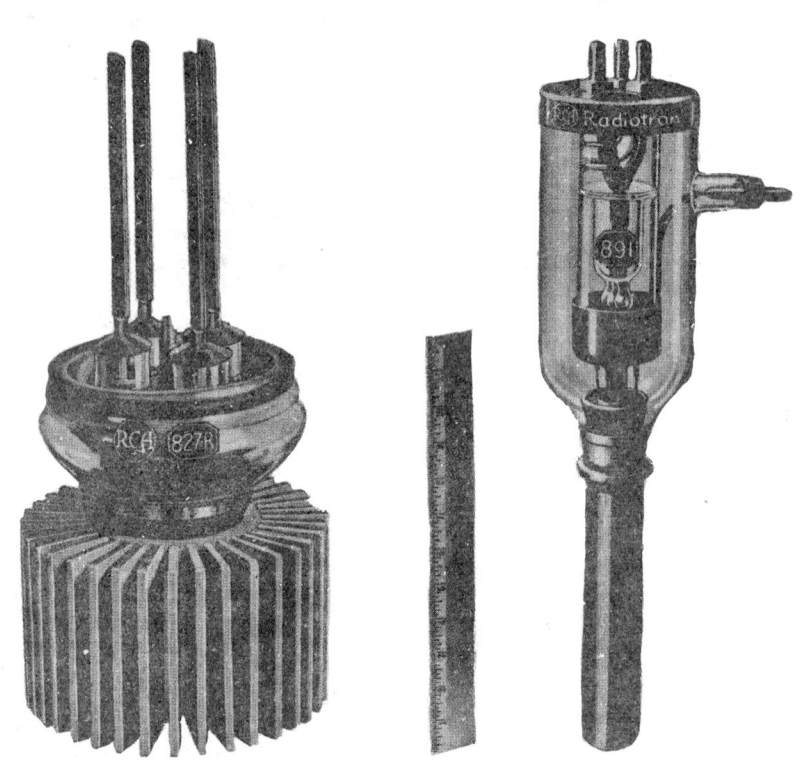

Fig. 4-19. Air-cooled power tube with radiating fins.

Fig. 4-20. Water-cooled power tube.

In the case of transmitting tubes, the manufacturer furnishes information giving the maximum allowable value of each of these limiting factors, and in order to insure maximum tube life it is important to consider each of them individually.

CATHODE-RAY TUBES, KINESCOPES

Electrons after emission from a small cathode can be focused into a narrow beam or pencil. This electron beam can then be deflected by electric or magnetic fields, and the utilization of this principle leads to the cathode-ray tube, or *kinescope* as it is called in television reception. It is used for display of radar information, and also as a laboratory tool to study the operation of electric circuits. A photograph of a kinescope is shown in Fig. 4-21, with a detailed drawing of an electric field deflection type of tube in Fig. 4-22.

Chapter 4   *Vacuum Tube and Transistor Principles*   153

In Fig. 4-22, K represents the cathode, with heater not shown; G serves as the grid to control the spot brightness on the screen S. Anode No. 1, F, is known as the focusing anode since by variation of its potential the electron beam may be sharply focused on the screen. Anode A is the second or

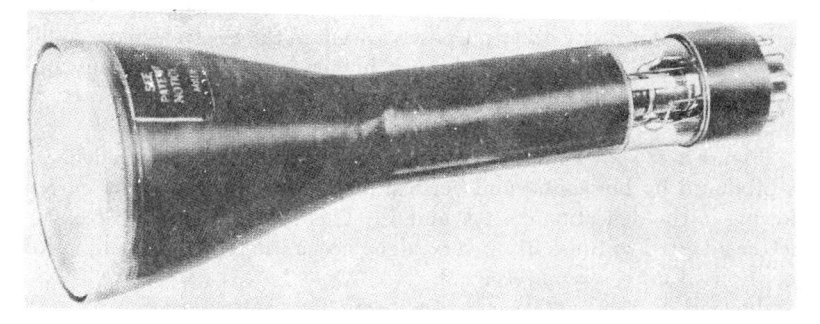

FIG. 4-21. Cathode-ray tube. (Courtesy Allen B. Du Mont)

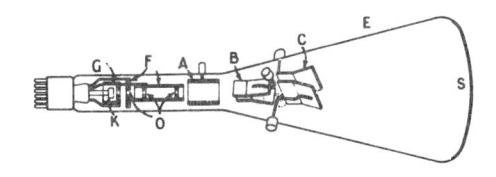

FIG. 4-22. Schematic arrangement of electrodes in a cathode-ray tube of the electrostatic-deflection type. (Courtesy RCA)

accelerating anode, B and C are two pairs of deflecting plates, S is a screen deposited on the inner surface of the glass envelope, and composed of a fluorescent substance which glows or gives off light when bombarded by high-velocity electrons.

In operation, electrons emitted by the cathode are accelerated by anodes F and A, and by virtue of the apertures in the various tube elements and the form of the electric field about F and A, the electrons are constricted into a narrow beam or pencil along the axis of the tube. The degree of concentration or focusing is controlled chiefly by the potential of F, the first anode. After passing through anode A, the electrons proceed at constant velocity to the screen, since there is no further accelerating field, and collide with screen S in a small spot at the center of the area, causing a glow to appear there.

If an alternating voltage is applied between the pair of plates B, the electrons in the beam will be attracted toward the plate which is positive at the moment, and repelled by the one which is negative. The beam is therefore deflected up and down as the voltage varies, and the luminous spot moves correspondingly on the screen. The other pair of plates, C, is arranged at right angles to the first, and any voltage across these plates will produce horizontal deflections, to the left or right depending on polarity.

This tube forms an extremely versatile tool for investigating electrical phenomena. It may be used for examining wave forms of currents or voltages, for comparison of frequencies, for obtaining volt-ampere characteristics, and for many other purposes. Because the electrons have so little mass, the beam will respond at much higher frequencies than any other indicating device available, and the range of applications is practically unlimited.

Figure 4-23 shows another type of cathode-ray tube, in which deflection is produced by horizontal and vertical magnetic fields set up by currents flowing in the deflecting coils X and Y. This type of deflection control is better adapted to tubes using very high accelerating voltages, and is commonly used in television receivers.

Instead of focusing the electron beam by electrostatic means as described above, it is possible to accomplish the same purpose by the use of a magnetic field directed along the tube axis. Such a field is easily produced

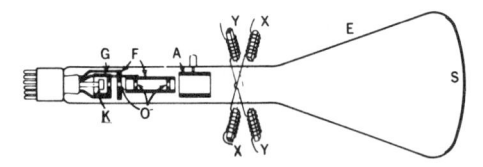

Fig. 4-23. Schematic arrangement of electrodes in a cathode-ray tube of the magnetic-deflection type. (Courtesy RCA)

by means of a coil surrounding the neck of the tube, approximately at the location of F in Fig. 4-23. The anode F may then be omitted, and this simplification is again advantageous with tubes using very high accelerating potentials.

A variety of fluorescent materials is available for the formation of the screen S. The color of the glow varies with the material, and this may be a matter of some importance. For oscillographic use, a screen giving a bright green color is usually employed, since this color is easily obtained and it has a good visual quality. For photographic use a deep blue glow is more useful, on account of its greater actinic power. By using a suitable mixture of phosphors, a glow approaching white in color may be obtained, and this is preferred for television screens, since the pictures present a more pleasing appearance.

When the electron beam moves away from a particular spot on the screen the glow disappears, but not instantaneously. Various fluorescent substances exhibit different rates of decay of brightness after the excitation is removed, and tubes are now available with either long-persistence or short-persistence screens. Those with long persistence usually glow in two

Chapter 4   *Vacuum Tube and Transistor Principles*   155

colors, a short-time blue flash followed by the long persistence of yellow or orange. These screens are useful for study of extremely brief phenomena, such as radar information. For general oscillographic use a screen with medium persistence is usually employed. A table of usual screen performance is given below:

CATHODE-RAY TUBE FLUORESCENT SCREENS

Phosphor Number	Color	Persistence	Application
P1	Green	Medium short	Oscillography
P2	Blue-green	Long	Oscillography
P4	White	Short	Television
P5	Blue	Very short	Fast photography
P7	Blue-white; then yellow	Very long	Radar screens
P11	Blue	Short	Photography
P12	Orange	Medium	Limited
P14	Blue, then orange	Long	Non-recurrent phenomena
P15	Blue-green, ultraviolet	Short	Flying-spot scanner

The ability of an electron stream to produce fluorescence, and thus to indicate its own position, is utilized in another class of tubes to indicate the magnitude of a voltage on one of the electrodes. These tubes are referred to as *electron-ray* or *magic-eye* tubes, and typical examples are the 6E5 and 6U5. These tubes have a fluorescent screen bombarded by electrons from the cathode, and a control electrode deflects the electrons to produce a wedge-shaped shadow in the general glow covering the screen. The angular width of this shadow depends on the potential of the control electrode, the usual available range being from 0° to 90°. The control electrode is attached to the plate of a triode unit contained in the tube envelope, and by using the triode as a d-c amplifier the device becomes quite sensitive as an indicator of voltages applied to the grid. This type of tube is customarily used as a tuning indicator on radio receivers, but it is also very useful as a null indicator in bridge measurements, and after calibrating the shadow angle it may even be used as a voltmeter. The 6U5 differs from the 6E5 in having a remote cutoff, and it can therefore handle a wider range of input signals.

ICONOSCOPES

A type of tube used in some television cameras also employs an electron gun and focused beam of electrons. The optical image is focused by a lens onto an image plate, consisting of a mosaic of tiny silver dots on a mica card backed by a metal plate. Each dot is photoelectrically sensitized and gives up a number of electrons proportional to the light intensity striking it and thus assumes a positive potential, charging the small capacitor formed between the tiny dot and the backing plate.

The electron beam, directed toward the mosaic, is deflected horizontally across the picture by a pair of deflecting plates, or by a controlled magnetic field. After each horizontal deflection or line, the beam is deflected downward by the width of two lines and the horizontal deflection repeated. After all the odd-numbered lines are scanned the operation is repeated on the even-numbered lines until the entire optical image has been covered. As the electron beam strikes each silver dot the capacity between dot and backing plate is discharged by addition of electrons to the dot, and a current flows to the backing plate proportional to the charge, and thus to the light intensity on that particular dot. A picture consists of a composite of tiny dots, scanned at a rate giving 30 complete pictures per second.

THE ORTHICON

A more sensitive camera tube is called the *image orthicon*. This employs the electron-beam-scanned mosaic image plate to which is added internal amplification through use of a secondary-emission multiplier.

When an electron beam strikes an electrode, each beam electron may cause to be emitted one or more secondary electrons which can then be attracted to a more positive electrode. By special sensitization the number of electrons emitted per incident beam electron may be increased to 8 or 10 each. If this ratio is called S, and the process is repeated over several successive stages, the total amplification of current can reach S^n, where n is the number of stages. For $S = 10$, and $n = 6$, the amplification is 1,000,000.

In the image orthicon, electrons are collected on each silver dot in accordance with its potential. The remaining uncollected electrons in the beam are reflected or turned back from the dot. This returning beam thus has density varying in proportion to the degree of darkness or lack of light on the particular dot, and the returning beam of electrons thus is varied in accordance with the picture.

FIG. 4-24. An image-orthicon tube. (Courtesy RCA)

The returning electron beam is caused by a magnetic field to strike a sensitized electrode near the cathode and to produce secondary electron emission. These electrons pass successively through a number of secondary emission stages, receiving secondary emission amplification, before the electrons strike a collector electrode and the current is taken out externally.

By use of this internal amplification, which proves to be less noisy than external pentode amplifiers, the sensitivity of the camera tube is much increased, permitting television pickup in locations of low light intensity. An image orthicon tube is shown in Fig. 4-24.

THE KLYSTRON

Still another tube employing an electron beam is the *klystron*. Electrons emitted from cathode K in Fig. 4-25 are accelerated by a positive voltage on anode A, and a beam of electrons passes through a hole in the anode. The input cavity or gap is supplied with a high-frequency voltage such that an alternating voltage exists between G_1, the entering grid, and G_2, the exit

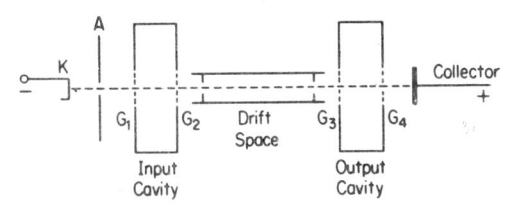

FIG. 4-25. Schematic of a klystron tube.

grid. Thus the electron beam is given small alternate accelerations and decelerations, or small variations in velocity, superimposed on its larger velocity along the tube axis.

The potentials, and the length of the drift space are adjusted so that in passing along the drift space the faster electrons will just catch up with a preceding slow group. Thus at the output of the drift space the electrons are then travelling in bunches or groups. In passing through grids G_3 and G_4 of the output cavity these bunches represent pulses of current, and part of the energy of the electron bunches is given up in exciting an oscillating current in this cavity. Finally the electron beam is caught by the collector. This process is known as *velocity modulation*.

Since only a small voltage is required to produce the initial velocity variation, and larger voltages are available at the output, the klystron is an amplifier. By returning part of the output back to the input cavity in proper phase the tube can also be made to oscillate or generate high frequencies.

THE MAGNETRON

A generator of very high power at wavelengths of a few centimeters is the *magnetron;* a cut-away view is shown in Fig. 4-26. In this tube a high voltage acts to accelerate electrons from a central cathode to a surrounding anode. Directed parallel to the cathode is a strong magnetic field which tends to curve the electron paths in a direction tangential to the cathode and anode surfaces. The anode is divided by pockets or cavities, each of which serves as a resonant circuit at the output frequencies, producing radio-frequency voltages between the anode fingers visible in Fig. 4-26.

FIG. 4-26. Cut-away view of a magnetron. (Courtesy Raytheon Mfg. Co.)

As a result of the d-c anode voltage, the magnetic field, and the radio-frequency voltages, the electron paths become bunched into rotating spokes as in Fig. 4-27, and these rotate in synchronism with the radio-frequency fields. Radio-frequency energy is extracted from the rotating electron spokes by the pockets or cavities, much as acoustic resonance is produced by blowing across the mouth of a bottle.

The magnetron is used as a power source for radar and other microwave systems. Both the klystron and magnetron are discussed at greater length in Chapter 18.

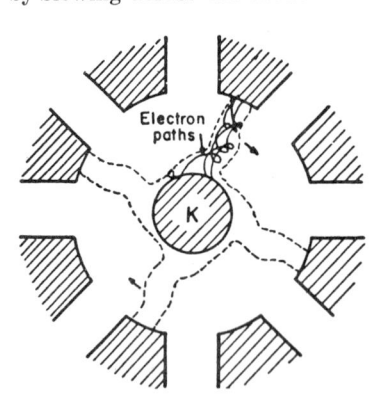

FIG. 4-27. Individual electron paths and the rotating electron "spokes" in the magnetron.

GAS TUBES

If a tube is not completely evacuated, or if a small amount of gas is intentionally introduced, some electrons in travelling toward the plate may collide with gas molecules. Such collisions may result in releasing electrons from their orbits within the gas atoms, and once free these electrons likewise are attracted to the plate. They may in turn collide with other gas molecules, and liberate more electrons. This process is known as *ionization*, and it may be cumulative in nature if enough gas is present so that a large number of collisions take place in the space between cathode and plate.

A definite amount of energy is required to dislodge an electron from an atom of any particular gas, and the voltage necessary to furnish this amount of energy is called the *ionization potential* for that gas. Typical values are as follows:

Gas	Ionization Potential (volts)
Argon	15.7
Neon	21.47
Helium	4.0
Mercury	10.38

The atom which has lost an electron, and therefore also a definite amount of negative electricity, has acquired a net positive charge. It is attracted to the cathode, but on account of its greater mass the atom (or ion) moves more slowly than an electron traveling toward the plate. It may, however, acquire considerable kinetic energy befcre completing its journey, and the impact of large numbers of such positively charged particles can have a destructive action on the cathode surface, particularly if this is of the coated type. This action is referred to as cathode bombardment; it may be quite important in such tubes as the mercury-vapor rectifiers considered in Chapter 5. Other gases frequently used are xenon, argon, and hydrogen.

SEMICONDUCTORS

Metals such as copper, gold, silver, or iron are excellent conductors of electricity whereas materials such as porcelain, polystyrene, rubber, or quartz are considered excellent insulators. Intermediate between both groups are the *semiconductor* materials, silicon, germanium, lead sulphide (galena), and various other metallic oxides and sulphides.

The high conductivity of the good conductors is due to a great number of relatively free electrons in the metallic crystal that provide ample numbers of charge carriers to give high currents under low applied voltages (low resistance). The extremely low conductivity of the insulators is due to the lack of available free electrons, all such charges being rigidly bound to the atoms of the material. Without charge carriers that are free almost no current will flow when an emf is applied.

Germanium or silicon seem the most useful semiconductor materials at present, but much work is being done to develop compounds of several metals which may offer similar or more desirable properties.

In semiconductors the number of free charge carriers is always severely limited and is a function of temperature or of impurities present. At absolute zero temperature pure germanium would be an insulator. As the temperature is raised some of the electrons are broken loose from their associated atoms by thermal vibration, and become available as charge carriers. In addition, wherever an electron has left its normal position in the interatomic bond, there is an electron vacancy or *hole* in the crystal framework and this appears as a fixed positive charge. Thus when an electron is released from a bond by thermal energy there is created an *electron-hole pair*, and this situation is symbolically represented in Fig. 4-28.

It is known that elements of the Fourth Group of the Periodic Table, as are silicon and germanium, have four *valence* electrons in their outermost shell of electrons surrounding the nucleus. It is these valence electrons

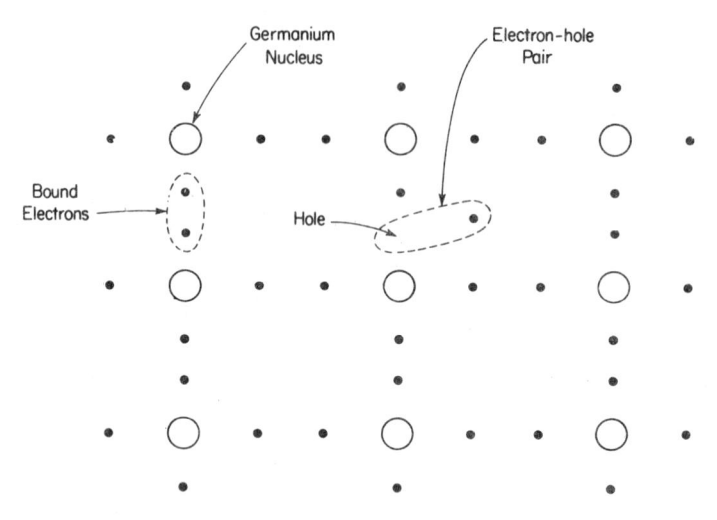

FIG. 4-28. Electron distribution in the lattice of a crystal of germanium.

which contact other atoms in forming crystals and determine the crystal shape or *lattice*. The four electrons from each atom of silicon or germanium in Fig. 4-28 are shared or bound in valence bonds by four other atoms, each of which in turn contributes one electron to the bond.

Upon the application of increased temperature a few of the electrons may be freed or the bonds broken and electron-hole pairs formed as shown, making possible *intrinsic conduction* by both positive and negative carriers, as will be described. It should be noted that the number of such pairs created at usual temperatures is very small compared to the total number of atoms in a crystal, so that the number of available charge carriers is limited but increases with temperature. Such materials then have a negative temperature coefficient of resistance.

While the hole is bound to an atom it is still attractive to electrons bonded to neighboring atoms, and under the influence of an applied electric field or emf one of these electrons will occasionally break loose and jump to a nearby hole, leaving its own hole behind. Another electron may jump into this second hole and thus the hole *appears* to move from atom to atom through the crystal. A major difference between the motion of a free electron and the motion of a hole is in their relative speeds or *mobilities*. Usually it is possible for an electron to travel about twice as fast as a hole, and this difference is important in some semiconductor devices. In any case, owing to the devious paths required in passing through the crystal, the charge velocities are much below those expected in vacuum tubes.

It is found that appreciable production of electron-hole pairs, giving rise to intrinsic conduction, occurs only above about 85°C for germanium or approximately 180°C for silicon. In most semiconductor devices this phenomenon is to be avoided so that operating temperatures are limited for devices currently in use. Intrinsic conduction tends to mask the more desired conduction effects to be described next.

If a means is employed to limit conduction largely to one kind of carrier, namely to electrons for N-type conduction, or to holes for P-type conduction, very interesting properties can be obtained in semiconductors. For instance, if extremely small quantities of a material such as arsenic, from the Fifth Group of the Periodic Table, are added to extremely pure germanium or silicon, it is found that the resulting material now has a few free electrons contributed or donated by the arsenic, and the arsenic is known as a *donor* and the resultant germanium as N-type.

The number of impurity atoms is so controlled as to be very small with respect to the number of germanium atoms, possibly only one in 10^{10} to 10^{14} and the arsenic atoms take positions in the crystal lattice without disturbing its fundamental pattern. However, the arsenic, from the Fifth Group of the Periodic Table, has five valence electrons, one more than is needed to complete the bonds with the surrounding germanium atoms. It is this extra electron which is then free or unbound, and is donated to the crystal. Note that the crystal is still electrically neutral, since the arsenic nucleus contains

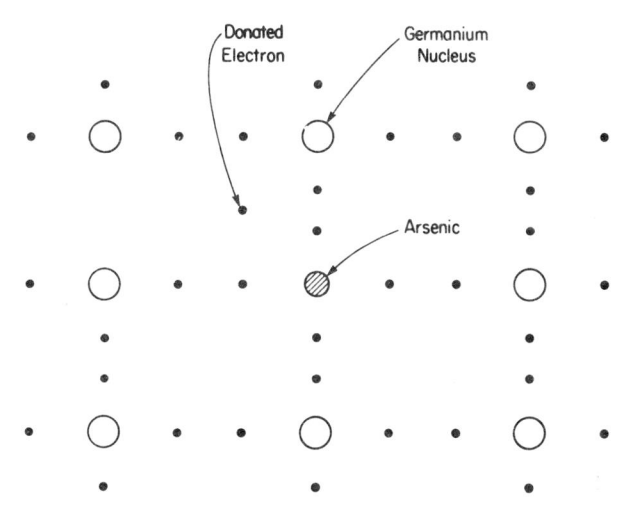

Fig. 4-29. Effect of N-type impurity in donating an electron to a region in a crystal.

a balancing positive charge for this electron. The excess electron is available to move as a current, and this is the important point. Conduction in such material is by electrons or is of N-type, and the situation is illustrated in Fig. 4-29 (N refers to a negative carrier).

If an element from the Third Group of the Periodic Table is added to the semiconductor as a controlled impurity, the situation becomes as in Fig. 4-30. Elements from the Third Group, such as boron or gallium, have only three valence electrons and can satisfy only three of the germanium bonds.

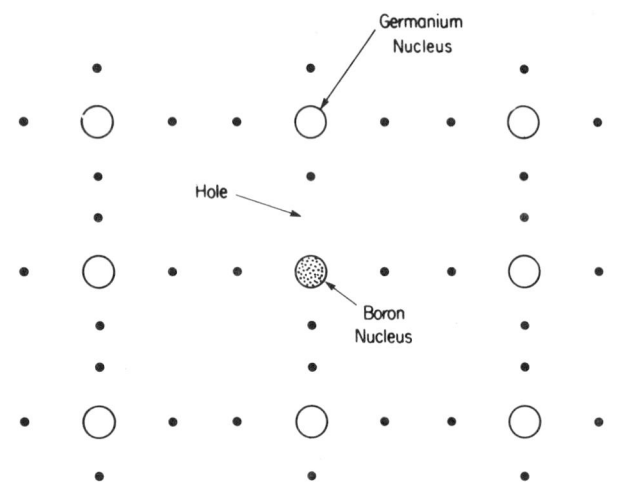

Fig. 4-30. Effect of introducing a boron impurity to add "holes" to a semiconductor lattice.

This leaves a built-in hole in the lattice for every impurity atom present, and makes conduction by holes as the majority carrier possible in P-type material (P refers to a positive carrier). Again the number of impurity atoms is very small. It is possible for the hole to migrate through the crystal under the forces given by an applied emf.

Semiconductor materials are prepared in very pure forms by various melting techniques under controlled atmospheres. The semiconductor is grown as a single crystal by dipping a small seed crystal into the melt and slowly withdrawing it under rotation. Single crystal material is desired to give more uniform conduction properties, since the presence of crystal boundaries leads to inconsistent characteristics. While the crystal is being pulled from the melt, controlled amounts of impurities may be added at proper times to give N or P material, or to give successive N or P layers as required for *junction* diodes and transistors.

Several other methods of forming junctions are illustrated in Fig. 4-31. The diffused junction is formed by welding a pellet of impurity (usually indium) onto the germanium, the junction being formed at the interface between the materials. In the surface-barrier type the impurity is plated onto the semiconductor and diffused into it by application of temperature. In the point-contact unit the junction is produced near the contact by *forming* or passing a héavy current through the point, causing a change in the base material near the point which apparently converts it to P material.

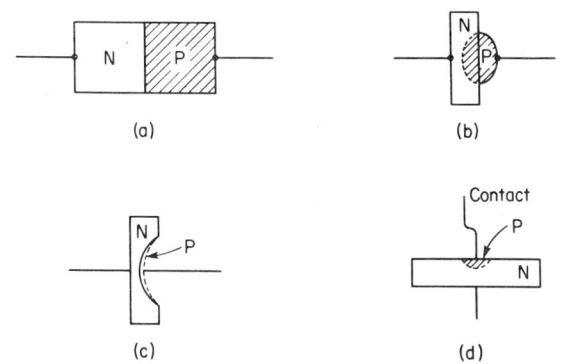

FIG. 4-31. Forms of junction: (a) grown; (b) diffused; (c) plated; (d) point-contact.

SEMICONDUCTOR DIODES

Junction units formed by growing crystals having alternate layers of N and P material, or by the other techniques of Fig. 4-31, make very desirable rectifiers or diodes at low frequencies. At the higher frequencies the internal capacity formed by the material interface shunts the rectifier and resort is had to the point-contact unit which remains usable up to some thousands of megacycles because of its low capacity.

Referring to (a), Fig. 4-32, an N-P junction without an applied voltage consists of two regions, one containing excess holes, the other excess or free electrons. These are spoken of as *majority carriers*, since in each case they far outnumber the carriers of opposite sign. Each region is electrically neutral when the total charges of all the atoms are considered.

When a potential is applied to make the P region positive, the situation is altered. Holes and electrons are forced toward the junction by the applied electric field, and electrons move across the junction and fill the adjacent holes. At the same time the positive emf is extracting electrons (injecting holes) at the P terminal and these holes migrate to the junction to replenish the hole supply at that point. The negative potential is also supplying electrons to the N region, and so a current flows as long as the emf is applied. This is the *low-resistance* or forward direction of current flow through the diode.

At (c) the diode is biased with the P region negative. Both holes and electrons are attracted toward the respective terminals and away from the junction region. This leaves the junction region devoid of charges and is the

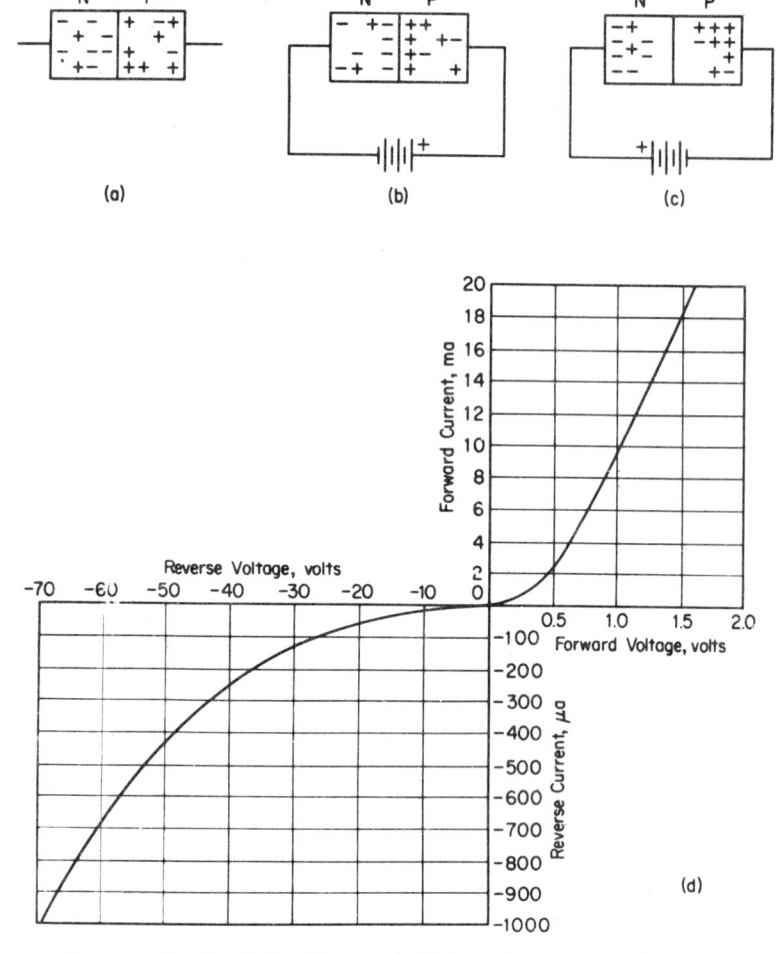

FIG. 4-32. Junction diodes: (a) normal; (b) forward biased; (c) back biased; (d) diode volt-ampere characteristic.

high-resistance or *back-biased* condition. There may be a very small current i_o under this condition because of the intrinsic conduction of the material. It is usually necessary that i_o be very small compared to any desired operating current, and since i_o is a sensitive temperature function the operating temperature may be limited.

In (d) is shown the sort of volt-ampere curve obtained from a diode. The different scales used in forward and reverse directions should be noted. The forward current is measured in milliamperes, whereas the reverse current is in microamperes, showing a very high ratio of reverse to forward resistance. The reverse resistance is limited at some high value of reverse voltage, at which breakdown occurs and a large current may flow. This is referred to as the *Zener limit*.

THE TRANSISTOR

It is possible to join two junction diodes back-to-back to form a semiconductor triode called a *transistor*. For this purpose a crystal is grown with a very thin (0.001-inch or less) P layer between two N layers of greater

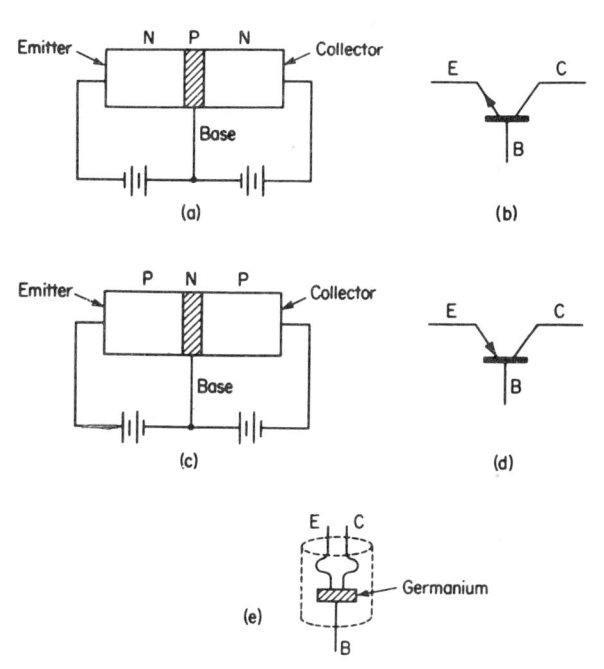

FIG. 4-33. Transistor types: (a) N-P-N junction unit; (b) symbol for (a); (c) P-N-P junction unit; (d) symbol for (c); (e) point-contact unit.

thickness, to produce an N-P-N unit. A crystal may also be grown with P-N-P layers to give similar performance, and these configurations are shown at (a) and (c), Fig. 4-33.

When the input diode formed by the left-hand N layer and the P layer of (a) is biased in the forward direction it constitutes a low resistance and electrons readily enter the hole-rich *base* region of P material. It is then said that the N layer emits electrons and it is named the *emitter*. Upon entering the P region some of the electrons combine with holes, but most come under the influence of the positively biased N region to the right and are attracted or collected by the *collector* or right-hand N terminal.

If the potential of the base with respect to the emitter is varied by a signal e_s as in Fig. 4-34, the flow of electrons will be modulated or varied in accordance with the varying base-to-emitter potential. Because the hole-rich P layer is so thin only a few of the electrons crossing the emitter-base junction are able to combine with holes in the base, and 95 to 99% of the emitted electrons reach the collector as output current. Hence the output or collector current is from 95 to 99% of the input or emitter current, and the base current, being the difference between these two, is from 1 to 5% of the emitter current.

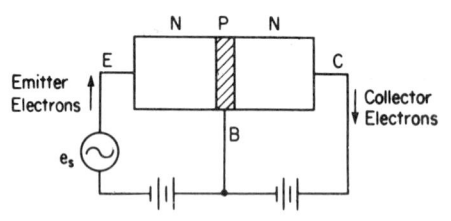

FIG. 4-34. Signal control of an N-P-N transistor circuit.

Although the ratio of output current to input current is less than unity, the device of Fig. 4-34 is nevertheless capable of both voltage and power amplification. Reference to Fig. 4-32(d) shows that a very small change in base-to-emitter voltage will, for this *forward-biased* junction, produce a large change in emitter current, with a correspondingly large change in collector current as explained previously. To obtain this same change in collector current by means of a voltage applied between base and collector would require for this *reverse-biased* junction a much larger voltage change, of the order of 1,000 times greater than the base-to-emitter voltage change. This implied voltage amplification may be realized in Fig. 4-34 by placing a high resistance of the order of 10,000 ohms in series with the collector-base lead. The changing collector current produced by signal e_s will develop a much larger output voltage across this high resistance load. Because the output voltage is greater than the input voltage (whereas the input and output currents are nearly equal) it is evident that the output power will also be greater than the input power.

Another way of viewing the voltage and power amplification achieved by the transistor is to note that emitter or input current flows in a low-resistance circuit, whereas the nearly equal collector or output current flows in a high-resistance circuit.

Current amplification a is defined as the ratio of the *change* in the collector current Δi_c to a given *change* in emitter current Δi_e which produces it, and for junction units this usually approximates 0.95 to 0.99.

It should be remembered that the *emitter* is always *forward-biased*, the *collector back-biased*. With this rule the indicated bias polarities of (c), Fig. 4-33, for the P-N-P type of unit become obvious. It may then be considered that the positive emitter injects holes into the N-type base region (or attracts electrons from the base). These holes attract negative electrons from the negatively biased collector, and cause a collector current of electrons to flow from collector to base. A few of the injected holes combine with electrons in the base region so that once again the ratio of collector to emitter currents will be of the order of 0.95 to 0.99.

The symbols for N-P-N and P-N-P junction transistors are shown in (b) and (d) respectively of Fig. 4-33. The direction of the arrow on the emitter terminal shows the *forward direction* or direction of easy (positive) current flow.

The point-contact type of unit is usually made with N-type germanium, and if it be assumed that P layers are formed near the contact points in the forming operation, then the unit takes on some of the properties of a P-N-P configuration. However, because of 0.001- to 0.002-inch spacing of the contacts, some of the holes injected by the emitter flow directly across to the collector and there create a positive space charge which is attractive to electrons. This increases the flow of electrons into the base from the negative collector. Because of the presence of this space charge, the collector current in a point-contact transistor may be several times that flowing in the emitter circuit, and values for α of two to three are to be expected.

QUESTIONS AND PROBLEMS

1. What are the relative advantages and disadvantages of tungsten, thoriated tungsten, and oxide-coated cathodes?

2. Why is it common practice to connect the return circuits to the center tap of the filament transformer in a-c heated filamentary tubes?

3. What limits the current flow in the lower regions of Fig. 4-3? In the upper regions of Fig. 4-3?

4. What three factors limit the output obtainable from a tube? What effect does each one of these have?

5. What is the fundamental structural difference between a diode and a triode? What is the physical effect which makes the triode operate as it does?

6. From the grid-plate characteristic curve for $E_b = 200$ v on Fig. 4-6, determine values of transconductance (g_m) for various values of plate current. Plot a curve of g_m against I_p.

7. Determine the a-c plate resistance (r_p) for the triode plate characteristics shown in Fig. 4-6, for a grid voltage of -5.0, and for two values of plate current.

8. From the family of plate characteristics for the 6J5 triode, Fig. 4-7, determine r_p, μ, and g_m for a quiescent operating condition of $E_g = -6.0$ and $E_b = 200$ v. Compare the value obtained for μ with the product of g_m and r_p.

9. On the plate characteristics for the 6J5 triode, construct load lines for a load resistance of 50,000 ohms and the following values of E_{bb}: 160, 240, and 320 v. Repeat this procedure for load resistances of 40,000 ohms and 20,000 ohms (note that all the load lines for a given value of load resistance are parallel to each other).

10. Construct a load line for 50,000 ohms load resistance through the quiescent point considered in Problem 8. Using this load line, determine the voltage amplification of the tube. Using the equivalent-circuit diagram and the tube constants determined in Problem 8, determine the voltage amplification again and compare it with the value obtained above.

11. Explain completely the shape of the tetrode plate characteristics, shown in Fig. 4-12, for plate voltages less than that indicated by point D.

12. Determine the values of R_s and C, in Fig. 4-14, required to obtain 100 v at the screen from a 250-v source, if the normal screen current is 0.8 ma, the lowest operating frequency is 100 cycles, and the effective a-c resistance in the tube between the screen and cathode is 70,000 ohms.

13. Trace the development of the multi-element high-vacuum tube from the triode through the tetrode and to the pentode. Give the reasons for and the effects of each added element.

14. Determine the voltage amplification for the 6AU6 tube operating through a quiescent point of $E_b = 300$ v and $E_g = -1.5$ for a load resistance of 50,000 ohms. Compare the amplification obtained here with that for the 6J5 tube of Problem 10.

15. In what different manners do the beam power tubes and the pentodes achieve the same results with regard to secondary emission?

16. Determine the pattern produced on the cathode-ray screen if a sine wave of voltage is supplied by the transformer secondary to the deflection plates of a cathode-ray tube, as shown in Fig. 4-35.

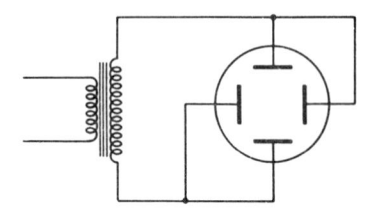

Fig. 4-35

17. Name and describe four electron tubes operating by use of electron beams.

18. From chemical considerations list the materials available as donor and acceptor elements for impurities in transistor materials.

19. Describe the rectifying action of a P-N junction.

20. Describe the amplifying action of a junction transistor.

CHAPTER 5

RECTIFIED POWER SUPPLIES

One of the important uses of diodes has already been referred to in Chapter 4 — that of providing direct currents and voltages from an a-c source. The tube in such service is known as a *rectifier*, and some of the properties of rectifiers and associated filter circuits will be considered in this chapter.

HALF-WAVE RECTIFIER

The simplest form of rectifier circuit is that shown in Fig. 5-1, and the wave form of the current in the load resistance is shown in Fig. 5-2. The tube permits current to flow when its plate is positive with respect to its cathode, but none flows when the plate is at a negative potential. If voltage

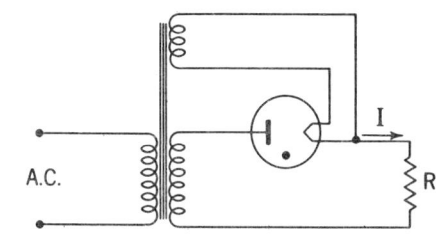

FIG. 5-1. Half-wave rectifier circuit with resistance load.

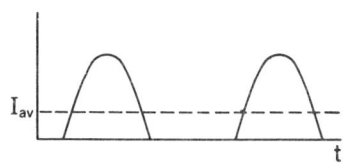

FIG. 5-2. Current flow in rectifier of Fig. 5-1.

drop within the tube is neglected, the current will consist of unidirectional pulses having the form of half sine waves, and the average current throughout the cycle, or d-c component, will be $1/\pi$ or 0.318 of the peak value. Since the peak anode current is limited for any particular tube by the electron emission of the cathode, the above relation determines the maximum d-c load current that can be supplied.

The maximum voltage across the tube occurs in the idle or nonconducting half cycle, and its value is equal to the peak value of the transformer secondary voltage. This peak inverse voltage must not exceed the voltage-breakdown rating of the tube, and thus the maximum load voltage which a given tube can supply is fixed.

FULL-WAVE RECTIFIER

The wave form of the rectified load current can be improved by making use of both positive and negative halves of the a-c cycle, and one way of doing this is shown in Fig. 5-3, with the corresponding wave form of load current in Fig. 5-4. The tube has two separate anodes and one cathode, and

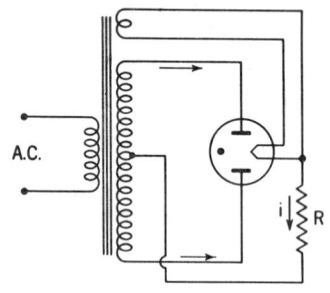

FIG. 5-3. Full-wave rectifier circuit with resistance load.

FIG. 5-4. Current flow in rectifier of Fig. 5-3.

typical tubes are the 5U4G and 5Y3G. The transformer secondary is provided with a center tap, and on tracing through the circuit it is seen that in each half of the cycle one half of the secondary winding and one rectifier anode carry current, the other being idle. The d-c component of current is $2/\pi$ or 0.636 of the peak value of current.

FILTER CIRCUITS

The output wave forms shown in Figs. 5-2 and 5-4 are entirely satisfactory for some applications, such as the operation of relays, battery charging, and so forth, but they are not smooth and continuous enough to

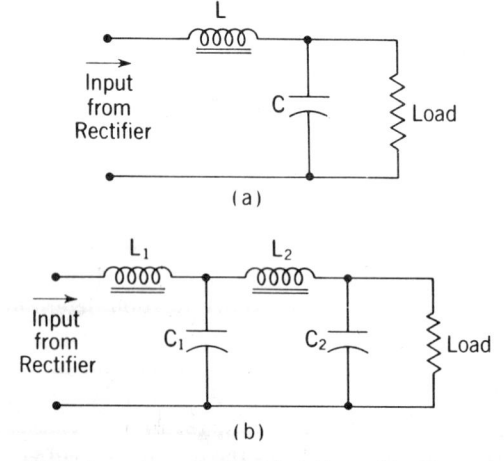

FIG. 5-5. Choke-input filter circuits, (a) single-section, (b) two-section.

be useful for B-voltage supply of amplifiers and radio receivers. Service of this sort requires that the supply voltage be almost pure d-c, with very little *ripple* superimposed upon it.

Smoothing of the rectified a-c voltage is accomplished by the use of filter circuits composed of inductance and capacitance, or resistance and capacitance. Figure 5-5 shows single-section and two-section filters of the choke-input type. The rectified current in flowing through the inductance L encounters a high reactance at ripple frequency, but very little resistance to the d-c component, and as a result the fluctuations are greatly reduced. The capacitor C in parallel with the load helps still more in this direction by absorbing most of the remaining fluctuations in current, since its reactance at ripple frequency is much less than the load resistance. If the reduction in ripple is still not sufficient, another section of filter may be added, as shown in Fig. 5-5 (b).

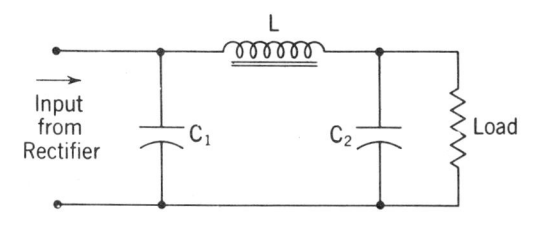

Fig. 5-6. Capacitor-input filter.

Another form of filter circuit is shown in Fig. 5-6. This is known as the capacitor-input filter, since the capacitor C_1 is supplied directly by the rectifier. In operation, the capacitor C_1 is charged to the peak voltage available from the rectifier, and this charge is withdrawn gradually by the load current. Fluctuations in current and voltage are smoothed out by L and C_2, as in the choke-input filter. No further current is supplied by the rectifier until its voltage is again higher than that remaining on C_1. This operation is shown in Fig. 5-7.

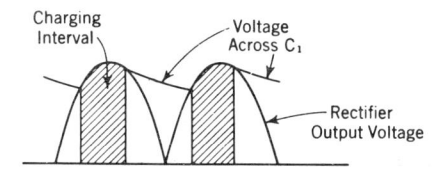

Fig. 5-7. Operation of capacitor-input filter.

It is found that the filtering effect of an inductance increases with increasing load current, whereas the filtering effect of a capacitor decreases with increasing load current. When inductance and capacity are combined in a filter circuit it is not surprising to find that the filtering effect and ripple present are more or less independent of load current. Actually the ripple present is found to be proportional to $1/LC$.

In comparing the two types of filters, it is seen that rectifier current flows continuously in the choke-input filter, whereas it flows for only a brief part of each cycle in the capacitor-input circuit. For the same d-c load current, the peak anode current in the rectifier will therefore be much larger with capacitor input than with choke input to the filter. Rectifiers for supplying large amounts of load current are commonly provided with filters using choke input, for this reason.

Another comparison between the two circuits concerns the voltage regulation, or the variation of output d-c voltage with load current. By referring to Fig. 5-7, it can be seen that if the load current is increased, the voltage across C_1 will fall more rapidly, and the average or d-c voltage will decrease. This effect is not present in the choke-input circuits of Fig. 5-5, and consequently the voltage regulation is better with choke input than with capacitor input.

It should be noted, however, that the input inductance loses its effect if the current through it is too small, and when this takes place the circuit behaves very much like that of Fig. 5-6. The approximate point at which this lower limit of load current is reached, in the case of a single-phase full-wave rectifier operating at 60 cycles, is where $L = R/1{,}200$. Here R is the d-c load resistance in ohms, and L is the inductance of the filter choke in henrys. Thus, if a rectifier is to furnish 400 ma at 2,000 v d-c, the apparent load resistance is

$$R = \frac{2{,}000}{0.40} = 5{,}000 \text{ ohms,}$$

and the minimum size of filter choke will be

$$L = \frac{5{,}000}{1{,}200} = 4.2 \text{ henrys.}$$

If, however, the load current is likely to fluctuate between 400 and 100 ma, this choke will not be suitable, for at the lower current the d-c load resistance has become 20,000 ohms, and the necessary inductance changes to

$$L = \frac{20{,}000}{1{,}200} = 16.7 \text{ henrys.}$$

A choke of this size or larger will provide satisfactory operation over the entire range of load currents. To reduce size and cost, so-called "swinging chokes," having large inductance at low current and considerably smaller inductance when saturated by larger d-c current values, are available for such service.

Still another comparison may be made between choke-input and capacitor-input filters. In the case of the filter with choke input, the d-c output voltage is approximately equal to the *average* rectified voltage of the supply. With capacitor input, however, the d-c voltage approximates the *peak* rectified voltage at light loads, and the amount of decrease with load depends on the size of the first capacitor, C_1. More output voltage is therefore available for the same a-c supply, especially if a large input capacitor is used, and for this reason most of the power supplies used in radio receivers and small amplifiers are of the capacitor-input type.

Filter circuits of resistance and capacitance were referred to earlier. A typical example of this type of filter has already been discussed in connection with the screen supply shown in Fig. 4-14, and for use with rectifiers the R-C filter is very practical and economical wherever the current drain is small, as in the case of cathode-ray oscillographs, photocells, vacuum-tube voltmeters, and so on.

RECTIFIER TUBES

Two broad classes of tubes are used in rectifier service, the high-vacuum and the gas-filled. In the high-vacuum tube, emission is controlled by space charge, and increase of current through the tube is accompanied by increase of anode potential; that is to say, voltage drop in the tube. In the gas-filled tube, ionization of the gas takes place, and consequently the voltage drop across the tube can never greatly exceed the ionization potential of the gas.

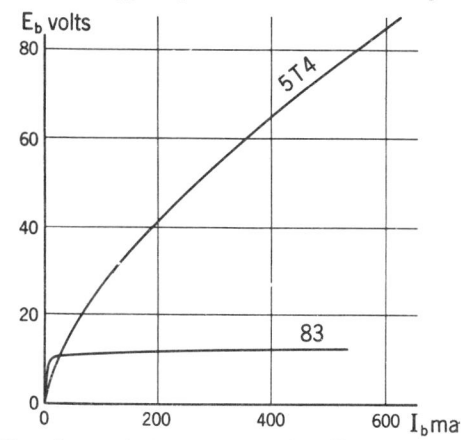

FIG. 5-8. Plate characteristics of two typical small rectifier tubes. Type 83 is a gas tube and type 5T4 a high-vacuum tube.

Most tubes of this type use mercury vapor as the gas, and for these the tube drop is approximately 15 v, independent of the current flowing. Figure 5-8 shows current-voltage characteristics of the 5T4, a high-vacuum rectifier, and of the 83, a mercury-vapor rectifier, which have approximately the same maximum ratings.

In rectifiers employing mercury-vapor tubes, it is important (except for the smaller types) that the filament or cathode be brought to full working temperature before the plate voltage is applied. If this is not done, electron emission will be limited and the voltage drop across the tube will be excessive. Mercury vapor tubes must also be given time to reach mercury operating temperature if sufficient gas pressure for proper operation is to be obtained. If the surrounding temperature is low external mercury heaters may be required. If either cathode temperature or mercury vapor pressure are low, the tube voltage drop will rise. Greater voltage drop causes increased bombardment of the cathode by positive ions as described in Chapter 4, and when the voltage drop exceeds about 24 in mercury vapor, this bombardment may be intense enough to cause disintegration of the cathode surface.

Mercury-vapor tubes should not be used with capacitor-input filters unless a protective resistance is placed in series with the input capacitor, to limit the peak charging current to a value within the tube peak current rating.

VOLTAGE-DOUBLER RECTIFIER

The cost and weight of the transformer shown in the rectifier of Fig. 5-3 may be considered excessive for some low-cost radio receivers, and the circuit of Fig. 5-9 used instead. This circuit produces a full-wave output, with no-load d-c voltage nearly twice the peak value of the a-c supply voltage.

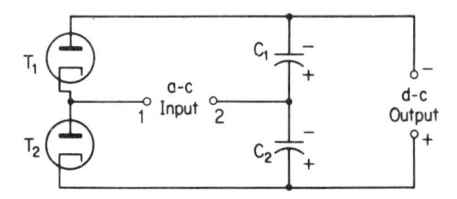

FIG. 5-9. Voltage-doubler rectifier circuit.

When a-c terminal 1 is positive, the anode of T_2 is positive and the tube conducts, placing the a-c line across capacitor C_2 and charging this capacitor to the peak voltage of this cycle of the a-c supply. In the next half cycle terminal 2 will be positive, diode T_1 will conduct, T_2 being nonconducting with negative anode, and the a-c supply will be connected across capacitor C_1, charging it to the peak of the a-c line voltage on this half cycle with the polarity indicated. As viewed from the load, the capacitors C_1 and C_2 are in series and the output voltage on open circuit is the sum of the voltages to which they have been charged.

The load current will partially discharge the capacitors and thereby lower the average output voltage. The larger the capacitors, the less the decrease in voltage and the better the voltage regulation. Capacitors as large as 40 μf are commonly used in this circuit.

As applied to radio receivers and other small electronic systems, the diodes T_1 and T_2 are commonly enclosed in a single envelope, and both cathodes heated by a series heater winding. Typical examples of such tubes are the 25Z5 and the 25Z6-GT. The circuit is also employed with selenium rectifiers as diodes, thus eliminating the need for heater power.

Using 16-μf capacitors, the circuit described will furnish a d-c output of 220 v, 60 ma, with an a-c input of 117 v rms.

VOLTAGE QUADRUPLERS AND MULTIPLIERS

The basic principle employed in the voltage-doubler rectifier of Fig. 5-9 was to charge each capacitor separately and then to discharge them in series, utilizing the sum of their separate voltages to supply the load. This principle can be extended to still higher multiples of the input voltage, and Fig. 5-10 shows a voltage quadrupler.

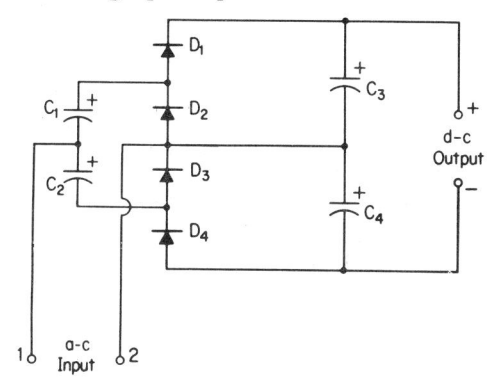

Fig. 5-10. Voltage-quadrupling circuit.

Here the diodes are represented by the symbol for a germanium or selenium rectifier, the arrow indicating the direction of easy current flow. To explain the operation of the system, assume that the a-c supply voltage is 115 v rms (162 v peak), and that the d-c output circuit or load is open.

It may be seen that C_1 and C_2 will each be charged through D_2 and D_3 to 162 v with polarities as shown, in the manner described in the previous section. However, with terminal 1 positive, there is another path from 1 through C_1, D_1, and C_3 back to terminal 2. In this path the supply voltage peak is 162 and the voltage on capacitor C_1 is 162 in the same direction, so that the voltage acting to charge C_3 is their sum or 324 v. Similarly C_4 will be charged to the same potential through D_4 and C_2 in the next half cycle when terminal 2 is positive. The d-c output voltage is the sum of the voltages on C_3 and C_4, namely 648 v, four times the peak value of the supply.

As in the case of the voltage-doubler, load current will reduce the output voltage below the open-circuit value, the amount of the drop depending on the size of the capacitors.

176   *Rectified Power Supplies*   Chapter 5

Figure 5-11 illustrates a further extension to produce a voltage multiplication of eight times the supply peak. It consists of two quadrupler circuits in series, and it is suggested that the student trace through the charging paths for the various capacitors, and determine the voltage to which each is charged.

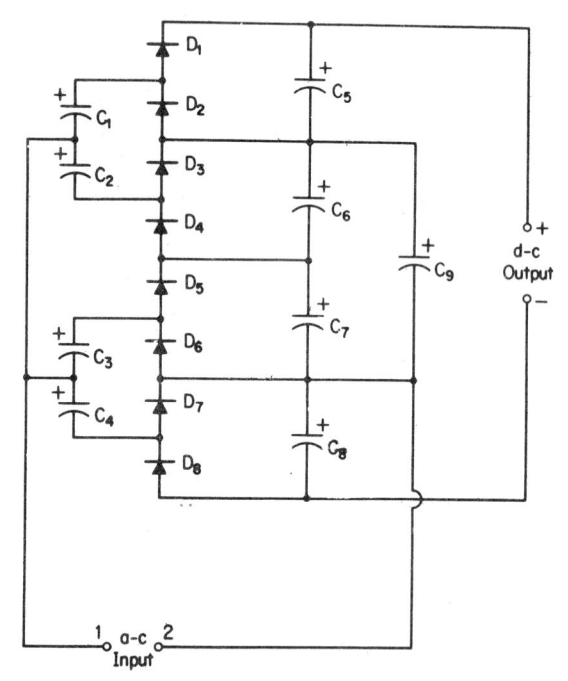

FIG. 5-11. Eight-times voltage multiplier rectifier.

The capacitor C_9 is a blocking capacitor which permits the line to supply the two units in parallel, without short-circuiting the d-c output voltage. This capacitor should be large enough so that its reactance is negligible compared to the resistance of the load. As an example, if $C_1 \ldots C_8$ is each 0.25 µf, and an output of 1,000 d-c v and 20 ma is required, C_9 might be 2 µf. This represents a 60 cycle reactance of 1,330 ohms, compared to a load resistance of 50,000 ohms.

BRIDGE RECTIFIER

The full wave rectifier circuit of Fig. 5-3 requires a center-tapped secondary winding on the transformer, and only half the winding is active at any instant. The *bridge rectifier* of Fig. 5-12 provides full-wave output with an untapped secondary winding of half as many turns compared with the circuit of Fig. 5-3, for the same d-c output voltage. However, four rectifiers, instead of two, are required.

Chapter 5   *Rectified Power Supplies*   177

By tracing the circuit, it can be seen that current flows to the load through D_2 and D_3 when terminal 1 of the input is positive, and through D_1 and D_4 when terminal 2 is positive. The current flows through the load in the same direction on both half cycles.

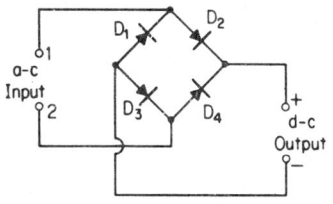

FIG. 5-12. Bridge-rectifier circuit.

If rectifiers with filamentary cathodes (not separately heated) are used, the filaments of D_2 and D_4 may be supplied from a common filament winding of the transformer. The filaments of D_1 and D_3, however, are not at the same d-c potential, and must be supplied by separate transformer windings.

REGULATED POWER SUPPLIES

Even though the output of a rectifier-filter be satisfactorily smooth and free from ripple, there may be some fluctuation of voltage caused by variations in the a-c supply, or in the load itself, and for some applications these fluctuations cannot be tolerated. Figure 5-13 shows one method of obtaining a stable value of d-c voltage, depending on the use of a glow-discharge tube such as the OD3/VR150. The anode potential of this tube remains nearly constant at 150 v over a range of currents up to 40 ma. The size of

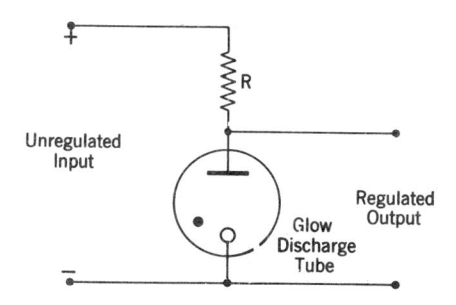

FIG. 5-13. Glow-tube rectified voltage supply.

the resistor R is chosen so as to provide the proper amount of voltage drop resulting from the sum of load current and current through the regulator tube. Thus, if the supply voltage or unregulated input is 250 v, and the load requires 20 ma at 150 v, a glow-tube current of 20 ma (less than 30 ma) should be chosen. The resistor R is then required to provide a voltage drop of 250-150 = 100 v, while carrying a total current of 40 ma. The required value for R is then 100/0.040 = 2,500 ohms. Fluctuations of the input voltage will vary the tube current, but the tube voltage and thus the load voltage and current, will remain nearly constant.

SERIES-TUBE REGULATORS

For improved control over wider current and voltage ranges, circuits similar to Fig. 5-14 are employed. The load current flows through the series 6AS7G tube (or tubes in parallel), and control is obtained by varying the internal resistance of this tube and thereby its voltage drop. A constant reference potential is obtained from a glow-tube voltage regulator. Control

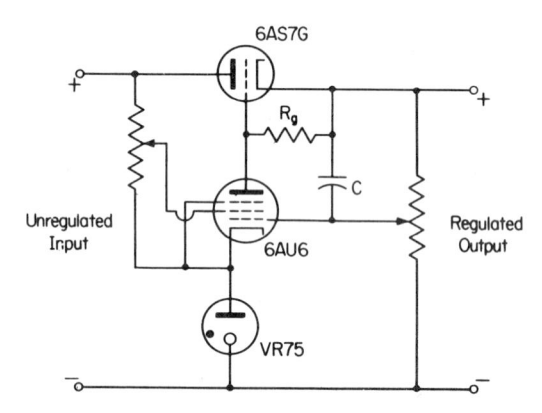

FIG. 5-14. Series-tube electronic voltage rectifier.

is achieved by comparing a portion of the output voltage to the constant voltage reference. If an unusual difference exists the amplifier tube adjusts the grid bias of the 6AS7G to a new value to return the output voltage to normal.

If the output voltage tends to rise, the grid potential of the 6AU6 will rise (cathode remaining constant at the glow-tube potential) and the 6AU6 plate current will increase. This current increase lowers the potential of the grid of the 6AS7G, raising its internal resistance and producing a greater drop for a given load current, thus restoring the voltage at the output terminals to near its original value.

Regulator circuits of this type are extremely effective over a considerable range of output currents, and output fluctuations are easily held below 0.02 v per milliampere. The regulator is also helpful in smoothing out ripple voltage present, and capacitor C is present to place a greater percentage of the ripple voltage on the grid of the 6AU6, thus aiding in eliminating ripple. The cost and weight of the filter components can sometimes be considerably reduced in comparison with an unregulated supply.

QUESTIONS AND PROBLEMS

1. What is the maximum inverse voltage applied to the tube in a half-wave rectifier circuit operating from a transformer with secondary voltage of 300 (rms)?

2. What is the maximum inverse voltage applied to the tube in a full-wave rectifier circuit operating from a transformer with a secondary voltage of 300 (rms) from the center tap to either of the outside leads?

3. Trace the operation of the voltage-doubler circuit through one complete cycle.

4. What are the relative advantages and disadvantages of choke- and capacitor-input filters?

5. Determine the minimum amount of inductance that the choke in a choke-input filter should have if it is to operate with a single-phase full-wave rectifier and is to supply a minimum current of 60 ma at 400 v d-c.

6. A 5-henry choke is available for the construction of a choke-input filter to be used with a single-phase full-wave rectifier. The rectifier supplies a load which may draw a current varying from a minimum of 10 ma to a maximum of 50 ma at a d-c voltage of 1,000. What is the maximum amount of bleeder resistance that should be used at the output of the filter?

7. What is the advantage of using a "swinging" choke in a choke-input filter?

8. What danger is incurred if the plate supply on a mercury-vapor rectifier tube is turned on before the cathode has reached its full operating temperature?

9. Explain completely the operation of a typical series-tube voltage-regulator circuit.

10. Explain the performance of the 8-times voltage-multiplier rectifier of Fig. 5-11, stating the voltages on all capacitors under open-circuit load conditions.

CHAPTER 6

TRANSMISSION AND RECORDING OF SOUND

NATURE OF SOUND

Sound consists of a pressure wave in air, traveling in all directions from the source of the sound, with a speed of approximately 1,100 ft per second in air under atmospheric pressure at sea level. Our ears recognize and respond to certain *frequencies* in these pressure waves, and such frequencies are said to comprise the range of audible sound. This range, in young people with good hearing, extends from approximately 20 cycles to approximately 20,000 cycles, but decreases with the age of the listener.

A sound wave is set up by the vibration of a material body in contact with the air. Thus, in the case of the human voice the vibration occurs in the vocal cords; in the violin the sound originates in the motion of the string but is transferred to the surrounding air by consequent vibration of the back and belly of the instrument, because these parts have much greater area of contact, or better coupling, with the surrounding air. Examination of other musical instruments and of other sources of sound reveals that the action is similar in each case — a solid object or area is set in vibration and the vibration is communicated by contact to the surrounding air. To accomplish the electrical reproduction of speech or music, as in a radio receiver, phonograph, or a public address system, the diaphragm or cone of a loud-speaker is set in vibration, and the transmission of this vibration to the surrounding air sets up sound waves which will, to the ear, constitute a more or less faithful copy of the waves set up by the original source.

When the vibrating surface, such as a loud speaker cone, moves outward some of the air close to the cone is compressed, and a compression wave is generated which travels outward in all directions. Immediately thereafter the cone moves backward and a rarefaction or reduced pressure wave is set up which follows the compression wave. These compressions and rarefactions may be considered as positive and negative halves of a sound wave radiating outward from the loud-speaker cone. The pressure variations for ordinary sounds are only a very tiny fraction of the total static air pressure; a sound which is barely audible may be produced by an excess pressure of $1/10^9$ of the static pressure, and a sound so intense as to cause a sensation of pain in the ear requires only an excess pressure of $1/10^4$ above the static value.

The *waveform* of the sound depends upon the source which produced it, and for most voices and musical instruments is quite complex. Just as in the case of electrical waves, the sound wave may be analyzed into a fundamental component and a succession of harmonics. In studying the properties of any electrical system for the transmission or reproduction of sound, each of the harmonic components present in the sound wave may be evaluated separately, and the total response will then be the combination of all these. This procedure is justified by the principle of superposition, as outlined at the close of Chapter 3.

DISTORTION

In the process of electrical transmission or reproduction of sound, the output waveform generally deviates from the original form in some degree, and this deviation is classed as distortion. There are three principal types of distortion, and these will be considered separately.

Frequency distortion. Every transmission system has a limited range of frequencies over which it can operate, and the response of the system within this range may be different at different frequencies. Two typical frequency-response curves are shown in Fig. 6-1. It is customary to plot the abscissa or horizontal axis in terms of the logarithm of frequency, since this gives equal space to each octave, and the ear hears in terms of octaves.

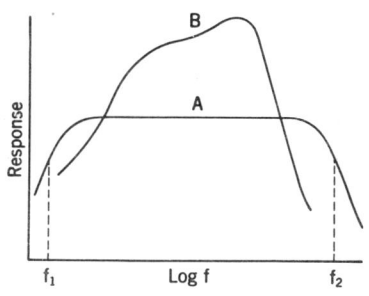

Fig. 6-1. Response curves illustrating frequency distortion.

It may be noted that curve A shows a much more uniform response over a range of frequencies than does curve B, and this might be a desirable feature of the response of A. On the other hand, if it were desired to select a group of frequencies out of a wider range, then a curve such as B might be desirable. In any case, variation of response with frequency is referred to as *frequency distortion*.

It is common to refer to the useful *bandwidth* of a system, and for a characteristic such as A, Fig. 6-1, this bandwidth is usually taken as extending between the indicated frequencies f_1 and f_2, at which the voltage response is 0.707 times the uniform response in the middle range. This means that the power output at these points, being proportional to voltage squared, is $\frac{1}{2}$ of the middle-range power output.

Nonlinear distortion. When a system or device has a curved characteristic, or relation between input and output, as is shown in Fig. 6-2, the system or device is said to be *nonlinear*. As may be seen from the figure, the output waveform corresponding to a pure sine wave input will be nonsinusoidal or distorted. Such action is called *nonlinear distortion*. New frequencies are present in the output which were not present in the input.

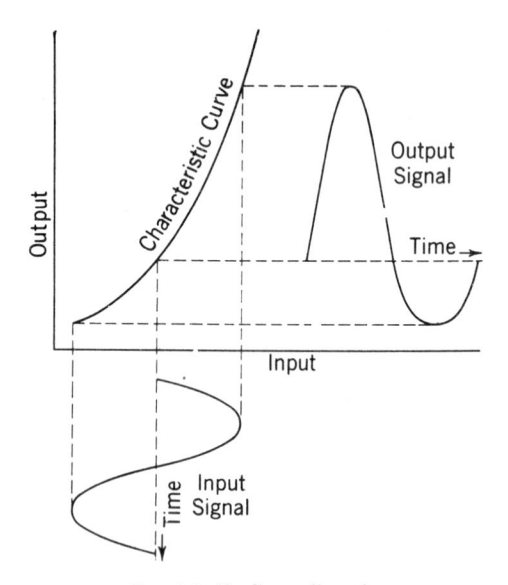

Fig. 6-2. Nonlinear distortion.

In the case illustrated, the new frequencies are harmonics in the output wave, and they are therefore all multiples of the original frequency. However, if the input signal contains more than one frequency, as is usually true, the output will also contain frequencies equal to the sums and differences of the input frequencies and of their harmonics. This latter result is called *intermodulation* distortion, and for pleasant music listening it should be kept small.

Delay or phase distortion. Signals passing through a transmission system always encounter a certain amount of delay, and if the delay time is different for different frequencies, the result will be an altered waveform. This is because harmonics in the output wave will appear at phase angles different, with respect to the fundamental, than those they occupy in the input wave, even though their amplitudes may be unaltered. This is known as *delay* or *phase distortion*.

HIGH-FIDELITY REQUIREMENTS

The male human voice employs frequencies from about 200 to 3,000 cycles, and most musical instruments employ a range covering not more than about 40 to 12,000 cycles for fundamental and important harmonics. So-called high-fidelity music reproducing systems employ equipment capable of much wider ranges, sometimes to 100,000 cycles with amplitude and intermodulation distortions below one or two per cent. These systems employ large amounts of negative feedback (as described in Chapter 7) to reduce the distortion in the audible range to such low values. The wide frequency range is obtained as a dividend from the use of feedback. It is also found that when large amounts of feedback are used, the gain and phase angle must be designed and controlled to frequencies far beyond the useful range, otherwise high-frequency oscillations may occur because the feedback has become positive at some frequency. These oscillations, even though beyond audibility, may overload the amplifier tubes causing distortion over portions of an output cycle, giving an effect of muddiness to certain tones.

LIMITED FREQUENCY RANGE EQUIPMENT

It has been found that amplifiers or ordinary telephone circuits, intended only for speech reproduction, can operate with high intelligibility when the frequency range is limited or reduced by filters to about 200 to 3,000 cycles. Female voices or music will not sound natural but the understandability of the transmitted speech will nevertheless be high. By restricting the transmitted frequency range the equipment cost can be lessened, and in radio transmission the frequency band required for the signal can be significantly reduced, causing less interference with adjacent signals.

MICROPHONES

A microphone is a device that transforms sound pressure into electrical energy. In most types of microphone the sound pressure acts upon a thin plate or diaphragm, setting it into vibration, and this mechanical motion is then utilized to produce electrical effects. The chief types are described below. The various designs are also built to incorporate desired directional pickup properties.

The carbon-grain microphone. One of the earliest microphones, and the type still most generally used in telephone practice, depends for its action on the fact that the electrical resistance between carbon granules in contact with each other varies with the contact pressure. Figure 6-3 shows a simplified sectional view of a single-cell or single-button carbon microphone, as is used in some telephone sets. A small brass cup contains two polished carbon disks, one fastened solidly in the cup and the other attached to the diaphragm. The space between the disks is partly filled with carbon granules, and as the diaphragm vibrates in response to the sound waves striking it, the varying pressure on the granules causes changes in the electrical resistance between the buttons. The microphone circuit is shown in the same figure, and from this it is seen that variation of microphone resistance will alter the current through the transformer primary, and so will set up induced voltages in the secondary.

By proper choice of diaphragm stiffness and mass, the moving system can be made to resonate near the middle of the speech range of frequencies. When this is done, the electrical output is large enough to operate a receiver over a considerable length of line without requiring amplification. The frequency response is then not very uniform, although it is entirely adequate for speech reproduction.

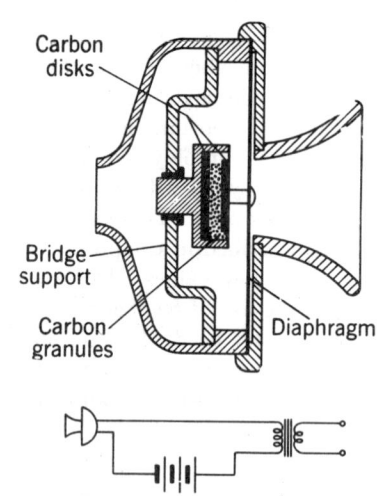

Fig. 6-3. Carbon grain microphone.

By using a very light diaphragm, tightly stretched, the frequency response is greatly improved, but at the expense of sensitivity. Carbon microphones with this type of construction, and having two *buttons*, or carbon cells, were used extensively in early broadcast practice.

The crystal microphone. Another type of microphone, widely used in public-address systems, depends for its action on the *piezoelectric effect* possessed by certain crystals, for example Rochelle salt. The term piezoelectric effect refers to the fact that when pressure is applied on the crystal in the proper direction, electrical potentials are produced between opposite faces of the crystal. The *sound-cell* type of microphone contains an assemblage of small crystals of this type, so connected that their piezoelectric potentials are in series. The sound falls on the crystals and vibrates them directly. The electrical output is quite small, but the frequency range and uniformity of response are excellent.

In another type of crystal microphone, a metal diaphragm is coupled mechanically to a crystal of Rochelle salt in such a way that vibration of the diaphragm causes a twisting of the crystal, and thereby the generation of a voltage at the terminals. This type has much greater output than the sound cell, but the frequency response is limited by the inertia and stiffness of the diaphragm and the associated driving members.

The crystal microphone has a high output impedance, making it well suited to direct connection to vacuum-tube amplifiers.

The dynamic microphone. Several types of microphones depend for their action upon the voltage induced in a conductor moving in a magnetic field. The *dynamic* or moving-coil microphone contains a small coil attached to a diaphragm, so arranged that when the diaphragm vibrates the coil moves back and forth in a radial magnetic field, and thus generates the output voltage. By careful design of the moving element, and by making use of air-chamber resonance, it is possible to obtain a nearly uniform response from 40 to 10,000 cycles. An incidental advantage is that the output impedance of the microphone is low, and the microphone cable is less sensitive to hum pickup than in the case of the crystal microphones. A transformer is required to couple such a low-impedance source to the high-impedance input of a vacuum tube grid.

Ribbon microphone. In this type the moving element is a very thin and flexible aluminum ribbon, upon which the sound waves act directly. It vibrates in a transverse magnetic field, and generates an electromotive force on the ribbon. The ribbon impedance is so low that a small step-up transformer is included in the microphone mounting to raise the impedance to a level suitable for transmission over a line. Most ribbon microphones respond to air-particle velocity in the sound wave, rather than to sound pressure, and they are referred to as *velocity microphones*. They can be made to have excellent frequency characteristics.

PHONOGRAPH PICKUPS

Any of the above basic principles suitable for microphones may also be employed to translate the motion of a phonograph needle into a varying voltage. However, desirable characteristics and design convenience have caused most phonograph pickups to employ either of two methods, the *piezoelectric crystal*, or an adaptation of the dynamic principle known as the *reluctance pickup*.

The crystal pickup. The piezoelectric crystal, usually Rochelle salt, may be so mounted that the motion transmitted from the record by the phonograph needle will cause a twist of the crystal and thereby the generation of a voltage proportional to the amplitude of the motion. Crystal pickups are high-impedance devices suitable for direct connection in the grid circuit of a vacuum tube and will develop one to two volts output from the average record.

Reluctance pickups. A voltage may be generated by moving a conductor relative to a magnetic flux, as in the dynamic microphone, or by moving or changing the magnetic flux relative to a coil. The reluctance pickup employs the latter principle, whereby motion of a tiny steel armature causes the distribution of magnetic flux from a permanent magnet to change between two fixed coils. This change of flux induces a voltage in the coils. The reluctance pickup is nominally high impedance and is usually directly connected to the grid circuit of a vacuum tube. The usual output level is about 10 mv.

Since a voltage is present only when the magnetic armature is moving and the flux is changing, the voltage magnitude is proportional to the rate of armature movement, and the output voltage is proportional to needle velocity and not to the amplitude of needle movement. For constant amplitude of record cut this would mean that the output voltage of such a pickup will fall with falling frequency. The condition of a fixed limiting amplitude is necessary to prevent interference between adjacent grooves. There is no problem at the higher frequencies where amplitude is small in any case, but at the lower frequencies this danger of overcutting places a definite limit on amplitude below some arbitrary frequency called the *turnover frequency*, and having a value in the range of 300 to 500 cycles.

At the same time emphasis (distortion) in favor of the high frequencies is now generally introduced to raise the level of the recorded sound with respect to the high-frequency needle scratch. This is called high-frequency pre-emphasis, and when the high frequencies are restored to proper balance with the low frequencies by the appropriate de-emphasis network, the needle scratch is also reduced by the same amount.

The result of these two deliberate and arbitrary distortions is to require frequency-weighting networks in record-reproducing pre-amplifiers, which

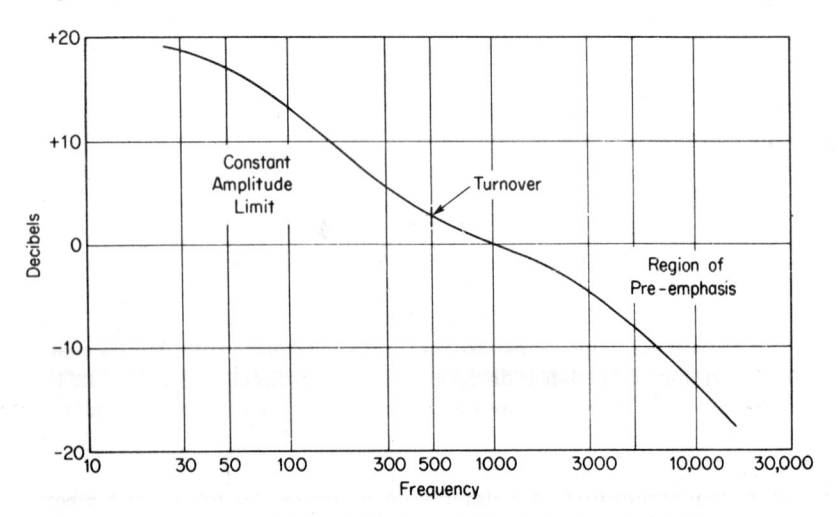

FIG. 6-4. Standard RIAA playback characteristics for preamplifier.

will raise the level of the low frequencies below turnover by the amount lost by the constant-amplitude characteristic, and will reduce the high frequencies by the standard amount of pre-emphasis. The standard RIAA curve for record playback from LP records is shown in Fig. 6-4. This is now an industry standard, although many other curves were used on older records.

TAPE RECORDING

Cellophane or paper tape, coated with a magnetic iron oxide, is employed for recording of sound or music. The tape is run past a gap in a magnetic circuit whose flux is caused to vary in accordance with the input signal. Some of the magnetism excited in the tape coating remains after passing over the gap. When the tape is rerun past the gap in the sound head this magnetic flux induces a voltage in the coil which may then be amplified and reproduced. A recording head is shown in Fig. 6-5.

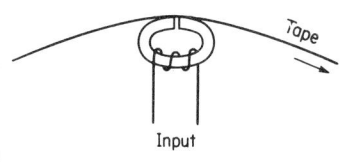

FIG. 6-5. Recording or pickoff head for tape recorder.

By using a super-audible frequency the recorded signal may be erased from the tape, and the tape reused.

REPRODUCERS

A reproducer is a device for converting electrical energy into sound. As in the case of microphones, this transformation usually involves an intermediate mechanical motion.

TELEPHONE RECEIVERS

The ordinary telephone receiver is the most commonly known acoustic device. A modified form, the watchcase type used in radio operators' headsets, is shown in section in Fig. 6-6. Two small coils are wound on soft-iron pole pieces, which are attached to the poles of a permanent magnet. The pole pieces attract the steel diaphragm with a steady pull caused by the permanent magnet, and with an alternating force set up by the voice currents flowing in the coils. The diaphragm is set into vibration, and sets up sound waves in the air in contact with it. The permanent magnet is necessary to avoid distortion in the output, since the diaphragm would be attracted *twice* in each cycle if only the a-c attraction were present.

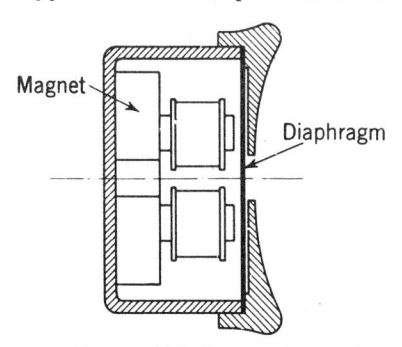

FIG. 6-6. Telephone receiver.

The ordinary receiver used with telephone instruments is wound for about 70 ohms resistance, and it has a definite resonance peak near 1,000 cycles, for the sake of sensitivity. By winding with many turns of fine wire, the sensitivity to weak currents can be greatly increased, and such receivers are very useful as indicators in a-c bridges and for radio communication systems.

LOUD-SPEAKERS

The commonest type of loud-speaker is shown schematically in Fig. 6-7. The moving coil, situated in a powerful radial magnetic field, carries the operating current. The reaction of the signal current with the magnetic field causes the coil to move back and forth along its axis. In this motion it carries with it the paper cone radiator. The cone is supported at its outer edge by a flexible suspension, and at least at the lower frequencies, it moves as a rigid piston, without appreciable bending or deformation. The result is a very effective transformation of the electrical input into sound energy radiated from the surface of the cone.

The radiation from the rear surface of the cone is opposite in phase to that from the front surface, and it is the function of the *baffle* shown in the figure to prevent these two effects from cancelling each other. The baffle will be effective at any frequency for which the distance from the front of the cone, around the edge of the baffle, to the rear edge of the cone, is greater than a half wave length of the sound. For example, at 100 cycles the wave length is $\lambda = 1,100/100 = 11$ ft (see Chapter 9) and the distance from front to back of the cone around the baffle edge should not be less than about 5.5 ft. Various means for meeting this requirement, employing curved baffles or horns, are used to return the rear radiation in phase with that from the speaker front. This is important because speaker efficiencies fall off at low frequencies, and by adding the radiation from the rear to that from the front an increased efficiency can be obtained to improve low-frequency response. One way in which this is done is to provide a planned path, usually involving a port or hole in the baffle, which will return radiation in the proper phase at some low frequency.

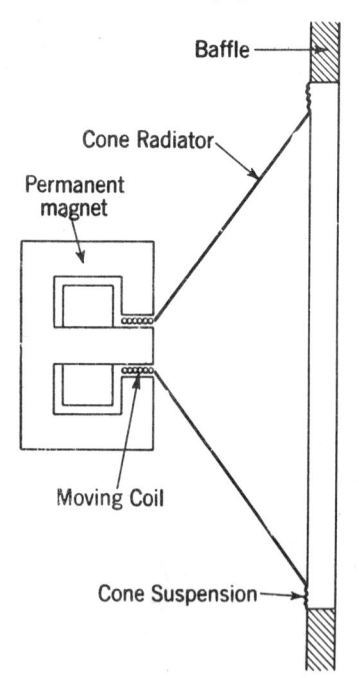

FIG. 6-7. Cone loudspeaker.

Cone speakers are also inefficient at high frequencies because of their large mass. This difficulty is often overcome in high-fidelity systems by employing two speakers, a cone and baffle such as described to provide the low frequencies up to possibly 1,000 to 1,600 cycles, and a small high frequency horn to carry the frequencies above. The amplifier output is divided into two frequency ranges by a dividing network or "crossover network" employing filter principles. Thus the low-frequency speaker or "woofer" receives only the low frequency band and the high-frequency speaker or "tweeter" receives only the higher frequencies. Some of the imperfections of the system are due to imperfect division or phase shift in the crossover network, but the net effect is still beneficial. Systems of unusual perfection may employ three or four speakers with a corresponding number of frequency bands.

TELEPHONE CIRCUITS

Two-way operation is essential for satisfactory telephone service, and many of the problems of the industry arise from this fact. A simplified circuit for obtaining two-way operation is shown in Fig. 6-8. This is called a local-battery system because a separate battery is required at each end of the line, and in such a telephone system each subscriber must have a

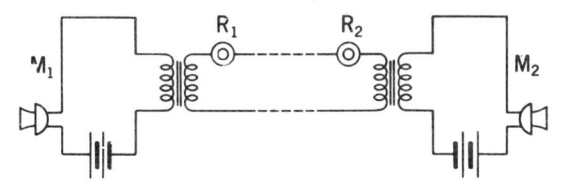

Fig. 6-8. Local-battery telephone circuit.

battery in his house equipment. The operation of the circuit is simple, since speaking into either microphone will set up voice currents in both receivers. The transformers isolate the direct current required for the operation of the microphone, and also improve the efficiency by stepping up the voltage and reducing the line current, or more nearly matching the microphone impedance to that of the line.

COMMON-BATTERY CIRCUITS

The obvious advantage of removing the batteries from the subscribers' premises to the central office led to the development of the *common-battery system*, which uses one large battery to supply microphone current to all subscribers' sets. This led to a new difficulty, however, in that the *voice currents* of all circuits in use flow through the same battery, and because of its internal impedance there is a possibility that some of these currents will find their way into other circuits. This effect is known as *cross talk*, and it may be avoided by use of a transformer called the *repeating coil*.

A repeating coil and its manner of use are shown in Fig. 6-9. It may be recognized as essentially a one-to-one transformer with primary and secondary windings split at the battery. Other repeating coils, connecting other pairs of subscribers, may be tied in at the points BC, $B'C'$, and when so connected will offer extremely high impedance to the flow of voice current from one channel to the other, but practically none to the flow of voice current in its own channel.

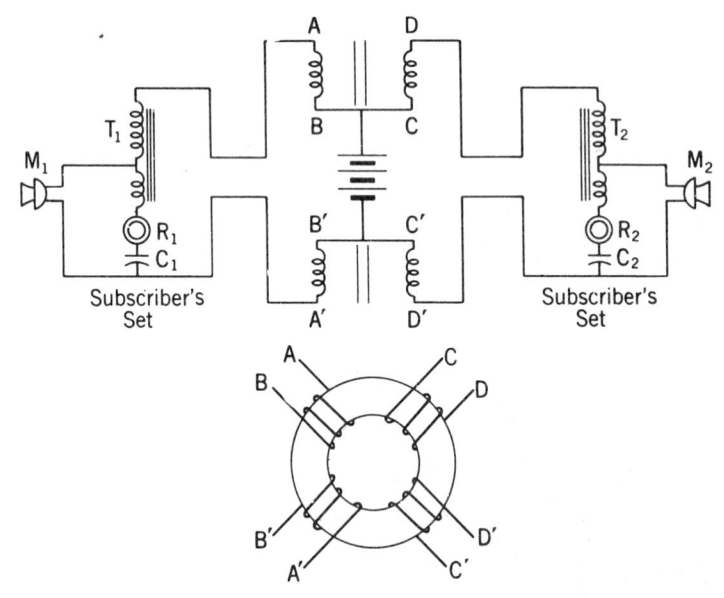

FIG. 6-9. Common-battery telephone circuit.

The talking circuit of the subscriber's set is different in the case of the common-battery telephone. Transformers T_1 and T_2, usually called *induction coils*, and capacitors C_1 and C_2, are connected as *auto-transformers*, that is, the lower portion of the winding serves as primary and also as part of the secondary to step up the a-c component of the microphone voltage.

Telephone systems require additional equipment for signaling the operator and subscribers, and for switching connections between subscribers. Their discussion is beyond the scope of this book.

TELEPHONE LINES

The transmission lines used in telephony are of two kinds, open-wire and cable. The open-wire lines are gradually being superseded by cable construction, either overhead or underground, for two principal reasons. In the first place, many more circuits can be accommodated — there is a standard cable containing more than 2,100 pairs of wires, and it is common practice to run 900 pairs in overhead cables on a single pole line; while a pole line carrying 50 pairs of open-wire construction would be a monstrosity. The other reason for preferring cable is that it affords much better protection against weather hazards and against electrical interference, both noise and cross talk.

THE DECIBEL

Losses and gains in telephone lines and other equipment are frequently stated in terms of a logarithmic unit, the decibel (discussed in Chapter 1), defined as

$$db = 10 \log \frac{W_2}{W_1},$$

where W_1 is the input power and W_2 the output power. If W_2 exceeds W_1, the value is positive and is spoken of as a *gain*. When W_1 exceeds W_2, the expression will be negative and indicates a *loss*.

The decibel is also used as a measure of the *amount* of power absorbed or furnished by any device, by comparing it with a standard power. This standard is referred to as *zero level*, and in telephone practice the power at zero level is 1 milliwatt. For example, if the power output of an amplifier is 4 w, its power level expressed in decibels is

$$\text{output} = 10 \log \frac{4}{0.001} = 10 \log 4{,}000 = 36 \text{ db}.$$

When so used, with 1 milliwatt as the zero or reference level, the unit is sometimes referred to as the *dbm*.

When used in the broadcast industry, and measured by instruments of specified dynamic characteristics, the readings in dbm are sometimes referred to as being in *volume units* or vu.

The decibel is a very important unit, and it is used extensively in all branches of electrical communications. The fact that it is logarithmic in nature makes it possible to obtain over-all effects, resulting from a combination of lines, amplifiers, and other equipment connected in sequence, by simply adding or subtracting their respective gains or losses. It is also true that the sensitivity of the ear is nearly a logarithmic function of sound intensity, and a change in sound level of 1 db is just barely perceptible to the average person, which shows further that the size of the unit is well chosen.

LOSSES IN TELEPHONE LINES

The smallest conductor used in standard open-wire telephone circuits is No. 12 B. & S. gauge, and a line of such construction will have a loss of approximately 0.06 db per mile. Cable circuits, on the other hand, make use of conductors not larger than No. 19, and a typical cable circuit with wire of this size will show a loss of about 1.0 db per mile. In this respect cable circuits are at a disadvantage as compared with open lines, especially for long distances, although recent improvements in the efficiency of telephone instruments have greatly extended the useful range of cable circuits.

REPEATERS

In long-distance lines losses are so high that satisfactory operation becomes impossible without the use of amplification. Amplifiers for telephone service are known as *repeaters*, and they must of course function in both directions along the line. Repeater stations are ordinarily installed at 50-mile intervals along the line, except in the newer wide-band transmission systems, for some of which repeater spacing is as little as 5 miles. Because of the total number of stages of amplification in a long-distance transmission, the performance requirements of telephone repeaters are very severe. Any distortion is cumulative, and any appreciable amount would soon result in unintelligible speech. The demands of high-quality program transmission for broadcast networks are even more exacting.

One method for obtaining two-way repeater service is shown in Fig 6-10. Two amplifiers are employed, one for each direction of transmission Feedback and oscillation are prevented by the *hybrid coils*, which are essentially three-winding transformers of balanced construction. A signal travelling from west to east encounters the first hybrid coil, where part of the power enters the west-east amplifier by way of the center taps on the

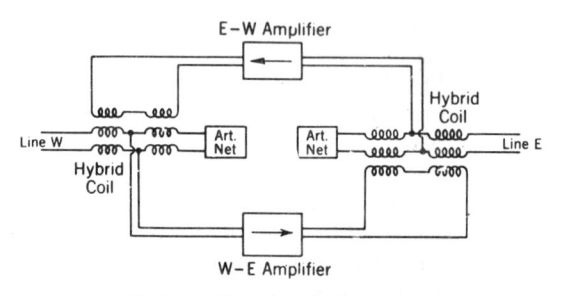

FIG. 6-10. Type 22 telephone repeater.

main windings. The amplified power is fed into the third winding of the other hybrid coil, and divides there into two equal portions, one passing out on the line to the east, the other into the artificial network. This network is constructed to have impedance properties equal to those of the line over the entire frequency band, so that the combination of line and artificial network is equivalent to a balanced bridge. As a result of this balance, none of the output of the west-east amplifier reaches the input terminals of the east-west amplifier, and feedback around the loop is avoided.

It will be noted that power is diverted and lost at both hybrid coils.

This loss is compensated for by raising the gain of the amplifier by an equivalent amount, which turns out to be 6 db in power.

QUESTIONS AND PROBLEMS

1. Determine the wave length in air for sound waves of the following frequencies: 20, 50, 100, 1,000, 10,000, and 20,000 cycles.

2. Explain the difference between frequency distortion, nonlinear distortion, and delay or phase distortion.

3. Why are the curves in Fig. 6-1 plotted to a logarithmic scale?

4. Explain the operation of a carbon-grain microphone.

5. Why is it essential to have a permanent magnet in a telephone receiver?

6. Explain the operation of a typical loudspeaker of the type shown in Fig. 6-7.

7. Explain the operation of the circuit in Fig. 6-8.

8. A telephone cable circuit using No. 19 conductors has an effective impedance (nearly pure resistance) of 450 ohms. If the power is being transmitted at a level of $+2.0$ db, find the values of voltage and current present on the line (1 milliwatt = zero level).

9. Repeaters similar to Fig. 6-10 are installed at 50-mile intervals in a No. 19 cable circuit having a loss of 1.0 db per mile. The input level to the repeater is 0.0 db, and the amplifier gain is adjusted so that the input to the next repeater is the same. Find the voltage at the output of the amplifier, if its load is 5,000 ohms (1 milliwatt = zero level).

10. The loss per mile in No. 19 cable is 1.0 db. If a 3,000 mile cable were used between New York and San Francisco without amplifiers, find the output power at San Francisco if 1 milliwatt is supplied at New York.

CHAPTER 7

AUDIO AND VIDEO AMPLIFIERS

FUNDAMENTALS OF AMPLIFIERS

In Chapter 6 it was shown how sound energy can be converted into electrical energy by means of a microphone, the electrical energy transmitted over wires, and then changed into sound by means of headphones or loudspeakers. Somewhere in this process it is generally necessary to amplify or increase the volume of the signal being transmitted. This amplification is done while the signal is in the electrical state by means of an audio amplifier or a repeater. Because energy cannot be created it is not possible to amplify the signal (that is, increase its energy) without taking energy from somewhere else. In an electrical amplifier the varying signal voltage is used to control the energy output of some other source such as a battery or power supply and this greater controlled energy becomes the output of the amplifier.

An *audio amplifier* is one suitable for amplifying signal voltages whose frequencies lie within the audible range. If the amplification is uniform for all frequencies within this band the amplifier is said to be *flat* over the range of frequencies considered. In general the amplification tends to drop off for both the very high and very low frequencies and special precautions must be taken in the design of the amplifier to prevent this. A modern broadcast amplifier is usually designed to be flat from 30 to 15,000 cycles. Audio amplifiers may use either vacuum tubes or transistors for the active or amplifying elements. Amplifiers using vacuum tubes will be considered first and then transistor amplifiers.

VACUUM-TUBE AMPLIFIERS

To obtain larger amplifications than can be secured with a single vacuum tube, an amplifier usually consists of several *stages*, the output of one stage being fed to the input of the next. An amplifier stage may be either a *voltage amplifier* or a *power amplifier*, and a complete audio amplifier usually consists of one or two stages of voltage amplification followed by a single power-amplifier stage.

A voltage amplifier is one designed to produce a large *voltage* amplification with very little power output. Such an amplifier could be used to furnish signal voltage to the grid of a succeeding stage, since when properly operated the grid circuit requires no appreciable power.

Chapter 7　　　　*Audio and Video Amplifiers*　　　　195

A power amplifier is designed primarily to supply a large amount of *power* to a loudspeaker or other power-actuated device. In this application the actual voltage step-up is of secondary importance and is usually sacrificed to improve the power-handling capacity of the stage.

CLASSIFICATION OF AMPLIFIERS

Audio amplifiers are also classified according to the operating conditions under which the tube works. The classifications in general use are Class A, Class AB, and Class B. A *Class A* amplifier is one in which the plate current flows continuously throughout the cycle of alternating voltage applied to the grid. The grid bias and alternating grid voltage are selected so that operation is confined to the linear or straight-line portion of the grid-voltage–plate-current characteristic curve. The shape of the output voltage wave will be similar to the input voltage wave on the grid of the tube.

A *Class AB* amplifier is one in which plate current flows for more than half but less than a complete electrical cycle.

A *Class B* amplifier is one in which the plate current flows for only one half of each cycle of the alternating grid voltage.

The subscript 1 may be used with the letter classification (for example, *Class A*$_1$) to indicate that the grid is not allowed to swing positive during

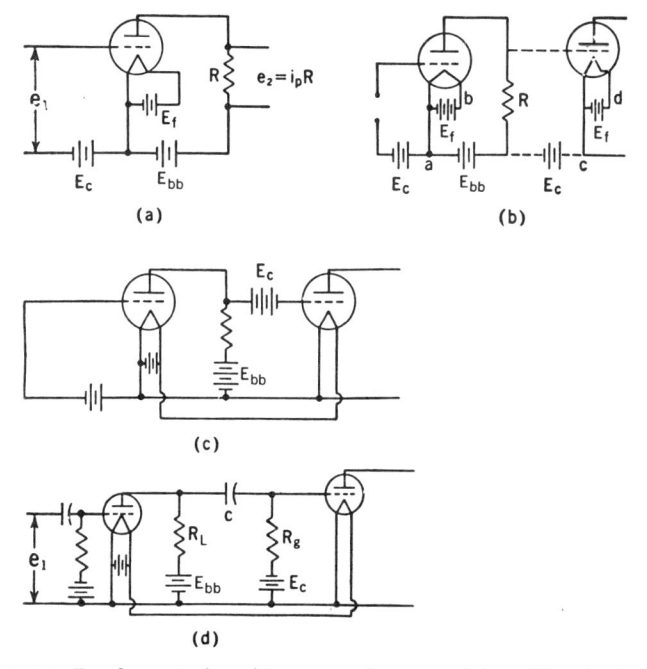

Fig. 7-1. Development of a resistance-capacitance coupled amplifier circuit.

any part of the cycle. The subscript 2 is used (for example, *Class AB*$_2$) to indicate that the grid does go positive for a fraction of the cycle.

These classifications are discussed more fully in the section on power amplifiers and in Chapter 11.

RESISTANCE-CAPACITANCE-COUPLED AMPLIFIER

It was seen in Chapter 4 that a voltage variation applied to the grid of a tube would produce a variation in its plate current. If this varying plate current is made to flow through a resistor R, as in Fig. 7-1(a), a varying voltage e_2 will be developed across the resistor, similar in all respects to the original voltage e_1, except that it may be many times larger. This larger voltage could then be applied to the grid of a second tube and still further amplification obtained. A practical circuit for connecting this second tube to the first is shown in Fig. 7-1(d). In order to see why each of the components R_p, C, and R_g is necessary for satisfactory performance, the operation of other simpler circuits will be considered.

Figure 7-1(b) represents the simplest form of coupling possible. The batteries E_f, E_{bb}, and E_c are necessary to maintain the correct *operating voltages* on the tubes. It was seen in Chapter 4 that the plate and grid voltages of a tube must be selected so that it operates on a linear or straight-line portion of its characteristic (E_g-I_p) curve. This selection is necessary if the output voltage e_2 is to be a faithful reproduction of the input voltage e_1. In practice it is usually desirable to supply power for heating the filaments of the tubes from a common battery or other source of supply. If this were done in Fig. 7-1(b) by connecting together points a and c and points b and d, it is evident that the battery voltages E_{bb} and E_c would be shorted.

The arrangement in Fig. 7-1(c) gets around this difficulty by connecting the negative of both filaments to the negative of the B battery. However, this arrangement places the B battery between the grid and filament of the second tube and so puts a large positive voltage on the grid instead of the small negative voltage usually required. Therefore a large negative voltage must be put in series with the grid of the second tube by the battery E_c. This circuit is sometimes used in amplifiers built for amplifying d-c voltages. It has the very serious disadvantage that the small d-c voltage on the second tube is obtained as the difference of two comparatively large voltages and so a small percentage change in either of the large voltages will produce a large percentage change in the voltage on the grid. For example, if the plate voltage E_b were 90 v and the required grid voltage on the next tube were $-4\frac{1}{2}$ v, the battery voltage E_c would have to be $-94\frac{1}{2}$ v. If now the voltage of E_b were to drop 5 v (that is, to 85) because of aging of the battery, the voltage on the grid would increase from $-4\frac{1}{2}$ to $-9\frac{1}{2}$, while if the voltage E_c had dropped 5 v the grid voltage would have changed from $-4\frac{1}{2}$ to $+\frac{1}{2}$. The result is that an amplifier coupled directly as in

Fig. 7-1(c) tends to be unstable and its operation may be affected too much by changes in supply voltages.

The connections of Fig. 7-1(d) are designed to overcome these and certain other difficulties. To prevent the voltage E_b from being applied to the grid of the next tube, a capacitor C, usually called a coupling capacitor because it couples the two stages together, is used. However, this leaves the grid disconnected from the filament as far as direct voltages and currents are concerned. This condition of *floating grid* must not be permitted because any electrons from the filament that reach the grid have no way of leaking off. This situation allows a large negative charge to build up on the grid and the tube becomes inoperative. For this reason the grid-leak resistor R_g is used and the correct bias voltage E_c is applied in series with it. It will be noticed that the same capacitor-resistor combination is used in the input of the first tube. This arrangement isolates the tube from any d-c voltage that may be present along with the signal voltage and ensures that the correct bias voltage will be applied to the grid of the tube.

Having arrived at the circuit of Fig. 7-1(d) as one that will be suitable as an amplifier, the next problem is to determine what values should be used for the resistors and capacitors. It is evident that each of these components has two separate functions to perform: the first is to assist in applying the correct d-c or operating potentials to the tube and the second is to provide the best conditions for the amplification of the alternating or signal voltage. Often the two functions call for widely different values and a compromise value must be used.

The plate resistor. When a tube is operating as a voltage amplifier the only requirement on the plate resistor R_p as far as the signal voltage is concerned is that the resistor be as large as possible. From the equivalent circuit of a triode shown in Chapter 4, it will be recalled that the tube acts like

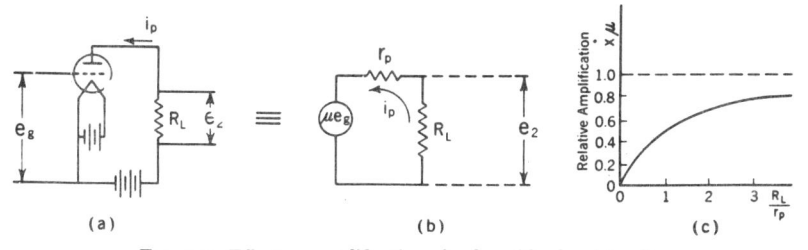

FIG. 7-2. Effect on amplification of value of load resistor R_L.

a generator having a voltage equal to μe_g in series with the plate resistance r_p of the tube. The way in which the voltage e_2 across R_L varies as the size of R_L is increased is shown in Fig. 7-2(c). R_L is shown as so many times r_p, and e_2 is shown as a fraction of μe_g. When the resistance of R_L is very small, nearly all of the voltage drop occurs across the resistance of the plate r_p, and the voltage e_2 is small; while if R_L is made very large compared with r_p, then nearly all of the total generator voltage (μe_g) will appear across R_L, and e_2 will be nearly equal to μe_g, which is the maximum value it can have. From this it appears that R_L should be made very large, say about a hundred times as large as r_p.

But now consider the d-c or operating voltage. The d-c voltage E_b on the plate of the tube will be the battery voltage E_{bb} minus the d-c voltage drop in the resistor. If R_L is made too large there is too great a voltage drop in this resistor and the resulting plate voltage will be too small. It would be possible to raise the plate voltage by increasing the battery voltage E_{bb}, but to do so might require several hundred to a thousand volts and this is not conveniently obtained. It is evident then that a smaller value of R_L would have to be used and a compromise value of 1 to 10 times r_p is generally chosen for a triode voltage amplifier, depending on whether r_p is high or low. For a power amplifier there are other considerations that dictate the value of R_L.

Grid-leak resistor. In Fig. 7-1(d) not all of the signal voltage across R_L will be applied to the grid of the second tube. The reactance of the capacitor C and the resistance R_g act together as a voltage divider and only that part of the voltage across R_g will be impressed on the grid. Hence, as far as the signal voltage is concerned, R_g should be as large as possible. When the circuit is properly operated there is no flow of grid current and so the resistor is not limited by the voltage-drop consideration as was the plate resistor. However, if the grid-leak resistor is made too large, the effect of a floating grid begins to appear, because any charge, either positive or negative, that collects on the grid cannot leak off fast enough. The actual size of the resistor required varies with the tube and usually lies in the range 0.1 to 10.0 megohms. The proper value is generally specified by the manufacturer of the tube.

The coupling capacitor. The function of the coupling capacitor is to keep the large positive voltage E_b off the grid and at the same time to offer minimum impedance to the signal voltage. This requirement indicates a large value for the capacitance, but again if C is made too large any charge that collects on the grid (and therefore charges this capacitor) will require too long a time to leak off through the grid-leak resistor. It is evident that this effect depends upon both the size of the capacitor and the size of the grid-leak resistor, since increasing either of them will increase the time required to discharge the capacitor, or the *time constant* of the circuit as it is generally called. The time constant is given by the product of $C \times R$ where C is in farads and R is in ohms. It represents the time in seconds required to discharge the capacitor to about one third (actually to 1/2.718) of the original charge. A typical value for a resistance-capacitance-coupled amplifier is 0.004 second, which could be obtained with a 1-megohm grid leak and a 0.004-µf capacitor or a 0.5-megohm leak and a 0.008-µf capacitor, or any other suitable combination.

Another practical consideration that enters into the selection of a coupling capacitor is that it should have small leakage, that is, a high leakage resistance of the order of hundreds or thousands of megohms. If the leakage resistance is small, an appreciable amount of direct current will flow through it. Then the plate resistor R_L, the capacitor leakage resistance, and the grid leak, all in series, will act as a voltage divider across the supply voltage E_{bb} and put a positive d-c potential on the grid. As an example, if $R_L = 0.1$ megohm, $R_g = 1.0$ megohm, and the leakage resistance is only 10 megohms, then for a plate supply voltage of 100 v there is a current of 0.009 ma through the circuit and this produces 9 v across the grid-leak resistor.

FREQUENCY RESPONSE OF A RESISTANCE-CAPACITANCE COUPLED AMPLIFIER

As was pointed out earlier, most applications of an audio amplifier require that it have fairly flat or uniform response over a wide range of frequencies. In other words, the amplifier must amplify all frequencies within this range approximately the same amount. The resistance-capacitance coupled amplifier is particularly suited to do this because its main impedances (the resistors) are independent of frequency over the audio range. At medium and high frequencies the reactance of the coupling capacitor is so small, compared with R_g, that it may be considered a short

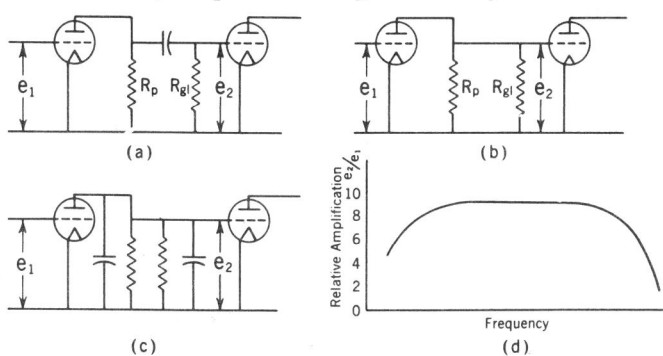

FIG. 7-3. Equivalent circuit for a resistance-capacitance coupled amplifier at (a) low, (b) medium, and (c) high frequencies, and (d) the type of response obtained.

circuit as far as the signal frequency is concerned. At very low frequencies, however, its reactance becomes large and some of the signal voltage appears across it as well as across the grid-leak resistor where it is wanted. For this reason the amplification will fall off at low frequencies. At very high frequencies the reactances of the tube capacitances and wiring capacitances become low compared with the resistances which they shunt (see Fig. 7-3), and so the amplification drops off. The tube capacitances involved are the plate-to-cathode capacitance of the first tube, which parallels R_L, and the grid-to-cathode resistance of the second tube, which parallels R_g. This is shown in Fig. 7-3 which also shows the equivalent circuits at low and high frequencies and the frequency-response curve of such an amplifier.

GAIN OF AN AUDIO AMPLIFIER

The voltage gain of an amplifier stage is given by the ratio of the signal voltage appearing across the grid of the second tube to the signal voltage on the grid of the first tube. It is quite easily determined over most of the frequency range, for which the equivalent circuit of Fig. 7-3(b) will apply. It has already been shown in Chapter 4 that the voltage E_p appearing across a resistor R in the plate circuit of a triode (Fig. 4-11) is given by

$$E_p = \frac{\mu E_s R}{r_p + R}.$$

In the case of Fig. 7-3(b), the resistance R consists of R_L and R_g in parallel. That is,

$$R = \frac{R_L R_g}{R_L + R_g}.$$

If the grid leak resistance R_g is very much larger than the plate resistor R_L, as is often the case, the effect of R_g in parallel with R_L will be small and R will be approximately equal to R_L. In this case the voltage amplification of the stage is the same as the voltage amplification of the tube, and is given by

$$\frac{e_2}{e_1} = \frac{\mu R_L}{r_p + R_L}.$$

It will be seen that as the plate resistor is made very large compared with r_p the plate resistance of the tube, the actual amplification approaches μ, the amplification factor of the tube — see Fig. 7-2(c). When R_L is 9 times r_p, the amplification will be $\frac{9}{10}$ of μ.

At low frequencies, where the exact circuit of Fig. 7-3(a) must be used it is a little more difficult to compute the gain. But if R_g is still very much larger than R_L, the gain will be given approximately by

$$\frac{R_g A}{\sqrt{R_g^2 + X_C^2}},$$

where

$$A = \frac{\mu R_L}{r_p + R_L},$$

and is the gain in the medium-frequency range. This is because R_g and

X_C act as a voltage divider across R_L, and the fraction of the total voltage that appears across R_g is given by

$$\frac{R_g}{\sqrt{R_g^2 + X_C^2}}.$$

X_C is the reactance of the coupling capacitor and is given by

$$X_C = \frac{1}{2\pi f C},$$

where f is the frequency (cycles per second) and C is the capacitance in farads. At that frequency which makes X_c equal to R_g, the amplification will have dropped to $1/\sqrt{2}$ times A, its value for the medium frequencies. Since power is proportional to the square of the voltage (across a fixed resistance), this means that the power will have dropped to one half, which corresponds to a drop of 3 db (see chart, Chapter 1). This is the *lower half-power frequency* of this amplifier stage and is considered as being the lower limit of frequency response where reasonable fidelity is required. Similarly the *upper half-power frequency* is that frequency at the upper end of the range where gain has dropped to $1/\sqrt{2}$ times A owing to the shunting effect of the tube and wiring capacitances. It can be calculated by considering the effect of these capacitances in parallel with the plate and grid resistors.

THE PENTODE AMPLIFIER

The circuit of a resistance-coupled pentode amplifier is shown in Fig. 7-4. Except for the connections to the two new grids to apply the proper d-c potentials to them, the circuit is similar to that of the triode. However, there are important differences in its operation, as was pointed out in

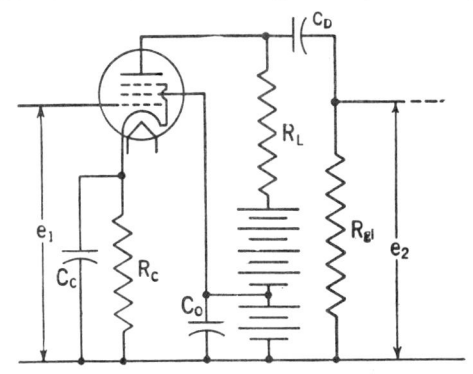

FIG. 7-4. Circuit of a resistance-capacitance coupled pentode amplifier.

Chapter 4. Because of the screening effect of the screen grid, the plate voltage has little influence on the plate current. This means that both the plate resistance r_p and the μ of the tube will be very high. The plate resistance is usually greater than a megohm, so that instead of using a plate resistor R_L having several times the value of r_p, it is now necessary to use one that is very much smaller than r_p, because its size is still limited by the d-c voltage drop across it. The gain of the stage will still be given by

$$\text{voltage gain} = \frac{e_2}{e_1} = \frac{\mu R}{r_p + R},$$

but now r_p will be much larger than R (R is even smaller than R_L), so it is possible to consider that $(r_p + R)$ is about the same as r_p and write

$$\frac{e_2}{e_1} = \frac{\mu}{r_p} R.$$

Now μ/r_p is equal to g_m, the mutual conductance or transconductance of the tube, so for a pentode the gain of a stage of amplification is given by

$$\text{voltage gain} = \frac{e_2}{e_1} = g_m R,$$

which is approximately equal to $g_m R_L$ if the grid-leak resistor is several times larger than the plate resistor R_L.

From this relation it is evident that the figure of merit or worth of a pentode as a resistance-capacitance coupled amplifier is its transconductance rather than its amplification factor. With the pentode as with the triode, the d-c voltage *on the plate* is the plate-supply voltage minus the voltage drop in the plate resistor. However, with the triode the plate current that flows is proportional to the plate voltage, so that if this plate voltage should be decreased owing to use of a larger plate resistor or lower plate-supply voltage, the plate current will decrease and so reduce the voltage drop in the plate resistor. In the case of the pentode there is no such compensating effect because the plate current depends upon the *screen voltage* and is almost independent of the plate voltage. For this reason, if a larger plate resistor is used or the plate-supply voltage is reduced, it is necessary to reduce the plate current by decreasing the screen voltage. If this is not done the plate voltage may be reduced to almost zero and the tube will cease to function as an amplifier.

MULTISTAGE AUDIO AMPLIFIERS

It is generally desirable to have more gain than can be obtained from a single stage and for this purpose two or more amplifier stages are connected in *cascade*. A typical circuit using two pentodes is shown in Fig. 7-5. The resistors R_L and R_g are the usual plate and grid-leak resistors, respectively, and the capacitors C are the coupling capacitors. Resistors R_D are voltage-dropping resistors to reduce the screen voltages below those used on the plates, and capacitors C_D are the necessary screen-resistor by-pass capacitors whose function has already been discussed in Chapter 4. Resistors R_C

Chapter 7   *Audio and Video Amplifiers*   203

are cathode resistors whose function is to furnish the required grid bias. Because the plate current must flow through these resistors in order to complete the path back to ground and the high voltage supply, there will be a positive voltage on the cathodes furnished by the IR drop across these resistors. Since the grids are operated at ground potential, the cathodes will

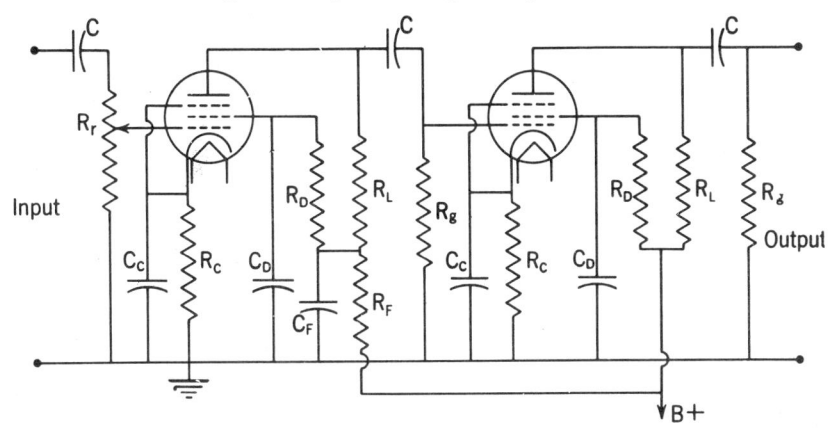

Fig. 7-5. A two-stage resistance-capacitance coupled pentode amplifier.

be more positive than the grids, or the grids will be negative with respect to the cathode, by the voltage drop across the cathode resistors. The required size of resistor is then given by

$$R_c = \frac{E_c}{I_p},$$

where E_c is the required bias and I_p is the d-c plate current flowing in that particular tube. The capacitors C_c are cathode by-pass capacitors whose function is to by-pass the signal currents around the cathode resistors. This low-impedance path is necessary, for otherwise the audio-frequency variations in the plate current would produce an audio-frequency voltage which would be introduced directly into the grid circuit.

The resistor and capacitor combination R_F-C_F constitutes what is known as a filter or decoupling circuit. One of the precautions that must be taken in constructing a multistage audio amplifier is to prevent the signal voltage in the output stage from getting back to the input circuit. If this *feedback* occurs, the signal which is fed back will be in such a direction that it either aids or opposes the original input signal. The first is called *regenerative* and the second *degenerative feedback*. With regenerative feedback the signal fed back from the output results in still more signal being fed into the input, and the signal strength may progressively increase until the amplifier reaches an oscillating or *singing* condition. This phenomenon is particularly troublesome with three-stage amplifiers or with two-stage amplifiers having a high gain, for then the output signal is very large compared with the input signal and it requires only a very small percentage of the output signal to be fed back to produce the unwanted oscillation. This feedback may occur in several ways, but in audio amplifiers the most common cause is the coupling between stages due to the common high-voltage supply. If batteries are used they will have a low internal resistance when they are new and no feedback difficulties should be experienced. However, as the batteries age, their internal resistance increases greatly. The alternating plate current of the last stage flows through this resistance, and the resulting alternating voltage is applied to the plate of the first tube and hence directly onto the grid of the second tube. This condition results in feedback.

If a power supply is used instead of batteries it must be well filtered, as outlined in Chapter 5. In this case the alternating plate current must flow through the output capacitor of the filter and so will produce a voltage drop across this capacitor. The capacitor is usually quite large, say 8 μf, so this voltage drop will be very small except at the low frequencies where the capacitor reactance $1/(2\pi fC)$ becomes appreciable.* (For example, an 8-μf capacitor has a reactance of only 20 ohms at 1,000 cycles, but this increases to 400 ohms at 50 cycles.) For this reason the frequency of the oscillation which occurs is generally quite low and gives rise to the sound known as *motor-boating*. This type of feedback can be eliminated by the filter circuit $C_F R_F$ (sometimes for high gain a two-section filter is required). The resistor and capacitor in series act like a voltage divider across the power supply with only the voltage across the capacitor being applied to the plate circuit of the first tube. If the capacitor reactance is small (that is, if the capacitor is large) and the resistor is large, this voltage will be only a small fraction of the original voltage appearing across the power supply.

HUM AND TUBE NOISE

When several stages of amplification are used, difficulty is often experienced with hum and other noise which is present with no signal applied to the input. This trouble occurs because in a high-gain amplifier even a

---

* For purposes of quickly estimating the effect of a capacitor it is well worth remembering that at 1,000 cycles a 1-μf capacitor has a reactance of 160 ohms approximately. From this it is possible to estimate without slide rule or pencil and paper the approximate reactance of any other size of capacitor at any audio frequency. For example, the reactance of a 0.1-μf capacitor at 50 cycles would be

$$160 \times \frac{1}{0.1} \times \frac{1,000}{50} = 160 \times 10 \times 20 = 32,000 \text{ ohms,}$$

while the reactance of an 8-μf capacitor at 100 cycles would be

$$160 \times \frac{1}{8} \times \frac{1,000}{100} = \frac{160}{8} \times 10 = 200 \text{ ohms.}$$

In the radio-frequency range it is useful to remember that 1,000 $\mu\mu f$ at 1,000 kc has 160 ohms reactance.

very small stray voltage picked up in the first stage may be amplified to a large signal at the output. Hum may originate from stray electromagnetic or electrostatic pickup from a-c power lines. This type of pickup can be eliminated by adequate shielding. Shielded tubes and grid leads are necessities for the first stages of a high-gain amplifier. A poorly filtered power supply is a common source of hum and for it the remedy is obvious. With a-c operation the filament leads carry alternating current, so they should be twisted together in order to reduce their magnetic field and should be kept as far as possible from grid and plate leads. (The corner of the chassis is a good place to run filament leads.) When direct current is used for the filament the grid return may be made to one side of the filament supply (usually the negative), but when alternating current is used to heat the filament the grid return should be made to the center tap of the filament transformer. If there is no center tap, a small center-tapped resistor may be connected across the filament leads and the grid return made to the center of this. Sometimes a small potentiometer is used and a screwdriver adjustment provided for setting to the position resulting in least hum.

Tube noise may also be a source of trouble in an audio amplifier. It may arise from mechanical vibration of the tube parts, in which case it is called *microphonic noise*. Replacing the offending tube with a nonmicrophonic tube will usually cure this source of trouble. Another form of tube noise appearing at very high amplifications is a hissing sound produced by the so-called *shot-effect*. This effect is due to the random motion of electrons in the tubes, in particular the first tube of the amplifier, because any noise occurring there is amplified by all the other stages. A similar type of noise, which places an upper limit on the amplification that can be obtained, is that known as *thermal agitation*. The electrons in any conductor or resistor are in constant motion back and forth, with an average velocity that increases with the temperature of the material. This random motion of charges produces fluctuating voltages in the conductor or resistor and these are amplified by the amplifier. The noise caused by thermal agitation in the grid resistor of the first tube is of most importance, because it receives the full amplification of the amplifier. The noise can be reduced by using a lower value of resistance but this will also decrease the signal gain and the signal-to-noise ratio will not be improved.

OVER-ALL GAIN AND FREQUENCY RESPONSE

The total gain of an amplifier consisting of several stages is the *product* of the gains of the individual stages when these gains are expressed as voltage ratios. For an amplifier having a voltage gain of 100 in each of the first two stages and 2 in the last stage, the over-all gain would be 100 × 100 × 2 = 20,000. Power gain is proportional to the *square* of voltage gain, so the overall power gain of the amplifier would be 10,000 × 10,000 × 4 = 400,000,000. When the gain is expressed in decibels the total gain is the *sum* of the gains of the individual stages. From Fig. 1-3 of Chapter 1 it will be found that the gain of the above amplifier would be 40 + 40 + 6 = 86 db. The over-all frequency response of an amplifier is also the product of the frequency response of the individual stages when this frequency response is expressed as a voltage ratio, as in Fig. 7-3(d). If the gain of an amplifier stage at 10,000 cycles were only half of the gain at some reference frequency such as 1,000 cycles, then the gain at 10,000 cycles of two such stages in cascade would be only $\frac{1}{2} \times \frac{1}{2} = \frac{1}{4}$ of the gain at the reference frequency. If the frequency response were expressed in decibels, a single stage of the above amplifier would be *down* 6 db (see Fig. 1-3) at 10,000 cycles below the gain at 1,000 cycles. Two similar stages in cascade would be *down* 6 + 6 = 12 db at 10,000 cycles. It is apparent that if the over-all frequency response of an amplifier is to be reasonably good, the frequency response of a single stage must be very good.

NONLINEAR DISTORTION IN AUDIO AMPLIFIERS

All of the three types of distortion discussed in Chapter 6 are found in audio amplifiers. The first of these, frequency distortion, has already been covered under resistance-capacitance coupled amplifiers, and the third, phase distortion, is not important in audio amplifiers. *Nonlinear* or *ampli-*

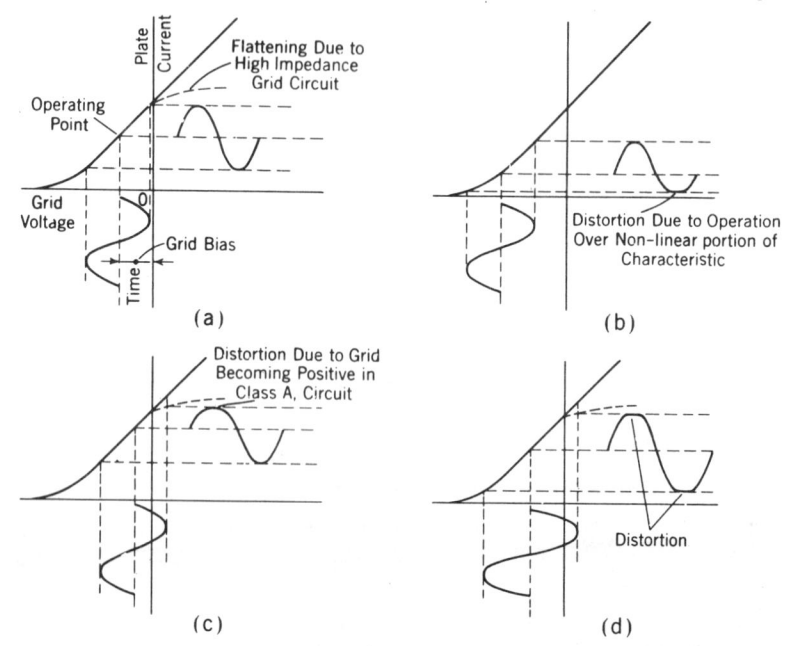

FIG. 7-6. Nonlinear distortion produced by incorrect operating conditions. (a) Correct operating conditions; (b) grid bias too large; (c) grid bias too small; (d) correct grid bias but grid driving voltage too high.

Chapter 7   **Audio and Video Amplifiers**   207

tude distortion usually appears in the output or power stage of an amplifier because that is where the signal is of greatest amplitude, but it may be produced in any of the other stages if incorrect operating conditions are used. Figure 7-6 shows possible ways in which nonlinear distortion may be introduced into an amplifier.

In Fig. 7-6(a) are shown the correct operating conditions with the alternating grid voltage swinging over the linear or straight-line portion of the characteristic curve. The result is a plate-current wave form that is a faithful reproduction of the grid voltage wave form.

In Fig. 7-6(b) is shown the effect of using too large a value for the d-c grid bias. Operation has now been carried into the nonlinear portion at the lower end of the E_g-I_p curve and the resulting plate-current wave will be flattened off on the lower peaks. This flattening results in the introduction of new frequencies (particularly second harmonic) in the output that were not present in the input.

In Fig. 7-6(c) is shown the effect of using too small a value of grid bias. This insufficiency allows the instantaneous grid voltage to swing positive over a fraction of a cycle. If the upper part of the characteristic curve is straight (solid-line portion) no particular harm will result, but if it curves sharply and flattens off (dotted curve) after passing into the positive grid-voltage region, the upper peaks of the plate-current wave will be cut off and serious distortion will result.

This dotted curve is the one that usually applies. It is obtained when the grid circuit has a high resistance, as was the case in the resistance-capacitance coupled amplifier. As long as the grid is negative it attracts no electrons and no grid current flows. However, when the grid goes positive it attracts some of the electrons being emitted by the filament, and grid current flows. If the grid circuit has zero or very small resistance, this flow of grid current has little effect on the grid voltage; but if this grid current has to flow through a high resistance (as would generally be the case), a large voltage drop occurs across the resistance and this drop reduces the grid voltage. This voltage reduction in turn reduces the plate current which would flow for positive grid voltages, and changes the characteristic curve, as shown by the dotted portion. If special low-impedance grid circuits are used it is possible to operate in the positive grid region without serious distortion, and it is sometimes desirable to do this because of the larger power outputs obtainable. Such operation is indicated by use of the subscript 2 after the letter, indicating the class of operation.

Figure 7-6(d) shows operation with the correct grid bias but with too large a signal voltage on the grid. In this case both the upper and lower peaks are flattened off and the distortion may become quite bad.

DYNAMIC CHARACTERISTICS OF A VACUUM-TUBE CIRCUIT

From the discussion it is apparent that the selection of the correct operating voltages for a tube is quite important if a large signal with low distortion is desired. To determine the correct operating voltages it is necessary to consider the action of a tube *in a circuit*. Figure 7-7 (curve *A*) shows a mutual-characteristic curve for a tube relating grid voltage to plate current. However, this curve cannot be used in plotting the operation of the tube in a circuit because it is for a *constant* plate voltage and makes no allowance for the fact that the actual plate voltage *decreases* as the plate current increases owing to the voltage drop in the plate resistor. The curve

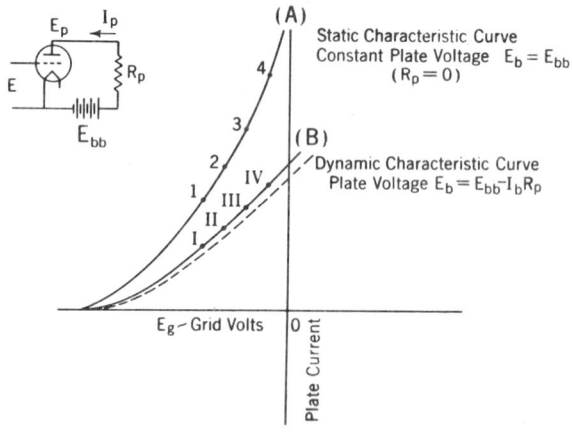

FIG. 7-7. Static and dynamic characteristic curves of a vacuum tube.

which takes this into account and which shows actual plate current against grid voltage when there is a resistor in the plate circuit is a lower curve, such as (*B*) in Fig. 7-7. This curve is the *dynamic* or operating characteristic mentioned in Chapter 4. It depends upon the size of the load or plate resistor as well as on the characteristics of the tube. This dynamic characteristic can be obtained very easily from the plate characteristics of the tube and the load line. Each intersection of a plate-characteristic curve with the load line shows the plate current that will flow for a given grid voltage E_c and for the *actual* plate voltage on the tube. Therefore, if these corresponding values of plate current and grid voltage are plotted on the diagram of Fig. 7-7, they will lie along a line such as *B*. This line is the dynamic characteristic curve for the value of load resistor used to construct the load line in Fig. 4-10. For a larger value of resistor, the load line of Fig. 4-10 would be less steep and the dynamic characteristic of Fig. 7-7 would also be less steep as indicated by the dotted curve *B*.

The dynamic characteristic obtained in this way can be used to plot the plate-current changes that will occur as the grid voltage is varied. This plot is shown in Fig. 7-8. If the dynamic characteristic is curved, then nonlinear distortion will occur and new frequencies (chiefly second harmonic) will be introduced into the output, as shown in Chapter 6. However, the dynamic characteristic will be straighter than the static characteristic and, in general, the higher the value of load resistance the straighter will be the dynamic curve. The amount of distortion, as indicated by the percentage of second harmonic present in the output wave, can be estimated roughly from the shape of the plate-current wave in Fig. 7-8. The solid line shows the actual plate current, which has a greater amplitude for the top half of the wave than for the lower half. Because of this difference in amplitudes the average or d-c current will increase slightly by the amount shown as a in the diagram. In addition there will be introduced a second harmonic

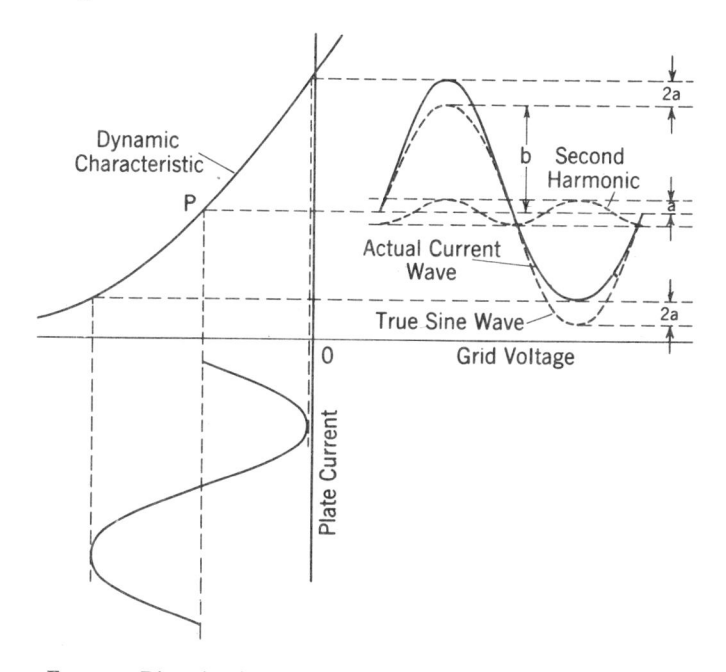

Fig. 7-8. Distortion introduced by a nonlinear dynamic characteristic.

component, shown dashed, which has an amplitude approximately equal to a. This second harmonic wave, together with the d-c increment, should add to the undistorted fundamental sine wave (also shown dashed and labeled "true sine wave") to produce the actual output wave. The percentage of second harmonic is given approximately by $(a/b)100$.

It is also possible to determine the second-harmonic percentage directly from a set of plate-current-plate-voltage curves such as those of Fig. 4-10. This second method is dealt with in most tube manuals.

SELECTION OF THE OPERATING POINT

Figure 7-8 shows that, in order to avoid distortion, the tube should be operated only over the straight-line portion of its characteristic and the lower curved portion of the dynamic curve should be avoided. Because the grid should not be allowed to swing positive, this requirement locates the operating region between the point where the curvature starts at the lower end of the curve and the point $E_c = 0$ at the upper end. The operating point P should then be midway between these limits. This same operating region can be located directly on the load line of Fig. 4-10. It extends along the load line from the curve $E_c = 0$ down to the region where the plate characteristics become curved.

AVOIDING NONLINEAR DISTORTION

To avoid excessive distortion in an amplifier stage certain precautions must be observed.

(1) The correct plate and grid voltages should be used so that operation takes place on the linear portion of the dynamic characteristic curve. The manufacturer usually specifies two or three sets of correct operating voltages, and where possible one of these should be used. It should be remembered that the voltage on the plate of the tube will be less than the supply voltage by the voltage drop in the plate resistor or coupling transformer. In the latter case the drop will be small, but in the former case it may be the larger part of the supply voltage.

(2) The load resistance should be of the right value. In the case of triodes this usually means about two or three times the plate resistance of the tube if maximum power output with small distortion is desired. In the case of pentodes the amount of distortion increases quite rapidly if the wrong load resistance is used. The correct value is usually specified by the manufacturer and this recommendation should be followed closely. With pentodes it should also be remembered that if the value of plate resistor is increased it is necessary to reduce the plate current by reducing the screen voltage in order that the voltage drop across the plate resistor should not be too great.

(3) Even with correct operating potentials and load resistance, serious distortion will occur if the signal voltage on the grid is too great. In this case both peaks of the output wave will be flattened off. The remedy is to reduce the input signal voltage by means of a potentiometer or volume control in the grid circuit.

TRANSFORMER-COUPLED AMPLIFIER

Another type of coupling commonly used in audio-frequency amplifiers is transformer coupling. Although its most general use is in coupling the output stage to the speaker or load, it may also be used as interstage coupling. Here it has the advantages of giving somewhat higher gain and eliminating the voltage drop that occurs in the plate resistor when resistance coupling is used. It has the disadvantages of greater cost, increased space requirement, and the possibility of a poorer frequency response. In a resistance-capacitance-coupled amplifier all the voltage gain is produced by the tubes and the maximum gain possible from a stage is given by the amplification factor μ of the tube in that stage. As was shown in an earlier section, the actual amplification is always less than μ by the factor

$$\frac{R}{R_L + R}.$$

With a transformer-coupled stage, however, the maximum possible gain is given by $\mu \times n$, where n is the step-up turns ratio of the transformer. For reasons to be given, n is generally limited to about 3, but even so with low-μ tubes this is an important factor and the early audio amplifiers were mostly transformer coupled. However, the introduction of pentodes and high-μ triodes has largely removed this advantage because it is now possible to obtain easily amplifications of the order of 100 times with a single resistance-coupled amplifier stage. For this reason transformer coupling is used mainly in the power amplifier stage or stages where certain other advantages make its use desirable.

Figure 7-9 shows a transformer-coupled amplifier, an approximate equivalent circuit, and a typical curve of amplification against frequency. The amplification is fairly uniform over the medium-frequency range from b to c but falls off at the low-frequency end (a to b) and again at the high-frequency end (d to e). The amplification represented by the flat portion b to c is approximately $\mu \times n$. For all low and medium frequencies

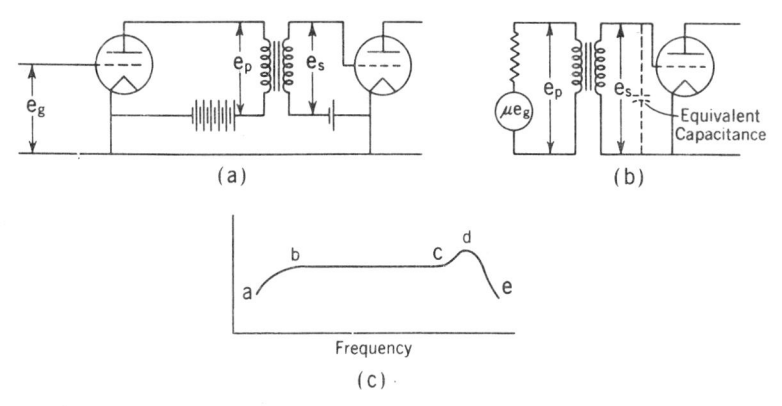

Fig. 7-9. (a) Transformer-coupled amplifier circuit; (b) approximate equivalent circuit; (c) typical frequency response curve.

up to the point c, the voltage e_s across the secondary of the transformer is n times the voltage e_p across the primary. For a given signal voltage e_g on the grid of the tube, the voltage e_p depends upon the ratio of the reactance of the primary winding of the transformer to the plate resistance r_p of the tube. It is very small when that reactance is much less than r_p and increases as the reactance increases, approaching a constant value μe_g as the reactance of the transformer primary becomes very large compared with r_p. This behavior is similar to the way the output voltage of a resistance-capacitance-coupled amplifier increases as the load resistor is increased and a curve similar to the curve of Fig. 7-2(c) would apply. This is shown in Fig. 7-10, where e_p is plotted against the ratio of the primary reactance X_p to the plate resistance r_p. The reactance of the primary winding depends upon the frequency and

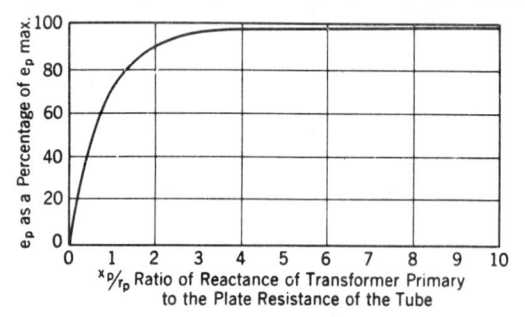

Fig. 7-10. Voltage e_p across the primary winding of a transformer as a function of the ratio of the reactance of the winding to the resistance of the source.

is given by $2\pi fL$ where L is the inductance of the winding, a constant, and f is the frequency. The curve of Fig. 7-10 is therefore also a picture of how e_p varies with the frequency. At a frequency which makes $2\pi fL = r_p$, e_p will be 0.707 times its maximum value μe_g. Above this frequency the amplification changes very little but below this frequency it falls off sharply.

It is still necessary to explain the hump at point d and the sharp drop in amplification at still higher frequencies. Both primary and secondary windings of the transformer consist of a large number of turns of wire and there is considerable capacitance between windings and between the individual turns of each winding. These capacitances can be represented approximately by an equivalent capacitance shunted across the secondary winding, as in Fig. 7-9(b). At low frequencies this capacitance has negligible effect because its reactance will be very high, but at high frequencies where its reactance becomes smaller it acts as a shunt to by-pass the signal and the amplification is lost. This effect is shown by the portion of the curve from d to e. At some point intermediate between the high-frequency and low-frequency cases, represented by the hump d, there is a frequency at which resonance can occur between this equivalent capacitance and the leakage reactance of the transformer. At this frequency the amplification may rise to quite large values, producing serious frequency distortion. This situation will be especially probable if the plate resistance r_p has a very low value.

Since the amplification of a transformer-coupled amplifier stage is given approximately by μn, it would appear possible to produce very high amplifications by the simple expedient of making the turns ratio extremely large. However, if this is done by winding on a large number of turns on the secondary, the equivalent capacitance across the secondary will be increased to such an extent that the high-frequency response of the transformer will drop off badly. On the other hand, if the turns ratio is increased by decreasing the number of primary turns, the primary reactance will be decreased and the low-frequency response will suffer. Usually a transformer ratio of 1:3 is about the greatest that can be used.

A transformer is generally designed to be used with a tube having a certain value of plate resistance. The effect of using tubes having plate resistances higher and lower than this proper value is shown in Fig. 7-11.

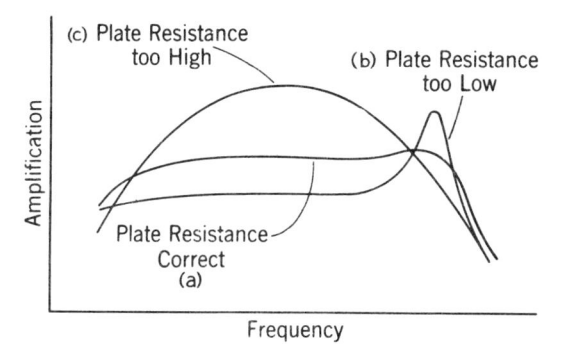

Fig. 7-11. Effects of different plate resistances on the frequency response of a transformer-coupled amplifier.

Curve (a) is for a tube having about the correct value of plate resistance. If the plate resistance of the tube is much lower than the proper value (curve b) there is danger that the hump in the amplification curve will rise to a high peak. If the plate resistance of the tube is much higher than the proper value, the primary reactance of the transformer at low frequencies will not be large enough compared with r_p, and so the low-frequency amplification will be reduced. At the high-frequency end the shunting effect of the equivalent capacitance will be greater and the high-frequency response will also be reduced. Because tubes with large plate resistances also have large amplification factors, the amplification in the middle-frequency range, still being given by μn, may be quite high and the resulting curve will be as shown in c. This explains why it is not feasible to use pentodes or high-μ triodes with a transformer to obtain very large amplifications. If for some reason it is desired to use a transformer with a tube having a high plate resistance, this can be done by shunting the primary winding of the transformer with a resistance equal to the plate resistance of the tube that should be used with the transformer. This shunt has the effect of reducing the amplification in the medium-frequency range and the resulting frequency response will be about the same as would be obtained with a tube having the correct plate resistance. The actual amplification that results will be proportional to the transconductance of the tube being used and will be independent of its amplification factor.

PUSH-PULL CIRCUITS

To get greater output than can be obtained from a single tube, two tubes are often connected in a *push-pull circuit* as shown in Fig. 7-12. At the instant the voltage on the grid of tube 1 is a positive maximum, the voltage on the grid of tube 2 will be a negative maximum. At this instant the plate current of tube 1 will be increasing and the plate current of tube 2 will be decreasing. These currents will produce voltages in the secondary of the transformer that are in the *same* direction, and so twice the power output

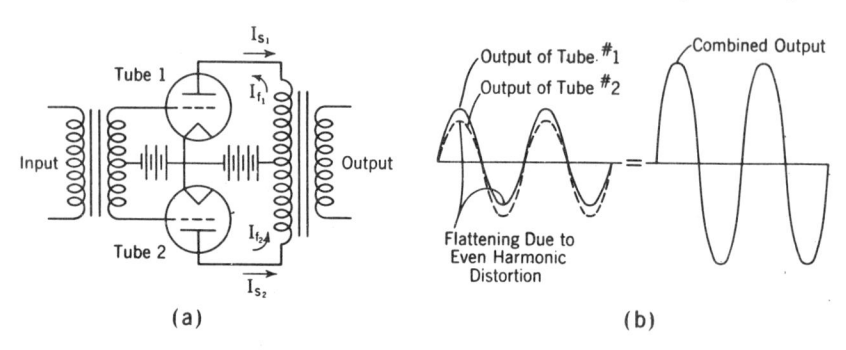

FIG. 7-12. Push-pull amplifier circuit and its effect in the reduction of even harmonic distortion.

can be obtained. The circuit has the advantage that distortion from the second and all even harmonics will be reduced, because for these harmonics the currents will flow in *opposite* directions through the transformer and so cancel out. Even harmonics give to the lower half of a wave a shape different from that of the upper half, as for example the flattening off of the lower half shown in Fig. 7-6, when too large a grid bias was used. In a push-pull circuit the top half of the current wave of one tube is added to the bottom half of the other, and a symmetrical wave results. This is shown in Fig. 7-12(b). Because twice the output of a single tube can be obtained with *less* distortion, it is possible to obtain more than twice the output with the *same* distortion given by a single tube. Push-pull operation also has the advantage that any a-c hum voltage from the power supply will produce currents in opposite directions in the primary of the transformer and no hum voltage will appear across the secondary. However, this advantage applies only to hum introduced in this stage, since hum introduced in preceding stages will be amplified just as is the signal voltage. The d-c *plate currents* to the tubes also flow in opposite directions in the transformer primary winding so that large plate currents may be used without danger of saturating the core.

POWER AMPLIFIERS

The final stage of an audio amplifier is usually a power-amplifier stage, because loudspeakers, modulation transformers, and most other devices which the amplified signal must drive require power (that is, *current* as well as voltage) for their operation. Class A amplifiers are used extensively as power amplifiers and the essential difference between their operation as power amplifiers and their operation as voltage amplifiers is in the magnitude of the load resistance and the plate voltage used. For voltage amplification the load resistance is made as high as possible to obtain maximum voltage gain; but when power output is the chief consideration the load resistance is reduced to a value that gives most power for a certain permissible value of distortion. With power amplifiers the highest possible plate voltage is used because the power output possible is directly dependent on the plate voltage.

To obtain larger power outputs for a given tube size and plate voltage, Class AB and Class B operations are used. The different amplifier classifications are shown in Fig. 7-13. In Class A_1 operation the grid-voltage swing is generally confined to the linear portion of the curve and the plate

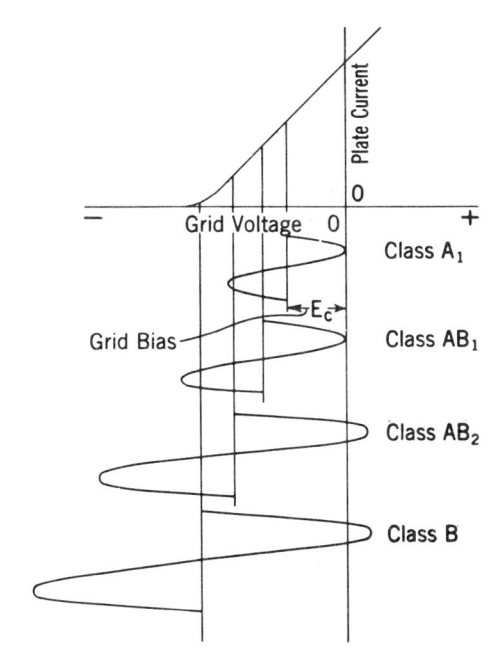

Fig. 7-13. Amplifier classifications.

current flows for the entire cycle. The grid never goes positive and hence the maximum grid swing permissible is limited to that shown in Fig. 7-13.

By increasing the grid bias E_c a larger swing of grid voltage can be obtained without the grid ever going positive; such operation is known as Class AB$_1$. In this case the negative peak of the grid-voltage swing carries the operation off the curve and the plate current is zero for a fraction of a cycle. This would produce excessive distortion with a single tube, but by using two tubes connected in push-pull the current waves are made to supplement each other as shown in Fig. 7-12(b); thus the distortion is kept within reasonable limits.

If the grid-voltage swing is increased still further with the same grid bias, a greater output can be obtained but the grid will be positive for a small portion of the cycle. Such operation is called Class AB$_2$. It requires a driving voltage fed from a low-impedance source, or grid-circuit nonlinear distortion of the type shown in Fig. 7-6 will result. Practically, this means that because the grid circuit draws current it requires *power* from the preceding stage. For this reason the preceding stage should be a Class A$_1$ amplifier feeding the grid circuit of the Class AB$_2$ stage through either a *one-to-one* or a *step-down* transformer.

Even greater power output can be obtained from a tube by operating it Class B, in which case the grid bias is increased almost to plate-current cut-off so that plate current flows only for the positive half of the grid-voltage swing. Because this type of operation is used chiefly to modulate the radio-frequency power of a transmitter, a more detailed discussion of it will be given in Chapter 11.

INVERSE FEEDBACK

Distortion in an audio amplifier can be reduced and the frequency response of the amplifier greatly improved by the use of inverse feedback (also called negative or degenerative feedback). This improvement in quality is obtained at the expense of the gain of the amplifier, but with modern high-μ tubes it is easy to obtain all the gain that may be desired. In an inverse-feedback circuit some of the output voltage is fed back to one of the preceding stages in a direction such that the voltage fed back opposes the original signal voltage at that point. If some distortion has occurred in the amplifier and the output contains new frequencies not originally present in the input, these new frequencies are now introduced into the input in a direction such as to tend to cancel those produced in the amplifier itself. Again, if the amplifier produces frequency distortion so that some frequencies are amplified more than others, these over-amplified frequencies produce a large signal in the input (owing to the feedback). This large signal opposes the original signal and the *actual input signal* at these frequencies will be small. A numerical example will help make this situation clear.

Suppose an amplifier without feedback requires 1 v input across *cd* to produce 100 v output (Fig. 7-14). A feedback circuit is then added, which feeds back 9 v to the input when the output is 100 v. If this 9 v is in the *opposite* direction to the original 1 v, it will now require 10 volts

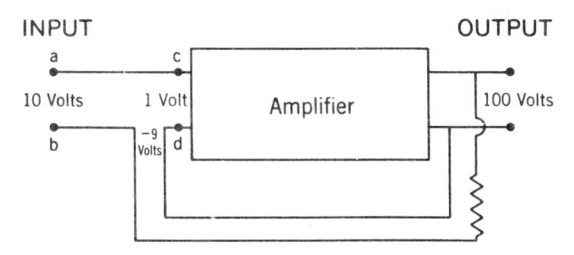

```
INPUT                                    OUTPUT
    a           c
10 Volts     1 Volt    Amplifier        100 Volts
    b        -9
            Volts  d
```

FIG. 7-14. Principle of inverse feedback.

$(1 + 9 = 10)$ at the input terminals *ab* to keep 1 v at *cd* and so maintain the 100 v output. The original gain of the amplifier without feedback was

$$\frac{\text{output volts}}{\text{input volts}} = \frac{100}{1} = 100.$$

The new gain with feedback added is

$$\frac{\text{output}}{\text{input}} = \frac{100}{10} = 10.$$

The first effect of feedback has thus been to reduce the gain of the amplifier. These results can be expressed mathematically by saying that the gain of the amplifier with feedback is given by

$$\text{gain} = \frac{A}{1 - \beta A},$$

where A is the amplification without feedback and β is the fraction of the output voltage that is fed back. In the above example $A = 100$ and $\beta = -9/100 = -0.09$, so that

$$\text{gain} = \frac{100}{1 - (100 \times -0.09)} = \frac{100}{1 + 9} = 10.$$

Now consider the effect on frequency response. Suppose the amplifier alone has considerable frequency distortion and amplifies a 1,000-cycle signal 100 times but amplifies a 3,000-cycle signal 200 times. For the same input signals the ratio of 3,000-cycle to 1,000-cycle signal in the output will be $200/100 = 2$, without feedback. When feedback is applied it will require for the 1,000-cycle signal 10 v at *ab* to produce 100 v at the output. (As before, 9 v of this will be used in overcoming the feedback voltage and the other volt will appear across *cd*.) However, for the 3,000-cycle signal it requires only $9\frac{1}{2}$ v at *ab* to produce 100 v at the output, because it requires 9 v to overcome the feedback voltage and only $\frac{1}{2}$ v across cd to produce 100 v at the output. Thus it is found that with feedback the amplifier gain for 1,000 cycles is $100/10 = 10$, but for 3,000 cycles it is $100/9.5 = 10.5$ approximately. If a 10-v 1,000-cycle signal is applied at ab the output will be $10 \times 10 = 100$ v, but if a 10-v 3,000-cycle signal is applied at ab the output will be $10 \times 10.5 = 105$ v. With feedback, then, for the same input signal the ratio of the output voltages at 3,000 cycles and 1,000 cycles will be $105/100 = 1.05$. Instead of having twice as much 3,000-cycle output as 1,000-cycle output (that is 100% greater) in comparison with the case without feedback, with feedback the 3,000-cycle output is only 5% greater than the 1,000-cycle output.

Of course the over-all gain has been reduced but this can be overcome by designing the original amplifier for very high gain.

Practical inverse-feedback circuits. Inverse feedback may be used over one, two, or three stages. That is, the voltage may be fed back to the same stage, to a preceding stage, or to the stage before the preceding stage. One- and two-stage feedback is fairly easy to apply, but three-stage feedback is more difficult because of the possibility of phase shift or delay in the amplifier. If this phase shift amounted to nearly a half cycle at any frequency the voltage fed back would aid rather than oppose the original

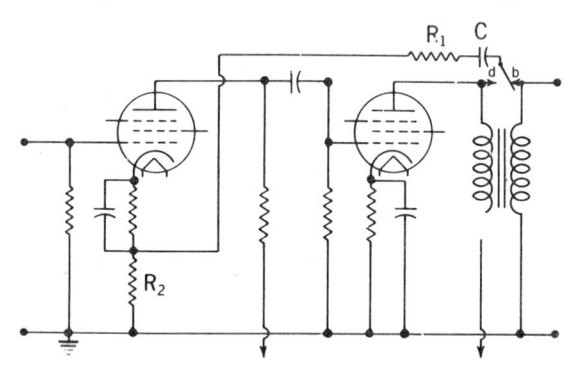

Fig. 7-15. Two-stage amplifier with inverse feedback.

signal and oscillation would result. Figure 7-15 shows a typical two-stage feedback amplifier. Voltage is picked off the output stage at either of points a or b and fed back through a blocking capacitor C and resistor R_1 to the cathode circuit of the first tube. The amount of feedback is set by adjusting the sizes of R_1 and R_2. If the voltage is picked off at a, any distortion introduced by the transformer is not compensated for by the feedback circuit. If the connection is made to point b, the feedback circuit will also compensate for frequency distortion introduced by the transformer. However, if the secondary winding is a low-impedance winding such as would be used to feed the voice coil of a loudspeaker, there may not be sufficient voltage available for feedback at the point b, and it will be necessary to make the connection to point a.

The big advantage of inverse feedback in an amplifier is that it makes the gain and frequency response of the amplifier almost independent of changes that may occur in tube characteristics due to variations in supply voltages, aging of the tubes, and so on. The gain and frequency response depend mainly on the feedback circuit, and because this is composed only of resistances and a capacitor, the characteristics will remain constant over a long period of time. For this reason inverse feedback is applied to a-c operated vacuum-tube voltmeters, d-c amplifiers, and other equipment that requires an amplifier of constant gain.

CATHODE-FOLLOWER AMPLIFIER

It is often necessary to connect an amplifier to a low-impedance load such as a coaxial cable. As explained in the section of this chapter on impedance matching, it is then desirable to match the output impedance of the amplifier to the load impedance. This matching can sometimes be done by means of a transformer, but the frequency range of transformers is restricted, and in addition, they are relatively expensive circuit components. Ordinarily resistance-coupled amplifier circuits have a high output impedance, but by a simple reconnection of the circuit it is possible to provide a low-impedance output suitable for direct connection to a low-impedance load. This reconnected circuit is known as a *cathode-follower* circuit.

The cathode-follower amplifier results from an arrangement of the input voltage, load resistor, and output voltage different from the basic voltage amplifier shown in Fig. 7-1(a). The effects of this arrangement are to give a voltage gain of less than one for the cathode follower, but a power gain much greater than one is common. As mentioned before, the cathode follower also has an outstanding feature which is very useful in broad-band video circuits; it provides a low output impedance suitable for feeding into coaxial cables having rather large values of shunting capacitance.

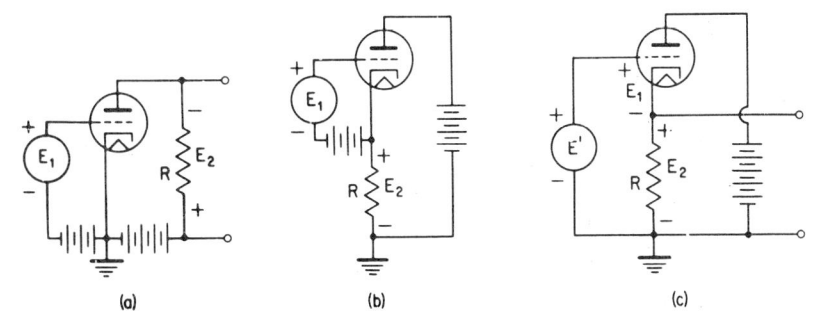

FIG. 7-16. Development of cathode-follower circuit.

Consider the circuit shown in Fig. 7-16(a). The output voltage E_2 is given in terms of the input voltage E_1 and the circuit constants as:

$$E_2 = E_1 \frac{\mu R}{r_p + R}.$$

The circuit of Fig. 7-16(b) has the relative positions of the load R and supply battery interchanged. The input signal voltage E_1 and the bias battery are still introduced between the cathode and grid. This circuit arrangement does not change the relationship between the input voltage E_1 and the output voltage E_2.

Consider the third figure where a new input voltage E' is to replace the input voltage E_1. Assume that the portion of voltage between cathode and grid is to remain equal to E_1. The new input voltage E' must be equal to the sum of E_1 and E_2. When the voltage gain of the third circuit is expressed in terms of the input voltage E' and the output voltage E_2 it will be noted that this gain must be less than one:

$$\frac{E_2}{E'} = \frac{E_2}{E_1 + E_2} = \frac{\mu R}{r_p + R + \mu R} < 1.$$

A reduction in the bias battery must be made to allow for the average voltage drop across the load R. In most circuits the average voltage is made just equal to the required bias voltage and thus self-bias only is used.

The output impedance of this cathode follower may be determined by use of the fact that the terminal voltage across any simple generator is reduced to one half when the applied load equals the internal resistance (see Table 7-1). If the load R in the cathode follower is made very large the output voltage would become very nearly equal to the input voltage E'. If the input voltage E' is now maintained constant and the value of the load R is reduced we may find what load gives an output voltage just $\frac{1}{2}$ of the input voltage.

$$\frac{E_2}{E'} = \frac{\mu R}{r_p + (\mu + 1)R} = \frac{1}{r_p/\mu R + (\mu + 1)/\mu}$$

$$\frac{E_2}{E'} \approx \frac{1}{2} \quad \text{when} \quad (\mu + 1)R = r_p$$

that is, when $R = r_p/(\mu + 1)$.

The output impedance of the cathode follower is found to be given by this figure:

$$Z_{\text{out}} = r_p/(\mu + 1) \approx 1/g_m,$$

where the symbol $\approx$ means approximately equals.

An actual cathode-follower circuit appears as in Fig. 7-16(c). In this circuit the load is connected across the resistance R, so that R is effectively in parallel with the output impedance cealculated by the equation above.

Chapter 7   Audio and Video Amplifiers

The effective output impedance of the cathode follower is therefore that of the combination of $r_p/\mu + 1$ and R in parallel and is

$$Z_{\text{out}} \text{ (effective)} = \frac{R r_p/(\mu + 1)}{R + r_p/(\mu + 1)} = \frac{r_p R}{r_p + (\mu + 1) R}.$$

For approximate calculations this exact formula may be replaced by the simpler expression

$$Z_{\text{out}} \approx \frac{1}{g_m}.$$

The output impedance of cathode-follower circuits generally lies between about 50 and 500 ohms. This low value is achieved because the output voltage changes very little for changes in the value of the load impedance. The properties of cathode followers can be illustrated by means of examples.

Example 1: Given a tube having a g_m of 4,000 micromhos, a mu of 80 and an r_p of 20,000 ohms. Let this tube be used in a cathode-follower circuit with a load of 1,000 ohms.

$$\text{Voltage gain is } \frac{E_2}{E'} = \frac{80 \times 1,000}{20,000 + 81 \times 1,000} = 0.792,$$

$$Z_{\text{out}} = \frac{20,000 \times 1,000}{20,000 + 81 \times 1,000} = 198 \text{ ohms};$$

or

$$Z_{\text{out}} \approx \frac{1,000,000}{4,000} = 250 \text{ ohms (approx.)}$$

Example 2: Given a tube having a g_m of 4,000 micromhos, a mu of 80 and an r_p of 20,000 ohms. Calculate what value of load R is to be used so that the output impedance will be 70 ohms. Find the voltage gain in this case.

$$Z_{\text{out}} = 70 = \frac{20,000 R}{20,000 + 81 R} \quad \text{giving} \quad R = 98 \text{ ohms.}$$

$$\text{Voltage gain} = \frac{80 \times 98}{20,000 + 81 \times 98} = 0.28.$$

Example 3: Given a tube having a mu of 20 and r_p equal to 10,000 ohms. If this tube is used in a cathode-follower circuit with a load of 2,000 ohms what is the voltage gain?

If the grid resistor is 100,000 ohms compare the power required in the input across this resistor for a 10-v input with the power developed in the load resistor.

$$\text{Voltage gain} = \frac{20 \times 2,000}{10,000 + 21 \times 2,000} = 0.77.$$

$$\text{Input power} = \frac{E^2}{R} = \frac{10^2}{10^5} = 1 \text{ milliwatt.}$$

$$\text{Output power} = \frac{7.7^2}{2,000} = 29.6 \text{ milliwatts.}$$

The power gain in this case is 29.6.

GROUNDED-GRID CIRCUIT

In Fig. 7-17(a) and (b) the usual vacuum-tube circuit of Fig. 7-16(a) and the cathode-follower circuit of Fig. 7-16(c) have been redrawn omitting the d-c bias supplies for clarity in studying the circuit action for signal voltages and currents. Examination of these circuits reveals that in the "ordinary" connection of Fig. 7-17(a) the cathode is at ground potential for

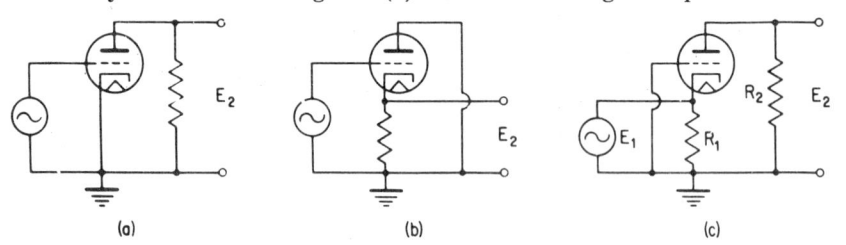

FIG. 7-17. Possible triode connections: (a) grounded-cathode or common-cathode; (b) grounded-plate or common-plate (cathode follower); (c) grounded-grid or common-grid.

signal voltages and is common to input and output circuits. Hence this connection is often termed a *grounded-cathode* or *common-cathode* circuit. In the cathode-follower circuit of Figure 7-17(b) the plate is at ground potential for signal voltages and is common to input and output circuits. Hence this connection is termed a *grounded-plate* or *common-plate* circuit.

There is one other possible connection as shown in Fig. 7-17(c). In this circuit the grid is operated at ground potential for signal voltages and is common to input and output circuits. Hence this connection is known as the *grounded-grid* or *common-grid* circuit.

Operation of the grounded-grid circuit can be analyzed in a manner similar to that used for the ordinary grounded-cathode circuit. In the grounded-grid circuit of Fig. 7-17(c) the input voltage affects the strength of the plate current in two ways. First the voltage E_1 between grid and cathode is amplified in the usual way by the amplification factor μ and appears in the plate circuit as an amplified voltage μE_1. However in this grounded-grid circuit the voltage E_1 is also introduced directly in series in the plate circuit, and with such a polarity to aid or increase the plate current. Hence the effective voltage in the plate circuit is now $\mu E_1 + E_1 = (\mu + 1) E_1$ instead of just μE_1 as in the ordinary grounded-cathode circuit. The plate current which results from this effective voltage must flow through r_p, R_2, and R_1 in series so the plate current will be

$$I_p = \frac{(\mu + 1)E_1}{r_p + R_1 + R_2}.$$

The output voltage E_2 will be

$$E_2 = I_p R_2 = \frac{(\mu + 1)E_1 R_2}{r_p + R_1 + R_2}.$$

When this expression is compared with the usual expression for the output voltage of the grounded-cathode circuit, i.e.,

$$E_2 = \frac{\mu E_1 R_2}{r_p + R_2},$$

it is observed that as far as the output is concerned the grounded-grid circuit using a tube with an amplification factor μ and plate resistance r_p acts like an ordinary grounded-cathode circuit having a slightly higher amplification factor $(\mu + 1)$ and a higher plate resistance $(r_p + R_1)$. However the input voltage E_1 is now applied across the effective resistance of R_1 and so power is absorbed from the generator, with a consequent reduction in gain. For this reason the grounded-grid circuit is seldom used at audio frequencies.

At higher frequencies the grounded-grid circuit does have some advantages which warrant its use. Because the grid is grounded, capacitive currents flowing through the grid-plate capacitance as a result of the output voltage do not flow through the input circuit as they do in the ordinary grounded-cathode connection. Hence there is no interaction between output and input circuits and the need for "neutralization" of radio-frequency amplifiers as discussed in Chapter 11 does not exist for grounded-grid circuits. Grounded-grid circuits are also used at much higher frequencies where special high-frequency tubes are designed for this type of operation.

PHASE INVERTER

Many amplifier circuits require a change from single amplifiers to push-pull amplifiers somewhere along the circuit. This change can be made in a number of different ways, one common method being the use of a transformer having two secondary windings in series. The voltages obtained from the outside leads will be 180° out of phase when each is referred to the mid-tap lead. Transformers are rather limited in frequency response, however, and are costly, and therefore other methods of obtaining two voltages equal in magnitude but differing by 180° phase will be considered.

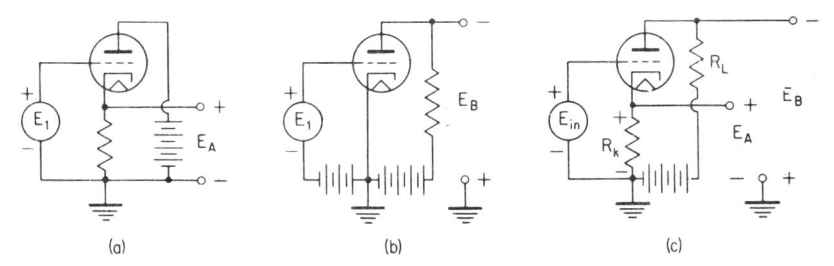

FIG. 7-18. Development of phase-inverter circuit.

The cathode-follower circuit presented a voltage gain of less than one but with zero degrees phase relationship between the input voltage and output voltage. The basic voltage amplifier circuit offers 180° phase shift between the input voltage and the output voltage and usually has a voltage gain greater than one. Let these two circuits be combined in the manner shown in Fig. 7-18.

It should be noted that the same a-c signal current flows through each of the load resistors in Fig. 7-18(c). The magnitude of a-c signal voltage developed across these loads must be equal, when $R_K = R_L$. It should also

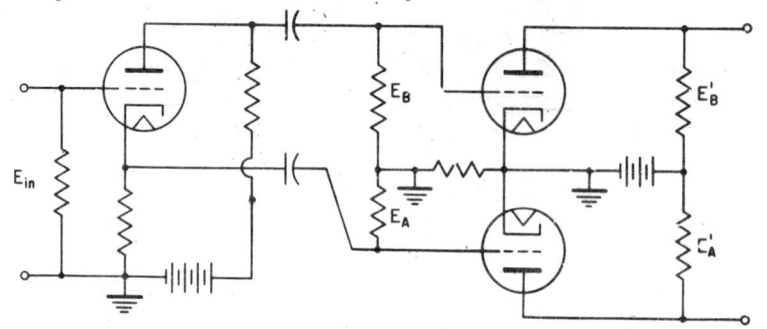

FIG. 7-19. Push-pull amplifier using phase-inverter circuit.

be noted that with respect to the ground reference point the currents flow in opposite directions in these load resistors. The direction of a-c signal voltage with respect to the ground reference point is therefore reversed. These two a-c voltages may then be coupled with coupling capacitors into the grids of a push-pull amplifier stage in the manner of Fig. 7-19.

This circuit has the advantage that only one tube is required for the task of changing from single to push-pull. Disadvantages of this circuit are the

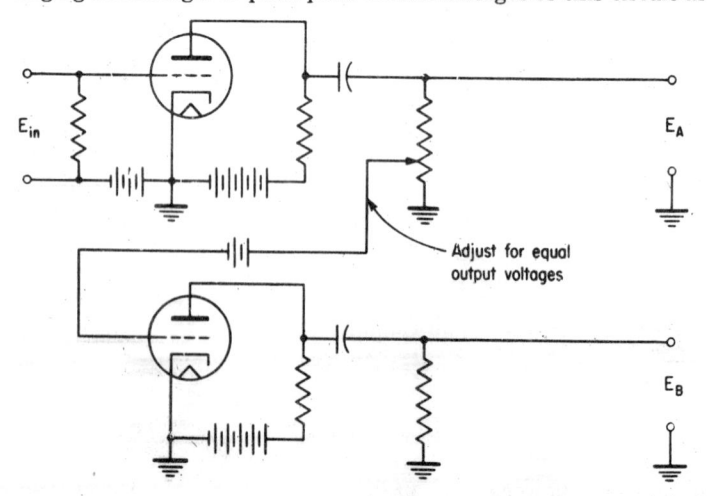

FIG. 7-20. Phase inverter circuit using two tubes to obtain better balance.

floating cathode and the lack of voltage gain from the input to each of the outputs. This circuit will remain balanced over a wide frequency range when the shunt capacitance on each output is correctly portioned (not equal because the output impedances from the two outputs are not equal).

A second method uses two tubes in the inverter circuit and offers some voltage gain. The circuit is given in Fig. 7-20 and consists of two voltage amplifier stages with a voltage-divider element.

The input signal is presented to the upper tube where it is amplified and inverted (shifted by 180°). The output of this amplifier is one of the two push-pull outputs as indicated, and also becomes the input to the lower inverter tube after passing through the voltage divider. The purpose of the voltage divider is to offset any voltage gain in the lower stage. The second push-pull output comes from this lower stage after a second inversion (180° shift). The voltage divider is adjusted until the two push-pull output voltages are equal in magnitude. The disadvantages of this circuit are in the extra tube and the fact that the output signals do not remain exactly 180° in phase or equal in magnitude over wide frequency ranges.

VIDEO AMPLIFIERS

For certain applications, especially in television, an amplifier is required which will pass a very wide band of frequencies, usually from about 20 cycles up to 4,500,000 cycles. Such amplifiers are called *video* amplifiers because they are used to amplify the picture signal in television as compared

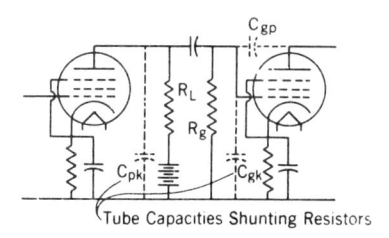

FIG. 7-21. Capacitances in a resistance-capacitance coupled amplifier.

to the sound signal which requires an *audio* amplifier. The chief difference between a video amplifier and an audio amplifier is that it must be designed to amplify high frequencies as well as the medium and low frequencies. It is not possible to build a transformer-coupled amplifier to cover such a wide range, so some form of resistance-capacitance coupling is generally used.

The amplification of an ordinary resistance-capacitance coupled amplifier drops off at the high frequencies because of the shunting effect of the tube and wiring capacitances. These capacitances are shown in Fig. 7-21. The shunting capacitance is comprised of C_{pk} of the amplifier tube, plus C_{gk} and an effective input capacitance equal to $(1 + \mu) C_{gp}$ of the next tube in the circuit. There is also a stray wiring capacitance C_W whose exact location is difficult to specify but is approximately parallel to C_{pk}. The upper half power frequency is determined at the frequency where the reactance of this total shunting capacitance equals the value of resistance in parallel with this capacitance. An equivalent circuit for the high frequency range of the R-C amplifier is given in Fig. 7-22.

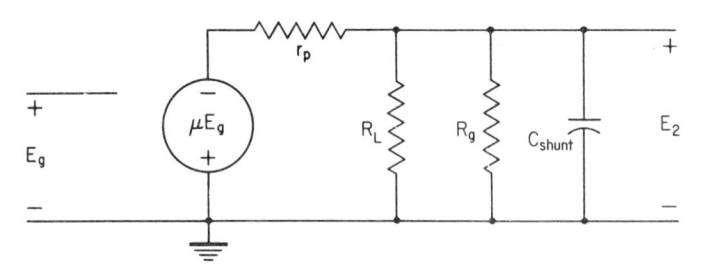

FIG. 7-22. Equivalent high frequency circuit of a resistance-capacitance coupled amplifier.

The upper frequency range can be extended by reducing the value of R_L, but the process has a practical limit because the gain of the amplifier would be reduced for these lower values of R_L.

The upper frequency range can also be extended by improving the ratio of tube g_m to tube shunting capacitances. Assume that a gain of 10 is required for a stage operating in the midband region. The gain of the amplifier stage is given approximately by the product of g_m times R_L when both R_g and r_p are much greater than R_L. Larger values of tube g_m would therefore permit smaller values to be used for R_L with the required midband gain of 10. If the increase in g_m is obtained while maintaining approximately constant the values of shunting capacitance, the high-frequency range would be increased by the use of lower R_L values. Conversely, if the shunting capacitance can be reduced while keeping the value of g_m about constant, the high-frequency range would be increased and the midband gain could be the required value of 10.

This ratio of tube g_m to tube-shunting capacitances has important control over the relative midband gain and band width which may be expected from any tube. Great progress has been made in the last decade in the design and production of new tubes having an increased ratio $g_m/(C_{pk} + C_{gk} + C_{gp})$. The use of these tubes has not only permitted greater gains for the same band width but has extended the upper frequency range as well. Many of these new tubes are possible because of the fine precision now being employed. A recent tube has 400 grid wires per inch; each wire is only 3/10 of a millimeter in diameter and is spaced only 1/400 of an inch from the cathode surface.

Chapter 7   *Audio and Video Amplifiers*   227

COMPENSATED VIDEO AMPLIFIERS

If the upper frequency range could be extended only by lowering the value of R_L there would not be much gain per stage in a very wide band amplifier. Attention must therefore be given to the cause of limiting the high-frequency range, the shunting capacitance. It will be noted that the effect of this capacitance is to reduce the effective load impedance at high

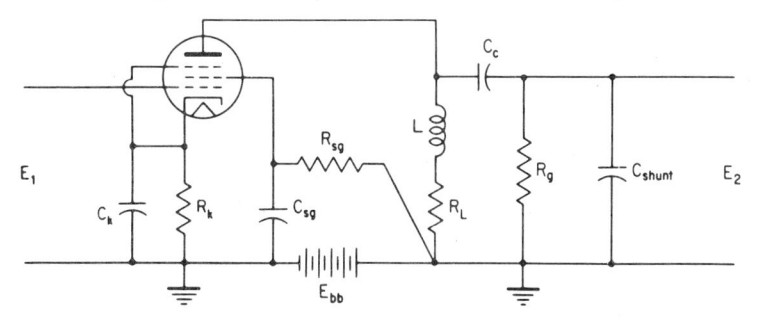

FIG. 7-23. Shunt-compensated video amplifier circuit.

frequencies. If a small inductance is placed in series with R_L the impedance of this branch can be made to increase in the same range of frequencies that the shunting effect of the capacitance branch becomes noticeable. The circuit is shown in Fig. 7-23.

By means of this device the effective load impedance presented to the tube can be maintained more nearly constant over a wider frequency range. The response obtained for a typical stage with increasing inductance values is given in Fig. 7-24. These curves show relative values. To the values shown there must be added the low frequency gain in decibels (20 log$_{10}$ $g_m R_p$). Curve I is for the uncompensated case while curve IV shows the effect of over-compensation, that is, of too large a value of inductance. It will be noted that with a suitable choice of L (curve III) the gain of the amplifier will be nearly constant for a considerable extension of the frequency range.

The required size of the inductance L can be computed from the formula

$$L = \frac{R_L Q}{\omega_2} \quad \text{where} \quad \omega_2 = 2\pi f_2 = \frac{1}{R_L C_{\text{shunt}}}$$

is 2π times the frequency at which the gain of the uncompensated amplifier is 3 db down (the 0.707 point).

Q is a factor which can be chosen to give the desired type of response. In the example for which the curves of Fig. 7-24 were drawn, R_L was chosen as 5,000 ohms and the shunt capacitance C_{shunt} was assumed to be 15.9 $\mu\mu$f, so that the half power frequency f_2 is 2 megacycles. Curve I is the uncompensated case for which Q, and therefore L, is equal to zero. Curve II is for $Q = 0.25$ corresponding to $L = 99$ microhenrys. Curve III has been

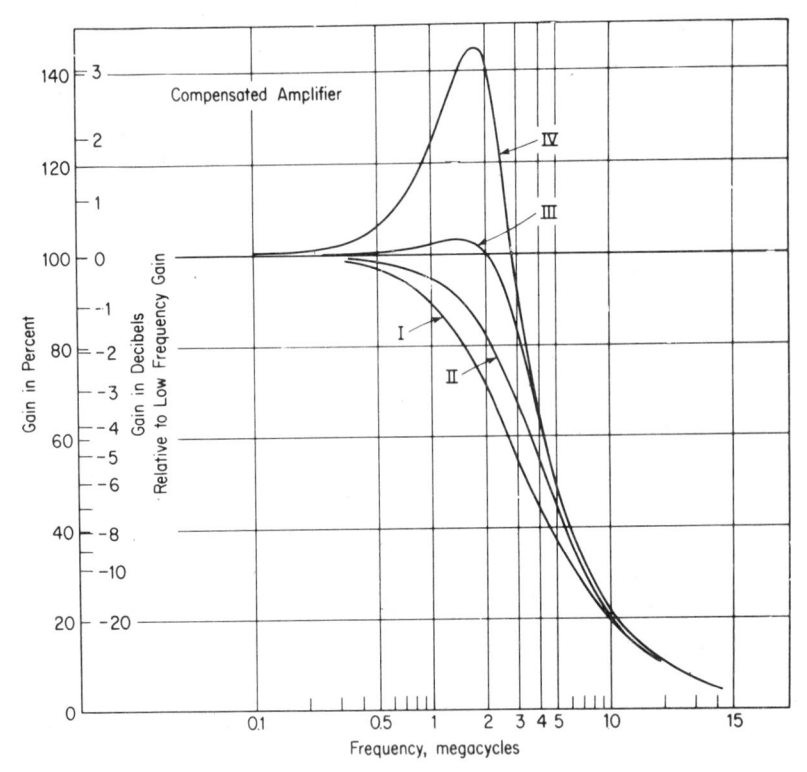

FIG. 7-24. Characteristic relative-gain curves of shunt-compensated video amplifier stages.

drawn for $Q = 0.5$ for which $L = 198$ microhenrys, and Curve IV for $Q = 1$ which corresponds to $L = 396$ microhenrys. Usual values for Q lie between 0.34 and 0.5. The case of $Q = 0.414$ is of interest because it represents the maximum compensation that can be employed without peaking the response. (Curve III shows a slight peaking and Curve IV a large amount of peaking).

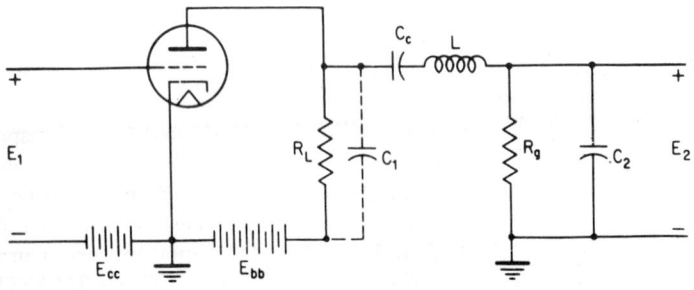

FIG. 7-25. Series-compensated video amplifier circuit.

A second form of high-frequency compensation is given in Fig. 7-25 and is known as the series circuit. The total shunting capacitance is divided into two sections on each side of the inductance. It is well known that the series resonant circuit can present across each element a voltage higher than the total series applied voltage (see Chapter 3). As the shunting effect of C_1 (which represents the stray capacitance) reduces the gain in the high-frequency range, the series circuit of L and C_2 coming into resonance may be used to increase the gain. In this manner the high-frequency range is extended beyond the half-power frequency of the uncompensated amplifier. The degree of this compensation depends not only on the value of L used but also upon the relative size of C_1 and C_2. Most amplifier circuits will have the larger capacitance in C_2 but this series compensation circuit works at its best when the larger capacitance is in C_1. In some cases a small capacitor is shunted across C_1 to improve the ratio and give better response even with the increase in total shunting capacitance.

Low-frequency compensation is sometimes used in these circuits. This may offer advantages in reducing the size of coupling capacitors and cathode bypass capacitors so that the net cost and performance is improved despite the introduction of additional components. The basic circuit used for this low-frequency compensation is given in Fig. 7-26. Loss of gain occurs at low frequencies because of the voltage divider action of X_c and R_g. If the gain of the stage up to the left side of this coupling capacitor is

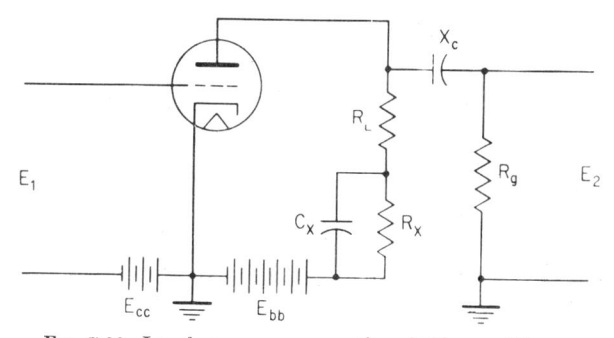

FIG. 7-26. Low frequency compensation of video amplifier.

increased over the same frequency range as that in which this voltage divider action is effective, the circuit can be compensated. An addition is made to the normal plate load R_L in the form of an impedance consisting of R_x and C_x in parallel. In the midfrequency range the reactance of the capacitor C_x is very low and the plate load is approximately R_L. In the lower frequency range the reactance of C_x increases and the gain of the circuit is therefore increased. By proper selection of the values of R_x and C_x the overall gain can be maintained at the lower frequencies.

In addition to the gain versus frequency response of a video amplifier the time delay or phase shift through the amplifier is also of importance. A complex wave such as a television signal contains a large number of frequency components. If these different components suffer different time delays the resultant wave shape will be distorted. It is important therefore, that the time delay through the amplifier be nearly independent of frequency. Because a tenth of a microsecond corresponds to a phase shift of 36° at 1 megacycle but 72° at 2 megacycles, it is evident that a phase shift proportional to frequency is the desired response. Video amplifiers are usually designed with a view to good delay response as well as good amplitude response. In the shunt-compensated circuit of Fig. 7-24 an inductance corresponding to a value of $Q = 0.34$ gives best delay response.

DIRECT-CURRENT AMPLIFIERS

Most amplifiers whose circuits are similar to those described above have poor low-frequency response and will not amplify d-c voltages at all. Since it is sometimes necessary to amplify d-c voltages, or a-c voltages of very low frequency, changes must be made in the circuit of the amplifier. Many special circuits have been developed which can amplify direct voltages. Such amplifiers are known as *direct-current amplifiers*. The most common type of d-c amplifier is the *direct-coupled amplifier*. Direct-coupled amplifiers are used in some oscilloscopes to amplify voltages before applying them to the plates of the cathode-ray tube and in vacuum-tube voltmeters to increase their sensitivity. One type of direct-coupled amplifier, known as the Loftin-White amplifier, has been used in radio sets as an a-c amplifier. A type of direct-coupled amplifier is also used in regulated power supplies (see Chapter 5).

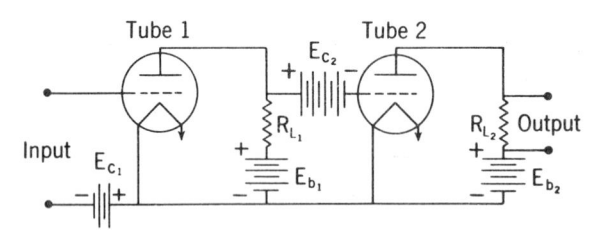

Fig. 7-27. D-c amplifier circuit.

It was pointed out earlier in this chapter that the drop in low-frequency response in a resistance-capacitance-coupled amplifier is mainly caused by the drop in voltage across the coupling capacitor between stages. The coupling capacitor blocks off d-c voltages completely so that they do not reach the grid of the tube in the next stage. Hence, if d-c voltages are to be amplified this capacitor must be removed entirely from the circuit. However, if it is removed the plate voltage of the previous stage applies a

Chapter 7 **Audio and Video Amplifiers** 231

high positive bias to the grid of the tube. To overcome this positive bias, the negative bias on the grid must be greatly increased, so that a large battery is required. The circuit in Fig. 7-27 shows an amplifier of this type. The first stage is the same as used in resistance-capacitance-coupled amplifiers, except that there is no capacitor in the input lead to the grid of the first tube. In the second stage, the bias battery E_{C_2} has to overcome a positive voltage equal to the supply voltage of the previous stage minus the d-c voltage drop in the plate resistor of this stage, and also has to supply the necessary negative bias for the tube. This circuit is an example of a direct-coupled amplifier, which could be used for amplifying either d-c or a-c voltages. This circuit is not often used because it requires such a large bias battery in the second stage.

In order to reduce the number of batteries required, the circuit can be changed as shown in Fig. 7-28. The plate resistor of the first tube and one side of the filament, as well as the plate resistor of the second tube, are connected to taps on one battery. The plate supply for the first stage is the voltage between taps a and c. If the filament of the tube in the second

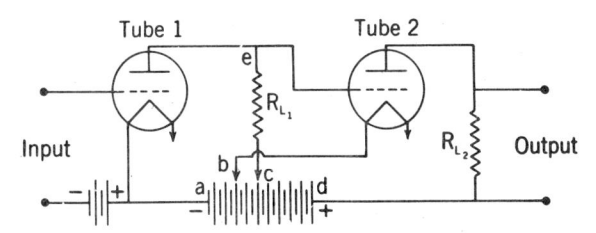

Fig. 7-28. D-c amplifier circuit using common d-c supply.

stage were connected to point c instead of b, there would be a *negative* bias on this stage due to the plate current in the first tube flowing in the resistor R_{L1}. This voltage is usually too large a negative bias for the second tube, so that part of it is bucked out by moving the filament lead to tap b. The plate voltage on the second stage is the voltage between taps b and d. In this circuit it should be noted that the filaments of the two tubes are not at the same potential, so that either separate filament supplies, or tubes with indirectly heated cathodes, must be used.

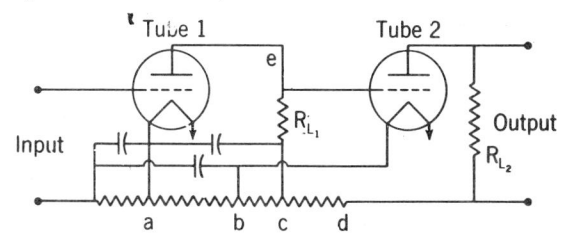

Fig. 7-29. Loftin-White direct-coupled amplifier.

Instead of using a tapped battery in the circuit of Fig. 7-28, a voltage divider can be used as shown in Fig. 7-29. This circuit is known as a Loftin-White direct-coupled amplifier. When a-c voltages are to be amplified, it is necessary to by-pass the various sections of the voltage divider with low-reactance capacitors. At the lowest frequency to be amplified these capacitors should have reactances that are low compared with the resistances they by-pass.

If more than two stages are used in a d-c amplifier, it becomes very difficult to make the amplifier stable. Any small changes in the voltages on the first tube are amplified to such an extent that it is hard to maintain the correct grid bias on the last tube in the amplifier. For this reason d-c amplifiers seldom have more than two stages.

PUBLIC-ADDRESS SYSTEMS

One of the important uses for an audio amplifier is the amplification of speech and music, either indoors or outdoors, so that the sound may be heard over a large area. A *public address system* (more briefly *p. a. system*) is the name given to the complete installation required for this application.

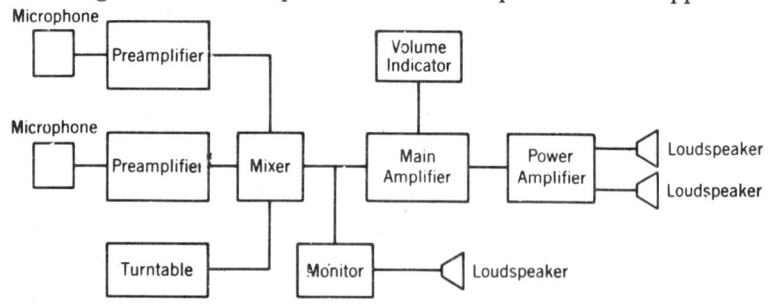

FIG. 7-30. Block diagram of a public-address system.

A public-address system includes all necessary amplifiers, microphones, volume controls, mixing systems, loudspeakers, record turntables, and sometimes volume indicators and monitoring systems. A block diagram of the main parts of a large public-address system is shown in Fig. 7-30. Not all of these elements are included in smaller public-address systems, which may consist of only one microphone input, one phonograph input, and a simple volume-control system followed by one amplifier to feed from one to three loudspeakers. Only very large installations use volume indicators and monitoring amplifiers. In these large installations, such as in a theater, each of the parts shown in the diagram occupies a separate chassis. The usual small public-address system contains the amplifiers, the mixer, and the power supply all on one chassis.

Preamplifiers. Most microphones now used have very low voltage outputs so that considerable amplification must be used with them to produce the power required by the loudspeakers. While any properly designed amplifier with enough gain can be used, it is usual to employ a *preamplifier* ahead of the main amplifier. A preamplifier is an amplifier of moderate gain (about 50 db for example) especially designed to have very low hum and noise levels in its output. This low-noise feature is essential since any hum or noise in the first stage will be amplified by succeeding stages and will interfere with the desired audio signal. The noise level in a preamplifier can be kept low by using good-quality components in the circuit, by making careful joints and connections, and by adequate shielding. A preamplifier is particularly susceptible to hum, so that care must be taken to keep it away from power transformers, filter chokes, and filament and power leads. A good way to reduce hum is to build the preamplifier on a separate chassis, but with proper care it is possible to include it on the chassis with the main amplifier.

Mixers, volume controls, and switching circuits are common sources of noise, so it is usual to place them in the circuit *after* the preamplifier. In this way, the microphone output is amplified enough by the preamplifier so that it overrides any noise introduced in these circuits. Noise from moving contacts in mixers and switching circuits is mainly caused by dirt, so that regular cleaning is necessary to keep down their noise level. They may be cleaned with carbon tetrachloride.

Mixers and volume controls. When several microphones and phonograph turntables are used at one time, it is necessary to be able to control the volumes from each of them separately, and to combine or *mix* their outputs. This is accomplished by means of a *mixing circuit* or *mixer*. While mixing circuits are ordinarily used to mix the outputs of several sources to obtain the most natural balance between the various sounds, sometimes they are used to change the natural levels to obtain special effects. A mixing circuit is usually followed by a *master volume control* or *attenuator* which changes the volume of all the signals after they have been mixed.

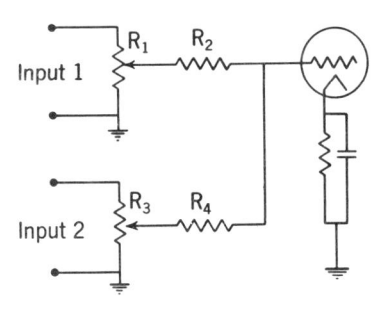

Fig. 7-31. Resistance mixing circuit.

One of the simplest mixing circuits, known as a *resistance mixer*, is shown in Fig. 7-31. In this circuit the inputs are applied to separate volume controls (variable resistors) and the variable arms of the controls connected together through the resistances R_2 and R_4. These resistances are necessary in the circuit since, if they were left out, turning one volume control to zero volume would short-circuit the output of the other. They must not be made too small, typical values being 0.25 to 0.5 megohms. The variable resistors R_1 and R_3 are usually 0.5 to 1.0 megohms. Resistance mixers of this type do not give completely independent control, since changing one control affects the volume of the other slightly. In most applications, however, the change is not noticeable, as it is usually less than 6 db.

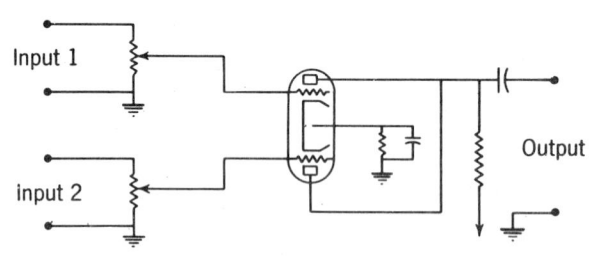

FIG. 7-32. Electronic mixing circuit.

Another simple type of mixing circuit is shown in Fig. 7-32, which is called an *electronic mixer*. In this circuit the two inputs are connected to the grids of a twin-triode tube and the two plates are in parallel, with a common plate resistor. An electronic mixer gives completely independent control of the two inputs. Figure 7-33 shows a slightly different type of electronic mixing circuit which is often used in preference to that of Fig. 7-31, as it has a higher gain.

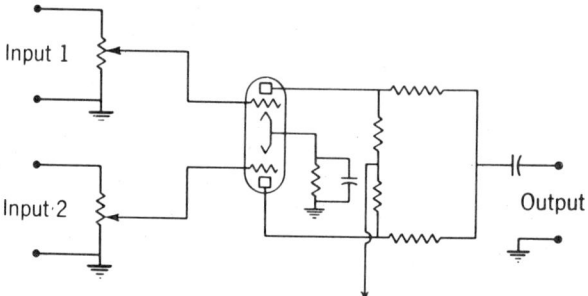

FIG. 7-33. Improved electronic mixing circuit.

Broadcasting studios and theaters employ mixing circuits which contain *constant-impedance* volume controls or attenuators. These constant-impedance volume controls are of three main types, called T-pads, H-pads, and L-pads, which are illustrated in Fig. 7-34. Their action may be seen by considering a T-pad as follows. The three resistances are made variable and are connected or ganged to a single control knob. R_1 and R_2 are made equal and vary together. As the control knob is turned, resistances R_1 and R_2 decrease and resistance R_3 increases by the correct amount to maintain the impedance looking into the pad at a constant value. But

Chapter 7   *Audio and Video Amplifiers*   235

as the knob is turned in this direction, the voltage drop across R_1 and R_2 decreases and that across R_3 increases, so that more and more of the input voltage appears at the output, which may be thought of as being in parallel with R_3 if R_2 is small. Thus a T-pad will change the volume in the circuit without changing the input impedance. The action of L- and H-pads may be analyzed similarly. T-pads and H-pads maintain constant impedance on both the input and output since they are symmetrical. The L-pad will maintain constant impedance in only one direction, that is, on the side containing the series arm.

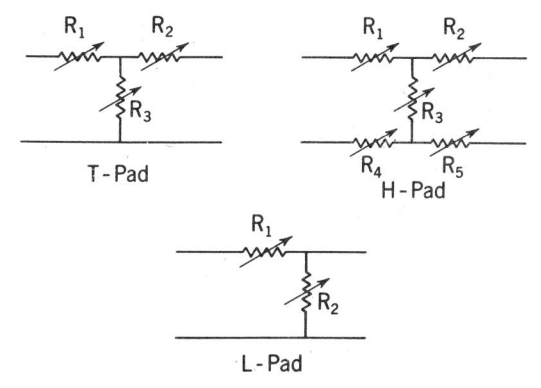

FIG. 7-34. Attenuators.

The way these pads can be connected in the circuit to form a mixer is shown in Fig. 7-35. This circuit uses two T-pads as mixers and one T-pad as a master volume control. Mixer 1 determines the volume of input 1, and mixer 2 that of input 2, while the master volume control gives control over the volume of the total signal after mixing. Since changing the

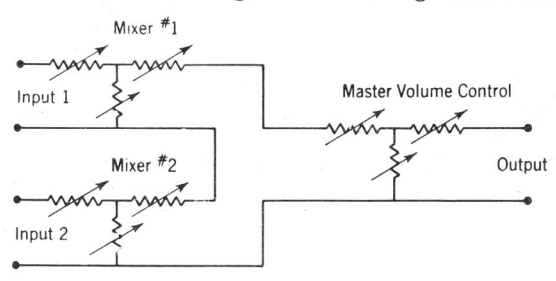

FIG. 7-35. Mixer system using T-pads.

setting of mixer 1 does not vary the impedance at its output terminals, there will be no change in the impedance connected to the output of mixer 2. Consequently there is no change in the voltage supplied by mixer 2 as mixer 1 is varied. This means that the settings of the mixer controls are

236　　　　　　　　Audio and Video Amplifiers　　　　　　Chapter 7

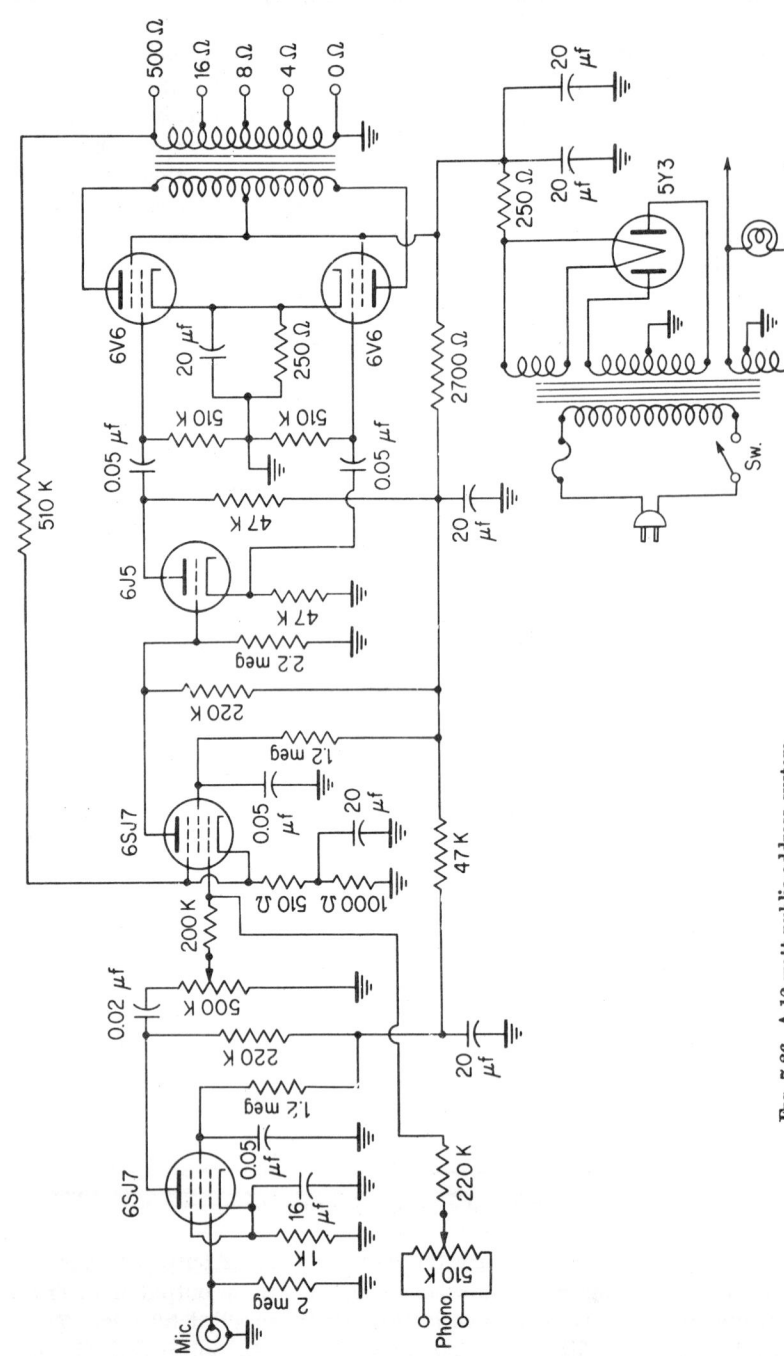

Fig. 7-36. A 12-watt public-address system.

independent of each other. In the circuit shown, the outputs of the two mixing pads are really connected in series. Other arrangements are possible in which the pads are connected in parallel. By using more pads it is possible to mix as many circuits as desired, while always having separate control on each circuit.

Public-address amplifier circuits. The purpose of the main amplifier is to amplify the output of the mixing circuit up to a level sufficient to drive the loudspeakers to the desired volume. A wide variety of circuits is employed in these amplifiers, the circuit used depending on a number of factors — such as power output required, type and number of input circuits, the quality or fidelity required, portability, and cost. Public-address amplifiers are usually rated in terms of their power output, which is the amount of power delivered to the load, with no more than 5% to 10% distortion. The power-output stage is ordinarily a Class A amplifier in low-power high-fidelity amplifiers, Class AB_1 or AB_2 in medium-power amplifiers, and Class B in the large amplifiers. Class AB amplifiers are very popular for public-address work because of their excellent power output at low cost. The remainder of the amplifier is designed to provide the necessary amplification between the output of the mixing circuit and the input to the power stage. Factors in the design of such amplifiers have already been considered earlier in the chapter.

A circuit diagram of a typical public-address amplifier is shown in Fig. 7-36. This amplifier has an output of 12 to 15 w, and is provided with one microphone input and one phonograph input. A resistance mixing circuit similar to that shown in Fig. 7-31 is used. The preamplifier for the microphone is a 6SJ7 connected as a pentode resistance-capacitance-type amplifier. The output of the mixer is amplified by a 6SJ7 pentode which is resistance-capacitance-coupled to the next stage, called the driver, which supplies grid excitation to the power-output stage. The 6J5 driver is connected as a phase-inverter in the manner of Fig. 7-19 to provide a balanced push-pull input for the push-pull power-output stage.

The power stage in the amplifier uses a pair of 6V6 tubes operated Class AB_1. The output transformer is provided with a tapped output winding in order to feed a 500-ohm line or to feed directly to a low-impedance loudspeaker. A connection from this output winding back to the cathode of the second 6SJ7 through an appropriate resistance provides negative feedback for reduction of distortion.

Impedance matching. In connecting together the various parts of the public-address amplifier system that have been described above, there are some precautions to be observed, one of which is *impedance matching*. When two pieces of equipment are connected together, their impedances are said to be matched when the impedance of the load is adjusted to the best value to absorb power from the source. Impedances are matched for two reasons: (1) to obtain maximum power output; (2) to reduce distortion.

To see why impedance matching gives maximum power output, consider the circuit shown in Fig. 7-37, which consists of a generator (such as a battery, a power amplifier, or any device capable of supplying electrical power) and a load resistance. All generators have some internal loss which can be considered as an *internal resistance*. In the figure, R_1 represents the internal resistance of the generator, R_2 the load resistance, and E the voltage of the generator. The problem is to determine what the size of R_2 must be to give maximum power output.

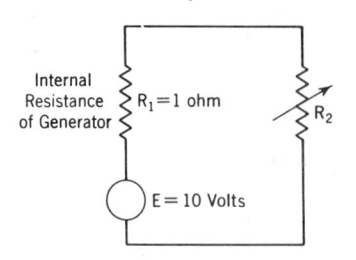

Fig. 7-37. Equivalent circuit of a generator connected to a load.

Suppose the internal resistance of the generator is 1 ohm and its voltage is 10 v. Then if various values of load resistance are connected to the generator, that is, if various values are assigned to R_2, the current can be calculated by Ohm's law and the power in the resistor R_2 determined. Table 7-1 shows the results of some calculations for various values of R_2.

TABLE 7-1

R_2 (ohms)	Current (amperes)	Voltage across R_2	Power in R_2 (watts)
0.1	9.07	0.907	8.20
0.5	6.67	3.33	22.2
1	5	5.00	25
4	2	8.00	16
100	0.099	9.99	0.998

An examination of this table shows that the maximum power that it is possible to take from the generator is 25 w, and it occurs with a load resistance of 1 ohm, that is, a load resistance equal to the internal resistance of the generator. It can be shown that no matter what the voltage or the internal resistance of the generator, the *maximum power output* is obtained when the load resistance equals the internal resistance of the generator, that is, when their impedances are matched. However, it should be noted that the *voltage* across the load resistance increases as the load resistance is increased, and is not a maximum when the impedances are matched. Hence if the maximum *voltage* across the load is required, the impedances should not be matched, but the load impedance should be as large as possible. When the source is a vacuum tube there is a definite load resistance which will give the maximum power with a particular allowable amount of distortion. This load is specified by the manufacturer.

In a long transmission line such as a telephone line, it is found that signals transmitted over it will be distorted unless the load resistance at its output end has a particular value, called the *characteristic impedance* of the line. The characteristic impedance of the line is the impedance that must be matched for maximum output and minimum distortion. If the terminal impedance is not equal to this characteristic impedance, it is found that some of the energy sent down the line is *reflected* when it strikes the impedance mismatch at the end. If the line is long, reflected energy is largely lost, and since the amount of reflection depends somewhat on frequency, different amounts of power will reach the load resistance depending on the frequency. This variation produces distortion of the transmitted signal, which can be avoided by matching impedances. The characteristic impedance of most telephone lines is about 500 ohms.

Another case where mismatched impedances may cause distortion is in circuits involving transformers and in long microphone cables. In these cases the distortion is not usually serious except for widely mismatched impedances.

The microphone cables ordinarily used are high-capacitance cables, so that if a long length of such a cable is used to connect a high-impedance microphone to the grid of the preamplifier, the capacitance of the cable which is in parallel with the high impedance of the microphone by-passes the higher frequencies. In order to reduce the loss of high-frequency response, it is usual to transform the high impedance of the microphone to a low impedance by means of a microphone transformer or by an amplifier at the microphone with a transformer or cathode follower output. The capacitance of the line has much less shunting effect on the low-impedance output, so that longer lines can be used without losing high-frequeucy response. Microphone cables as well as transmission lines are designed to have low characteristic impedances, since low-impedance lines pick up less hum and noise than high-impedance lines.

Operation of public-address systems. In the operation of public-address systems it is often necessary to have the microphones and loud-speakers in the same room or auditorium. Such arrangements cause considerable difficulty owing to *acoustic feedback* from the loudspeaker to the microphone. Any noise in the room picked up by the microphone is amplified and fed to the loudspeaker, which reproduces the noises louder than before. This louder noise repeats the same cycle and thus the sound builds up to the limit of the amplifier. The result is a howl at a frequency for which the over-all system has most gain. Feedback can be avoided by taking suitable precautions to limit the amount of sound from the loud-speaker that can be picked up by the microphone. Methods for doing this include the use of directional loudspeakers such as horns, curtains or drapes on the walls to reduce reflections, directional microphones, and careful control of the volume. Manipulation of the tone control on the amplifier will sometimes help to stop acoustic feedback.

Where several loudspeakers are to be used, care must be taken to see that the audience will not be able to hear the sound from two or more loudspeakers at different distances with about the same volume, otherwise echoes will interfere with clarity of sound. When only one microphone and one or two loudspeakers are employed, a good arrangement is to put the loudspeakers above and slightly forward of the microphone. This makes the sound from the loudspeakers reach the audience at about the same time as the direct sound from the speaker and thus avoids echoes.

Since the characteristics of rooms vary so widely it is difficult to lay down specific rules for placing microphones and loudspeakers. Each particular job requires special consideration, so that either experience or trial and error must be used to determine the correct placements.

TRANSISTOR AUDIO AMPLIFIERS

Most operations utilizing vacuum tubes can also be performed with transistors. Particularly at audio frequencies, transistors possess many practical advantages which will often dictate their use instead of vacuum tubes, although they have certain shortcomings as well. Their small size, light weight, long life, ruggedness and freedom from microphonics are important advantages, and their high efficiency and low power requirements make battery operation feasible and economical. On the debit side,

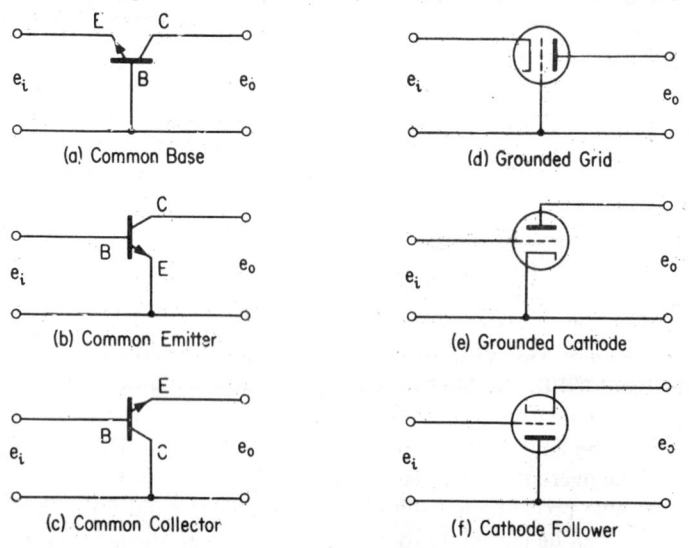

Fig. 7-38. Analogous transistor and vacuum-tube circuits.

transistors are quite temperature-sensitive and careful circuit design is required to prevent extreme variations in gain and to allow interchangeability of transistors.

At audio frequencies junction transistors are used almost exclusively. Compared with the point-contact type, junction transistors have higher gain, better efficiency, lower distortion, greater stability, and lower noise. Their poorer high-frequency performance which limits their use at radio frequencies is not a serious handicap at audio frequencies.

BASIC TRANSISTOR CIRCUITS

Three basic circuits can be used with transistors, and these are shown in Fig. 7-38 at (a), (b), and (c). If it be considered that the emitter is analogous to the cathode, the base to the grid, and the collector to the anode of a triode vacuum tube, then the similarities of the basic transistor circuits to those of the vacuum tube are readily apparent. The common-base (or grounded-base) circuit corresponds to a grounded grid circuit, the common-emitter to a grounded cathode, and the common-collector to a grounded plate (cathode follower). As with its vacuum-tube counterpart the common emitter is the most frequently used circuit. In Fig. 7-38 an N-P-N unit has been indicated, and d-c biasing must be applied accordingly. For a P-N-P unit opposite bias polarities would be required. As in the case of the analogous tube circuits, the characteristics of the three basic transistor circuits differ, and so they will be studied separately.

The operating conditions and quiescent or Q point for a transistor are specified in terms of currents, in contrast to a vacuum tube whose operation is usually described in terms of voltages. Thus in a transistor there will be three currents, i_e the emitter current, i_b the base current, and i_c the collector current. Unlike the tube, there will always be input current in the transistor, so that often its input circuit will have an impedance which is low compared to that of the vacuum tube.

Because of the presence of an input current the performance analysis and equivalent circuit of the transistor will be more complex than in the case of the vacuum tube. With the above currents there will also be potentials on the emitter, base, or collector called v_e, v_b, and v_c as measured to one of the three electrodes chosen as the reference or common element. Arbitrary choice of this common element then leads to the three fundamental transistor circuits of Fig. 7-38.

EQUIVALENT FOR THE COMMON-BASE TRANSISTOR CIRCUIT

Figure 7-39 shows the equivalent circuit for a transistor in the common-base circuit. Internal resistances r_e, r_b, and r_c are known as the emitter, base, and collector resistances respectively. The generator $r_m i_e$ in the collector lead represents the transfer emf or inherent amplifying property of the transistor, as did the μe_g generator for the vacuum tube. It may be noted that the output of this generator is dependent on the input *current* in the transistor, whereas it is related to the input *voltage* in the vacuum tube.

For small signal rms inputs the currents and voltages have effective values given by I_e, I_b, I_c, V_e, V_b, and V_c. Using small a-c signals of low frequency it is then possible to determine the values of r_e, r_b, r_c, and r_m from measurements conducted on the transistor. As is usual in network experi-

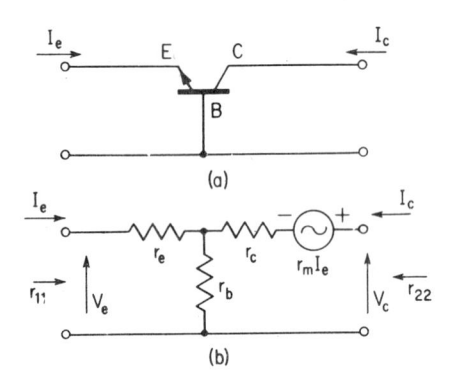

FIG. 7-39. Equivalent circuit for the common-base transistor circuit.

ments, measurements may be taken on the equivalent circuit (or on the transistor itself) with the input applied to the 1,1 or emitter-base terminals with the output or collector terminals 2,2 open or shorted, or input may be applied to the collector-base terminals with the emitter terminals open or shorted.

With input on terminals 1,1 and with the collector-base or output terminals 2,2 open, it can be seen that the *transistor input impedance* is

$$r_{1,1} = r_e + r_b.$$

Reversing the situation with the emitter circuit open, the *transistor output impedance* is

$$r_{2,2} = r_c + r_b.$$

It is also possible to measure transfer impedances, or ratios of voltage produced in one circuit to a current flowing in a second circuit. For instance the *forward transfer resistance* is defined as the ratio of the voltage V_c which appears at the open collector terminals to the current I_e flowing in the emitter circuit. It can be seen that with the collector open the voltage V_c will then be

$$V_c = r_b I_e + r_m I_e$$

or the forward transfer resistance is

$$r_{2,1} = \frac{V_c}{I_e} = r_b + r_m.$$

A *reverse transfer resistance* can also be defined as the voltage V_e appearing at the open emitter terminals when a current I_c flows in the collector circuit. Then

$$V_e = r_b I_c$$

and the reverse transfer resistance is

$$r_{1,2} = \frac{V_e}{I_c} = r_b$$

in the common-base circuit. This quantity has no analogous term in the vacuum-tube circuit at low frequencies.

From the above relations or transistor measurements it is possible to find that

$$r_e = r_{1,1} - r_{1,2}, \quad r_c = r_{2,2} - r_{1,2}, \quad r_b = r_{1,2}.$$

These relations contribute a physical significance for the transistor parameters in terms of easily made measurements. This view may be further supported in the case of r_m, the mutual transfer resistance, by noting that the potential transferred from the input circuit is $(r_b + r_m) I_e$, and that under collector short-circuit conditions in the output circuit the voltage relations must lead to

$$(r_m + r_b)I_e + (r_c + r_b)I_c = 0$$

from which

$$-\frac{I_c}{I_e} = \alpha = \frac{r_m + r_b}{r_c + r_b} = \frac{r_{2,1}}{r_{2,2}}.$$

(The negative sign results from the choice of current directions shown in the figure). For the junction transistor it is found that r_b usually has a value of the order of 100 ohms and so is very small compared with r_m or r_c which may approximate a megohm. Therefore

$$\alpha \approx \frac{r_m}{r_c}, \quad \text{or} \quad r_m \approx \alpha r_c,$$

and this allows computation of r_m.

In all of the above it has been assumed that the input a-c signals are small, so that the transistor characteristics may be assumed linear. In addition, as the applied frequency is raised all these resistive terms become somewhat reactive or become impedances, owing to internal transistor capacitances and delays in charge motion. However, in the usual operating ranges of audio amplifiers the resistive approximation is found to yield reasonable results.

A transistor normally transmits in both directions, while in a vacuum tube at low frequencies conduction in the plate circuit has no effect in the grid circuit. When the transistor is connected into a circuit as in (a), Fig. 7-

40, the input impedance Z_in seen by the generator at the 1,1 terminals may be affected by the value of load Z_L, and likewise the output impedance Z_out of the circuit as seen at 2,2 may be affected by the source impedance Z_s.

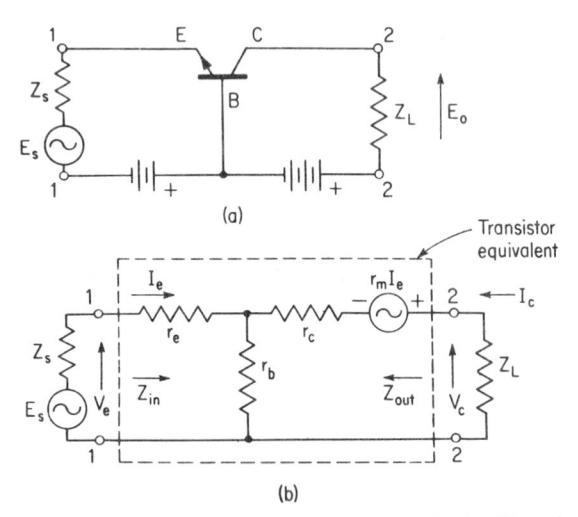

Fig. 7-40. (a) N-P-N transistor in the common-base circuit; (b) equivalent circuit for (a).

By observation of the equivalent circuit in (b), two circuit equations can be written for sinusoidal currents and voltages I_e, I_c, V_e, and V_c, with the base common, as

$$V_e = (r_e + r_b)I_e + r_b I_c$$
$$0 = (r_b + r_m)I_e + (r_c + r_b + Z_L)I_c.$$

The input impedance $Z_\text{in} = V_e/I_e$ and this can be solved for, giving

$$Z_\text{in} = r_e + \frac{r_b(r_c - r_m + Z_L)}{r_c + r_b + Z_L}.$$

Because of the effects of shunt capacitances the load Z_L will usually be small with respect to r_c and to $r_c - r_m = r_c(1-\alpha)$, as is usual with a pentode tube. The input resistance of a common base circuit then reduces to an approximation:

$$Z_\text{in} \approx r_e + r_b(1-\alpha).$$

The output impedance $Z_\text{out} = V_c/I_c$ and this can be found as

$$Z_\text{out} = r_c - r_b\left(\frac{r_m - r_e - Z_s}{r_e + r_b + Z_s}\right).$$

Usually the second term on the right is small and so as an approximation for the common-base circuit input:

$$Z_\text{out} = r_c.$$

Chapter 7   **Audio and Video Amplifiers**   245

The current gain $A_i = I_c/I_e$ can also be found as

$$A_i = \frac{-(r_b + r_m)}{r_c + r_b + Z_L}$$

and for usual parameter magnitudes this approaches

$$A_i \approx -\alpha$$

because $r_b \ll r_m$, $r_b < Z_L \ll r_c$, and $r_m = \alpha r_c$.

The voltage gain $A_v = V_c/V_e$ can be calculated, and after use of $r_b \ll r_m$ becomes

$$A_v \approx \frac{\alpha Z_L}{r_e + r_b(1 - \alpha)}$$

and no phase reversal is involved.

Justification for the above assumptions is found in the parameter values of a typical N-P-N junction transistor, where $\alpha = 0.97$, $r_e = 40$ ohms, $r_b = 100$ ohms, $r_c = 1.5$ megohms, and therefore $r_m = 1.46$ megohms. When employed in a typical amplifier with $Z_s = R_s = 100$ ohms, and $Z_L = R_L = 0.1$ megohm, gain values of $A_i = -0.97$ and $A_v = 2{,}260$ are obtained. The power gain can then be computed for the resistive source and load as

power gain $= A_i A_v = 0.97 \times 2{,}260 = 2{,}190 = 33.4$ db.

The input impedance will be 43 ohms and the output impedance will be 1.5 megohms. Such a simple amplifier circuit for an N-P-N unit appears in Fig. 7-41.

If the magnitude of the transistor resistances r_e, r_b, and r_c is known in advance, an indication of the *approximate* performance of a transistor in the common-base circuit of Fig. 7-40(a) can also be inferred from a direct inspection of the equivalent circuit of Fig. 7-40(b). With a small

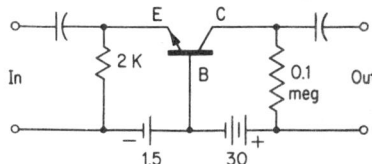

FIG. 7-41. Common-base operation in a practical circuit.

r_b of about 100 ohms and a very large r_c of 1.5 megohms, the effect of the secondary circuit on the input impedance is neglected, and Z_{in} consists of r_e in series with the *effective* resistance of r_b. To determine the effect of r_b on the input impedance recall that the base current I_b which flows through r_b represents the difference between the emitter current I_e and a nearly equal collector current $I_c = \alpha I_e$, so that $I_b = I_e(1 - \alpha)$. [With the current directions shown in Fig. 7-40(b), $I_b = I_e + I_c$, but $I_c = -\alpha I_e$, so that $I_b = I_e(1 - \alpha)$.] Because the base current which flows through r_b is smaller than the current I_e through r_e by a factor of $(1 - \alpha)$, and because the voltage across r_b also will be correspondingly reduced, the effect of r_b on the input impedance will be reduced by the factor $(1 - \alpha)$. Therefore,

$$Z_{in} \approx r_e + r_b(1 - \alpha).$$

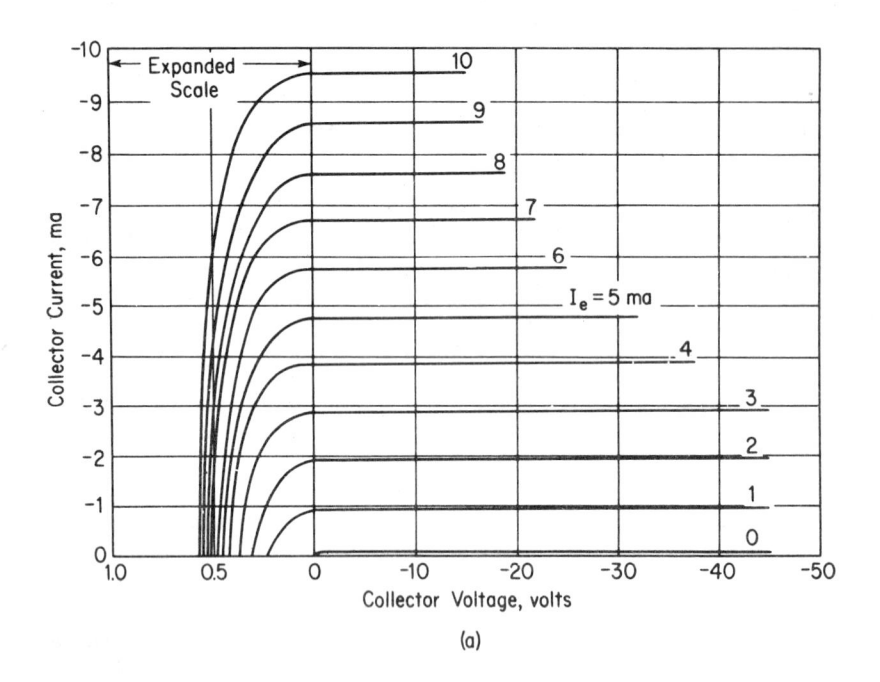

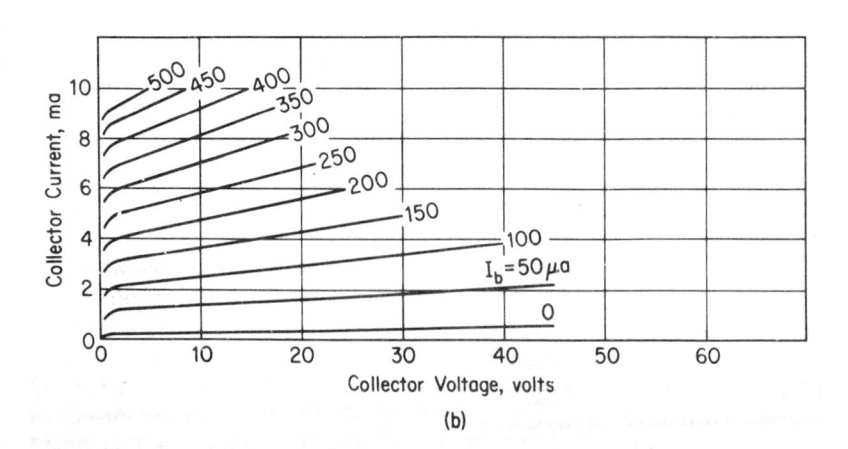

FIG. 7-42. Collector characteristics of a typical transistor (type 2N44). (a) Common-base connection; (b) common-emitter connection.

Looking back into the circuit at terminals 2, 2, the effect of r_b and the primary circuit can be neglected, so that the output impedance will be

$$Z_{\text{out}} \approx r_c.$$

The current gain, defined as $A_i = \dfrac{I_c}{I_e}$ will be approximately $-\alpha$.

The voltage gain will be the ratio of the voltage developed across the load impedance Z_L by the output current I_c to the input voltage V_e which is equal to $I_e Z_{\text{in}}$. Therefore,

$$A_v = \frac{-I_c Z_L}{I_e Z_{\text{in}}} = \frac{\alpha Z_L}{Z_{\text{in}}} \approx \frac{\alpha Z_L}{r_e + r_b(1 - \alpha)}.$$

The power gain is

$$A_v A_i = \frac{\alpha^2 Z_L}{Z_{\text{in}}}.$$

Summarizing these results, it is seen that for a typical transistor connected in a common-base circuit, the input impedance will be low, of the order of 50 ohms; the output impedance will be very high, in the megohms; the current gain will be slightly less than unity; the voltage gain can be fairly high and is proportional to the load impedance, which in practice is limited to about 100,000 ohms; the output voltage has the *same* phase as the input voltage; the power gain is medium, 30 db being an average value.

GRAPHIC CHARACTERISTICS

Characteristic curves can be drawn for the transistor as for the vacuum tube, and Fig. 7-42(a) shows a family of collector-current characteristics for a junction transistor, each curve drawn for a constant value of emitter current, and with collector voltage measured to the base in the common-base circuit of Fig. 7-38(a). Part (b) of Fig. 7-42 shows the collector current characteristics for the common-emitter circuit. In this circuit the input current is I_b rather than I_e, and the curves are drawn for fixed values of I_b, with the collector voltage measured to the emitter in the common-emitter circuit of Fig. 7-38(b).

As was the case for the vacuum tube, the reciprocal of the slope of these curves yields the value of an important transistor parameter for small-signal operation. The reciprocal of the slope of (a), Fig. 7-42 is called the output resistance $r_{2,2}$ in the common-base circuit. This is approximately equal to the collector resistance r_c and to the transistor output impedance previously discussed from the basis of the equivalent circuit. The curves of Fig. 7-42(b) yield the same quantity for use in the common-emitter circuit. From the flatness of the curves it can be seen that $r_{2,2}$ will be large, possibly greater than a megohm.

Other curve families may be drawn, such as those relating emitter current and voltage, emitter voltage and collector current, collector voltage and emitter current, and similar relations to base current. Because of the multiplicity of these relations the graphical characteristics lose some of the value attributed to them in the vacuum tube. They do find use in graphical load-line analysis for large signal performance exactly as in the case of the vacuum tube. Particular attention must be paid in the location of the Q point to avoid exceeding dissipation ratings.

THE COMMON-EMITTER CIRCUIT

Figure 7-38(b) shows the common-emitter circuit to be similar to the vacuum-tube grounded-cathode circuit, and because of the high available gain the common-emitter form is much employed. A common-emitter circuit is shown in Fig. 7-43(a) and an equivalent circuit for (a) is shown in Fig. 7-43(b). It will be observed that the input terminals to r_e and r_b are re-

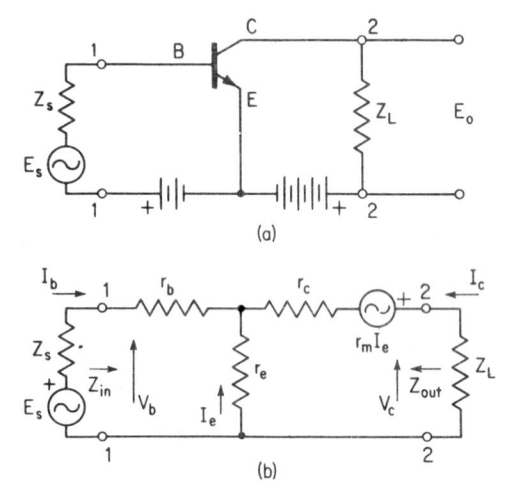

FIG. 7-43. (a) Common emitter circuit; (b) equivalent circuit for (a).

versed from the common-base circuit, introducing a 180° phase shift. Also the input current now is I_b rather than I_e. With these differences in mind, the approximate performance of the common-emitter circuit relative to the common-base circuit can be determined. For a given I_e, the input current being I_b, is smaller by the factor $(1 - \alpha)$. The current amplification can therefore be expected to be larger by the reciprocal of this factor, so that

$$A_i \approx \frac{\alpha}{1 - \alpha}.$$

[This ratio $\alpha/(1 - \alpha)$ is usually called β and is often listed in transistor manuals rather than α.] On the other hand, to produce the same I_e as before, approximately the same voltage (with a sign reversal) between terminals 1,1 will be required; and because the output voltage is I_cZ_L as before, the expression for voltage gain remains as before. That is,

$$A_v \approx \frac{-\alpha Z_L}{r_e + r_b(1-\alpha)}.$$

The output voltage has a phase opposite to that of the input voltage as with a grounded-cathode vacuum-tube circuit. The input impedance, being input voltage divided by input current, will be increased by the factor $1/(1-\alpha)$ compared with the common-base circuit, so that

$$Z_\text{in} \approx r_b + \frac{r_e}{(1-\alpha)}.$$

The output impedance will be still high, although reduced below its value for the common-base connection. The power gain being A_iA_v will be greater than for the common-base connection by the factor $1/(1-\alpha)$.

Using a transistor having the parameters previously given, but now connected in a common-emitter circuit with $Z_s = R_s = 400$ ohms, $Z_L = R_L = 20{,}000$ ohms, the performance will be $Z_\text{in} = 1{,}430$ ohms, $A_i = 32.3$, $A_v = -450$, power gain $= A_iA_v = 14{,}500 = 47.3$ db.

The performance of a transistor in a common-emitter circuit can be summarized as follows: The input resistance is higher than for the common-base connection, usually ranging between 500 and 1,500 ohms. The output resistance depends upon the generator impedance; although lower than for the common-base circuit, it is still high, ranging from about 50,000 ohms to 2 megohms. Although the possible voltage gain is about the same as for the common-base circuit, the value of load impedance used and resulting voltage gain is usually less in order to maintain stability. However, the common-emitter circuit also has current gain, so that the power gain is quite high, values between 35 db and 50 db being common. These values compare favorably with those obtainable from a pentode. This can be achieved in a circuit such as that of Fig. 7-44 with a plate supply voltage of only 3 to 10 v, compared with several hundred volts for the pentode.

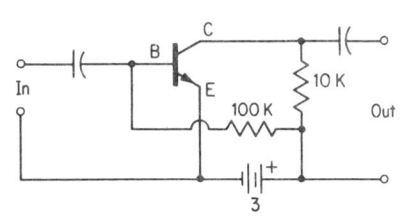

Fig. 7-44. Typical common-emitter amplifier using an N-P-N transistor.

COMMON-COLLECTOR CIRCUIT

A common-collector circuit (biased for an N-P-N transistor) is shown in Fig. 7-45(a) with an equivalent circuit in Fig. 7-45(b). In this case the output current is I_e and the input current is I_b. Since these currents are related in magnitude by $I_b \approx I_e(1 - \alpha)$, the current gain will be

$$A_i \approx \frac{-1}{1 - \alpha}$$

where the negative sign is chosen to agree with the current directions shown in Fig. 7-45(b). If it is assumed that the impedance of the shunt arm of the equivalent circuit is sufficiently high that it can be neglected, then the input

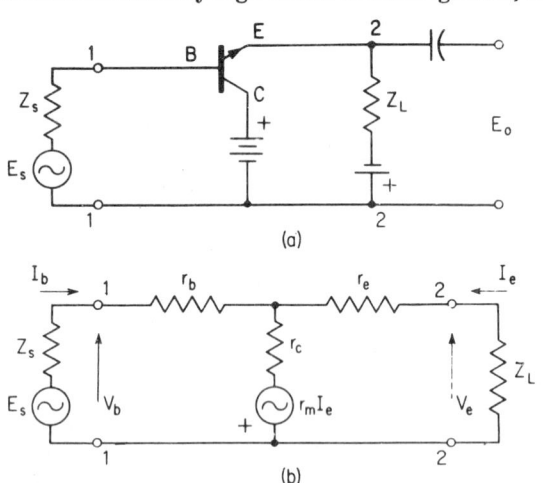

Fig. 7-45. (a) Common-collector circuit; (b) equivalent circuit for (a).

impedance will be given approximately as r_b in series with the effective resistance of r_e and Z_L. Because the current through these latter impedances is larger than the current through r_b by the factor $1/(1 - \alpha)$, the input impedance will be

$$Z_{in} \approx r_b + \frac{r_e + Z_L}{1 - \alpha} \approx r_b + \frac{Z_L}{1 - \alpha} \quad \text{for } Z_L \gg r_e.$$

Similarly, the output impedance will be

$$Z_{out} \approx r_e + (r_b + Z_s)(1 - \alpha).$$

The voltage gain is

$$A_v = \frac{-I_e Z_L}{I_b Z_{in}} \approx \frac{Z_L}{r_b(1 - \alpha) + Z_L} = \frac{1}{1 + (r_b/Z_L)(1 - \alpha)}.$$

It is always less than unity, but approaches unity for $Z_L \gg r_b(1 - \alpha)$. There is no phase reversal, the output voltage being in-phase with the input voltage.

The power gain is

$$A_v A_i \approx \frac{1}{1 - \alpha} \cdot \frac{1}{1 + (r_b/Z_L)(1 - \alpha)} \approx \frac{1}{1 - \alpha} \quad \text{for } Z_L \lessgtr r_b.$$

For the typical transistor previously considered and connected now in a common-collector circuit with $R_s = 400$ ohms and $R_L = 10,000$ ohms, the approximate performance will be $Z_{in} = 330,000$ ohms, $Z_{out} = 53$ ohms, $A_i = -33$, $A_v \approx 1$, and power gain $= 33 = 15$ db. Because the common-collector circuit is often used to match a high-impedance generator to a low-impedance load, more usual values for generator and load impedance are, for example, $R_s = 17,000$ ohms, $R_L = 500$ ohms. For these values $Z_{in} \approx 17,000$ ohms and $Z_{out} \approx 500$ ohms, so that impedance matching has been obtained.

Summarizing results, the common-collector circuit is seen to have performance characteristics similar to the cathode-follower tube circuit as was predicted. The circuit has current gain, but the voltage gain is less than unity. The resulting power gain is low, ranging from 10 to 20 db. The input impedance depends upon the load impedance Z_L, and the output impedance depends upon the generator impedance Z_s. For normally used values of generator and load impedances the input impedance is high, being $Z_{in} \approx (Z_L/(1 - \alpha))$, whereas the output impedance is low, being $Z_{out} \approx Z_s(1 - \alpha)$.

BIAS CIRCUITS FOR TRANSISTORS

In addition to its higher gain, the common-emitter circuit has the advantage that both input and output circuits can be biased from a single battery. Figure 7-46 shows several methods for furnishing proper bias to the base and collector from a common supply. In Fig. 7-46(a) an N-P-N transistor has a positive d-c voltage on the collector given by $V_C = V_{CC} -$

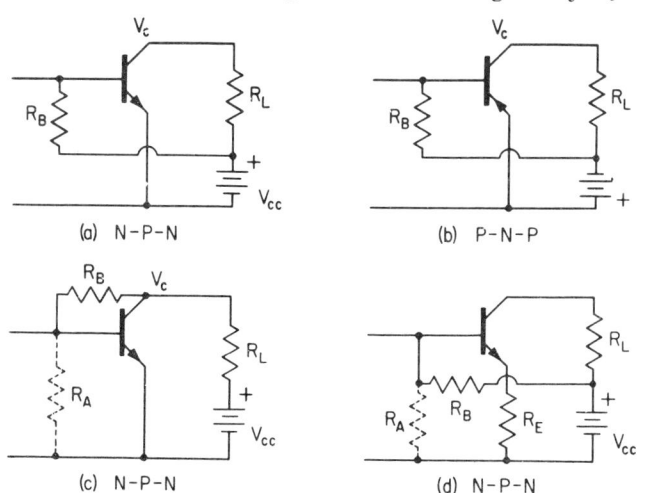

FIG. 7-46. Transistor bias circuits. (a) Fixed bias with N-P-N transistor; (b) fixed bias with P-N-P transistor; (c) voltage-feedback self-bias; (d) current-feedback self-bias.

$I_C R_L$, where I_C is the d-c collector current. It has a smaller positive d-c voltage on the base given by $V_B = V_{CC} - I_B R_B$, where I_B is the d-c base current flowing through the base bias resistor R_B. Actually the important independent parameter is the base *current*, rather than base voltage. Because, in this forward or low-resistance direction, the voltage between base and emitter is always very small (less than 0.1 v), the base current is given to a good approximation by $I_B = V_{CC}/R_B$, which makes it possible to calculate the required value of R_B if the desired base bias current is known from characteristic curves. For many transistor types and commonly used supply voltages the required value of R_B ranges from 100,000 ohms to 1 megohm, with 250,000 ohms being a common value.

In Fig. 7-46(a), positive current flows in at the collector and base, and out at the emitter (electrons move in the opposite direction); the base is slightly positive with respect to the emitter and the collector is more positive, so that *with respect to the base* the emitter is negative and the collector is positive. Following the discussion of Chapter 4, these are the correct polarities and directions for an N-P-N transistor. For a P-N-P transistor all bias polarities and current directions must be reversed, and this is achieved simply by reversing the polarity of the supply battery V_{CC} as indicated in Fig. 7-46(b). It is important that correct battery polarity be used with either type of transistor in order to avoid transistor burn-out. The circuits of Figs. 7-46(a) and (b) are called *fixed-bias* circuits in contrast with the *self-bias* circuits of Figs. 7-46(c) and (d).

One of the problems associated with transistor circuits is that of establishing and maintaining the correct operating point. Transistor characteristics may show considerable variation, even among transistors of a given type number, and these characteristics may change with temperature and age. The bias circuit of Fig. 7-46(c) has an advantage over that of Fig. 7-46(a) in that it tends to be self-stabilizing. The voltage source providing the bias current through R_B is in this case the collector voltage V_C rather than the supply voltage V_{CC}. When for some reason or other the collector current tends to be higher than normal, voltage V_C will be low owing to the larger $I_C R_L$ voltage drop. Therefore the base input current will be less, which is in the direction to decrease the collector current. Correspondingly, a lower than normal collector current raises V_C, which increases I_B and hence tends to increase I_C. Therefore this self-bias circuit tends to stabilize the operating point.

The connection of Fig. 7-46(c), wherein the bias current is varied according to the variations in collector voltage V_C, is known as a *voltage-feedback* connection. Stabilization of the operating point can also be attained in the circuit connection of Fig. 7-46(d), which is called *current-feedback*. Here a change in emitter current I_E changes the voltage drop $I_C R_E$ across a resistor in the emitter circuit; and so changes the potential of the emitter relative to the base. The resultant change in base current is in the correct direction to tend to stabilize the operating point.

In addition to feedback of the d-c bias potentials and currents, both self-bias circuits of Fig. 7-46(c) and (d) produce negative feedback at signal frequencies. This signal-frequency feedback can be avoided by by-passing the feedback resistors with shunt capacitors which provide a low-impedance path for the signal frequencies. However a reasonable amount of negative feedback is usually desirable, so all or a portion of the feedback resistors are often left unbypassed.

In practical bias circuits an additional resistor R_A, shown dotted in Fig. 7-46(c) and (d), is sometimes connected in the circuit as shown. By adjusting the size of R_A with respect to R_B an independent control of the value of bias current results, so that optimum stabilization with correct bias current can be obtained.

MULTISTAGE TRANSISTOR AMPLIFIERS

When more gain is desired than can be obtained with a single transistor stage, several stages can be connected in cascade using either resistance-capacitance or transformer coupling. Resistance-capacitance or R-C coupling has the advantages of compactness, light weight, and relatively low cost. Transformer coupling has the advantages of higher gain, less loss, lower d-c supply voltage, and better operating point stability. The disadvantages of transformer coupling are the greater size, weight, and cost of the transformer. Both types of coupling find practical application.

With transformer coupling, any of the three basic configurations, common-base, common-emitter, or common-collector can be used. As demonstrated in the typical examples given previously, the order of gain to be expected would be 30 db for the common-base, 45 db for the common-emitter, and 15 db for the common-collector connection. Although other considerations may sometimes dictate the use of common-base or common-collector configurations, the greater gain of the common-emitter configuration makes it the preferred connection for cascaded transformer-coupled stages. However with R-C coupled stages in cascade between equal generator and load impedances the common-emitter connection is the *only* configuration capable of producing gain and so is the only connection used (except for input and output stages where the other types are sometimes used for their impedance-matching capabilities).

That common-base stages connected in cascade cannot produce gain (when connected between equal generator and load impedances) is easily demonstrated. As stated in Chapter 4, in the common-base connection the output or collector current I_c is always slightly less than the input or emitter current I_e (that is, $I_c \approx I_e$). However the current flow in the output circuit is at a much higher voltage level and impedance level than in the input circuit. This means that to produce a given current change, say one milliampere, in output current I_c requires a relatively high voltage change in the output circuit. This same change in I_c can be produced by changing I_e with a relatively small voltage change in the input circuit. It follows that a high impedance Z_L can be placed in the output circuit and there will be developed across it a back-voltage $I_c Z_L$ which can be much higher than the input voltage $I_e Z_{in}$. Hence voltage gain and power gain are achieved with the common-base connection when the load impedance is higher than the input impedance.

When common-base stages are connected in cascade as in Fig. 7-47 the situation is different. The input impedance of one stage becomes the load impedance for the previous stage, so input and load impedances are now equal. Therefore the output voltage of the first stage, $I_{c_1} R_{in_2} = I_{c_1} R_{in_1}$, is

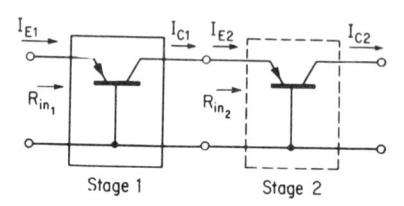

Fig. 7-47. Common-base stages in cascade. (Provisions for obtaining correct bias not shown.)

slightly less than the input voltage $I_{e_1} R_{in_1}$. The output power $I_c^2 R_{in}$ is also slightly less than the input power $I_e^2 R_{in}$. Thus the voltage gain, power gain, and current gain are all less than unity for cascaded common-base circuits, which do not use transformers.

In any practical version of Fig. 7-47 it would be necessary to use coupling resistors and capacitors in order to apply the correct bias potentials to the transistor. These resistors are effectively in parallel with the input and output impedances of the transistor and their only effect, as far as signal-frequency power is concerned, is to add to the losses and so decrease the gain still further.

In a somewhat similar fashion it is possible to demonstrate that R-C coupled common-collector stages are incapable of producing gain. For the common-collector stage it will be recalled that the output voltage is always slightly less than the input voltage; that is the voltage gain is always less than unity. Nevertheless power gain is possible when the load resistance is *less* than the input resistance, because the power input is V_{in}^2/R_{in} and the power output is V_{out}^2/R_L. However in cascaded circuits (without transformers) where R_L is equal to R_{in} the power gain will always be less than unity.

From the previous discussion it can be concluded that the only configuration suitable for cascaded R-C coupled stages is the common-emitter, which has both voltage and current gain. These considerations do not rule out the use of the other configurations for input and output stages. Here the impedance-matching capabilities of these other configurations combined with the ability to produce reasonable amounts of gain when properly matched, often dictate their use with generator or loads which have very high or very low impedances. Moreover, mixed combinations of the configurations can result in useful amplifier circuits with appreciable gain. For example, a common-base circuit followed by a common-collector circuit is useful for connecting a low-impedance generator to a low-impedance load. The common-base circuit feeding into the high input impedance of the common-collector can produce about 30 db of gain. The common-collector circuit matching between the high output impedance of the common-base and the low load impedance can yield about another 15 db. Even so, it will be found that the total gain produced by the different possible combinations will nearly always be considerably less than that obtainable from the same number of cascaded common-emitter stages. The practical considerations in coupling together common-emitter stages are covered in the next two sections.

MULTISTAGE R-C COUPLED TRANSISTOR AMPLIFIERS

Figure 7-48 shows a simple resistance-capacitance coupled common-emitter transistor audio amplifier. As was done with vacuum-tube amplifiers, it is instructive to consider the factors involved in choice of values for the various resistors and capacitors. Bias resistor R_B is selected to provide the correct bias current to the base. Its value depends on the transistor characteristics and supply voltage used, but 0.1 to 0.5 megohms is a common range. Value of load resistor R_L involves a compromise between

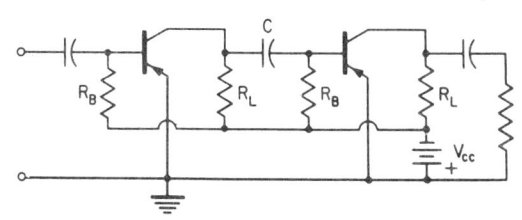

FIG. 7-48. Resistance-capacitance coupled common-emitter audio amplifier using P-N-P transistors.

amplifier gain and size of the supply voltage. Increasing R_L increases the gain that can be obtained but also increases the supply voltage V_{CC} required to yield specified values of collector voltage and current. A compromise value of 10,000 ohms is typical for R_L. In operation the input resistance of the second stage is effectively in parallel with R_L for the first stage at signal frequencies. For a common-emitter amplifier this input resistance is low, of the order of 1,000-2,000 ohms, so the actual load impedance for stage one is much lower than the value of resistor R_L, and of course the gain obtainable is also correspondingly lower. In the example on page 249 a gain of about 47 db was calculated for a typical common-emitter transistor amplifier with

$R_L = 20,000$ ohms. When this load resistance is shunted by the input resistance of the following stage (say 2,000 ohms), this gain figure will be reduced by 8 to 10 db. (The voltage gain is reduced by about 10 db but the current gain actually increases slightly.) In addition, it is necessary to allow from 2 to 4 db for losses in the coupling network. Therefore a resulting gain of about 35 db per stage might be expected.

The size of the coupling capacitor C must be chosen to have a negligible reactance at mid-frequencies compared with the resistance R_L. The low-frequency response depends upon the size of this capacitor reactance in comparison with the output resistance of stage 1 and the input resistance of stage 2. Because the coupling capacitor is effectively in series with the output resistance R_{out_1} of the preceding stage and the input resistance R_{in_2} of the succeeding stage, the low frequency response depends upon the value of $X_c = 1/2\pi fC$ compared with $(R_{out_1} + R_{in_2})$. The response will be down 3 db from the midband gain at the frequency which makes these two quantities equal in magnitude. The output resistance R_{out_1} consists of R_L in parallel with the (usually) much higher output resistance of the transistor, so is approximately equal to R_L. Because R_{in_2} is usually considerably smaller than R_L, good low-frequency response of the amplifier depends mainly on keeping X_c small compared with R_L. A value of 1 µf for C is usually adequate. Fortunately with the low voltage supply ordinarily used with transistors, low-voltage miniature-type capacitors are usually satisfactory in this application.

TRANSFORMER-COUPLED TRANSISTOR AMPLIFIERS

Resistance-capacitance coupling has three disadvantages which are not present with transformer-coupled transistor audio amplifiers. It was seen that when two stages are cascaded, using R-C coupling, the input impedance of the second stage effectively becomes the load impedance for the first stage. With the common-emitter circuit this input impedance is quite low,

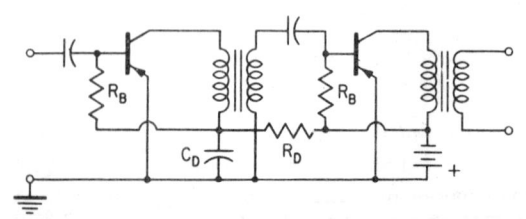

FIG. 7-49. Transformer-coupled audio amplifier using common-emitter connection.

so that the effective load resistance for the first stage is low, and the resulting gain of the first stage may be some 10 to 15 db lower than it would be with a properly matched load impedance. In addition, the efficiency of the R-C amplifier is low because of the power loss in the coupling resistors.

Finally, the d-c supply voltage must be higher than would otherwise be necessary because of the d-c voltage drop in the coupling resistors. All of these shortcomings can be avoided by use of transformer coupling. Of course audio transformers will add to cost, size, and weight, but the development of miniature transformers for use with transistors has decreased the importance of the size and weight considerations.

Figure 7-49 illustrates a simple transformer-coupled transistor audio amplifier. The transformer is a step-down transformer designed to match the high output impedance of stage 1 to the relatively low input impedance of stage 2. A transformer designed to match 20,000 ohms on the primary side to 1,000 ohms on the secondary side will be about right for many transistors.

Also shown in Fig. 7-49 is a *decoupling filter* consisting of the resistor R_d and capacitor C_d. The purpose of this filter is to prevent the amplified signals of a high-level stage from being fed back to the low-level input stage. Such feedback is likely to occur in battery-operated amplifiers using a common battery supply for all stages. The resistance of a battery is low when new, but it increases as the battery ages. A large signal current flowing in the output stage then produces an appreciable signal-frequency voltage across the battery resistance, and this voltage is impressed on the input stage. The feedback may cause serious distortion and even oscillation. In order for the filter to be effective, the reactance of the capacitor C_d should be small compared with the resistance R_d at the low-frequency end of the band. Too large a value for R_d will result in too great a d-c voltage drop across it, so the capacitor C_d must be fairly large. Common values are 10 to 30 μf for C_d, and from 500 to 10,000 ohms for R_d, depending upon the drop in supply voltage that can be tolerated.

PUSH-PULL AND COMPLEMENTARY SYMMETRY TRANSISTOR CIRCUITS

As with vacuum-tube amplifiers, push-pull circuits are often used for the output stages of transistor audio amplifiers as illustrated in Fig. 7-50. When operated Class A, the push-pull circuit offers the advantages of double the output power of a single-ended stage with a great reduction of

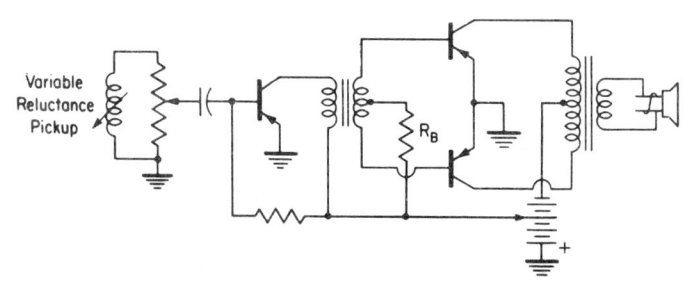

FIG. 7-50. Push-pull common-emitter circuit used in a transistor phonograph amplifier.

second-harmonic distortion. The push-pull circuit can also be operated Class B with the transistors biased at or near output current cut-off. Each transistor of the push-pull pair then conducts for one-half cycle of the signal voltage. Class B operation makes possible larger power outputs from transistors having given power dissipation capabilities. Under maximum signal conditions the power conversion efficiency is high, of the order of 65 per cent. In addition, when no input signal is present the idling power is very low because of the bias adjustment to near current cut-off conditions. Therefore Class B operation provides large power output with a minimum drain on the power supply — an important consideration for battery-operated amplifiers. However Class B operation also requires careful design and proper adjustment of the bias if severe distortion at the cross-over point (from one transistor to the other) is to be avoided. The distortion can

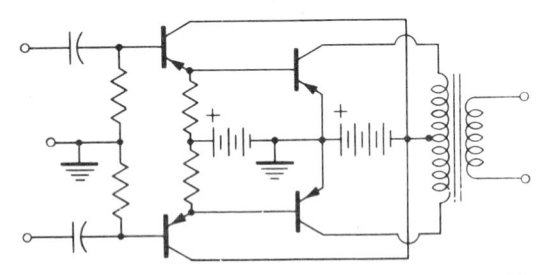

Fig. 7-51. Class B common-emitter push-pull output stage with a low-impedance common-collector driver stage.

be minimized by use of a driver stage having a low-impedance output. This low impedance can be obtained by use of a suitable step-down transformer to couple the driver stage to the push-pull stage. An alternative arrangement shown in Fig. 7-51 makes use of a push-pull common-collector stage with its low output impedance to drive the Class B output stage.

A very useful transistor circuit that has no vacuum-tube counterpart is the *complementary symmetry* circuit, one form of which is depicted in

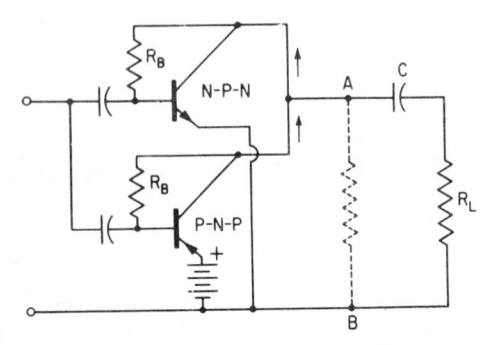

Fig. 7-52. Push-pull transistor amplifier with complementary symmetry.

Fig. 7-52. This circuit provides all the advantages of push-pull action, but both input and output are single-ended so that neither phase-inverter nor transformer is required. The complementary symmetry circuit depends upon the symmetrical properties of N-P-N and P-N-P junction transistors. As seen in Fig. 7-52, the N-P-N and P-N-P units are in series with the battery and with each other as far as the d-c energizing current is concerned, but both their inputs and their outputs are in parallel for the a-c signal voltages. With no signal applied at the common inputs the positive d-c current *in* at the collector of the N-P-N unit is just equal to the positive d-c current *out* at the collector of the P-N-P unit so that no current would flow through a d-c load resistor connected between points A and B. Therefore this d-c load resistor can be dispensed with, and in this sense each transistor forms the d-c load for the other. Now consider results when an a-c signal is applied at the common inputs. As the base voltages (with respect to the emitters) are made more positive, the positive current *into* the collector of the N-P-N transistor increases and the positive current *out* of the P-N-P unit decreases so that the *difference* current would flow through a load resistor connected between A and B. This action should be compared with the operation of a vacuum-tube push-pull circuit. In such a circuit, by use of a phase inverter or push-pull input transformer, the grid of tube 2 is made to go more negative as the grid of tube 1 goes more positive. Therefore positive current into the plate of tube 1 increases, and positive current *into* the plate of tube 2 decreases. In order to obtain the *difference* between these currents it is necessary to use a push-pull output transformer in which the plate currents flow in opposite directions through the primary winding. In the complementary symmetry transistor circuit this difference current is obtained directly without use of the output transformer. Phase opposition for the input voltages is not necessary for complementary symmetry transistors because a positive input voltage applied to the base of an N-P-N unit acts like a negative input voltage applied to a P-N-P unit and vice versa.

For an a-c input voltage applied in Fig. 7-52, the difference current is also a-c, so it can be made to flow through capacitor C and load resistor R_L. A d-c load resistor between A and B is therefore still unnecessary, and it may be eliminated with consequent elimination of the losses which normally occur in it. Efficiency is therefore higher. In addition, all the advantages of push-pull operation are obtained with complementary symmetry without the inconveniences usually associated with balanced operation. By adjusting the bias to near current cut-off each transistor will conduct for one half of the cycle and Class B operation can be obtained.

When the load impedance to which the output power is to be delivered is the low-impedance voice-coil of a loudspeaker it is desirable that the output impedance of the output stage also be quite low. This result can be achieved through use of the complementary symmetry common-collector stage shown in Fig. 7-53. Here the load impedance is connected between emitter and ground as is standard with the common-collector circuit. The bias resistors, R_1, R_2 and R_3, form a voltage divider across the supply

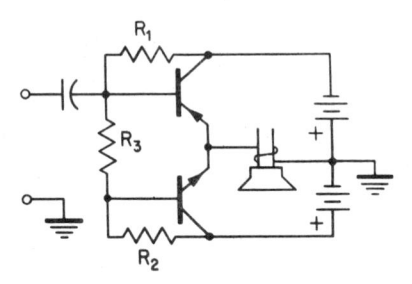

Fig. 7-53. Common-collector complementary symmetry Class B stage.

voltage, and can be selected to provide nearly distortionless Class B operation. Resistor R_3 is low enough to cause negligible unbalance between the two transistor circuits.

QUESTIONS AND PROBLEMS

1. The energy supplied to the speaker of a radio is many times larger than the energy received at the antenna. From where does this extra energy come?

2. What is meant by a *flat response* for an audio amplifier? Is such a response desirable?

3. What determines the operating class of an amplifier? Distinguish between Class A, Class AB, and Class B operation.

4. Why aren't amplifier tubes hooked in cascade with the plate of one tube connected directly to the grid of the next tube?

5. What is meant by the *operating voltages* of a tube?

6. What is the function of the grid-leak resistor? What is the *time constant* of a coupling capacitor of 1 µf together with a grid-leak resistor of 1 megohm? Is this a suitable value for use in an amplifier?

7. How does an increase in plate resistor size increase the amplification of a triode amplifier? Why is not the resistor made extremely large in practical amplifiers?

8. How would an increase in the size of the coupling capacitor of a resistance-coupled amplifier affect the voltage gain of the amplifier at high frequencies and at low frequencies? How would an increase in the value of the grid-leak resistor affect the voltage gain at high frequencies and at low frequencies? Should the plate resistor be large or small compared to the grid-leak resistor of the next tube?

9. What characteristic of a pentode determines its worth in a resistance-capacitance coupled voltage amplifier? How does the value of the load resistance affect the voltage amplification?

Chapter 7   *Audio and Video Amplifiers*   261

10. Why is the capacitor C_D placed in the circuit of Fig. 7-5? How is the grid bias obtained in this circuit? What components prevent feedback through the common plate supply?

11. If the gain in decibels of each stage of an amplifier is known, how can the over-all gain be obtained? What is meant by the expression "down 3 decibels at 5,000 cycles"?

12. Explain the various ways in which amplitude distortion may be produced. How does the dynamic I_p-E_g characteristic differ from the static characteristic? What is the *load line*, and how is it used in determining the dynamic characteristic?

13. Study Fig. 7-8 and devise an experimental method for testing for the presence of second-harmonic distortion. (How does the average value of the plate current change with grid excitation?)

14. Explain the salient features of the frequency response of a transformer-coupled amplifier. How does this response depend upon the plate resistance of the tube?

15. Draw a diagram of a simple resistance-capacitance-coupled triode amplifier showing all of the necessary bias resistors and by-pass capacitors for power-pack operation.

16. Draw a circuit diagram of a simple resistance-capacitance-coupled pentode amplifier showing all the necessary bias resistors and by-pass capacitors for power-pack operation.

17. Draw a circuit diagram of a push-pull amplifier. How do the output, distortion, allowable grid swing, and power-supply hum for a push-pull amplifier compare with those of a single triode stage?

18. Arrange Class A_1, AB_1, AB_2, and B amplifiers in the order of their power output and again in the order of the distortion they produce. Which classes of operation demand large input power?

19. What class of operation would give the largest undistorted power amplification in push-pull operation?

20. What is feedback, regenerative feedback, degenerative or inverse feedback? Explain how inverse feedback reduces frequency distortion.

21. What special characteristic must a video amplifier have? Why isn't transformer coupling used in video amplifiers?

22. What are the limiting factors to the high-frequency response of the circuit in Fig. 7-21? How does the circuit of Fig. 7-23 extend the frequency response over that of Fig. 7-21?

23. Does an increase in the load resistance of a triode amplifier increase or decrease the amplitude distortion? Does this apply to pentode amplifiers? How would the voltage gain be affected in these two cases?

24. In your own words explain the amplifying properties of a junction transistor when connected (a) as a common base stage, (b) as a common-emitter stage, (c) as a common-collector stage.

25. Show the correct bias polarities for N-P-N and P-N-P transistors when connected in common-emitter circuits.

26. Draw the circuit diagram for a possible two-stage common-emitter R-C coupled audio amplifier, showing typical values for resistors, capacitors and supply voltages.

27. Discuss the relative merits and disadvantages of resistance-capacitance and transformer coupling in transistor audio amplifiers.

28. Explain clearly how a complementary-symmetry transistor amplifier gives a push-pull action, even though inputs and outputs are connected in parallel.

29. For a junction transistor what is meant by the factor α? the factor β? If $\alpha = 0.97$ for a junction transistor what is the value of β?

CHAPTER 8

PULSE AND SWITCHING CIRCUITS

Nonsinusoidal waveforms, such as pulses, sawtooth waves, square waves, and the like are of great importance in electronic equipment. In their generation the vacuum tubes or transistors are ordinarily operated as very fast switches rather than as amplifiers, and distortion is usually desired, rather than avoided. A considerable variety of circuits are employed for various functions.

SIMPLE PULSE-FORMING CIRCUITS

The simple circuits of Fig. 8-1 are available for producing an output which is proportional to the rate-of-change or slope (derivative) of an input wave. This operation is illustrated in Fig. 8-2 (a) and (b), where a

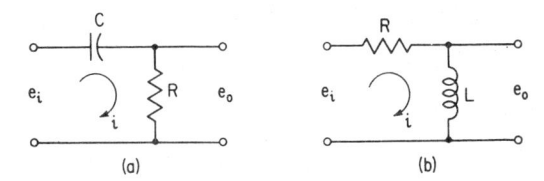

FIG. 8-1. Networks for producing an input proportional to the rate of change of e_i.

wave of sawtooth form is distorted by such a rate-of-change circuit as to produce a rectangular set of pulses, positive where the input has an upward slope, negative where the input has a downward slope.

This action is further demonstrated in (c) and (d), Fig. 8-2, where a square wave is applied to the circuit and sharp pulses obtained only at the instants of rise and drop. Theoretically the heights of these pulses should be infinite because the slope of the square waves is infinite, but circuit limitations, including the fact that no voltage can ever rise at infinite speed in a practical circuit of R, L, or C, causes the response to be more accurately represented by Fig. 8-3.

Such circuits are frequently employed to sort out waveforms having a given slope, or to reshape distorted pulses, usually in combination with tubes or transistors.

Assume that in (a) of Fig. 8-1 the reactance of capacitor C is large, or C is small, such that the current i is largely determined by C and not by R. If

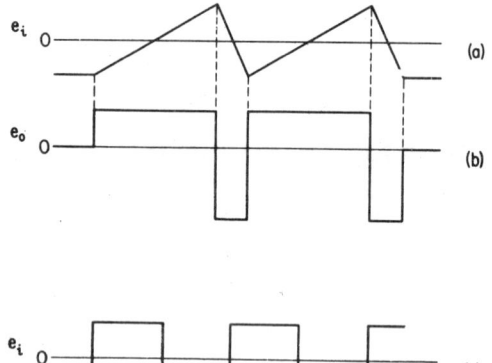

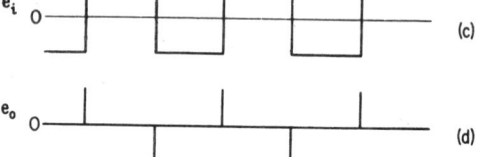

Fig. 8-2. (a) Input wave; (b) output proportional to the rate-of-change of the input wave; (c) and (d) same for a square-wave input.

this current is to be fixed largely by the reactance of C, then R must be small, or

$$\frac{1}{\omega C} > R, \quad \frac{1}{2\pi f} > RC, \quad \text{and} \quad \left(\frac{1}{f} = T\right) > 2\pi RC,$$

so that a more general statement of the condition on the circuit is that the product RC (time constant) is small with respect to the period T of the applied wave.

The charge on a capacitor is proportional to the applied voltage, and therefore the rate of change of charge, or the current, is proportional to the rate of change of the applied voltage. In this case the output $e_\text{out} = iR$, so that we can conclude that under the assumption involving $1/\omega C$, the output

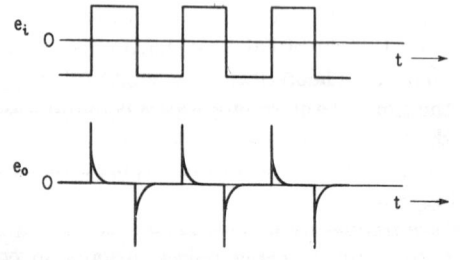

Fig. 8-3. Actual performance of a circuit of the form of Fig. 8-1(a).

voltage will be proportional to the rate of change, or the derivative, of the input voltage, and the circuit will perform as shown in Fig. 8-2.

If a 1,000-cycle square wave is applied, then the product RC must be small with respect to 1,000 microseconds, which is the period of the square wave.

A possible value would make $RC = 50 \times 10^{-6}$, obtained if $R = 100,000$ ohms and $C = 500$ μμf. Then RC would have a value equal to 1/20 of the wave period. The name *time constant* for the RC product is logical if we refer to the last equation above. Here we find RC associated in the same equation with time, and it is actually found that the units of RC are seconds, where R is in ohms and C in farads. With R in ohms and C in μf the units of time become microseconds. Physically the value of time given by RC represents the time required to charge C through resistance R to 63% of its final value, and thus the time constant of a circuit is an index figure for the circuit, measuring the relative slowness with which charge builds up in it. The charge will build up to 98% of its full value in four time constants.

The circuit of Fig. 8-1(b) will also respond to the rate of change of input because the voltage across an inductor is proportional to the rate of change of current. Because a pure inductance cannot be obtained, the circuit performance does not approach the ideal as well as that of the RC circuit, and the RL version is not often used.

The circuit of Fig. 8-1(a) is that of the RC coupling for amplifiers and in such an application usually no distortion or rate-of-change action is desired. It is reasonable to ask under what conditions the circuit ceases to distort and becomes a suitable coupling circuit. This can be easily determined.

If the reactance of C is assumed negligible with respect to R, then

$$i = \frac{e_i}{R} \quad \text{and} \quad e_o = e_i$$

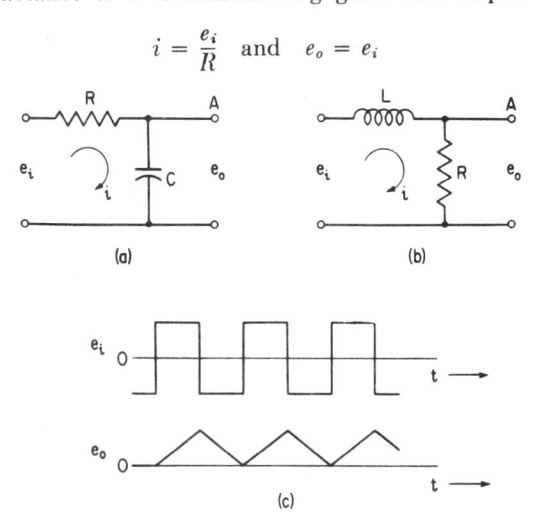

Fig. 8-4. (a) and (b), shaping circuits giving an output proportional to a summation of the current; (c) operation of (a).

or the circuit is distortionless as desired for amplifier coupling. For small reactance C must be large, and so the RC time constant must be large with respect to the time of a cycle of the lowest frequency to be passed. For rate-of-change operation the time constant RC must be small; for distortionless coupling RC must be large.

The circuits of Fig. 8-4 may be used to obtain an output voltage proportional to the charge on the capacitor, or to the integral or summation of the input wave. This is a form of wave distortion which is at times useful, particularly in separating long and short pulses, as in television synchronizing circuits.

For instance, in (a) the voltage across C is proportional to the charge q on the capacitor. The charge q is equal to a summation of all the current received by the capacitor in an interval. If the reactance of C is small, or C is large, then the value of i is fixed by resistance R and

$$i = \frac{e_i}{R}.$$

Thus the charge on the capacitor, and the output voltage is proportional to a summation of the current received by C. This implies a large value for RC, the time constant.

Performance might be as indicated in (c) for an input square wave. The output starts at zero and builds up at a uniform rate while the input voltage is constant or charge is being added to C at a constant rate. When the input reverses the charge starts to be removed or built up in the opposite direction and so an output triangular wave is obtained. The amplitude of the output variation is, of course, proportional to the time length of the input pulse, long pulses giving greater output voltage. Thus a means is provided to sort out long and short pulses, applying them to different circuits, as is necessary in television with the line and frame synchronizing pulses.

The circuit of (b) can be operated in similar fashion but because of the limitations imposed by the inherent resistance of the inductor, the operation of the circuit of (a) is more satisfactory.

CLAMPING CIRCUITS OR THE D-C RESTORER

It may be desired to insert a d-c component or a zero axis into a wave at a particular level, after the wave has passed through an RC amplifier and lost its own d-c axis. This operation can be performed by either a vacuum tube or a crystal diode in circuits such as in Fig. 8-5. The diode should have an internal forward resistance which is small with respect to R, and this condition is most easily met with crystals.

The RC circuit should have a long time constant, so that C will hold its charge over several cycles. In (a) when the signal is negative the diode circuit is open and the signal is transmitted to the output. However, if the wave were to start positive as at the instant W of Fig. 8-5(c) the diode conducts and connects point A to ground potential. The full applied voltage appears across C, and it charges to this potential quickly and remains there during the interval W-X. At X the input falls but the poten-

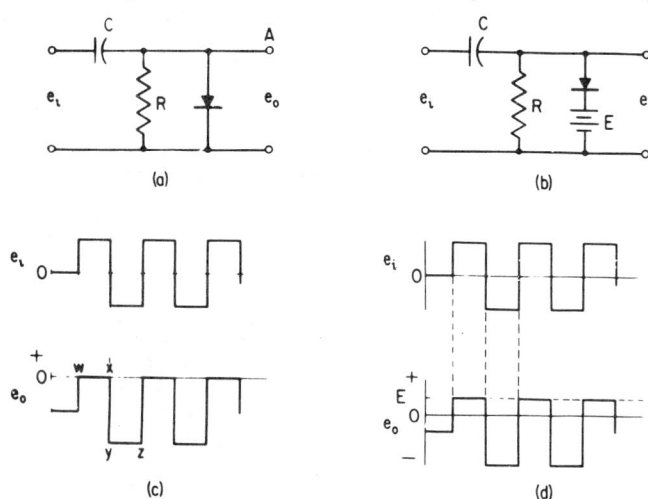

Fig. 8-5. (a) Clamping at zero; (b) clamping at $+E$ level; (c) effect of (a); (d) effect of circuit in (b).

tial across C cannot change since it cannot discharge rapidly through a large R, thus the output falls the full potential to Y. If RC is sufficiently large to prevent appreciable capacitor discharge during the interval Y-Z the output stays constant, or follows the input, once more going to zero when the input swings positive. Thus the output wave has its peaks clamped at zero, or a d-c axis is inserted at a known level.

In (b) the diode is biased with cathode positive so that the input wave has to go above zero by the amount of the bias E before the diode conducts and operates the clamp.

Since the grid-cathode path of a triode also constitutes a diode, the location of a similar RC long-time-constant circuit in the grid circuit of a triode will introduce a clamping action, followed by the triode gain. This is

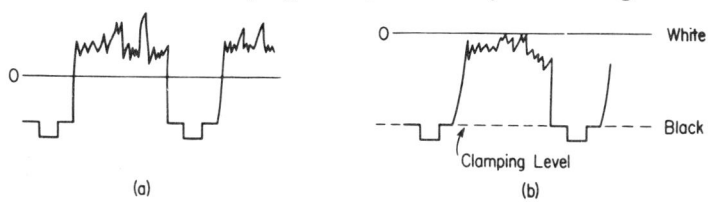

Fig. 8-6. (a) Television signal as received through amplifier; (b) signal clamped at black level.

the manner of operation of the grid leak detector in simple radio receivers, the reintroduced "d-c" component then actually carrying the audio signal modulation.

Clamping is also employed to fix the d-c level at a voltage equivalent to black in television receivers. Without this action the overall picture tone would depend upon the signal strength, weak signals having no tones other than shades of gray.

CLIPPING

Unwanted portions of waveforms may be clipped off or eliminated in another form of diode circuit in Fig. 8-7. If it is found desirable to eliminate all positive portions of the applied wave above a certain level E, the circuit

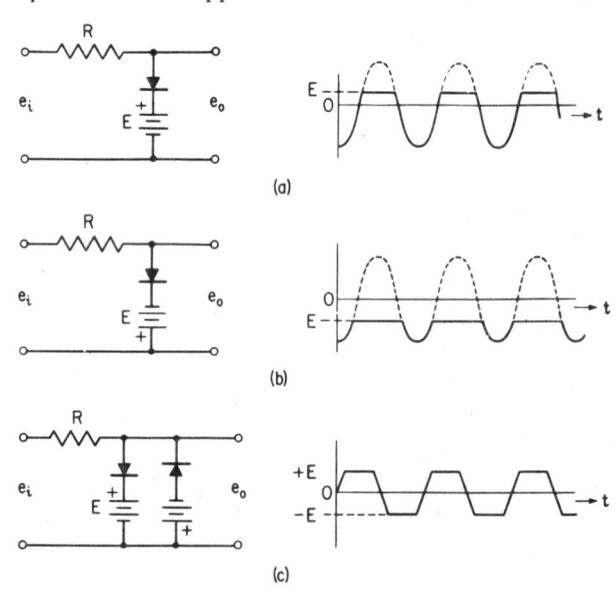

FIG. 8-7. Circuits for clipping at levels other than zero.

of (a) may be used. As soon as the input exceeds E the diode has a positive anode and conducts, reducing or shorting the output down to the fixed value of E. When the wave swings negative the diode opens, allowing the input wave to be passed to the output.

Resistance R should be large with respect to the diode forward resistance, and crystal diodes are preferred. In cases where the circuit is placed in the output of a pentode the resistor R may be eliminated, the high plate resistance of the pentode sufficing. Reversal of the series voltage allows the wave to be clipped at negative levels as in (b). Use of two diodes and two voltages allows a wave to be clipped on both halves, and if this is carried out

Chapter 8   *Pulse and Switching Circuits*   269

at a low point on a sine wave the result is a fair approximation to a square wave.

The triode clipper of Fig. 8-8 performs by grid clipping, using the grid and cathode of the triode as a diode. It will clip on the positive portions of the wave, limiting the positive excursion to zero, and clip the input at the

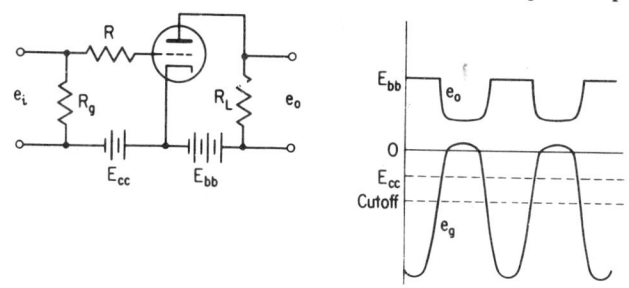

Fig. 8-8. Triode clipping.

other extreme through driving the tube beyond cutoff. The bias E_{cc} should be set half way between zero and cutoff bias to give symmetrical clipping, and R should be large with respect to the grid-cathode resistance which may be considered as falling to a few thousand ohms.

OSCILLOGRAPH SWEEP VOLTAGES

In the cathode-ray oscillograph it is desired to cause the electron beam to plot input wave forms against time on the screen. Thus a pulse chain or a sine wave will appear in true form for study of circuit action. For this purpose the wave to be studied is applied to the so-called *vertical* or *y*-axis circuit, and a voltage which varies uniformly with time is applied to the *horizontal* or *x* axis of the tube. An *x-y* plot is then obtained on the oscilloscope screen, in which the waveform to be studied appears plotted against time. The uniformly increasing voltage applied to the *x* axis is called a *sweep* voltage and appears as a sawtooth in form.

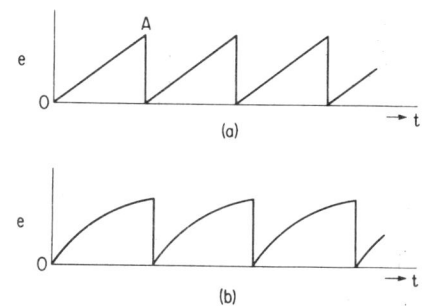

Fig. 8-9. Sawtooth wave: (a) perfect linearity; (b) effect of nonlinear action.

Such a sawtooth wave appears in Fig. 8-9(a), and when applied to the x-deflection circuit of the oscilloscope will cause the electron spot to progress uniformly across the screen. At time A the spot will stop and the electron beam and the trace on the screen will jump back to the starting point, the cycle then repeating. If the sawtooth wave is not perfectly linear on its rise as in Fig. 8-9(b), as a result of design difficulties, the plotted wave may appear bunched at one end, that is, the scale of time will be distorted.

A commonly used generator of such waveforms appears in Fig. 8-10. This circuit employs a gas triode or *thyratron*, the main feature of which is that it does not conduct at all until a certain minimum anode potential is reached. At potentials above this critical breakdown value the tube carries the full current permitted by the circuit and continues to do so until the applied potential falls to a very low value. The anode voltage at which breakdown occurs can be controlled and set by the grid voltage E_{cc}, more negative values causing higher breakdown potential.

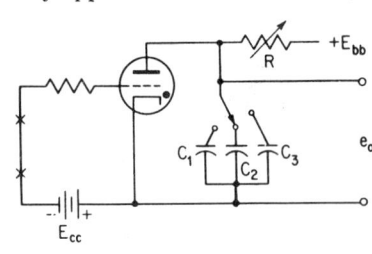

Fig. 8-10. Simple sweep voltage generator.

Anode potential is applied through R and is equal to the voltage on capacitor C_2, or whichever one of several sizes is chosen. The capacitor begins to charge and its voltage increases, thus starting the sloping portion of the sawtooth. At point A of Fig. 8-9 the critical anode voltage of the thyratron is reached, the tube fires, producing practically a short circuit between its plate and cathode, and almost instantly discharges the capacitor bringing the sweep voltage practically to zero and extinguishing the thyratron. The thyratron cannot again fire or conduct until the critical voltage is reached, so the capacitor once more starts to charge and the sweep cycle is repeated.

Several capacitors and a variable resistor R are available to vary the charging rate and to change the number of sweeps or sawteeth per second as desired to suit the voltages being observed on the vertical or y-axis oscillograph plates.

Actually the voltage across a charging capacitor varies exponentially with time in accordance with $E_{bb}\,\epsilon^{-t/RC}$, and only at values of e_{out} small with respect to E_{bb} or at the beginning of charge, is the sawtooth wave approximately linear. As a result only small maximum voltages are available, and these must usually be amplified to provide sufficient sweep voltage for an oscillograph. Forcing the charging action to higher output voltages will result in nonlinear variation of the sweep voltage, as in Fig. 8-9(b), and nonlinear variation of time across the oscillograph screen.

The gas tube cannot turn on and off too rapidly, and such circuits are usually limited to frequencies of 20,000 to 40,000 per second. Vacuum-tube

Pulse and Switching Circuits

circuits employing similar operating principles are available for higher frequencies.

SYNCHRONIZATION OF THE SWEEP

An input signal may be applied to the grid of the thyratron at x,x of Fig. 8-10, to lock the operation of the sweep circuit to the frequency of the wave to be observed on the cathode-ray screen.

The synchronizing input voltage will change by small amounts the voltage e_{out} at which the thyratron conducts or fires. This gives a slight correcting action on the timing of the sawtooth wave, and insures that the sweep always stops and starts at the same point, usually the positive peak, on the grid wave. The circuit may also be synchronized at a slower rate, so that e_{out} approaches firing value only on the peak of every second or third or every nth input wave, giving sweep timing such as to display a number of cycles of input on the screen at once.

FURTHER SWEEP CIRCUITS

If the current available for charging the capacitor were constant, the rate of charge would be constant and the variation of capacitor potential would be linear with time. Constant current can be approximately supplied by a pentode tube in series with R, since the current through a pentode is almost independent of anode potential.

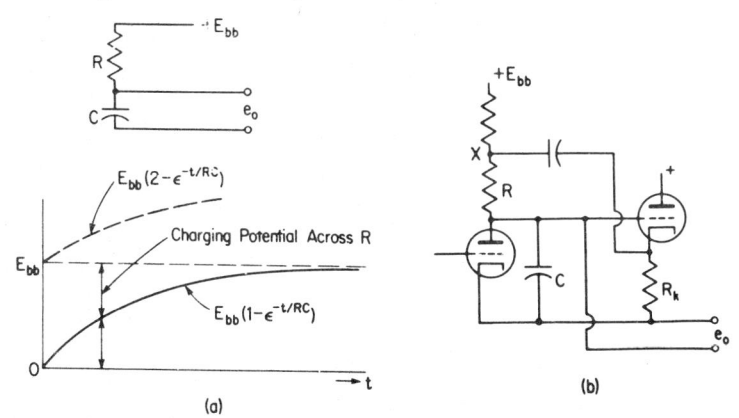

Fig. 8-11. (a) Capacitor charging circuit and charging curve; (b) bootstrap sweep circuit.

Another circuit for obtaining a more linear sweep voltage is the *bootstrap*. Figure 8-11(a) shows that the falling off of charging current to the capacitor at high e_c values is due to the fact that the charging potential across R varies as $E_{bb}\,\epsilon^{-t/RC}$. In the bootstrap circuit a cathode follower is used to provide feedback from the output to the capacitor, and the capacitor voltage becomes a more linear function of time. The sweep output voltage still will differ slightly from linearity because the cathode-follower gain is not quite unity. Considerable improvement in linearity of a sweep can be obtained, however.

Circuits must also be designed to furnish linear sweep currents for use with magnetically deflected cathode-ray tubes. They ordinarily employ an inductor as in Fig. 8-12, which may be the deflection coil on the cathode-ray

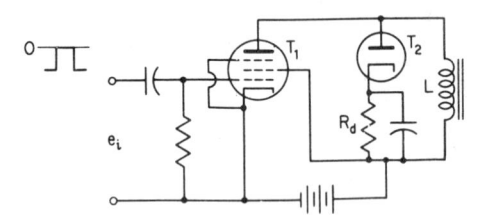

Fig. 8-12. Circuit for developing linear sweep current.

tube. Exponential build-up of current through this inductor is obtained, and by using as the time interval only that time in which the current remains small, fair linearity is obtained, much as with the electric sweep circuit. The current in the circuit will be given by

$$i = \frac{E_{bb}}{R_L + r_b}\left[1 - \epsilon^{-(R_L+r_b)t/L}\right]$$

where r_b is the d-c resistance of the tube used, in the operating range. Then if L is made large compared to the sum of the inductor resistance R_L and r_b, the sweep current will be essentially linear for about the first 5 per cent of the rise time. This states that the time constant $L/(R_L + r_b)$ should be large with respect to the time of a sweep.

To stop the sweep the tube T_1 is driven suddenly to cutoff by the input square wave. During the conduction interval the inductive voltage in the coil has been in such a direction as to maintain the diode T_2 nonconducting. When T_1 goes to cutoff the inductive voltage in the coil reverses as the current in L starts to fall, and T_2 conducts, shorting the coil. This allows the energy stored in the magnetic field of L to be quickly dissipated in the resistance of T_2 and R_d, and the current is quickly brought to zero. The resistor R_d is so chosen that no oscillations will occur and the current in the deflection coil will be quickly damped out, without at the same time exceeding the peak current ratings on T_2.

The input will usually be a square wave derived and in step with an oscillator controlled by the signal to be observed on the CRO. This control is from the line sync pulse in the case of a television receiver. This control insures that the sweep current is in synchronization with the picture to be observed.

Chapter 8 **Pulse and Switching Circuits** 273

THE BLOCKING OSCILLATOR

The blocking oscillator circuit of Fig. 8-13 is used to generate short pulses precisely spaced in time. The inductance L_1 is tuned by C, which may be a small actual capacitor or may be the distributed capacity of L_1. This low-C circuit should also have considerable resistance provided by winding the coil with fine wire, so that the tuned circuit will have a low Q. There should be close coupling between L_1 and L_2, so that a small change in plate current can induce a very large voltage in the grid coil. The values of R_g and C_g should also be large, or the time constant $C_g R_g$ should be long with respect to the length of the expected pulse. In fact, the spacing between successive pulses will be approximately equal to the value of the time constant.

As the voltage induced in L_1 drives the grid positive, C_g charges due to electrons reaching the positive grid. A large swing of output voltage is obtained, since the plate current first is driven up to a large maximum value by the positive voltage induced in L_1, and then is driven to cutoff as C_g charges negatively due to the attracted electrons.

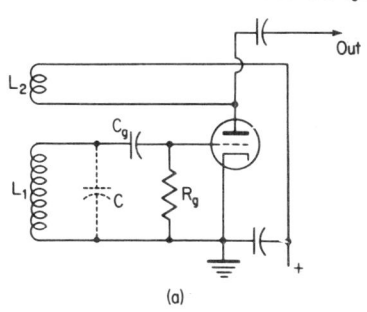

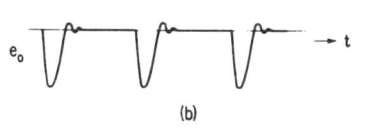

FIG. 8-13. (a) Blocking oscillator circuit; (b) output pulses.

Because of the large R_g value these electrons are unable to leak off the grid rapidly. The charging action is fast, and the time of the pulse is therefore short, because C_g charges through the grid-cathode resistance of about 1,000 ohms. After cutoff the tuned circuit is effectively isolated and a small oscillation may occur in the tuned circuit as in (b), but it is quickly damped by the low Q of the tuned circuit.

The grid remains negative beyond cutoff or the tube is blocked for quite a period, until C_g can discharge through the large R_g value. As soon as C_g discharges up to the cutoff voltage, the plate current starts to flow; this increasing plate current induces a positive grid voltage which further increases the plate current, and a second very fast rise pulse is begun as in (b), Fig. 8-13. Then C_g charges; the tube is once more cut off and the cycle is repeated.

The interval between pulses is a function of the $C_g R_g$ time constant. A clipper may follow the oscillator to take off only the tops of the negative pulses for use as precisely spaced timing pulses or marker signals.

Intermittent oscillation, blocking, or "squegging" may be accidentally obtained in other oscillator circuits and is usually an indication of a $C_g R_g$ time constant which is undesirably large.

ECCLES-JORDAN TRIGGER CIRCUIT

The trigger or flip-flop circuit is generally used in most electronic computers and counters. The basic circuit of Fig. 8-14 has two stable states or conditions, either with T_1 conducting and T_2 cut off, or with T_1 cut off and T_2 conducting. The circuit is unstable under any other condition and will drive itself to one state or the other. The voltages at points A and B go up or down almost instantaneously, and these voltages can be used to control switching or counting operations.

The operation may be explained by first assuming that T_1 and T_2 have exactly equal plate currents. Some slight variation in emission or other erratic effect may momentarily increase the current in T_1. This effect reduces the voltage at point A which in turn makes more negative the potential of the grid of T_2. However, this change in grid voltage on T_2 lowers its plate current, which in turn raises the potential at B. Raising or making more positive the voltage at B means that the voltage on the grid of T_1 is raised or made more positive, and thus the plate current of T_1 is further increased. Following this action step by step shows that any change from equality in either plate current will be followed by a succession of events which quickly lead to one tube being cut off, the other being fully conducting. Thus equality of plate currents is an unstable condition.

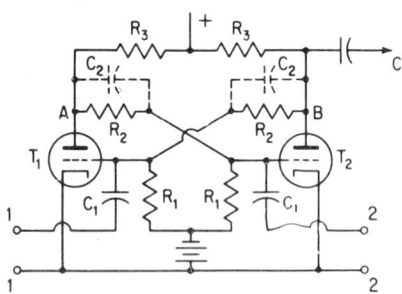

FIG. 8-14. Eccles-Jordan trigger circuit or flip-flop.

If T_1 is conducting and T_2 nonconducting in a stable condition, then application of a negative-to-grid pulse at terminal 1, sufficiently large to drive the plate current of T_1 just slightly past the balanced current condition, will cause the circuit to switch or trigger to its second condition, with T_2 conducting and T_1 cut off. A negative pulse on the grid of T_2 would then cause the circuit to trigger and reverse again.

Capacitors C_2 are usually added to speed up the response on short pulses, giving quick transfer of charge to the grid-cathode capacity of the tubes. Otherwise the internal tube capacity would have to charge through R_2 and this might take a little time and a short pulse might not operate the circuit. With capacitors C_2 in place it is possible to trigger the circuit with pulses of a microsecond or less. The time constants R_2C_2 should be large with respect to the time length of a pulse, so that the pulse voltage is fully effective at the other tube's grid. For usual dual triodes C_2 may be of the order of 50 to 100 $\mu\mu f$.

The circuit may also employ pentodes by using the R_2C_2 circuits to control the voltages on the screens. This leaves the control grids free for the triggering pulses, and then it is found that both grids may be triggered in parallel with negative pulses. The initial differences of voltage on the C_2 capacities when one tube was on and one off are sufficient to unbalance the circuit action and produce triggering, even though both tubes are driven together by the same negative grid pulse.

It takes some very short time after pulsing to transfer conduction from one tube to the other, and if a second pulse were to arrive during this interval it would not be detected. It is then said that the two pulses were too close together or could not be *resolved*.

One use of the circuit is in reshaping received or distorted pulses. Across A-B there appears a square voltage wave for each received pulse, regardless of exact input-pulse form. The output square wave may be passed through a circuit of the form of Fig. 8-1 to give a positive and a negative pulse. These waves may then be passed through a rectifier and one polarity eliminated, giving one new properly shaped pulse for each distorted input pulse. Received noise has also been eliminated in the process.

Successive pulses of one polarity can initiate the switching action in the circuit of Fig. 8-15. If a sufficiently large negative pulse is applied at the input terminal to the two anodes, the anode voltage may be driven down sufficiently so that both tubes will be cut off and the potentials at A and B made equal, regardless of their initial condition. The capacitors C_2 and C_2', however, were not equally charged, because if T_1 was on and T_2 off before the triggering pulse, then initially the voltage across C_2 was much less than across C_2' because the potential of A was lower than that of B. This means that C_2 will discharge faster than C_2', and the grid of T_2 will reach conducting level before the grid of T_1. It will then be found that a switching transition takes place, with T_2 on, T_1 off. Thus the triggering operation is easily performed with negative pulses on the common anode lead. Positive pulses have very little effect and are not ordinarily used.

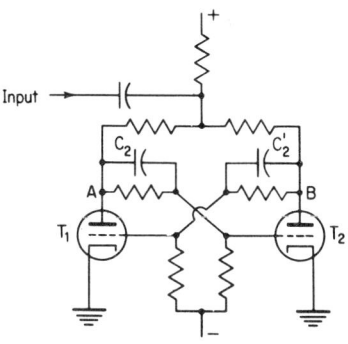

FIG. 8-15. Means of anode triggering.

SCALING CIRCUITS

In Fig. 8-16, if T_1 is off, T_2 on, T_3 off, T_4 on, then the application of a negative pulse to the input will cause T_1 to turn on, T_2 off. This will transmit a positive pulse to T_3 and T_4 which is ineffective. A second negative pulse on the input will turn off T_1 and turn on T_2, causing a negative pulse at C. Thus every second input pulse will cause an output negative pulse at C, and T_1 and T_2 may be called a *scale-of-two* circuit.

The negative pulse transmitted at C by the second input pulse will cause T_3 to turn on, T_4 off, and give a positive output pulse at D. The neon bulb T_b across the T_4 plate resistor will likewise be off.

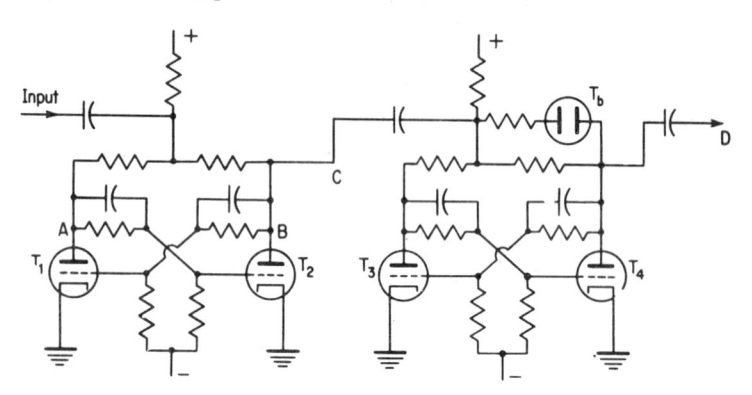

Fig. 8-16. Scale-of-four circuit.

It can be reasoned that at the fourth input pulse a negative pulse will appear at C, T_3 will turn off and T_4 on, a negative pulse will appear at D, and T_b will light. Thus T_b will light for every fourth input pulse, and the four tubes are called a *scale-of-four* circuit.

Additional tubes may be added, another scale-of-two making the overall one for a scale-of-eight, and so on. By ingenious use of feedback paths, causing some tubes to trigger out of order, it is possible to convert a scale-of-sixteen to a scale-of-ten or a decade circuit, and these have considerable application in counting devices operating at very high speed.

MULTIVIBRATORS

If one resistor R_2 is removed from an Eccles-Jordan trigger circuit, and a bias added, then the circuit becomes that of a *one-shot* or *monostable* multivibrator. If both resistors R_2 are removed, leaving the C_2 capacitors, then a *free-running* or *oscillating* multivibrator is obtained, so named because of the large number of harmonics in its output.

These circuits are shown in Fig. 8-17. In (a) the one-shot circuit has only one stable condition, namely that of T_1 cut off and T_2 conducting. A positive pulse applied to T_1 will cause the conduction to transfer to T_1 with T_2 off, but after a transient period in which C_2 discharges through R_1, the circuit will return to its initial stable condition and await another pulse.

A positive input pulse makes T_1 conducting, causing the potential of point X to drop, possibly by several hundred volts. Since C_2 cannot discharge quickly the voltage across it stays constant. If X went from $+300$

to +25 v, or a drop of 275 v, then point Z initially at cathode potential must also drop 275 v to keep the voltage across C_2 constant at the instant of switching T_1 on. This means that the grid of T_2 goes to -275 v and the tube cuts off. Capacitor C_2 begins to discharge and Z rises toward zero. When it reaches cutoff, T_2 conducts, lowering the potential at Y and the grid of T_1, and the circuit quickly triggers back to the initial condition with T_2 on, T_1 off.

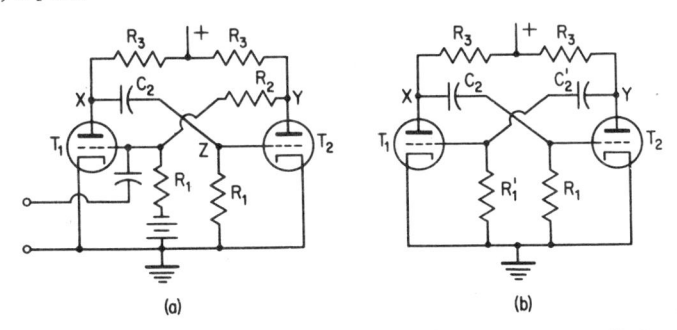

Fig. 8-17. (a) One-shot multivibrator; (b) free-running or oscillating multivibrator.

The waveforms of Fig. 8-18 illustrate the action, and show the sort of output waveforms available at the anode connection of T_2. If the grid of T_2 is returned to a positive bias voltage, more precise operation will be obtained since e_{c2} approaches cutoff steeply and triggers T_1 more precisely.

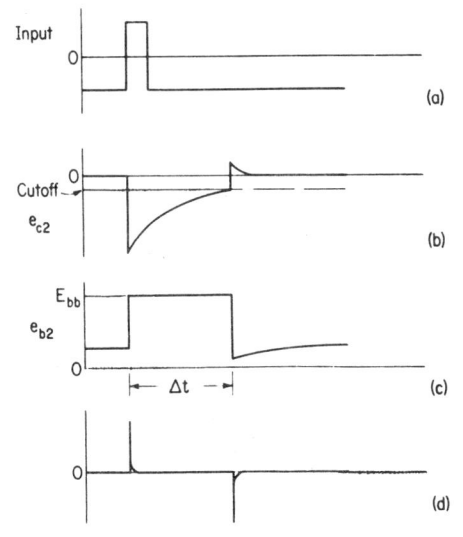

Fig. 8-18. (a) Input pulse; (b) grid voltage on T_2; (c) plate voltage on T_2; (d) shaped output of T_2.

By passing the output of T_2 through a shaping circuit of the form of Fig. 8-1 a pair of very sharp pulses can be obtained. The negative pulse is accurately delayed Δt seconds behind the input pulse, and the system may be used as a delay circuit. The circuit may also be used to renew or reshape distorted input pulses.

The free-running form at (b), Fig. 8-17, has no stable condition; conduction simply oscillates from one tube to the other, the frequency of oscillation being determined by the time constants $C_2 R_1$ and $C_2' R_1'$. These need not be equal, in which case unequal or unsymmetrical periods of conduction are obtained for the two tubes.

If the circuit were ever stable, both grids would be at zero and the plate currents would be equal. However, some tiny change would occur in the plate current of one tube, and if this were T_1 and its plate current increased, the potential at X would fall. This would carry down the grid of T_2 by the same amount, since C_2 cannot instantaneously change its potential; the plate current of T_2 would drop, the voltage at Y would rise, raising the grid voltage on T_1 and adding to the increase in its plate current. This action would be cumulative, leading to the condition of T_1 conducting, T_2 off.

As time passes the charges on C_2 and C_2' readjust by discharge through R_1 and R_1'; the voltage on the grid of T_2 rises above cutoff; current flows in T_2 lowering the potential of Y and the grid potential of T_1 through C_2' and reducing the current in T_1. This action continues until the currents just pass equality and then the circuit once more triggers to T_2 on, T_1 off. The cycle is again repeated, giving an oscillation.

Since grid current flows in the tube with a positive grid, its capacitor discharges faster through the tube grid resistance than can the other capacitor through R_1 or R_1'. As a result it is the time constant of the circuit associated with the conducting tube that determines the time of delay before the next triggering action. Thus unequal time constants will give unequal delays and unsymmetrical waveforms.

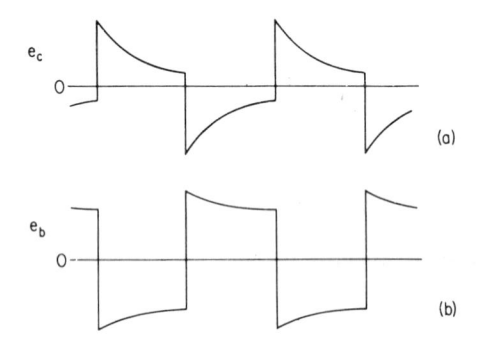

Fig. 8-19. (a) Grid-to-grid voltage waveform for oscillating multivibrator; (b) plate-to-plate voltage waveform in circuit of (a).

Waveforms as in Fig. 8-19 are obtained across both grids, or across both plates. These are rich in harmonics, and the circuit is often used to generate high-order harmonics. If the plate-to-plate wave is clipped the circuit becomes an excellent source of accurate square waves.

SYNCHRONIZATION OF THE MULTIVIBRATOR

It is often desired that a free-running multivibrator run in step, at some submultiple of a certain input frequency. This may be desired to divide input frequencies by integral factors as in a frequency standard. There a 1-megacycle crystal standard-frequency oscillator may be divided successively by factors of 10 with locked or synchronized multivibrators, to give output frequencies of 100 kc, 10 kc, 1 kc, and 100 cycles, all accurate to the same percent precision as the original 1-megacycle standard.

The synchronizing input may be introduced into the grid returns as in (a), Fig. 8-20, where it adds to the grid voltages obtained from the $C_2 R_1$ circuits. This addition is shown in (b) as a wavy variation of grid voltage

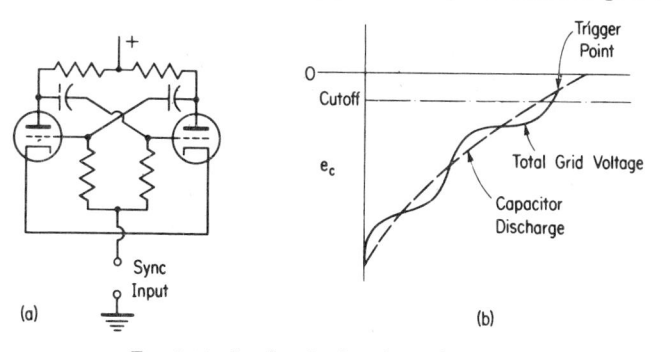

FIG. 8-20. Synchronization of a multivibrator.

e_c. Stable operation requires that the grid voltage reach the triggering point when the synchronizing voltage is going positive. If the $C_2 R_1$ action tends to trigger the circuit early, the added synchronizing voltage holds off or delays the action. If the capacitor discharge action were to attempt to trigger the circuit late, the positive part of the synchronizing signal would trigger the circuit a little early, and stability and exact lock-in would be obtained. As shown, the circuit would be locked with a count-down factor of four (two cycles for triggering each tube).

TRANSISTOR TRIGGERS

A transistor equivalent of the free-running multivibrator appears in

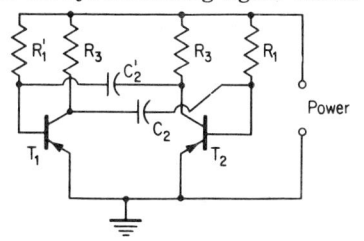

FIG. 8-21. Free-running transistor multivibrator.

Fig. 8-21. The resistors and capacitors are labelled as in the tube circuits, and since the base is analogous to the grid, it should be possible to compare the functions of the transistors and tubes.

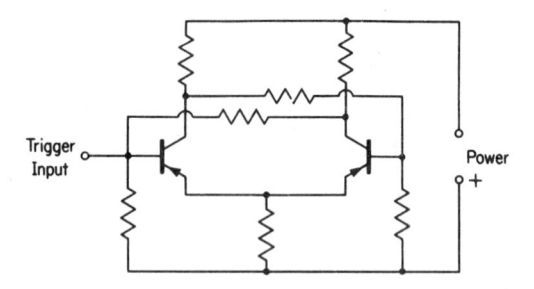

FIG. 8-22. Transistor flip-flop.

Flip-flop and single-shot versions of the circuit are easily attained by altering the appropriate R_1C_2 and $R'_cC'_2$ circuits. One form of a flip-flop appears in Fig. 8-22.

QUESTIONS AND PROBLEMS

1. Prove that if inductor L is lossless the circuit of (b), Fig. 8-1, will give an output voltage proportional to the rate of change of e_i. What assumption is needed?

2. If L is lossless, prove that the circuit of (b), Fig. 8-4, will give an output voltage similar to that of (a). What assumption is needed concerning element values?

3. Describe how clamping a television signal at the black level gives satisfactory picture contrast.

4. How is clipping used to obtain the sync signals in a television receiver?

5. Devise a circuit for clipping off all of a sine wave voltage except the top 10% of the positive half wave, leaving small bumps as the output.

6. Explain how the e_{out} wave of Fig. 8-6 can be further improved and squared by adding another triode clipper.

7. If the sawtooth wave of Fig. 8-9 is used as a cathode-ray sweep voltage, and if the time of one sweep O-A corresponds to the time of two cycles of a sine wave applied to the vertical CRO plates, plot accurately the picture to be expected on the screen.

8. Show what happens to the picture of Problem 7 if the sweep is nonlinear as in (b), Fig. 8-9.

9. Explain how a sweep voltage generator is synchronized or locked onto a signal.

10. Devise a clipping circuit for the wave of Fig. 8-13 (b), to give an output consisting of accurately timed marker pulses.

11. If neon lamps T_b are added to all required flip-flops, draw up a schedule showing which lamps will be lighted after each successive input pulse in a scale-of-sixteen counter.

12. Suggest values for C_2, C_2', R_1, R_1' of Fig. 8-16 which will give a symmetrical output wave at a frequency in the neighborhood of 10 kc.

13. Explain the action of the circuit of an oscillating multivibrator in terms of the waveforms of Fig. 8-19.

CHAPTER 9

ELECTROMAGNETIC WAVES

NATURE OF WAVES IN ANY MEDIUM

Water waves, sound waves, radio waves — apparently widely different phenomena — have certain characteristics in common.

Each type of wave provides a means for transferring energy. It is probable that nearly all energy is transmitted by means of wave motion. When a large steamer plows through the water and sets up waves that rock a small boat half a mile away it is easy to see that the energy required to do the rocking was transmitted by means of waves. When speech is heard across a room and energy from a radio station is received a thousand miles away it is not difficult to believe that the energy was transmitted by means of waves. However, when a man pushes on one end of a steel bar and the other end pushes against an object he is moving with it, it is not so obvious that the energy has been transmitted by wave motion. But if the vibrating cone of a loudspeaker were alternately pushing and pulling at one end of the bar, the wave motion which carried the vibration to the other end of the bar would become more apparent. In this case, if the alternations were rapid enough, it might be found that when the front end of the bar was pushing forward the other end might already be pulling backward. This would be because of the time taken for the wave to travel down the bar. This same time is required when the man pushes the bar, but it is so small that its existence is not generally realized with this type of motion.

Figure 9-1 is a representation of wave motion. Figure 9-1(a) might represent a cross-section of a water wave at a particular instant. If the wave is moving from left to right, (b) would be a picture of the wave an instant later. It will be seen that the crest of the wave which was at position 1 in (a) has moved over to position 2 in (b). Part (c) shows the same wave at a still later instant, at which time the crest has moved to position 3. The rate at which the wave is moving from left to right is called the *velocity of the wave*. Next consider the motion of the point *a*, which might be a cork floating on the water or a particle of the water itself. At the instant represented by (a) the particle *a* is on the crest of the wave. An instant later the crest has moved on and the particle has dropped down, as shown in (b). Still later it occupies the position shown in (c). While the motion of the *wave* has been continuously forward, the motion of a *particle* such as a has been up and down along a vertical line such as 9. The maximum distance either side of the line fg traversed by the particle is called the *amplitude* of the wave. This is shown as the length h in (a) and corresponds to the definition of the amplitude of a sine wave given in Chapter 3.

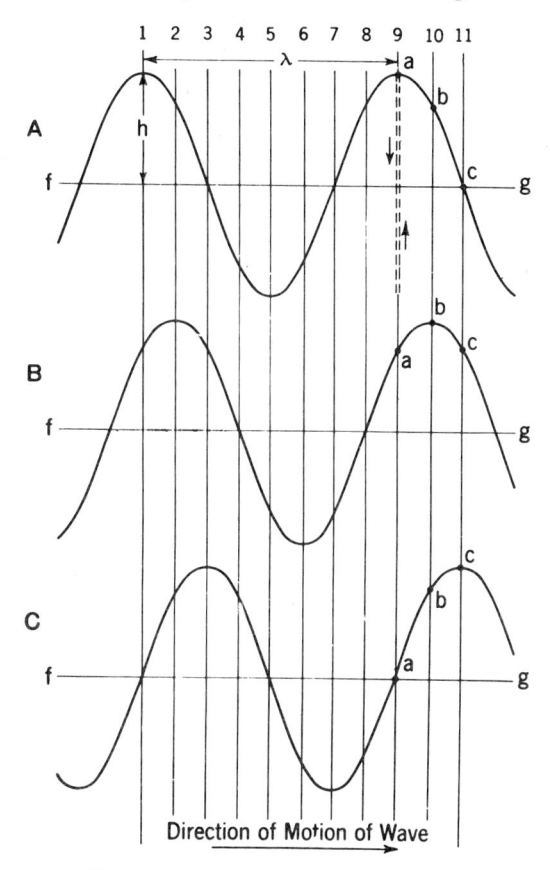

Fig. 9-1. Representation of wave motion.

The distance between successive crests of the wave is called a *wavelength* and is represented by the Greek letter λ (lambda). Of course this is also the distance between successive troughs or any two corresponding points on successive waves.

The number of oscillations per second made by a particle such as a is known as the *frequency*, and is designated by the letter f.* If the waves were being generated by moving a board up and down in the water the frequency would depend upon the number of times per second the board

* The symbol f for frequency should not be confused with the abbreviation f for farad.

was moved up and down. That is, *the frequency depends upon the source.* On the other hand, the speed with which the waves traveled outwards would be independent of how rapidly the board was moved up and down and would depend only upon the properties of the medium — in this case upon the properties of water. If some other liquid such as oil or alcohol were used, the velocity of the waves would be different. *The velocity of propagation depends only upon the medium* and is independent of the source of the waves.

The frequency with which the particle a moves up and down along the line 9 is also the frequency with which the waves are going past the line, for each time a reaches a top peak in its journey up and down, the crest of a wave is going past.

For a given frequency f and velocity of propagation V, the wavelength λ is fixed and is given by

$$\lambda = \frac{V}{f}.$$

λ is usually expressed in meters, V in meters per second, and the frequency f in cycles per second.

This can also be written as

$$V = \lambda f,$$

which states that the velocity with which the wave is moving is equal to the length of a wave times the number of waves per second passing a given point.

The velocity of sound in air is about 344 m per second, so that a 344-cycle oscillation would produce a wave 1 m long. The velocity of electromagnetic waves is 300,000,000 m per second and with them it requires an oscillation frequency of 300 megacycles to produce a wave 1 m long.

TRANSVERSE AND LONGITUDINAL WAVES

The waves pictured in Fig. 9-1 are known as transverse waves because the motion of the particle is at right angles to the direction of motion of the wave. That is, the wave is moving forward, left to right, and the particle is moving up and down. With sound waves in air, on the other hand, the particle motion is back and forth in the direction in which the wave is moving. Such a wave is called a *longitudinal* wave. It is illustrated in Fig. 9-2. The density of the lines represents the pressure in that region.

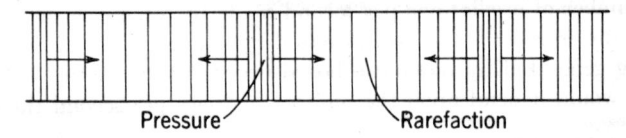

FIG. 9-2. Longitudinal wave motion.

Pressure maxima and minima correspond to the crests and troughs of the wave of Fig. 9-1. The wave of pressure is moving from left to right and the arrows indicate the direction in which the particles are moving. The particles move in both directions from a pressure maximum in toward a pressure minimum. This particle movement results in a pressure maximum being formed where a pressure minimum existed a moment before; and in this manner the wave moves on. The individual particles, however, only oscillate back and forth in a manner similar to the up-and-down oscillation of the particle in the case of transverse waves.

PHASE IN WAVE MOTION

In Fig. 9-1, if attention is concentrated upon the motion of two particles, say a and c, it will be noticed that c does exactly what a does but at a later time. In A, the particle a is at the crest and the particle c is half way between trough and crest but on the way up. In B, a is on the way down and c is still on the way up. In C, a is half way down and c has just reached the crest. The motions could be followed through a complete cycle and it would be found that the movement of c was similar to that of a but coming after it by a constant interval of time. The particle c is said to *lag a* in *phase*. There is a *phase difference* between a and c and this difference may be expressed as a fraction of a cycle. In this case c lags a (or a leads c) by a quarter of a cycle. A complete cycle consists of 360°, so that a quarter of a cycle is 90°. Phase difference is usually expressed in degrees, as, for example, c lags a by 90°.

Phase difference is of importance when more than one wave is involved. If two waves of the same frequency are moving together in the same direction they will combine to give a resulting wave which will be the sum of the

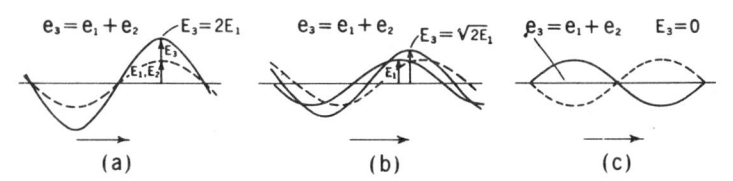

Fig. 9-3. Addition of sine waves: (a) in phase, (b) 90° phase difference, (c) 180° phase difference.

two waves. If two equal waves have the *same phase*, that is, if corresponding particles on the two waves reach the crest at the same instant, the amplitude of the resulting wave will be double that of one wave. This is shown in Fig. 9-3(a). However, if the waves have 90° phase difference as shown in Fig. 9-3(b), the amplitude of the resulting wave is only 1.414 (or $\sqrt{2}$) times that of a single wave. This result can be proven by adding the two waves together point by point. Figure 9-3(c) shows the special case where the waves differ in phase by 180° In this case, because the particle motion due to one wave is exactly equal and opposite that due to the other at all points and all times, the amplitude of the resulting wave is zero. That is, the waves have canceled each other. Complete cancellation can occur only if the two waves are of equal amplitude. This case will be of particular interest in the study of directional antenna arrays.

Although the amplitude of the resulting wave can be obtained as above by plotting the waves and adding them point by point, it can be obtained much more rapidly using a simple geometrical construction. This construction is shown in Fig. 9-4 for the three cases just discussed. The procedure is to lay off from an origin or point O two lengths proportional

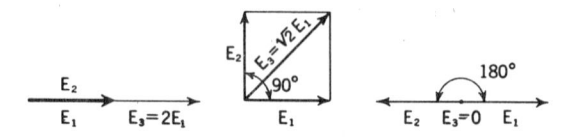

Fig. 9-4. Vector addition of sine waves.

to the amplitudes of the two waves being added. The lengths are drawn with an angle between them equal to the phase difference. For the three cases being considered the phase differences were 0°, 90°, and 180°. With the lengths laid off the parallelogram is completed and the length of the diagonal from O will be proportional to the amplitude of the resulting wave. It will be seen that this method gives the same answers in the three cases considered as did the point-by-point addition.

REFLECTED WAVES AND STANDING WAVES

When a moving wave such as a water wave strikes a boundary such as a solid wall, the original wave is abruptly halted but a reflected wave is set up which travels in the direction opposite to that of the original wave. This combination is shown in Fig. 9-5, where the incident wave traveling from left to right is shown by the solid curve and the reflected wave traveling from right to left is shown by the dotted curve. This gives rise to two waves existing simultaneously at the same place, so they can be added together to show the resultant just as were the waves of Fig. 9-3. However, these waves are traveling in opposite directions so that it is not surprising to find a result different from that obtained in Fig. 9-3. Parts A to I of Fig. 9-5 show the original wave at successive instants of time. It is traveling forward, left to right, as was the wave of Fig. 9-3. The reflected wave is also shown at these same instants. It is traveling backward, right to left. It will be noted that the reflected wave is just the mirror image of the continuation of the original wave beyond the boundary. The resultant of the two waves is shown by the dashed wave in parts A to I. Figure 9-5J shows all the resultant waves of parts A to I plotted on top of one another, that is, J shows the resultant over a complete cycle. A striking feature of this

Chapter 9　　　　　　***Electromagnetic Waves***　　　　　　**287**

is that at some points the resultant is *always* zero. This resultant occurs at one quarter of a wave length from the boundary, again at three quarters of a wavelength and every odd quarter wavelength from the end. Such

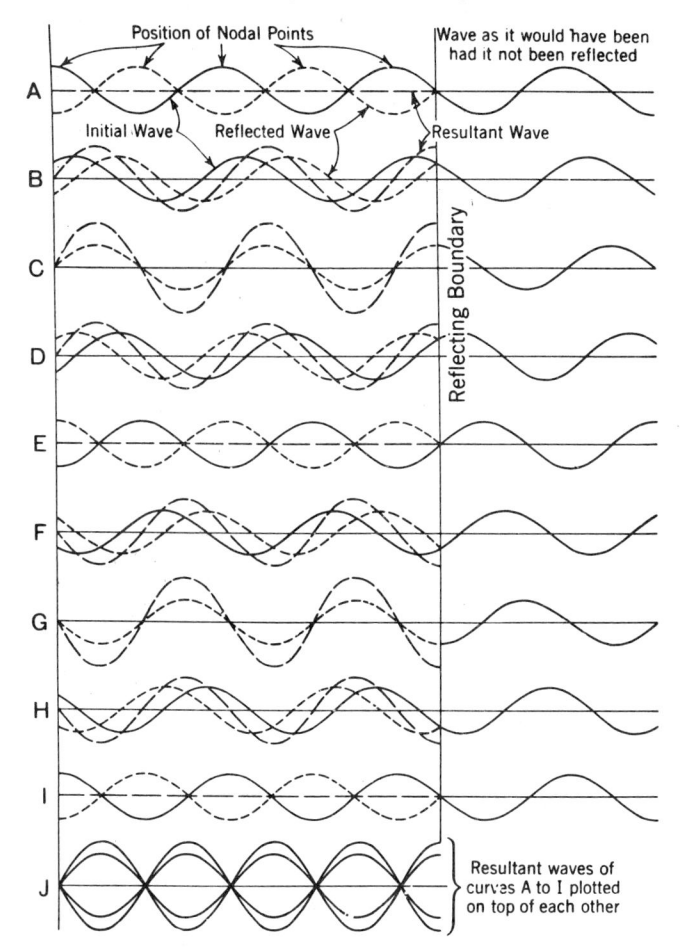

FIG. 9-5. Reflection of a wave at a boundary, showing addition of initial and reflected waves to give standing waves.

points are called *nodes*. It will also be noticed that the resultant reaches its largest values at the boundary and at points distant from the boundary by multiples of one-half wavelength. These points of maximum amplitude are called *loops* or *antinodes*.

Figure 9-6 shows a more detailed picture of the resultant wave for the successive instants of time A through I over a complete cycle. It will be seen that although particles are everywhere in motion (except at the nodes) the wave seems to be *standing still* as there is no forward motion of the crests as was the case with the wave of Fig. 9-3. For this reason such a resultant wave is called a *stationary wave* or *standing wave* in contrast to the *traveling* or *progressive* wave of Fig. 9-3.

Fig. 9-6. Resultant waves of Fig. 9-5 over a complete cycle.

An effective illustration of such wave motion can be obtained with a length of rope. If the far end of the rope is left free and the rope is jerked up and down at the near end, a wave of motion will be seen to travel down the rope. If the far end of the rope is then fixed solidly to a wall or other support and the near end of the rope is jerked vigorously, a wave will be seen to travel down the rope, be reflected at the boundary, and return to the sending end. The next step is to send a continuous wave motion down the rope by continuously moving the near end of the rope up and down. Interference between the incident and reflected wave will become evident, and if the frequency of the up-and-down motion is varied it will be found possible to produce standing waves on the rope. In this case the *nodes* will be at the end and half-wavelength distances from the end and the loops will come at the odd quarter-wavelength points. Whether a node or a loop appears at the reflecting boundary depends upon the *boundary conditions*. In the case of the rope fixed solidly at the far end, the boundary conditions were such that there could be no motion at this point — that is, it must be a nodal point.

An understanding of the ideas of wave motion, that is, traveling waves and standing waves, is of great assistance in the study of antennas and their feeder systems.

ELECTROMAGNETIC WAVES ON WIRES

When a pair of parallel wires is used to connect a battery or generator to a load there will be a voltage V between the wires and a current I flowing through them as shown in Fig. 9-7(a). Because of the voltage between the wires there will be an *electric field* about them which will everywhere have

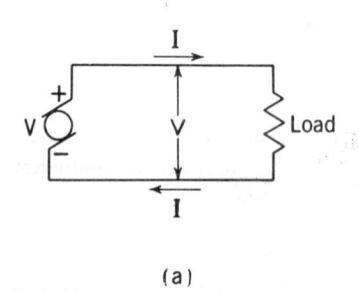

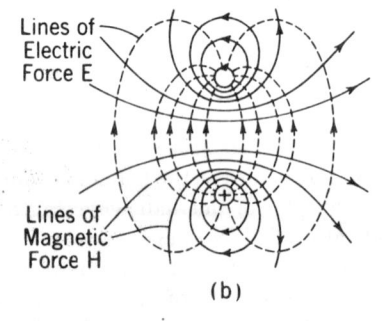

Fig. 9-7. Electric and magnetic fields about a pair of parallel wires.

the direction shown by the *lines of electric force*. The density of these lines in any region is proportional to the *strength of the electric field E* (volts per meter) in that region [Fig. 9-7(b)]. The electric field is strong near and between the wires and becomes weaker the further away from the wires one goes. It will be noticed that the lines terminate on the wires at right angles to their surfaces.

Because of the current I flowing through the wires there will be a magnetic field H surrounding them, as shown. The density of these lines is proportional to the *magnetic field strength* and their direction indicates the direction of the magnetic field. The lines of magnetic field strength H are everywhere at right angles to the lines of the electric field E.

The arrows on both the electric and magnetic field lines indicate the directions of the fields which correspond to the directions of voltage and current shown in Fig. 9-7(a). If the generator terminals were reversed so that the top line became negative and current flowed from right to left in the top line, the directions of both electric and magnetic fields would be the reverse of those indicated by the arrows. If the d-c generator were replaced by an a-c generator so that voltage and current on the wires were alternating, the electric and magnetic fields would also be alternating; that is, their positive and negative directions would change with the reversal of voltage and current. Moreover, at the instant that the voltage and current were zero the electric and magnetic fields would also be zero, so that with a-c operation the electric and magnetic fields are continually being built up and then collapsed back again to zero.

SOUND WAVES AND ELECTROMAGNETIC WAVES

It will be interesting and instructive to compare the transmission of sound waves in a speaking tube with electromagnetic waves along a pair of parallel wires. In Fig. 9-8 the sound tube is shown with an oscillating piston or diaphragm at the sending end S and a flexible diaphragm at the receiving end R. When the piston is moved back and forth quite slowly, the pressure is the same all along the tube and the diaphragm R moves in and out with the driving piston S. R and S are then *in phase*. Actually there is a small time interval between the maximum forward position of S, position 1, and the maximum forward position of R, position 3. The time interval is so small compared with the time for a complete oscillation (a period) that it can be neglected in this case. However, as the piston is speeded up to a higher frequency, the time for an oscillation becomes small and the time interval required for the pressure produced at S to reach the diaphragm at R becomes important. When the piston is vibrating fast enough it will be possible to have it back at position 2 by the time R has reached position 3. In this case R would be lagging S by one half cycle or 180°. R and S are then said to be 180° out of phase. As the frequency is increased still further, the driving piston may have moved from 1 to 2 and back to 1 again by the time the pressure has reached R to move it to position 3. In this case R and S would move in and out together, but R would lag S by a complete cycle or 360°. For the case shown in Fig. 9-8 the piston has made two complete movements in and out before the disturbance has

(a) Transmission of sound waves along a sound tube

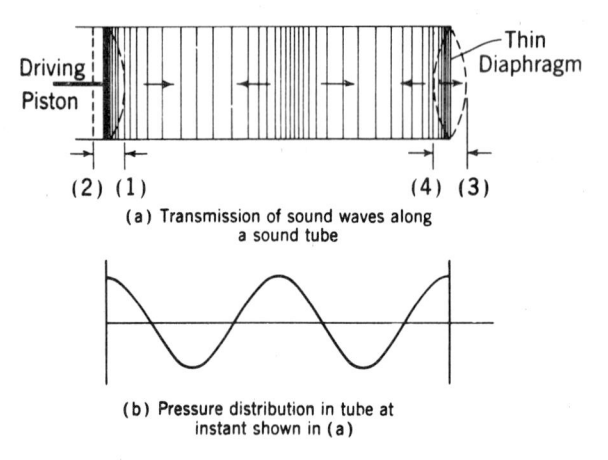

(b) Pressure distribution in tube at instant shown in (a)

Fig. 9-8. Transmission of sound waves along a tube, showing the instantaneous pressure distribution.

reached R, and R lags S by two complete cycles or 720°. Figure 9-8(b) shows the pressure which would exist along the tube at the instant that S and R are in their most forward positions. It is evident that *for this frequency* the tube is just two wavelengths long.

It is important to note the differences that exist between the slow and rapid operation of the driving piston. When the frequency of oscillation was low so that the corresponding wave length $\lambda = V/f$ was very large compared with the length of the tube, the pressure in the tube was everywhere the same and the direction of motion of the air particles was the same in all parts of the tube. However, when the frequency was increased so that the wavelength was reduced to the same order of magnitude as the length of the tube, these conditions changed. The pressure was different in different parts of the tube and the particle velocity was forward in some places and backward in others. Quite similar effects will be observed with electric waves.

Figure 9-9 illustrates the transmission of electric energy along a pair of parallel wires, or a transmission line as it is generally called in communication work. As in the case of sound in a tube, as long as the alternations of the generator are slow so that $\lambda = V/f$ is large compared with the length of the line, the voltage will be practically the same all along the line (resistance drop considered negligible) and the current will be in the same direction along the line; that is, left to right on the top wire and right to left on the bottom, or vice versa. This situation is just the familiar 60-cycle case, because the wavelength at a frequency of 60 cycles is

$$\lambda = \frac{300,000,000}{60} = 5,000,000 \text{ m} = 3,100 \text{ miles,}$$

and this is long compared with any transmission line that might be used.

However, when the frequency is increased to very high values such as are used in radio work, the corresponding wavelength becomes small and even short transmission lines may be several wavelengths long. Figure 9-9 shows the case for which the frequency has been increased until the wave

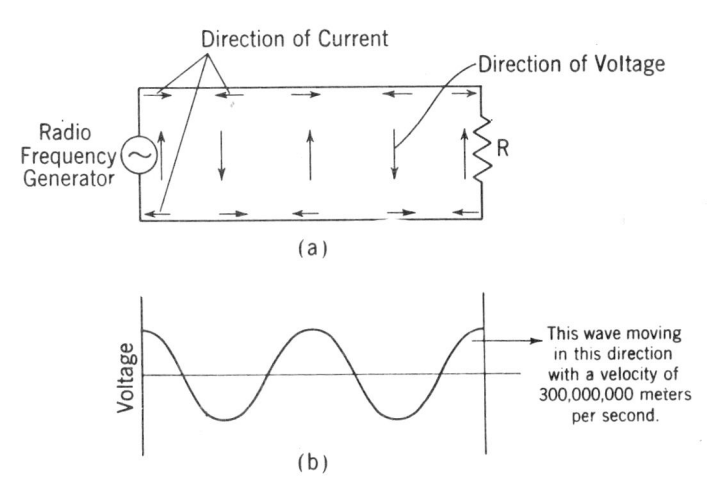

Fig. 9-9. Voltage and current on a transmission line at the instant of maximum generator voltage.

length is just one-half the length of the line or the line is two wavelengths long. The instantaneous direction of current is shown in Fig. 9-9(a) and the instantaneous voltage distribution is shown in Fig. 9-9(b). The similarity to particle velocity and pressure in the sound tube will be evident. No longer is the voltage between the wires constant along the line, for as a voltage maximum leaves the generator and starts on its journey down the line the generator voltage changes and goes through two complete alternations (in the case shown) before the first voltage maximum reaches the load at the end of the line. Similarly, the direction of current from the generator changes four times (two complete cycles) while a particular current crest is traveling down the line.

The idea of current leaving one terminal of the generator, going around the circuit and returning to the other terminal now becomes rather awkward, and it is better to think of it as positive current leaving one terminal and negative current leaving the other, and these two current mates traveling down the line together. Of course, when the alternator voltage and current reverse, the terminals from which the positive and negative currents leave will reverse and the currents along the line will be as shown in Fig. 9-9.

So far the transmission of energy from the generator to the load has been considered in terms of voltages and currents along the line. As was seen earlier, corresponding to these voltages and currents are electric and magnetic fields that surround the wires and travel down the line with their respective voltage and current mates. Actually, then, the energy is conveyed from generator to load *by these fields* through the space surrounding the wires, and the wires themselves merely *guide* the energy to its destination. This guidance corresponds to the acoustic case, where the sound tube serves merely as a guide, the sound energy being conveyed by the motion of the air in the tube. For the transmission of electromagnetic waves along wires it is immaterial whether one considers voltages and currents or electric and magnetic fields, but for the transmission of electromagnetic waves in space *where there are no wires*, consideration of the electric and magnetic fields is necessary.

STANDING WAVES

If the end of the sound tube of Fig. 9-8 is closed by a solid plate instead of the flexible diaphragm, the pressure wave is reflected from it instead of being absorbed. The reflected wave travels back down the tube, interfering with the incident wave and producing standing waves as in Figs. 9-5 and 9-6. Because the end of the pipe is solidly closed, the pressure can build up to a maximum and there is a pressure loop at the end and a pressure node one quarter of a wavelength from the end as in Fig. 9-6. The particle velocity, however, must be zero at the end because the solid plate is immovable, and the layer of air next to the plate must also have zero velocity. The distribution of particle velocity is therefore as shown in Fig. 9-10, with a node at the end and at half wavelength from the end. Figure 9-10(b) is an alternative representation of Fig. 9-10(a). The only difference between two

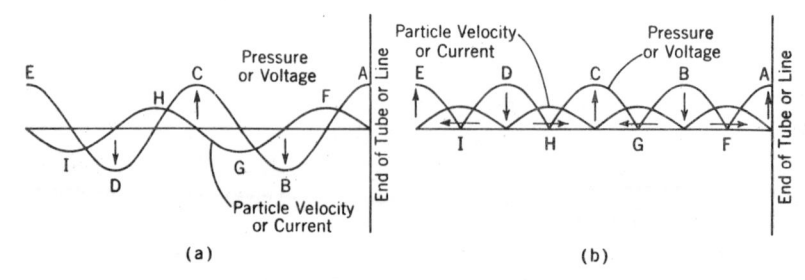

FIG. 9-10. (a) Standing waves of pressure and particle velocity in a closed-end tube, or voltage and current on an open-ended line; (b) alternative representation, not considering phase differences between adjacent loops.

adjacent crests such as B and C is that they have 180° phase difference, the one being negative when the other is positive and vice versa. Because a pressure meter (in the case of sound) and a voltmeter (in the case of electricity) cannot measure *phase*, the pressure or voltage indicated by such instruments would be as shown in Fig. 9-10(b).

If the transmission line of Fig. 9-9 is open at the end instead of being terminated in a resistance, standing waves of voltage and current are set up as in the case of the sound tube. With an open-ended line the current must be zero at the end but the voltage can go to a maximum, so that the voltage and current distribution are as indicated in Fig. 9-10. If, on the other hand, the line is shorted instead of being left open, reflection also occurs but in this case the voltage must be zero at the end (because of the short circuit) and the current can go to a maximum. For this case the voltage and current distributions shown in Fig. 9-10 are interchanged.

CHARACTERISTIC IMPEDANCE Z_0

In Chapter 7 mention was made of the desirability of terminating a transmission line in its characteristic impedance. The *characteristic impedance* of a transmission line is that value of terminating impedance which will produce no reflections. Its value depends upon the geometry of the line, that is upon the conductor size and spacing, and also upon the line losses and the dielectric material of the line. For a line that has low losses the characteristic impedance is nearly a pure resistance R_0, and is given by the formula

$$Z_0 = R_0 = \sqrt{\frac{L}{C}} \text{ ohms}$$

where L and C are the inductance and capacitance of a unit length of the line. For a parallel-wire line with an air dielectric the value of this characteristic resistance can be obtained from the formula

$$Z_0 = R_0 = 276 \log_{10} \frac{b}{a} \text{ ohms}$$

where a is the radius of the wires and b is their spacing, center-to-center.

Because there are no reflected waves when a line is terminated in Z_0 (read Z-zero) the effect is just the same as though the line were not terminated but were extended to infinity. Also, since the effect on a certain length of line of adding an infinitely long extension is the same as terminating it in Z_0, the *input* impedance of the infinitely long extension must be Z_0. Thus the characteristic impedance of a line is also its input impedance when the line is infinitely long, or when there are no reflected waves on it. Now any impedance is the ratio of a voltage to a current, and the input impedance of a line is the ratio of the input voltage to the input current. When a line is terminated in its characteristic impedance and has no reflections, the magnitudes of the voltage and current (as read by a voltmeter and ammeter, respectively) are almost constant along the length of the line if the line has small losses. When the line losses are not negligible, the voltage and current will both decrease exponentially along the line (which is terminated in Z_0) but their *ratio* will remain constant and equal to Z_0. Magnitudes of voltage and current along properly terminated lines having no loss and having

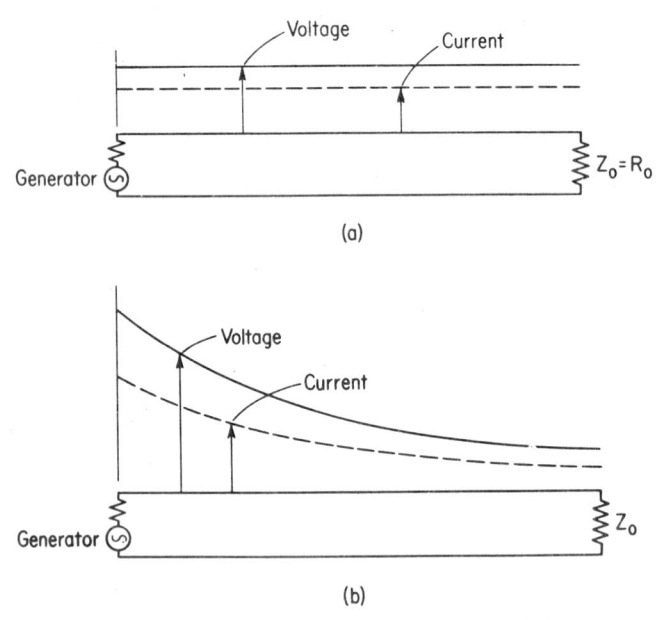

Fig. 9-11. Magnitude of voltage and current along a transmission line that is terminated in its characteristic impedance (a) when the line has negligible loss, (b) when the line has appreciable loss.

appreciable loss are shown in Fig. 9-11(a) and (b). Many transmission lines used at very high and ultra high frequencies are designed to have quite small losses so that the distributions shown in Fig. 9-11(a) are often a good approximation.

PHASE OF VOLTAGE AND CURRENT

Figure 9-11(a) shows only the *magnitudes* of voltage and current along a lossless line that is terminated in its characteristic impedance. The phases of both voltage and current change at a uniform rate along the line. That is, the voltage and current at successive points along the line reach their maxima at successive instants of time as was indicated in the sketch of wave motion (Fig. 9-1). Such a wave, the crest of which advances at a uniform rate along the line, is known as a *traveling wave*. Thus, a line terminated in its characteristic impedance always carries traveling waves of voltage and current. In contrast, when a line is terminated in either an open circuit or a short circuit, standing waves of the type illustrated in Fig. 9-5J result.

STANDING WAVE RATIO

When a line is terminated in some impedance other than Z_0, a short circuit, or an open circuit, the wave motion that results can be thought of as a combination of traveling and standing waves. The ratio of the maximum voltage (or current) to minimum voltage (or current) is a direct measure of the "mismatch" of the termination. This voltage ratio or current ratio (the two are always equal) is known as the *standing wave ratio*. When the line is properly terminated in its characteristic impedance, the standing wave ratio is unity — that is, there are no standing waves. For any other value of terminating resistance R, the standing wave ratio is given by

$$\frac{V_{max}}{V_{min}} = \frac{I_{max}}{I_{min}} = \frac{R}{R_0} \quad \text{(for } R > R_0\text{)};$$

$$\frac{V_{max}}{V_{min}} = \frac{I_{max}}{I_{min}} = \frac{R_0}{R} \quad \text{(for } R < R_0\text{)}.$$

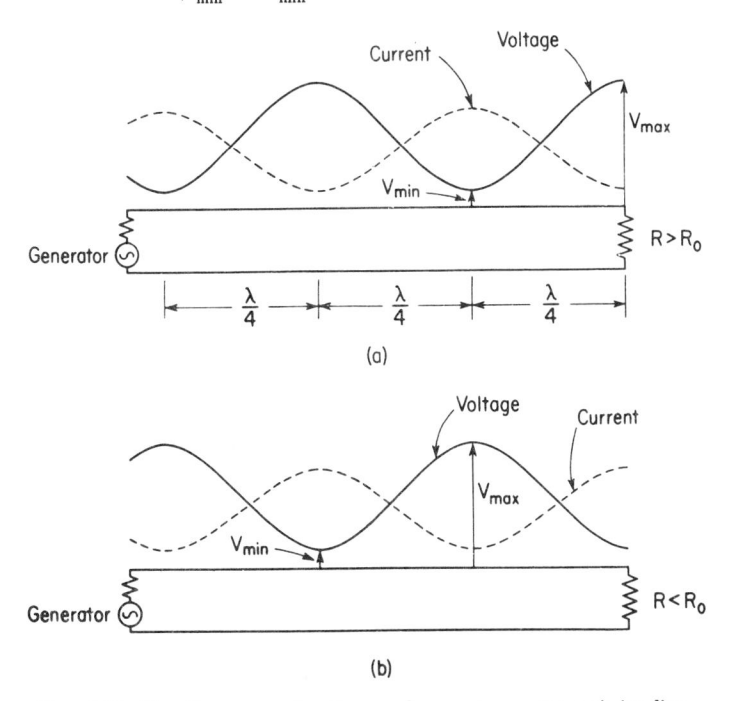

Fig. 9-12. Standing waves of voltage and current on a transmission line that is terminated in a resistance R which is (a) greater than the characteristic resistance R_0, (b) less than R_0.

A low-loss line having a resistive characteristic impedance has been assumed. For the first case where the resistance load R is greater than R_0 the voltage maxima will occur at the termination and at points that are multiples of half-wavelength from the end, as in the open-circuited case. The current *minima* will occur at these same points, and the current maxima will occur at odd multiples of a quarter wavelength from the termination. For the second case, where R is less than R_0, the situation is just reversed, with a current maximum at the termination and a voltage maximum a quarter of a wavelength from the termination. These voltage and current distributions are shown in Fig. 9-12.

It will be noted that the standing wave ratio becomes very large when the terminating resistance is either very small or very large compared with the characteristic resistance R_0 of the line.

If the line is terminated in an impedance that is not a pure resistance standing waves will still occur, but the maxima or minima will be displaced

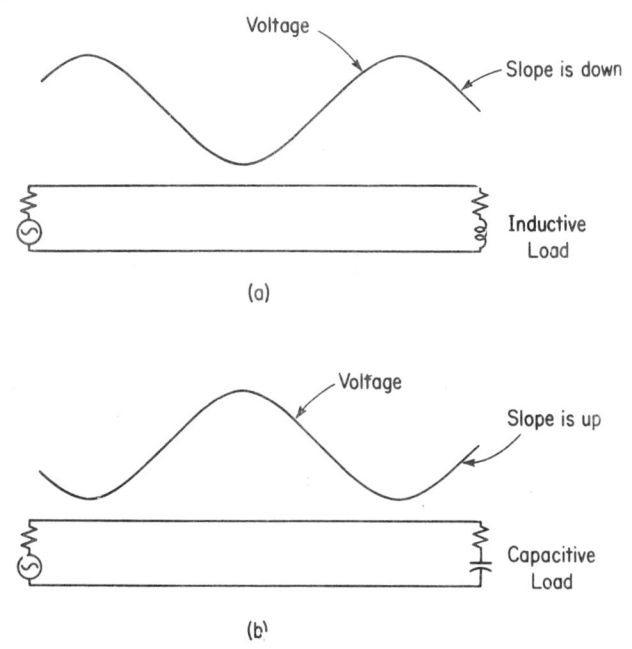

Fig. 9-13. Standing wave of voltage on a line which is terminated in a load having reactance.

along the line to left or right depending on the nature of the terminating impedance. If the terminating impedance has a reactance that is inductive the voltage curve will slope downward toward the termination, whereas if the terminating impedance is capacitive the voltage curve will slope up toward the termination. These results are indicated in Fig. 9-13.

COAXIAL TRANSMISSION LINES

In addition to the parallel-wire transmission lines which are used extensively at power and audio frequencies (power and telephone lines), transmission lines may also have the form of a *coaxial* or *concentric* cable. This form, indicated in Fig. 9-14 has several advantages and is very much used at the higher radio frequencies. Its chief advantage is that, when it is properly connected to its generator and load, through balance transformers if necessary, all the fields are confined to the region between the two conductors and so are completely isolated from the electric and

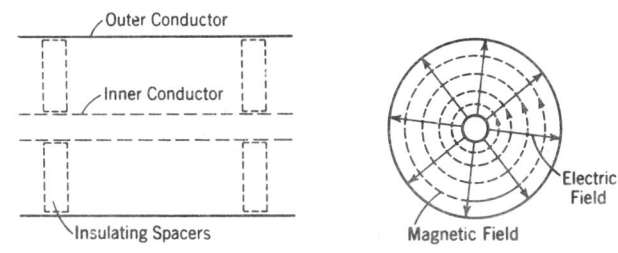

FIG. 9-14. Concentric transmission line showing electric and magnetic fields.

magnetic fields or voltages and currents of adjacent lines. Thus several cables carrying currents of the same or different frequencies may be grouped together without significant interaction. Coaxial cables are extensively used for conveying television programs from one location to another.

Figure 9-14 shows the distribution of electric and magnetic fields within a coaxial line. The direction of the electric field is radial, and that of the magnetic field is circular about the inner conductor. Equal and opposite currents flow in the inner and outer conductors. Because of the phenomenon known as "skin effect" at high radio frequencies, the current is concentrated near the surface of the conductor. In a coaxial cable the current is concentrated near the *outer* surface of the inner conductor and the *inner* surface of the outer conductor, adjacent to the associated electric and magnetic fields. The characteristic impedance of a coaxial line is given by the same formula as for parallel wire transmission lines,

$$Z_0 = \sqrt{\frac{L}{C}} \text{ ohms.}$$

For an air dielectric between the conductors, the characteristic impedance of coaxial lines can be computed from the formula

$$Z_0 = 138 \log_{10} \frac{b}{a} \text{ ohms}$$

where now a is the outside radius of the inner conductor and b is the inside radius of the outer conductor. When the space between conductors is filled with a solid dielectric having a dielectric constant greater than unity, the capacitance between the conductors is increased and the characteristic impedance of the line is decreased in proportion to the square root of the capacitance. Most flexible cables, both parallel-wire and coaxial, use a pliable solid dielectric and so have a characteristic impedance that is lower than the value for the corresponding air dielectric line. Flexible coaxial cables usually have a characteristic impedance of about 50 ohms, whereas for air dielectric coaxial cables it ranges between about 50 and 75 ohms. Air dielectric parallel-wire lines have a characteristic impedance of the order of 300 to 600 ohms. For the flexible solid dielectric parallel-wire lines (twin-lead ribbon) in common use at radio frequencies (particularly for TV reception) the characteristic impedance is often made to be 300, 150, or 75 ohms.

WAVEGUIDES

In addition to their transmission along parallel-wire and coaxial transmission lines, electromagnetic waves may also be guided along the inside of hollow conductors. Such guiding systems are called *waveguides*. Although the transmission lines just considered will guide waves of all frequencies, waveguides will transmit waves only when the frequencies are very high. Frequencies above the audible range are called *radio frequencies*, and for convenience in discussion the radio frequency spectrum has been divided up into the following bands, each of which covers a ten to one frequency range. Starting at the low-frequency end these bands are:

LF	(low frequencies)	30 to 300 kc
MF	(medium frequencies)	300 kc to 3 mc
HF	(high frequencies)	3 to 30 mc
VHF	(very high frequencies)	30 to 300 mc
UHF	(ultrahigh frequencies)	300 to 3,000 mc
SHF	(superhigh frequencies)	3,000 to 30,000 mc

For the waveguide type of transmission discussed in this section only the uhf and shf bands are ordinarily of interest.

Two common types of waveguides are the rectangular guide of Fig. 9-15(a), and the circular guide of Fig. 9-15(b). Wave transmission in wave-

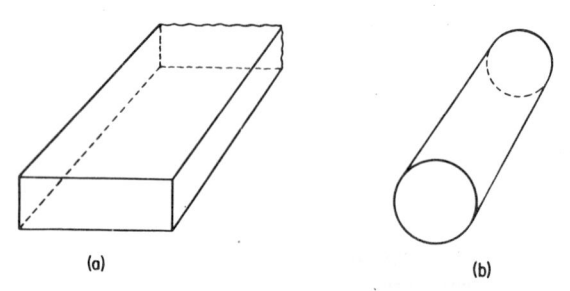

Fig. 9-15. Rectangular and circular waveguides.

guides is similar in many respects to that which occurs on an ordinary coaxial transmission line, but there are some differences as well. Some of these similarities and differences are illustrated in Fig. 9-16, which shows the field configuration within a coaxial line and that within a circular waveguide for one particular type of wave. In the case of the transmission line the lines of electric flux terminate on equal and opposite charges on the inner and outer conductors. In the waveguide of Fig. 9-16(b) there is no inner conductor and the flux lines turn back to terminate on charges of

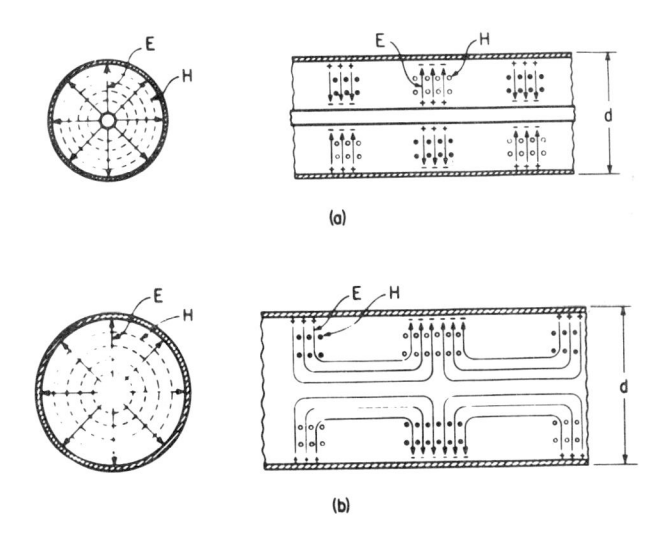

Fig. 9-16. Electric and magnetic field configurations (a) in a coaxial line, and (b) for one type of wave ($TM_{0,1}$) in a circular waveguide.

opposite sign on the same (outer) conductor as shown. The configurations of the magnetic flux lines are similar for the two cases. Although there is not much difference between the two types of transmission illustrated in Fig. 9-16 there is one difference that is of great practical importance. The transmission between the two conductors of the coaxial line will take place at any frequency right down to zero or dc. In other words the diameter d of the coaxial line can be any fraction of a wavelength. For the waveguide of Fig. 9-16(b) however, transmission will take place only when the diameter d is of the order of a half wavelength or greater. Because a wavelength at 60 cycles is about 3,000 miles it is apparent that the wave guide type of transmission is impractical at power, audio, and the lower radio frequencies. On the other hand at uhf and shf this type of transmission is quite feasible. For example, at 3,000 megacycles a wavelength is only 10 cm or about 4 in., and suitable waveguides have quite reasonable dimensions. Because of the absence of the inner conductor (where most of the loss occurs in a coaxial line) the efficiency of waveguide transmission is quite high and it becomes the preferred method at shf.

The type of wave illustrated in Fig. 9-16(b) is only one of a large number of wave types or *modes* that can be propagated within hollow conductors. The most commonly used modes for guides of rectangular and circular

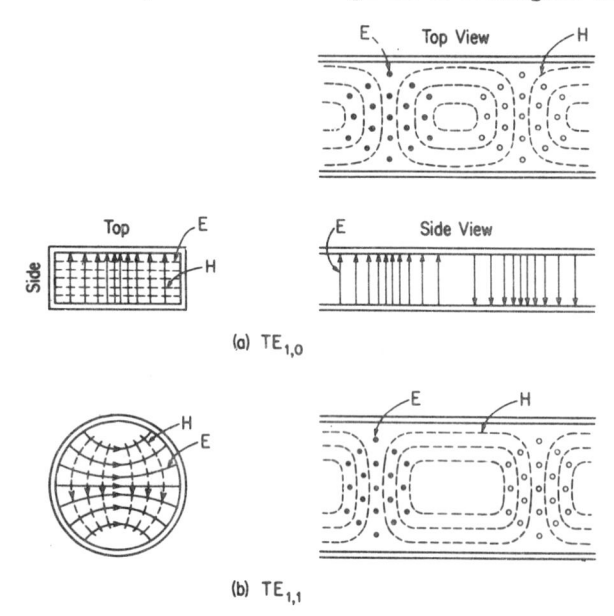

Fig. 9-17. (a) $TE_{1,0}$ mode in a rectangular waveguide. (b) $TE_{1,1}$ mode in a circular waveguide.

shapes are shown in Fig. 9-17. For the rectangular guide, the wave shown is the $TE_{1,0}$ mode, whereas for the circular guide the $TE_{1,1}$ wave is shown. The wave illustrated in the circular guide of Fig. 9-16(b) was the $TM_{0,1}$ mode.

The designations used for the various modes have the following significance. All modes which propagate within an ordinary single-conductor waveguide (as contrasted to a two-conductor transmission line) are either TM or TE waves. TM is the abbreviation for *transverse magnetic* and means that lines of magnetic field lie entirely in a transverse plane perpendicular to the axis as in Fig. 9-16(b), and therefore there is no axial component of H, as there is for example in Fig. 9-17(a) or (b). Similarly a TE wave is *transverse electric* with no component of electric field in the axial direction. Two examples of transverse electric fields are shown in Fig. 9-17. The number subscripts in the mode designations for rectangular guide indicate the number of half sine-wave loops of field variation that exist in the x and y directions respectively when the axis of the guide is taken as the z

direction. By convention the x direction is parallel to the broad face of the guide and the y direction is parallel to the narrow face. Therefore a $TE_{1,0}$ wave is a transverse electric wave which has one half sine-wave loop of variation in the x direction (along the broad face) and zero loops of variation in the y direction. That is, the field is constant or uniform along the y axis, but along the x axis the field varies sinusoidally from zero at one wall to a maximum at the center and down to zero at the other wall. For circular guides the notation is similar except that the numeral subscripts refer to the number of loops of field variation in the circumferential and radial directions respectively. Thus the $TM_{0,1}$ wave illustrated in Fig. 9-16(b) has no variation in the circumferential direction, but has one loop of variation in the radial direction.

WAVES IN THREE DIMENSIONS

If the sound tube of Fig. 9-8, the transmission line of Fig. 9-9, or the waveguides of Fig. 9-15 are left open, a certain amount of the energy which they are guiding escapes or is radiated from the end. The energy spreads out in three dimensions in space as illustrated in Fig. 9-18. Because the same amount of energy is spreading out through surfaces of ever-increasing

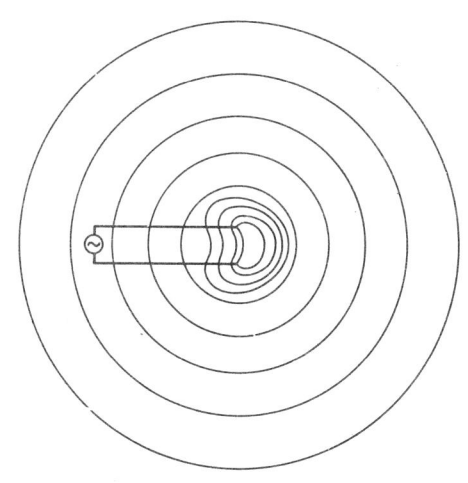

FIG. 9-18. Radiation from the open end of a transmission line, sound tube, or waveguide.

size, the energy that flows *through a given area* decreases as the distance from the open end increases. This statement means that the pressure or electric-field strength, as the case may be, decreases with increasing distance. It is found that both pressure and electric-field strength are inversely proportional to r, the distance from the source.

Because a radio wave released into space becomes weaker as the distance from the transmitter increases, it is important that as much energy as possible be radiated from the source. In the case of the sound tube, radiation from the end can be increased by opening the end out into a horn. The effect of this enlargement is to set a larger volume of air into motion and so increase the amount of energy radiated. Of course this extra radiated energy is furnished by the driving piston; the effect of the horn is to increase the back pressure on the piston and so make it do more work. In this respect the horn is like a transformer or impedance-matching device because it matches the low "impedance" of the air to the relatively high impedance of the driving mechanism.

(a) Sound Horns

(b) Electromagnetic Horns

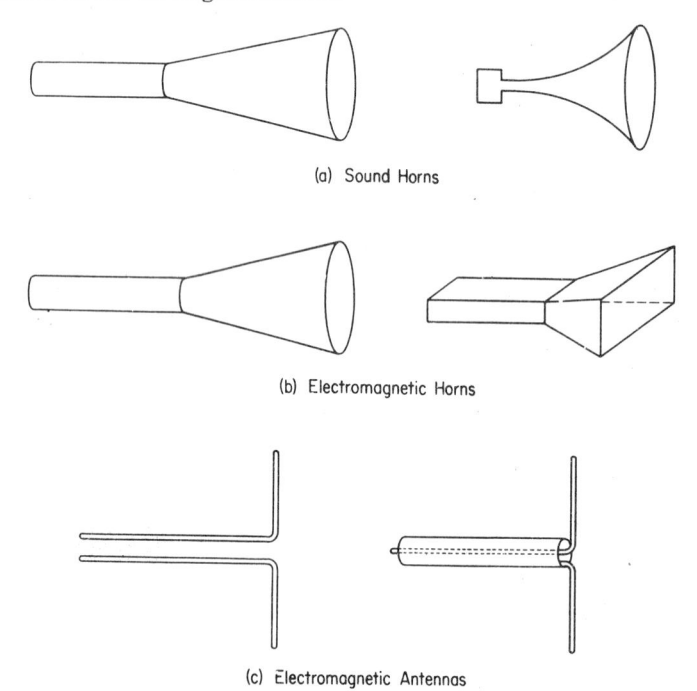

(c) Electromagnetic Antennas

FIG. 9-19. Acoustic and electromagnetic radiators.

Figure 9-19(a) shows two common forms of sound radiators. In the first of these the sound tube has been opened out into a simple horn or megaphone; the second is the well-known exponential horn.

In an analogous manner a circular or rectangular waveguide can be opened out into an *electromagnetic horn* as illustrated in Fig. 9-19(b). Here again the primary function of the horn is to match the "impedance of free space" to that of the waveguide, and so cause more of the energy to be radiated.

The end of a transmission line can also be opened out in a similar manner as indicated in Fig. 9-19(c), where the ends of the line have been turned back to form an *electro-magnetic radiator* or *antenna*. The current distribution and directions of the corresponding magnetic field are shown in Fig. 9-20(b) and (c); the voltage distribution and corresponding electric

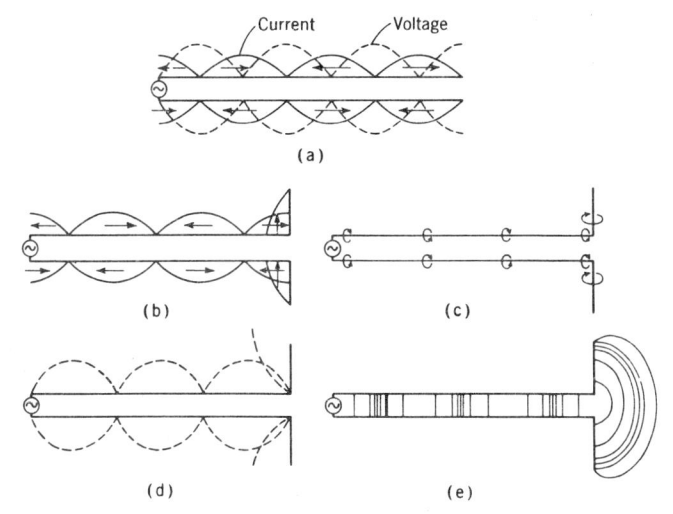

Fig. 9-20. End of a transmission line opened out to increase radiation: (a) line before opening out; (b) opened-out line showing current distribution; (c) corresponding magnetic field; (d) voltage distribution; (e) corresponding electric field.

field configuration are shown in Fig. 9-20(d) and (e). As in the case of the horn, the opening out of the line into an antenna increases the radiated energy, and this energy is taken from the generator. Here again, the antenna acts as an impedance-matching device to couple or match the transmission line to "free space."

DIMENSIONS OF AN ANTENNA

In order to radiate energy effectively and efficiently a radiator should have dimensions at least of the order of one-quarter to one-half wavelength. For the case of an acoustic radiator the area of the opening should be of the order of one-half wavelength square so that a reasonable volume (measured in cubic wavelengths) of air is set in motion with nearly the same phase. For an antenna, the requirement that its length be of the order of one-half wavelength can be explained in terms of Fig. 9-20(b). Upon examining the currents in the turned-back or radiating portions of the line it will be found that they are in the *same* direction; likewise, the corresponding magnetic and electric fields in space will be in the same direction and will reinforce each other to produce relatively strong fields. This reinforcement in turn means a large amount of energy radiated. The longer the antenna, up to one wavelength, the more of this current there will be in the radiating portion and the stronger will be the fields. When each of the turned-back portions is longer than one-half wavelength so that the total antenna length is greater than a wavelength, there will be included on the antenna portions of other current loops in which current is flowing in the reverse direction (Fig. 9-21). These reverse currents change the directions in which

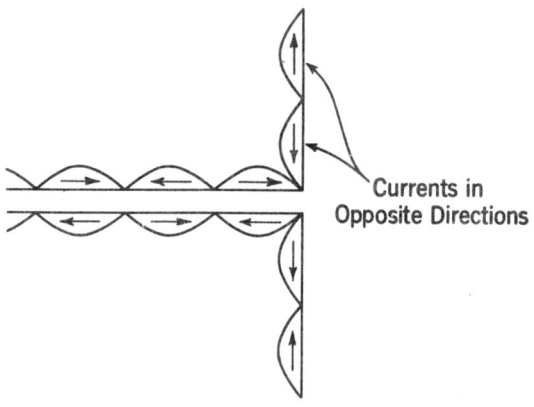

Fig. 9-21. Antenna longer than one wavelength, showing canceling current loops.

maximum energy is radiated so that the antenna becomes a *directional* antenna (see Chapter 19). As will be seen later, there are certain advantages to making an antenna just one-half wavelength long, and this is a very commonly used length.

The proper dimensioning of an acoustic radiator or loudspeaker is quite difficult when it is desired to radiate the entire audio frequency band, because whereas an 8 in. diameter horn or loudspeaker is satisfactory at 1,000 cycles, it is much too small at 100 cycles and too large at 10,000 cycles. Fortunately the proper dimensioning of an electro-magnetic radiator or antenna is usually much simpler. This is because, as will be explained in the next chapter, the audio frequencies are not transmitted directly, but instead are translated to a small band of radio frequencies centered about a "carrier" frequency. Hence if an antenna is dimensioned to be one-half wavelength long at the carrier frequency it will be nearly one-half wavelength long for each of the frequencies in the entire band being transmitted.

RADIATION RESISTANCE

When a transmission line is left open or is shorted at the end, the electromagnetic wave is reflected and sent back up the line so that there is no net flow of energy along the line (except a small amount to supply resistance losses). However, as the end of the line is opened out into a radiator some of the energy is radiated into space instead of being reflected back, and energy is taken from the generator. As far as the generator is concerned, then, the antenna is like a resistor absorbing power at the end of the line. The particular value of resistance which would absorb the same amount of power as the antenna is called the *radiation resistance* of the antenna. The power absorbed by a resistor is I^2R; the power absorbed, and therefore radiated, by an antenna is I^2R_a, where R_a is the radiation resistance of the antenna and I is the current flowing in at the feed point.

When the antenna is of the order of one-half wavelength long the power radiated for a given current is large and the radiation resistance has a reasonably large value. However if the antenna is short compared with a half wavelength, the power radiated for a given current will be very small and hence the radiation resistance will be quite low. Under these conditions, because there is always some ohmic loss in the antenna, the efficiency will also in general be quite low — hence the desirability of making the length of an antenna of the order of one-half wavelength.

MECHANISM OF RADIATION

So far, nothing has been said as to how the electromagnetic waves manage to leave the wires and travel on in free space where there are no

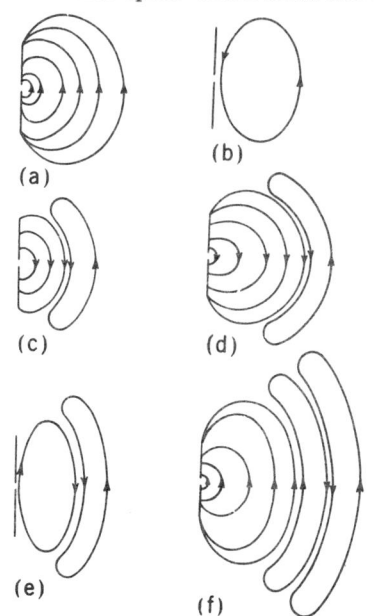

Fig. 9-22. Electric-field distribution about an antenna at successive instants over a complete electrical cycle.

wires and therefore no currents and no charges. The mechanism of radiation is complicated and any simple picture can but suggest the manner in which it occurs. Figure 9-22(a) indicates the electric-field distribution about an antenna at the instant that the voltage between the two halves of the antenna is a maximum. As the voltage goes to zero the charges upon which the lines of electric force end flow towards the center (the antenna current) and the lines contract or collapse back to zero. However, it takes a certain length of time for them to travel outward and back in again (they move with the speed of light); if, therefore, the voltage alternations on the antenna are very rapid, the voltage may have reached zero and be building up in the opposite direction before some of the outer-most lines have collapsed. This condition is illustrated in Fig. 9-22(b) for a particular line. This line then becomes detached and is pushed out into space by the new set of lines expanding outward with the increasing voltage on the second half cycle. This process continues and results in loops of electric force or electric strain moving out into space with the velocity of light. Along with these lines of electric force, and at right angles to them, is a magnetic field or lines of magnetic force also moving outward with the speed of light. These lines of magnetic force form circles about the antenna, the circles increasing in diameter as the field moves outward.

When electromagnetic waves are guided along wires, there exist both a moving electric field corresponding to the moving charges (electrons) and a moving magnetic field which is generally considered as being produced by the current or moving charges. In turn, the moving or collapsing magnetic field generates a voltage which corresponds to the electric field. It was Clerk Maxwell who pointed out that the intermediate step of charges and currents is not necessary and that a *changing electric field is equivalent to a current*, which he called displacement current, and so can produce a magnetic field directly. In a similar way a moving or changing magnetic field generates an electric field.

In this manner the propagation of electromagnetic fields in regions where there are no conductors is explained.

DIRECTION OF THE ELECTRIC AND MAGNETIC FIELDS

From the above it will be evident that the direction of the electric field at any point remote from an antenna is at right angles to a line from the antenna to the point and lies in the plane through the antenna and the point. This arrangement is shown in Fig. 9-23, where the plane through the antenna and the point is the plane of the paper. The direction of the magnetic field is perpendicular to this plane and therefore perpendicular to the electric field. The strengths of both the electric and magnetic fields decrease with distance, being inversely proportional to r, the distance from the antenna. Also the field is stronger at points along lines perpendicular to the antenna (such as long the line OA) than in other directions making a smaller angle with the antenna (such as along OP). This relation is expressed by saying that the field strength is proportional to $\sin \theta$ where θ is the angle between the antenna and the direction of radiation. When θ is zero, that is, when the point considered is directly above the antenna, the field strength is zero. The above relation is true for very short antennas,

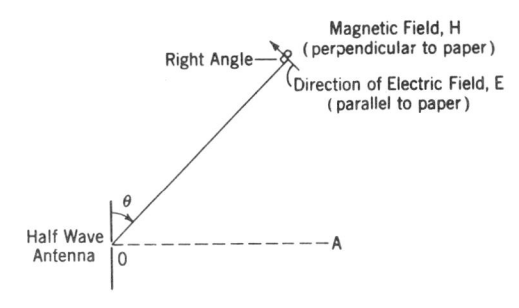

FIG. 9-23. Direction of the electric field about an antenna.

considerably less than a half wave long. Longer antennas become *directional* and may radiate much more at certain angles than at others.

THE RECEIVING ANTENNA

If the generator or transmitter at the end of the transmission line in Fig. 9-20 were replaced with a receiving set, the antenna would be a *receiving* antenna (Fig. 9-24). As the electromagnetic waves radiated from a distant transmitting antenna speed by, the electric field induces a voltage in the receiving antenna. This voltage alternates in direction at the same fre-

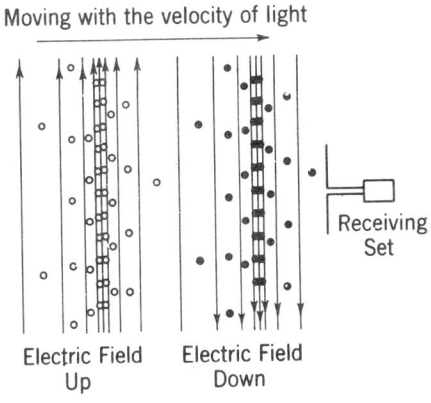

FIG. 9-24. Reception of an electromagnetic wave by a receiving antenna.

quency as the voltage in the transmitting antenna. It is also possible to consider that it is the lines of magnetic force moving past the antenna which generate a voltage in it. Either point of view is acceptable and both lead to the same answer. However, in calculating the voltage actually induced in the antenna, the voltage should be considered as the result of the electric field *or* the magnetic field, but not both, since these are but two different ways of viewing the same phenomenon.

QUESTIONS AND PROBLEMS

1. In connection with wave motion, define the following: frequency, wavelength, velocity of the wave, phase difference, nodes and loops, transverse and longitudinal waves.

2. Calculate the wavelength (a) of a 1,000-cycle sound wave; (b) of a 1,000-cycle radio wave. *Answer:* (a) 34.4 cm, (b) 200 km.

3. What are the differences between a standing wave and a traveling wave? Explain fully.

4. Show how the voltage (as measured by a voltmeter) varies along a line having zero resistance carrying (a) a traveling wave; (b) a standing wave.

5. Find the resultant of two waves, each of amplitude A, which differ in phase by 60°, by (a) plotting and adding instantaneous values, (b) vector addition. Repeat for phase differences of 90°, 120°, 180°, 270°.

6. Figure 9-5 shows a wave reflected at a boundary *without change of phase*. This could be a voltage wave reflected from the open end of a line or a current wave reflected from the short-circuited end of a line. Redraw Fig. 9-5 to show a wave reflected with 180° phase reversal. This would represent a voltage wave reflected from a shorted end or a current wave reflected from the open end of a line.

7. What is meant by the radiation resistance of an antenna?

8. Would one expect to receive much signal on a horizontal antenna when the transmitting antenna is vertical? Explain in terms of the directions of the electric and magnetic fields about a vertical transmitting antenna.

9. For antennas that are shorter than one-half wavelength the radiation resistance varies approximately as the square of the length of the antenna. If an antenna which is one-quarter wavelength long has a radiation resistance of 12 ohms, what would be its radiation resistance if it were only 1/20 of a wavelength long?

10. What would be the standing wave ratio on a lossless transmission line that is terminated in (a) an open circuit, (b) a short circuit?

CHAPTER 10

TRANSMISSION AND RECEPTION OF SIGNALS BY RADIO

ELECTROMAGNETIC WAVES AS MESSAGE CARRIERS

To convey a message from one point to another a message carrier is required. Electromagnetic waves, commonly called *radio waves*, are ideal message carriers for transmitting information through space, where other carriers would be impractical or uneconomical, as in the case of messages between moving vehicles and aircraft. The space medium for transmission is unlimited and efficient, the speed of transmission from the transmitting antenna to the receiving antenna being approximately 186,000 miles per second.

Throughout the radio industry the term "carrier" is universally understood in this sense, and the word is commonly used in the literature by writers. For services utilizing more than one carrier, as in television, separate identification is employed, such as the "picture" or "video" carrier, and the "sound" or "aural" carrier.

MODULATION

In Chapter 9 we discussed how electromagnetic waves are produced, and in Chapters 19 and 20 we shall deal with radio wave propagation and antennas. In this chapter we discuss the process of modulation, by which the message is attached to an electromagnetic-wave message carrier.

Radio waves transmitted continuously, with each cycle an exact duplicate of the other, will carry no information. To convey information, the carrier must have superimposed on it a sequence of changes which can be detected at the receiving point. In other words, the carrier must, in some way, be *modulated*.

AMPLITUDE MODULATION

One form of modulation is interruption of the carrier by turning it on and off at intervals corresponding to the dots and dashes of the telegraph code. In this case the amplitude of the carrier changes from zero to its full value. Changing the amplitude of the carrier is called *amplitude modulation*.

Any variation of the carrier amplitude from its otherwise continuously repetitive state may be detected by suitable circuits at a distant receiver. By causing the carrier amplitude to vary continuously, duplicating the complex variations of speech or music, exact detection and reproduction of sound is possible at a receiver.

Amplitude modulation is utilized in many services such as television, standard broadcasting, aids to navigation, telemetering, radar, and facsimile. Although the message content may vary widely for these, the mechanism of combining the message and the message carrier at the sending terminal is basically the same. In amplitude modulation the amplitude of the waves is varied, duplicating faithfully the fluctuations of the message. At the receiver these variations are detected, or *demodulated*. Although the more precise terms are *demodulation* for the process and *demodulator* for the device, the terms *detection* and *detector* are widely used.

Following reception and demodulation at the receiver (the process of removing the message from the carrier) the carrier is of no further use and is discarded.

In amplitude modulation the frequency of the carrier remains unchanged at all times, modulated or unmodulated, and only the amplitude is changed in accordance with the modulating signal. However, the frequency instead may be changed in accordance with a modulating signal, and this is called *frequency modulation*.

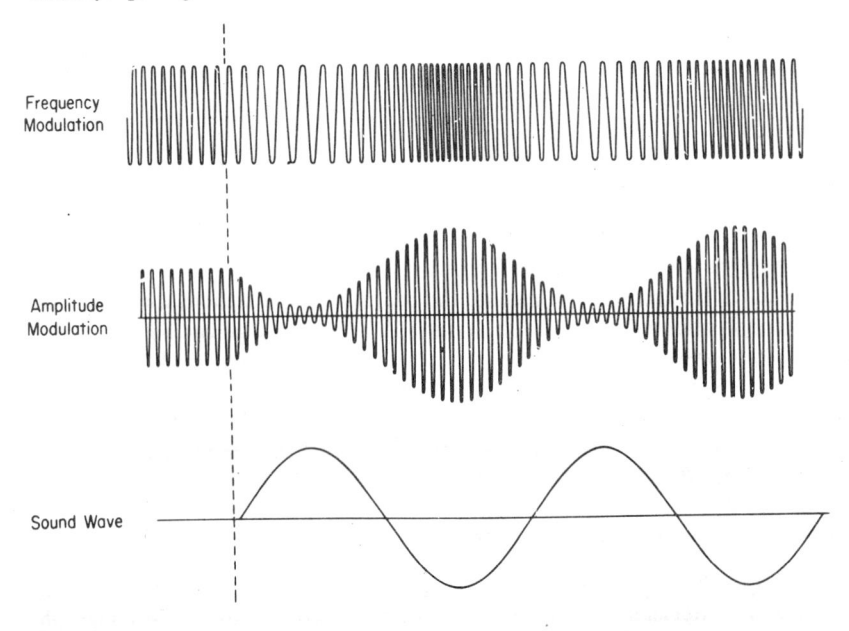

Fig. 10-1. Illustration of amplitude and frequency modulation.

Chapter 10   *Transmission and Reception of Signals by Radio*   311

FREQUENCY MODULATION

In a frequency-modulated system the amplitude of the carrier remains unchanged at all times, and the frequency is made to fluctuate symmetrically above and below the average value of the carrier. For example, a carrier frequency of 1,000 kc may be caused to swing between 925 and 1,075 kc, or any other amount chosen in accordance with the signal voltage. In frequency modulation the deviation of the carrier frequency from its average value is proportional to the instantaneous strength of the modulating signal.

At the bottom of Fig. 10-1 is shown an electrical wave at the frequency of sound, such as 1,000 cycles; in the center, a carrier which is amplitude-modulated by the 1,000-cycle wave; and at the top, a carrier which is frequency-modulated by the 1,000-cycle wave.

PHASE MODULATION

Other methods of modulation are employed which are related to amplitude or frequency modulation. For example, in phase modulation the phase of the carrier is alternately advanced and retarded in amounts proportional to the voltage of the message wave. As explained in Chapter 3, an alternating wave may be represented by a line or phasor. The carrier phasor velocity is accelerated and decelerated about the normal unmodulated phasor velocity. This action also produces frequency modulation; the higher the modulating frequency, the faster the phasor must move, because the time-per-message frequency cycle is shorter. The instantaneous frequency of the carrier is determined by the instantaneous carrier phasor velocity. Therefore, in phase modulation, frequency modulation is produced, but the amount is proportional to the modulating frequency as well as to the amplitude of the signal.

A phase modulator could be converted to a frequency modulator by using an audio input equalizer having attenuation inversely proportional to the modulating frequency. Certain types of frequency modulators employ phase modulation with the audio input equalizer added to produce the frequency-modulation characteristic. With a fixed sine-wave tone as modulator, it is normally impossible to tell at the receiving point whether frequency or phase modulation is in use.

PULSE MODULATION

Pulse modulation is definitely related to amplitude modulation and may also be related to frequency modulation. This system is employed in some types of microwave relay systems.

In this system the carrier may be modulated with short rectangular pulses. In the unmodulated condition these pulses may be evenly spaced and identical. During modulation they may be modified in amplitude, in width, or in relative position with respect to each other, or combinations of these changes may be employed. In general, pulse modulation is confined to microwave relay systems in which a multiplicity of messages are transmitted simultaneously and adequate bandwidth is available.

More details about modulation will be provided in following chapters. The various systems are described here as briefly as possible to accentuate the importance of modulation of the carrier by the message signals and the fact that there are a variety of methods, each having advantages in one or more of the radio communications systems.

THE COMPONENTS OF A RADIO COMMUNICATIONS SYSTEM

The essential components of a radio communications system are listed below and illustrated in Fig. 10-2.

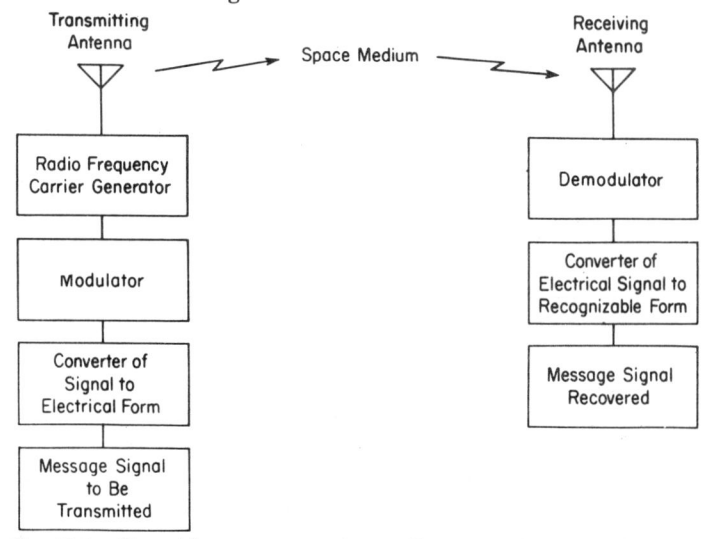

FIG. 10-2. Essential components of a radio communications system.

1. A device which converts the original signal into an electric current which is a replica of the message.
2. A generator of radio-frequency carrier current.
3. A modulator which attaches the message to the carrier.
4. An antenna which converts the modulated radio frequency currents to electromagnetic waves.
5. A medium through which the electromagnetic waves may travel to the receiving antenna.
6. A receiving antenna which converts the electromagnetic wave energy back to modulated radio frequency currents.
7. A demodulator which separates the signal from the carrier and restores it to its original form.

8. A loudspeaker or other device which converts the electrical signal to a form which can be used, for example, to sound energy.

The modulator which attaches the message to the carrier may take many forms. It may be a telegraph key, a continuously variable speech-facsimile-television type of amplitude control, or it may consist of other than amplitude-modulated types, so long as it accomplishes the purpose of attaching the message to the carrier by causing a change in its otherwise unvarying repetitive state of wave motion.

The transmitting antenna in its simplest form is an elevated conductor tuned to resonance at the carrier frequency, and to which the output of the transmitter is connected. Currents flowing in the antenna produce radio waves, as explained in Chapter 9. The radio waves flow freely through space from the transmitting antenna to the receiving antenna, attenuated only by intervening solid objects, water, or ionized gaseous layers in the atmosphere. The elevated receiving antenna intercepts the radio waves, causing a flow of currents to the receiver through interconnecting wires or transmission lines.

The demodulator, an essential part of a radio receiver, removes the message from the carrier and restores it to its original form. The simple demodulator for amplitude-modulated systems is a rectifier of radio frequency currents. Figure 10-1(b) shows an amplitude-modulated radio frequency current. The center line is the axis. Above the axis the radio frequency current flows in one direction, and below the axis it flows in the

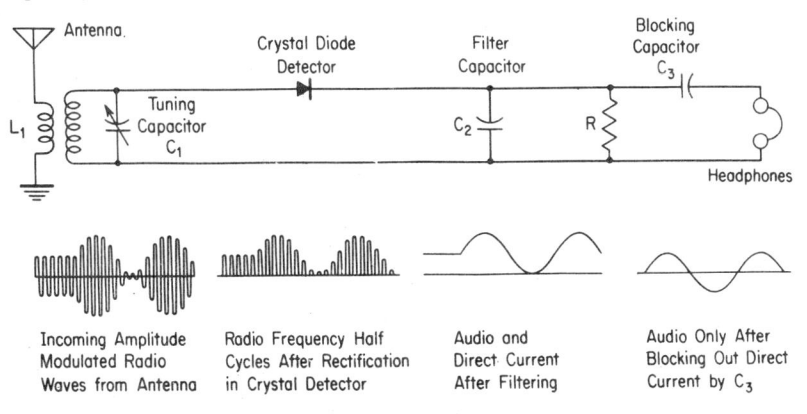

FIG. 10-3. Graphic illustration of demodulation process in crystal diode receiver.

opposite direction. Insertion of a device to stop the flow in one direction produces pulsating direct currents at carrier frequency, which vary in amplitude in accordance with the message. Now if these pulsating, unidirectional, amplitude-modulated currents are used to charge a capacitor in a circuit, as shown in Fig. 10-3, the carrier frequency pulsations charge the capacitor C_2 to a voltage proportional to the voltage of the envelope or maximum value of the electrical wave. This envelope is proportional, in an amplitude-modulated wave, to the strength of the signal at any instant.

TYPES OF SERVICE

Radio communication is employed in a great variety of services throughout the world, and these services may be grouped as described below. Radio communication other than microwave systems (whose transmission can be confined to narrow paths) is not used or authorized where communication by wire would serve as well; otherwise the medium would become overcrowded, and services in which radio communications are indispensable would be limited.

In the *fixed services* the transmitting and receiving stations remain in fixed geographical locations. Examples are transoceanic telegraphy, radiophoto, teleprinter, telephony, and facsimile. Primarily, the fixed services are used for communications over very long distances, usually over water.

In the *mobile services* communication is carried on between vehicles in motion or between vehicles and immobile stations built to communicate with them.

In *broadcasting* the message is sent from one point for general reception anywhere within a large geographical area.

FORMS OF TRANSMISSION

Information can be transmitted on the radio carrier in a variety of ways as described below.

Radiotelegraphy. The message may be in the form of dots and dashes of the telegraph code. The simplest means of creating the dots and dashes is to interrupt the carrier by a telegraph key, as shown in Fig. 10-4. The letter "A" is formed by sending first a dot and then a dash. The letter "N" is formed by sending first a dash and then a dot, the reverse of "A". In ship-to-ship and ship-to-shore radio telegraphy the key is ordinarily

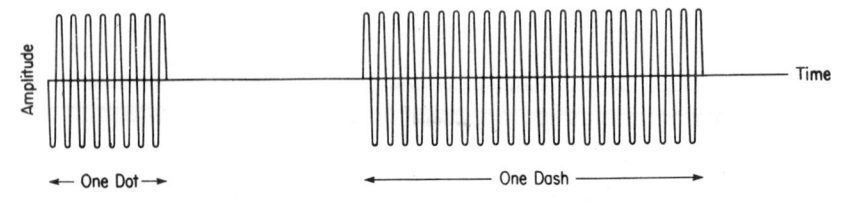

FIG. 10-4. Continuous wave telegraphy accomplished by keying carrier on and off as dots and dashes of telegraph code. Dot and dash shown form letter "A."

operated by hand at the rate of about 20 words per minute. In high-speed transoceanic circuits the messages are transcribed to paper tape, by perforating it, and the tape is fed into machines which key the transmitter at speeds of about 300 words per minute. At the receiver the process is reversed. This type of transmission is called *continuous wave telegraphy*.

In manual transmission and reception the receiver operator ordinarily listens to the incoming signals through earphones. The international distress signal ship frequency is 500 kc. When this continuous wave is demodulated, impulses are produced at the rate of 500,000 per second, far above the range of audibility. In order to produce a tone which the operator can hear, another wave of say 501,000 cycles is introduced in the receiver, prior to demodulation, by a local oscillator. The local incoming signal and the local oscillator combine to produce the difference frequency of 1,000 cycles to which the headphones respond. Continuous wave receivers contain these separate local oscillators as standard equipment, and means of frequency control are included to provide the preferred headphone tone frequency.

A carrier also may be modulated by an audio-frequency tone which is keyed on and off. This eliminates the need for the special tone-producing local oscillator at the receiver. However, this form of transmission, called *interrupted continuous wave telegraphy*, offers little advantage and is seldom used.

A more common form of high-speed telegraphy is called *carrier shift telegraphy*. As the name implies, the frequency of the carrier is shifted in accordance with the signal dots. Hence it is a form of frequency modulation.

Teleprinter. In services which handle a large amount of message traffic, manual telegraph transmission and reception is too slow and expensive, and so machines are used. The code may consist of combinations of perhaps five dots and spaces. The message is transcribed to a paper tape on a manually operated machine resembling a typewriter, which perforates the tape with holes corresponding to the dots and spaces. These tapes are then fed into a machine at high speed, and the dot perforations are converted into the electrical signals which key the transmitters. At the distant receiver the dot combinations automatically select the proper electric typewriter key in sequence, and the message is typed directly onto another length of paper tape.

Radiotelephony. For this method of transmission the information may be in the form of speech, music, or any other sound to which the human ear responds. The waveform of speech and music is very complex, including simultaneously a great many individual sounds of different frequencies and amplitudes. Differences in the combinations occur from instant to instant. For true reproduction of music all frequencies from about 30 to 15,000 cycles must be transmitted through the system equally well, without the addition of sounds not present originally. When such a complex wave modulates a carrier, the carrier amplitude envelope follows the complex wave motion instant by instant, fluctuating above and below its axis. Figure 10-1 shows in the center the carrier envelope, amplitude-modulated by a simple sine wave. Figure 10-5 shows how the envelope may appear over a very short interval of time when a complex wave is being transmitted. This is a 24-millisecond interval of an announcement going to the broadcast network from Radio City.

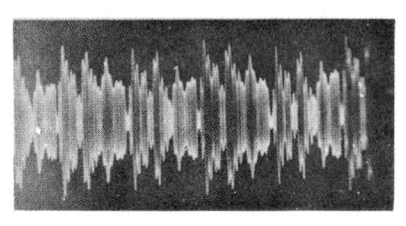

FIG. 10-5. A 24-millisecond interval of a speech modulated wave.

Radio facsimile. The information transmitted by facsimile may consist of printed matter, drawings, photographs, etc. The figure to be reproduced is scanned by a minute beam of light which is reflected to a photoelectric cell with a brightness proportional to the picture brightness. The photo-electric cell in turn produces a fluctuating electric current which modulates the carrier, and the brightness-modulated carrier conveys the signals to the receiver. There the picture may be reconstructed by exposing a sensitized piece of paper, element by element, in synchronism.

Television. The conventional television system involves two separate transmitters, one utilizing frequency-modulated radio telephony for the sound, the other conveying picture information, at a very fast rate, picture element by element except for periodic interruptions when synchronizing pulses are conveyed. A composite television wave form differs from a sound wave principally by its tremendously wide frequency range. For picture transmission much more information must be transmitted in a given time. In the United States the standard television frequency range is about 4,200,000 cycles. It is essential that television signals be transmitted through the system with unusually small tolerances with respect to uniform amplification, time delay, and amplitude linearity for all frequencies throughout this vast frequency range. The tolerances are the most rigorous of any form of transmission. Amplitude modulation is employed for broadcasting the picture signals.

Radar. Radar is used primarily to determine the presence and position of moving objects such as ships and airplanes. The radio carrier is transmitted in short bursts, or pulses, with relatively long intervals between, and is directed as a narrow beam in the desired direction. The carrier waves or pulses scan the skies sector by sector. When they strike a distant object

Chapter 10   *Transmission and Reception of Signals by Radio*   317

such as an airplane some of the wave energy is reflected back to a receiver near the transmitter, which measures the time traveled out and back. Since it is known that radio waves travel 186,300 miles per second, the distance of the reflecting object may be determined. Its direction in space from the radar station is found by observing the maximum reflected signal intensity of the radar beam. Conventionally, both distance and direction are displayed on a cathode-ray oscillograph tube at the receiving location.

Radio aids to navigation. Radio is utilized to a great extent in navigation. Ships or aircraft employ a variety of radio systems for blind landings, exact measurement of altitude, accurate determination of position in space and of distance and direction from other radio stations or aircraft, maintenance of a fixed course, automatic flight control, and the location of storm clouds.

Telemetering. Radio telemetering is the process of automatically conveying precise information for metering purposes from one point to another. Examples of radio telemetering systems are the remote indication of water levels in reservoirs, rivers, etc., and transmission of scientific data from rockets or aircraft under flight test.

Radio control. Radio control of distant mechanisms is very useful because of the flexibility, speed, and reliability of the technique. It may be employed in the operation of unmanned aircraft and rockets, control of public utility switching and distribution, operation of boats, control of telephone system switching, and control of motor and railroad vehicles.

SCOPE OF RADIO COMMUNICATIONS

About 3,000 standard broadcasting stations, 600 television stations, and 600 FM broadcasting stations are in operation in the United States. Radio and television reception has become a part of the daily lives of our citizens. Numerically, however, the approximately 4,000 familiar stations constitute only a very small fraction of the total number of FCC radio authorizations.

The Federal Communications Commission has approximately 1,900,000 listed radio authorizations, covering sixty different kinds of radio services. It has issued authorizations for radio transmitters in various important services in the numbers shown below:

Marine	70,000	Public safety	245,000
Aviation	60,000	Amateur	160,000
Industrial	325,000	Common carrier	2,000
Land transportation	307,000		

Over 1,500,000 radio operator's licenses and permits are outstanding, of which over one million are for commercial operation. By far the largest number of authorizations have been made to the Safety and Special Radio Services, which include the armed forces, public services of all kinds, and private businesses. Because of the scope and importance of these radio services, a more complete description of them is given in an appendix, which includes information regarding applications for authorization, regulatory matters, and so forth.

SIDEBAND FREQUENCIES

When the carrier is modulated for message transmission, frequencies higher and lower than the frequency of the carrier are produced. Since they are distributed over a finite portion of the spectrum on each side of the carrier frequency, they are called *side frequencies* and are referred to collectively as *sidebands*. These sidebands contain all of the message information, and without them no message could be transmitted.

An unmodulated carrier has continuously repetitive waves, all identical. When amplitude-modulated, these waves fluctuate above and below a

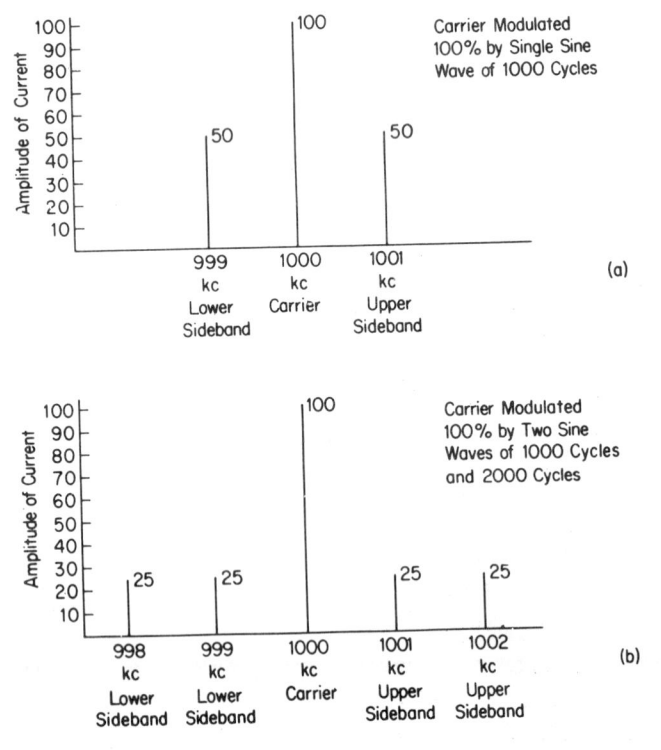

Fig. 10-6. Sideband and carrier frequencies and amplitude relationships for 1000-kc carrier modulated by (a) 1,000 cycles, and (b) 1,000 cycles and 2,000 cycles of equal voltage.

fixed axis. The average amplitude does not change. For example, a 1,000-kc carrier modulated by one kilocycle will have an average constant amplitude but will contain fluctuations at the rate of 1 kc, as shown in Fig. 10-1. During modulation additional power is transmitted, representing the message, which is all contained in new sidebands. These sidebands appear identically above and below the carrier and correspond in frequency to the sum and difference, respectively, of each component of the modulating signal and the carrier frequency. This is shown in Fig. 10-6(a). For a 1-kc modulating frequency the side frequencies are at 999 and 1,001 kc. With two modulating frequencies of 1 kc and 2 kc, the side frequencies become 998, 999, 1,001 and 1,002 kc, as shown in Fig. 10-6(b).

The sum of the current or voltage amplitudes of all sidebands cannot exceed the amplitude of the carrier. For example, if the transmitter were fully modulated by a single sine wave, the amplitude of the upper and lower sidebands each would be one half of the carrier amplitude. Since the power is proportional to the square of the voltage or current amplitude, the power in each sideband would be one fourth of the carrier power. Thus the total power of the two sidebands would be half of the carrier power.

If two equal sine waves were used for modulating, the amplitude of each of the four sidebands would be one quarter of the carrier amplitude, the power in each would be one sixteenth, and the total sideband power would then be four sixteenths or one quarter of the carrier power.

Figure 10-6(a) shows that the sidebands are produced both above and below the carrier, and that they are identical. It is not necessary in all cases to transmit both sidebands, since they are identical, and if one sideband were eliminated and one were left intact the message could still be recovered by demodulation at a receiver.

SINGLE AND VESTIGIAL SIDEBAND TRANSMISSION

After an amplitude modulated carrier is produced, one of the sidebands can be eliminated prior to transmitting. This is called *single sideband transmission*. It is used to improve system efficiency in some radio services such as transoceanic radio telephony. In other services, such as television, one complete set of sidebands is transmitted, those at frequencies above the carrier, while the other set of sidebands below the carrier is transmitted only in part, i.e., those closest to the carrier. This is called *vestigial sideband transmission*.

The use of single sideband transmission in transoceanic telephony improves the transmission efficiency and also reduces the frequency spread required to accommodate the sidebands. Vestigial sideband transmission in television is used only for the latter purpose. In each type of transmission the modulation originally produces both sets of sidebands. Thereafter, by technical processes, one set is wholly or partially eliminated before transmission.

Single or vestigial sideband systems are not used in all services because of added cost and technical complexity required of the transmitting and receiving apparatus, or for other reasons mentioned briefly later in this chapter.

SUPPRESSED CARRIER TRANSMISSION

It has been explained that if the carrier alone were transmitted, without any sidebands, no message would be present because the message content is in the accompanying sidebands. The sidebands are produced only when modulation of the carrier takes place. Since both the sidebands lower in frequency and those higher in frequency than the carrier contain the essential message information, for demodulation of the message at the receiving point only the carrier and one set of sidebands are essential. A carrier is necessary at the transmitter and also at the receiver, but it is not necessary to transmit and receive the same carrier. The carrier may be suppressed at the transmitter before transmission and replaced at the receiver by a locally generated carrier of the same frequency. So long as one set of sidebands is transmitted and received and is combined at the receiver with a carrier identical to the one suppressed, the requirements for message transmission and reception are satisfied. This technique is termed *single sideband suppressed carrier transmission.*

ADVANTAGES OF SINGLE-SIDEBAND SUPPRESSED CARRIER TRANSMISSION

The advantages of single-sideband suppressed carrier transmission compared with conventional double sideband and carrier transmission are given below.

1 The elimination of one set of sidebands and the carrier reduces the required transmission bandwidth to that of only the one set of sidebands. For a speech communication circuit of 3,000 cycles, the single-sideband suppressed carrier system requires a radio frequency bandwidth of only 3,000 cycles. In a conventional double-sideband and carrier system both sets of sidebands require a radio-frequency bandwidth of 6,000 cycles. The single-sideband system relaxes the tolerances on the transmitter bandwidth and conserves radio-frequency channel width.

2. If the modulation, single-sideband suppression, and carrier elimination are accomplished in the transmitter at low power, the final high-power amplifier will be required to amplify only the power in one set of sidebands. The overall system is at least four times as efficient as the conventional system in transmitting message intelligence power because, with the carrier eliminated, the sideband power level may be increased by that amount within the rating of the transmitter power tubes.

3. The absence of radio-frequency carrier waves eliminates carrier-frequency heterodyne beat-note interference at receiving points.

4. A substantial degree of privacy of communication is achieved, because reception of only a single set of sidebands produces no intelligible signal in a receiver until a local carrier of the correct frequency is generated and combined with it prior to demodulation.

5. The narrower frequency band of a single-sideband suppressed carrier signal makes it possible to narrow the receiver pass band and thus reduce the amount of noise and interference received.

Single-sideband suppressed carrier transmission is employed increasingly in commercial and military radiotelegraph, telephone, and teleprinter systems. It is primarily limited to such systems because local carriers produced at the receiver must be precisely controlled in frequency. For example, it is never used in broadcast transmission, because of the need for simple receivers for this service.

CONCEPT OF THE RADIO CHANNEL

It is apparent, then, that radio communication systems, to carry the message sideband frequencies as well as the carrier frequency, require an irreducible amount of frequency spread which depends upon the highest modulating frequency to be transmitted. In other words, a pathway of a definite width is required, commonly referred to as a *channel*.

A radio channel may be visualized as a private pathway from which all other traffic is excluded (within a certain geographic area). This pathway is composed of a segment of the frequency spectrum. As an example, television Channel 4 extends from 66 to 72 megacycles. Channel 3 extends from 60 to 66 megacycles. Each of the television channels is assigned exclusively for the use of one station over the largest geographical area it may reasonably serve, and all other uses are prohibited within specified distances to prevent interference.

The width of a channel is established only after a thorough study of the minimum width which could be used satisfactorily to transmit the carrier and sidebands required.

PRACTICAL CONSIDERATIONS OF CHANNEL WIDTH

All radio services must share the space medium, which is severely crowded. To fit as many channels as possible into it, each channel width must be limited. Since sidebands of the highest message frequency must be transmitted, the message bandwidth must be limited to the minimum necessary to meet the requirements. For that reason in the mobile services utilizing speech, the audio frequency bandwidth has been restricted to 3,000 cycles.

In standard broadcasting it would be desirable to transmit a bandwidth of 15,000 cycles and have a much wider channel than the 10 kc which is used. But to do so would reduce the number of channels which could be fitted into the block of frequencies available from 540 to 1,620 kc. Therefore, a system was adopted in which the fidelity of transmission is subordinated in favor of more channels and hence more stations.

FM broadcasting was created to provide a more perfect system of sound broadcasting, but to do so it was necessary to use a different part of the frequency spectrum, 88 to 108 megacycles, and adopt a much wider channel. In an FM system, as will be explained in Chapter 14, amplitude-modulated noise may be greatly reduced in the receiver if the carrier frequency deviation above and below the axis is made large during modulation. A good compromise between channel width and noise suppression is a carrier frequency deviation about five times as great as the highest modulating frequency. For a 15,000-cycle audio bandwidth the deviation then becomes 75,000 cycles each side of the axis. This spans 150,000 cycles. To accommodate sidebands and provide a small guard band, the channel width of 200 kc was adopted for FM broadcasting. A deviation of plus or minus 75 kc is thus considered to represent 100% modulation for these system standards.

In television, technical standards of transmission required unusual development and compromise to compress the channel width to 6 megacycles. A wider channel would have seriously restricted the number of channels and stations which could have been authorized.

In other services similar practical considerations have been and are being dealt with to assure maximum utilization of the frequency spectrum.

RADIO FREQUENCY SPECTRUM

It is possible to produce electromagnetic waves at any frequency from a fraction of a cycle up to thousands of millions of cycles. The most useful radio frequency spectrum extends from about 13,000 cycles to 10,000 megacycles. At frequencies below 13 kc a channel may become very wide compared with the carrier frequency, creating severe problems of circuitry and utilization, and the dimensions of efficient transmitting antennas become impractically great.

The propagation of radio waves varies widely at different frequencies throughout the radio spectrum, as will be explained in Chapter 19. This provides a choice of frequencies best suited for various types of communications. For example, frequencies between about 5 and 25 megacycles are suitable for communications over long distances, but poor for radar, television, and many other purposes. Frequencies from about 50 megacycles upwards for several hundred megacycles are best for television because the propagation of such radio waves is effective over about 75 miles, with comparative freedom from long-distance interference, and also because the wide channel of 6 megacycles may be satisfactorily transmitted and received without the prohibitive problems of band-pass circuitry of lower carrier frequencies.

Chapter 10   *Transmission and Reception of Signals by Radio*

Frequencies from 40 to 500 megacycles are readily adaptable for communication to and from mobile units because propagation is satisfactory without prohibitive long-range interference.

Frequencies from about 500 kc to 1,600 kc are adaptable for standard broadcasting because both local and rural service may be provided effectively without prohibitive interference from very long distances.

For many applications of radar, frequencies of thousands of megacycles are required so that moderate sized antennas of great directivity and power gain can facilitate propagation characteristics similar to light.

SHARED USE OF THE RADIO FREQUENCY SPECTRUM

There are tens of thousands of radio transmitters throughout the world. Each transmitter employs a carrier of some specific frequency. Obviously, if all of these transmitters were to operate on the same carrier frequency to which all receivers responded, the interference would be intolerable. Successful radio communication requires that the individual transmitters and receivers be able to operate with little interference from other transmitters. This is achieved by the use of different carrier frequencies and by designing receivers that select only the carrier frequency desired.

The demand for radio communications has been so great that it has been increasingly difficult to find space in the radio frequency spectrum in which to make carrier frequency assignments. To provide for a maximum number of stations it is essential that adjacent carriers be as close in frequency as possible to permit communication without interfering with receivers tuned to other transmitters.

The 5- to 25-megacycle range is most suitable for communication over distances of 1,000 to 12,000 miles, but these frequencies must serve all nations of the world and must be restricted to services for which they are uniquely suitable. Because of the extreme ranges at which interference may be caused, frequency sharing is limited. The services seeking or using this part of the spectrum include transoceanic telegraphy, international broadcasting, facsimile, teletypewriter, standard frequency transmission, military, toll telephony, and navigational aids.

The growth of standard broadcasting service has always been limited by the lack of frequency space, requiring the sharing of channels, the limitation of maximum transmitting power, minimum geographical separation of co-channel and adjacent channel stations, and the use of directional antennas to minimize interference.

REGULATION OF RADIO

Although all nations of the world have a sovereign right to the free use of all parts of the frequency spectrum, they have had to draw up international treaties and agreements to divide the radio frequency spectrum into segments assigned to specific services throughout the world. Thus, for example, the ships of all nations share certain frequencies, are able to standardize types of equipment, share common calling and traffic-handling frequencies, and sail any seas without interference from local services. Such agreement vastly simplifies the problems of handling traffic and distress calls, standardizes technical and operating details and procedures such as channel width, carrier frequency drift tolerances, and maximum powers, and vastly simplifies the problems of administration and regulation.

Each nation has its own regulatory body which exercises authority and control over radio in its own country. Acting within the scope of international treaties and agreements, it assigns frequencies, provides policing, establishes technical rules and standards, controls common carrier rates and practices, and protects and safeguards the public interest within its purview.

In the United States the Federal Communications Commission is responsible to the Congress for the regulation of all nongovernmental radio of the United States, Alaska, and Hawaii. The FCC determines who shall be authorized to operate a radio transmitter of any description for either domestic or international communication and has authority to impose penalties for misuse under powers delegated by the Congress.

With limited exceptions for certain types of low-power equipment, anyone desiring to operate a radio station must apply for a Construction Permit, giving a complete description of the facilities proposed and the nature of the operation intended. If two or more competing Construction Permit applications are received for a single channel, a public hearing may be held at which each applicant presents reasons why he should receive the permit. Upon receipt of a permit and satisfactory completion of construction, a license application must be filed, and upon authorization operation may be started.

The Federal Communications Commission participates in negotiating regional treaties and agreements for the mutual advantage of the United States and its neighboring countries. Such agreements are in effect or pending with Cuba, Mexico, Canada, and the Caribbean island countries covering the use of television and standard broadcasting channels.

RADIO TRANSMITTERS

Radio transmitters are basically simple, consisting of two principal parts, one for producing the carrier and one for modulating it. For simplicity the first may be described as the radio frequency section and the second as the modulating section.

Radio frequency section. The radio frequency section is fundamentally a radio frequency oscillator used to produce an oscillating current at the proper frequency. To maintain the carrier frequency at its specified value, such oscillators normally contain special design features such as a quartz crystal as a precise resonant circuit element, and operate at low power.

Chapter 10   *Transmission and Reception of Signals by Radio*   325

Amplifiers are needed, then, to increase the power to the level desired for satisfactory communication. Transmitters contain a required number of radio frequency amplifiers; there may be 5 for a 50-kw broadcast transmitter. Since the power delivered by a crystal oscillator may be of the order of one-half watt, there would need to be a power amplification of 100,000 times for normal operation.

Modulating section. Since the electrical message signal to be transmitted is originated at a very low level, amplifiers are required to facilitate modulation. In a conventional 50-kw broadcast transmitter, the input power level of the audio frequency signal would be approximately 10 mw. The power amplification required in the modulating section from the input terminals to the output of the last stage, the modulator, would be about 3,500,000. This would normally require about four audio amplifier stages.

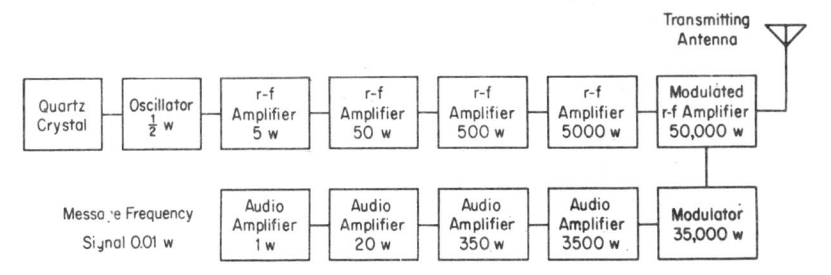

```
                                                              Transmitting
                                                              Antenna
         ┌─────────┐  ┌──────┐   ┌──────────┐  ┌──────────┐  ┌──────────┐  ┌──────────┐  ┌──────────────┐
Quartz   │Oscillator│  │ r-f  │   │   r-f    │  │   r-f    │  │   r-f    │  │ Modulated │
Crystal  │   ½ w    │──│Ampli-│───│Amplifier │──│Amplifier │──│Amplifier │──│r-f Amplifier│
         │          │  │fier  │   │   50 w   │  │  500 w   │  │  5000 w  │  │  50,000 w  │
         └─────────┘  │ 5 w  │   └──────────┘  └──────────┘  └──────────┘  └──────────────┘
                                                                                  │
Message Frequency    ┌──────┐   ┌──────────┐  ┌──────────┐  ┌──────────┐   ┌──────────┐
Signal 0.01 w        │Audio │   │  Audio   │  │  Audio   │  │  Audio   │   │ Modulator│
                     │Ampli-│───│Amplifier │──│Amplifier │──│Amplifier │───│ 35,000 w │
                     │fier  │   │   20 w   │  │  350 w   │  │  3500 w  │   └──────────┘
                     │ 1 w  │   └──────────┘  └──────────┘  └──────────┘
                     └──────┘
```

FIG. 10-7. Composition of 50-kw Class B modulated broadcast transmitter.

In a frequency-modulated transmitter the modulating section differs considerably from that of an amplitude modulated transmitter, but it performs the same function of amplifying the input audio frequency signal and producing the modulation of the carrier.

Figure 10-7 shows, in block diagram, the sections discussed.

SIMPLE RADIO RECEIVERS

The receiving antenna delivers to the input circuit of a receiver modulated radio-frequency transmissions, perhaps from a multiplicity of stations on different frequencies. Therefore, before demodulation the desired transmission must be selected. This is accomplished by tuned selective circuits. In complex receivers selectivity may be carried on in more than one section, but almost always it is included prior to demodulation.

Frequently the desired transmissions are so very weak when received that without amplification they would be very difficult or impossible to detect. Therefore, it is common practice to include radio frequency amplification before demodulation. In practice the selectivity and radio frequency amplification are obtained jointly in radio frequency amplifiers with tuned input and output circuits.

After the received modulated carrier signal has been segregated by the tuned circuits and amplified to a usable level, it is demodulated. The demodulated signal usually is at too low a level to achieve its ultimate purpose, such as operating a loudspeaker, a facsimile printer, a television picture tube, etc., so additional amplification is applied. It is conventional to include a volume control for precise power adjustment in combination with this amplification.

The "tuned r-f" receiver, described and illustrated in Fig. 10-8, is the simplest basic type. For some applications it is satisfactory; however, radio-frequency amplifiers provide only limited amplification at the higher radio frequencies. With them the carefully controlled and shaped selectivity

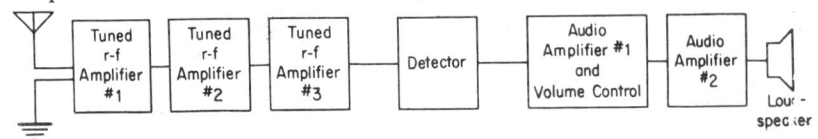

Fig. 10-8. Tuned radio-frequency type receiver. Tuning controls for r-f amplifiers 1, 2, 3 conventionally combined mechanically (but not electrically) for convenient single-dial tuning control. This type receiver not widely used.

characteristics required on television channels could not be attained economically or practicably, and over a wide range of frequencies the selectivity characteristics of the tuneable circuits vary too widely.

To overcome these disadvantages a more complex receiver is very widely used, called the *superheterodyne*.

SUPERHETERODYNE RECEIVERS

The basically simple superheterodyne receiver circuits are employed in almost every form of receiver used in broadcasting, television and communications. As indicated previously, the tuned r-f type of receiver has certain disadvantages. At the higher frequencies, tubes do not function as well as at lower frequencies. Also, when tuned selective circuits must vary to respond to any selected channel in a wide band of frequencies, the selectivity changes over the tuneable range. In addition, where it is necessary to carefully shape the bandpass characteristics to give uniform response over a wide range of sideband frequencies, as in television, it is difficult and often impossible to do so and at the same time obtain sharp cutoff at the sides of the band. These disadvantages could be avoided if the carrier frequency and its sidebands were converted to a lower frequency at which tubes and circuits perform better.

Use of a so-called intermediate radio frequency (i-f) amplifier which operates in a lower frequency range can substantially solve the problems of bandpass selectivity and amplification. This i-f amplifier may be perma-

Chapter 10 *Transmission and Reception of Signals by Radio* 327

nently tuned to amplify a fixed band of frequencies allocated to a particular service. It is then necessary to convert any incoming desired transmission from its original carrier and sideband frequencies to match the frequency band of the i-f amplifier. This is accomplished easily by the heterodyning process which will be explained more fully in the next chapter. A tuneable oscillator is added to the receiver to produce the frequency band to which the i-f amplifier responds. Also, a mixing tube, often called the *first detector*, must be added. Conventionally the detector which demodulates the output of the i-f amplifier is called the *second detector*. Because the i-f amplifier functions at a frequency intermediate between the incoming carrier and the final message frequency, it is called an *intermediate frequency amplifier*.

The important parts of such a receiver designed for broadcast reception are shown in Fig. 10-9. A 1,000-kc carrier is indicated as being received. From a local oscillator built into the receiver a second frequency is introduced into the input circuit together with the incoming carrier. This second

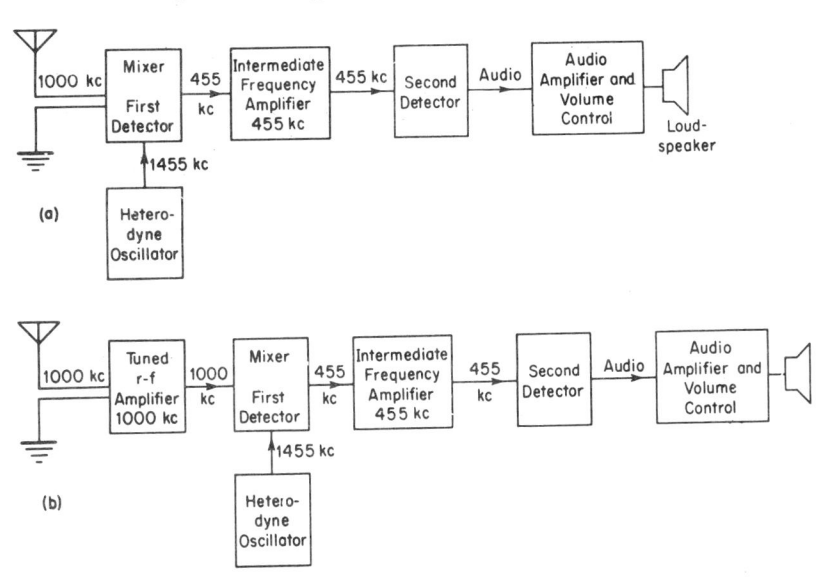

Fig. 10-9. (a) Superheterodyne receiver with antenna circuit directly connected to mixer (first detector); (b) superheterodyne receiver with tuned r-f stage preceding mixer (first detector).

frequency is made to differ from the incoming carrier frequency by 455 kc. The first detector, called a mixer tube, mixes these two frequencies and demodulates the combination to produce a new carrier of 455 kc, with the same sidebands as the original. In the 455-kc i-f amplifier, amplification, selectivity, and band shaping are accomplished. Thereafter demodulation is accomplished in a conventional manner.

When a carrier of other than 1,000 kc is desired, the local oscillator frequency is adjusted so that the difference between carrier and oscillator again produces 455 kc for the intermediate frequency amplifier.

Advantages of extra sensitivity and selectivity in a superheterodyne receiver may be gained by using radio-frequency amplification before mixing, and such a receiver is shown in Fig. 10-9(b).

PROBLEMS OF RECEIVER SELECTIVITY

If the bandpass circuits of receivers were ideal, they would accept without attenuation all frequencies in the channel desired, and reject with infinite attenuation all frequencies outside of that channel. Since this ideal is not approachable at a reasonable cost, adjacent channel interference is highly possible. Knowing this, the radio regulatory bodies have, in some cases, had to standardize on channel widths containing a "guard band" to increase the frequency separation between stations, and thus minimize interference. In other cases, it has been necessary to assign only alternate channels in an area, such as for FM broadcasting, leaving the intervening channels unused. Figure 10-10(a) shows the selectivity characteristic of a conspicuously poor broadcast receiver. Figure 10-10(b) corre-

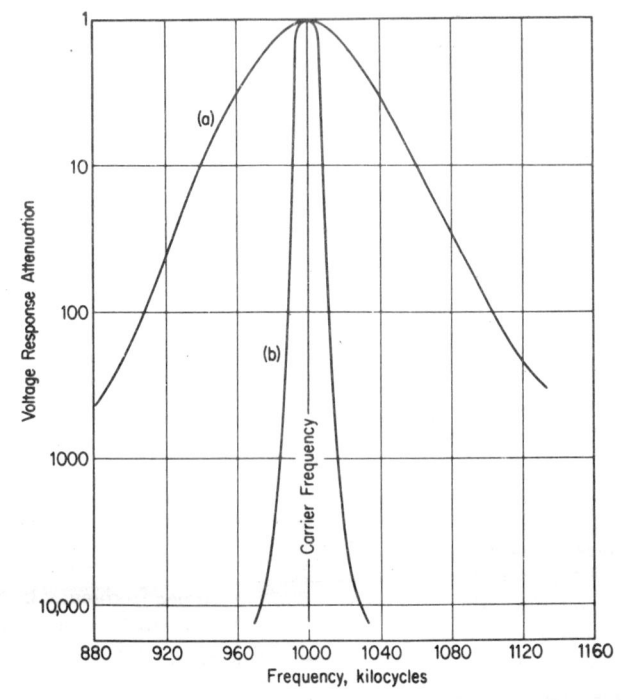

FIG. 10-10. Selectivity curve of broadcast receivers. (a) Simple tuned r-f; (b) modern superheterodyne.

sponds to a good modern broadcast receiver, in which a compromise was made between selectivity and band-pass characteristics. Chapter 13 will deal further with selectivity.

As technology improves, receiver selectivity is being increased and channel widths in some cases are being narrowed correspondingly. In one service, channel widths are being cut in half. By thus doubling the number of channels available for the service, the efficiency of spectrum utilization is being increased and the public need for additional stations may be more adequately met.

Adjacent channel response of different selectivities is shown in Fig. 10-10. Receiver (a) would be very vulnerable to interference and useless in most American cities having local broadcasting channels separated by 40 or

ROCKY POINT, L.I.
TRANSMITTING PLANT

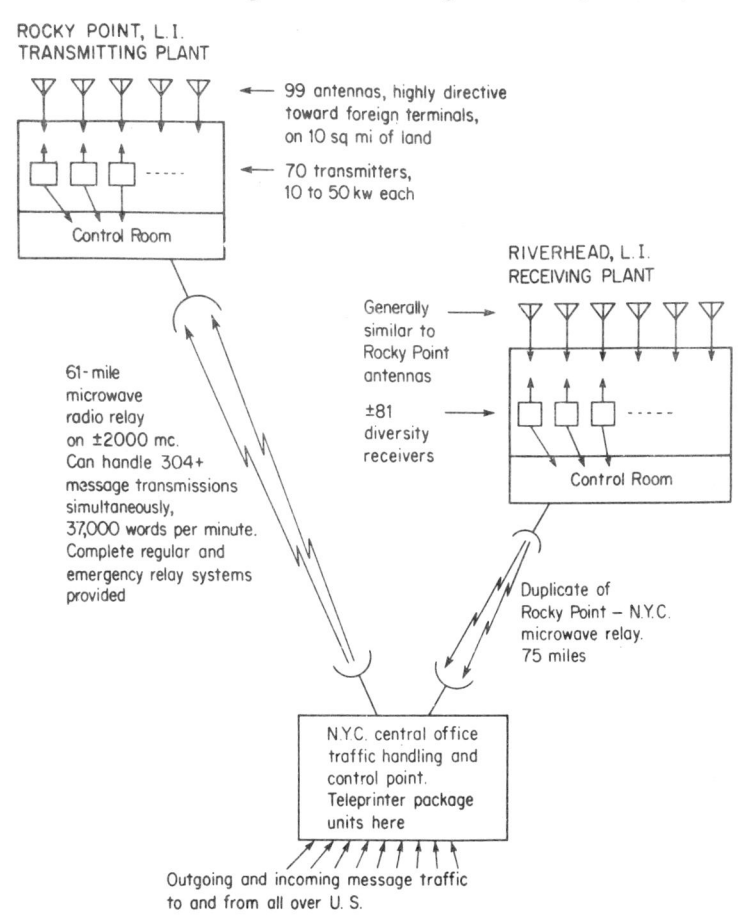

← 99 antennas, highly directive toward foreign terminals, on 10 sq mi of land

← 70 transmitters, 10 to 50 kw each

Control Room

RIVERHEAD, L.I.
RECEIVING PLANT

Generally similar to Rocky Point antennas

±81 diversity receivers

Control Room

61-mile microwave radio relay on ±2000 mc. Can handle 304+ message transmissions simultaneously, 37,000 words per minute. Complete regular and emergency relay systems provided

Duplicate of Rocky Point – N.Y.C. microwave relay. 75 miles

N.Y.C. central office traffic handling and control point. Teleprinter package units here

Outgoing and incoming message traffic to and from all over U.S.

FIG. 10-11. Layout of modern point-to-point radio station.

50 kc. As many as seven stations could be received simultaneously. The selectivity curve of receiver (b), representing about the best produced, would be able to select local stations within 30 kc. However, because of variations the regulatory body does not make local assignments closer than 40 to 50 kc.

EXAMPLE OF A MODERN OVERSEAS RADIO COMMUNICATIONS SYSTEM

A modern overseas communications system is a highly developed and mechanized complex of widely separated elements but nevertheless is basically simple. For reception of messages from overseas a single receiving plant requires scores of receiving antennas and as many as 70 complete receiving sets. A large building and many hundreds of acres of open flat land are also required. In addition, the receiving plant must be geographically separated from the transmitting plant to avoid interference from the very powerful transmissions, and must be isolated from heavily populated areas where radio interference is heavy. Such plants are connected to a central office in a large city such as New York by microwave relay over which incoming or outgoing messages are relayed by direct electrical continuity. (See Fig. 10-11.)

FIG. 10-12. Teleprinter used in overseas message transmission and reception. Tape contains perforations and type. (Courtesy RCA Communications, Inc.)

Chapter 10   *Transmission and Reception of Signals by Radio*   331

Messages filed anywhere in the United States for transmission to Europe or Latin America are forwarded by wire line to a central radio office in New York City. They are transcribed on paper tape by perforations in a machine resembling a typewriter. An example of a transcribed message on perforated tape is shown on Fig. 10-12 which shows a teleprinter unit used in transmission and reception. The perforations on the tape are in a code resembling the telegraph code, each letter, number, etc. being perforated separately in sequence. The tape, when run through the teleprinter, causes corresponding electrical currents to flow and these key the transmitter. Many messages may be transcribed on a single tape. When it is full it is ready for insertion in the transmitting system. A number of tapes may be inserted side by side in parallel "channels." Each message is automatically coded with a serial number when special perforations indicate the end of the message. When one tape has been transmitted the system automatically starts the "ready" tape in the adjacent channel. In transmission the coded characters actuate an adjacent teleprinter which reproduces the message being transmitted as a record and to indicate that the machine is functioning properly. The message signals are transmitted to a vast high-power transmitting plant on Long Island about 75 miles distant. In New York a large number of central office "package set" transmitter units are in

Fig. 10-13. Teleprinter package set for transmission, reception, monitoring, and recording. (Courtesy RCA Communications, Inc.)

use simultaneously, each feeding its own distant transmitter, antenna, and radio channel. Figure 10-13 shows a section of the central office in New York City.

Each vertical group of compartments contains a "package unit." At the top shelf is a typing reperforator which automatically monitors and records all overseas traffic transmitted by the "package set." At the middle shelf is a similar typing reperforator which constitutes the receiving end of the circuit. The lower shelf contains transmitter-distributors.

At the 75-mile distant transmitting plant the messages arriving from New York terminate in a master control room shown in Fig. 10-14. Here each message channel from New York is "patched" to the proper trans-

FIG. 10-14. Master control room at overseas radio transmitting plant, Rocky Point, Long Island. (Courtesy RCA Communications, Inc.)

mitter for the country of destination. There are more than 50 high-power transmitters in the adjoining halls; Figure 10-15 shows a 50-kw transmitter.

In the hundreds of acres of land surrounding the transmitter building there are scores of large highly directional transmitting antennas, each "aimed" toward some overseas receiving station. Another switching system is provided so that the RF outputs of the individual transmitters may be switched to the inputs of the individual antennas desired in setting up a circuit.

Chapter 10   *Transmission and Reception of Signals by Radio*   333

At the receiving plant special receiving sets are employed, and these will be described in Chapter 13.

Traffic is handled in teleprinter transmission at a speed of about 60 words per minute. Although somewhat slower than conventional code transmission, the handling is minimized and the overall cost is substantially

FIG. 10-15. Fifty-kilowatt radio transmitter at overseas transmitting plant, Rocky Point, Long Island. (Courtesy RCA Communications, Inc.)

lower. This system also handles radio photo, facsimile, and audio frequency traffic. Audio traffic includes program transmission service for which a portion of the channels are available for relaying broadcast programs to and from foreign countries. An additional service is provided by which teletypewriter services may be established directly from the office of a subscriber to another office in a foreign country, passing by direct electrical patched connection through all intermediate wire lines, central offices, transmitting and receiving plants, etc.

The system described, the East Coast facilities of RCA, is one of the more elaborate of several in operation. For handling traffic to the Orient and the Pacific similar plants are located in California. Such systems also are operated by other communications companies and by the U. S. Government agencies, in the United States and various other parts of the world. Overseas toll telephony is handled over similar facilities of the American

334   Transmission and Reception of Signals by Radio   Chapter 10

Telephone and Telegraph Company. The radio relay system which connects New York, Riverhead, and Rocky Point is described in Chapter 21.

EXAMPLE OF A POLICE RADIO SYSTEM

Police radio systems are used extensively in the United States. A typical system is shown diagrammatically in Fig. 10-16. The base station

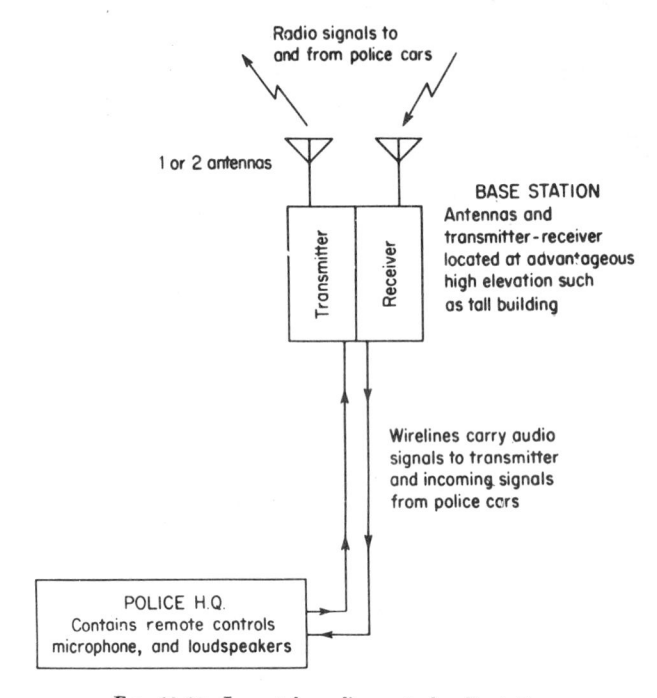

FIG. 10-16. Layout for police central radio station.

transmitter-receiver is illustrated in Fig. 10-17. The transmitters and receivers employ frequency modulation. The base transmitter has 75 w of r-f output and normally operates on a single frequency for all transmissions. The companion receivers and transmitters are mounted in the same cabinets for the base station and also for the patrol cars. Further descriptions of the apparatus are given in Chapters 12 and 13.

The base station antennas are located on a high building or tower so that the radio waves to and from the cars are not impeded. The transmitter-receiver unit is located nearby so that it may be connected efficiently by short lengths of r-f transmission line not exceeding a few hundred feet. The base transmitter-receiver unit is connected to the control point at police headquarters by telephone lines. The control point at police headquarters often is located in the same building but in other cases may be as much as

Chapter 10 *Transmission and Reception of Signals by Radio* 335

Fig. 10-17. Base station receiver: 75-watt transmitter used in land mobile service such as police. (Courtesy Du Mont Laboratories, Inc.)

several miles distant. No manual attendance is required at the base-station; operation is by remote control.

Figure 10-18 shows a typical police headquarters control unit with microphones, loudspeaker, and remote-control box. Figure 10-19 shows a patrol car transmitter-receiver unit, with the loudspeaker built in and the microphone connected by a cord. The front section which contains the control apparatus is located on the automobile dashboard. This may be physically removed from the transmitter-receiver cabinet at the rear if the latter is mounted under the seat or in the baggage compartment, as is usually done. Power is obtained from the car battery and car generator which is made specially large when radio apparatus is to be used. All transmissions from the car are normally conducted on a single carrier frequency, and no tuning at the car or the base station is required after the initial installation. This type of system is used for many types of land mobile operations.

FIG. 10-18. Police headquarters radio control point. Unattended fixed tuned transmitter-receiver controlled from here. (Courtesy Du Mont Laboratories, Inc.)

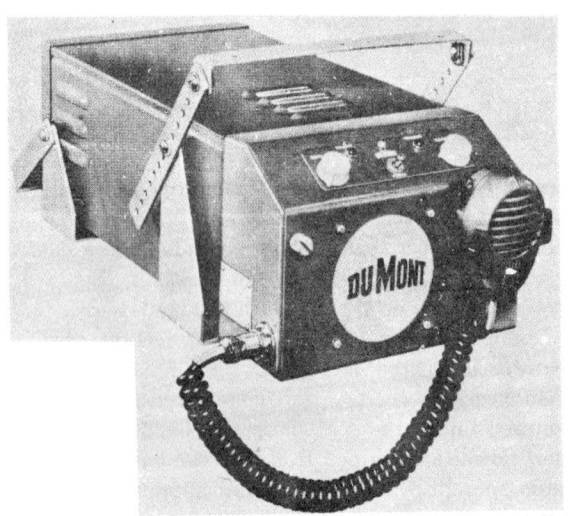

FIG. 10-19. Automobile transmitter-receiver for land mobile service such as police. Forward control head detaches when transmitter-receiver section at rear is mounted in trunk or under seat. Transmitter 75 w. Frequency modulated used in this service. (Courtesy Du Mont Laboratories, Inc.)

QUESTIONS AND PROBLEMS

1. What advantages has radio as a means of communication?
2. Under what conditions are radio communications used to best advantage?
3. Name the different forms of transmission.
4. What is a carrier and what function does it perform?
5. What is modulation?
6. Name the basic parts of a radio communications system and describe their functions.
7. Describe an amplitude modulated radio transmitter and explain the function of each part.
8. What is a radio channel?
9. Describe a simple receiver and explain the function of each part.
10. Describe the superheterodyne type of receiver and explain how it functions.
11. Explain the advantages of the superheterodyne type of receiver.
12. What are sidebands?
13. Explain how sidebands are produced.
14. Why is selectivity important in radio receivers?
15. Why must frequency allocations be carefully planned and rigidly controlled?
16. What body in the United States is responsible for the control of the non-military radio services?
17. What are the approximate limits of the useful radio frequency spectrum?
18. Explain the basic differences between amplitude and frequency modulation.
19. What is the speed of travel of a radio wave?

CHAPTER 11

AMPLITUDE-MODULATION DETECTORS, RADIO-FREQUENCY AMPLIFIERS AND MODULATORS

Chapter 10 described briefly the elements of a radio communications system, modulation and demodulation at the transmitter and receiver respectively, and the function of message-frequency and radio-frequency amplifiers. Chapter 11 will explain these processes and devices in greater detail.

DETECTION OR DEMODULATION

It was explained that demodulation or detection is the process of removing the message from the radio frequency message carrier and restoring it to its original form. There are several different types of detectors, but in amplitude modulation systems they all function in the same general way. Three steps are required, consisting of *rectification, filtering,* and *separating out the message*.

Rectification. Figure 11-1(a) shows an unmodulated carrier starting at Point 1 and continuing to Point 2. Modulation then begins and the carrier voltage is caused to fluctuate above and below its axis forming a carrier envelope corresponding to the wave shape of the message signal.

Chapter 5 explained that when alternating currents are applied to the terminals of a rectifier they pass freely in one direction but are cut off or blocked in the opposite direction. The first step in detection is to apply the currents of Fig. 11-1(a) to a rectifier and produce unidirectional pulses at carrier frequency. Since, from this process, a useful voltage must be developed, a resistor R is added in series with the rectifier. The circuit is shown in Fig. 11-1(b). Now there will be produced across R a succession of half-cycle unidirectional current pulses corresponding to the upper half of the waveform shown in Fig. 11-1(a). The r-f signal has now been rectified and the first step of detection is completed to produce the rectified signal shown in Fig. 11-1(c).

Chapter 11　　*Radio-Frequency Amplifiers and Modulators*　　**339**

Filtering out the pulses. The second step in detection is to eliminate the r-f pulses and obtain a smooth flow of current that varies at audio frequency corresponding to the envelope of the carrier as shown in Fig. 11-1(a). Capacitor C_1 is now connected across R [Fig. 11-1(d)]. The addition of capacitor C_1 smooths out or filters the pulses, thereby producing a continuous voltage across R which will vary in accordance with the envelope of

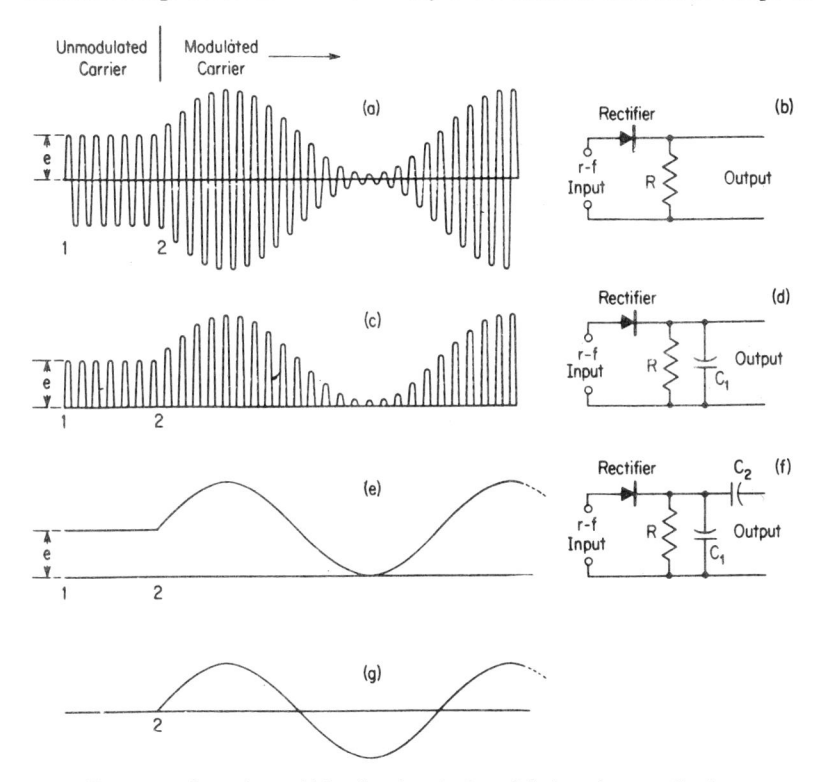

Fig. 11-1. Operations which take place in demodulation of an amplitude-modulated signal.

the input signal. The operation is as follows. When the first part of a half-wave pulse is impressed upon the filter RC_1, the voltage starts to rise and it continues to do so until the pulse reaches maximum amplitude. As the voltage of the half-wave pulse starts to decrease, the voltage across R would normally decrease with it. However, capacitor C_1 now has been charged up. It now discharges slowly through R, maintaining current through R which can change only slowly, i.e., at an audio frequency rate. The current, flowing through R, thereby produces the audio voltage across it as shown in Fig. 11-1(e). Thus, the pulses have been converted to a current which flows smoothly and continuously. When the input signal is unmodulated, a constant d-c voltage is produced across R, as shown by (e) of Fig. 11-1. During the time when the carrier is modulated, the voltage across R follows the modulation voltage. Since the modulation is a slowly varying quantity, R is able to discharge C at a sufficient rate so that the output voltage across R can vary in accordance with the modulation. The current variation is the same as that of the envelope of the modulated wave shown on Fig. 11-1(a).

Restoring the message signal. The waveform of Fig. 11-1(e) during modulation is a replica of the waveform of the original message signal, with one qualification. It is a direct current modulated by the message signal. To recover the original message signal, the modulated direct current may be converted to alternating current. This is accomplished by adding to the circuit the capacitor C_2 as shown on Fig. 11-1(f). C_2 is an a-f by-pass capacitor which blocks the flow of direct current but permits free passage of alternating current. The insertion of C_2 now produces at the output terminals an a-f voltage which is a faithful replica of that used to modulate the distant transmitter. Detection is now complete. The final result is shown in Fig. 11-1(g).

REQUIREMENTS FOR RECTIFIERS

A rectifier is a device which is an effective conductor of electricity in only one direction between its terminals. Because radio communications utilize such high frequencies for message carriers, radio rectifiers must meet certain special minimum requirements. For example, the capacitance between terminals must be low so that the alternating r-f currents do not flow around the device instead of through it. The dimensions must be small so that the distributed capacitance is minimized and compact equipment may be fabricated. The rectifying properties must be satisfactory even at the extremely low voltages which must be rectified, and the operating characteristics should be stable over a long useful life of many thousands of hours.

Three principal types of rectifiers are used in radio detectors: crystal diodes, vacuum tube diodes, and vacuum tube triodes.

CRYSTAL DETECTORS

Many types of crystals possess rectifying properties which make them very satisfactory for use in detectors. In the early days of radio, carborundum, silicon, galena, and various pyrites were extensively used, in some cases with polarizing voltages. In commercial systems carborundum was often preferred because of its ruggedness, stability of adjustment, and sensitivity. Galena possessed excellent properties for sensitivity to weak signals but was poor with respect to ruggedness and stability. The scientific knowledge of crystalline structure and the mechanisms of crystal rectification were not well understood, and the development of excellent vacuum tubes caused crystal rectifiers to be limited in their use until World War II. At that time vastly expanded new operations on frequencies of hundreds or thousands of megacycles produced problems of radio rectification which could not be met satisfactorily with vacuum tubes because of the effect of even the lowest realizable values of capacitance and the finite time of electron transit at such frequencies. Crystals were superior in these respects and came into wide use. In the following years research in solid state physics vastly expanded the scientific knowledge of crystals; they are now grown to carefully controlled and planned degrees of purity or impurity.

Crystal detectors are now extensively used in broadcast and television receivers and other amplitude modulation services.

Crystal detectors have some substantial advantages over vacuum tubes. They are very small, have only two simple connections, contain no filaments which must be heated and which at intervals burn out, are comparable in sensitivity with a diode tube, are very cheap, have very low capacity, may be made almost invulnerable to shock and damage, and require no power for their operation.

Modern crystal rectifiers commonly contain germanium or silicon because they can be manufactured easily, have good operational stability, and are efficient. The simplest crystal rectifiers contain a crystal of about one eighth of an inch on a side, approximating a cube with one face imbedded in a soft metal base plate forming one of the terminals. The other terminal may be a small point which bears on the crystal. The unit normally is enclosed in a small air-tight moisture-proof capsule.

The diagram of a simple crystal detector circuit was shown in Fig. 10-3.

DIODE TUBE DETECTORS

Diode detectors are incapable of amplifying an incoming signal, but they are almost universally used because of their ability to reproduce faithfully the message signal waveform, and because they are simple and inexpensive. Any required amplification is easily obtainable in associated amplifier stages. A diode detector may be either a crystal or a vacuum tube that consists of a cathode and a plate. It is conventional to use a loading resistor of from one-half to 1 megohm so as not to affect the i-f

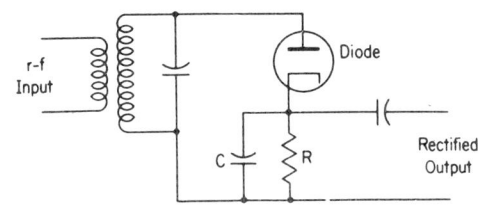

FIG. 11-2. Diode vacuum tube detector.

selectivity. The most widely used amplitude modulation detector is the diode vacuum tube (see Fig. 11-2). The circuit of a diode vacuum tube detector is the same as for a crystal diode detector with the exception that the vacuum tube circuit must provide for cathode heating connections.

The characteristics of an ideal diode rectifier are shown in Fig. 11-3. The input-output linearity shown may be closely approximated by good circuit design, including a high value of loading resistance. A correspond-

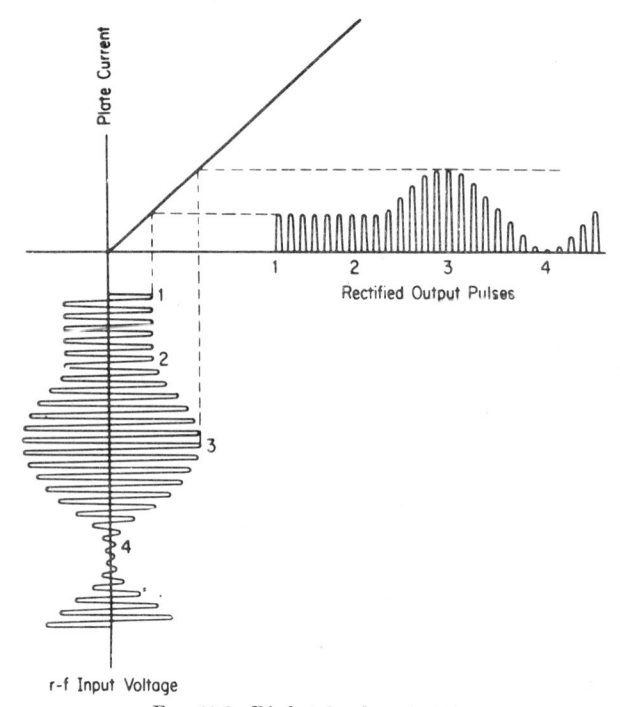

Fig. 11-3. Diode tube characteristic.

ing curve for a crystal diode would be very similar. Under ideal conditions a diode detector will reproduce exactly the modulation envelope, and even under practical conditions the output will contain very little distortion so long as the diode load impedance to modulation frequency currents is a very large fraction of the load impedance to direct currents. If the ratio is undesirably small, the voltage developed across the load at modulation frequencies is limited. A cause of distortion from this source is the use in the output circuit of too large capacitors across the load resistor. Since d-c does not flow through capacitors and audio frequency does, the ratio of impedance will always be less then unity.

A diode may present an input impedance for a carrier different from the input impedance to sideband frequencies. The input impedance, being a

Chapter 11   *Radio-Frequency Amplifiers and Modulators*   343

finite value, causes a load to be placed on the circuit from which input power is obtained. This causes the input voltage to be somewhat lower than would be the case if the diode were not operating. The degree is dependent upon the ratio of the driving circuit impedance to the diode input impedance and the ratio may differ for carrier and sidebands, in effect changing the percentage of modulation.

The resulting limitations may differ in degree for different sideband frequencies and may cause modulation peaks to be flattened selectively. Such difficulties may be minimized or avoided by careful design.

TRIODE DETECTORS

Triode vacuum tubes may be used in detector circuits. The methods of detection form two general classes, plate-circuit detection and grid-circuit detection. There are two types of plate-circuit detectors, the linear plate detector and the square-law plate detector. In all types of triode detection the triode tube also acts as an amplifier, producing gain in signal.

Plate-circuit detection-linear. Figure 11-4 shows a triode connected as an amplifier with a resistance output R. The grid is biased to the cutoff

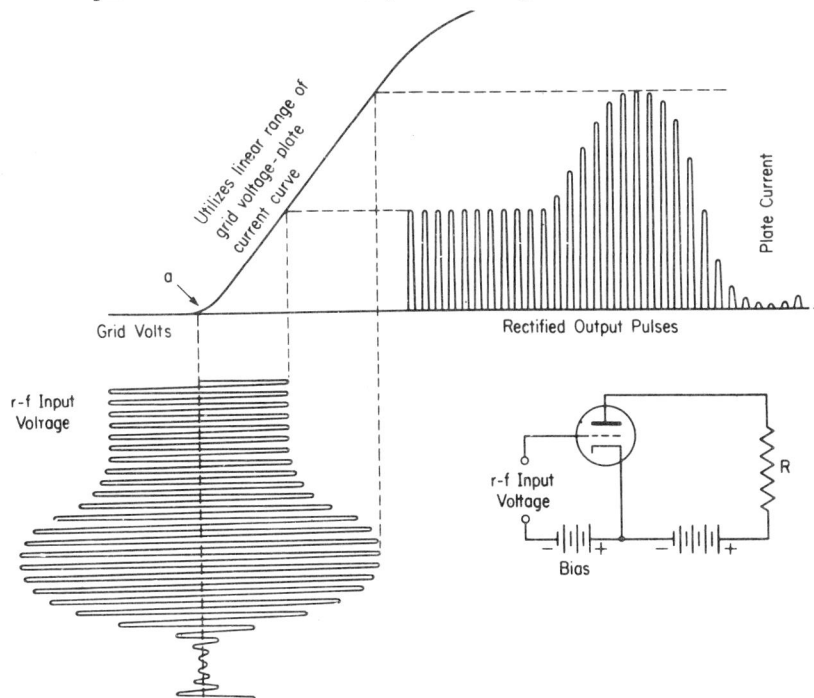

Fig. 11-4. Linear triode rectifier.

point by the negative bias voltage. A radio-frequency signal impressed on the grid will produce half-cycle pulses of radio frequency through R. The envelope of the half-cycle pulses in R will be the same as the envelope of the positive half of the input voltage envelope. This process in effect is rectification, and if the half-cycle pulses are put through a filtering circuit the full process of detection has been performed. Because the tube is connected as an amplifier, there will also be amplification between input and output.

A linear plate-circuit detector diagram is shown in Fig. 11-5. The triode is biased to the cutoff point. The signal input voltage is impressed on the grid through the tuning circuit L_2 and C_1. The output load circuit of the

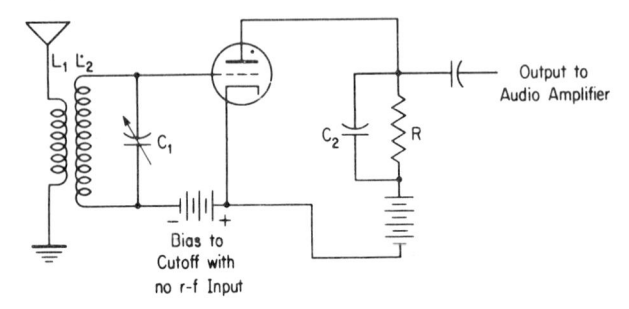

FIG. 11-5. Linear plate circuit detector.

tube is composed of resistance R shunted by capacitor C_2, which form the filter for the detector action. Separation of the audio frequency is attained by the output capacitor. The linear portion of the characteristic curve must be used or there is distortion. The curvature in the characteristic at point a in Fig. 11-4 introduces distortion, which is relatively small when the maximum extent of the characteristic curve is utilized.

Plate-circuit detection-square-law. Square-law plate detection was used in the early days of radio in spite of its large distortion because of a lack of good r-f amplifiers. With modern tubes the signal input voltages may easily be built up by radio amplifiers enough to operate properly a diode detector or a linear plate detector. Figure 11-6 shows the lower end of the characteristic curve of a triode used in square-law detection. This portion is in the vicinity of a in Fig. 11-4. A small signal input voltage is impressed on the grid. This causes a plate current to flow. As may be seen from Fig. 11-6, the curvature of the characteristic causes distortion of the output current. Curve d indicates the average current of the r-f waves. Owing to the curvature in the characteristic, the envelope of d is not a true reproduction of the envelope of the input signal. The output signal pulses may be put through a filter, producing a varying current as shown by d. The audio-frequency component may then be separated by means of a blocking capacitor.

Chapter 11   *Radio-Frequency Amplifiers and Modulators*   345

The circuit diagrams of the linear and the square-law detectors are similar. The essential difference between these two types is that the driving input voltage is high enough in a linear detector to produce operation over the straight and linear portion of the curve, whereas in the square-law detector it is not and operation is confined largely to the curved portion of

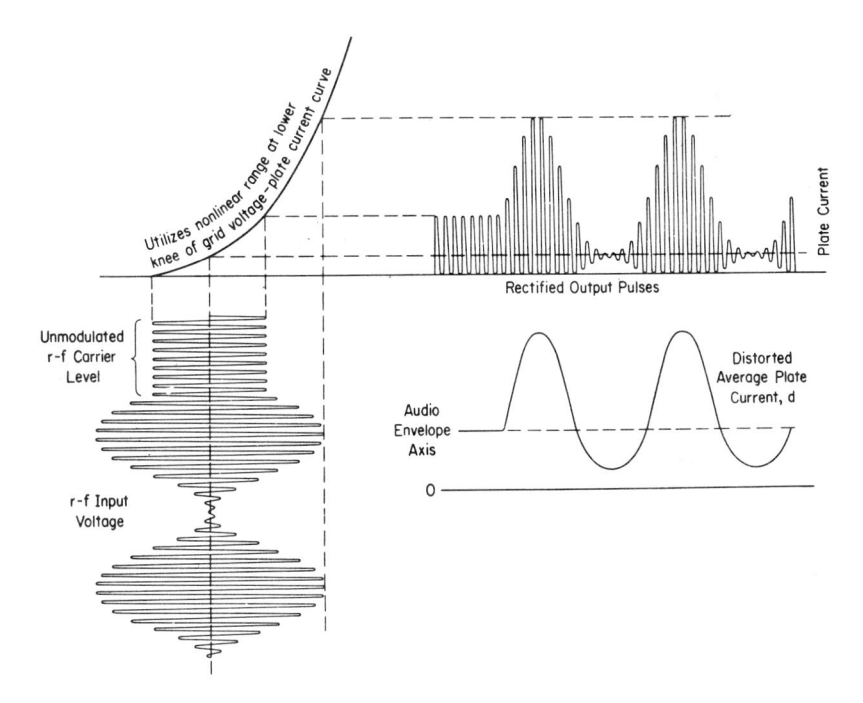

FIG. 11-6. Square-law plate detection. This distorted output is typical of small-input triode square-law detectors.

the characteristic curve. This may be seen clearly by comparing Figs. 11-4 and 11-6 which differ only in the range of operation on the grid voltage-plate current curve. The square-law detector is a *weak signal device*. It may become a linear detector by increasing the r-f input voltage a large amount to utilize the long linear part of the curve. Hence the linear detector is a *strong signal device*.

The sensitivity of square-law detectors is always lower than that of a corresponding circuit with sufficient input signal to give linear detection. This is evident from the fact that the slope of the curve which relates r-f input to instantaneous audio output is steepest in the linear portion, as illustrated in Fig. 11-7. Three portions of the curve are sampled. In each case the input voltage change is equal. The output current change is seen to be much greater where the curve is steepest.

Grid-circuit detection. In the triode detectors described so far, rectification has taken place in the plate circuit, but it also may take place in the grid circuit. Figure 11-8 shows that the input circuit composed of inductance L_2 and capacitor C_1, the filter circuit composed of resistor R and

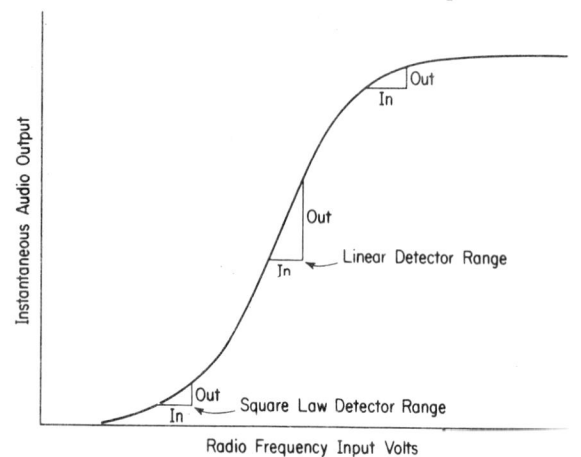

Fig. 11-7. Input-output relations of r-f detectors.

capacitor C_2 and the grid and cathode of the tube, together comprise a diode detector circuit such as that shown in Fig. 11-2. There are only two differences. One is that a grid is used in Fig. 11-8, whereas a plate is indicated in Fig. 11-2. The other is that an output plate circuit is added in Fig. 11-8.

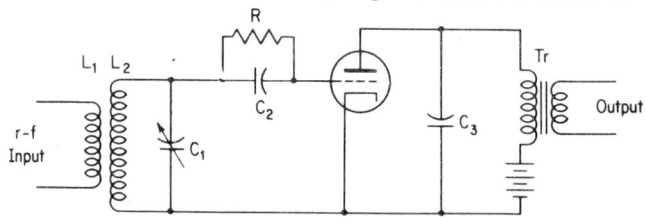

Fig. 11-8. Grid circuit detection.

The rectification takes place in the grid circuit causing a varying current to flow through R. The d-c component is such that the voltage across R is negative at the grid end and bias voltage is therefore automatically supplied to tube T so that it may function as an amplifier. The audio-frequency component of the varying voltage across R is impressed through inductance L_2 across the grid and cathode of the tube. The inductance L_2 is small enough so that at audio frequencies it has negligible impedance. The audio-frequency component across the grid and cathode is amplified by the tube, producing an audio frequency that is transformed from the plate circuit to the output terminals through the audio transformer T_r. Part of the radio-frequency signal voltage appearing across the tuned input circuit is impressed upon the grid through resistor R and capacitor C_2. This is also amplified by the tube. By-pass capacitor C_3 is provided to shunt the radio frequency in the output circuit to ground.

Grid-circuit detection is characterized by high sensitivity and high output audio-frequency signal for a given radio-frequency input, but in general is more susceptible to distortion.

Plate and grid-circuit triode detectors have been replaced almost entirely by diode detectors because the signal input voltage may easily be built up in radio amplifiers to a value high enough to operate a diode properly.

HETERODYNE DETECTION OF CONTINUOUS WAVES

In Chapter 10 reference was made to heterodyne detection in superheterodyne receivers and in reception of continuous waves in telegraphy. A clear understanding of heterodyne detection is important because of its very wide use. Heterodyne detectors are often referred to as *mixers*, particularly in connection with superheterodyne receivers.

Figure 11-9(b) shows an incoming signal E_s. It is desired to change the frequency, shown as 1,000 kc, to some other frequency which may be either an audio or radio frequency. It will be assumed that the incoming 1,000-kc continuous wave is keyed on and off to produce telegraphic dots and dashes. The detected output of such a signal cannot be heard because it contains no audio frequency modulation identifiable in the headphones. It is possible to create a 1,000-cycle audible tone during the time the wave trains are coming in as dots and dashes of the telegraphic code. This is accomplished, as explained briefly in Chapter 10, by inserting a frequency of 1,001 kc, from a separate local oscillator, a frequency differing from the incoming signal by the desired 1,000 cycles. These two wave trains are shown as Fig. 11-9(a) and 11-9(b). The 1,001-kc signal is inserted at the detector input where the incoming signal also appears.

If two separate radio-frequency signals are applied to a rectifier there will be a net voltage representing the sum of the two separate voltages. If the two voltages occur at identically the same frequency, the output will be at the same frequency. But what occurs if the frequencies differ? At one instant the signals are assumed to be in the same direction simultaneously and they will then add. But with one signal rising and falling at a different repetition rate from the other one, there will also occur instants when currents will flow in opposite directions and cancel. The sequence of addition and cancellation obviously will occur at a rate corresponding to the frequency difference between the two voltages. This produces the result shown on Fig. 11-9(c).

It has been shown that when rectification takes place the waveform of the output audio voltage is a duplicate of the radio frequency envelope at the input. The continuous combination of two sine-wave radio-frequency currents of different frequencies, produces a new sine wave component in a rectifier at the difference frequency.

The heterodyne process is almost universally used in such services as ship-to-shore telegraphy in which continuous wave carriers are keyed on and off to make dots and dashes, with no other modulation. This form of

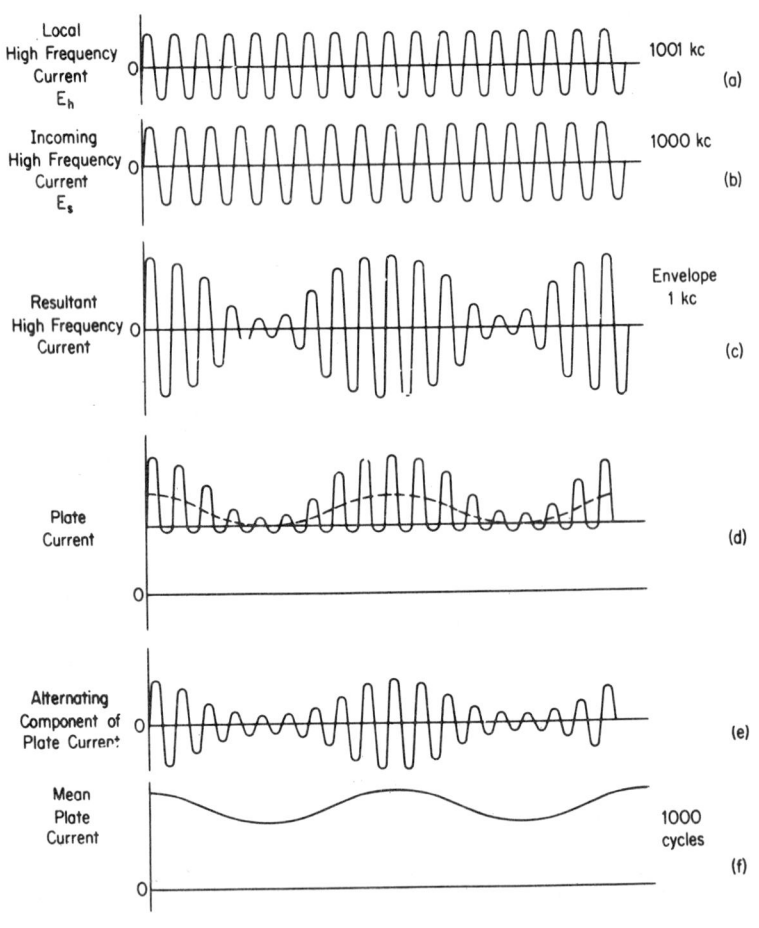

Fig. 11-9. Detection by heterodyne process.

transmission and reception is most efficient, requires only simple apparatus and a minimum of sideband space for a channel. The heterodyne process in the receiver is a vital part of the system because of its flexibility, simplicity, efficiency, and because the heterodyne oscillator produces an audible output only when a carrier is being received. The reason for the latter will be clear if it is recalled that a single CW carrier, upon detection, does not produce an

Chapter 11   *Radio-Frequency Amplifiers and Modulators*   349

audible beat note. The local CW heterodyne oscillator may be considered to be a single side frequency produced locally and the audible note results from the envelope of the carrier and side frequency.

If a superheterodyne receiver were used for CW reception, one beating oscillator would be utilized to heterodyne the carrier frequency down to the intermediate amplifier frequency. Another beating oscillator would also be required to produce an audible headphone signal. In this case the second oscillator injection, separated by some audio frequency such as 1,000 cycles, could be at the input to the second detector where it would combine with the intermediate frequency to produce a 1,000-cycle headphone tone.

The heterodyne process may be utilized to reduce the frequency of a signal to any desired amount, by proper adjustment of the frequency of the *beating*, or *heterodyne oscillator*. If the signal is modulated, the output at the new lower frequency will have the same modulation, either amplitude or frequency, as the original signal. In the design of UHF television receivers, for example, the beating oscillator is used to reduce to about 42 megacycles any frequency between 470 and 890 megacycles, by changing the tuning of the beating oscillator frequency between 512 and 932 megacycles. In long-lines telephony and microwave radio relaying, radio frequency carriers and subcarriers are very extensively employed, and the positioning of these carriers and subcarriers is accomplished by the heterodyne process. It is also used in instrumentation. Sometimes heterodyne frequency shifting is used in two or more stages. It is one of the most useful and widely employed techniques in radio communications.

DETECTION OF INTERRUPTED CONTINUOUS-WAVE SIGNALS

Since interrupted continuous-wave signals are amplitude modulated by audible signals at the transmitter, they may be received on any of the conventional types of amplitude modulation detectors. The fact that the original audible tone used for modulation is reproduced at the detector output makes it unnecessary to use a beating oscillator to produce an audible headphone signal.

RADIO FREQUENCY AMPLIFIERS

Radio frequency amplifiers are employed in transmitters and receivers and many other devices for amplifying r-f power, current, or voltage. A large variety of tubes have been specially designed for the purpose, ranging in size from a tiny UHF receiver amplifier tube about the size of a pencil eraser to huge transmitter amplifier tubes which are five or six feet long, require large amounts of forced cooling by water, and handle hundreds of kilowatts of power. The operation of all such tubes, regardless of size, is fundamentally the same when conventional triodes, tetrodes, or pentodes are used. Tubes used for audio frequency amplification may be used for radio amplification in many cases. Radio amplifier tubes especially designed for the purpose have short internal connections to minimize distributed inductance and capacitance, have the lowest interelectrode

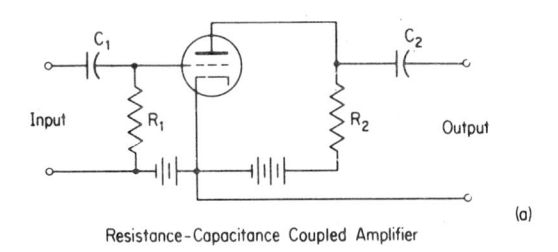

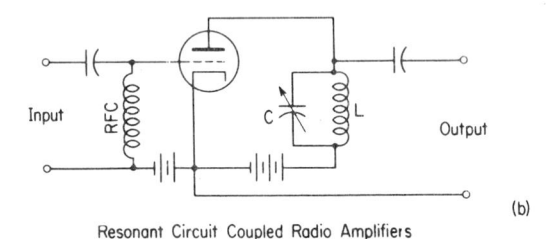

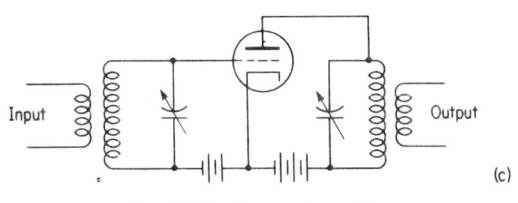

FIG. 11-10. Types of amplifiers.

capacitance possible to minimize internal feedback, and may include screen grids.

It is pertinent to note the essential difference between audio and radio amplifiers. Fig. 11-10 shows circuits for each type.

DIFFERENCE BETWEEN RADIO AND AUDIO AMPLIFIERS

The resistance-capacitance coupled amplifier is conventional and has been described in Chapter 7. It will amplify a wide range of audio frequencies and also will function at radio frequencies, although with increasing frequency the tube capacitance shunting R_1 increasingly reduces the gain. Obviously such an amplifier does not respond to one frequency to the exclusion of others. It contributes nothing to receiver selectivity. The plate circuit efficiency is very low because the plate current I^2R loss in R_2 is large. This alone would make it unsuitable for power amplification. The loading and coupling elements, while suitable for low-power audio amplification, are not suitable for radio amplification. The principal difference between audio and radio amplifier circuits lies in the loading and coupling elements.

Figure 11-10(b) shows a radio amplifier having a tuned circuit for loading and coupling. We now have an adjustable frequency-selective load, the power loss in the plate coupling resistor has been eliminated, and the plate supply voltage may be lower. This follows because the plate rectifier voltage is applied in full to the tube plate, instead of being divided between the tube and the coupling resistor R_2. To achieve rated plate voltage at the plate the rectifier must supply the rated voltage and, in addition, the d-c voltage across the coupling resistor R_2, when using resistance coupling as shown in Fig. 11-10(a)

Figure 11-10(c) shows a radio amplifier with tuned circuits in both the grid and plate circuits. Where radio amplifiers are employed in cascade, the circuit of Fig. 11-10(b) is approximately equivalent to that of Fig. 11-10 (c) because the plate tuning circuit of the preceding stage adds selectivity in the same manner as does the input tuning circuit of Fig. 11-10(c).

Chapter 3 showed that a parallel tuned circuit has a relatively high impedance at the resonant frequency and that the impedance decreases rapidly on either side of the resonant frequency. The resonant impedance may be made very high and may be adjusted to the optimum value for either high-power transfer efficiency or higher voltage amplification and selectivity. Since the impedance of the tuned circuit decreases rapidly on either side of resonance, the voltage developed at the terminals will be reduced correspondingly. Selectivity is thereby achieved.

In receivers the performance features most desired in radio amplifiers are frequency selectivity, high gain, and ease of tuning. In transmitters the emphasis is on plate efficiency and high gain, with selectivity and ease of tuning usually secondary.

Resistance-capacitance coupled amplifiers are very effective for high quality audio amplification at low power. It has been shown that the plate efficiency is low, but for that application the amount of power involved is so small that it is unimportant. This condition is unique to that application. In radio amplifier applications other conditions unique to them dictate the use of resonant circuit coupling and loading.

RESONANT COUPLING FOR IMPEDANCE MATCHING

A resonant coupling circuit can act as an impedance-matching circuit. It is possible to adjust the impedance to any desired value. This adjustment is particularly important in transmitters, where large amounts of power are used and efficiency is important. The impedance of a resonant coupling circuit can readily be adjusted for the optimum load impedance of the amplifier, thereby attaining maximum output for a given allowable loss.

Figure 11-11(a) shows a parallel resonant circuit consisting of a capacitor C in parallel with inductance L, which is in series with resistance R.

At terminals Z-Z, the impedance Z ideally is a pure resistance at resonance. The circuit shown in Fig. 11-11(a) has the approximate constants for resonance at 1,000 kc. At 1,000 kc, when R is zero, Z is infinite;

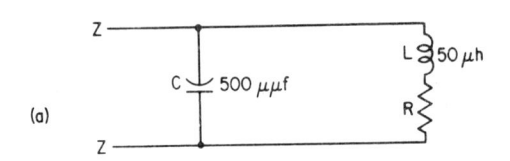

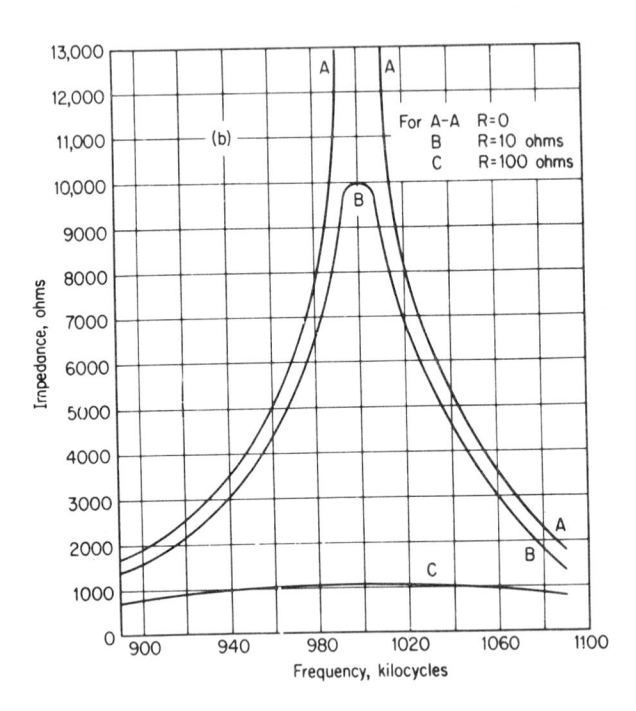

Fig. 11-11. Effect of tuned circuit on load impedance.

when R is 10 ohms, Z is 10,000 ohms; and when R is 100 ohms, Z is 1,060 ohms. R may be selected so that the impedance Z will be whatever value is the proper load impedance for the amplifier. As an example, the 10-ohm load has been transformed to 10,000 ohms. The use of parallel resonant circuits for impedance transforming was more fully discussed in Chapter 3.

A parallel resonant circuit may be adjusted to present the proper load impedance to a tube operating at a specific carrier frequency, such as 1,000 kc. In radiotelephony, however, the amplifier must also amplify the sidebands properly.

Referring to Fig. 11-11(a), assume that the circuit is resonant under three conditions, namely R equals 0, 10 ohms, and 100 ohms. Also assume that there is no appreciable resistance in C and that L is a pure inductance and has no resistance. The impedance Z over a band of frequencies under the three conditions is shown.

A tuned circuit such as is shown in Fig. 11-11(a) appears as a pure resistance at its terminals only at the resonant frequency. At other frequencies the impedance Z is partly reactive.

Curve A shows that when R equals zero the impedance Z rises to infinity at the resonant frequency of 1,000 kc. Curve B shows Z when the resistance is 10 ohms. If the highest modulation audio frequency is 10,000 cycles, the band of frequencies to be transmitted is from 990 to 1,010 kc. Curve B shows that the impedance varies somewhat over this range. A radio amplifier can work satisfactorily with this difference if simple precautions are taken. Curve C of Fig. 11-11(b) is a plot of Z when R is equal to 100 ohms. Over the band of frequencies from 990 to 1,010 kc the impedance is substantially equal at all frequencies. However, it is to be noted that the impedance Z is much lower than that usually required for the plate load of a tube. If an attempt is made to make the impedance characteristic too flat, the impedance at resonance is then too low as a coupling device for a tube. Amplifiers built for wide pass bands use more complex circuits for coupling.

HARMONIC SUPPRESSION

Vacuum tubes generate harmonics that produce undesirable radiations when coupled to an antenna if precautions are not taken to suppress them. The second harmonic of 1,000 kc is 2,000 kc, the third harmonic is 3,000 kc, and so forth. If the harmonics were not suppressed, but were allowed to radiate, they could and sometimes do cause interference to other services which employ those frequencies for communication. The curves of Fig. 11-11(b) do not extend to the harmonic frequencies, but it is apparent that the impedance at such frequencies would be low, the harmonic voltages developed would be correspondingly low, and that the circuit would discriminate against harmonic energy transfer. However, this discrimination is not adequate alone for suppressing transmitter output harmonics.

It is an inherent characteristic of heavily loaded radio amplifiers to generate harmonics because their operation occurs over a large range of the grid voltage-plate current characteristic curve. Under such conditions the curve is not a straight line. Figure 11-12(a) shows such a curve with bends near the plate current cutoff and the plate current saturation regions. The ideal curve, which may be approached but is never fully realized, is shown as a straight dashed line. It may be seen that the plate current in practice falls below the ideal curve at low values and rises above it at higher values. Figure 11-12(b) shows the component of plate current produced by this nonlinearity. For the fundamental frequency the plate current for one-half

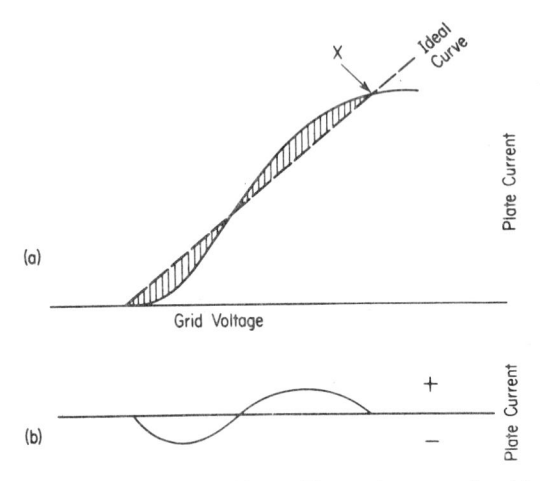

FIG. 11-12. Illustration of how r-f harmonics are produced in r-f amplifiers.

cycle will rise from zero to maximum at Point X and then decline, producing a single plate-current pulse. But the nonlinear component during this interval will, starting at zero, go negative and then positive before it reaches Point X.

While the fundamental plate-current pulse is providing power for one-half cycle of r-f current in the output tuned circuit, the nonlinear component is providing a negative exciting pulse followed by a positive exciting pulse, producing r-f current of twice the fundamental frequency. Since the nonlinear component of plate current is not a symmetrical sine wave it must contain not only second harmonic components but also higher order components representing higher order harmonics.

Methods of reducing harmonic currents in transmitter output circuits will be explained in sections to follow.

RADIO AMPLIFIER CIRCUITS

Basically most classes of radio amplifiers are alike. A simple radio amplifier is shown in Fig. 11-13. Tube T works into the resonant circuit composed of capacitor C_1 and inductor L.

The signal input voltage is impressed on the grid. Bias voltage reaches the grid through the radio-frequency choke RFC which prevents the input signal voltage from being shorted through the bias supply. The radio-frequency by-pass capacitor C_2 is of such a value that points X-X are at substantially the same radio-frequency potential without shorting the plate voltage. The resonant output-coupling circuit is composed of L and capacitor C_1. So far as radio frequency is concerned, the coupling circuit is connected directly between the cathode and the plate of the tube. Blocking capacitor C_3 is of such a capacitance that the output is effectively connected

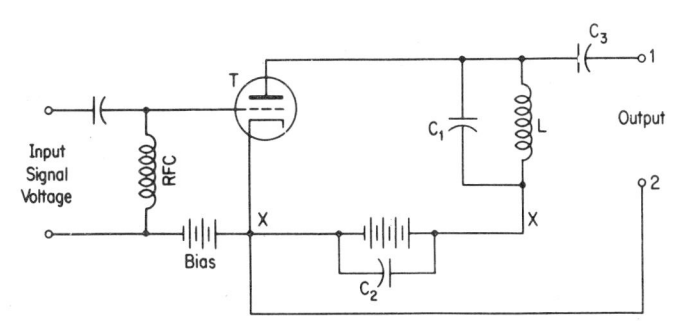

FIG. 11-13. Simple triode r-f amplifier.

to one end of the resonant circuit. C_3 prevents the plate voltage from appearing across the output terminals. Output terminal 2 is effectively connected to the other end of the resonant circuit through capacitor C_2. Whatever radio-frequency voltage develops across $C_1 L$ also appears across the output terminals 1, 2.

There are many ways of coupling the signal input voltage into an amplifier and of coupling the output to the load. There are also many methods for connecting the bias voltage to the grid, and for supplying plate voltage to the tube. The resonant output circuit or tank circuit, as this is often called, takes many forms.

AMPLIFIER INPUT CIRCUITS

Five different input circuits for impressing the input signal voltage on the grid of a radio amplifier are shown in Fig. 11-14. L and C_1 in Fig. 11-14(a) represent the output circuit of a driving amplifier. The radio-frequency input voltage is impressed upon the grid through radio-frequency blocking capacitor C_2. R-f choke prevents the bias supply from shorting the radio-frequency voltage between the grid and cathode. The blocking capacitor C_2 prevents the plate voltage of the preceding amplifier from being impressed on the grid.

C_1 and L represent the input tuning circuit to the radio amplifier in Fig. 11-14(b). R is the cathode bias resistor. C_3 is a radio-frequency bypass around the cathode resistor, therefore effectively placing all of the input voltage directly across the grid and cathode. In Fig. 11-14(c) capacitor C_1 and inductor L represent the output circuit of a preceding amplifier. L induces a radio-frequency voltage in L_2. Capacitor C_3 by-passes the radio-frequency voltage directly to the cathode, thereby impressing all of the voltage from L_2 across the grid and cathode. The circuit in Fig. 11-14(d) is often employed when the output circuit of a tube in a transmitter is not adjacent to the tube which follows it. This is called link coupling. The impedance of the link circuit is kept low. The circuit shown in Fig. 11-14(e)

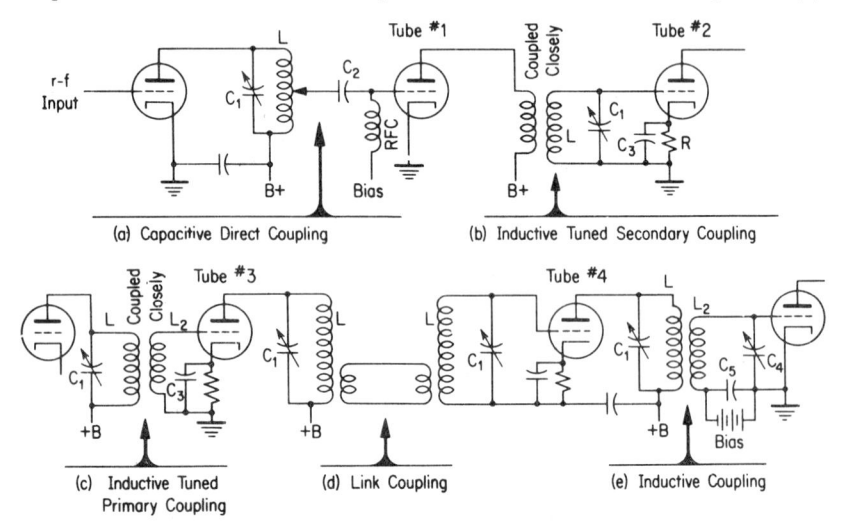

FIG. 11-14. Five different input coupling circuits for radio amplifiers.

may be used where it is desirable to use a separate bias supply and still be able to ground one end of the tuning capacitor. As may be seen, capacitor C_4 is grounded at one end. The capacitor C_5 is large enough so that it is a short circuit for radio frequency. It prevents the bias voltage from being shorted. The bias voltage is fed to the grid through inductor L_2.

SINGLE-ENDED OUTPUT CIRCUITS

There are many types of output circuits that may be used with radio amplifiers. The fundamentals of most of these are illustrated in Fig. 11-15. In Fig. 11-15(a), L and C_1 comprise the tuned output circuit. The plate supply is fed to the plate through inductor L. C_2 is an r-f by-pass capacitor effectively connecting one end of the output circuit to the cathode. C_3 allows passage of the output radio frequency but stops the plate voltage from being impressed upon the output terminals. The plate supply is said to be series fed because the plate voltage is fed in series through the tuning inductance L to the plate.

Figure 11-15(b) illustrates parallel feed of the plate voltage. The plate voltage is fed to the plate through an r-f choke. As may be seen, the plate voltage and the output circuit are, in effect, in parallel. R-f choke prevents the radio energy from passing into the plate supply, and capacitor C_3 prevents d-c plate voltage from passing into the tuning circuit. Inductors L_1

Chapter 11   *Radio-Frequency Amplifiers and Modulators*   357

and L_2 form a radio-frequency transformer. The output of the amplifier is inductively fed from L_1 to L_2.

A somewhat different type of output circuit is shown in Fig. 11-15(c). Tuning is accomplished by variable capacitor C_1. Inductors L_1 and L_2

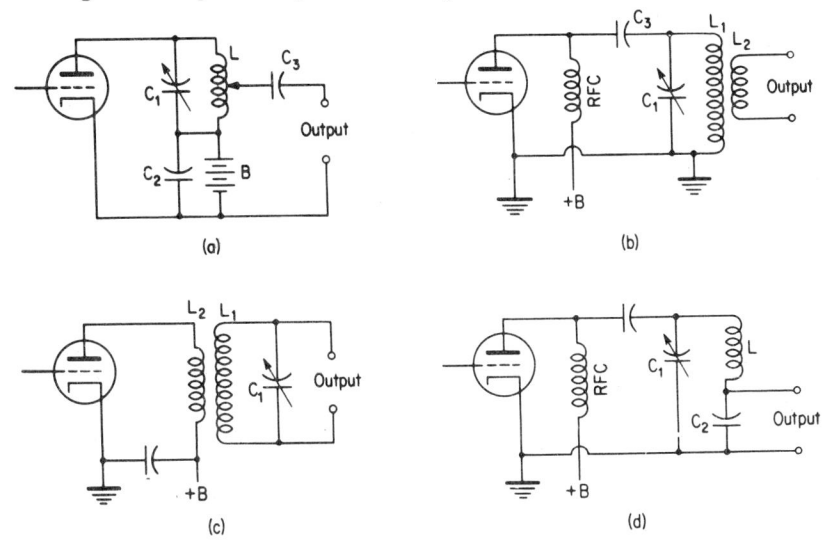

FIG. 11-15. Four types of output circuits for single-ended radio amplifiers for transmitters and receivers.

form a radio-frequency transformer. Therefore, the tuning circuit is reflected into the inductor L_2 and is effectively in series with the plate of the tube.

An output circuit used extensively in transmitters is shown in Fig. 11-15(d). Inductor L and capacitors C_1 and C_2 form the output tuning circuit. In this circuit the output of the amplifier is obtained from across capacitor C_2. This type of circuit helps to reduce harmonics.

Output circuits such as those in Fig. 11-15(b) and (d) are used extensively in transmitters. The output circuit shown in Fig. 11-15(c) is extensively used in receivers. Variations of Fig. 11-15(a) are also used in receivers.

GRID-BIAS VOLTAGE SUPPLIES

Bias voltage may be supplied to an amplifier by a number of different methods. Several of these are shown in Fig. 11-16. In each of the figures shown, L_1 and L_2 represent the input and output circuits respectively.

Figure 11-16(a) shows a battery in use as a grid-bias supply. This type of supply is used when external bias is required in portable battery-operated receivers. It is seldom used in amplifiers deriving their power from a 60-cycle a-c source because the bias voltage can be supplied readily by other means. Cathode biasing is shown in Fig. 11-16(b). R is in series with the cathode of the tube. The plate current returns to the cathode through R and therefore there is a d-c voltage drop across the resistor.

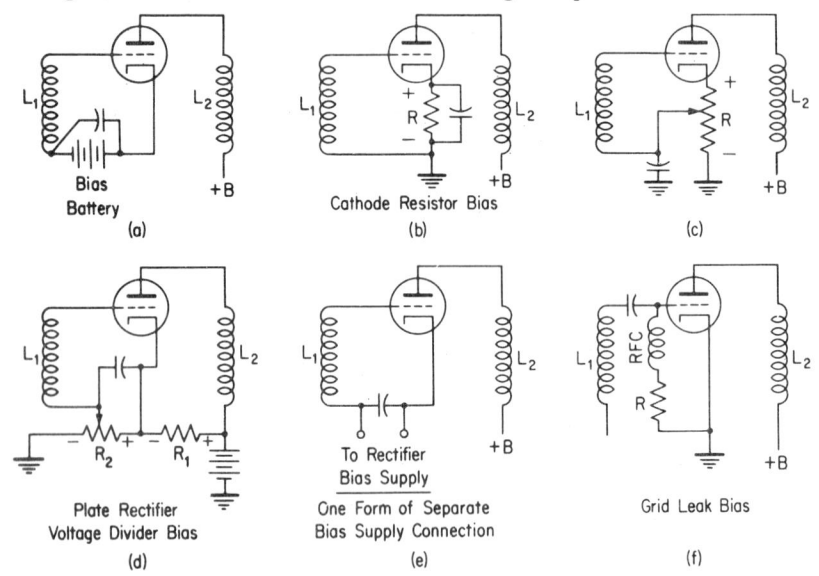

Fig. 11-16. Methods of obtaining grid bias.

(a) Bias Battery
(b) Cathode Resistor Bias
(c)
(d) Plate Rectifier Voltage Divider Bias
(e) One Form of Separate Bias Supply Connection
(f) Grid Leak Bias

The polarity is positive at the cathode end and negative at the other end. By connecting the grid return lead as shown in Fig. 11-16(b), the grid is biased negative in relation to the cathode. The voltage across R is equal to the plate current multiplied by the resistance. By using the proper value of resistance, the proper bias voltage is obtained for the amplifier.

Figure 11-16(c) shows cathode biasing except that the grid return is made to the arm of a potentiometer so that the bias voltage to the tube may be varied at will. This type of biasing is not generally used but is convenient in experimental and test equipment.

Figure 11-16(d) shows biasing by means of a voltage divider across the plate supply. The divider is composed of resistor R_1 in series with a potentiometer R_2. The flow of current through the divider is such that negative bias is supplied to the grid.

Biasing shown in Fig. 11-16(b) is extensively used in receiver radio amplifiers, low-powered audio amplifiers, and in transmitters to some extent. The circuit of Fig. 11-16(c), as explained, is a variation of that in Fig. 11-16(b) and has a limited use. Biasing as shown in Fig. 11-16(d) is principally limited to low-powered amplifiers. However, it is not in such general use in receivers as is the biasing shown in Fig. 11-16(b).

Chapter 11   *Radio-Frequency Amplifiers and Modulators*   359

Figure 11-16(e) shows a method of biasing extensively used in transmitters wherein a separate rectifier is used to supply the voltage.

Grid-leak biasing is shown in Fig. 16(f). This is used fairly extensively in transmitter radio amplifiers. When the radio-frequency grid input voltage exceeds the bias voltage, an amplifier draws grid current. As shown in Fig. 11-16(f), the grid current returns to the cathode through radio-frequency choke RFC and resistor R. Current through R produces a voltage drop across R and this biases the grid. Grid-leak bias has the disadvantage that if the signal input voltage to the grid is removed the bias on the tube is removed. Because there is no flow of current through R the tube will draw an abnormally high plate current and it may be damaged. To avoid this possibility, sometimes a fixed bias supply is inserted between R and ground for part of the bias voltage. The voltage drop across R contributes the rest. The fixed bias is high enough to limit the plate current to a safe value in the event that the signal input voltage to the grid is removed.

RADIO AMPLIFIER CLASSIFICATIONS

Radio frequency amplifiers may be adjusted to serve specific functions. The types of operation are classified for convenience in identification and reference. These classifications are explained below; they are similar to those for audio amplifiers as discussed in Chapter 7, but are repeated here for further development.

Class A amplifier. In a Class A amplifier the grid bias and the input signal voltage are such that throughout the entire cycle of the alternating input signal voltage, plate current flows continuously. This type of amplifier is most useful when maximum voltage amplification of the input signal is desired. The efficiency is low in terms of useful power output compared with plate d-c power input but is highest of any class of amplifier in the ratio of output signal power compared with input signal power.

Class B amplifier. In a Class B amplifier the grid bias and the input signal voltage are such that plate current flows only during approximately one half of the cycle of the alternating input signal. This type of amplifier operates at much higher plate efficiency than does a Class A amplifier. The grid bias is adjusted so that plate current is approximately zero when no input signal is present. This type of amplifier is usually employed in push-pull to obtain linear amplification of audio frequency or modulated radio frequency signals at high efficiency and tolerable distortion.

Class C amplifier. In a Class C amplifier the grid bias and input signal voltages are such that plate current flows during substantially less than half of the cycle of the input signal voltage. The grid bias is appreciably higher than necessary to cut off plate current in the absence of input signal voltage, usually twice as great or more.

Class C amplifiers are not linear and cannot be used to amplify audio or modulated radio frequency signals. The grid input signal is usually high enough so that a further increase does not produce any increase in plate circuit output. This type of amplifier produces the highest plate efficiency of any type and is superior as a plate modulated radio amplifier. It also produces the greatest amount of radio frequency output from a tube. It is very widely used in the radio frequency stages of radio transmitters.

APPLICATION NOTES

Class A amplifiers are employed for both audio and radio frequency amplification, the principal difference being in the type of coupling elements used to obtain the proper impedances. In radio amplifiers tuned circuits are commonly used to adjust impedances and obtain frequency selectivity.

Class B amplifiers are also used for both audio and radio frequency but not to the same degree as Class A amplifiers. In audio use push-pull operation is required because a single-sided amplifier amplifies only the positive voltage swings of an audio signal. However, single-sided radio amplifiers may be used with reasonable linearity because the missing negative swings of voltage are reconstructed in the plate tank circuit which has oscillating currents. The high efficiency of Class C amplifiers is obtained as a result of the high bias and large driving voltage. With this adjustment grid current and plate current are caused to flow during a shorter time interval than the full positive half cycle of driving voltage. A very large pulse of plate current flows for a relatively short time, reducing the tube heating, as will be explained later in this chapter. This increased efficiency makes it possible to raise the operating power level before the rated tube heat dissipation is exceeded.

Class B amplifiers are commonly used as the *modulators* to provide the large amounts of audio power required in amplitude-modulated transmitters when the final stage is the one modulated, in which case the stage is always a Class C amplifier. Class B amplifiers have been extensively used as linear radio amplifiers of modulated radio frequency power but this use is diminishing in favor of new and more efficient arrangements.

Reference is often made to voltage amplifiers as a class apart from power amplifiers. The connotations of these terms in radio amplifiers correspond with those of audio amplifiers and were explained in Chapter 7.

CLASS A RADIO AMPLIFIERS

Class A amplification is characterized by large amplification of the input signal, low *plate circuit efficiency*, and a sacrifice of maximum achievable power output level in favor of faithful reproduction of the input signal. This class amplifier has unique advantages where very low input signal levels are available, as in receivers, maximum possible amplification is

Chapter 11   *Radio-Frequency Amplifiers and Modulators*   361

desired, and maximum achievable output power level is not required. The principal differences between Class A, B, and C amplifiers are those of voltage settings.

The radio amplifier shown in Fig. 11-13 may operate as a Class A, B, or C amplifier, as desired, by adjusting the bias voltage, the plate voltage, and the signal input voltage to the proper values. The circuit diagrams for the three types of radio amplifiers usually are identical.

Figure 11-17 shows the grid voltage-plate current relationship for a Class A radio frequency amplifier. The d-c grid bias is adjusted so that the zero signal plate current is in the center of a straight portion of the curve,

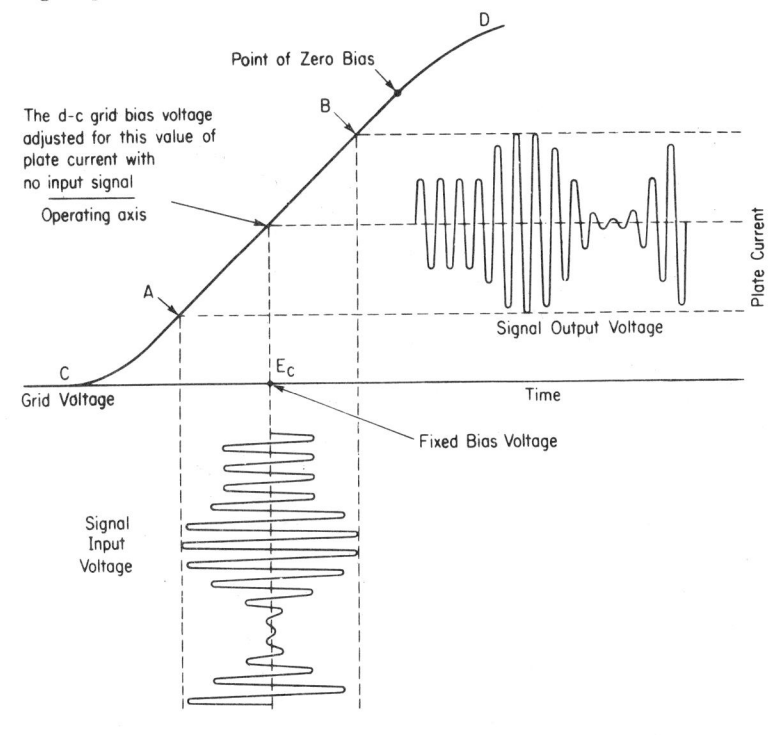

FIG. 11-17. Relation between grid voltage and plate current for Class A amplification. It applies to either radio or audio amplifiers.

which is required for faithful reproduction of the input signal. The plate current excursion, shown as points A and B, is limited to the portion of the curve where a change of signal input voltage produces an identical change in output signal voltage. The reproduction of the envelope wave shape then is faithful. The d-c grid bias voltage is halfway between points A and B and the plate current no-signal value is also halfway between points A and B.

The input signal voltage excursion is fully utilized to cause an amplified voltage to appear across the load, and the amplification takes place over the steepest portion of the curve as shown in Fig. 11-17. Therefore, the input signal amplification is the greatest that could be produced by the tube. In terms of output-to-input signal power amplification, the Class A amplifier is more efficient than either the Class B or Class C types because the steepest portion of the curve is utilized and because the full voltage range of the input signal is employed to control the plate current. This expression for efficiency should not be confused with the one for plate circuit efficiency in which the ratio of d-c plate input power and signal output power is used. In a Class A amplifier the plate circuit efficiency is lower than that of either a Class B or Class C amplifier because only a limited portion of the maximum possible plate current excursion is used.

Class A amplifiers are employed in receivers to achieve the largest possible undistorted amplification per stage. This minimizes the number of stages and therefore cheapens and simplifies the design. The penalty of low plate circuit efficiency in receivers is unimportant because the power thus lost in small receiving tubes is not significant. In transmitters employing large power tubes, plate circuit efficiency is very important.

Class A amplifiers may employ triodes, pentodes, or screen grid tubes. Pentodes are very widely used because they will produce very high voltage amplification. Great advances have been made in the development of tubes for receiver radio amplification. The signal voltages obtained from receiving antennas are conventionally expressed in terms of microvolts. Since the power is so extremely low the tubes required to amplify it may be very small in size and power ratings. As new classes of receiving tubes have been developed for radio amplification, they have been characterized by smaller size, lower supply voltage ratings, higher amplification, greater freedom from self-oscillation, and efficient operation on increasingly higher carrier frequencies. They have been described as "pencil" tubes, "miniature" tubes, and "microminiature" tubes. The latter in physical size are about equal to a lead pencil eraser, have inter-electrode spacings measured in millionths of an inch, and mark a new frontier in amplification at carrier frequencies up to about 1,000 megacycles. Such tubes are most near the ideal for small power radio amplification, but naturally have little value in transmitters where higher powers are involved.

Class A radio amplifiers, as stated, may be quite free of distortion of modulated carrier waves, which is required. However, for unmodulated carrier amplification, such as for C.W. telegraphy, linearity is not important. The input signal voltage and plate current excursions could extend over the complete range C to D of the curve shown in Fig. 11-17 and thus produce more power as Class B or Class C amplifiers. For practical reasons, however, in receivers this offers no worthwhile advantage and is used little, if at all.

Chapter 11　*Radio-Frequency Amplifiers and Modulators*　363

AMPLITUDE DISTORTION IN RADIO FREQUENCY AMPLIFIERS

Amplitude distortion has been discussed in preceding chapters, particularly with reference to audio amplifiers. But such distortion may also occur in radio amplifiers because of unequal amplification of the envelope of a modulated wave. Any significant change in modulation-envelope wave shape caused by nonlinear distortion adds new signal frequencies, changes the amplitudes of others, and produces a different sound in the receiver loudspeaker. The reproduced sound may and usually does become very discordant and unpleasant. Figure 11-18(a) shows a sine-wave modulated carrier relatively free of distortion. Figure 11-18(b) shows the same wave after nonlinear distortion intentionally was caused by improper adjustment of the transmitter through which it passed. This original envelope wave shape was a 50-cycle sine wave but it now has been distorted and new envelope frequencies have been added. They can be heard in headphones, and can be seen and measured with suitable instruments.

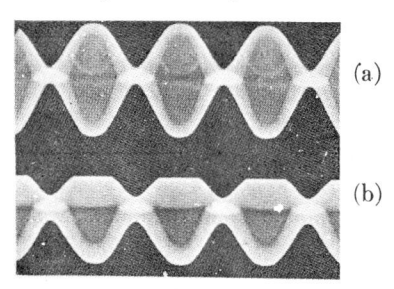

FIG. 11-18. Distortion of modulated carrier envelope by severe amplitude nonlinearity in r-f amplifier.

Frequencies in the original complex wave shape may also be changed in amplitude. If the change in this direction is substantial, it may be detected by ear. But it will not normally produce unpleasant and discordant results. The characteristic unpleasant sound of amplitude distortion is produced primarily because new frequencies are added. They are referred to as distortion products.

IMPROPER OPERATION

Improper operation of a Class A amplifier produces distortion of the modulation envelope. Figure 11-17 illustrated the condition of proper operation. Figure 11-19 illustrates how nonlinear distortion of the modulation envelope is produced when the d-c grid bias is too high. Operation in the upward direction from the axis to point B is still linear, but in the downward direction to point A it occurs over the curved portion of the operating characteristic curve, limiting follows and the modulation envelope is badly distorted. Reducing the bias to the value shown on Fig. 11-17 would produce proper undistorted operation.

Improper operation of a Class A amplifier also may occur if the grid bias is too low. Figures 11-17 and 11-19 differ only in the shifting of the operating axis from the center of the curve, where it operated properly, to the lower knee of the curve where distortion was produced. If it were shifted to the curved upper knee, limiting would occur over the upward swings of modulation. Another type of distortion would also enter, as described below.

When the voltage on the grid of a tube becomes positive a current flows through the impedance of the input circuit, causing a voltage drop that changes the bias point of the amplifier. It is therefore necessary, if

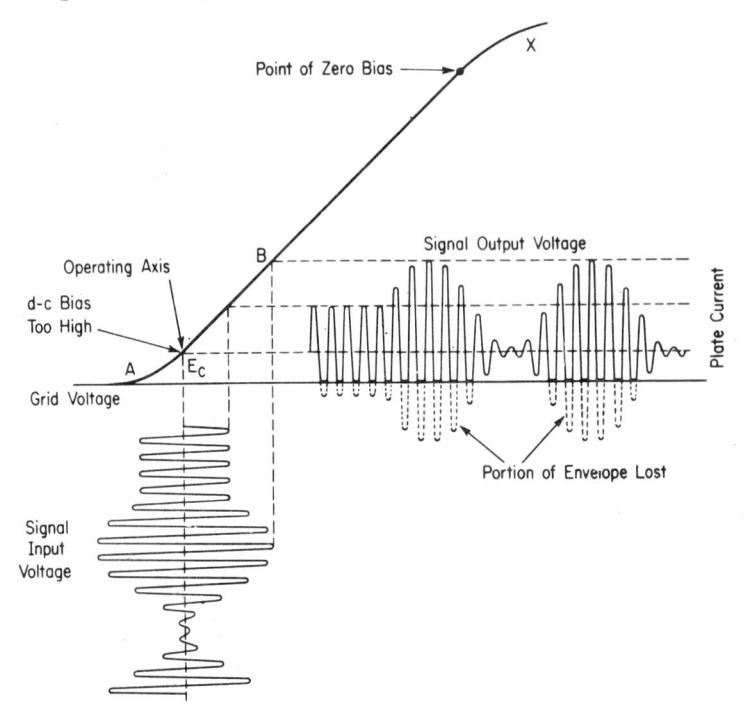

Fig. 11-19. Relation between grid voltage and plate current in Class A amplifier when d-c bias is too high and nonlinear distortion occurs.

grid current is drawn, to make the resistance of the input circuit low. When it is low, a comparatively high input power is required to produce the voltage necessary on the grid to obtain the desired output. Class A amplifiers are used to obtain high voltage amplification, and therefore high-impedance input and output circuits are used. Owing to the high-impedance input circuit, grid current should not be allowed to flow in a Class A amplifier. Operation where grid current is drawn is shown in Fig. 11-20. The bias voltage E_c is exceeded by the peak signal input during part of the time, and therefore the grid draws current in that interval. The grid current flowing in the internal impedance of the input circuit causes the terminal voltage to be less than the open circuit voltage, so that the actual voltage applied to the grid is distorted during this part of the cycle as shown on Fig. 11-20. Operation between A and C is linear, and therefore the input voltage is faithfully reproduced over that portion. From C to B grid current flows, causing the reproduction to be nonlinear, as indicated by the envelope of the modulated wave.

Distortion in a Class A amplifier will be indicated if a d-c ammeter in the plate circuit changes during operation. Although the plate current fluctuates above and below its axis, the positive and negative fluctuations

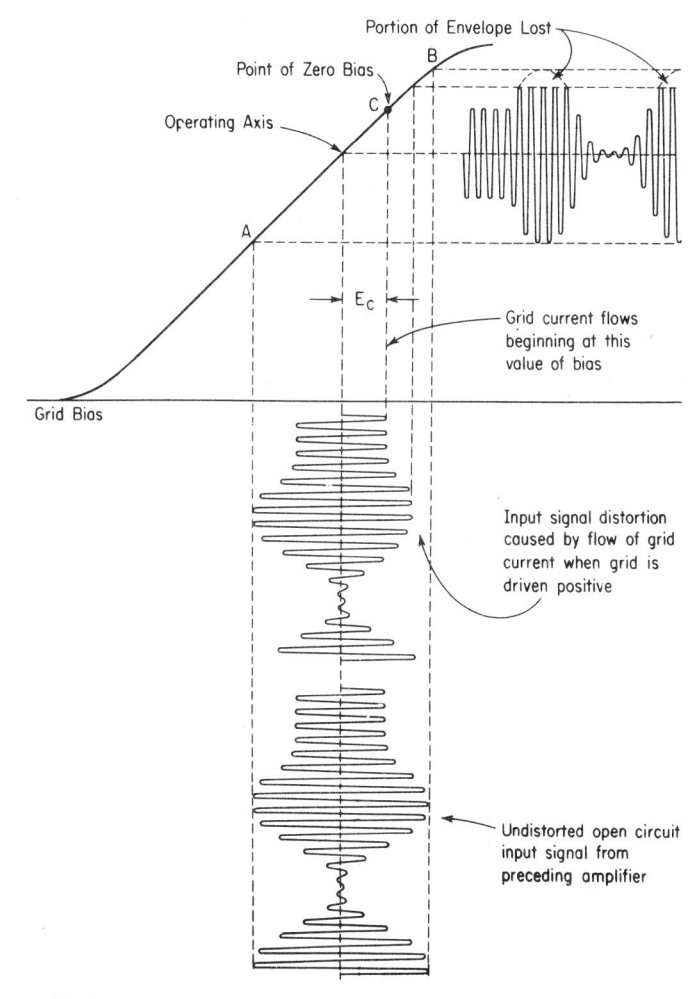

Fig. 11-20. Relation between grid voltage and plate current in Class A amplifier when grid is driven positive, impedance presented to preceding driver stage is reduced, and grid input voltage is caused to drop.

are equal for proper operation and the average current is constant. Nonlinear distortion will cause the plate current to attain greater amplitude in one direction than the other and produce a change in average current.

HEAT DISSIPATION LIMITS IN POWER TUBES

Power is required to heat the cathode of a tube. Because plate circuit efficiency is less than 100%, part of the plate power input is wasted in the form of heat. Power in Class B and Class C amplifiers is also dissipated in the grid circuit. This heat can destroy power tubes if not confined within a safe heat-dissipation rating. For this reason it has been customary to rate tubes in terms of allowable watts or kilowatts of plate heat dissipation. So long as this is not exceeded there is a wide range of permissible adjustments.

The heat dissipated in small power tubes may require no forced air or water cooling. In much larger sizes both may be required, water for the plate cooling and forced air for the grid and seals. In some television power tubes water cooling of the grid has been utilized also to permit handling of larger amounts of power in small spaces. Water-cooled power tubes are clamped in water jackets. Water is forced over the plate and pumped to evaporation coolers or other devices for reducing the temperature and then again is piped to the jacket. In one obsolete type of 50-kw broadcast transmitter 100 kw of plate heat is required to be drawn off the power output stage. Pure water is required. Contamination with salts and other impurities reduces the resistance of the water path to ground so much that the 10,000- to 18,000-v plate voltages cause heavy d-c currents to flow. Plates of air-cooled power tubes are equipped with cooling fins which channel the cooling air around the anode and associated fins.

NEUTRALIZATION

The grid-to-plate capacity of a triode tube is very small, and therefore at audio frequencies the impedance of this capacitance is extremely high. At radio frequencies, however, an appreciable amount of energy from the plate circuit may feed back to the grid circuit through this capacitance. Feedback from the plate circuit to the grid circuit may cause the tube to start oscillating which interferes with normal operation as an amplifier. It is possible to counteract the energy feedback from the plate to the grid. This process is called *neutralization*. In Fig. 11-21(a), C_{pg} represents the interelement plate-to-grid capacitance. Energy from the output of the tube is fed from plate to grid through C_{pg}. This feedback usually causes oscillation. The feedback through C_{pg} may be neutralized as shown schematically in Fig. 11-21(a). The tuning capacitor for the output circuit is really two capacitors in series, allowing the ground return to be made on the center of the two. The voltage from N to M is across the plate and cathode. This feeds voltage back to the grid through capacitance C_{pg}. The voltage

Chapter 11   *Radio-Frequency Amplifiers and Modulators*   367

from M to O is opposite in polarity or 180° out of phase with the voltage across MN. The voltage across MO causes a current to flow through the neutralizing capacitor C_n to the grid. If the capacitance of C_n is made equal to the plate-to-grid capacitance C_{pg}, the same voltage is fed to the grid

(a) Plate Neutralization

(b) Plate Neutralization

(c) Plate Neutralization

(d) Grid Neutralization

(e) Direct Plate-to-Grid Neutralization

(f) Cross Neutralization

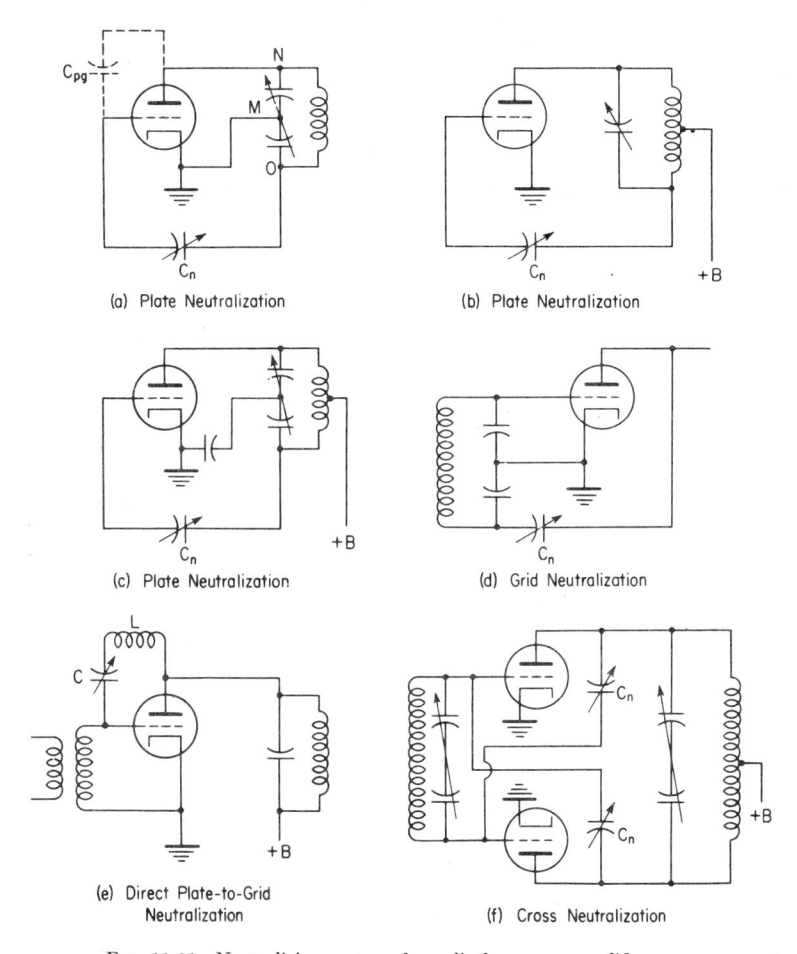

Fig. 11-21. Neutralizing systems for radio-frequency amplifiers.

through C_n as is fed through C_{pg}. Inasmuch as the resultant voltages on the grid are equal and of opposite phase, they cancel out at the grid, thus preventing oscillation.

Neutralization may be accomplished in many ways. Figure 11-21(b) and (c) shows two practical forms of plate neutralization. C_n in both cases is the neutralizing capacitor. In practice the neutralizing capacitor is made variable and of a capacitance in excess of the plate-to-grid capacitance.

The reason for this higher capacitance is that stray capacitances between wiring and pieces of equipment in the circuit will require a neutralizing capacitance higher than the plate-to-grid capacitance. The type of neutralization shown in Fig. 11-21(a), (b), and (c) is called *plate neutralization* because energy is fed from the plate circuit back to the grid.

Grid neutralization may also be employed. This is illustrated in Fig. 11-21(d). The operation is identical with plate neutralization except that voltage is fed from the grid circuit to the plate. C_n is the neutralizing capacitor.

Figure 11-21(e) illustrates a type of neutralization which is often used in transmitters. C_{pg} represents the plate-to-grid capacitance of the tube. Capacitor C and inductor L represent the neutralizing circuit. Inductor L, of the proper value, is connected from plate to grid. In practice, a capacitor C is inserted in series with inductor L so that the plate voltage is not impressed on the grid. The current fed from the plate to the grid through L is out of phase with the current fed through C_{pg}. If L is of the proper value the resultant voltages on the grid will be equal and out of phase. The voltage fed back through C_{pg} is neutralized on the grid by the voltage fed through L.

Another type of neutralization is illustrated in Fig. 11-21(f). In push-pull the neutralization is usually what is called *cross neutralization*. The two capacitors in Fig. 11-21(f) labeled C_n are the neutralization capacitors. As may be seen, they cross-feed from the plate of one tube to the grid of the other.

CLASS A PENTODE RADIO AMPLIFIER

Pentode tubes have a number of characteristics that make them ideal for use as voltage radio amplifiers. The principal advantages are large voltage or power amplification per stage, and freedom from the need for neutralization. The elements in a pentode are the filament or cathode, control grid, screen grid, suppressor grid, and plate. The suppressor grid is located between the plate and the other elements. When connected to the cathode or grounded, it behaves as an electrostatic shield between the plate and the control grid. It also minimizes the effects of secondary emission from the plate.

Figure 11-22 shows the diagram of a pentode radio amplifier. G_3 is the suppressor grid. There is an interelement capacitance between the plate and the suppressor grid. As may be seen, because of the radio-frequency by-pass capacitor C_4 this capacitance is virtually across the output inductance L_3 and thus forms part of the output circuit. There also is an interelement capacitance between suppressor grid G_3 and control grid G_1. This capacitance is virtually across the grid input circuit and therefore becomes part of the input tuning circuit. The two capacitances are independent of each other owing to the fact that the suppressor grid is connected to the cathode. The capacitance from the plate to the control grid

Chapter 11 *Radio-Frequency Amplifiers and Modulators* 369

G_1 is, for all practical purposes, not present. The residual grid-to-plate capacitance of a typical pentode tube is about 0.008 $\mu\mu f$. If the suppressor grid and the screen grid were removed, the capacitance would be several

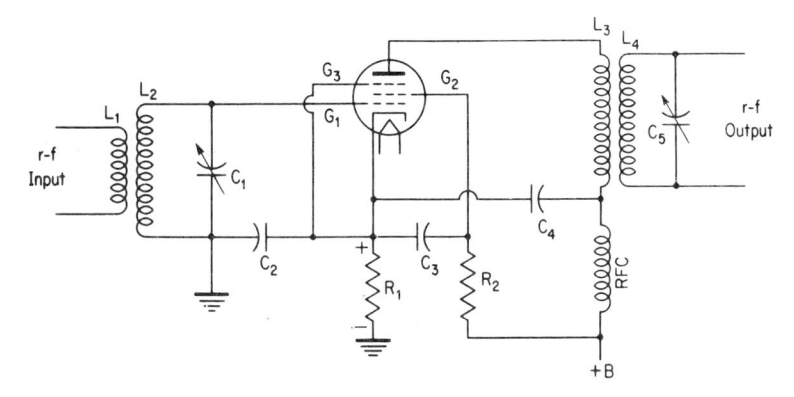

Fig. 11-22. Typical pentode radio-frequency amplifier applicable to either receivers or transmitters. Receiver would preferably use Class A adjustments.

micromicrofarads. The screen grid reduces the plate-to-grid capacitance so that it cannot feed enough energy back from the plate to the grid circuit to cause oscillation. The screen grid, accordingly, eliminates the need for neutralization.

Capacitive or inductive coupling between the plate and grid circuits of a radio amplifier, if of sufficient value, will cause oscillation. Referring to Fig. 11-22, if the output circuit coils L_3 and L_4 are placed in inductive relation to the input circuit inductors L_1 and L_2, there may be feedback causing oscillation. Oscillation could also be caused by running the leads from the grid and plate too close to each other. Other circuit components of the input and output circuits, if placed in proximity to each other, may cause enough feedback to make the tube oscillate.

The fundamentals of the circuit are the same in a receiver or a transmitter. An indirectly heated cathode is shown in Fig. 11-22. Use of this type of cathode is common practice in receivers. For high power in transmitters a directly heated cathode is usually preferred. The input resonant circuit to the tube is composed of L_2 and C_1. The signal input voltage impressed on L_1 is induced into L_2. L_1 and L_2 comprise a radio-frequency transformer. L_4 and C_5 inductively couple into the plate circuit through the transformer formed by L_4 and L_3. This sequence constitutes the output resonant circuit of the amplifier. The plate of the tube receives its current through an r-f choke and L_3. The r-f choke stops the radio-frequency energy from passing through to the source of the plate voltage. The need for this blocking will be discussed in detail in a later paragraph. Screen grid G_2

receives its voltage through resistor R_2. Screen grids usually operate at a voltage somewhat lower than the plate voltage and they also draw a small amount of current. Screen-grid current passing through resistor R_2 causes a voltage drop that effectively reduces the voltage on the screen grid G_2. C_3 is an r-f by-pass from the screen grid to the cathode. R_2 and C_3 form a filter to keep the radio-frequency from passing through to the plate supply. R_1, connected between the ground and the cathode, is a *cathode resistor*. Current drawn by the plate and the screen grid G_2 passes through R_1 to the cathode, causing a voltage drop in R_1. The voltage across R_1 has the polarity shown in Fig. 11-22. Control grid G_1 is connected to the negative end of R_1 through inductor L_2. Owing to the polarity of the voltage drop in R_1 the control grid is negative in respect to the cathode. By proper selection of resistor value, the bias on the control grid may be set at the proper operating value. C_2 is a by-pass capacitor and serves two purposes. It is a short circuit for radio frequency, and therefore the input resonant circuit is effectively connected across the cathode and control grid. Capacitor C_2 also effectively by-passes the radio frequency around resistor R_1.

Figure 11-22 shows a practical circuit of a radio amplifier. Many circuit variations are possible. The class of the amplifier, the service that it is to perform, and the power govern the specific circuit that is used and the values of the component parts.

SHIELDING TO PREVENT SELF-OSCILLATION

Shielding is resorted to in order to prevent oscillation. The pentode tube itself is usually placed in a metal shield. In receivers this shield may be a small metal cylinder that fits over the tube. Many types of tubes are made with a metal envelope instead of glass and thus are sufficiently self-shielded. The input and output often need to be shielded. The tuning inductors often are fully shielded, and limited shielding is sometimes used around the variable tuning capacitors. It is also common practice in some applications to shield the lead from the input tuning circuit to the grid of the tube. In general it is important that the circuits of one amplifier be shielded from the circuits of others.

In triode amplifiers it is not so important as in other types to keep the various elements of an amplifier stage isolated from each other. However, it is advisable to follow the same general practice. In order to isolate the stages of amplification from each other, each stage is sometimes shielded. Shielding the tuning inductors is usually sufficient in receivers and low-powered transmitters. In high-powered transmitters, in addition to shielding the individual tuning circuit inductors, each amplifier stage may be shielded completely.

SHIELDING TO PREVENT RADIATION

Radio frequency amplifiers, where radio frequency currents flow through open wires, can produce radio waves capable of causing severe

Chapter 11   *Radio-Frequency Amplifiers and Modulators*   371

interference to other services. As an example, television receivers contain amplifiers for horizontal synchronizing pulses which, approximating square-wave shapes, are very rich in 15,750-cycle harmonics. These harmonics extend up through the standard broadcasting band from 540 kc to 1,620 kc and unless confined in the receiver may cause severe interference to broadcast reception in localized residential areas. As another example, amplifiers and oscillators of r-f currents often produce strong harmonic radiation which can produce long-distance interference to other services operating on these higher harmonic frequencies.

Radiation from circulating currents in the transmitter frame is called *frame radiation*, and it has caused problems as the radio frequency spectrum has become crowded with new services. Increasingly it has become necessary to provide elaborate shielding of radio-frequency amplifiers, particularly those carrying substantial amounts of power. This shielding ordinarily must completely enclose all amplifier and oscillator stages carrying substantial r-f currents, including the openings for control devices and meters. In addition, the r-f currents conventionally are filtered from all wiring leading out of the amplifier compartments, except those leading to the antenna. Double shielding in some instances has been necessary to adequately confine the fields at harmonic or spurious frequencies.

TETRODE AMPLIFIERS

Tetrode or screen-grid tubes contain a filament or cathode, control grid, screen grid, and plate. Screen-grid tubes may be used in radio amplifiers. The pentode tube has displaced the screen-grid tube in most receiving-set applications. The screen grid in a tetrode tube functions to reduce the plate-to-control-grid capacity. In addition to the screen grid, the pentode tube also has a suppressor grid. The plate-to-grid capacitance in a pentode is less than in a tetrode. Therefore an amplifier using a pentode is less liable to oscillate, and is usually preferred in radio amplifiers.

FILTERING D-C SUPPLY CIRCUITS TO RADIO AMPLIFIERS

In previous paragraphs mention was made of *filtering circuits* in plate and screen-grid leads from plate-supply sources, the necessity for isolating various circuit components from each other, and the importance of isolating cascaded amplifiers also from each other.

A series of amplifier stages is usually operated from the same plate and bias supplies. If cascaded stages of amplification are properly shielded it is still possible for feedback to occur through the plate or grid power supplies. A common way to prevent it is to use filters. Referring to Fig. 11-22, r-f choke and capacitor C_4 form a filter which prevents energy in the plate-circuit from feeding into the plate supply, where it in turn could feed into the plate circuit of a preceding tube and from there back to the grid. In some cases, this feedback could be enough to cause oscillation. R-f choke is a high impedance to the radio frequency. C_4 presents a very low impedance and therefore by-passes the radio frequency energy around r-f choke and the plate supply. In this manner the output is prevented from feeding into the plate power supply. In some cases a resistor may be substituted for r-f choke. Resistance R_2 and capacitor C_3 form a filter for the screen grid G_2 in the circuit of Fig. 11-22.

Figure 11-23 shows four different types of power-supply filters. These are arbitrarily labeled types 1 to 4. Type 1 is the same as the filter composed of r-f choke and C_4 in Fig. 11-22. Type 3 is the same as the filter

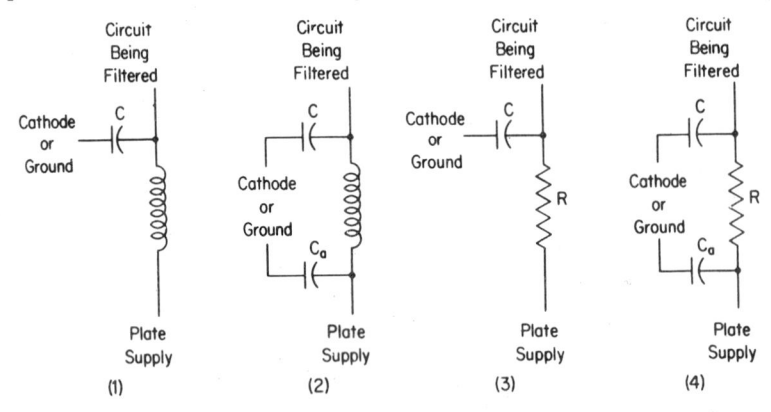

Fig. 11-23. Methods of filtering radio- or audio-frequency currents out of leads to power supply systems to prevent feedback through common power supply and to keep r-f out of rectifier tubes.

composed of R_2 and C_3 in Fig. 11-22. Type 2 and type 4 are the same as types 1 and 3 except that an additional capacitor C_a has been added. In most cases there is a large capacitor across the output of the d-c supply circuit and therefore C_a is unnecessary.

In transmitters, there is another reason why r-f supply circuit filters are important. Mercury-vapor rectifying tubes are in general use for changing 60-cycle alternating current into direct current for plate and bias power supplies. Radio frequency currents affect the operation of mercury-vapor rectifying tubes, and therefore it is important that such currents be kept from them.

AMPLIFIER TUBE COMPLEMENTS

The output power of radio transmitters varies from a fraction of a watt up to several hundred kilowatts. To be economical, the power capacity of the output tubes should not be much higher than required for the operating power of the transmitter.

It is impractical for tube manufacturers to produce a special tube for every use. Therefore it becomes necessary to use two or more tubes in the output amplifier in some cases.

Chapter 11   *Radio-Frequency Amplifiers and Modulators*   **373**

Use of combinations of two or more tubes in transmitter amplifiers has made it feasible to standardize to some degree the power capability of vacuum tubes. A multitude of different power outputs is obtained by proper combination of a limited number of vacuum tubes of various power ratings. When two or more tubes are used in an amplifier, they may be operated either in *parallel* or in *push-pull*, and must be similar.

PARALLEL OPERATION OF POWER TUBES

R-f amplifier tubes operate effectively when connected with their respective elements in parallel, as shown by Fig. 11-24, if certain conditions are observed. The tubes must be of the same type, and even so minor differences exist within the rating tolerances which may cause the loads to

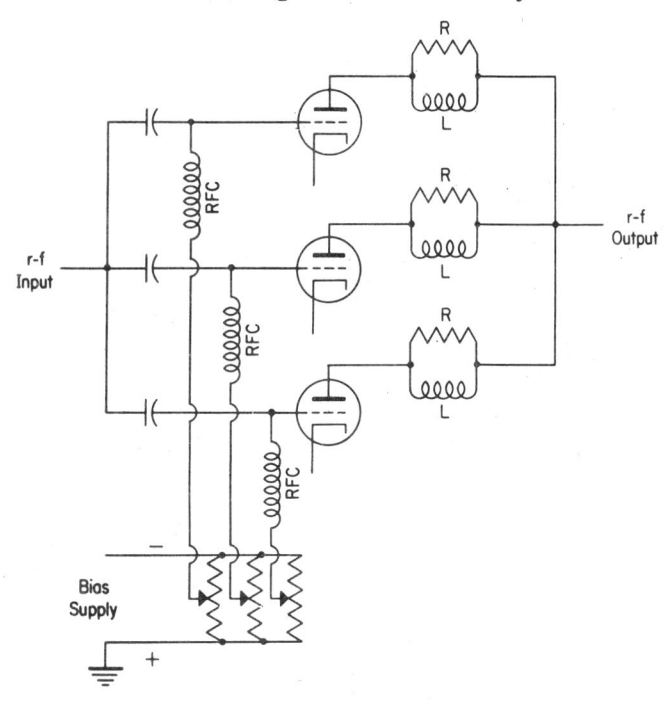

FIG. 11-24. Parallel operation of radio-frequency amplifier tubes.

be somewhat unequal. When operated close to maximum rating it is often desirable to select tubes which match closely in operating characteristics. The selection may be made by comparing the grid and plate voltages and marking the serial numbers of tubes which are most nearly identical in performance. In high-power amplifiers separate grid bias control also may be provided to insure equal division of plate power input and output. (See Fig. 11-24.)

Parallel-tube amplifiers are more vulnerable to parasitic oscillations than single-tube amplifiers because of the greater complexity of the r-f wiring and the greater number of distributed circuit elements such as tubes, resistors, capacitors, etc. It is not an inherent characteristic of parallel-tube amplifiers to oscillate but, when they do, simple remedies may suffice, such as the insertion in the plate circuits of damping circuits LR such as are shown in Fig. 11-24. The inductors L may be small at 1,000 kc and consist of 10 or 15 turns of wire on a two-inch diameter, shunted with a resistor of about 35 ohms. These damping circuits may be very effective at the high parasitic frequency without appreciably affecting performance at the desired frequency.

Since parallel oscillations are produced by output voltage being fed back to the grid circuit it is necessary that the grid leads be short. One method of making them so is to arrange the tubes in a circle as shown in Fig. 11-25. This usually provides the shortest possible connections and is a very

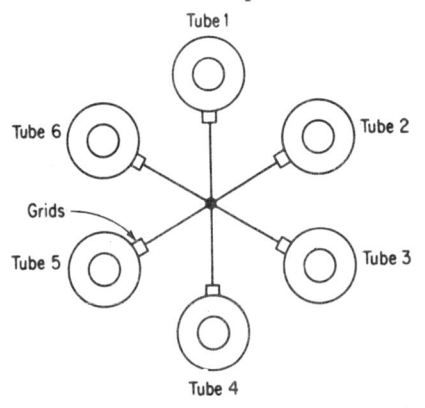

Fig. 11-25. Ring arrangement of parallel connected power tubes.

advantageous arrangement, particularly when the grid terminal is brought out at some distance from the plate terminal, as for instance from opposite ends. Both grid connections and plate connections are then short and direct.

PUSH-PULL OPERATION OF POWER TUBES

The second method of using multiple tubes in a radio amplifier is to operate them in push-pull. This requires an even number of tubes. Push-pull operation has several advantages over parallel operation.

Cross neutralization of a push-pull amplifier does not complicate the circuits as much as does neutralization of a single-ended amplifier. Two tubes in push-pull operation will give a slightly higher power output than will the same two tubes in parallel. Also, the even harmonics produced by the tubes are canceled out provided the circuits are carefully balanced. A

Chapter 11   *Radio-Frequency Amplifiers and Modulators*   375

circuit diagram of a push-pull triode radio amplifier is shown in Fig. 11-26(a). L_2 and C_1 comprise the resonant input circuit. Input tuning capacitor C_1 is a double-section assembly. Each section is in effect a capacitor by itself, connected in series so that the midtap may be grounded to form a balanced input circuit and so that the voltage impressed on the

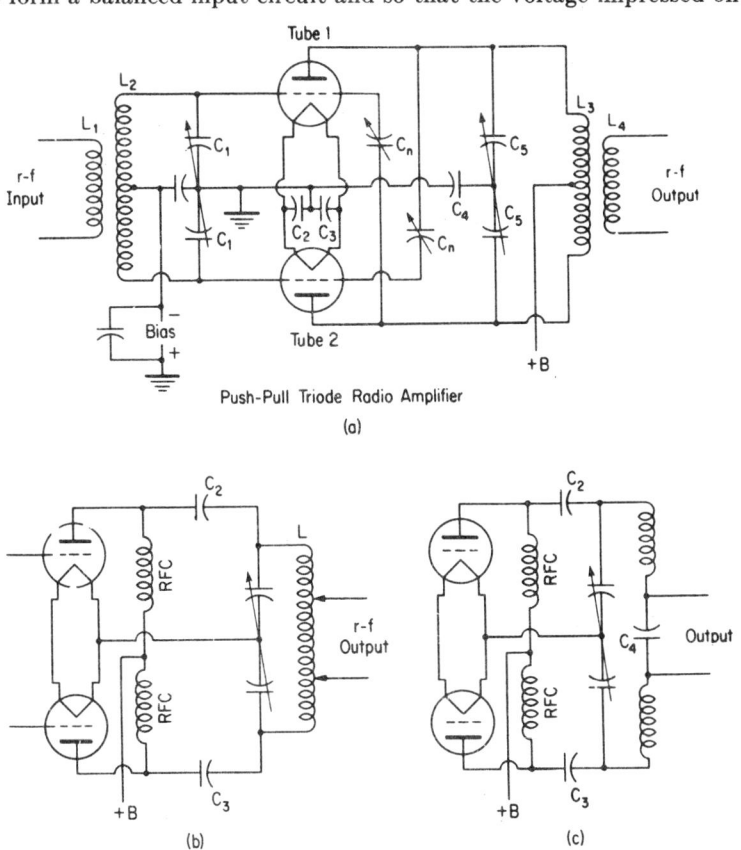

FIG. 11-26. Push-pull r-f amplifiers.

grid of each tube shall be the same. The output circuit consists of inductors L_3 and L_4 and capacitor C_5 and is similar to the input circuit. Capacitor C_5 is connected from the midtap of the tuning capacitor to ground, thus giving a balanced output circuit so that each tube works into the same impedance. The two capacitors labeled C_n are for neutralizing and are connected for cross neutralization.

Slightly different output connections are illustrated in Fig. 11-26(b) and (c). In (b) the output is directly coupled by tapping the plate tank coil in the center. This requires a different form of plate power feed so r-f chokes are added for r-f isolation between plate and power supply. The circuit has been simplified to emphasize the principal feature.

In Fig. 11-26(c) the circuit corresponds to that of Fig. 11-26(b) except that the plate tank coil is split in the center and a coupling capacitor, C_4, is added. Since C_4 has a low impedance at harmonic frequencies, compared with the carrier frequency, there is less harmonic voltage produced across the output terminals and consequently harmonics are reduced in relative amplitude. Tubes may be used in parallel on each side of a push-pull circuit.

HIGH-LEVEL AND LOW-LEVEL MODULATION

A transmitter in which the last radio stage is plate modulated performs the modulation function at high power level. Consequently such modulation is referred to as *high-level modulation*.

Often it is more economical or more satisfactory to perform the modulation function in a low power stage and amplify the modulated carrier in linear amplifiers. This is referred to as *low-level modulation*.

By their nature Class C amplifiers cannot amplify linearly an amplitude modulated wave, as will be explained. Such linear amplification requires that the grid and plate circuits be so adjusted that the grid input r-f voltage is faithfully reproduced in the output circuits. For this purpose the adjustments of the Class A or Class B amplifiers are required.

CLASS B RADIO AMPLIFIERS

Class B radio amplifiers are used principally in transmitters employing low-level modulation, where they have the advantage of much higher plate efficiency than Class A amplifiers and much better linearity than Class C amplifiers. Because push-pull amplifiers substantially cancel out even harmonics, Class B radio amplifiers are normally designed for push-pull operation. A Class B amplifier always produces a small amount of distortion at the lower extremity of the grid voltage plate current curve because there is an area where the curve is not linear. By using tubes in push-pull, the distortion which results from one tube is largely cancelled out by the other. For explanation of the operation reference is made to Fig. 11-27. Complete plate current cutoff would not be produced unless a disproportionately high bias were applied because of curvature near the cutoff point, Point A. A small amount of plate current will always flow and therefore is not adequately modulated to achieve perfect linearity.

Figure 11-27(a) shows a single-tube Class B amplifier operating curve. For simplicity it will be assumed that the curve A, B, C is straight and linear. During the time interval 1 to 2, unmodulated signal voltage is impressed on the grid. Owing to the operation at the cutoff point, the negative half of each cycle of the signal voltage impressed on the grid theoretically produces no current in the plate circuit. The result is that half-cycle pulses of plate current are made to flow as shown in interval 1 to 2. These pulses are impressed on the resonant output circuit. A resonant circuit has the property that if part of a cycle is impressed upon it, the pulse excites the resonant circuit and produces a full wave. It is possible, therefore, but not practical to operate Class B amplifiers single-ended. Figure 11-27(b)

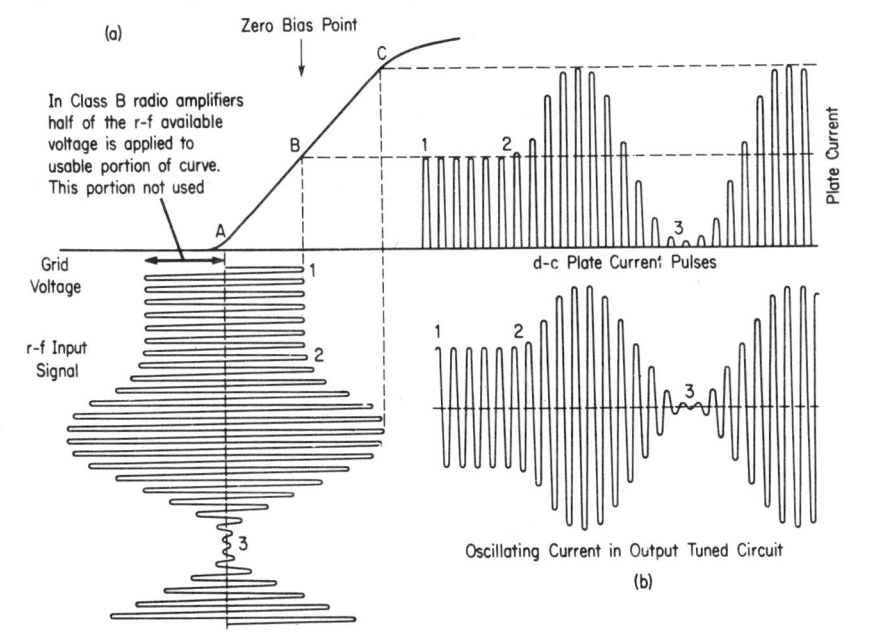

FIG. 11-27. Current flow in Class B r-f amplifier.

represents the instantaneous tank current circulating in the output resonant circuit. Time interval 1 to 2 corresponds to interval 1 to 2 on Fig. 11-27(a) and represents the full-sine-wave current flowing in the tank circuit as a result of the application of the succession of half-wave pulses of power from the plate.

Class B radio amplifiers may be operated over the characteristic curve extending up to where the top slopes off. The operating portion of the curve as shown in Fig. 11-27(a) is from A up to where the curvature starts at point C. To amplify faithfully a 100% modulated radio signal, the signal input voltage without modulation must be half of AC or up to point B. During the interval 2 to 3 the carrier is 100% modulated by an audio frequency. As may be seen from Fig. 11-27(a), the part of the signal input voltage to the left of the operating axis theoretically has no effect on the instantaneous plate current i_p.

The plate circuit pulse train and the tank circuit oscillating current have an envelope which ideally duplicates the input r-f envelope, as shown between 2 and 3 in Fig. 11-27(b).

As explained previously, Class B radio amplifiers conventionally are made for push-pull operation to help overcome practical imperfections of secondary importance inherent in single-ended Class B amplifiers.

In Fig. 11-27(a), curve ABC does not increase ideally in the downward-modulation direction because of curvature, so in a single-ended amplifier there is limiting which causes distortion of the envelope. But in a push-pull amplifier, when this tube is limiting, the opposite tube is at the positive upward extreme of modulation and is not limiting. Envelope distortion is thereby reduced. In push-pull operation one tube produces pulses in the plate circuit as shown in Fig. 11-27(a). During the time this tube is not producing pulses the second tube does so, and therefore the plate tank circuit receives pulses at each half cycle, one from each tube. The even harmonics which are undesirable in transmission are at the same time greatly reduced, because they represent currents of about equal amplitude but opposing polarity.

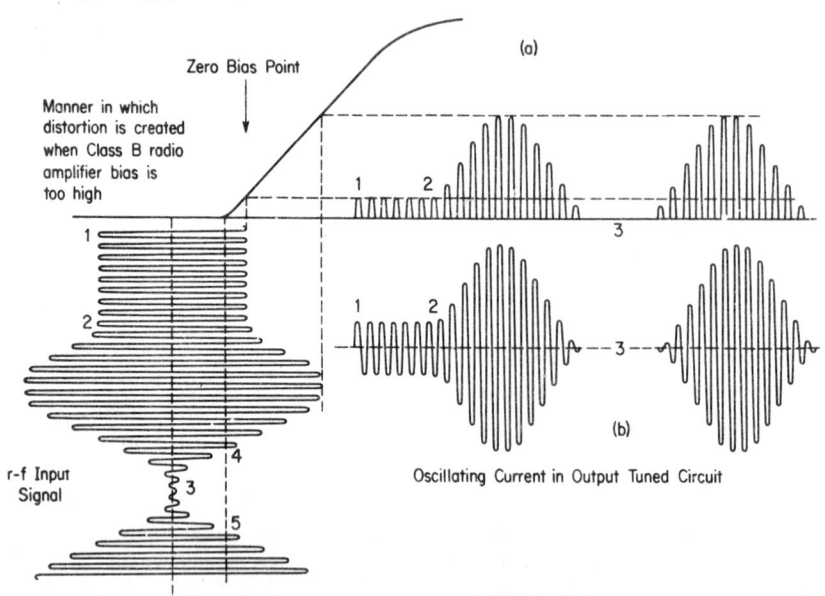

Fig. 11-28. Current flow in Class C r-f amplifier excited by a modulated wave.

As may be seen from Fig. 11-27(a), the signal input voltage at 100% modulation exceeds the bias voltage and therefore the grid goes positive. For this reason the impedance of the grid input circuit must be low. Therefore a higher input power is required for a Class B amplifier than is required for a Class A radio amplifier where the input impedance is high. A plate efficiency as high as 35% is obtainable from Class B radio amplifiers.

Figure 11-28 shows the effect produced when the bias is too high. During modulation the signal input voltage between points 4 and 5 is such that there is complete cutoff of the plate current, and therefore the envelope of the radio-frequency pulses in the plate circuit is not a sine wave. Part of the full envelope has been cut off. [See Fig. 11-28(b).]

The single-ended amplifier circuits shown so far and the push-pull radio amplifier shown in Fig. 11-26 could be operated as Class B radio amplifiers if the constants of the circuits are properly selected and the voltages are properly adjusted.

CLASS B MODULATORS

Class B audio amplifiers may operate at efficiencies as high as 50% to 70% for maximum output. They are employed extensively in high-power high-level modulation because of the simplicity, efficiency, and high quality obtainable. In plate modulation the entire plate input power to the modulated stage must be modulated. This includes both that part which is converted to useful output and the part which is dissipated in losses. For a 50-kw telephone transmitter having a modulated amplifier plate efficiency of 75%, the amount of plate input power which must be modulated is 67 kw. To modulate the input power fully, one half as much audio frequency power may be required, so the modulator of the 50-kw transmitter must be capable of supplying 33.5 kw. A Class A amplifier, with its inherent low plate efficiency, would be very large and expensive.

Figure 11-29 shows a circuit diagram of a Class B audio push-pull modulator. The operation is simple. Conventionally the output is transformer-coupled to the load consisting of the modulated amplifier circuits.

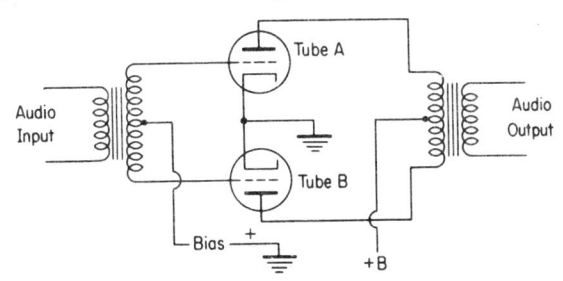

FIG. 11-29. Class B audio-frequency modulator.

There is one very important difference between Class B radio and Class B audio amplifiers. In the radio amplifier, half-wave driving pulses are produced. These excite the tuned circuit which, having oscillating r-f currents, reproduces the missing half wave. This does not occur in an audio amplifier. Therefore, Class B audio amplifiers for faithful reproduction of an input signal are always required to operate push-pull. The circuit diagram of Fig. 11-29 differs in no way from the circuit of a Class A push-pull amplifier. The only differences are those of voltages, currents, and impedances which a circuit diagram does not show.

In Class B audio modulators or amplifiers, voltages on a hypothetical positive side of the signal voltage axis are amplified by one tube and on the hypothetical negative side by the other tube so that the full wave is reproduced. One tube is idle while the other one works, and they constantly alternate to produce in combination the full wave.

Figure 11-30 shows the grid-voltage-plate-current relation of two tubes used in Class B audio amplification. Bias voltage E_c fixes plate current at the approximate cutoff point of the characteristic curve.

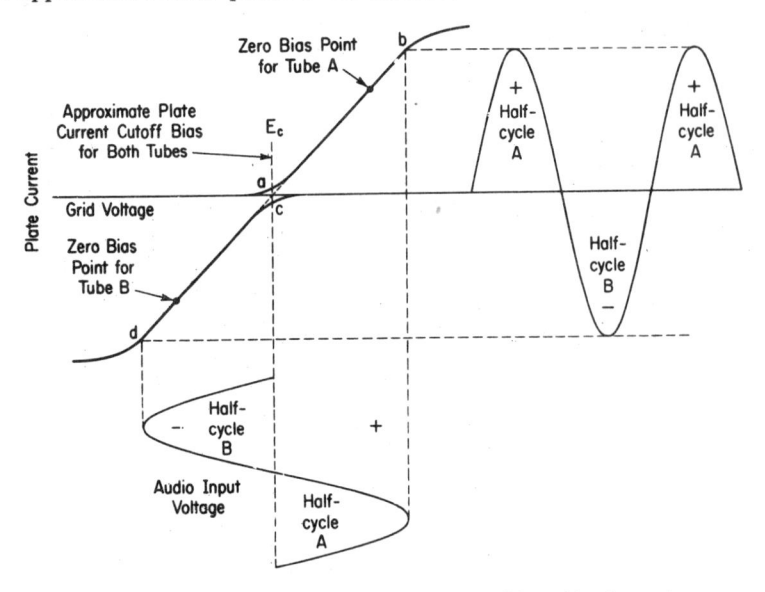

Fig. 11-30. Class B audio modulator or amplifier-grid-voltage-plate-current relations.

The characteristic curve of tube A is shown in the upper portion of the diagram. The characteristic curve of tube B is shown turned in such a way that the two characteristic curves form a straight line. Both tubes are biased to almost the cutoff point so that there is little plate current flowing when there is no signal input voltage. During the half cycle labeled A the plate current of tube B remains very low or goes to zero. In tube A, during the half cycle A, the total grid voltage (the sum of the bias voltage and the signal input voltage) becomes more positive, and operation takes place over the portion of the characteristic curve of tube A from a to b. A half-cycle pulse of plate current is produced. During the half cycle B of grid input, the grid of tube A becomes more negative, and therefore the plate-current flow from tube A remains very low or goes to zero. The total voltage on the grid of tube B, however, becomes more positive and operates over the portion of the characteristic curve of tube B from c to d. This causes a half-cycle pulse of plate current to flow. Half cycles supplied by tube A and tube B form a full-wave reproduction of the input signal voltage.

The grids in Class B audio amplifiers may go positive, and therefore draw grid current, so it is necessary to use a low-impedance input. This requires a relatively high audio frequency grid input power to produce the necessary undistorted grid driving voltage. For proper operation it is necessary that the bias voltage shall not change when the input signal is applied. Without an input signal no current is drawn from the bias supply. However, as soon as the input signal becomes high enough so that the total voltage on the grid is positive, current is drawn by the grid. In order that the bias voltage may be held constant the regulation of the bias-voltage supply must be good.

Pulses of current are drawn from the plate supply by the tubes. The amount varies from near zero to the total plate current at the crest of the output audio-frequency wave. This range is large. As in the grid circuit, it is necessary to maintain the plate voltage nearly constant. It is therefore necessary to have a plate supply with good regulation for a Class B audio amplifier. To operate properly, Class B audio-frequency amplifiers call for careful design to meet the rigid requirements.

Class B audio amplifiers using triode tubes have a relatively low power amplification. Pentode tubes produce a higher power amplification but have disadvantages as will be discussed in a later paragraph. The low power-amplification gain of a class B audio amplifier using triode tubes is more than offset by the high power output and high plate efficiency for applications where a large amount of audio power is required.

CLASS C RADIO AMPLIFIERS

Class C radio amplifiers have two outstanding advantages for radio transmitters: with them the greatest possible amount of r-f output power may be achieved, and they operate with the highest plate efficiency of any class of amplifier. It follows that they are very widely used in amplifying small amounts of power up to desired carrier power levels and in producing large amounts of power most efficiently. They are always used in plate modulated stages.

The plate current excursion encompasses the whole range from nearly zero to almost full saturation during each positive half cycle of driving signal voltage. Therefore, during most of the time that plate current flows it is employed in producing output power. There is no zero-signal plate current. When r-f excitation is lost the plate input power drops to zero.

The r-f voltage produced across the plate tank circuit is proportional to the amplitude of the plate current pulses that produce and sustain it. The greater the amplitude of these pulses, the higher becomes the r-f voltage. It is not necessary that the pulses be half cycles in duration, but only that they occur at the correct time. If they were of full amplitude but very short duration, the plate tank voltage would still receive a powerful stimulus. Because of this narrower pulse a high efficiency may be achieved.

When no current flows between plate and cathode, no plate power is dissipated. Power is lost, because the tube is not 100% efficient, but the loss occurs only during the intervals when plate current flows to produce a pulse of power to the tuned output load circuit. If this plate current flow is produced at the most opportune time and cut off at other times, the efficiency is increased.

The plate power input to the amplifier at any instant is the product of the instantaneous plate voltage and the current flowing at that instant. Part of this power appears in the load and the remainder appears in the tube. The portion appearing in the tube is the product of the instantaneous plate current and the instantaneous voltage drop between plate and cathode. The portion appearing across the load at the same instant is the product of the plate current and the instantaneous voltage drop across the load. If the flow of plate current is restricted to those intervals when the drop across the load is highest, and correspondingly across the tube is lowest, high efficiency may be achieved.

Oscillating current flows in the output-tuned circuit. The r-f voltage across the load is shown on Fig. 11-31(a). It may be seen that the voltage varies constantly, reaching alternate high and low values with respect to the d-c plate voltage. At the time when the r-f voltage is greatest and of opposite polarity to the d-c plate supply voltage, the net voltage across the plate-cathode is very low. If the plate current is introduced at that time the power lost in heating the tube, represented by the product of the plate current and plate-cathode voltage drop, is lowest. Referring to Fig. 11-31 (a), it may be seen that the pulse of plate current, to occur at the most opportune time (when the plate voltage is lowest), must be of much shorter duration than one-half cycle. Otherwise the plate cathode voltage and the corresponding plate power loss increases.

The manner in which the short plate current pulse is created from a sine-wave r-f input voltage is shown in Fig. 11-31(b). It shows that the d-c grid bias is much higher than required to cut off the plate current in the absence of r-f driving voltage. When a "positive" half cycle of r-f is applied to the grid, as illustrated, only the effective portion shown in black is of sufficiently high voltage to cause plate current to flow. That portion occupies a much smaller interval of time than the full half-cycle 180°. Thus the narrow pulse of plate current is permitted to flow only at the most advantageous time, when the voltage drop between plate and cathode is lowest. The

Chapter 11 *Radio-Frequency Amplifiers and Modulators* 383

width of the plate current pulse for best efficiency may be as low as one third of the full 180° half cycle. It is controlled to the desired width by adjustment of d-c bias and r-f grid-driving voltages. Narrowest pulses are produced with very high d-c bias and correspondingly high r-f input voltage.

There is a practical limit to the r-f driving voltage which may be applied safely to the grid. Figure 11-31(a) shows that the d-c voltage between plate

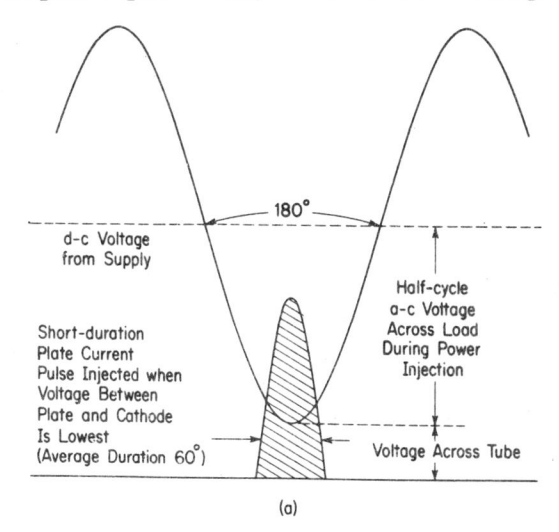

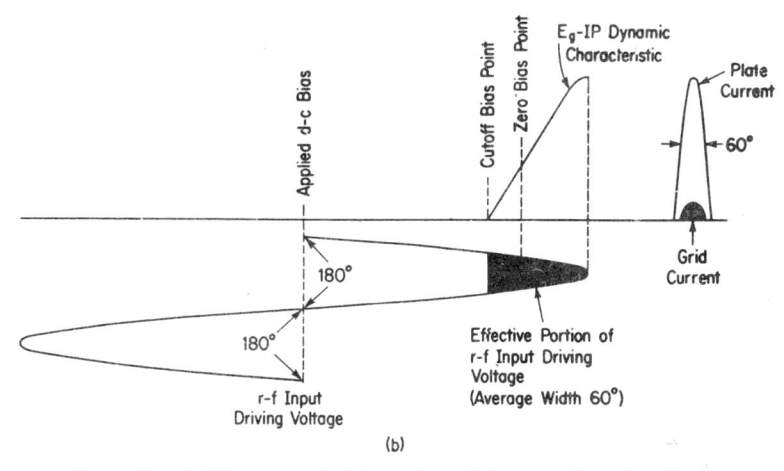

Fig. 11-31. (a) Manner in which high plate efficiency is achieved in Class C amplifiers; (b) manner in which short duration plate current pulses are created in Class C amplifiers.

and cathode drops to a small value during the time the r-f driving voltage is maximum. Suppose the grid voltage became substantially more positive than the instantaneous plate voltage. Then the electrons would no longer be attracted to the plate but would instead flow to the more positive grid. The grid current would rise sharply, perhaps enough to cause destruction through heating. Ordinarily tubes used for Class C service have maximum ratings for all important voltages and currents, including grid current. In designing Class C amplifiers the instantaneous r-f grid voltage is not permitted to exceed the instantaneous minimum plate voltage. Where maximum plate efficiency is desired, the instantaneous grid voltage may be almost as large but the grid driving power required increases rapidly. Where low grid driving power is an important consideration, the r-f grid voltage peak is kept substantially below the minimum plate voltage, the duration of the grid driving voltage pulses is extended, the plate efficiency is reduced, and the power output may be somewhat greater.

Class C amplifiers may be biased by means of a grid leak as explained earlier in this chapter. The value of the grid leak resistance required may be calculated by estimating the proper value of grid current from the published tube ratings and dividing it into the grid bias voltage desired, in accordance with Ohm's law. The heat dissipation in the grid leak would equal $I_g^2 R$.

The maximum voltage across the output tuned circuit is given by

$$E_0 = E_b - E_{p\,min}$$

where E_b is the plate d-c supply.

$E_{p\,min}$ is the minimum drop from plate to cathode. The power output to the tuned circuit is given by

$$P_{out} = \frac{E_0 I_p}{2}.$$

The power input to the grid is

$$P_g = \frac{E_g I_g}{2}.$$

The power amplification is

$$P_{amp} = \frac{E_0 I_p}{E_g I_g}.$$

Class C radio amplifiers of large size may achieve plate efficiencies of 80% without notable difficulty and 90% in extreme cases.

Class C amplifiers are commonly used as plate-modulated output stages in broadcast and many other types of transmitters. Modulation may be accomplished very satisfactorily and efficiently by causing the plate supply voltage to be varied in accordance with a modulating signal. Conventionally, the voltage to the amplifier is changed by connecting a modulator in series with the plate supply source. When the peak modulator a-c output voltage equals the d-c supply voltage 100% modulation may be achieved. On negative half-cycle peaks the modulator voltage is equal and of opposite polarity to the d-c supply voltage, they cancel, no voltage is applied to the plate, and the instantaneous output of the amplifier becomes zero, for 100% minimum modulation. During positive half-cycle peaks the modulator voltage adds to the d-c supply voltage, the voltage applied to the plate is therefore twice normal, the plate current becomes approximately twice normal as a result, and the instantaneous power output is thus quadrupled. Plate modulation of Class C amplifiers is described more completely in Chapter 12.

BALANCED MODULATORS

Chapter 10 explained the performance and advantages of suppressed carrier transmission. The production of sidebands without the carrier is

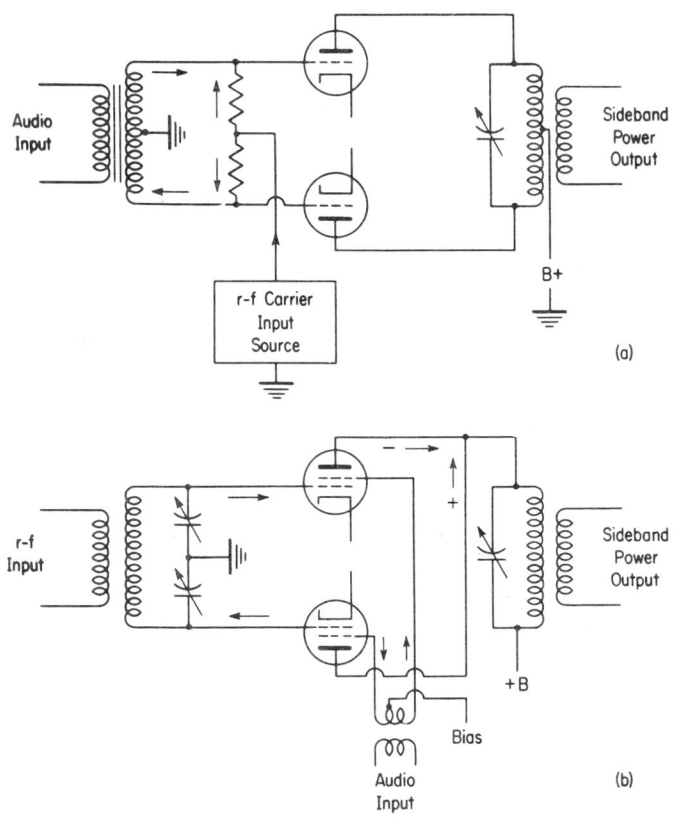

Fig. 11-32. Balanced modulators for producing sideband power with carrier suppressed.

accomplished in balanced modulators. Figure 11-32(a) shows a simple type of balanced modulator.

Audio frequency is applied to the grids in push-pull from the audio input transformer. The plate circuit is also connected push-pull. However, the r-f carrier is inserted at the center of resistor R and goes to each grid in what is called "push-push." The result of this unique connection is to produce only sidebands in the tuned output r-f circuit. The push-push carrier voltage on the grids causes plate currents to flow into the output-tuned circuit equal in amplitude and opposite in phase, so that no net r-f voltage appears across the tuned circuit. However, when audio voltage in push-pull is applied to the grids, there is produced sideband power in the plate circuit at frequencies representing the sum and the difference of the audio and radio frequencies applied to the grid.

Figure 11-32(b) shows a balanced modulator using screen grid tubes. In this case the r-f input is in push-pull on the control grids, the audio input is in push-pull on the screen grids, and the plate circuits are connected in parallel. R-f plate voltage from the tubes is equal in amplitude but opposite in phase, so again no r-f carrier voltage appears across the tuned circuit.

PENTODE RADIO AMPLIFIERS CLASS B AND CLASS C

Pentode tubes, capable of a power output of several hundred watts, are available for use in transmitters. In addition to having a large voltage amplification factor, they also provide high power amplification. Pentode tubes are in extensive use in transmitters in stages previous to the modulated amplifier.

WIDE-BAND TELEVISION INTERMEDIATE FREQUENCY AMPLIFIERS

Unique requirements for wide-band radio amplifiers are met in television receivers. As explained more fully in other chapters of this book, the receiver must be designed to pass a wide band of frequencies of several megacycles and then cut off very sharply above the carrier, and provide a specified slope below the carrier.

The following table shows in the column at the left the Channel 4 frequency specifications.

CHANNEL 4 Frequencies (megacycles)	HETERODYNE OSCILLATOR Frequency (megacycles)	I-F AMPLIFIER Frequencies
Lower limit of channel..........;..66	113	47
Picture carrier..................67.25	113	45.75
Color subcarrier................71.05	113	41.95
Sound carrier..................71.75	113	41.25
Upper limit of channel..........72	113	41

Chapter 11   *Radio-Frequency Amplifiers and Modulators*   **387**

The center column gives the Channel 4 heterodyne oscillator frequency used to beat the carrier frequencies to the intermediate frequencies. At the right are shown the intermediate frequencies. Figure 11-33 shows the desired intermediate frequency-amplifier voltage-amplification characteristics for finest detail reception. Sufficiently wide band characteristics are

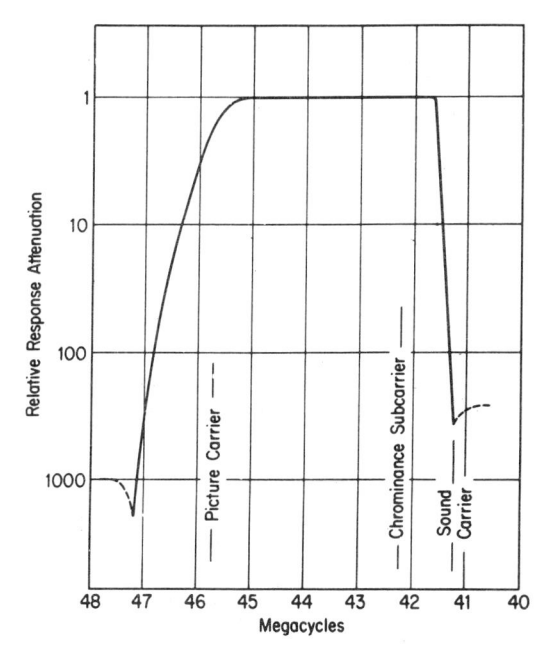

FIG. 11-33. Ideal response of an intermediate-frequency amplifier for television receiver. Sound takeoff separate.

not practicably attainable with the simple types of tuned i-f amplifiers in which several stages in cascade employ circuits tuned to the same frequency. Several methods of obtaining the desired results are used. One method consists of using several stages, each tuned to a slightly different frequency and staggered to span the desired wide-band of frequencies. This is called *stagger tuning*. For example, three amplifiers in cascade could be tuned respectively to 42, 43.5, and 45 megacycles. Since there are three stagger-tuned stages, such a combination is referred to as a *stagger-tuned triple*. By the judicious design of individual circuits the response curves would overlap and produce uniform response.

Another method employed is the use of over-coupled double-tuned circuits. Chapter 3 illustrated how different amounts of mutual coupling between tuned circuits may be employed to control bandwidth and how it may be additionally controlled by adding resistors.

Another method is the use of complex filter networks as coupling sections between stages.

Stagger tuning may be utilized in the coupling circuits of a single i-f amplifier stage, or may be employed between successive stages to obtain wide-band amplification, but in either case shaping of the envelope at the edges of the passband requires the use of supplementary trap circuits.

An i-f amplifier may consist of a first stage tuned near one end of the passband, a second stage tuned near the other end of the passband, and a third stage tuned near the center to provide the required overall bandwidth, to which there would be added special trap circuits to accomplish the envelope shaping at the upper and lower edges of the passband. The circuitry is complex and permits much flexibility in obtaining the desired results. The superheterodyne type of receiver is ideal for television reception because it is possible to select the best i-f frequency band to provide most easily the precisely shaped envelope characteristics required, and because the one i-f amplifier functions equally well for all channels.

Figure 11-34 shows one form of television intermediate-frequency amplifier. The shaping of the overall bandpass characteristic is accomplished in several stages. The output circuit of the mixer tube is a broad band circuit

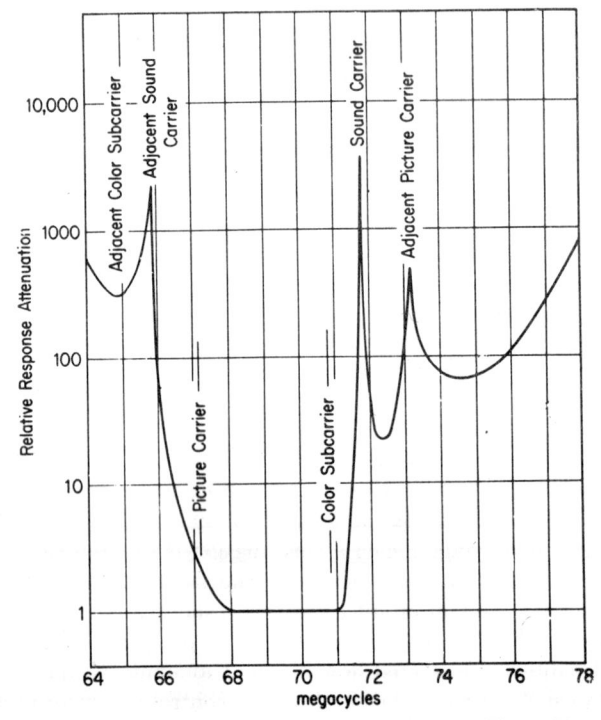

FIG. 11-35. Response of TV i-f amplifier shown in Fig. 11-34.

A to which is coupled a 41.25-megacycle trap circuit. This absorbs energy at the specified frequency. Circuit *B* is a 39.75-megacycle trap for attenuation of that and adjacent frequencies. Circuit *C* is another series-tuned trap which attenuates frequencies of 47.25 megacycles and adjacent frequencies. Circuit *D* is a second 41.25 megacycle trap which absorbs energy at that and adjacent frequencies from the first i-f transformer. Circuit *E* is the first of the "stagger" triple tuned circuits for maximum response at 45 megacycles. Circuit *F* is the second stagger-tuned circuit in the triple and has maximum response at 42 megacycles. Circuit *G* is the last of the stagger triple circuits and is tuned to 44 megacycles. This circuit feeds the second detector where the intermediate frequency is converted to the video frequency which was originally used to modulate the transmitter. The overall response at Channel 4 obtainable with this amplifier is shown on **Fig. 11-35.** Note that attenuation is plotted in this curve in contrast with transmission in Fig. 11-33, and hence the two curves are inverted with respect to each other.

DOHERTY HIGH-EFFICIENCY RADIO AMPLIFIER

As stated previously, the plate circuit efficiency of Class B radio frequency amplifiers is limited to about 35% if linear amplification is to be accomplished. In high-powered transmitters, such as are employed in broadcasting, this low efficiency is undesirable because 65% of the plate input power is dissipated in heating the tubes. This would require larger and more expensive amplifier tubes, more elaborate and costly cooling systems to remove from the tubes the heat produced by the unused plate input d-c power, a rectifier of correspondingly high rating and cost, larger primary substation and control equipment, and a substantially greater cost for the larger power consumption.

The Doherty type of linear radio-frequency amplifier overcomes these disadvantages by producing average plate circuit efficiencies of 60 to 65%

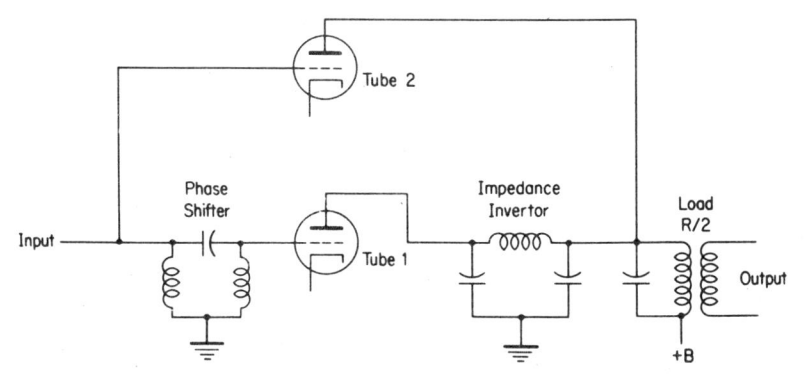

FIG. 11-36. Simplified diagram of Doherty amplifier.

independent of the degree of modulation. It is accomplished by dividing the load between two tubes in a unique circuit such as that shown in Fig. 11-36.

It is important to bear in mind that maximum power output is obtained from a tube when its impedance is matched by the load impedance, but that maximum plate efficiency, with lower power output, is obtained when the load impedance is high in comparison with the apparent output impedance of the tube. Therefore, a variation in the apparent impedance of the load can be used as a means of varying the output power while keeping the efficiency fairly constant. This is the method used in the Doherty amplifier to achieve high overall efficiency.

The key to the operation of this amplifier is the impedance inverter shown in Fig. 11-36. Properly proportioned inductive and capacitive components create a network having the properties of a quarter-wave transmission line in which the phase of the voltage is retarded 90° regardless of the terminating resistance and in which the input impedance is *inversely* proportional to the terminating resistance. In other words, the load resistance seesaws. When it goes up on one side, it goes down on the other. The impedance inversion is the action desired. The phase retardation is not desired but may be corrected as will be explained. Impedance R on one side of the inverter produces R also on the opposite side; $2R$ on one side produces $\frac{1}{2} R$ on the opposite side, etc. The equation for such performance is:

$$Z_s = \frac{Z_L^2}{Z_R}$$

where Z_S is the sending end impedance; Z_R is the receiving end impedance; Z_L is the network characteristic impedance.

As shown in Fig. 11-36, tubes 1 and 2 have their plates connected to a common load, but between tube 1 and the load there is the impedance inverter which causes lowered impedances on either terminal to appear as increased impedances on its opposite terminal. As stated earlier, this device causes incidental and undesired phase shift of 90° between the plate and the load. To cause the phases of the currents at the common outputs to match and add directly at the load, an opposing compensating 90° phase shifting network is connected ahead of the grid circuit of tube 1. The phase shifter network is similar to the impedance inverter except that the inductors are replaced by capacitors and the capacitors are replaced by inductors to produce a compensating phase advance of 90°.

As shown in Fig. 11-36, the common load impedance is made $R/2$, which is the desirable load impedance for maximum output for two tubes operating in parallel in conventional manner. Up to about half of maximum modulated transmitter output level, tube 2 is biased beyond cutoff, is inactive and has no effect on tube 1. Note that the impedance inverting network on the load side is terminated in $R/2$ but that the action of the

Chapter 11   *Radio-Frequency Amplifiers and Modulators*   **391**

inverter causes the impedance at the tube 1 side to be $2R$, which is substantially higher than the apparent tube output impedance and which, therefore, achieves higher operating efficiency.

The function of tube 1 is to deliver up to approximately half of maximum modulated output. Above about half power, tube 2 goes into operation automatically and adds the remaining additional power required. When tube 2 goes into operation, tube 1 is just beginning to saturate.

At low modulation levels, tube 1 operates very efficiently, with tube 2 inactive. At higher modulation levels tube 2 goes into operation and starts delivering power to load $R/2$. As it does so the voltage across $R/2$ rises. This increases the apparent load impedance at the load end of the impedance inverter, which in turn causes this increase to appear as a decrease at tube 1 plate. As the load decreases from $2R$ to R at that point, tube 1 can and does deliver more power. Increasing the modulation to full value causes tubes 1 and 2 to operate into the optimum impedance for maximum power output, $R/2$, and to divide the load equally.

The key to the efficiency of the Doherty amplifier is that over the full range of modulation the tubes operate with an average efficiency of about 60%, whereas in the conventional linear amplifier maximum efficiency is attained only at maximum modulation levels, with a much lower average.

The Doherty amplifier has had somewhat limited application for high-quality telephony, such as broadcasting, because of greater complexity in establishing and maintaining the adjustments within desirable tolerances.

CHIRIEX HIGH-EFFICIENCY R-F MODULATING SYSTEM

In all the radio amplifiers discussed so far in this chapter the amplitude modulation was produced by creating amplitude variations in grid or plate

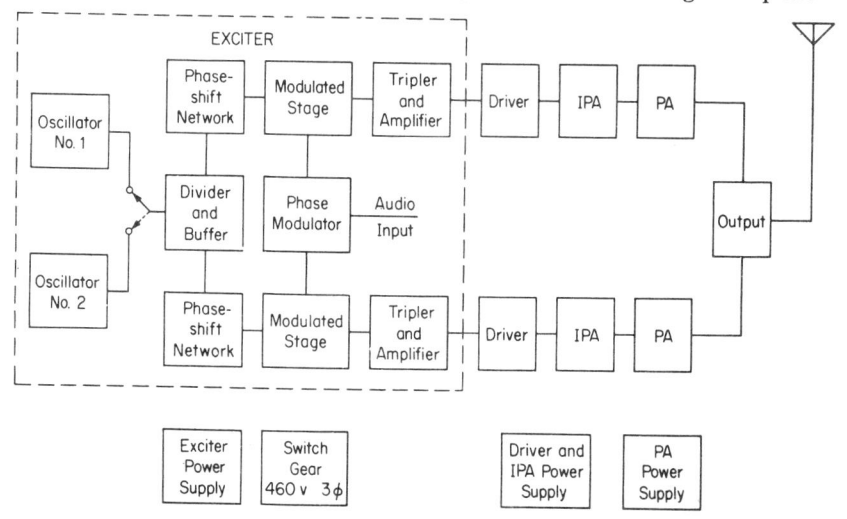

Fig. 11-37. Chiriex high-efficiency modulating system.

voltages. A system invented by H. Chiriex operates on a different principle and makes possible the smallest and most efficient amplitude-modulated transmitters known in the high-power field. Modulation is accomplished in a push-pull radio frequency amplifier by "out phasing" the grid voltages to produce a large amount of amplitude-modulated output power by means of a small amount of phase-modulated controlling power. A simplified diagram of such a transmitter is shown in Fig. 11-37.

The principle of operation is simple. If the grid input voltages of the push-pull power amplifier tubes are in phase opposition, at 180°, the output voltages are also in phase opposition, and the combined output power is zero, which is 100% negative modulation. If the grid input voltages are in phase, the output voltages are also in phase and the outputs of tubes 1 and 2 add to produce 100% positive modulation. The unmodulated grid voltage phase axes are adjusted to an intermediate value so that the phases of the two grids swing equal and opposite amounts to produce plate-modulated output power.

The Chiriex modulating system has the advantage that the output tubes operate very efficiently as Class C amplifiers and they do not require bulky

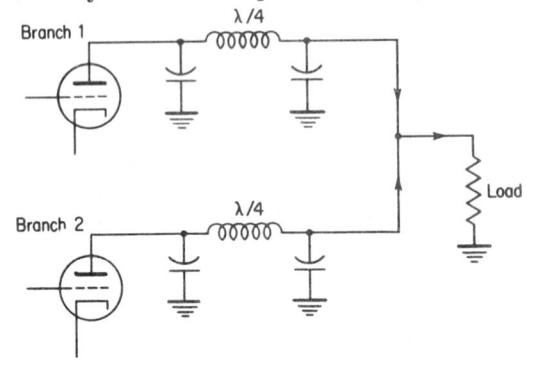

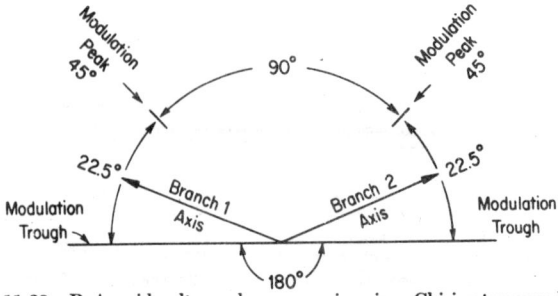

Fig. 11-38. P. A. grid voltage phase excursions in a Chiriex type modulating system.

and expensive modulating apparatus or expensive linear amplifiers. Figure 11-37 shows oscillators 1 and 2. One is used and the other is kept in reserve for emergency use. They produce oscillations at the carrier frequency. They feed a divider-buffer stage which divides the carrier frequency to one third, separates it into two equal parts, and feeds them to two channels differing in phase by 180°. Each branch starts with a phase-shifting network which causes the original 180° phase difference to become 135°. The signals then go to identical phase modulator stages where the phase modulation is accomplished. The two signals having a phase difference of 135° are caused to have fluctuations in phase corresponding to the message frequency envelope, and for 100% modulation the total excursions are plus and minus 7.5° from the axis, in opposite directions. This produces between the two branches differences in phase which vary from 120 to 150° The phase-modulated signal is then multiplied in frequency three times to the carrier level. In so doing the individual vector excursions of plus and minus 7.5° are also tripled to plus and minus 22.5° and the overall phase differences in the two branches then become 90 and 180°. Figure 11-38 shows the phasor relationships at the grids of the final radio amplifiers where the phase modulation is converted to amplitude modulation. Figure 11-38 also shows the output coupling networks by which the push-pull tubes combine their outputs into a common load.

The reduction of the carrier frequency to one-third prior to phase modulation was performed to achieve improved linearity in the modulation process. Having thus performed the modulation under optimum conditions, the tripling process expanded the phase modulation excursion to the desired amount at the same time that the carrier frequency was restored.

PROBLEMS OF DESIGNING RADIO-FREQUENCY AMPLIFIERS AT THE HIGHER FREQUENCIES

Neutralization of r-f amplifiers may be viewed as a penalty paid in cost and circuit complexity in exchange for stable operation and more flexibility in the use of vacuum tubes. It is a very useful tool and functions satisfactorily over a wide range of carrier frequencies. But as the radio industry has developed new services at increasingly higher carrier frequencies, the technical problems which have required major effort have included methods of producing high power. The problem has been one of designing satisfactory power tubes, or finding superior methods of using existing designs. A transmitter is designed around the power tubes which enable its performance.

The problems primarily are those of reducing interelectrode capacitances, building in shielding between output and input circuits, maintaining small mechanical dimensions, removing the heat generated in small spaces, and minimizing stray capacitance and inductance. At high frequencies, for example, the inductance and capacitance required for resonance of the output-tuned circuit become very small. The problem of stray inductance and capacitance is explained below. Any piece of apparatus and any interconnecting wire has a capacitance to ground. They also have capacitance from one to the other. These are called *stray capacitances*. They are illustrated in Fig. 11-39. In addition, any piece of interconnecting wire has inductance. The connecting wire of any piece of apparatus to any other

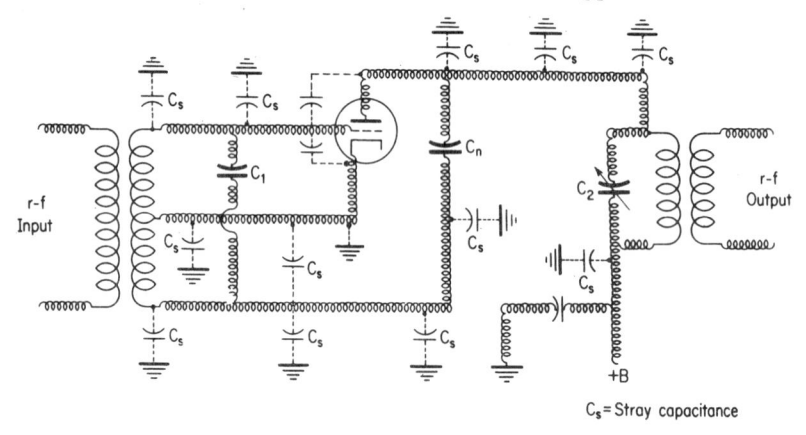

Fig. 11-39. Stray capacitances and inductances as they, in effect, would appear in a circuit diagram for a high-frequency radio amplifier.

piece may be viewed as a continuous inductance connecting them. At low carrier frequencies these stray capacitances and inductances are insignificant compared with the large values used in the circuit apparatus components and may be disregarded at the carrier frequency. They often form random tuned circuits which produce parasitic oscillations at very much higher frequencies, but this is incidental to the main point under consideration.

When the inductance and capacitance become so small for resonance at ultra-high frequencies that the stray inductances and capacitances are an appreciable fraction of their values, control is lost at the carrier frequency. The equation for inductive reactance of a wire, $2\pi fL$, indicates that the reactance increases directly with frequency. At high frequencies of, for example, 25 megacycles, the reactance of a short length of straight wire becomes appreciable. The capacitances of tuning capacitors are measured in millionths of a microfarad. The stray capacitances may be comparably large. A reduction of the value of the strays, and methods of minimizing their effects, becomes necessary by eliminating apparatus components where possible, shortening interconnections, perhaps increasing the wire diameter to reduce the inductance, making the assembly more compact, etc.

When the carrier frequency becomes high enough, the stray capacitances may exceed the capacitance required for tuning the plate tank circuit.

Chapter 11 *Radio-Frequency Amplifiers and Modulators* 395

Before reaching that point the strays may make neutralization critical, unsatisfactory, or impossible. If the neutralizing capacitor could be eliminated, it would be a step toward the desired simplicity of design. Fortunately, a circuit has been developed by which this may be accomplished. It consists of grounding the control grid as described in the section which follows.

GROUNDED GRID RADIO AMPLIFIERS

Operation of high-power radio amplifiers at very high radio frequencies often may be facilitated by grounding the grid instead of the cathode as shown in Fig. 11-40. This avoids the need for neutralization and is of particular advantage where uniform amplification with good stability is

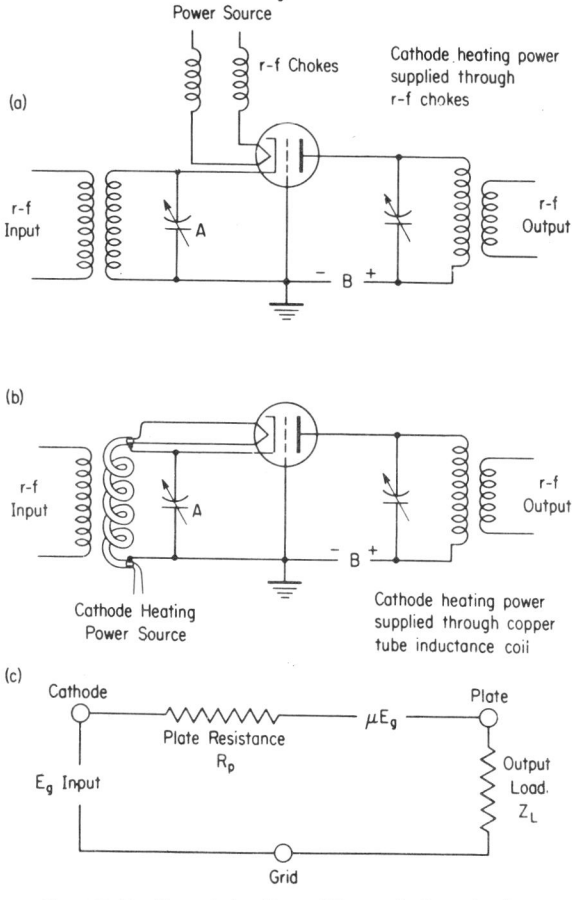

FIG. 11-40. Grounded grid amplifier equivalent circuit.

essential over a broad band of frequencies. The grounded grid acts as a shield between input and output circuits. Ordinarily grounded grid operation is used with triodes at frequencies so high that neutralization of triodes is difficult or undesirable, or for economy in design.

In the grounded grid circuit the input r-f voltage is applied between grid and cathode in the usual manner except that the cathode must have a substantial r-f impedance to ground to avoid shorting the input r-f voltage. This high impedance is normally attained by the use of a tuned circuit or r-f chokes which also carry the cathode heating current to the cathode, as shown. In grounded grid operation the grid voltage remains fixed at zero and the cathode voltage is caused to vary with respect to it, the reverse of the usual procedure.

Figure 11-40 shows that when the cathode is driven positive with respect to the zero voltage grid, the driving voltage is in series with the external circuit connecting the plate and cathode. The plate current flows through the plate to cathode circuit in phase with the input current. Therefore, input driving power appears in the output circuit. In a conventional amplifier the tube amplification factor is designated as μ and is an index of the degree of voltage amplification. In a grounded grid circuit the input and output r-f voltages add to produce plate current. This causes the circuit to function as though the amplification factor were $(\mu + 1)$.

In a conventional amplifier, capacitive currents flowing from plate to grid may produce oscillations because the power is fed back to the control element. In the grounded grid circuit the grid ideally is maintained at ground potential. R-f plate current flowing to the grid may pass to ground. The feedback from the output to the input circuit is minimized, the interaction is very small, and neutralization is not required.

Grounded grid circuits are common in the output stages of high-power high-frequency transmitters because conventional circuits have the disadvantages that:

1. The accumulated capacitances of the tube, the neutralizing system, and other conventional components are relatively large. They are in effect shunt capacitances across the output-tuned circuit. At very high frequencies they become so large that the output tank circuit may no longer be tuned, or the tuning range becomes so limited as to be unsatisfactory.

2. At high frequencies the neutralizing circuit resonates near the fundamental frequency of operation and no longer is controllable or effective.

3. Neutralizing capacitors must be built to withstand high voltage and must be located in confined spaces which complicate design and make for instability.

4. Operation of pentodes, tetrodes, and neutralized triodes becomes increasingly unstable and inefficient when the operating frequency is in-

creased to the point where short apparatus connections become highly reactive inductances, and where screen grids cannot be kept at ground potential. Neutralizing can be accomplished over a decreasing frequency range as the frequency is increased, until finally the range disappears completely.

The tube functions as a very stable series booster. The voltage amplification is:

$$E_\text{amp} = \frac{E_g \, (\mu + 1)}{R_p + Z_1}$$

where R_p = plate resistance and μ = amplification factor.

Grounded grid amplifiers may be operated in Class A, B, or C service. They require more input power than conventional amplifiers but, as stated, most of the input power is utilized as useful output, not lost. The driver power required is the product of the signal voltage times the signal current at the input. The output power is the product of the plate signal current and the combined input and output signal voltages.

When the cathode is driven it must be maintained at a high impedance above ground to avoid shorting the input circuit. But power at the same time must be delivered to the cathode to heat it.

Two methods of delivering cathode heating power are shown in Fig. 11-40. Figure 11-40(a) shows cathode current fed through separate chokes which have high r-f impedance. Figure 11-40(b) shows an input tuning inductor constructed of copper tubing through which the heating wires are carried. Figure 11-40(c) shows the equivalent circuit of a grounded grid amplifier. The alternating voltage appearing in the output load is the sum of the input voltage and the component produced by amplification because the input, tube, and output are all in series.

CATHODE COUPLED AMPLIFIERS

A cathode coupled circuit may be employed where a very low output impedance is desired. It is characterized by a load between cathode and ground as shown in Fig. 11-41. The advantages are:

1. Output impedances of very low values may be obtained for feeding very low impedance lines such as coaxial cables.

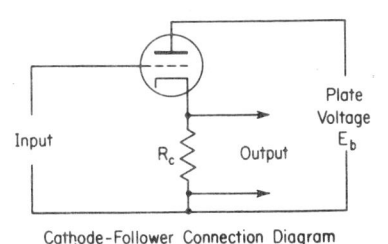

Cathode-Follower Connection Diagram

Cathode-Follower Equivalent Circuit

FIG. 11-41. Cathode follower circuits.

2. The input capacitance is very low which is important when very wide band transmission is desired.

3. The circuit is highly degenerative which reduces amplitude distortion and frequency selectivity.

4. The stability against oscillations is grea

The disadvantage is that the gain is very low, actually less than unity. The voltage amplification for the circuit shown is:

$$E_{\text{amp}} = \frac{\mu R_c}{r_p + (\mu + 1)R_c}$$

where r_p = plate resistance and μ = amplification factor.

This circuit may be thought of as an electronic transformer which produces a very low impedance output without the prohibitive limitations of a conventional wire-wound transformer in very wide band operation.

Figure 11-41 shows that the output voltage across R_c is also in the grid circuit and that the phases of the two voltages produce degeneration. In other words, the input voltage is reduced by the output voltage. It is for this reason that the voltage amplification cannot equal unity.

Cathode-coupled circuits are extensively used in television transmitting systems in which a very wide band of frequencies must be accommodated with very little *phase shift*, frequency selectivity, or amplitude nonlinearity, and which must employ low impedance coaxial cables instead of ordinary wiring for passing video signals. In a video system, frequency-selective phase shift causes some frequency components to be accelerated or decelerated more than others in the arrival time at the terminal. Thus the high-frequency and low-frequency components of a complex wave arrive at slightly different times at the receiving kinescope and misregister, causing very undesirable edge effects on pictures.

Uniform amplification of all frequencies is equally important as will be evident from the consideration of complex television wave shapes. They consist of combinations of fundamental and harmonic frequencies. These must combine in both amplitude and phase to reproduce faithfully the original wave. If some frequencies are attenuated more than others, streaking and smear may appear on the picture because the complex wave shapes are then not accurately reconstructed.

The characteristic high input impedance and low output impedance make cathode coupling a very valuable technique in designing television pulse generators and amplifiers. The fact that the output voltage rises and falls in the same polarity with the input voltage, and follows it, is the reason for cathode-coupled stages often being called cathode followers. In most conventional amplifiers the polarity of the input voltage is the reverse of the output voltage. The polarity of television video circuits is especially important because wave shapes usually are not symmetrical.

Chapter 11   *Radio-Frequency Amplifiers and Modulators*   **399**

GRID-MODULATED AMPLIFIERS

Modulation may be accomplished in the grid circuit of a tube and in some applications, such as television transmitters, this has distinct advantages. For example, in plate modulation, such as was described in the preceding section, the modulation transformers, reactors, etc., are necessarily large and are characterized by high distributed capacitance and leakage inductance which limit the modulation bandwidth and introduce undesirable random resonant conditions. Therefore, it is desirable to avoid their use. A small and simple modulating system is desired in which the modulating section is not required to operate at very high power. Instead

(a) Control Grid Modulation   (b) Suppressor Grid Modulation

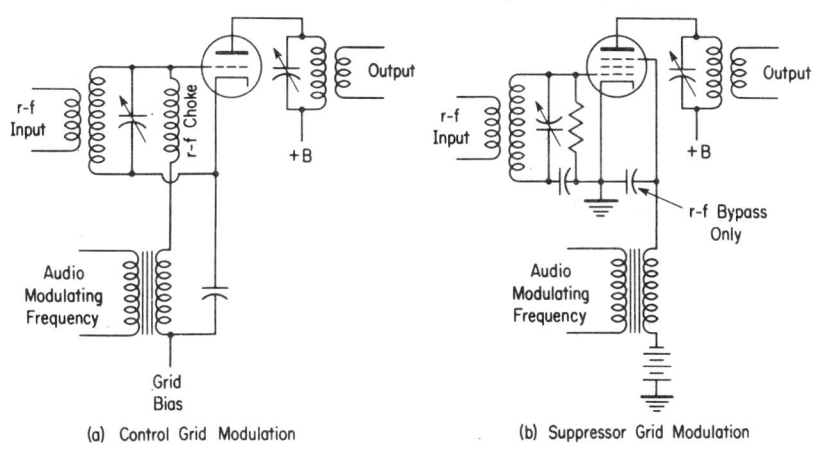

FIG. 11-42. Grid modulation systems.

of modulating the plate voltage, the grid voltage may be modulated. This is common practice in television. A grid-modulated triode amplifier is shown in Fig. 11-42(a). The grid voltage-plate current relationship is shown in Fig. 11-43.

In order to realize maximum plate efficiency a grid-modulated amplifier is biased substantially higher than required for plate cutoff, the Class C condition. Then the plate current flows in pulses of a duration substantially shorter than one-half cycle, as was explained for Class C operation, the plate heat dissipation is reduced, and the efficiency is correspondingly increased. In grid modulation the radio frequency driving power input is then adjusted so that in the unmodulated condition the peaks of plate current extend to the center of the linear range, halfway between cutoff in the downward direction, and saturation in the upward direction. This is similar to the plate current condition described for a linear radio amplifier which has a low plate efficiency. During instants of maximum positive modulation the tube operates as an efficient Class C amplifier driven almost to the point of saturation.

400   *Amplitude-Modulation Detectors,*   Chapter 11

Normal adjustment of the grid-modulated amplifier provides for the signal modulating voltage to drive the r-f plate current upward almost to saturation during positive maximum modulation peaks and to reduce it very close to the cutoff point on maximum negative peaks. The plate

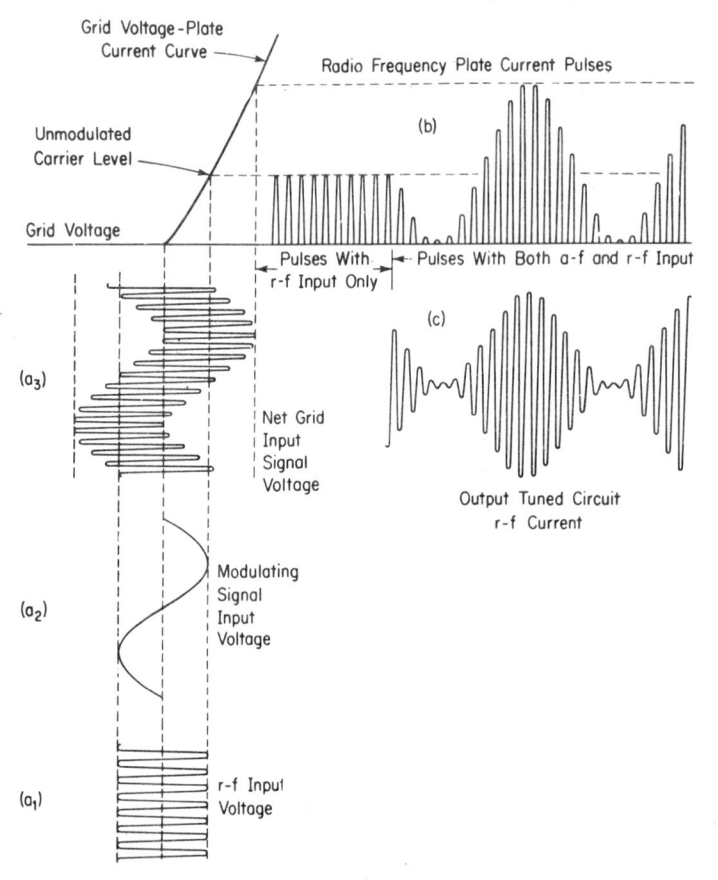

FIG. 11-43. Relationship between r-f and signal input voltages in grid-modulated radio amplifier.

efficiency in the absence of modulation is about half that of a Class C amplifier. Averaged over several cycles of full modulation by a sine-wave signal voltage the plate circuit efficiency is approximately 50%. Since the r-f plate current and plate voltage of a grid-modulated amplifier reach a maximum of about half those of a conventional Class C amplifier, the power output is limited to about one fourth. In the interest of obtaining maximum linearity of modulation over the full modulation range, grid-modulated amplifiers may be adjusted for best performance in that respect at the expense of plate efficiency.

Chapter 11 *Radio-Frequency Amplifiers and Modulators* 401

The circuit diagram of Fig. 11-42(a) shows that the modulating frequency input voltage is in series with the grid bias and alternately adds to it and subtracts from it, producing modulation of the bias. Figure 11-43 shows the input and output voltage and current relationships, Fig. 11-43 (a)$_1$ the constant amplitude r-f input voltage, Fig. 11-43(a)$_2$ the modulating signal input voltage, Fig. 11-43(a)$_3$ the combination of these voltages, Fig. 11-43(b) shows the plate current pulses which are produced, and Fig. 11-43(c) shows the output-tuned-circuit r-f current.

A screen-grid tube of the pentode type may be modulated in the suppressor grid circuit, as shown in Fig. 11-42(b). The connections are conventional for a radio amplifier except that the modulating signal voltage is inserted in series with the suppressor grid connection. A high order of linearity is not achieved.

Grid modulators require little driving power, particularly when they are not operated above the point where grid current flows, but the range over which linear modulation occurs is somewhat limited. They are used where efficiency may be sacrificed in favor of more important characteristics, such as wideband transmission for television.

PLATE MODULATION OF SCREEN-GRID TUBES

Screen-grid tubes of the pentode type may be used as plate-modulated radio-frequency Class C amplifiers. In order to plate-modulate a screen-grid tube properly, it is necessary to modulate the screen-grid voltage also. The screen grid is connected to the plate-supply circuit through its dropping resistor. The plate-supply voltage is obtained through the secondary of the modulator output audio transformer and therefore the screen grid and the plate voltage are modulated simultaneously. This type of modulation may be used where the highest fidelity is not required.

PARASITIC OSCILLATIONS

In constructing high-power amplifiers difficulties are sometimes caused by *parasitic oscillations*, particularly when a number of tubes are connected in parallel. Such oscillations ordinarily occur at frequencies much higher than the operating frequency, are not expected, often first become evident when highly abnormal operation takes place, and may be quite difficult to identify. They waste power, make adjustment at the desired frequency difficult or impossible, and may produce such heavy circulating oscillatory currents that the glass envelope of the tube is melted or the seals are destroyed. Parasitic oscillations can be detected, and ordinarily proper amplifier operation cannot be achieved until they are eliminated.

The voltages produced may in some cases be so great that long flashovers and destruction of circuit elements occur in high-power amplifiers. Such oscillations are usually the result of the existence in both the grid and plate circuits of tuned circuits of very high frequency, which are resonant to the same frequency, for which the neutralizing system is not balanced and is therefore ineffective. These very high frequency tuned circuits ordinarily consist of random inductance and capacitance having no planned relationship to the tuning system for which the amplifier was designed. When they occur the feedback from plate to grid through the tube or the neutralizing system easily may be more than sufficient to produce *tuned-grid-tuned-plate oscillations*. The preferred correction is to identify the unknown tuned circuits and eliminate them by changes in wiring or substitution of circuit elements without affecting the performance on the desired frequency.

Parasitic frequency oscillations result from conventional types of tube operation, but are created by accident, and are usually difficult or impossible to predict. They are very often encountered in the construction of high-power amplifiers.

Figure 11-44(a) shows a radio-frequency amplifier in which accidental combinations of circuit elements produced tuned-grid tuned-plate oscillations. The parasitic frequency-tuned grid circuit included the grid bias choke and three capacitors, one of them being half of the input tuning capacitor, another the bias supply by-pass capacitor, and the third the blocking capacitor which confined bias to the grid.

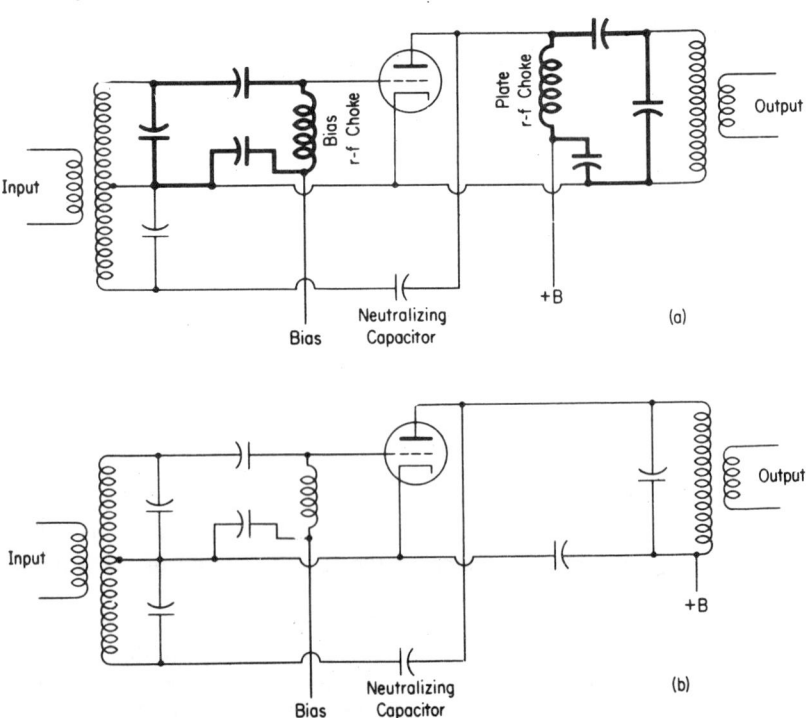

FIG. 11-44. Tuned-grid tuned-plate parasitic oscillation.

The parasitic frequency-tuned plate circuit included the plate current supply choke, output tuning capacitor, plate supply by-pass capacitor, and a plate-blocking capacitor which confined plate rectifier power to the tube. The parasitic tuned circuits are drawn heavily to identify them. In this instance, as usual, the neutralizing system was balanced and functioned at the desired frequency but was unbalanced and ineffective at the parasitic frequency. Correction was achieved by modifying the plate circuitry to eliminate the parasitic tuned circuit as shown in Fig. 11-44(b).

QUESTIONS AND PROBLEMS

1. Identify three types of detectors and explain their characteristics.

2. Describe heterodyne detection of CW signals.

3. What functions do the first and second detectors perform in a superheterodyne receiver?

4. Explain the three steps in demodulation of a telephone modulated carrier.

5. Describe the principal differences between radio- and audio-frequency amplifiers.

6. What is an intermediate-frequency amplifier?

7. Describe three methods of providing grid-bias voltage to an r-f amplifier.

8. Name and describe three classes of r-f amplifiers.

9. In what way does a Class B r-f amplifier differ from a Class C r-f amplifier?

10. What is neutralization and what is its purpose?

11. (a) How is a single-ended r-f amplifier neutralized? (b) How is a push-pull r-f amplifier neutralized?

12. What are the advantages of pentode r-f amplifiers?

13. What purposes does shielding serve in r-f transmitters?

14. In cascaded radio amplifiers, why is it desirable to filter the grid bias and plate supply leads?

15. (a) What difficulties may arise in operating r-f tubes in parallel? (b) How may the difficulties be overcome?

16. Which class of r-f amplifier is best suited for a plate-modulated stage? Why?

17. Why is it ordinarily not necessary to neutralize tetrode or pentode amplifiers?

18. Why are Class A amplifiers used for radio-frequency amplification in receivers?

19. What methods are used in television receivers to achieve the required bandpass characteristics?

20. (a) Describe the advantages of the Doherty high-efficiency amplifier. (b) Explain its performance.

21. (a) Describe the advantages of the Chiriex type of r-f modulating system. (b) Describe its performance.

22. What is low-level modulation? What is high-level modulation?

23. What is a parasitic oscillation?

24. Explain the circuit of a grounded grid amplifier.

25. Under what conditions is a cathode coupled stage of advantage?

26. What unique problems are encountered in the design of high frequency power amplifiers that are not encountered at low frequencies?

27. What is a balanced modulator? How does it operate?

CHAPTER 12

AMPLITUDE MODULATION RADIO TRANSMITTERS

The function of the radio transmitter is to accept a signal, such as an audio frequency, from a pair of terminals and provide every intermediate step of processing required to feed an antenna with a modulated carrier of the specified power and frequency. The entire process is governed by many factors and many specifications must be met to insure satisfactory transmission and conformance with the Rules and Standards of the Federal Communications Commission. If a transmitter is to be designed properly it is necessary first to list carefully and completely every function it is to perform, every requirement that is to be met, and every permissible tolerance.

INFLUENCE OF TRANSMITTER FUNCTION ON DESIGN

A radio transmitter designed for ship telegraphy is relatively simple. It would be of limited power of perhaps 1,000 w, be equipped only for continuous wave telegraph transmission, and would normally use as its source of primary power the ship's generator. A continuous-wave high-power transmitter for transoceanic telegraphy would have corresponding requirements to meet, unique to that service. In the broadcasting service a more complex form of modulation and other differing requirements are involved. In the section which follows, the most important specifications of a typical transmitter are considered to illustrate the factors which govern design and performance. A 10-kw broadcast transmitter was selected because the relative complexity provides more area for discussion of performance features.

TYPICAL TRANSMITTER PERFORMANCE REQUIREMENTS

AUDIO AMPLIFICATION

For broadcasting, the message signal is speech, music, and other sounds. There must be provided enough audio amplification to permit full 100% modulation of the carrier with a reasonable reserve for minor amplifier gain variations resulting from differences in individual tube characteristics, etc.

In order to know how much audio amplification will be required, it is necessary to know what audio power level will be available at the input terminals. If it is established that the input level will be 10 mw of power, it is then possible to specify safely 1 mw to include a reasonable reserve. There is thus established one important specification: It will be possible to obtain 100% modulation when one milliwatt of audio power is available at the input.

INPUT IMPEDANCE

The input audio level is now known. But it is also necessary to know the proper terminating impedance to provide a match for the circuit preceding the input terminals, to insure against audio-frequency selectivity and power loss. If it is established that the driving circuit can be either 150 or 600 ohms, it is then possible to design the input circuit to be adjustable to either, ordinarily by tapping the primary of the audio input transformer or changing resistor combinations. Thus there is established another specification: The input circuit will be designed to be either 150 or 600 ohms.

AUDIO-FREQUENCY RESPONSE

Standard AM broadcasting is a high-quality service, but as a service it is not capable of the high fidelity of FM broadcasting because of inadequate channel width and severe selectivity requirements, as explained in Chapter 10. It is necessary to decide what audio bandwidth the station is to transmit. A 30 to 15,000-cycle wide band transmission is expensive and cannot be generally used on broadcast receivers. The average receiver responds from about 50 to 7,500 cycles so it is reasonable to impose a requirement that all frequencies in this range shall be transmitted without more than 10% variation in amplification (about 1 db). The best few receivers respond to a range of about 30 to 10,000 cycles, so for them it is reasonable to transmit this wider range with a somewhat looser tolerance, about 15% (about 1.5 db). This amount of variation may be sensed reliably by a listener only under back-and-forth comparisons of the two levels with unvarying tone modulations, and is ordinarily most difficult to sense when the different levels are comprised of speech or music which fluctuate continuously. Thus, there may be established another specification: that the transmitter shall amplify all frequencies from 50 to 7,500 cycles without a plus or minus difference of more than 10%, and for 30 to 10,000 cycles not more than 15%.

AUDIO DISTORTION

In low-quality telephone systems considerable distortion is permissible and is tolerated to keep down costs. This is not the case in broadcast transmission, but it becomes increasingly expensive in tubes and apparatus as perfect undistorted linear transmission design is approached. A compromise of 2.5% audio distortion is reasonable. Thus there is established another specification: that at any percentage of modulation up to 100% the audio distortion at any frequency or combination of frequencies shall not exceed 2.5%. This represents a level which is ordinarily unnoticeable but which could not be greatly exceeded without noticeably unpleasant effects.

TRANSMITTER NOISE

It would be ideal to have a transmitter which processed a signal without introducing any extraneous noise such as a-c hum from tube cathodes, etc., but perfection is not attainable. An unnecessarily elaborate noise-suppression design imposes expensive design innovations. Transmitter hum or noise which is audible at a receiver is undesirable. What degree of suppression is adequate but not unnecessarily expensive? It is known that human ears are more sensitive to some frequencies than others, and this may be taken into account in specifying tolerable hum or noise levels. For simplicity it will be considered that all hum or noise shall be suppressed below the 100% modulation level by not less than 60 db (1,000 times in voltage). Thus there is established another specification: that unweighted transmitter noise shall be at least 60 db below full modulation level.

MODULATION SYSTEM

Should the transmitter employ high-level or low-level modulation? Should it employ the Chiriex type, the Doherty type or the grid-modulated type amplifier? Or should it utilize a Class B modulator and a high-level Class C modulated amplifier? This decision is made by considering the unique demands of a given output power rating, in this case 10 kw. An analysis of costs showed the last type of system to be most desirable. Thus there is established another specification: that the transmitter shall employ audio Class B modulation of a Class C radio power output stage.

CARRIER FREQUENCY

What carrier frequency should the transmitter be designed for? Since the standard broadcast band spans 540 to 1,620 kc, it would be desirable to have only one design which could be made to operate anywhere within that range, rather than have many designs, each for a single frequency or restricted band. This is feasible without prohibitive extra cost. Therefore it is possible to fix another specification: that the transmitter shall be capable of meeting all performance specifications when adjusted to any channel between 540 kc and 1,620 kc.

CARRIER FREQUENCY STABILITY

The Federal Communications Commission requires that in the standard broadcasting service the carrier frequency may not at any time deviate more than plus or minus 20 cycles from its assigned frequency. The purpose of this regulation was to confine carrier-beat frequencies from interfering stations to very low and relatively inaudible frequencies. Thus it is necessary to consider two phases of frequency inaccuracy. First is the initial adjustment of carrier frequency. This is important, because if there is a fixed inaccurate setting of 15 cycles off zero, a movement of only 5 cycles is permissible in that direction. Second is the drift representing imperfect freedom from the influence of temperature, voltage, etc., in the frequency-determining circuits. This drift changes with time.

Modern quartz-crystal frequency-determining circuits will remain fixed within one or two cycles per million over very long periods, and small trimmer capacitors may be employed in oscillating crystal circuits to adjust the frequency exactly on the assigned value. Therefore, meeting the FCC specification of 20 cycles is not a serious problem. Actually, a specification of plus or minus 5 cycles is reasonable, so it may be adopted. An inexpensive quartz crystal is ground for the specified frequency and inserted in the carrier-frequency-determining circuit by a plug-in arrangement.

TYPE OF OUTPUT CIRCUIT

The transmitter must feed some kind of a load. In radio circuits it is important to know if the load must be balanced (equal impedances to ground on the two sides) or unbalanced (one side grounded). Modern transmitters and also transmission lines employ to advantage grounded circuits in matching impedances and in minimizing spurious radiations. These circuit elements are expensive, so this specification is important. If it is assumed that the antenna and the transmission line to it are conventional, it is possible to fix the specification that the output circuit shall be grounded. Experience has shown this to be desirable.

OUTPUT IMPEDANCE

There has been specified an unbalanced output circuit but what should be its impedance? This is another important specification because the circuit elements are expensive and they must function properly. The performance of the station depends upon proper matching of the transmitter to its load.

Upon consideration of a wealth of information about transmission lines and antennas, it is known that an adjustable range of about 40 to 250 ohms will meet any condition likely to be encountered in the broadcasting service, so the specification may be added that the transmitter output circuits shall be adjustable to feed any load in that range.

CARRIER AMPLITUDE SHIFT

In an amplitude plate-modulated transmitter ideally the *average* carrier power does not change. However, imperfect voltage regulation of circuits, principally power supplies, may cause the average carrier level to drop during periods of heavy modulation. This may be sufficient to introduce distortion because modulation is then unnaturally limited. Therefore, a maximum degree of carrier shift should be specified to protect against nonlinear modulation. Experience has shown that 5% is reasonable, so there may be added another specification that carrier shift between 0 and 100% modulation shall not exceed 5%.

POWER RATING

It is, of course, necessary to specify the carrier power rating of the transmitter. In this instance 10 kw was specified.

MAXIMUM AMBIENT TEMPERATURE

Heat is dissipated in a radio transmitter because all of the power used to operate it is not converted to useful output power. This heat may cause the temperature in a poorly ventilated transmitter room to rise to a point where the apparatus components become hot beyond their ratings and fail during operation. It is therefore necessary to specify a limit to the safe ambient room temperature. Experience has shown that 45°C is reasonable, so there may be added the specification that the maximum ambient temperature in the operating area shall not exceed 45°C.

POWER SUPPLY

The transmitter must obtain primary power from a public utility supplier or some other source for the various rectifiers, cathode heating circuits, blowers, etc. Should it be single phase or three phase? What should be the voltage? If it does not conform with normally available power supplies, an expensive converter will be required, the transmitter will require expensive rebuilding, or a special substation will be necessary.

It is known that for transmitters of 10-kw rating there are substantial advantages to the use of three-phase input power. There are also advantages in cost in not requiring high-voltage circuit breakers, switches, and wiring. Therefore, a commercially available three-phase low-voltage source is preferable. Thus there shall be specified 208/230 v, three phase, 60 cycles for maximum advantage and convenience.

POWER CONSUMPTION

Switching, metering, and protective devices will be required at the point where power circuits are terminated in the transmitter room or building. But for what amount of power should they be designed? No purpose would be served by providing such devices to process 5,000 kw if the load were to be only 25 kw. Furthermore, the cost would be far greater than necessary. A circuit breaker must be furnished of a capacity great enough to safely interrupt a full short circuit on the power line. It will be far smaller on a low-voltage line than on a high-voltage line. Prior to installing equipment the proposed primary load must be known at least approximately so that the power line may be tied in with the transmitter. The load drawn on the 10-kw broadcast transmitter will vary during modulation because Class B audio modulation is to be employed. In this case it may be calculated that the power drawn would be 20 kw in the absence of modulation, 23 kw during average modulation, and 33 kw during sustained full modulation, with a power factor of 85%. But this is not sufficient information. The primary load changes with modulation. If the voltage also changes, the transmitter does not obtain in full measure the power required with increasing modulation, and carrier shift and modulation distortion result. Hence there is needed another specification that the permissible variation of power-line voltage shall not exceed 5% owing to poor voltage regulation.

The load drawn by the transmitter does not represent the full consumption of the station because of the loads for light, heat, miscellaneous power, antenna-tower marker lights, low-level audio monitoring and control apparatus, etc. Therefore, the loads are estimated, at about 50 kw maximum. It is now possible to specify that the source of primary power must provide 50 kw with not more than 5% voltage variation. This information will now be available for the public utility company and the correct switching, metering, and protective devices may be selected.

OTHER SPECIFICATIONS

Many other specifications are established, such as frame dimensions, which are important with respect to the size of elevator access doors, corridors, etc., or the weights of the transmitter units for which the supporting floors must be designed, or the types and numbers of tubes which must be in operation and carried as spares. But the discussion of the most pertinent technical specifications given above is sufficient for this introduction.

CRYSTAL CARRIER FREQUENCY OSCILLATORS

It has been explained in Chapters 10 and 11 that the demand for radio channels has compelled the regulatory bodies to impose close tolerances on radio channel widths and also on carrier frequency drift from the assigned values. With respect to the latter, when a broadcast station carrier frequency drifts, the sidebands move correspondingly. Even though the carrier may remain in the channel, the sidebands may be caused to move outside and cause interference to other stations. Also, when a carrier on the same channel from a distant point comes in and causes interference to a local station, it is first evident as a heterodyne beat note when the carriers are not at exactly the same frequency. If the beat note is of only a few cycles per second, neither the audio amplifier, the loudspeaker, nor the ear

Chapter 12   *Amplitude Modulation Radio Transmitters*   411

respond well because they are very inefficient at such frequencies. But a beat note of 500 or 1,000 cycles, which meet a much better response, is very evident and troublesome. Hence, the FCC has minimized the beat note type of broadcast interference by imposing the requirement that broadcast stations must not exceed a frequency inaccuracy of 20 cycles from the assigned carrier frequency. Two stations drifting in opposite directions from the assigned frequencies thus can produce a maximum heterodyne note of only 40 cycles.

One of the highest broadcast carrier frequencies is 1,600,000 cycles. The 20-cycle tolerance is equivalent to a drift of only thirteen ten thousandths of one per cent. Stated differently, it is only one part in 80,000. Fortunately, this requirement not only can be met, but it was originally possible safely to impose it because of the adaptation of crystal control to oscillator frequency-determining circuits. Tolerances of accuracy similar in percentage but not identical are imposed on practically all radio services to conserve spectrum space and minimize interference. As stated previously, it is possible with commercial type crystal oscillators to meet a much closer tolerance of 5 parts per million or better, corresponding to only three ten thousandths of one per cent, or one part in 250,000. Such accuracy is obtainable only because of the use of crystals. Such frequency-determining circuits are almost universally used in all radio services. The services as we know them could not have attained equal development, use, and efficiency without them. The accuracy of 5 parts per million is attained at practical commercial cost, but much greater accuracy is possible at greater cost. It has been demonstrated that more expensive crystal oscillators may be produced which will function within less than one part per billion over a 24-hour period and one part per billion over a 30-day period.

THE SIMPLE VACUUM TUBE OSCILLATOR

A vacuum tube amplifier can be made to oscillate because energy from the output can be returned to the grid circuit, in the correct phase to keep feeding upon itself. The losses and useful output energy consume power which is drawn from the plate power supply. Consider the circuit shown in Fig. 12-1(a). When the switch in the grid circuit is closed, a sudden change in plate current to a higher value will occur. This will cause a disturbance in the resonant circuit *LC*. A voltage will be built up across *L* and oscillating current will be produced in *LC*. The plate current which was stable before the grid switch was closed will become stable at a new value. The field across the coil will produce a few oscillations in the tuned circuit *LC* but they will die out quickly because the power will be used up in heating the resistance of the circuit and no new power is added. But suppose some of this oscillating energy, before it died out, were to be coupled back to the grid as shown in Fig. 12-1(b)? The plate oscillating power fed back to the grid will then be amplified in the plate circuit and become stronger. The cycle repeats. The part fed back to the grid becomes increasingly stronger and the cycle builds up until the tube is oscillating at maximum possible amplitude. In other words, it is saturated. It is important to remember

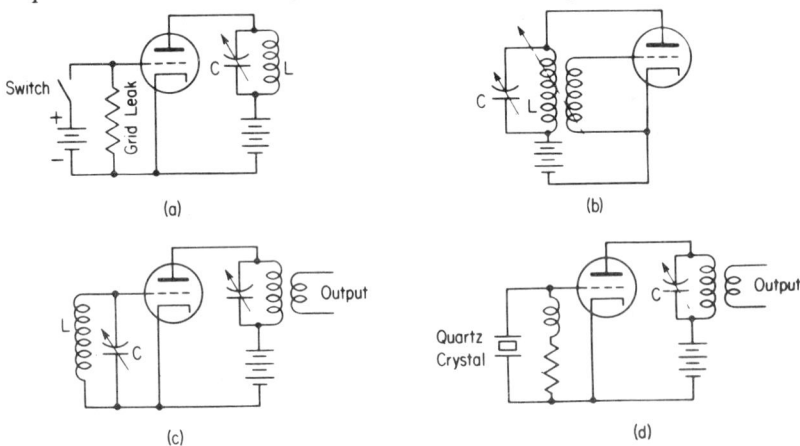

FIG. 12-1. Evolution of triode crystal oscillator.

that if the phase of the power fed back to the grid circuit is incorrect it will cause the plate oscillations to diminish, and not increase, and no oscillations can be sustained.

Now it is seen that feedback is involved in maintaining oscillations. There are many methods of accomplishing feedback but they all accomplish the same purpose. It will be sufficient to describe the most common type of oscillator circuit of interest, namely the *tuned-grid tuned-plate* oscillator, shown in Fig. 12-1(c). Here the feedback is accomplished through the internal capacitance of the tube, primarily the plate to grid capacitance. If this is sufficiently high, as it is in triodes, an interesting situation develops.

If the resonant circuit LC is tuned to a frequency slightly lower than that of the grid-tuned circuit, oscillations will be sustained because the energy fed back is in correct phase. But if the LC circuit is tuned to a higher frequency than that of the grid, degeneration occurs and oscillations are quenched, rather than aided.

Naturally, the frequency is determined by the tuned circuits. But these circuits have appreciable resistance compared with their reactance, at best an appreciable fraction of one per cent (values of Q of a few hundred). They may be as good coils and capacitors as can be made, but they still have a moderate ratio of reactance to resistance. It may be shown that the higher the ratio of reactance to resistance in the frequency-determining resonant circuit, the less will the frequency be affected by external factors which may change. The resistance of coils changes with temperature. Tuning capacitors change with temperature also. Good as they may be, they are not good enough for modern carrier-frequency-determining oscillators. To answer this need quartz crystals are far better. A quartz crystal in an oscillator circuit functions as an extremely stable tuned circuit of very high reactance and very low resistance. When properly connected, the piezoelectric effect will produce vibrations and radio-frequency voltages of the great frequency stability required and readily obtainable by no other means.

QUARTZ CRYSTALS

Quartz crystallizes in hexagonal prisms in several varieties such as amethyst, rose quartz, and cairngorm and is cut to create jewelry, lenses, prisms, etc. It is hard, generally almost without color, and somewhat resembles glass. It has a very high melting point and is little affected by corrosive substances. It is very plentiful on earth, but crystals of sufficiently good quality for radio use are found in quantity only in a few places, such as Brazil. Crystals vary in size from a diameter of a fraction of an inch up to about twelve inches. A typical crystal is shown in Fig. 12-2.

THE PIEZOELECTRIC EFFECT

Quartz crystals have the property, under favorable conditions, of developing an electrical charge when they are compressed. Correspondingly, they change dimensions slightly when a polarizing voltage is applied on opposite faces. This is called the piezoelectric effect. It has been employed to create ultrasonic waves in water by applying an alternating voltage across the faces of flat sections of crystal. This piezoelectric coupling between voltages across the crystal and mechanical forces within it results in the reflection of the mechanical constants of stiffness and mass into an electrical equivalent

Fig. 12-2. Section of quartz crystal as it is found in its natural state.

of a circuit which is resonant at one frequency and anti-resonant at another. The equivalent Q of this electrical circuit may run into many thousands, and hence its stability as an oscillator will be very great.

For radio use, selected crystals are sawed into slabs which may approximate one square inch of area and one-eighth inch in thickness. The crystalline structure is such that a carefully prepared crystal will display the piezoelectric effect vigorously at a single frequency which is remarkably free of variation. The frequency is determined by the dimensions of the crystal slab and the section of the mother crystal from which it was cut. In manufacture the slabs are sawed out and carefully ground to the proper uniform thickness to give the piezoelectric effect at the frequency desired. The final stages of grinding require great care, the use of very fine abrasives, and frequent measurements of the crystal frequency made in an oscillator. If the slab is ground too thin, the time and effort is wasted because the frequency will be too high.

One axis extensively used is about 35.5° from an optical axis, and crystals cut from it may be dimensioned for frequencies from about 500 kc to about 10,000 kc. The frequency at which the piezoelectric effect occurs is determined by the thickness and is

$$F_{KC} = \frac{1675}{\text{thickness in millimeters}}.$$

In seeking highly stable frequency-determining circuits for radio transmitters the quartz crystal was found to be very satisfactory. Quartz oscillators comprise in most respects the world's best secondary frequency standards. (The *primary* frequency standard is the rotation of the earth.) For radio use they are simple, cheap, and reliable. Figure 12-1(d) shows a quartz crystal triode oscillator with the crystal substituted for the grid-tuned circuit of Fig. 12-1(c). As the capacitance of C is increased from its lowest value, a point is reached at which the circuit starts to oscillate as evidenced by a reduction in plate current. As the capacitance is increased, the oscillations become stronger as evidenced by diminishing plate current, until they attain maximum amplitude. This point represents a plate tuning adjustment at which the resonant frequency of LC is slightly lower than the frequency of vibration of the quartz slab. A further increase in capacitance causes the oscillations rather abruptly to stop because the phase of the feedback voltage from plate to grid is no longer regenerative.

When pentode tubes are used as crystal oscillators, the plate-to-grid capacitance ordinarily is not high enough to provide the feedback and an external capacitor is added.

Quartz crystals are not completely invulnerable to changes in frequency caused by changes in temperature. Therefore, where close frequency tolerances are required, the crystal holder is enclosed in a special housing including a heater element thermostatically controlled to maintain the crystal at uniform temperature. Crystal oscillators include tubes or transistors. Changes in tube voltages slightly affect the frequency, so voltage variations are often guarded against. Changes in the plate tuning affect the frequency because they are reflected back to the grid circuit through the plate-grid capacity. It is quite common to use a plate tuning system with a non-variable capacitor, and often only a coil is used which is self-resonant with its distributed capacitance. It is quite common to use circuit elements carefully selected for a low temperature coefficient to minimize any variable that affects the frequency of the system. The frequency is affected by the

Chapter 12   *Amplitude Modulation Radio Transmitters*   415

loading of the oscillator plate circuit so it is common practice to load the oscillator very lightly and follow it with a specially designed amplifier which causes no reflected effects. This is usually referred to as a *buffer amplifier*. In the operation of a crystal oscillator a grid current will flow through the r-f choke and the grid-leak resistor R. There is also a radio-frequency voltage across the crystal and a radio-frequency current is caused to flow through the crystal. If the current is too high the crystal is liable to crack, and therefore the power output of a crystal oscillator tube must be limited. The power output, using triode tubes, may be a watt or less and is seldom more than 5 w.

Pentode and tetrode oscillators load the crystal more lightly than do triodes and somewhat more output is obtainable, so such tubes are preferred. A tetrode crystal oscillator is shown in Fig. 12-3. The crystal is connected between the control grid and ground and across it is connected a

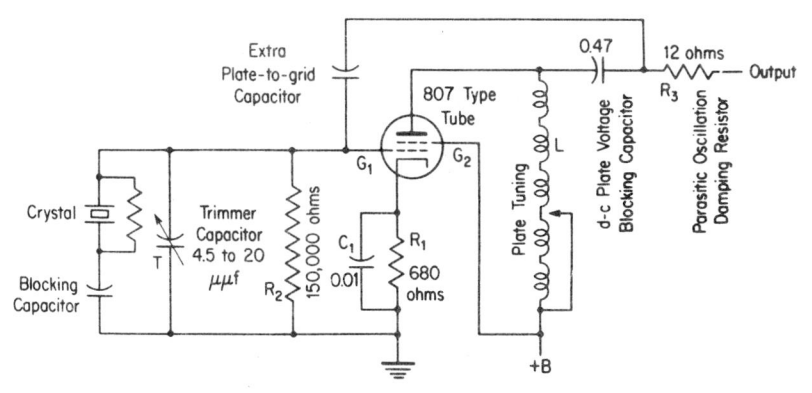

Fig. 12-3. Crystal-controlled tetrode oscillator.

150,000 ohm resistor R_2 to provide bias voltage. Protective bias voltage is also obtained from R_1 and C_1 at all times, even when oscillations are not present, which is not the case with bias produced by grid leak R_2. The output circuit includes an inductor L which is tapped so that the proper load may be obtained for the frequency at which the oscillator will operate. A variable plate tuning capacitor is not included because tuning, once adjusted by selection of the proper tap on L, does not require further change. The tuning of L is accomplished by selecting the proper inductance to resonate with its distributed capacitance.

The plate-to-grid capacitance of a tetrode tube is so small that external capacitance is added as shown on Fig. 12-3. A variable trimmer capacitor is connected across the crystal so that exact fine adjustment of carrier frequency may be made conveniently at the time of installation in the field. A d-c blocking capacitor is connected in series with the crystal to eliminate bias voltage from its terminals.   In series with the output lead a 12-ohm resistor R_3 is connected to guard against parasitic frequency oscillations.

A pentode oscillator would differ little from the tetrode circuit of Fig. 12-3 except for the addition of the suppressor grid and its connection to the cathode. The pentode, having lower grid-to-plate capacitance, almost always requires external feedback whereas the tetrode in some applications may not.

A photograph of a crystal oscillator identical to the one shown in Fig. 12-3 is shown in Fig. 12-4. The crystal and constant-temperature control device are mounted in the plug-in sealed can at the upper left. The plate

Fig. 12-4. Quartz crystal oscillator assembly. (Courtesy RCA)

loading inductor with taps is shown at the right. The crystal trimming capacitor is located in the center underneath the panel, and the adjustment is made with a screwdriver by turning the shaft shown with the screwdriver slot. The entire assembly is mounted in a metal shielding can with only the upper portion of the tube and crystal housing protruding for easy replacement, as will be shown later.

Figure 12-5 shows the internal construction of the quartz crystal assembly within the plug-in metal housing shown on Fig. 12-4. The unit at the left contains a crystal cut from an optical axis known as the AT axis which has unique advantages over a frequency range from 325 to 3,000 kc. This is commonly used for oscillators in broadcast transmitters. The

Chapter 12   *Amplitude Modulation Radio Transmitters*   417

assembly second from the left was designed for television transmitters in which the quartz crystal frequency is multiplied in frequency multiplier stages to obtain the carrier frequency. The assembly in the center employs the AT or another cut known as the BT cut for frequencies between 1.8 and

Fig. 12-5. Representative types of quartz crystal assemblies used in radio transmitters. (Courtesy RCA)

8.5 megacycles. This unit contains a means of varying the separation between the quartz crystal electrodes to control the frequency of oscillation over a range of plus or minus 0.03 to 0.04%. The assembly fourth from the left was designed for crystal cuts known as the CT or DT cuts which have unique advantages over a frequency range of from 70 kc to 350 kc.

The assembly at the right contains a unique circular quartz slab cut for use in high-frequency communications equipment operating in the range of 15 to 50 megacycles. In Fig. 12-5 the 14-w heater unit, used for maintaining constant temperature, was omitted to give a better view of the assemblies.

The highest frequency at which quartz crystals respond well at their natural resonant frequency is about 16 megacycles. However, specially prepared plates will respond at harmonic frequencies of 3, 5, 7, and 9 times the natural frequency. There are few practical uses for those above 3 times. The circular slabs for 15-50 megacycles are of the harmonic type.

BUFFER AMPLIFIERS

As stated previously, a crystal oscillator must operate under constant conditions for maximum frequency stability. In order that it may operate in a manner independent of external effects, it is customary to use special care in the design of the following amplifier to minimize the reflected-back effects of tuning, loading, heating, external electromagnetic or electrostatic fields, etc. In other words, the following amplifier not only amplifies but also acts as a buffer against influence from other r-f stages. It is commonly referred to as the *buffer amplifier*.

Because of the small feedback capacitance from plate to grid in pentodes and tetrodes, the effects of plate circuit conditions are minimized in the grid circuits. For that reason they are preferred as buffer amplifiers over triodes which require neutralization to be comparable in that respect. Pentodes and tetrodes are preferable for other reasons, such as lower input capacitance, greater output, and freedom from the need for neutralization to prevent self-oscillation.

It is particularly desirable to use a buffer amplifier between a crystal oscillator and a modulated amplifier stage because of the large fluctuations in voltages and currents inherent in the modulation process. In large transmitters it is customary to include a complete spare crystal oscillator for emergency use because of the vital need for adequate frequency control and the relatively low cost. In earlier years, transmitters often were provided with complete duplicate units including not only the crystal oscillator but also one or two buffer amplifier stages, but with more recent knowledge of quartz crystal cutting and improved and more varied types of tubes the practice is being abandoned in favor of only spare oscillators.

Figure 12-6 shows a typical buffer amplifier employing two tetrodes operating in parallel to give the desired output power. Resistors R_1 are 47 ohms and are provided to suppress parasitic oscillations. Resistors R_3 are of equal value and with the small inductors L_3 are provided for the

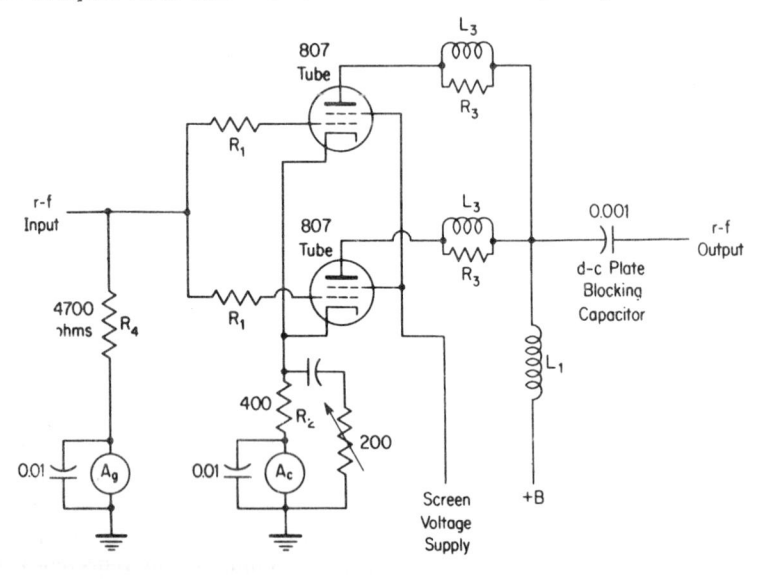

FIG. 12-6. Typical buffer amplifier utilizing type 807 tubes.

same purpose. Ammeter A_c is for measuring plate current (and screen current), knowledge of which is highly desirable in tuning and checking performance. The cathode circuit is a very convenient location for measurement of these currents because the ammeter is at ground potential and

Chapter 12   *Amplitude Modulation Radio Transmitters*   419

does not require high voltage insulation from ground as when it is located in the plate circuit. Ammeter A_G is for measuring grid current, knowledge of which is very desirable when tuning and checking performance. The resistor R_4 is to prevent shorting the grids. The capacitors around the ammeters are to keep radio-frequency currents from flowing into the d-c ammeters. Inductor L_1 is of high impedance.

DRIVER AMPLIFIERS

The requirements unique to crystal oscillators and buffer amplifiers have been explained in the order in which these stages appear in the radio-frequency section of a transmitter. The next stage or stages prior to the modulated stage are the intermediate stages and the driver stage. The intermediate stages of a broadcast transmitter are of straight-forward radio-frequency amplifier design and may utilize triodes, tetrodes, or pentodes depending upon personal preference or the availability of standard types of tubes rated to handle the power and frequency.

In selecting an intermediate amplifier tube it is desirable to consider the grid and plate d-c voltages necessary for various adequately rated types to keep to a minimum the number of special rectifiers which may be necessary. Where an adequately rated amplifier tube type is available which would operate at the same voltage as other stages, it would normally be preferable over other adequately rated types which would require additional and special rectifiers. The selection of tubes requires knowledge of the radio-frequency power available to drive the grids. If the tube is too small it will saturate at much lower driving power than its purpose dictates. Driving power will be wasted and limited power output will be delivered. If it is too

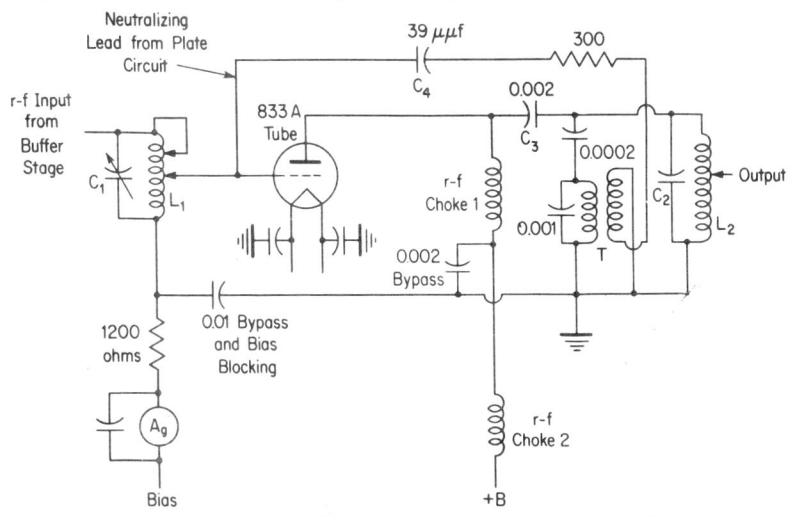

FIG. 12-7. Radio-frequency driver amplifier.

large the driving power will not be adequate to drive it into the efficient operating range. Tubes also are rated in terms of maximum frequencies at which they may be operated. A tube designed for broadcast frequency operation may have mechanical dimensions and other characteristics which inhibit its use at frequencies used for television, radar, etc. Therefore, large tubes usually are down-rated with increasing frequency to a point where they are inoperative or prohibitively inefficient at some specified high cutoff frequency.

Figure 12-7 shows a typical diagram of a driver amplifier such as would be used to excite the modulated final radio amplifier stage in a high-level modulated broadcast transmitter. L_1 and C_1 constitute the input tuned circuit. Adjustment of r-f grid driving power is accomplished by tuning this circuit to resonance and adjusting the tap from the grid. Since the lower end of L_1 is at ground potential at r-f, and maximum voltage above ground is at the top on the diagram, the tap is moved up to increase excitation.

R-f choke 1 is an inductor which passes d-c to the plate from the rectifier supply but which does not permit the passage of r-f current. T is an r-f transformer provided to obtain conveniently the r-f power to be fed back to the grid for neutralization. The secondary connection is shown reversed to emphasize the need for correct phasing of the feedback voltage. Reversing this connection would cause oscillation, not prevent it. L_2 and C_2 constitute the output tuned circuit. R-f choke 2 is for prevention of passage of r-f current into the rectifier, the importance of which was explained in Chapter 11. The excitation for the grid of the following amplifier is adjusted by tuning the output tuning circuit L_2C_2 and adjusting the r-f output tap upward or downward on L_2. C_3 blocks d-c plate voltage from flowing to ground through the output circuits. C_4 prevents grid bias voltage from flowing to ground through the secondary of T.

A driver stage differs from any other intermediate amplifier stage principally in that it must deliver adequate power at all times to the grid of the modulated amplifier, and in the process must often meet special requirements of good voltage regulation under varying load conditions. The requirements for driving a Class C modulated amplifier are simpler than for grid-modulated amplifiers in which, in some transmitters, power is wasted in absorption resistors as the price for improved voltage regulation. For example, if a modulated amplifier grid required a widely varying amount of driving power up to 1,000 w of r-f power, a driver amplifier of only 1,000-w output ordinarily would not provide constant voltage. But if a 5,000-w amplifier delivered 4,000 w to a resistor across which the grids were connected, the grid load variation up to 1,000 w is only a fraction of the total 5,000 w delivered, and the voltage would remain much more constant. The requirements for a driver bear the same consideration as an intermediate amplifier with the added requirement that it may need to function as a constant voltage source of power under varying load conditions.

Chapter 12 *Amplitude Modulation Radio Transmitters* 421

AMPLITUDE-MODULATED RADIO-FREQUENCY AMPLIFIERS

Figure 12-8 shows a Class C radio amplifier and Class B audio modulator. The input to the Class C amplifier is through a parasitic frequency damping circuit to the grids of three tubes operating in parallel. Neutralization of

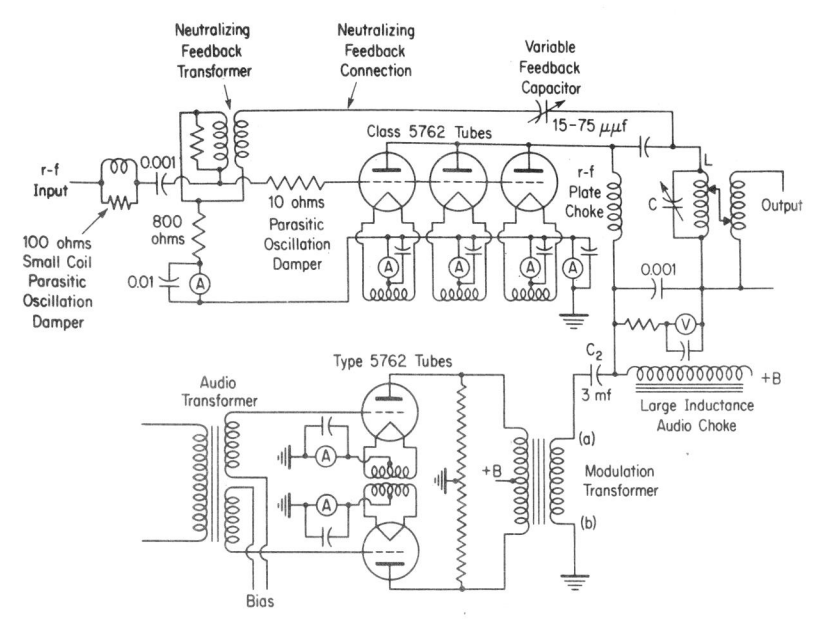

FIG. 12-8. Amplitude-modulated radio-frequency amplifier.

the amplifier is accomplished by feeding plate voltage back to the grid through a variable neutralizing capacitor and a radio-frequency transformer as shown and so connected as to produce degenerative phase of feedback voltage. Radio-frequency output is obtained through a network including L and C so designed as to minimize harmonic radiation. The Class C amplifier is conventional except for the neutralization transformer and conforms with descriptions of such amplifiers given in Chapter 11. The Class B audio modulator likewise is conventional.

Plate power is fed to the r-f amplifier through the audio modulator choke and plate r-f choke and is blocked from the secondary of the modulation transformer by a 0.3-mf capacitor C_2 as shown. Modulation occurs through variation of the plate supply voltage. This is accomplished by the modulators. The power output of the modulators appears across the secondary of the modulation transformer at terminals a and b. Capacitor C_2 is very large and permits free passage of audio-frequency power, but blocks d-c plate power. The audio modulator choke is required to prevent an audio-frequency short circuit of the modulation transformer output. This choke permits free passage of d-c but blocks audio-frequency currents. An examination of this circuit will show that for audio frequency the output of the modulator is in series with the d-c plate supply to the r-f amplifier. During the positive half cycle of audio frequency the modulator output audio voltage rises and adds to the d-c supply voltage. At full modulation it is equal, and the two add to double the d-c voltage insofar as the r-f amplifier is concerned. When the plate voltage is doubled the current is also caused to be doubled, and thus at the peak it results in the r-f power output being quadrupled. During negative half cycles of audio output the audio voltage subtracts from the d-c supply voltage and the net power on negative peaks goes to zero in the absence of any voltage at the r-f power amplifier plate. Thus the output of the r-f stage varies from zero to four times carrier power at full modulation peaks. The simplified equivalent circuit of the plate-modulated system shown is illustrated in Fig. 12-9

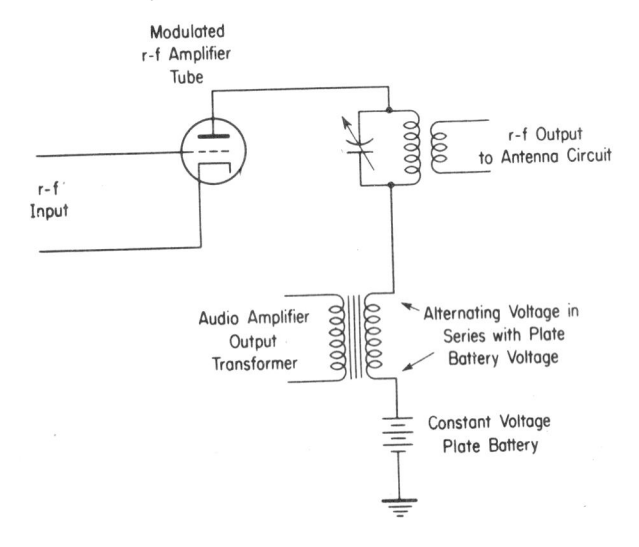

FIG. 12-9. Equivalent circuit of plate-modulated r-f amplifier.

which omits all circuit elements not needed to visualize the operation. The audio-frequency transformer adds at audio frequency plus and minus values of plate voltage to the unvarying battery voltage.

There are many methods of accomplishing amplitude modulation but they fall into the two general classifications of *grid modulation* and *plate modulation*. Plate-modulated r-f amplifiers may have plate efficiencies of as much as 80%. Since grid-modulated amplifiers have but a fraction of that efficiency, plate modulation is preferred in the absence of any unique reason to choose otherwise, as in television transmitters.

In a plate-modulated radio amplifier the relationship of modulating wave, plate voltage, and carrier envelope are shown in Fig. 12-10. Figure

12-10(a) shows, starting at the left at point zero, a condition where the modulating wave voltage has not yet been applied and there is no modulation. In this condition the radio amplifier d-c plate voltage is 5,000 v as shown in Fig. 12-10(b) and unmodulated carrier waves are produced as shown by (c).

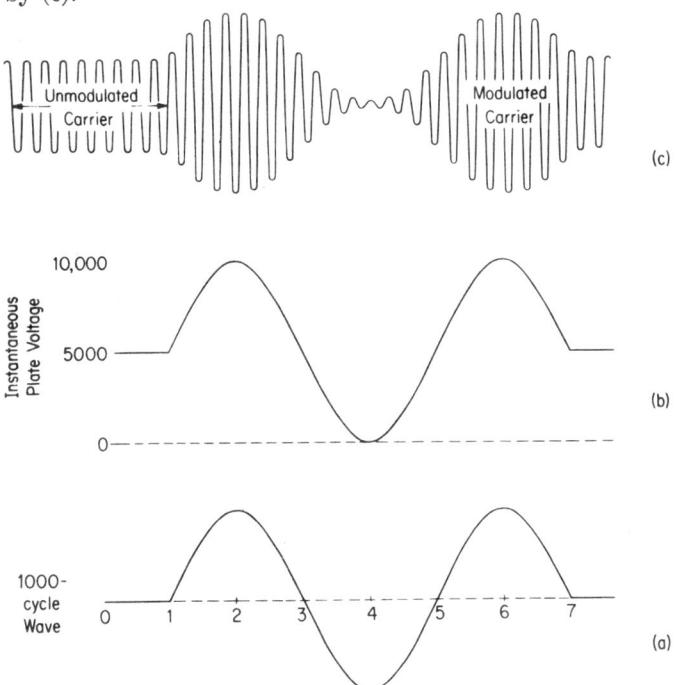

FIG. 12-10. Relationship between modulating wave, plate voltage, and carrier envelope in plate modulation.

At Point 1 on Fig. 12-10(c) a 1,000-cycle modulating wave appears. At Point 2 its voltage has increased to maximum in what shall be called the positive direction. This has caused the plate voltage to double for 100% modulation, as shown in Fig. 12-10(b), and has also caused the carrier voltage to double as shown in (c). The increase of plate voltage from 5,000 to 10,000 v is due to the addition of audio voltage in series with the plate supply rectifier voltage. The added 5,000 v is at audio frequency. At Point 3 the modulating voltage has dropped to zero and is about to reverse to what we shall call the negative direction. But at the zero point the modulator is producing no voltage to add to the rectifier voltage, so momentarily the carrier returns to the unmodulated level.

Upon full reversal at Point 4, the modulator again produces momentarily an additional 5,000 v, but this time it is negative with respect to the rectifier 5,000 v and the two voltages cancel as shown by Fig. 12-10(b). Without any plate voltage the radio amplifier output momentarily drops to zero, as shown at (c). The cycle then starts to repeat. At Point 7 the modulating voltage is removed, and, until it is applied again, the unmodulated condition continues.

For 100% modulation the plate voltage varies from zero to double its unmodulated value. When the plate voltage is doubled, the current is also caused to be doubled so the power momentarily is quadrupled. Therefore, a plate modulated radio amplifier must be capable of producing peaks of power four times as great as the unmodulated value. Also, for this power to be obtained, the radio-frequency grid-driving voltage must be sufficiently high to drive the plate current to the peak value.

NEUTRALIZING OF R-F AMPLIFIERS

Figure 12-8 shows a neutralized power amplifier, including a variable neutralizing capacitor. One of the most important steps in preparing a transmitter for operation is the tuning and neutralization of the r-f amplifiers.

It was explained that the grid-plate capacitance of a triode permits r-f current to flow from the plate to the grid-cathode capacitance, creating feedback from the output to the input circuit, and thus producing self-oscillation. It has been explained that neutralization consists of feeding back an equal and oppositely phased current to the grid-cathode capacitance, through a separate circuit, so that the resultant voltage at the grid is zero, or so small that the feedback is reduced to insignificant proportions.

Since, in a single side amplifier, a neutralizing capacitor connected directly from plate to grid would produce a voltage of the same phase as that of the grid-to-plate capacitance, a method must be found to produce a neutralizing voltage of opposite phase. This is accomplished by connecting an inductor in series with the neutralizing capacitor or using a transformer as shown on Fig. 12-8. The inductor or transformer alone could provide the required neutralizing voltage, but it is convenient to use a variable series capacitor to block out the d-c plate voltage and provide the adjustable r-f feedback voltage desired.

Figure 12-11 shows the circuit of Fig. 12-8 reduced to simplified form. Neutralization is accomplished as follows:

1. Connect cathode-ray oscillograph to the output tuned circuit to indicate the r-f voltage developed across it. This connection should be made through a capacitor of about 10 $\mu\mu$f, of sufficiently high voltage rating to protect the apparatus and the operator from the d-c plate voltage.

2. Apply low plate voltage to the tube. It should be low enough to protect circuit components against self-oscillation or improper adjustment.

3. Apply r-f voltage to the grid and tune the output tank circuit for maximum developed r-f voltage.

4. Adjust the neutralizing capacitor for minimum output voltage. Part of the output voltage is produced by power feeding from the grid to the output through the grid-plate capacitance. Feeding an equal and opposite voltage through the neutralizing circuit will *reduce* the output voltage. Minimum output voltage indicates that the desired neutralization is achieved. This indication may be obtained with the plate voltage removed, if desired.

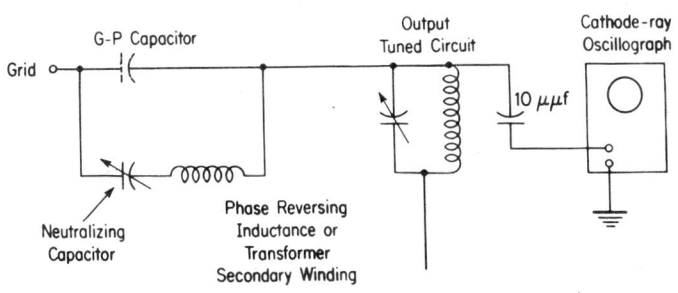

FIG. 12-11. Circuit elements of particular significance during neutralization of circuit shown in Fig. 12-8.

5. Alternately make fine adjustments in the output tuning and the neutralizing capacitor until neither control has appreciable reaction on the other.

6. Check the neutralization by observing the plate current of the preceding tube while the output tuning is varied widely on the stage being neutralized. The reaction should be very slight when neutralization is correct because the output circuit reactance is effectively isolated from the input.

The final adjustment consists of connecting the load to the amplifier. If the load is substantially non-reactive, the optimum amplifier output tuning adjustment will change little as the proper degree of loading is achieved.

In push-pull r-f amplifiers, the desired reverse-phase neutralizing voltage is obtained conveniently by connecting the neutralizing capacitor from its respective grid to the plate of the opposite tube, as has been explained. No separate phase reversing inductances are required.

POWER SUPPLY REGULATION

The various types of modulating systems have been described in Chapters 11 and 12. The performance of these systems in the manner desired depends upon the sources from which power is drawn. For example, it may be assumed that when a bias voltage is chosen it will always remain constant at the chosen value. But if it does not, then the operation of the amplifier is modified. If the change in bias is small and unvarying, little or no harm may result. But suppose it changes during modulation owing to conditions caused in the modulating process. If the bias is caused to increase, plate current flow is reduced when it should not be. Nonlinear amplification and distortion result.

INVERSE FEEDBACK IN TRANSMITTERS

Chapter 7 described and explained inverse feedback which is employed to reduce distortion and improve the frequency response in amplifiers. It was shown that if new frequencies appear in the output of an amplifier which were not present in the input signal, a portion of the output may be fed back to the input in reverse phase to provide a cancelling voltage.

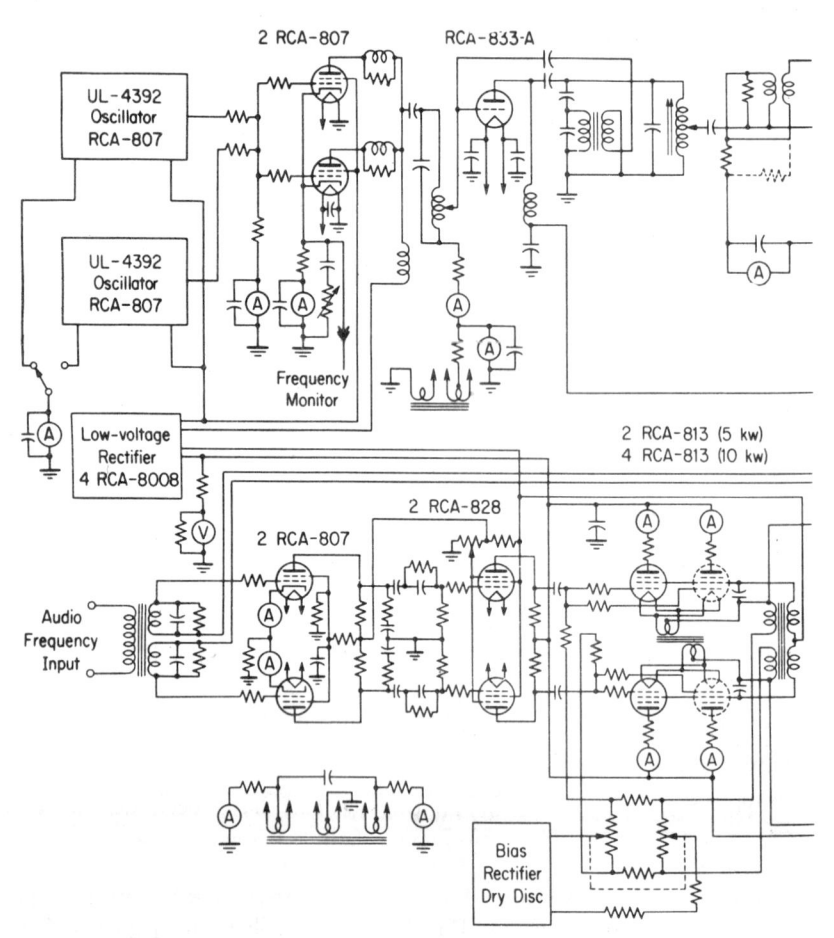

Fig. 12-12. Schematic circuit of amplitude-modulated broadcast transmitter.

Chapter 12   *Amplitude Modulation Radio Transmitters*   427

Since a portion of the desired signal also is fed back and causes cancellation, the gain of the amplifier must be increased to compensate for that loss. Feedback also cancels undesired noise frequencies introduced in an amplifier from filament lighting a-c, imperfect rectifier hum filtering, etc.

Inverse feedback is generally employed in broadcast transmitters. If a small amount of power from the output of the transmitter is demodulated, the transmitter may be considered to be an amplifier and inverse feedback applied in conventional manner. This is *overall feedback*. In high-level Class B modulated transmitters, feedback conveniently may be used from the modulator output back to the input terminals. An example of modulator output to input inverse feedback is shown on Fig. 12-12. It may be seen

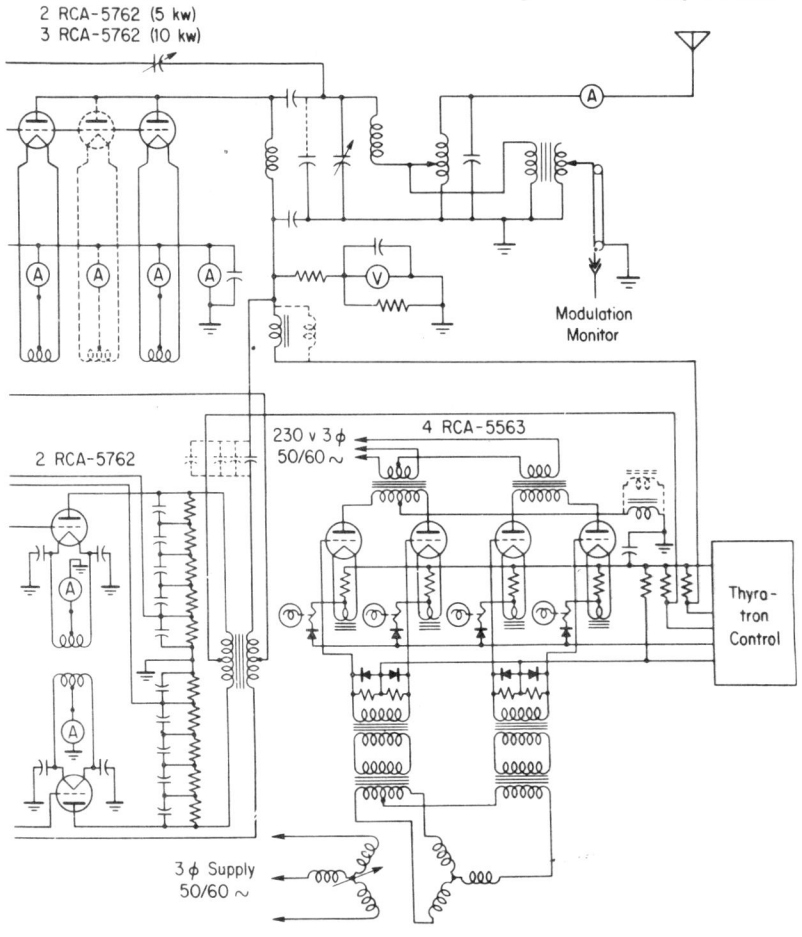

FIG. 12-12. (cont.)

that a resistor-capacitor combination is connected across the primary of the modulation transformer in a series arrangement with the center grounded. Feedback audio voltage is taken from the innermost two resistors and fed back to the center of the a-f input transformer in reverse phase.

SCHEMATIC DIAGRAM OF A 10-KW TRANSMITTER

Earlier in this chapter, in schematic diagrams, Fig. 12-3 showed a typical crystal oscillator circuit (Figs. 12-4 and 12-5 showed photographs of one), Fig. 12-6 showed a typical buffer amplifier, Fig. 12-7 showed a typical driver r-f stage, and Fig. 12-8 showed a typical Class B modulator and high-level modulated Class C r-f amplifier. Figure 12-12 combines these schematic diagrams in a somewhat simplified overall schematic diagram of a complete transmitter. The radio-frequency section is shown in the upper portion of the diagram and the audio-frequency section in the lower portion. The radio-frequency section has been described and explained stage by stage and the modulator was similarly treated. The transmitter described is a modern commercial 10-kw broadcast transmitter, the RCA type BTA-10H. To complete the description the audio-frequency section will be described.

FIRST AUDIO STAGE

The first audio-amplifier stage at the lower left is a conventional push-pull resistance-capacitance coupled amplifier employing two tetrodes, type 807. The input circuit utilizes an audio-frequency transformer to match the impedance of the input line to the grids, obtain a large step-up in audio voltage on the grids, provide an entry for the inverse feedback voltage from the modulator output, and give an input circuit in which both sides are equally above ground potential, balanced. Every phase of such amplifiers was described in Chapter 7. The combined plate and screen currents are metered in the cathode circuit as shown. The resistors in the grid leads are of low value to damp out parasitic oscillations. Plate voltage is obtained from the voltage divider resistor which also supplies screen grid voltage to the following stage. This divider is shown immediately above the upper 828 tube on Fig. 12-12. In parallel with each 12,000-ohm plate loading resistor there is a series circuit to ground consisting of a 2,000-ohm resistor and a 1,500-μf capacitor. The capacitor presents an impedance which rises inversely with audio frequency. Therefore, the load on the tube is such as to provide increasing amplification inversely with frequency. It will provide compensation for losses at low frequency elsewhere in the audio system.

SECOND AUDIO AMPLIFIER

The second audio amplifier is a conventional push-pull resistance coupled stage employing two type 828 pentode tubes. Immediately next to each grid there is a 470-ohm parasitic oscillation damping resistor. In series with each grid there is also a resistor and capacitor in parallel, located on the diagram between the parasitic resistor and the 807 plate blocking capacitor. The parallel circuit utilizes a 0.1-mf capacitor and a 0.22-megohm resistor which in combination contribute correction of the high-frequency response of the audio system. The plate loading resistors are 12,500 ohms each.

THIRD AUDIO STAGE

The third audio amplifier stage, which drives the modulator grids, is unique. It is cathode-coupled. Cathode coupling was explained in Chapter 7. Four type 813 pentode tubes are employed, two in parallel on each side of a push-pull circuit. The suppressor grids are connected directly to the cathodes. The most unique feature of this stage concerns the cathode and screen grid circuit. The output of this third audio stage is taken from the cathodes directly to the modulator grids. The secondary of the transformer was provided for the purpose of making the 813 screen grid voltages rise and fall at audio frequency instead of remaining at a fixed d-c voltage. This improves the linearity of the amplifier over the operating range. The audio voltage is in series with a d-c voltage which establishes the proper axis around which the audio voltage fluctuates.

On the 10-kw transmitter being described, inverse feedback is employed from the modulator output back to the first audio stage input. Figure 12-12 shows across the modulation transformer primary ten resistors in series, five on each side of a grounded center. Each resistor has across it a small capacitor. This constitutes a voltage divider of a total resistance of 40.2 megohms. This divider is tapped across the two innermost center resistors of 0.1 megohm each to provide a small audio voltage. Each side of the line has the same voltage to ground, or is balanced to ground. This line is led back to the center of the audio input transformer, as shown on Fig. 12-12, to accomplish overall audio system feedback. If overall transmitter feedback instead were employed, a small amount of r-f would have been extracted from the transmitter output, demodulated to audio frequency and applied to the input circuit.

POWER SUPPLIES

Bias voltage for the audio driver is provided by a rectifier unit employing selenium diode rectifiers. Selenium cells have the property of conducting very much more effectively in one direction than the other and with proper circuitry function in the same manner as tube rectifiers. Their use ordinarily is confined in radio transmitters to low-power rectifiers.

The low plate and screen voltages for the oscillators, buffer, driver r-f amplifier, and the first three audio stages are provided by a low voltage rectifier utilizing four 8008 rectifier tubes.

The circuit diagram is shown in Fig. 12-13. The anodes of tubes 3 and 4 are connected together and go to ground through an overload relay coil.

For 825 v the circuit is from ground through tube 3 and thence through the left half of the high-voltage winding to the center tap and out to the filter, for one half cycle. For the other half cycle tube 4 and the right half of the high-voltage winding are working. Thus full wave rectification is produced.

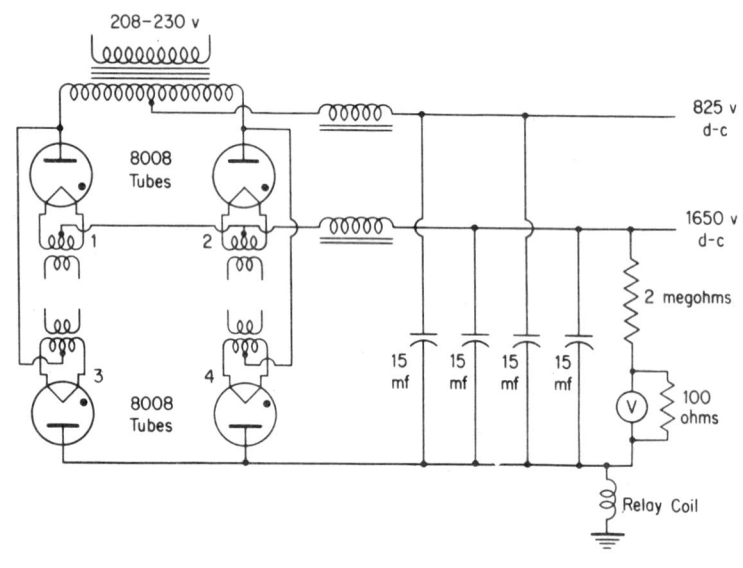

Fig. 12-13. Rectifier for low voltage plates and screens for oscillators, buffer, driver amplifier, and audio stages of 10-kw broadcast transmitter.

For 1,625 v the circuit is from ground through tube 3, thence from left to right through the full high-voltage winding and thence through tube 2 to the filter, for one half cycle. For the other half cycle the circuit is through tube 4, thence from right to left through the full high-voltage winding, thence through tube one to the filter. For the higher voltage full wave rectification is also produced.

The high voltage source for the r-f power amplifier and modulator provides up to 5,000 v from a rectifier assembly utilizing four type 5563-A thyratron tubes as shown at the lower right of Fig. 12-12.

A thyratron is a gas-filled grid controlled tube designed for rectification. The grids are ordinarily biased to about minus 100 v which prevents conduction with any amount of applied plate voltage. However, with positive plate voltage present conduction may be started by overcoming the bias with a higher sharp voltage pulse. Once started, conduction in these gas-filled tubes can be stopped only by eliminating positive plate voltage. The grid control is lost because of ionization of the gas. Positive plate voltage automatically is removed when each positive half cycle of a-c voltage is completed. On the next positive half cycle the grid again is in control and

Chapter 12   *Amplitude Modulation Radio Transmitters*   431

prevents plate current flow until again a short sharp positive voltage pulse eliminates its control. By controlling the time of application of the short positive pulses, the flow of current can be made to start at any desired part of the half cycle. This characteristic is used to advantage to control the voltage output of the rectifier. For maximum voltage, current flow is triggered on at the beginning of positive half cycles and continues until completion. For lower voltage output the triggering action is withheld until part of the half cycle is completed. For very low voltage the triggering action would be withheld until the half cycle were nearly completed so only a small amount of current would flow. The short firing pulses are of about 200 v and are obtained from two "pulse" transformers identified in the rectifier section of Fig. 12-12 by the resistors and metallic rectifiers across

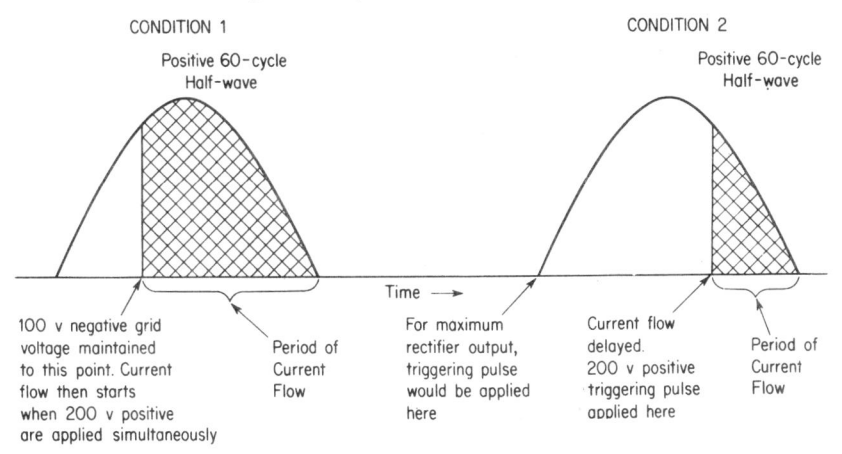

CONDITION 1   CONDITION 2
Positive 60-cycle   Positive 60-cycle
Half-wave   Half-wave

Time →

100 v negative grid voltage maintained to this point. Current flow then starts when 200 v positive are applied simultaneously | Period of Current Flow | For maximum rectifier output, triggering pulse would be applied here | Current flow delayed. 200 v positive triggering pulse applied here | Period of Current Flow

FIG. 12-14. Method of controlling output power of thyratron rectifier.

the secondaries. They convert 60-cycle sine waves to 60-cycle pulses or peaks. Figure 12-14 illustrates the method of rectifier output control.

The unique properties of thyratrons are also used for overload protection. Assume the triggering pulses are recurring regularly, ignite the rectifier tubes, and cause current to flow normally. In the event of a sudden accident, causing a short circuit, it is important to interrupt current flow as quickly as possible to prevent destruction. This is accomplished by introducing on the thyratron grids still a third voltage of minus 300 v which overcomes the 200-v triggering pulses. If a short circuit occurred in the middle of a 60-cycle positive half wave, 300 v would be applied to the grids. After the end of the half cycle the 300-v negative bias would prevent further conduction despite the continuing triggering pulses which are initiated independently.

Often a short circuit is momentary. To prevent noticeable transmitter outages under such conditions a supplementary control is added to the negative 300-v circuit which causes the voltage to be applied for a short interval and then removed. If the short no longer exists, operation is automatically restored. If it still exists, the sequence is repeated twice. If it still exists, the 300-v bias then automatically remains and manual restoration is required. This is called *automatic recycling*.

Grid voltage controls for the thyratrons provide a means of timing the application of the triggering pulses at any point desired on the positive half cycle of plate voltage.

FIG. 12-15. RCA type BTA 10-H 10-kw broadcast transmitter. (Courtesy RCA)

Figure 12-15 shows a photograph of the RCA 10-kw transmitter described. The left-hand cabinet, the modulator unit, contains at the bottom the modulation transformer, above it the type 5762 modulator tubes, above them the first and second audio-amplifier tubes, and at the top the third audio-amplifier type 813 tubes.

The second cabinet from the left, the power amplifier units, contains the r-f power amplifier with its three type 5762 tubes, neutralizing transformer, and neutralizing capacitor.

The third cabinet, the exciter unit, contains the air blower for cooling the modulator and power amplifier tubes. The duct system carries across

Chapter 12   *Amplitude Modulation Radio Transmitters*   433

cabinets 1 and 2 under the shelves, as may be seen. Above the blower are the four type 8008 low-voltage rectifier tubes. Above these tubes are the regular and emergency quartz crystal oscillator assemblies mounted in their metal housings. These are the units shown unenclosed in Fig. 12-4. Above the oscillators are the two type 807 buffer amplifier tubes, and at the top is the type 833-A driver amplifier tube.

The cabinet at the right, the rectifier and control unit, contains the main rectifier and the control devices. The lower section houses the thyratron

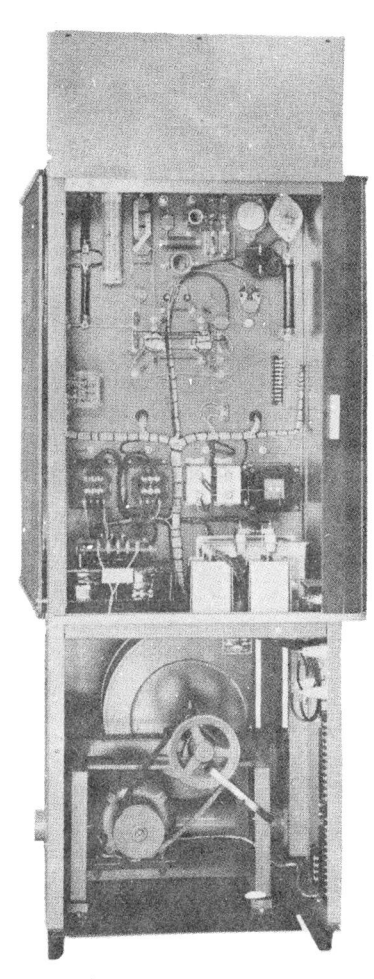

FIG. 12-16. Rectifier and control unit, rear view, BTA 10-H 10-kw transmitter. (Courtesy RCA)

FIG. 12-17. Power amplifier, rear view, BTA 10-H transmitter. (Courtesy RCA)

rectifier transformers. The rectifier tubes are in the rear at the top as shown on the rear view in Fig. 12-16.

Figure 12-17 shows the rear view of the power amplifier cabinet. The coil at the left mounted on the sidewall is the plate supply r-f choke. The capacitor above the shelf in the center is the plate blocking capacitor. The coils in the center and top are for plate tuning and loading. The capacitors

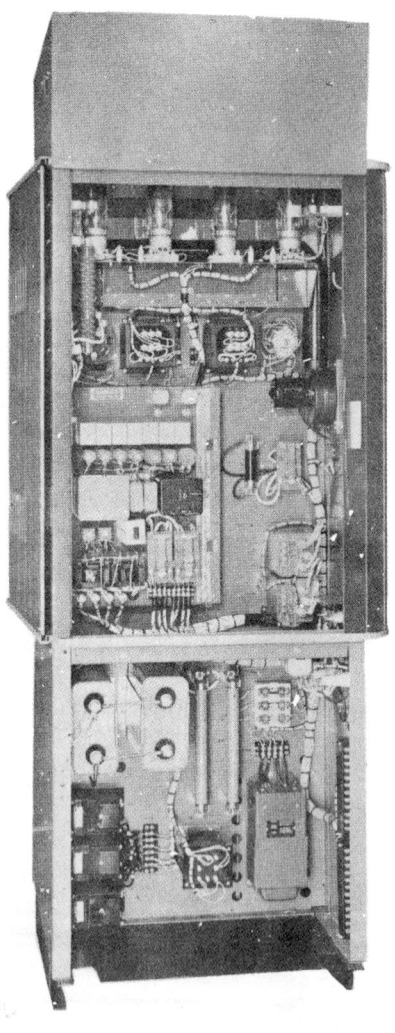

Fig. 12-18. Exciter panel, rear view, BTA 10-H transmitter. (Courtesy RCA)

on each side of the white plate blocking capacitor are of the vacuum type and are for output circuit plate tuning. The unit above the round hole is the variable plate tuning capacitor. The transformers in the lower compartment are for tube filament lighting.

Chapter 12 *Amplitude Modulation Radio Transmitters* 435

Figure 12-18 shows the rear view of the r-f exciter panel. Above the shelf are the components of the low-voltage 8008 tube rectifier, voltage dividers, etc. At the top are the radio-frequency components for the buffer and driver amplifiers.

SAFETY INTERLOCKS

The photographs of apparatus are shown with the doors open to illustrate internal design and construction. However, it is standard practice on this and nearly all other transmitters to provide automatic door switches which remove dangerous voltages when a door which closes them in is opened. Every precaution is taken to prevent persons from being injured accidentally. In addition, it is common practice to impose safety rules and penalties for violations.

SHIELDING

Figure 12-19 illustrates the method of shielding circuits to prevent undesired coupling from electromagnetic or electrostatic fields or external

FIG. 12-19. Illustration of shielding in a high-power transmitter. (Courtesy RCA)

radiation from them. The rear panels were removed for the photograph but, when added, completely enclose the individual compartments shown. The circuits shown are harmonic filter sections in an RCA 50-kw transmitter employing the Chiriex system of modulation. It is referred to as "ampliphase" modulation because it produces amplitude modulation from phase modulation. The glass-enclosed units are vacuum capacitors, consisting of metal electrodes enclosed in a vacuum to increase the rating and reduce the size.

TELEGRAPH TRANSMITTERS

A transmitter designed for CW telegraphy could be very similar to the radio-frequency system of the broadcast transmitter which has been described. The principal differences would be the elimination of the audio-frequency amplifying and modulating system and the addition of a keying system. The crystal oscillator and also possibly a buffer stage would operate continuously in order to eliminate any variation in operating conditions which would cause the frequency to change. In one of the low-power r-f stages a means would be provided for keying the carrier on and off, corresponding to the telegraph code characters. Conventionally the grid bias on the keyed amplifier tube would be made very high so that plate current is completely cut off. Then when the key is depressed this very high bias would be reduced so that normal plate current flowed. All of the following r-f amplifier stages would be biased below plate current cut-off so that no plate current flowed except when the telegraph key was depressed and r-f excitation came through to drive the amplifiers to normal output.

Since telegraph transmitters used on ships and for transoceanic service are required to change carrier frequency at times to take advantage of optimum wave-propagation conditions, and utilize channels on which there is a minimum amount of interference, they normally are provided with several crystals which may be switched into the circuit one at a time to obtain the carrier frequency desired. When the frequency is changed by a substantial amount, the amplifier circuits must be retuned for efficient operation, and this often is accomplished by a ganged switch which actually is a multiplicity of switches actuated by one long shaft. Other transmitters may use remote controlled relays for the switching function. On very high power transmitters manual retuning has been used but the trend is toward relay and motor-actuated switches.

KEYING METHODS

Figure 12-20 shows two methods of keying an r-f amplifier stage. Figure 12-20(a) simply keys the cathode return. When the key and thus the cathode lead is open, no plate current flows. Figure 12-20(b) shows a system utilizing a special *keying tube*. When the key is closed, the keying tube is biased beyond cut-off by high negative bias and no plate current flows

Chapter 12   *Amplitude Modulation Radio Transmitters*   **437**

through the keying tube. Plate current flows to r-f amplifier number 1 through R. The plate supply voltage is made high enough to allow for the loss of voltage in R and still provide normal plate voltage for the r-f amplifier.

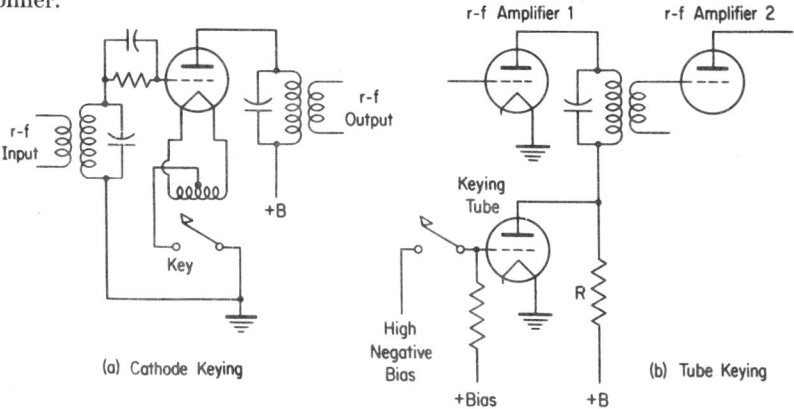

Fig. 12-20. Two methods o keying r-f amplifiers for CW telegraphy.

When the key is open, positive bias is applied to the keying tube grid and heavy plate current starts to flow through the keying tube. It has to flow through R. It greatly increases the voltage drop across R. This reduces the plate voltage to r-f amplifier tube 1 to the point where it is no longer able to drive amplifier number 2 above plate current cut-off. Thus, keying is accomplished.

FREQUENCY MULTIPLIERS

Quartz crystals perform effectively at frequencies up to about 15,000 kc when cut and processed properly. However, dimensional and other limitations prevent effective operation at much higher frequencies. Therefore, some method of generating higher carrier frequencies is essential. It is conventionally accomplished by cutting a crystal for a submultiple of the carrier and multiplying the crystal frequency the required number of times. For example, it may be desired to transmit on 30,000 kc. Crystals operate very advantageously in the range of 5,000 kc. If 5,000 kc were selected for the crystal frequency, it would need six times multiplication to provide the desired 30,000-kc carrier frequency. This frequency multiplication is accomplished conveniently and efficiently in *frequency multiplier* amplifiers.

Class C amplifiers are driven hard with short but heavy plate current pulses flowing during each positive half cycle. Operation extends far beyond the linear portion of the grid voltage-plate current curve. This causes substantial distortion which produces multiples of the driving frequency, or harmonics. In such an amplifier the plate-tuned circuit can be adjusted not to the fundamental but to a harmonic frequency, and r-f oscillating current will flow at the harmonic frequency. When so adjusted, the impedance at the harmonic frequency may be made optimum. At the fundamental it will be very low and little power will be absorbed. It is possible to use fundamental and harmonic tuned circuits in series and absorb power at both frequencies at a loss of efficiency, but the need to do so seldom arises.

Frequency multipliers may be used to produce higher frequency multiplication than 2 or 3, but at the higher orders the output decreases rapidly. Therefore, doubling or tripling is most often used, several times in cascaded multipliers when necessary. The 5,000-kc crystal frequency referred to would be doubled and the output of the doubler would then be tripled, or vice versa, to attain the 30,000 kc desired. The schematic circuit diagrams would be like those of conventional Class C amplifiers, the difference being in the use of higher biases and r-f driving voltages on the grid, and the tuning of the plate circuit to a harmonic frequency.

The plate efficiency in doubler stages will not be quite as high as in fundamental amplifiers with optimum adjustments, but in low power stages the moderate loss is of little consequence. Where frequency multiplication is accomplished in higher power stages, caution must be exercised not to exceed the heat dissipation ratings which otherwise might be exceeded because of the lower efficiency. Where 70% efficiency is obtainable at the fundamental frequency, 60% might be obtained at the harmonic frequency.

In television transmitters the frequency multiplication between crystals and carrier frequencies may range from 6 to over 80 because the channel frequencies are from 55.25 to 889.75 megacycles.

TRANSMITTER POWERS

Radio transmitters of conventional circuitry range in power from a minute fraction of a watt to over a thousand kilowatts. For instructions to the infantryman, for communicating with television camera crews within a studio, and for other purposes tiny fractional-watt units are carried in the pocket or the hand. For automobile communications transmitter-receiver combinations of about 2 cubic feet in size and a few watts are used by the tens of thousands at from 40 to 500 megacycles. Above certain sizes all steamships, numbering in the thousands, have radio communication employing transmitters ranging from about 50 to several thousand watts at frequencies from about 425 kc to 18,000 kc. Shore stations maintained for communications with ships use powers as high as 20 kw and may have a multiplicity of transmitters for use on several different frequency bands. For extremely reliable long distance communications, huge transmitters with powers up to 1,000 kw are used by the military at frequencies as low as 13 kc. Thousands of aircraft carry several transmitters and receivers usually of a few tens of watts and used between 108 and 152 mc. International broadcasting stations use powers up to 500 kw at frequencies from about 5,000 kc to 22,000 kc. Transoceanic common-carrier communications companies and the government services use hundreds of transmitters of from 5 to 50 kw in the same region. Over 100,000 radio amateurs in the United States use transmitters of powers up to 1,000 w at frequencies between the broadcast band and the microwave region. Private yachts by the thousands use transmitters of from 5 to 50 w for communications around 2,300 kc. In the United States about 3,000 broadcast stations use powers from 250 w to 50 kw between 540 and 1,620 kc, and about 500 television stations use powers from 100 w to 100 kw between 54 and 890 megacycles.

The services enumerated do not by a wide margin constitute the complete list but serve to emphasize the tremendous usage of radio transmitters and receivers for communications. Omitted completely are navigational aids, radar, radio relays, special military applications, telemetering, guidance systems, standard frequency transmissions, and many others.

SINGLE-SIDEBAND SUPPRESSED CARRIER SYSTEM

It was explained in Chapter 10 that for certain types of radio communications systems substantial advantages accrue by suppressing the carrier and one set of sidebands prior to transmission. The advantages were listed and evaluated.

Fig. 12-21. Sixty-watt telephone or telegraph transmitter designed for suppressed carrier single sideband or other forms of transmission. (Courtesy Radiomarine Corp.)

Figure 12-21 is a photograph of a single sideband, 60 w, suppressed carrier transmitter-receiver designed for telegraph or telephone transmission. It may also be operated as single-sideband-with-carrier equipment to make it compatible with conventional amplitude-modulated systems. It may be operated at any one of four pretuned channels between 3 and 15 megacycles. Figure 12-22 shows a photograph of the r-f chassis at the left and the power supply-audio chassis at the right.

A block diagram of the transmitter-receiver is shown on Fig. 12-23, the transmitter being at the top and the receiver at the bottom.

FIG. 12-22. Internal construction of SSB transmitter. (Courtesy RCA)

The microphone output is amplified in the microphone amplifier. For simplicity in the following explanation it will be assumed that the audio input is a 1,000-cycle tone. From the microphone amplifier the 1,000-cycle tone goes to a speech clipper which compresses the amplitude range of speech to increase its *average* level. The tone then goes to a balanced modulator which obtains 250-kc input from a crystal oscillator. From the balanced modulator there are produced 249- and 251-kc sidebands without a carrier. They go to a mechanical filter. This device is so tuned that it passes the 251-kc sideband but eliminates the 249-kc sideband.

There is now present a single 251-kc sideband without a carrier. This must be increased in frequency to the desired carrier frequency. It is ac-

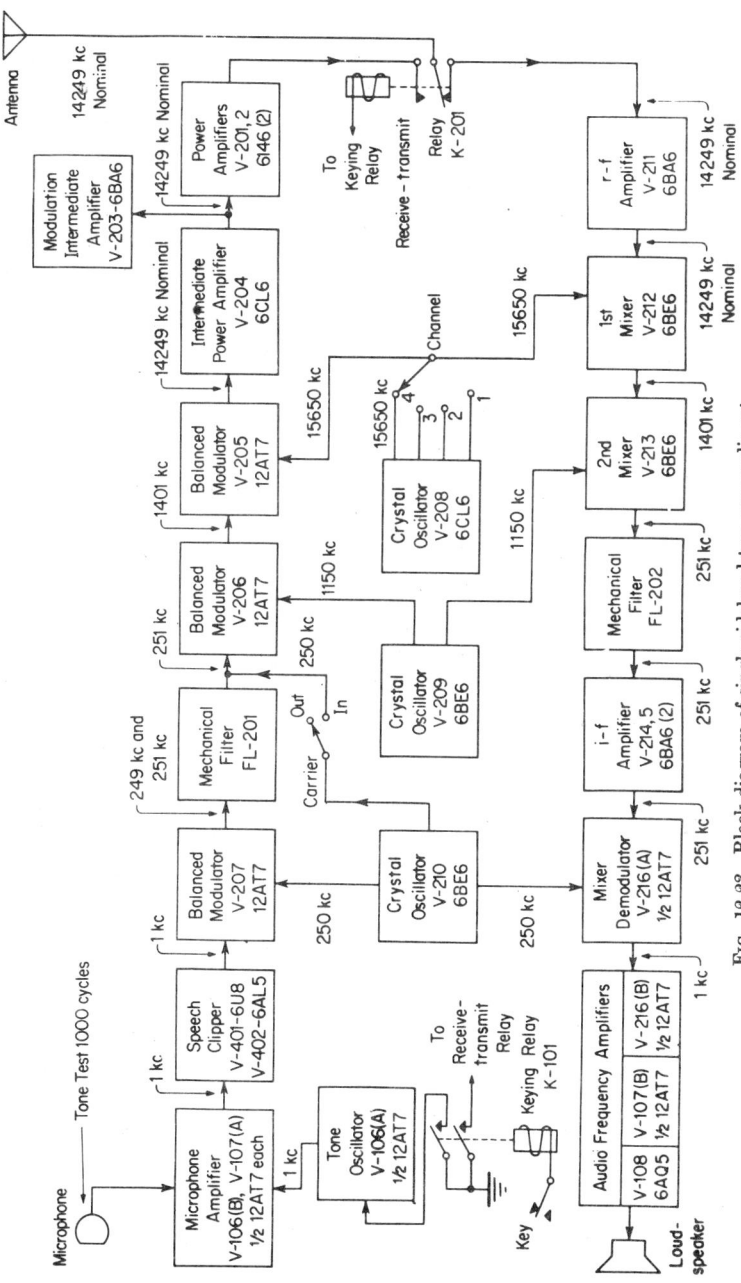

Chapter 12 *Amplitude Modulation Radio Transmitters* 441

FIG. 12-23. Block diagram of single-sideband two-way radio set.

complished in two following balanced modulators which heterodyne the frequency up in two stages to 14,249 kc, the carrier frequency as shown. Balanced modulators are used to heterodyne the two input frequencies and obtain a third difference frequency. The final operation is amplification of the 14,249-kc output from the final balanced modulator, to raise it to the desired carrier power. This is accomplished in two power amplifiers as shown in Fig. 12-23.

If it is desired to transmit a carrier with the sideband, to facilitate reception in a conventional receiver, the carrier of 250 kc is combined with the single 251-kc sideband at the input to the second balanced modulator, as shown in Fig. 12-23. The two frequencies will now be processed together in the following stages to produce the desired carrier and sideband frequencies.

Stations using this type transmitter for back-and-forth communication normally operate on the same carrier frequency because it is optimum for use in both directions. Incoming received signals go to an r-f amplifier. Then, using the same crystal oscillators used in heterodyning the transmitter frequency, the process is reversed and the incoming frequency is heterodyned *down* to 251 kc at which point the carrier of 250 kc is inserted, it is demodulated, and is fed to the audio amplifiers and loudspeaker. These operations are shown clearly in Fig. 12-23.

Because of the absence of carrier frequency beat-note interference with the suppressed carrier single-sideband system, it is possible for a number of distant stations using this system to operate on a "party line" basis, talking with each other freely and within limits simultaneously on the same channel frequencies.

It may be asked why the frequency was raised from 251 kc to 14,249 kc by heterodyning rather than using frequency multiplier stages of conventional type. One reason is that if the frequencies of a 250-kc carrier and a 251-kc sideband are multiplied five times they become 1,250 and 1,255. The sideband separation from the carrier is no longer 1 kc but has become 5 kc. This does not occur in the heterodyning process. If 1,000 kc is added to 250 and 251 kc there are produced 1,250 and 1,251 kc, and the sideband-carrier frequency relationship is not destroyed. But even more important, heterodyning may be accomplished without loss of linearity.

The specifications for the SSB (single sideband) system described are of considerable interest because they relate in a unique manner to other material in this chapter. They are shown below.

GENERAL DESCRIPTION AND TECHNICAL SUMMARY OF SSB SPECIFICATIONS

GENERAL

Channels: Four.

Type of operation: Simplex ("push-to-talk" telephone, or telegraph).

Frequency range:
    Channels 1 and 2: 3.0–6.7 megacycles
    Channels 3 and 4: 6.7–15.0 megacycles

Chapter 12 *Amplitude Modulation Radio Transmitters*

Antenna required: Resistance: 10–80 ohms. Capacitance: 300 μμf (min.). Single wire not to exceed one-quarter wavelength at highest channel frequency.

Crystals required: One 250-kc, one 1,150 kc, four carrier crystals (one per channel). Channel crystals must be 1,400 kc higher in frequency than the desired operating frequency. The same crystal serves both transmitter and receiver.

Emission:
Phone: Single sideband suppressed carrier or single sideband with carrier. Telegraph: Single sideband keyed tone.

Reception: Single sideband suppressed carrier; single sideband with carrier; keyed tone; single sideband keyed tone.

Keying speed: 30 words per minute, manual (break-in) operation; 60 words per minute, teleprinter operation.

TRANSMITTER

Power output: 60 w.

Frequency stability: ±0.0005%

Transmitted sideband: Lower.

Unwanted sideband suppression: 50 db.

Carrier suppression: 50 db.

Harmonic suppression: 56 db.

Audio input: (a) Single button carbon microphone from local handset or from up to three remote positions. (b) −6 dbm in 600-ohm line for full transmitter output.

Audio fidelity: ±2 db, 350–3,000 cycles.

Amount of speech clipping: 20 db.

Transmitted sideband distortion: Single tone, full power output, no clipping, 2.5% at 1,000 cycles.

Two-tone test: Distortion products, −26 db.

RECEIVER

Sensitivity: Better than 1 microvolt for 50 milliwatts output with 6-db signal-to-noise ratio.

Selectivity: Determined by mechanical filter characteristics: 3.2 kc nominal bandwidth for 6-db attenuation; 5.4-kc bandwidth for 60-db attenuation.

Audio fidelity: 2 db, 350–3,000 cycles.

Audio output: (a) 2 w maximum in speaker. (b) With 50-mw output in loudspeaker, audio level in 600 ohm line is −7 dbm.

Audio distortion: 2.5% (1,000 cycles at 50 milliwatts output).

Two-tone test: Distortion products, −26 db.

EXAMPLE OF A HIGH-POWER INTERNATIONAL BROADCASTING INSTALLATION

Because of the long distances to a variety of overseas target areas, international broadcasting stations employ very high power, highly directive antennas, frequently change operating frequencies, and require a large

444　*Amplitude Modulation Radio Transmitters*　Chapter 12

Fig. 12-24. Control section of 100-kw international broadcasting transmitter at Dixon, Cal., station serving Pacific areas in Far East.

Fig. 12-25. Transmitter-antenna switching frame at Dixon plant.

Chapter 12 *Amplitude Modulation Radio Transmitters* 445

number of antennas. Such a station owned by the U. S. Government and operated by the National Broadcasting Company is located at Dixon, California.

It employs 5 transmitters ranging in power from 50 kw to 200 kw. A photograph of a 100-kw transmitter control section is shown in Fig. 12-24. Incoming programs over telephone lines are routed through a centralized audio control room to the desired transmitter which feeds its power to one of many directional antennas. Since many transmitters and many antennas are used, and the combinations change frequently, an outdoor antenna-transmitter switching system is required (see Fig. 12-25). R-f lines from each of the transmitters lead to an outdoor swinging switch arm which is moved to the fixed position of whichever antenna switch is desired. Lines leading to the frame in all directions except the left are from huge antennas which fully occupy a square mile of land. Any transmitter may be connected to any antenna through the switch frame shown. One of the many high-power-gain antennas is shown in Fig. 12-26. It focusses the wave energy into a narrow beam directed toward a specific target area in the Far East.

As mentioned previously in this chapter, heat must be removed from power tubes. Figure 12-27 shows the cooling room of the Dixon plant. Above the catwalk at the left are air filters which remove dirt from the air circulating system. Cooling is accomplished by pumping the tube water through cooling pipes located in the forced-air circulating system.

FIG. 12-26. High-gain directional antenna at Dixon international station.

FIG. 12-27. Tube cooling room at Dixon international station.

The principle of focussing power into sharp radio beams is referred to elsewhere in this book. The principle of power gain is fully explained. The effective power in the beam from one of these huge antennas is the product of the transmitter output in kilowatts times the power gain and is as much as 20,000 kw.

R-F TRANSMISSION LINES

R-f power must be conveyed from one part of a transmitting system to another, such as from stage to stage or from the output stage to the antenna. At frequencies up to roughly 30 megacycles open wire lines are efficient and relatively inexpensive for this purpose. Often coaxial tube lines are preferred.

OPEN WIRE LINES

The simplest transmission line for transferring power from one point to another is a parallel pair of wires very similar to those seen on telephone and power-line poles. A pair of such parallel wires has distributed inductance, resistance, and capacitance. The inductance and capacitance vary with the spacing between wires and with the diameters of the wires. Each line has a characteristic impedance of its own which depends upon these factors, as explained in Chapter 9.

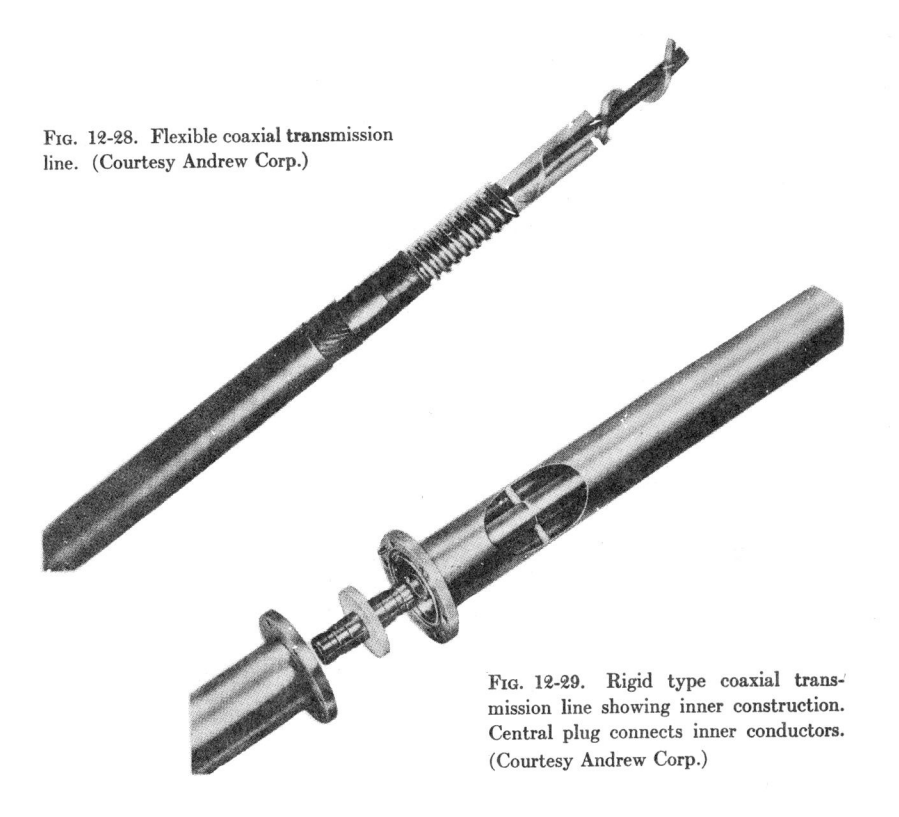

Fig. 12-28. Flexible coaxial transmission line. (Courtesy Andrew Corp.)

Fig. 12-29. Rigid type coaxial transmission line showing inner construction. Central plug connects inner conductors. (Courtesy Andrew Corp.)

Fig. 12-30. Method of routing coaxial line from transmitter to antenna. (Courtesy Andrew Corp.)

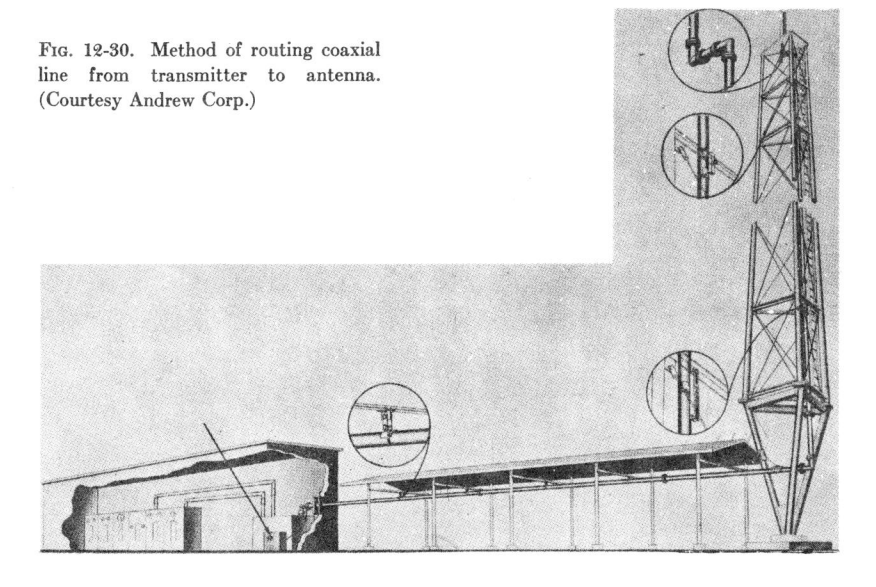

Fig. 12-31. Type equipment used to couple transmission lines to broadcast antennas. (Courtesy Andrew Corp.)

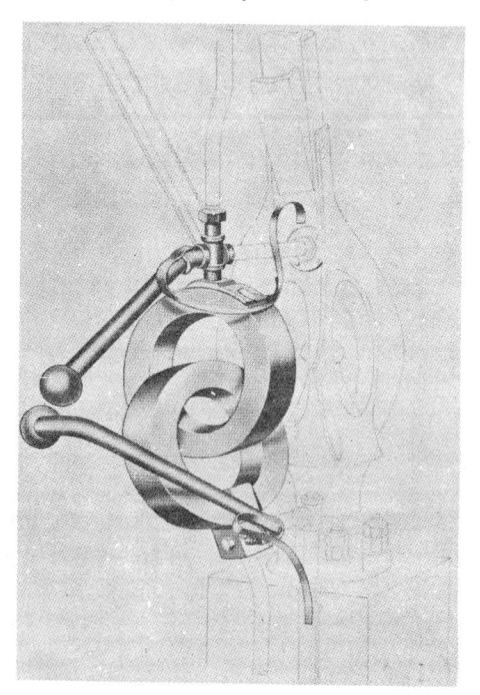

Fig. 12-32. Tower lighting isolation transformer. (Courtesy Andrew Corp.)

Chapter 12   *Amplitude Modulation Radio Transmitters*   449

COAXIAL TRANSMISSION LINES

Coaxial lines may be used for any frequency but above about 500 megacycles the losses become large owing to the copper and insulator dielectric losses. A coaxial line consists of one copper wire or tube forming the inner conductor located in the center of the larger copper tube and insulated from it. Figure 12-28 shows a flexible type of coaxial line and Fig. 12-29 shows the larger rigid type. The current flows on the inner conductor and on the inner surface of the surrounding tube. Since r-f currents flow on the surface of conductors and do not penetrate more than a few thousandths of an inch, the outside of the tube carries no current and is normally grounded. Figure 12-30 shows how a coaxial line is routed to connect a transmitter to an elevated antenna.

Coaxial lines used for transmitters normally have a characteristic impedance of about 50 ohms. They are manufactured in diameters up to $6\frac{1}{8}$ inches, and matching elbows, bends, gas seals, mechanical hangers, impedance transformer sections, etc., are made for each size. Such lines are universally used for VHF television, FM, mobile, and many other types of stations.

Coaxial lines for transmitters are made gas tight and filled with dry nitrogen or dry air to keep out rain and moisture. The pressure may be from 5 to 20 lb per square inch.

ANTENNA TUNING AND MATCHING

Figure 12-31 shows the type of equipment used in standard broadcasting stations to match the impedance of a coaxial or open wire line to one antenna or to several antennas of a directional array. In this service the steel towers are insulated from ground and function as the r-f current-carrying radiating element.

TOWER LIGHTING

Figure 12-32 shows a 60-cycle transformer used on insulated towers to deliver power to aviation obstruction warning lights, sleet melting resistors, etc. This transformer consists of air-core primary and secondary windings specially designed to provide adequate coupling and power transfer with large separations between them, as shown. The transformer is protected against flash overs by an air gap which is also shown. At the rear of the transformer is the tower leg insulator on the ground side of which the primary of the transformer is secured and on the upper side of which the secondary is secured.

QUESTIONS AND PROBLEMS

1. Describe important performance specifications for a broadcast transmitter.
2. What is the piezoelectric effect?

3. Describe the operation of a vacuum tube oscillator.
4. What is the function of a buffer amplifier?
5. How is plate modulation accomplished?
6. How is grid modulation accomplished?
7. What is overall inverse feedback? How does it operate?
8. How is carrier frequency stability obtained?
9. What is high-level modulation? Low-level modulation?
10. What is carrier shift? Why is it undesirable?
11. What is power supply regulation? Why is it important?
12. How does a CW telegraph transmitter differ from a broadcast transmitter?
13. How would you generate a 700-megacycle carrier of high frequency stability?
14. How is a CW telegraph transmitter keyed?
15. How does a frequency multiplier operate?
16. What form of modulation is most common in broadcast transmitters? Why?
17. What adjustments give optimum efficiency in Class C amplifiers?
18. How would you neutralize an r-f amplifier?

CHAPTER 13

AMPLITUDE MODULATION RADIO RECEIVERS

The purpose of a radio receiver is to process a modulated radio-frequency signal obtained from a receiving antenna and deliver to the output a reproduction of the original message signal, without more than tolerable addition of noise or interference or loss of character, and with sufficient amplitude to fulfill the system requirements. Sensitivity to weak signals usually is an important characteristic.

SENSITIVITY

The sensitivity is a measure of how weak a signal may be processed. In rating receivers in this respect, it is customary to establish some fixed receiver output power required for proper functioning and then rate the receiver sensitivity in terms of the lowest radio-frequency input voltage which will produce that amount of output. For example, it has been standard practice in home broadcast receivers to use one-half watt of audio power as the normal output when the incoming signal is modulated normally. If the sensitivity of the receiver is such that 50 microvolts of r-f signal at the input terminals are required to produce 0.5 w audio power at the output terminals, the sensitivity rating would be considered "50 microvolts sensitivity." This 50-microvolt rating is typical of a great many broadcast receivers which are intermediate between the best and the poorest in this respect. It is adequate for local and also fairly strong distant signals but falls short of the maximum sensitivity achievable.

The sensitivity depends upon the amount of amplification built into a receiver, including r-f amplification ahead of the detector, intermediate frequency amplification in superheterodyne receivers, and audio amplification following final demodulation.

A broadcast receiver built for reception of local stations need not be very sensitive. On the other hand a commercial receiver of the type used on ships, for transoceanic communication, international broadcast reception, military communications, etc., will employ far greater amplification to facilitate reception of the weakest signals it is possible to receive through natural and miscellaneous man-made noise interference. Such a receiver may have a sensitivity of one tenth of a microvolt, for example.

NOISE

How much sensitivity may be used? If there were no natural or man-made noise interference, there would be almost no limit. Very high power in transmitters would not be required because the weakest signals from very low power could be received merely by adding more and more receiver amplification as required. Unfortunately, nature has interposed barriers and fixed limits which technology has been unable to circumvent.

STATIC

Clouds contain electrical charges with respect to each other and also the earth. Under summer storm conditions these charges reach tremendous voltages which cause violent discharges evidenced as lightning strokes which may be between clouds, or clouds and the earth. This flow of current, often in the thousands of amperes, causes radio waves to be produced which may travel for hundreds or thousands of miles with sufficient intensity to interfere with radio communications in distant undisturbed areas.

The lowest radio frequencies of these waves depend upon the length of the lightning discharge paths. Since these paths often are large fractions of a mile, the ionized stroke path may be viewed as an antenna of about equivalent length producing heavy interference at long wavelengths. Lightning strokes often follow several paths, including branches, and produce complex frequency combinations. In addition, the sudden pulses of lightning current vary widely in character and duration. As in all pulse wave shapes, both fundamental and higher frequencies are present so it follows that interference from lightning discharges, called *static*, usually covers a wide band of frequencies extending from the lowest fundamental frequency upward, in diminishing amplitude. Stated in terms of frequency, natural static is most severe at the low radio frequencies, decreasing in amplitude with increasing frequency until above about 50 megacycles where it is of little consequence. At the low radio frequencies it is at times extremely severe and is never completely absent because of the good propagation of these frequencies over long distances from the tropical latitudes. Long-distance propagation also occurs on the medium and high frequencies, but is much more subject to diurnal and other frequency effects.

THERMAL AGITATION NOISE

Unfortunately there is another type of natural noise which exists at all radio frequencies. It is caused by the random movement of electrons in all electrical conductors, owing to thermal agitation of the atoms of which the conductor is composed. When the temperature of a wire is raised this agitation increases. The movement of electrons produces random differences of voltages across the ends of the wire. It should be borne in mind that the flow of current through a wire is the result of the flow of electrons. These random voltages caused by thermal agitation occur at all radio frequencies, quite uniformly throughout the spectrum. Therefore, the amount of noise which passes through an amplifier from this source depends upon the bandwidth of the amplifier. A narrow-band amplifier passes noise frequencies of only a narrow spectrum and therefore the amount of noise is limited. A wide-band system accepts more noise contributed from added portions of the spectrum, and the noise level is higher. For this reason receiver bandwidths adequate only for the carrier and sidebands are used to minimize noise. The noise, as heard from a loudspeaker, is best described as a hissing sound. On a television picture it appears as "snow."

The open circuit noise voltage across a wire is easily calculated. It is:

$$E = \sqrt{4KTRB}$$

where K is a constant equal to 1.37×10^{-23}; T is absolute temperature in Kelvin (about 300° for a living room); R is the resistance in ohms; B is the bandwidth in cycles.

For a 300° temperature this reduces to:

$$E_{microvolts} = 0.13 \sqrt{R \times B_{megacycles}}.$$

Thus we may calculate that with a wire of 100 ohms resistance and a broadcast receiver bandwidth of 10,000 cycles, the thermal agitation noise is about 0.13 microvolts. There would be no point in making the sensitivity of a receiver with an input impedance of 100 ohms appreciably greater than that necessary in order to receive a signal of 0.13 microvolts because such greater sensitivity would merely amplify the noise through which signals could not penetrate.

For a television system with a 100-ohm input impedance and a bandwidth of 4 megacycles, the noise threshold would be about 2.6 microvolts.

TUBE NOISE

Another type of natural noise limits usable receiver sensitivity. It is due to the fact that electrons from the cathode of a tube are emitted at an irregular rate and there is an added irregular flow between plates and screen grids. The first is often called the "shot effect."

An indirect method of calculating shot noise voltage is necessary. It consists of assuming that a tube has an "equivalent resistance" and that if it were connected from grid to cathode of a hypothetical "noiseless tube," at room temperature it would cause the same amount of noise output as thermal noise. Values of "equivalent resistance" for typical tubes are shown below.

Triode amplifiers	200–2,300
Triode mixers	900–6,000
Pentode amplifiers	700–7,000
Pentode mixers	2,800–35,000

A few computations, using the above values for R in the last equation, will show that for the broadcast receiver the shot noise threshold caused by a pentode may be as much as one-half microvolt. For the television system it may be several microvolts.

The receiving antenna, transmission line, and receiver input circuits all contribute noise. When integrated with the "shot noise" in the first tube, the noise threshold is very real and apparently inescapable. It establishes the limit of receiver sensitivity which can be used to advantage. Further improvement in the sensitivity of an overall receiving station must then be accomplished by building large and expensive receiving antennas which deliver more signal to the receiver terminals. This is standard practice at important commercial receiving stations. The large antennas are highly directive and serve two principal functions. First, they are sensitive to signal interference from only one direction, thus reducing external disturbances. Second, they intercept from the arriving electromagnetic field a large amount of r-f signal power which more effectively overrides the thermal agitation and shot noise produced in the receiving equipment. At the low radio frequencies, and to a lesser extent at the medium and high frequencies, most of the static noise comes from the tropics. It follows that antennas may be used to discriminate against noise of tropical origin if they are unusually sensitive to signals arriving from other directions.

RECEIVER NOISE FACTOR

The performance of a radio receiver, like that of any other piece of apparatus or machinery, satisfactorily and conveniently may be described by specifications, provided that the terms of the specifications are commonly used and understood. *Receiver noise factor* is such a specification.

It has been shown that thermal agitation noise in the input is an absolute natural limit to the useful sensitivity of a receiver. By common usage through accepted industry standards, insofar as internal noise is concerned, an ideal receiver is one in which the only source of noise is the thermal agitation noise in the antenna, with which the receiver is intended to operate. If the noise level exceeds that value, it departs from the ideal. The degree to which it departs is expressed as a factor, the receiver noise factor. It is the ratio of the noise power in the output of the receiver to the noise power which would be present in the output of an ideal receiver where the only internal noise source corresponded to the resistance of the driving source for which it is designed. It is commonly expressed in decibels (db).

It was shown that a receiver with a 100-ohm input impedance had 0.13 v of thermal agitation noise at the input. Receiver input circuits include, in addition to the antenna, other sources of internal noise, such as impedances separate from the antenna, shot noise, etc. These other sources of noise are added to the antenna thermal agitation noise. If the noise power is ten times as great as the antenna thermal agitation noise, the receiver noise factor is 10 db.

Chapter 13 *Amplitude Modulation Radio Receivers* 455

This is a very important specification because in designing a communications system, such as a microwave relay, a radar set, a radio broadcasting system, etc., it determines the minimum signal intensity required at the receiving point to overcome the receiver noise.

To determine the conditions which must be met to provide this minimum required field intensity the other components of the overall system must be evaluated, such as the propagation characteristics between the transmitting and receiving antennas, the antenna separations, the transmitter power, the desired ultimate signal-to-noise ratio from the overall system, and the efficiency of the transmitting and receiving antennas. It is significant to note that if the *noise factor* is 10 db the transmitter power must be ten times as great as would be needed for an ideal receiver to overcome the receiver noise. If the noise factor were 20 db, it would be 100 times as great. Because of the large cost of transmitting stations, the difference between a 10 db and a 20 db noise factor could impose an extra transmitter cost of hundreds of thousands of dollars. The factor under discussion assumes critical importance in designing a chain of microwave relay stations because the consequences of relatively small miscalculation are compounded when many relay stations are used in cascade. In sections to follow in this book there will be several references to receiver noise factor because, while at first glance it may not seem very important, it is a vital part of an overall system specification.

Since the thermal noise voltage developed in an impedance depends upon its temperature, and it is desirable to maintain simplicity in specifications, a temperature of 290°K is used in expressing and in measuring the noise factor. This is about average room temperature.

MEASURING RECEIVER NOISE FACTOR

A receiver specification as important as the noise factor must be measurable with accuracy. It may be accomplished in various ways. One is illustrated in Fig. 13-1. It utilizes a calibrated diode source of noise for comparison.

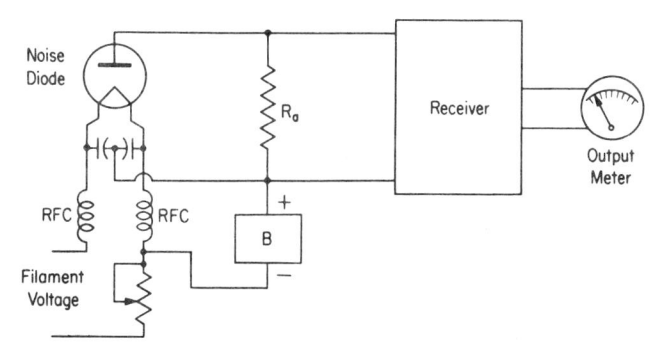

FIG. 13-1. Circuit for measurement of receiver noise factor.

A diode tube will provide noise voltage proportional to the current flowing through it so long as the plate voltage is sufficiently high to draw to the plate all the electrons emitted by the cathode, and thus prevent space-charge. The current through the diode is adjusted by controlling the cathode temperature. The cathode is usually a tungsten filament. Such noise diodes are specially made and are accurately calibrated for filament current-noise output voltage. The impedance of the noise diode is so high compared with R_a that it does not influence the accuracy of the measurements.

The measurement procedure is simple. With the diode deactivated, the noise level is noted on the output meter. The voltage gain of the receiver is then reduced by 3 db. The diode then is activated and the noise output is increased until the output meter returns to its original reading. Since this last operation doubled the output power by doubling the input power, the noise contributed by the diode equals the receiver-antenna noise, and the noise power may be read directly from the diode noise calibration. R_a is the resistance of the source for which the input circuit is designed, usually the radiation resistance of the antenna, or the characteristic impedance of the transmission line feeding the receiver. An equivalent resistance is substituted for the antenna when making the measurement. It may be a conventional type resistor. The output measuring meter is not critical as to response characteristics and may consist of a microammeter connected in series with the receiver diode detector to measure average detector current. The receiver gain control must be accurate since the accuracy of the measurement depends directly upon knowing that the original power output level is reduced to one half. The 3-db reduction of gain is equal to a 50% reduction in power. The thermal noise of R_a alone may be calculated and this is compared with the measured noise to derive the ratio which is called the noise factor.

RECEIVER HUM NOISE

It is convenient and economical to use a-c for heating the cathodes of receiving tubes, and is standard practice. At receiver sensitivities limited by the factors described in the preceding sections, hum "cross talk" into the amplifiers is ordinarily suppressed to satisfactory levels. Similarly, suppression of hum from high-voltage rectifiers can be controlled. The extent to which hum suppression is carried depends upon the purpose for which the receiver is built. In some small, simple, and very inexpensive broadcast receivers a relatively large hum amplitude is permitted up to the loudspeaker terminals. The loudspeaker response is made so low at the hum frequencies that they are tolerably attenuated. In receivers built for high-fidelity reception of speech and music the hum frequencies must be suppressed at the source by more expensive and careful design and selection of

Chapter 13   *Amplitude Modulation Radio Receivers*   457

tubes, components and circuitry, and in some cases by the type of inverse feedback described in Chapters 7 and 12.

AUTOMATIC VOLUME CONTROL

When tuning in stations, a person normally desires that the loudness of the speech, music, code signals, etc., be maintained at a pleasing level. Hand-operated volume controls on the audio amplifier are provided on broadcast and television receivers so that the adjustment may be made easily. But when tuning from station to station the received carrier voltages vary over a very wide range, which would make it necessary to readjust the volume for each station unless automatic adjustment were possible. Variations caused in sky-wave transmission are explained in Chapter 19. The signal input voltage to an automobile receiver will vary over wide limits depending upon the location of the vehicle. It is desirable in these situations that the signal coming from the loudspeaker remain substantially constant even though the input signals at various frequencies may vary greatly in intensity. Receivers incorporate *automatic volume control* circuits that make it possible to obtain substantially the same audio-frequency output with large differences in radio-frequency signal input. The voltage amplification of radio amplifiers varies with the voltages on the screen grid and the suppressor grid and with the bias voltage on the control grid. The automatic volume-control circuit functions to vary one or more of these voltages so as automatically to change the amplification of the radio or i-f amplifiers, or both. The voltage to accomplish this is obtained from the detector. It is easier to obtain the control voltage from a diode detector than it is from other types of detectors, and this is one reason why diode detectors are extensively employed in receivers.

The circuit diagram of a diode detector is shown in Fig. 13-2. Resistor R and the audio output capacitor comprise the output circuit of the detector.

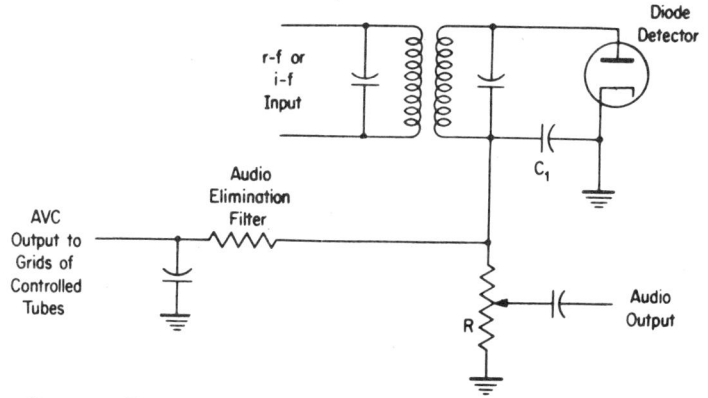

Fig. 13-2. How automatic volume control voltage is obtained from diode detector.

458   *Amplitude Modulation Radio Receivers*   Chapter 1

As was explained in Chapter 11, when a radio-frequency signal input voltage is impressed upon the diode detector, a d-c voltage is produced across the output resistor R. As the radio-frequency signal voltage is increased, the d-c voltage across R is increased. The negative d-c voltage across R is fed back to the bias leads of the radio or i-f amplifiers and is there employed to vary the amplification. As the signal input voltage to the diode detector increases, the d-c voltage across R increases, thereby reducing the amplification of the radio amplifiers. By proper adjustment, the audio-frequency output of the diode detector can be maintained at approximately the same intensity despite large changes in radio-frequency signal input voltages to the receiver. While Fig. 13-2 shows a single diode used for both detection and automatic volume control, often separate diodes are used for this purpose, or separate plates with a common cathode in one tube envelope.

AVC systems are built to provide control of gain with variation of average received r-f signal intensity. This control should not apply to audio-frequency fluctuations since this would smooth out the amplitude variations of the incoming carrier which constitute the modulation. Therefore, AVC systems include a simple resistance-capacity filter in the AVC control bus which eliminates audio-frequency components from the AVC voltage applied to the grids of the controlled tubes. This filter is shown on Fig. 13-2.

AUTOMATIC VOLUME CONTROL FIGURE OF MERIT

In receiver design and rating it is common engineering practice to apply a figure of merit which indicates the range of receiver r-f input voltages over which the AVC system will maintain the audio output level constant. Since some variation within this range would be permissible, the term "constant" must be defined, understood, and commonly used. In home broadcast receivers the AVC figure of merit is the number of decibels below an r-f input of 100,000 microvolts over which the audio output will not change more than 10 db, or about 3 to 1. For broadcast receivers designed for use in automobiles the r-f input reference level is 5,000 microvolts, instead of 100,000.

For home receivers the AVC figure of merit ranges from about 35 db to about 60 db, depending upon the cost and quality of the receiver. This means that in the case of the 60-db receiver the r-f input voltage may vary over a range of 1,000 to 1 without the audio output level changing more than about 3 to 1, so long as the r-f input voltage does not exceed 100 mv. For the 35-db receiver the r-f input voltage may vary about 60 to 1 without a variation of audio output level of more than about 3 to 1, so long as the r-f input level does not exceed 5,000 microvolts, or 5 μv. The lower figure of 5,000 microvolts for automobile receivers reflects the lower r-f voltage obtained from short automobile antennas and the nature of the receiving conditions.

Most radio receivers, including the inexpensive portable types, include AVC.

TUNED R-F RECEIVERS

This type receiver has limited use where tuning at will to any one of a large number of channels is desired. It has been generally described in Chapters 10 and 11. It has advantages mainly where only one channel, or a few closely separated channels, are to be received. The disadvantage of the tuned r-f receiver is that the band shaping and selectivity must be obtained in variable tuned circuits which differ in tuning characteristics over the tuning range. The impedance at the low-frequency end of the scale will be relatively high compared with the high end of the scale, tuning will be sharper, and the pass band will be correspondingly narrower. For broadcast use the change in selectivity and pass band is highly disadvantageous. At the high frequencies where the *per cent* separation between assigned carrier frequencies is lower, it is difficult to obtain adequate selectivity against adjacent channels. At the low frequencies where tuning is sharp the narrow pass bands may cut off essential sideband frequencies. In such receivers it may be desirable to mount all of the tuning capacitors on one long shaft. This imposes the requirement that a multiplicity of tuned circuits must be aligned accurately in order to tune to a common frequency over the full range. Inaccuracy in only one stage causes staggered tuning and appreciably reduces the sensitivity and selectivity. This problem is minimized for fixed frequency reception. The advantage in the superheterodyne of employing a fixed tuned intermediate frequency amplifier to obtain band shaping and constant selectivity for all channels does not exist in the tuned r-f receiver.

Tuned r-f receivers were used widely in the early years of broadcasting. Individual tuning dials for a multiplicity of radio amplifier stages were common, making tuning very cumbersome and difficult. When searching for weak signals, the extremely close tuning required in each stage for maximum amplification was difficult or impossible to attain initially because, until a station was heard, there was no way of knowing if one or more stages were mistuned and insensitive. Later, combining all tuning elements on a common shaft with only one dial interposed problems of tracking. Tuned r-f receivers are used principally in services on fixed frequency operation using carrier frequencies favorable for direct efficient amplification.

THE SUPERHETERODYNE RECEIVER

For utmost performance and flexibility the superheterodyne receiver has supplanted other types almost completely for either amplitude or frequency modulation reception. The advantages have been explained. More detailed discussion of its performance follows.

RECEIVER OSCILLATOR RADIATION

The heterodyne or local oscillator voltage is combined with the incoming carrier at the input of the mixer (first detector). This local oscillator voltage, if present at the antenna lead-in connection, can cause current to flow to the antenna and be radiated as electromagnetic waves, thus functioning as a low-power transmitter. The energy may be sufficient, small as it is, to cause interference to other receivers in the same or other services within an

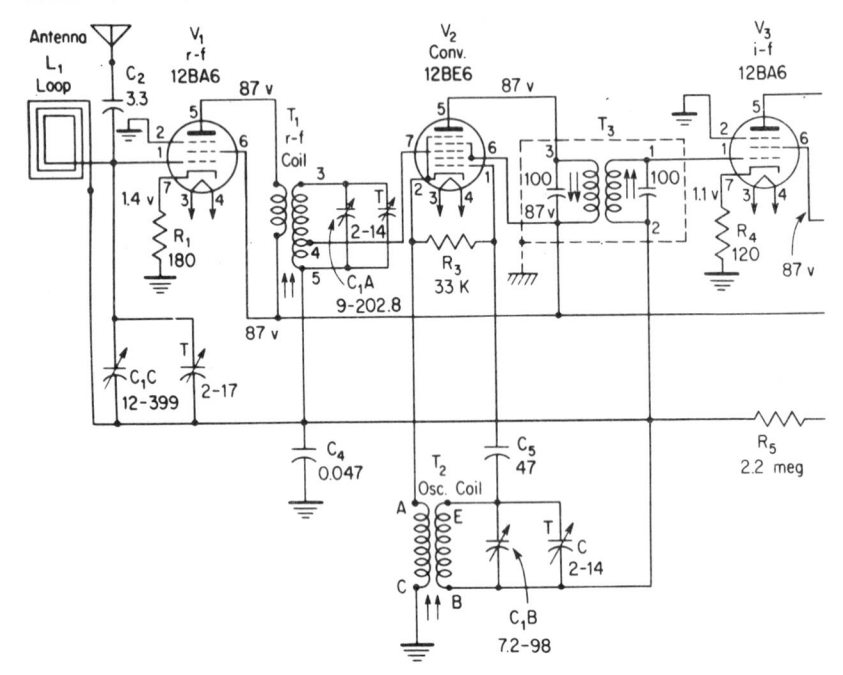

FIG. 13-3. Typical superheterodyne broadcast receiver.

area of roughly one-half square mile. In recent years the problem has required action, including cooperative endeavors, by industry associations and the FCC. Permissible tolerances have been established for the amount of receiver oscillator radiation and as the years go by these tolerances are tightened.

Two methods are usually employed in combination to suppress the radiation. One consists of using one or two tuned r-f amplifiers between the antenna and the mixer tube. These amplifiers are not required to tune sharply because oscillator radiation occurs on frequencies far from the immediately adjacent channels. In television receivers it is over 40 megacycles removed from the channel being received. The second method of reducing oscillator radiation consists of shielding the circuits which carry oscillator currents and which may radiate power directly from the receiver.

Chapter 13   *Amplitude Modulation Radio Receivers*   461

SPURIOUS RESPONSE

If two undesired signals are impressed on the first detector of a superheterodyne, the i-f amplifier ordinarily will reject them. However, if the two undesired frequencies happen to differ by the receiver intermediate frequency and reach the mixer, the mixer will pass into the i-f amplifier an interfering signal. Also harmonics of the local oscillator may combine with undesired signals, or even with harmonics of stray undesired signals to give

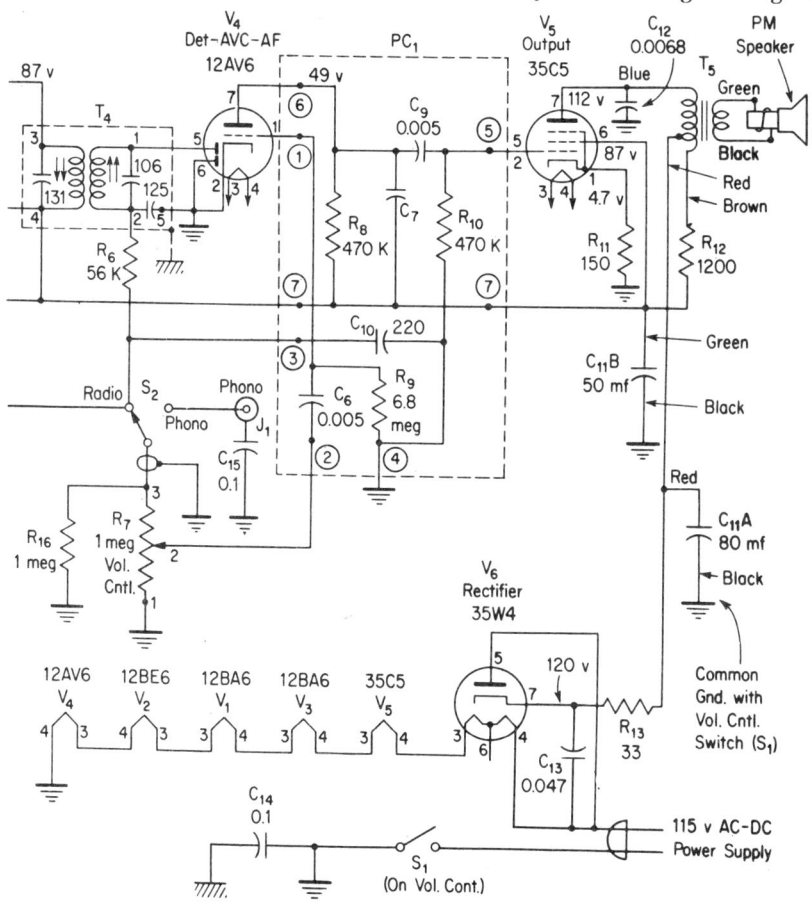

Fig. 13-3 (cont.)

the intermediate frequency. Selectivity ahead of the mixer is the only method of avoiding this form of random interference. The i-f amplifier cannot distinguish between a desired and an undesired signal if they are of the intermediate frequency. Here again is a case where the r-f amplifier ahead of the mixer will reject such incoming signals without being required to tune sharply.

IMAGE FREQUENCY RESPONSE

Another form of interference is avoided by using selectivity ahead of the mixer. In Chapter 10 the action of a superheterodyne receiver was illustrated by assuming that an i-f amplifier frequency centered on 455 kc, and that there was an incoming carrier of 1,000 kc. The heterodyne oscillator frequency of 1,455 kc was mixed with the 1,000 kc incoming carrier to produce the i-f of 455 kc. But suppose there were a strong carrier on 1,910 kc? This would mix with the heterodyne oscillator frequency of 1,455 kc and also produce 455 kc which the i-f amplifier would accept. This upper frequency is called the *image frequency* and is unique to superheterodyne receivers.

While tuned r-f amplifiers are effective in suppressing image response, spurious signals, and interference, another method is often used in receivers built for utmost performance. It consists of using two or more i-f amplifiers of different frequencies. An illustration will follow. In such receivers the first i-f is considerably higher than the second one. Such a receiver is referred to as a *double superheterodyne*, or triple, etc.

EXAMPLE OF SUPERHETERODYNE

Many superheterodyne receiver requirements are met with simple designs. For portable broadcast reception they may include one tuned r-f amplifier stage, only one or two i-f amplifier stages, and limited audio-frequency amplification. They may be constructed to operate from either a-c or d-c power supply systems, usually with some sacrifice in the amount of undistorted audio-frequency output power obtainable. The type of receiver most commonly used for broadcast reception is illustrated by the circuit diagram shown on Fig. 13-3. It is of the simple design referred to. It may be expanded to include additional radio, i-f, and audio stages, but such expansion basically amounts only to adding "more of the same" and incorporating refinements in detail.

The receiver illustrated includes a built-in loop antenna of small dimensions for use where sufficiently strong signals are available for satisfactory reception. The loop consists of a coil of wire of as large a diameter as the receiver housing can accommodate, and is tuned to resonance by a variable shunt capacitor shown on the diagram as C_1C. The loop antenna, being of small dimensions, does not have induced in it, by the radio field, a very large voltage; but for urban and suburban use, where strong signals are available, it is very convenient and satisfactory. For more remote areas a larger antenna may be necessary for noise-free reception, and for such conditions provision is made for its connection to one side of the loop, as shown. Across the loop tuning capacitor a small trimmer capacitor is provided so that the antenna tuning capacitor adjustments may be made to coincide and track with the other tuning adjustments made simultaneously by other variable tuning capacitors mounted on a common shaft.

The first tube is a pentode radio-frequency amplifier which is coupled to the first detector grid through an r-f transformer T_1. Plate tuning is accomplished with a variable capacitor C_1A in the secondary circuit and in this case also a small trimmer capacitor is connected in parallel to facilitate initial adjustment of the tuning controls. It may be seen that the input to the grid of the following first detector is taken from a tap on the tuned circuit. By proper selection of the tapping point the loading of the tuned circuit is adjusted to provide the optimum tuning characteristic and amplifier tube performance. Tapping across an increased number of turns would increase undesirably the effect of the internal detector-tube grid capacitance and resistance because a larger impedance would then be shunted.

The first detector is a pentagrid converter which functions both as the heterodyne oscillator and the first detector. In the oscillator circuit the feedback coil of the oscillation transformer is connected between cathode and ground, as shown on Fig. 13-3. The oscillator-tuned circuit is connected between the first grid, the oscillator grid, and ground through capacitors C_5 and C_4. The oscillator grid bias is obtained from grid leak R_3 which is connected between the grid and the cathode. The screen grid functions as the oscillator plate. An examination of this circuit will show that the oscillation transformer primary in the cathode-to-ground circuit is in effect in the plate circuit and provides plate-to-grid feedback voltage to sustain oscillations. The oscillator tuning is accomplished with variable capacitor C_1B which is shunted by a small trimmer capacitor to facilitate initial receiver alignment of tuned circuits.

The three variable tuning capacitors C_1A, C_1B, and C_1C are rigidly mounted on a common shaft so that tuning is accomplished with only one tuning knob.

The signal input is applied to the control grid as shown. In the electron path from cathode to plate, the electrons are subjected to control both by the signal voltage and the oscillator voltage so that in the plate output circuit the sum and difference frequencies appear. The desired 455 kc is selected by the permanently tuned output transformer which feeds the following i-f stage.

The i-f stage is a pentode with cathode resistor bias and operates in the conventional manner. The output circuit is an i-f transformer tuned to 455 kc similar to the one in the preceding stage. However, the secondary in this case drives the diode detector section of a multipurpose-detector—automatic-volume-control—audio-frequency-amplifier tube. Rectified signal output from the diode circuit passes through resistors. The audio output and AVC voltages are developed across R_7 and R_{16}. The rectified d-c voltage so developed acts to provide AVC negative bias to the grids of the r-f amplifier, the first detector, and the i-f amplifier. With strong input r-f signals the diode current increases, the AVC bias becomes more negative, and the amplification of the controlled stages decreases automatically to maintain a fixed output level. Over the effective control range of the AVC system strong and weak stations automatically cause the receiver gain to change so that a uniform audio output level is maintained.

Audio output from the diode is obtained from R_7, a potentiometer which provides for manual audio volume control to adjust the loudspeaker acoustic output to the loudness desired. The volume control feeds the control grid of the multi-purpose tube through capacitor C_6 as shown in Fig. 13-3. R_9 is the grid-to-ground resistor. The plate of the multi-purpose tube audio output is resistance-capacitance coupled to the control grid of the pentode audio output stage. This stage employs cathode resistance bias through R_{11}.

The power for cathode heating is applied to all tube cathode heaters in series, including the rectifier tube, as may be seen. Thus the voltage across all of these heaters in series is equal to the normal power line voltage of 115 v. No transformer is utilized, so either a-c or d-c may be used for heating. The heaters of these tubes are specially designed to operate at rather high voltage so that they may be operated efficiently in this manner, without series voltage dropping resistors. If low-voltage heaters were employed, a series resistor would be required to reduce the voltage applied to the heaters, imposing added cost and inefficient waste of power in heating the resistor. It may be noted that the order in which the heaters of the various stages are connected is V_4, V_2, V_1, V_3, and V_5. There is a reason for this order.

When heaters are connected in series, the voltage between the heater and cathode of a tube depends upon its order in the string. For example, the heater at the grounded end of the string will have only a very low heating voltage to the cathode. But the heater at the other end of the string will have a voltage of about 115 v to the cathode. The order of the heaters depends upon the manufacturer's safe rating of heater-to-cathode voltage for individual tubes and also the relative susceptibility to a-c hum cross-over into the various tube circuits.

Plate voltage for the receiver of Fig. 13-3 is obtained from a half-wave rectifier. One side of the power line is connected to ground. The other side is connected to the rectifier plate. In the case of an a-c supply rectified d-c is stored in filter capacitor $C_{11}A$ and C_{13} which, with resistor R_{13}, comprise the smoothing-out filter to the audio output amplifier. An additional smoothing filter network R_{12} and $C_{11}B$ is used thereafter to supply all other stages of the receiver. This plate supply employs no plate transformer. It may be observed that this circuit will function on either a-c or d-c. Since both the heater and plate supplies will do so, and no other power is needed, this receiver will operate equally well on either supply.

Chapter 13 *Amplitude Modulation Radio Receivers* 465

AC-DC receivers are relatively sensitive to interfering r-f voltages which flow along power lines, because of the more direct path in the power circuits of the receiver through which external voltages can enter, compared with receivers employing a-c transformers in the power supply circuits. For example, voltages are produced in the operation of fluorescent lamps which are of both audio frequency and radio frequency. These voltages cause currents to flow along power circuits to which the lamps are connected and enter the receiver through the power line cord. These currents may flow freely to the cathode heaters and by electrostatic coupling or resistance leakage may introduce these voltages into the cathode circuits. The use of cathode heating transformers introduces impedance to the flow of such interfering currents to the heaters and substantially reduces this susceptibility to power-line interference. Similar attenuation of power-line interference may be accomplished by inserting low-pass filters in series with the power line connection.

The audio output tube is coupled to its load, the loudspeaker, through an audio output transformer which provides the correct high-impedance load for the tube and the correct low impedance, 6 to 15 ohms, for the loudspeaker.

The receiver shown on Fig. 13-3 is provided with a switch which connects the audio-frequency manually operated volume control either to the diode detector output or to a phonograph jack mounted on the rear of the receiver chassis. This makes it possible to connect an external electrical phonograph pickup to the audio-amplifier input and use the receiver amplifier and loudspeaker for record reproduction.

The filter for eliminating audio-frequency fluctuations from the AVC voltage applied to the grids is shown on Fig. 13-3. It consists of resistor R_5 and capacitor C_4.

MULTIPLE FUNCTIONS IN RADIO RECEIVERS — SELECTIVITY

Receivers are often required to perform a variety of functions. It is prohibitively difficult to build tuned r-f receivers to incorporate the flexibility required. It is relatively simple to do so in the superheterodyne type of receiver. In communications type receivers it may be desirable to receive code signals, music, speech, facsimile, or other forms of modulation, and single-sideband suppressed carrier signals, for which differing pass bands and other features are necessary. This can be accomplished by constructing the intermediate-frequency amplifier to have variable passband characteristics and other features, changeable at will by built-in controls. For the ultimate in selectivity for CW reception, such receivers may include an extremely sharp filter which may be switched into the i-f circuit at will. Figure 13-4 shows an example of curves of a receiver i-f amplifier having a choice of band-pass characteristics which may be selected at will by a switch. For additional flexibility multi-purpose receivers may include also manual r-f gain control to maintain constant output with a wider range of r-f input voltages. Such receivers may also include a CW beat oscillator which may be switched into the i-f system to facilitate reception of continuous wave signals. These oscillators may have

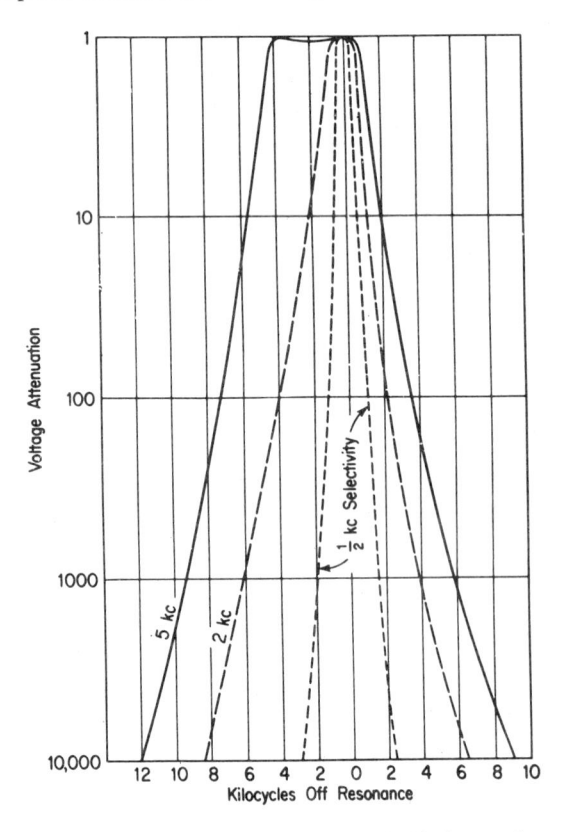

Fig. 13-4. Example of variable selectivity obtainable in superheterodyne intermediate-frequency amplifier for best reception of various types of transmissions.

a variable control to provide the beat-note pitch most needed or pleasing. In addition, a meter may be incorporated in such receivers to provide a measure of received signal voltage, often expressed both in volts and decibels. Such de luxe receivers may have more than adequate sensitivity to "reach down" to the noise level, may cover a tremendous frequency range from 540 kc or lower to 60 megacycles or higher, include special dial and control "band spread" for extremely fine tuning, accurate dial frequency calibration, specially developed input circuits to minimize shot noise, and noise-suppression circuits. The frequency ranges covered may encompass standard broadcasting from 540 to 1,620 kc, marine telephony, international broadcasting, telegraphy, telephony and facsimile, navigational aids, mobile, public safety, amateur, military, and other services throughout the world.

MULTI-BAND RECEIVERS

In the communications services, such as ship-to-shore, broadcast reception, and miscellaneous other applications, it is often desirable or necessary to be able to receive signals in two or more widely separated frequency bands. For example, communications from ship to shore may most effectively be conducted within hundreds of miles at frequencies between 500 and 700 kc. But at very long distances frequencies spaced at intervals between about 5 and 20 megacycles are necessary. In broadcast reception domestic stations operate on frequencies between 540 and 1,620 kc. But other broadcasting, for long-distance reception in foreign countries, is conducted on frequencies between about 6 and 21 megacycles.

Superheterodyne receivers may be built conveniently for such multi-band reception because the i-f and audio systems require no change. The r-f tuned circuits, consisting of the heterodyne oscillator and the r-f circuits preceding the first detector are the only ones requiring special attention. These ordinarily are limited in number to two or three, the antenna circuit tuning, the oscillator tuning, and possibly the tuning of an r-f stage. It is conventional in multi-band receivers to provide a set of separate tuning coils for each of the different bands to be received, and switch in the set required by means of separate switches mounted on a common shaft. The variable tuning capacitors for each position ordinarily are not switched but are retained to tune whichever set of coils is switched into use.

The use of multiple coil systems imposes a few problems which are solved without undue difficulty. For example, multi-band receivers are provided with multiple parallel dials which are closely calibrated in frequency for each band. These could be difficult to build accurately if trimmer capacitors were not provided for each coil, so that the fine tuning thus afforded in each circuit made it possible initially in assembly and test to adjust the frequency and resonant points to coincide with the frequencies indicated by the dial pointer. Such trimmer adjustments are made initially band-by-band at a reference frequency after which the tracking and calibration remains satisfactory at all frequencies in the bands.

RECEIVERS UTILIZING TRANSISTORS

Transistors are well adapted to use in radio receivers. In many respects they perform like vacuum tubes. Figure 13-5 shows a comparison of a tube and a transistor amplifier which may operate at either radio or audio frequencies, depending upon the input and output circuits. The emitter is the source of charged particles and is the counterpart of the vacuum tube

468 Amplitude Modulation Radio Receivers Chapter 13

cathode, but it emits positive "holes," instead of electrons. The transistor base is biased to control the flow of charges through the transistor and is the counterpart of the vacuum tube grid. The collector is the counterpart of the vacuum tube plate because it causes the output current to flow to the load circuit through the transistor.

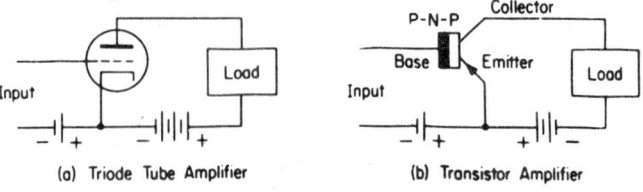

Fig. 13-5. Comparison of vacuum tube and transistor amplifier.

In the P-N-P transistor shown, the base requires a positive power supply voltage with respect to both the emitter and collector. The negative voltage bias on the base is low compared with that of the collector.

For the N-P-N transistor the polarity of these voltages is reversed. The input impedance of a transistor is low, about 600 ohms, compared with perhaps a megohm for a tube. The transistor output collector impedance is high.

Figure 13-6(a) shows a simple receiver employing a crystal diode detector and a P-N-P transistor amplifier. Modulated r-f is rectified in the diode. Capacitor C has substantially no impedance to r-f but at audio

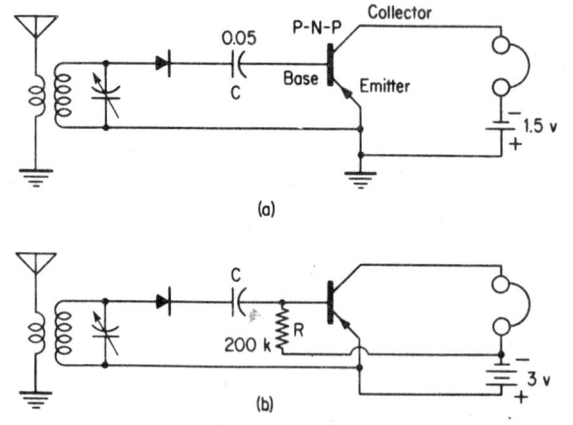

Fig. 13-6. Elementary transistor receiver.

frequency it has considerable impedance. Its charge varies with the modulation envelope, varies the bias of the transistor, and controls the flow of collector current to the headphones. The polarity of the diode is such that the base voltage is always negative with respect to the emitter.

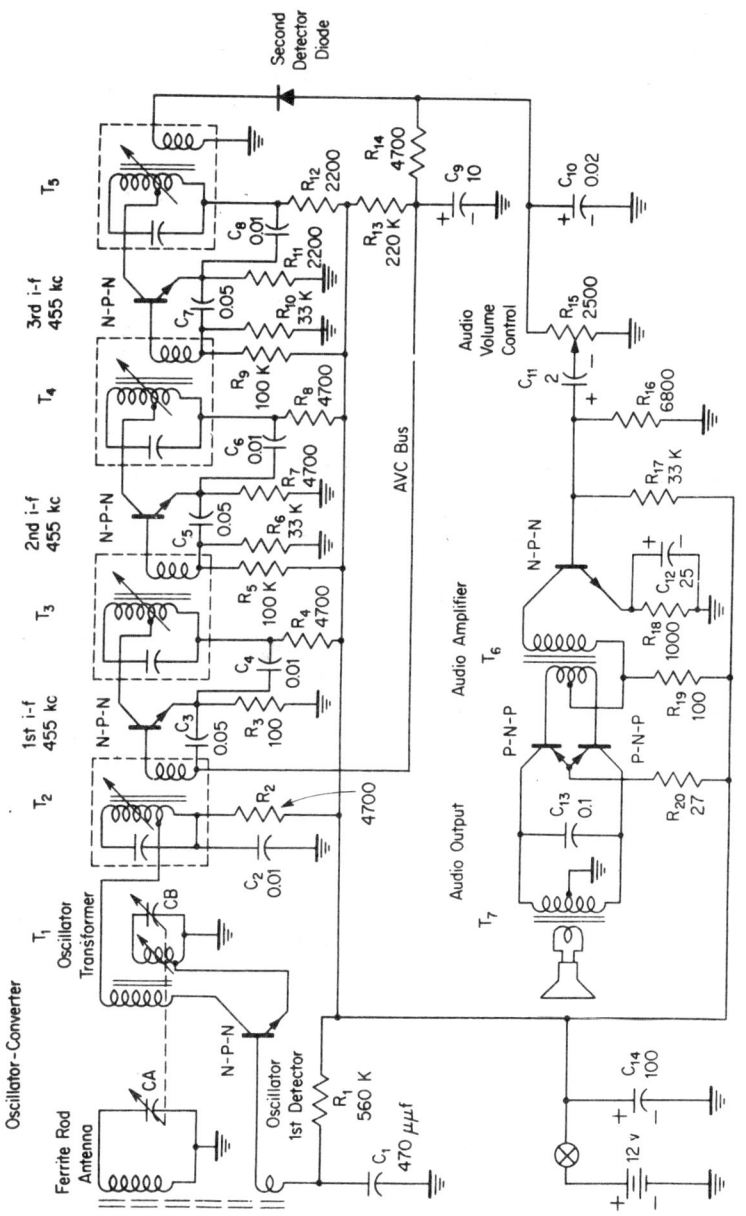

Chapter 13   *Amplitude Modulation Radio Receivers*   469

FIG. 18-7. Transistor radio receiver. (Courtesy Hazeltine Corp.)

Figure 13-6(b) shows a similar circuit with base bias provided through a resistor R of 200,000 ohms.

A TRANSISTOR SUPERHETERODYNE RECEIVER

Transistors have a number of inherent advantages over vacuum tubes for radio receivers. They are very small, are relatively immune to vibration and shock which produce microphonic spurious voltages or mechanical damage, require no cathode heating battery, produce little heat, and usually operate from a single small battery of low voltage. These characteristics of transistors make them particularly attractive for portable receivers.

A circuit diagram of a transistor receiver is shown in Fig. 13-7.

The antenna shown consists of a tuneable antenna coil about 8 inches long which is caused to resonate by a variable capacitor at the desired carrier frequency. This antenna coil has a ferrite core with high magnetic permeability which increases the inductance so that it may be built to resonate at broadcast frequencies with a small number of turns. Ferrite and powdered iron cores were specially developed for radio-frequency transformers and employ finely powdered ferromagnetic materials to minimize hysteresis and eddy current losses at radio frequencies. They are used widely in i-f transformers and similar applications where small size is desirable and are often used for tuning by being moved with respect to the windings to vary the effective permeability and inductance. In portable receivers ferrite or powdered iron core rod antennas normally are made long enough to utilize the full length of the receiver housing in order to intercept as much of the radio-frequency field as possible. The powder for the core is composed of ferromagnetic granules with a diameter of roughly 10 microns. They are combined with a binder, compressed and molded to produce the desired shape. For transformers, the cores may have maximum dimensions of a small fraction of an inch. For portable antennas they may be about a half-inch in diameter and as much as a foot long. In Fig. 13-7 the antenna variable tuning capacitor, C_A, is conventional and has a maximum value of about 200 $\mu\mu f$. The antenna output for the receiver is obtained from a coupling coil which is also wound around the ferrite core, as indicated on the figure, and coupled to the antenna coil.

The first transistor performs two functions. It is the heterodyne oscillator and also the first detector. The oscillator circuit is of the "grounded base" type, and the oscillations are produced by feeding power back from the collector to the emitter through the oscillation transformer T_1. Owing to the long transit time of current carriers passing from the emitter to the collector in the transistor shown, the voltage fed back from the collector to the emitter is delayed so that the phase relationship may be incorrect for producing oscillations reliably. To correct for this, capacitor C_1 is inserted in the base-to-ground circuit to produce a capacitive reactance which causes the current to lead the phase and thus produce compensation. The antenna voltage is applied to the base of the first transistor.

The output of the oscillator-first detector is 455 kc, the intermediate frequency used in this receiver. Amplification takes place in three i-f stages employing N-P-N transistors which produce a voltage amplification of about 10 per stage, or 1,000 altogether. The i-f transformers have high-Q tuned primaries which are tapped as shown properly to load the transistors which have about 10,000 ohms output impedance. The secondaries of the i-f transformers are untuned and match the 500-ohm input impedances of the following stages. In the type transistors used the feedback capacitance is low between collector and base and neutralization was not necessary to prevent oscillation in the i-f stages of this receiver. However, internal capacitance in a transistor cannot be ignored. If sufficiently high it may cause oscillations in a manner corresponding to the action in a vacuum tube r-f amplifier.

This receiver employs automatic volume control through the AVC bus to the first i-f amplifier base as shown on Fig. 13-7. The control voltage is negative and causes the collector current to be reduced, and the gain correspondingly to be reduced. The collector circuits are decoupled from the battery and the input circuits by resistance-capacitance filters R_4 and C_4 in i-f stage 1, R_8 and C_6 in stage 2, and R_{12} and C_8 in stage 3. The base bias is obtained from voltage dividers R_5-R_6 in stage 2 and R_9-R_{10} in stage 3. It may be seen that the positive 12 v from the battery will cause a current to flow in series through R_5 and R_6 to ground, and through R_9 and R_{10} to ground. By tapping between them, the desired fraction of 12 v is obtained for base bias.

The second detector is a point contact type of crystal diode which has as its load a 2,500-ohm audio-frequency gain control, as shown. C_{10} is an i-f by-pass capacitor. The automatic volume control voltage is obtained for the first i-f stage through R_{14} and C_9.

Audio voltage to the base of the first audio amplifier is obtained through C_{11}. Base bias to the first audio stage is obtained from voltage divider R_{16}-R_{17} which functions like the ones described for the i-f stages. The output from the collector goes to an interstage transformer T_6 which is designed to load the audio stage with 25,000 ohms.

The output stage is connected push-pull with the emitters common. R_{19} makes possible a low forward bias voltage on the base, and its purpose is to reduce the push-pull audio crossover distortion.

The AVC figure of merit for this receiver is 36 db.

The receiver described will produce the standard portable receiver audio output power of 50 mw when a 1,000 kc carrier field intensity of 475 microvolts per meter is available to excite the ferrite loop. The corresponding selectivity provides attenuation of about 100 to 1 at plus and minus 10 kc from the carrier frequency. The superheterodyne image rejection is about 200 to 1 and rejection of 455 kc incoming carrrier signals is about 100 to 1. It may be observed that this receiver requires only a single 12-v battery for its power supply.

MUTING SYSTEM

When a receiver is tuned to a station the automatic volume control system reduces the r-f gain to a point where the proper audio output level is obtained. What happens in a sensitive receiver when tuning between stations? The full gain of the receiver comes into play because the signal which prompted the AVC action is missing. The input noise is then amplified to a very high level which is annoying. This noise which is received when there is no r-f signal present may be reduced by a simple "muting," or "squelch," circuit, as shown in Fig. 13-8.

Detection occurs in a conventional diode detector having a load resistor R_1 and an r-f bypass capacitor C_1. The output drives an audio amplifier, as shown in Fig. 13-8. It is conventional in every respect except for the grid

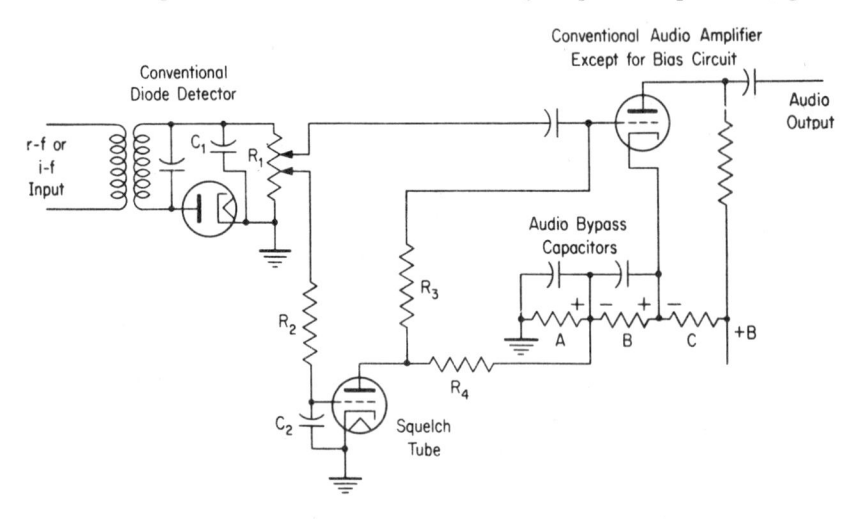

Fig. 13-8. Receiver noise muting circuit.

bias circuit. When only noise is present the audio amplifier bias is caused, by the squelch tube, to rise to the level where plate current is cut off and the tube becomes inoperative. This action is controlled by the amount of plate current drawn by the squelch tube.

When there is an r-f input signal voltage on the diode, a d-c voltage is developed across R_1 which furnishes bias to the squelch tube. In the squelch-tube grid circuit a filter, consisting of R_2C_2, smooths out the audio-frequency voltage fluctuations to provide steady d-c bias voltage. The circuit constants are adjusted so that when there is r-f input to the diode the bias voltage on the squelch tube will be high enough to stop its flow of plate current. Under these conditions, no current flows through R_4 and no voltage is produced across it. Then the audio-amplifier bias voltage will consist only of the normal bias voltage produced across section B of the voltage divider. What happens if the diode r-f input signal is removed?

In the absence of diode r-f input, no voltage is produced across R_1 and therefore there is no bias on the squelch tube. It automatically draws a large plate current which in turn produces a large voltage across its plate loading resistor R_4. Now the bias on the audio amplifier is the sum of the original bias plus the new R_4 voltage in series with it. This causes the audio amplifier plate current to be cut off and the tube made inoperative.

Squelch circuits such as this are used in mobile communications receivers. Normally the transmitter to which the receiver is tuned is inactive, and if the noise were not muted the receiver would produce a very high noise level for long periods. With muting, the receiver is silent until the carrier is received, at which time the receiver operates normally.

DIVERSITY RECEIVING SYSTEM

In Chapter 10 an overseas radio communications system was described. A more complete description of the receivers follows.

Radio waves used for overseas communications have carrier frequencies of from 5 to about 25 megacycles because such frequencies are propagated through the ionosphere over great distances, as will be described in Chapter 19. Since the ionosphere is not a stable area the signal intensities may fluctuate, or "fade," over a wide amplitude range, which reduces the circuit reliability. At times the signals drop below usable values. This may be caused by multiple reflections from the ionosphere which cause two or more waves to arrive over different paths of different lengths and differ in phase and amplitude. These waves, all from the same transmitter, do not differ in frequency but may change rapidly in phase from moment to moment as the relative path lengths change. That difference may cause random cancellations or additions in the receiver. Since this is so the fading characteristics will differ at locations separated by only a moderate distance, sometimes to the extent that there may be little or no instant-to-instant correlation. From a multiplicity of receivers at separated locations at least one of them at a time may receive a usable signal while for the others the signal may be unusable. In other words, if the antenna locations are duplicated in different locations, or *diversified*, and the outputs are all available to choose from, the circuit reliability will be improved. This type of reception is tremendously helpful in overseas telegraph and teleprinter communications and, in circuits where optimum performance is desired, it is indispensable. It is called *space diversity reception*.

It has been found by research and experience that three antennas spaced from 1,000 to 1,500 feet on a triangle represent about the optimum number and configuration. Each antenna is connected to its own individual superheterodyne receiver. All three are tuned to the same frequency and are substantially independent up to the output circuits. At that point electronic selecting circuits function to provide optimum performance, greatly

474 Amplitude Modulation Radio Receivers Chapter 13

improving the reliability. At the Riverhead transoceanic receiving station described in Chapter 10, about 81 diversity receiving systems are available; a large fraction of them are always in service. Figure 13-9 shows one of the message-traffic diveristy receivers at Riverhead. Incoming transmission

FIG. 13-9. Three-channel diversity receiver used for maximum performance in overseas reception of message traffic. (Courtesy RCA Communications)

lines from the receiving antennas may be seen at the upper right corner of the figure. They terminate on a switching panel. Diversity reception is not as advantageous for speech reception as it is for code because of selective fading which introduces distortion and undesirable effects.

SELECTIVE FADING

When an overseas radio-telephone signal consisting of carrier and sidebands, or sidebands alone, is transmitted, there may be several different ionospheric paths of differing lengths for it to traverse to the receiver. When two or more incoming multipath signals of random phase and amplitude combine, not only do the carriers fluctuate, but also the sidebands, often differently. The result may be a very different overall combination of sideband frequencies and amplitudes. In addition, a transmitted carrier

Chapter 13   *Amplitude Modulation Radio Receivers*   475

may have its amplitude reduced at a different rate than its sidebands so that a detected signal is in effect heavily over-modulated and therefore distorted. This is called *selective fading*, a term commonly used to describe fading at different rates of carrier and sidebands. It may often be observed in skywave reception of standard broadcasting stations at night, at distances of about 100 miles or more, when skywave and groundwave signals are received simultaneously

TRANSOCEANIC RECEIVERS

Each of the three channels of the diversity transoceanic receiver consists of a triple or quadruple conversion superheterodyne receiver designed for maximum performance and custom-built for this particular message

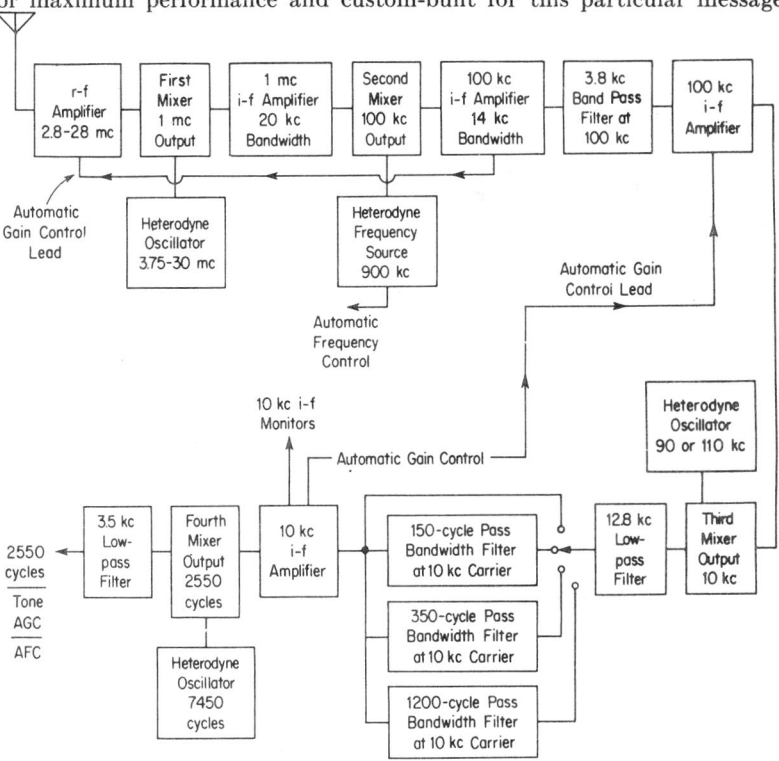

Fig. 13-10. Maximum performance message traffic receiver.

traffic service. Figure 13-10 shows a block diagram of one channel, with many details omitted for clarity. The operations in this receiver are described in sequence.

1. The incoming signals are amplified at carrier frequency. Figure 13-9 shows six large tuning controls at the tops of the two racks at the left. The

three at the top left each independently amplify frequencies from 2.8 to 9 megacycles in an identical manner. The three at the right each independently amplify frequencies from 9 to 28 megacycles in an identical manner. Because optimum performance is desired the overall frequency range is split into two parts for more efficient amplification of either part, and only one set of three grouped units is used at any time. Each of the three grouped units feeds one of the three identical receivers independently.

2. The carrier is heterodyned down to 1 megacycle and this is amplified in an amplifier with 20-kc bandwidth, centered on 1 megacycle.

3. The 1 mc is heterodyned down to 100 kc and this is amplified in an amplifier with a pass band of 14 kc centered on 100 kc.

4. The 100 kc is passed through a band pass filter centered at 100 kc which reduces the pass band from 14 kc to 3.8 kc and this is amplified.

5. The 100 kc is heterodyned down to 10 kc and this passes through a filter which eliminates any sideband or other frequencies which lie above 12.8 kc.

6. The 10 kc is next switched into one of four circuits to provide the minimum frequency pass band required for the carrier and sidebands of the particular form of message signal being received. For 75 words per minute on-off keying, the 150-cycle narrow band pass filter would be adequate and would be used. For 150 words per minute or "2 channel multiplex" the 300-cycle band pass filter would be used because the sideband frequency spread would require it. For 300 words per minute on-off keying, or "4 channel multiplex" or more, the 1,200-cycle bandpass filter would be used. For greater bandwidth requirements the filter by-pass position shown would be used.

7. The last conversion shown heterodynes the 10 kc down to 2,550 cycles, which is applicable to carrier shift keying.

This receiver has a rated "noise figure" at 28 megacycles of 6 db, which is very good. The superheterodyne "image ratio" at 28 megacycles is 100,000. The receiver described was not designed for telephone reception but is readily convertible. All oscillators and automatic gain control circuits are common to all three receivers.

A receiver such as this could be adapted to single-sideband suppressed carrier reception by the addition of a "local carrier" of the correct frequency, or diversion of one of the existing heterodyne oscillators to the purpose, and the use of a conventional detector. The missing carrier could be added in any of the i-f amplifiers and the detector added to follow it, so long as the band pass were 3 kc or more to accommodate the single sideband frequencies.

QUESTIONS AND PROBLEMS

1. What types of noise limit the useful sensitivity of a receiver?
2. How is receiver sensitivity rated?
3. Describe the principal types of radio receivers. What are their advantages and disadvantages?
4. Describe the operation of a superheterodyne receiver.
5. Describe the operation of automatic volume control. Why is it used?
6. What is an image frequency? How is it suppressed?
7. Describe a multi-band receiver.
8. What useful functions does an r-f amplifier perform in a superheterodyne receiver?
9. Why is double conversion sometimes used in superheterodyne receivers? What is it?
10. What is muting? Describe its operation.
11. Describe the advantages of transistors in receivers.
12. What is the meaning of "receiver noise factor"?
13. Why is receiver oscillator radiation undesirable? How is it suppressed?

CHAPTER 14

FREQUENCY MODULATION

GENERAL PRINCIPLES

Frequency modulation (FM) was proposed as a means of improving the signal-to-noise ratio of a radio system and making possible the transmitting of a wider range of audio frequencies than is permitted with the conventional amplitude-modulation system in the standard broadcast band. In all types of carrier transmission, music, speech, or other information is impressed upon a high-frequency carrier by changing the carrier in some respect. The process of amplitude modulation (AM) has already been discussed in detail. This chapter deals with a second method of changing the radio-carrier wave, that is, by changing the *frequency* (or *phase*) to correspond to the information or signal being transmitted. The amount of change depends upon the amplitude of the signal wave representing the information.

The fundamental distinction between amplitude and frequency modulation was discussed in Chapter 10 and illustrated in Fig. 10-1.

In a frequency-modulation system an increase in the amount of modulation means that the station occupies more of the assigned radio frequencies than when there is no modulation. If this process goes too far, the station being modulated interferes with other stations by trespassing upon their assigned frequency ranges, and so the frequency swing is limited by law. For frequency-modulation broadcasting the limit is 200 kc, or a deviation of 100 kc on either side of the center or assigned station frequency. In practice stations try not to swing the entire 200 kc, but use a total of 150 kc, or a 75-kc deviation on each side of the center frequency. In television sound channels (where frequency modulation is also used) 100% modulation is defined as a deviation of 25 kc above and below the center frequency of the sound carrier. The reduced frequency swing makes the tuning of the television set less critical than it might be otherwise.

The *modulation index* is defined as the frequency deviation divided by the audio frequency of the modulating wave when both audio and carrier are sine waves. The *frequency deviation* is the peak (greatest) difference between the instantaneous frequency of the modulated wave and the carrier frequency. For example, if the frequency deviation is 75 kc and the audio frequency transmitted is 15,000 cycles (15 kc) the modulation index is

Chapter 14 **Frequency Modulation** **479**

75/15 = 5. This ratio for the highest audio frequency and maximum frequency deviation is also called the *deviation ratio*. If a single audio frequency of 1,500 cycles (1.5 kc) is broadcast with full modulation, the modulation index is 75/1.5 = 50, ten times as great as before. For an audio frequency of 150 cycles (0.15 kc) the modulation index is 500, and so on. The modulation index is large for low audio frequencies and small for high audio frequencies of the same intensity.

The modulation index also varies with the loudness of the sound striking the microphone. When the maximum intensity of the sound occurs in the example above, the frequency deviation is 75 kc. If the sound decreases to one half of this intensity the frequency deviation is half as much as before and the modulation index for a 15-kc signal drops to half its former value, or 37.5/15 = 2.5. If there is no sound at the microphone the frequency deviation is zero and no sound is heard at the receiver from the modulated wave. To sum up, for any given audio frequency the modulation index varies directly with the signal level (that is, with the audio-frequency voltage); for any given signal level the modulation index varies inversely as the audio frequency; both of these effects may occur at the same time.

A suitable network of circuit elements is connected in the audio-frequency channel to make the voltage at high audio frequencies somewhat greater than it would be otherwise; this is called *pre-emphasis* of the high audio frequencies and tends to equalize the modulation index as well as to aid in the reduction of noise in the FM channel.

If the modulation index does not vary inversely as the audio frequency but remains constant for a given sound level, the modulation is what is called *phase modulation*. At a single audio frequency there is no way to distinguish between frequency modulation and phase modulation. Phase modulation is used in some FM transmitters with pre-emphasis in the audio channel so that the net result is an FM output.

Nearly any wave which is not a pure sine wave can be analyzed into components which are true sine waves. As an example, the blunt-topped wave shown in Fig. 14-1(a) can be constructed from a pure sine wave of large amplitude and another pure sine wave of three times the frequency of the first and having much smaller amplitude. The component wave is

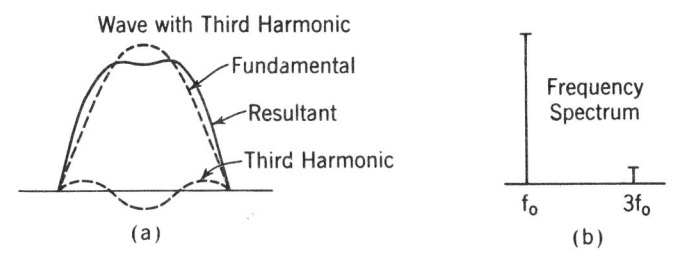

Fig. 14-1. Distorted wave and components.

drawn with dashed lines. This wave may also be pictured by a frequency-spectrum display as indicated in Fig. 14-1(b). In this spectrum the height of the line indicates the peak amplitude of each component and the horizontal distance is proportional to their separation in terms of frequency. The amplitude of the third harmonic is one-fifth that of the fundamental and the distance between them is $2f_0$, where f_0 represents the frequency of the fundamental (or lower frequency) wave.

In a frequency-modulation system the wave is alternately squeezed and stretched so that the shape of each cycle is no longer that of a sine wave but is slimmer or fatter than it would be if unchanged. Figure 14-2(a) shows one of the squeezed cycles and Fig. 14-2(b) gives the result of mathematically analyzing a frequency-modulated wave into sine-wave components and plotting the amplitudes of these components on a frequency-spectrum

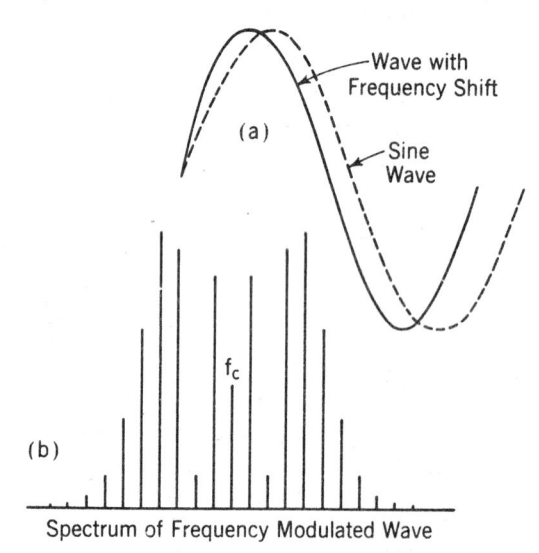

Fig. 14-2. Distorted wave and frequency-modulation spectrum.

diagram. The analysis requires the use of Bessel functions and so will not be given here. The center line, marked f_0, indicates the center or undeviated frequency, and the height of this line shows the amplitude of the center-frequency component under specified conditions. The other vertical lines denote frequencies which are called *side frequencies* (see Chapter 10). In FM there are many of these for each component in the signal, instead of only two as was the case in AM. With a large modulation index the number of important side frequencies is greater than with a small modulation index. The examples cited show that when the audio frequency is low the modulation index is large and vice versa. The side frequencies are spaced out from the center frequency at intervals which are proportional to the audio-frequency components; that is, each of the side frequencies set up by a 100-cycle tone is separated from its nearest neighbor by 100 cycles, the side frequencies of a 1,000-cycle tone are 1,000 cycles distant from each other, and so forth. When there are a large number of important side frequencies they lie closer to the center frequency than when the modulation index is small with a small number of side frequencies lying farther apart. In general the side-frequency amplitudes decrease as they lie farther from the center frequency and are negligible for values beyond the maximum frequency swing. The result is that the station will stay within the assigned band of frequencies and the side frequencies will not cause interference outside the station channel.

Noise in the output of a radio set may be broadly defined as any sound which was not present at the microphone. In case of message service, often called "communication service," this definition includes any disturbance which interferes with understanding the message coming over the air. As has been discussed, noise arises from many sources such as atmospherics, lightning, magnetic storms, diathermy machines, interfering radio stations, and so on — even from the random noise of electrons in motion in circuit components within the set itself. These disturbances are like radio signals in character, appear at all radio frequencies, and cannot be tuned out.

To overcome the effects of noise, the receiver must be able to separate the desired signal from the noise. Some of the methods intended to effect this are the use of high power at the transmitter, increase of modulation, selective circuits at the receiver, and directional antenna systems. Each of these methods has certain drawbacks.

In an amplitude-modulation system the interfering signal or noise adds directly or cumulatively to the desired signal and the result in the output may be serious interference if the undesired signal is even as small as 1% of the desired signal. With the frequency-modulation system, on the other hand, the addition of the noise and the desired signal is quite different and a condition results which is much more favorable to the reduction of noise. This fundamental difference has been proven both by mathematical analysis and by experiment. In frequency modulation, when two signals are added and the modulation index of the desired signal is fairly large while the undesired signal (either noise or interfering radio transmission) is less than half the amplitude of the desired signal, the modulation of their sum is determined almost entirely by the larger or desired transmission. In fact, no appreciable disturbing effect is noticed in the output of the receiver under these circumstances. This is a great improvement over the amplitude-modulation system. It is often called the *capture effect*. The same effect leads to "blanketing" of a weak station by a stonger one with the same center frequency, which is sometimes an advantage and occasionally a disadvantage.

The reduction of noise in an FM system may be explained with the aid of Fig. 14-3. The dark area in Fig. 14-3(a) represents the combination of noise and desired signal up to the highest audio-frequency component to be transmitted — say 15 kc. Noise is distributed over this entire frequency range and comes into the receiver equally as well as the program so that it can be heard in the loudspeaker output. It is assumed here that the pass

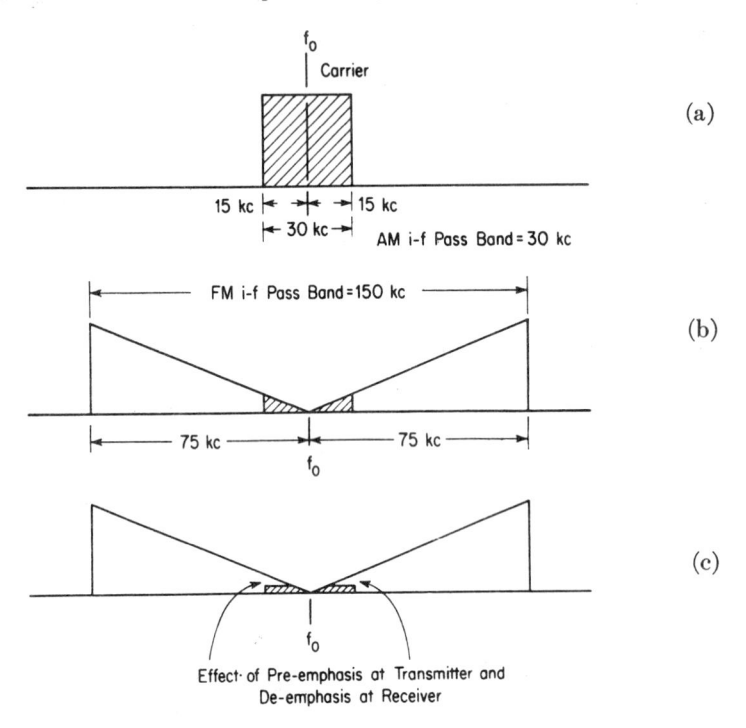

FIG. 14-3. Triangular noise spectrum with FM signals.

band of the i-f amplifier is 30 kc, or twice that of the highest audio frequency This is not likely to be the case in the standard broadcast band. Frequencies outside the pass band of the i-f amplifier do not come through the amplifier and so are suppressed.

The i-f amplifier in the FM set has a broader pass band — up to 150 kc ideally. However, because of the modulation index in the FM system being very great at low audio frequencies [near f_0 in Fig. 14-3(b)] noise is reduced in this portion of the spectrum. At increasing audio frequencies the modulation index decreases and the discrimination against noise is less, as indicated by the triangular portion of the diagram. At the upper limit of the pass band ($\pm$ 75 kc) the noise has the same effect as in an AM system. The a-f amplifier will accept and amplify frequency components only up to

15 kc, and within this range the noise level is low as indicated by the shading on the diagram. The improvement in signal-to-noise ratio between the two systems may be as great as 20 db. The de-emphasis circuit effects an additional improvement as indicated in Fig. 14-3(c) by cutting down both noise and signal at the receiver. Since the high frequencies are pre-emphasized at the transmitter, the noise components are attenuated proportionately more than the signal components. This improves the signal-to-noise ratio.

OTHER FACTORS IN FREQUENCY-MODULATION SYSTEMS

The only portion of the radio-frequency spectrum where wide-band channels (200 kc) are available to accommodate FM stations is in the VHF domain. The region assigned to frequency-modulation broadcasting is from 88 to 108 megacycles.

The fact that the range of signals at these frequencies is about one to three times the distance from the transmitting antenna to the horizon (or less than 100 miles in most instances) means that the same carrier frequency may be used by stations separated by intervals of some 200 miles without causing serious interference. In communication or emergency work the same frequency may be used by stations much closer together or by a whole group of stations; in some cases the capture effect of a strong carrier might be undesirable if it concealed the fact that a weaker station was on the air.

If two amplitude-modulation stations operate with carrier frequencies which are very nearly the same and signals from both stations appear in a radio receiver, a squeal or heterodyne is set up. This squeal often prevents the reception of a desired station and may be caused by an interfering station a thousand miles or more away, whose signal strength is very small compared with that of the desired station. Heterodyne effects between frequency modulation stations on the same or nearly the same frequency are eliminated by the limited distance covered by the signal and by the capture effect of the stronger of the two signals.

In broadcasting entertainment programs, the narrow AM channel has always been a barrier against transmitting the higher audio frequencies which give brilliance and life to music and speech. Frequency modulation, with a top audio frequency in commercial transmitters of 15,000 cycles, removes this limitation to high fidelity.

In amplitude-modulation systems the power output under conditions of 100% modulation must average 50% more than the power required with no modulation, and reaches peaks of power which are four times the unmodulated power. This radio-frequency power must be supplied by changes in the efficiency of the linear radio-frequency amplifiers or by the audio-frequency high-level modulator stage. With large stations this situation may lead to great engineering difficulties. In frequency-modulation systems, on the contrary, the modulator may be a receiving-type tube and the output stage, operating efficiently as a Class C amplifier, delivers the same amount of power to the antenna at all times. Changes in the modulation simply change the distribution of the power between the carrier and the various side frequencies.

Because of the noise-reducing properties, high fidelity, and limited interfering range of frequency-modulated systems, FM is often used in radio-relay circuits which supplement or replace wire-transmission lines or program circuits. Frequency-modulation transmissions may have several services multiplexed — such as a combination of sound and facsimile, or two sound channels to give binaural reproduction of sound, or combinations of television picture and sound signals. In the last case the picture uses amplitude modulation and the sound frequency modulation.

FREQUENCY-MODULATION SYSTEMS FOR COMMUNICATION SERVICES

The primary purpose of communication or message service is to convey information; tone quality and similar factors are of minor importance. For example, a restricted audio-frequency range of, say, 250 to 3,000 cycles is found to be adequate. With this compressed audio-frequency range the frequency deviation may be considerably reduced and stations assigned to narrower channels, which permits the simultaneous operation of more stations in a given area. The deviation ratio may still be large enough to retain the noise-reducing properties of frequency modulation; a top frequency of 3,000 cycles and a deviation ratio of 5 means a frequency swing of only 15 kc on either side of the carrier frequency or a channel width of 30 kc. As a matter of fact, experiments have shown that near the limit of the useful range of frequency-modulation signals for emergency service it is better to use a small deviation ratio, which still further reduces the channel width.

The useful sensitivity of any receiver is limited ultimately by the noise level. Because of their low noise level, frequency-modulation receivers are well suited for communication service. On the other hand, it is important that FM receivers have high gain to operate the limiter circuit (to be described later) when a limiter circuit is used. When no carrier is being received the frequency-modulation receiver is open to any random noise and static from the air, and its high gain and broad-band amplifiers make it noisier than a comparable AM receiver. For communication work, where the carrier is switched on only when needed for messages, the continuous noise of the receiver becomes very annoying. Some sort of effective squelch circuit is often included to prevent sound from coming from the speaker when no carrier is applied to the antenna of the set.

A system called "FM-FM" is used in telemetering information from guided missiles and pilotless aircraft as well as other similar applications. An electron-tube oscillator circuit operating at a low frequency (possibly in the a-f range) may be frequency controlled over a small frequency inter-

Chapter 14 *Frequency Modulation* 485

val by a barometer element, such as the pressure-sensing part of an aneroid barometer. An increase in atmospheric pressure may increase the frequency of the oscillator, for example, and a decrease could make the frequency lower. This amounts to frequency modulation of the oscillator output in response to the changes in the surrounding atmospheric pressure. The output of this oscillator in turn is used to frequency modulate an r-f transmitter, hence the name FM-FM. By the use of suitable sensing elements other information pertaining to temperature, humidity, acceleration, and other variables may be sent back to a recording or control station.

FREQUENCY-MODULATION TRANSMITTERS

The operation of many of the circuits in a frequency-modulation transmitter is identical with the corresponding circuits in an amplitude-modulation transmitter. There are, however, certain circuits which are unique to the frequency-modulation system and these will be considered in some detail.

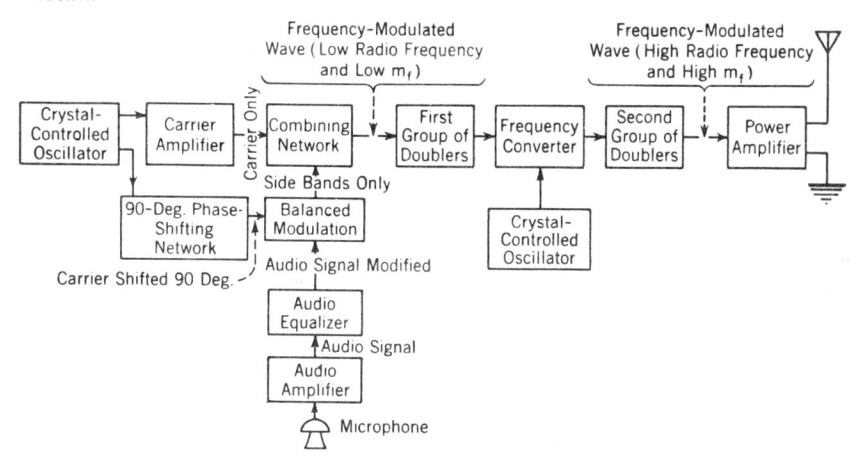

FIG. 14-4. Block diagram of Armstrong frequency-modulation transmitter system.

Two methods of obtaining frequency-modulated waves are in present use; these are (1) the Armstrong system, and (2) the reactance-tube or Crosby system. Other methods are possible and may be put to limited use.

In the Armstrong system of frequency modulation, the stages shown in the block diagram, Fig. 14-4, are used. This transmitter has an oscillator whose frequency is held at a constant value by means of a quartz crystal at about 200 kc. This constant-frequency wave is sent through two channels; one channel consists of a radio-frequency amplifier which builds up the amplitude of the wave, and the other channel includes a phase-shifting network and a balanced modulator. The balanced modulator is similar in most respects to the standard modulated stage used in amplitude-modulation transmitters, whose output can be analyzed into a carrier and a pair of side frequencies for each audio-frequency component in the signal from the microphone. However, in this case the carrier is not wanted and is removed by the action of this modulator. The phase-shifting network is needed because when the sidebands set up by this balanced modulator are added to the original carrier after a 90-degree phase shift one type of frequency modulation is produced. The deviation ratio is small and must be multiplied to reach a useful amount.

The output of the combining network is fed into a series of frequency-doubling amplifier stages. These consist of Class C r-f amplifiers whose plate circuit is tuned to double the frequency of that to which the grid circuit is tuned. The r-f signal applied to the grid is thus multiplied by two and the deviation ratio is also multiplied by two.

In frequency-modulation transmitters for broadcasting services it is found that a continuation of this multiplying process does not give a sufficiently large deviation ratio before the carrier becomes too high to be used. A frequency converter is then used to lower the carrier frequency without affecting the deviation ratio. This converter, which is of the same type as used in a superheterodyne receiver, is supplied with a fixed frequency from another crystal-controlled oscillator.

The output of the converter is passed through more doubler stages further to increase the deviation ratio and carrier frequency until the correct carrier frequency is attained. All of these operations can be carried out with receiver-type electron tubes. The output, with high carrier frequency and high deviation ratio, is then applied to as many Class C power-amplifying stages as are necessary to build up the power output of the transmitter to any amount desired.

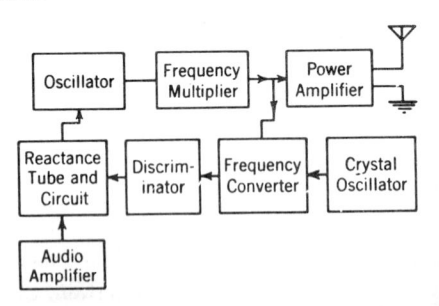

Fig. 14-5. Reactance-tube or Crosby system frequency-modulation transmitter.

In this system the modulation index is directly proportional to the audio-signal voltage. The audio frequency supplied to the modulator must be passed through an audio-equalizer network which makes the modulation index inversely proportional to the audio frequency by making the audio amplitude proportional to the audio frequency. That is, if a constant amplitude were maintained at the input to the system while sweeping through the audio spectrum, the output amplitude would increase directly in proportion to the increase in audio frequency. For instance, if the output were 1 v at 100 cycles, it would be 2 v at 200 cycles, and so on.

The reactance-tube system is shown in block diagram Fig. 14-5.

In this system the oscillator is not crystal-controlled but is self-excited. Another electron tube, the reactance tube, is connected in parallel with the tank circuit of the oscillator. The reactance tube is supplied with the usual plate and cathode-bias voltages as shown in Fig. 14-6. In addition,

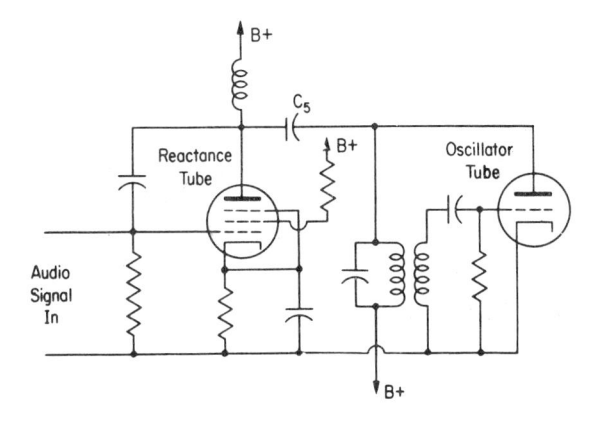

FIG. 14-6. Circuit diagram of oscillator and reactance tube.

its plate is supplied with alternating voltage from the tank circuit through capacitor C_5. By means of a suitable circuit, as shown, this tube is caused to act like a reactor (inductor or capacitor) and its reactance is varied in accordance with the audio frequency. The frequency of the oscillator is also changed because of the changing reactance connected across its tank circuit, and a frequency-modulated r-f wave appears at its output.

This carrier with frequency modulation is passed through a number of frequency doublers or multiplier stages to increase the frequency and deviation ratio. Finally, power amplifiers step up the power and feed the signal into the antenna.

Some means must be provided to keep the transmitter upon its assigned carrier or center frequency. A self-excited oscillator, such as that used here, does not hold closely enough to an assigned frequency to be satisfactory for this service. Stability may be obtained by comparing the output of the transmitter with a standard, crystal-controlled oscillator and feeding back a suitable correcting voltage from a frequency converter and discriminator (to be described later) to the grid of the reactance tube and thus keep the self-excited oscillator on the correct frequency.

FREQUENCY-MODULATION RECEIVERS

The typical frequency-modulation receiver is a superheterodyne similar to that used for AM in many respects but having certain special features. Among the latter are (1) a *limiter* to remove any amplitude modulation of the FM wave, including noise, appearing along with the desired FM signal and (2) a *discriminator* circuit to change the frequency variations into audio-frequency voltages. In some receivers these two items are combined in (3) a *ratio-detector* circuit. In entertainment receivers, the audio-frequency channel and especially the loudspeaker should be of the highest quality to take advantage of the high fidelity possible in FM systems. This high quality may require the use of special tweeters for the high frequencies, woofers for the low frequencies, and perhaps also a middle-frequency speaker as well.

The pre-emphasis at the transmitter to increase the modulation index at the higher audio frequencies must be corrected with a *de-emphasis* circuit in the receiver. This circuit to restore the correct balance of tone may consist

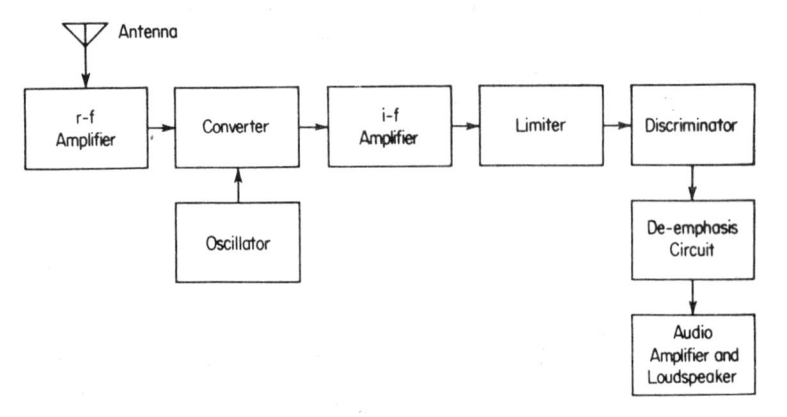

FIG. 14-7. Block diagram of typical FM receiver.

of a resistor and capacitor in series whose time constant is about 75 microseconds (such as a 75,000-ohm resistor and a 0.001-μf capacitor).

A block diagram of a typical frequency-modulation receiver is shown in Fig. 14-7.

The circuit for a typical limiter stage and its action are shown in Fig. 14-8. The limiter tube operates with a low plate voltage (40 to 80 v) so that it will reach its maximum output with a relatively low applied grid voltage. The voltage to drive the limiter is obtained from an i-f amplifier, usually of at least two stages. The circuit operates like a Class C amplifier

Chapter 14   **Frequency Modulation**   **489**

and even though large signals are applied to the input of the receiver, the output of the limiter remains nearly constant. Increasing the grid voltage increases the grid bias because of resistor R_2, and the bias voltage is maintained by capacitor C_1 until the input level changes markedly. Amplitude

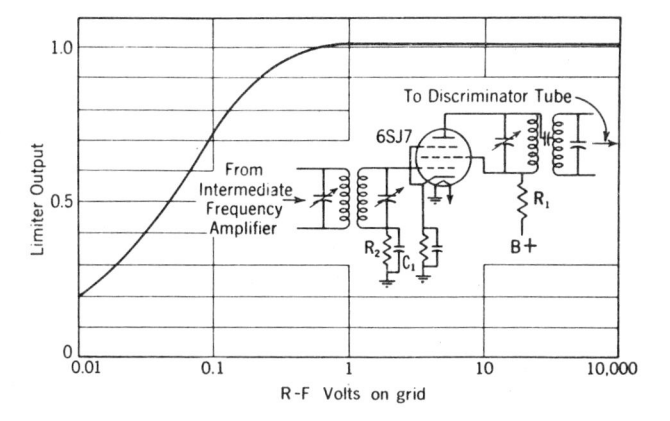

FIG. 14-8. Circuit of limiter stage and curve showing limiter action.

modulation is wiped off the signal no matter what its source; the limiter does not interfere with the frequency modulation because its action does not affect the instantaneous frequency of the wave.

The circuit diagram of a typical FM detector (discriminator) is shown in Fig. 14-9. When a voltage appears at the plate of the limiter stage the same voltage (E_1) is set up across inductor L_3; capacitors C_3 and C_5 are selected

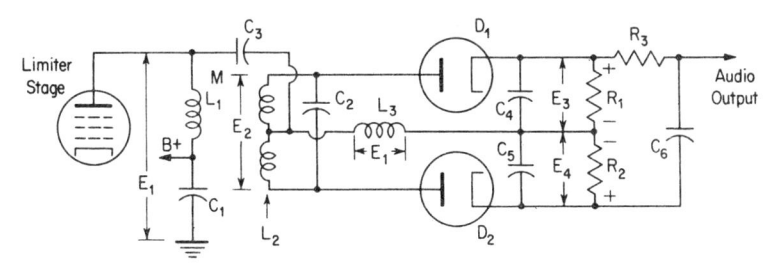

FIG. 14-9. Frequency-modulation detector or discriminator.

to have low reactance to the i-f carrier used. At the same time a voltage is induced in the secondary of the i-f transformer and half appears as $E_2/2$. Each diode has the sum of the voltage $E_1 + E_2/2$ applied between its plate and cathode and the result, after rectification, is smoothed by the filtering action of R_1C_4 and R_2C_5, respectively. The voltage between the a-f output terminal and ground is the difference of the output of the two diodes which are connected in opposing polarity. When a signal is applied of exactly the frequency to which the i-f transformer is tuned the voltages across the two diodes are equal and no a-f voltage appears at the output. The voltages E_1 and E_2 are 90° out of phase (because of the action of a tuned transformer) and the whole system is balanced. If, however, the frequency is changed by a small amount, E_1 and E_2 will no longer be 90° out of phase (again, because of tuned-transformer action) and the voltages developed across the diodes will differ. When the frequency varies at an a-f rate, a varying (a-f) voltage will be present at the output of the discriminator. Resistor R_3 and C_6 make up the deemphasis circuit. A typical curve showing the relation of the applied frequency to the resulting audio voltage is given in Fig. 14-10.

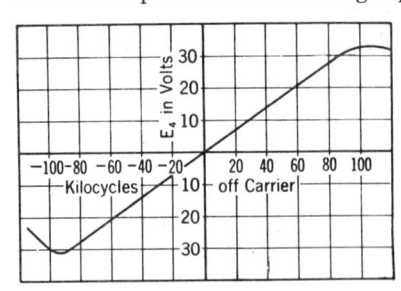

Fig. 14-10. Frequency variation plotted against audio-frequency output in a discriminator.

A circuit which responds to the ratio of the diode outputs and ignores amplitude modulation is shown in Fig. 14-11 and is called a *ratio detector*. No limiter is required and the preceding stage can operate at full gain. Up to the diodes this circuit is identical with that of the discriminator; the

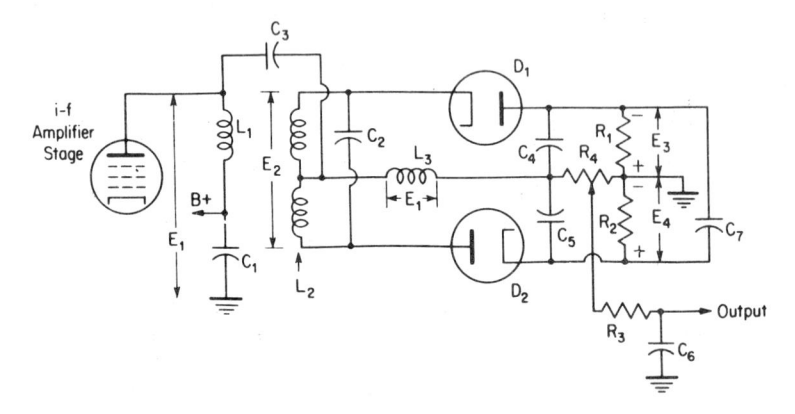

Fig. 14-11. Frequency-modulation ratio detector circuit.

diodes are connected in series aiding instead of opposing, and the a-f output is taken from a different point in the circuit. In addition a voltage-stabilizing capacitor C_7 with a time constant (in parallel with $R_1 + R_2$) of about 0 2 second holds the sum of the rectified voltages across R_1 and R_2 a constant as long as the amplitude of the incoming FM carrier is a constant. The ratio of the voltages developed across these resistors constitutes the a-f output. This ratio depends upon the frequency only and not the amplitude.

Chapter 14   **Frequency Modulation**   491

An increase in amplitude, for example, increases the total voltage but does not change the ratio of the voltages across the resistors. When the carrier frequency changes the voltages across R_1 and R_2 change and as a result an equalizing current flows through resistor R_4. The voltage drop resulting across this resistor is the a-f output. Resistor R_3 and capacitor C_3 make up the de-emphasis circuit.

Another method of converting frequency-modulation signals is to apply the incoming signal to a circuit tuned to a slightly different frequency from that of the signal, as shown in Fig. 14-12. A higher frequency causes an increase in the current in the tuned circuit and a lower frequency a corre-

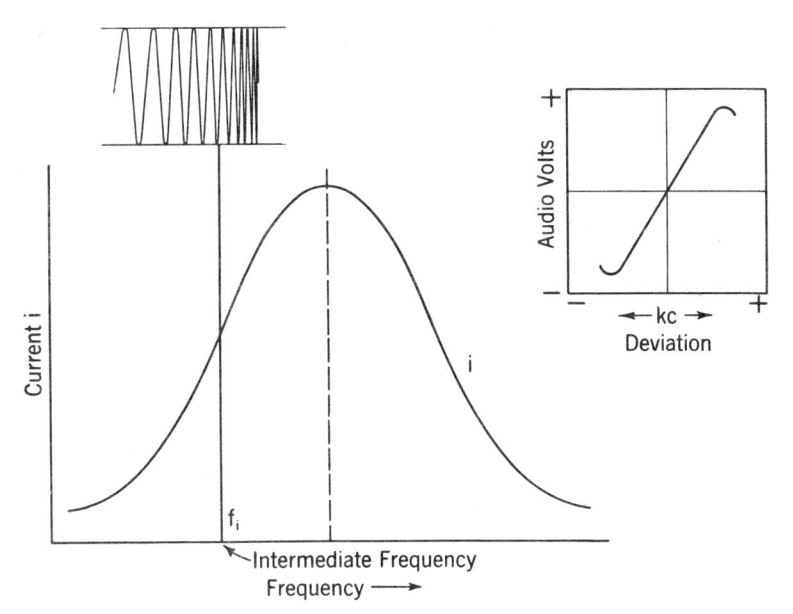

FIG. 14-12. Off-resonance tuning to convert frequency-modulation signals into amplitude-modulation signals.

sponding decrease in current. This current change converts the variation in frequency into a variation in the amplitude of a current, which may then be passed on to a standard amplitude-modulation type of detector. This effect explains the fact that frequency-modulation signals may be brought in with amplitude-modulation receivers under certain conditions. The tone quality is usually very poor because of the nonlinear character of the tuned circuit response.

TYPICAL FREQUENCY-MODULATION RECEIVER

Figure 14-13 shows the diagram of a representative circuit for a frequency-modulation receiver. Nine tubes are used: r-f amplifier and mixer, two i-f amplifiers and a limiter, a discriminator, a-f amplifier, output stage, automatic-frequency-control tube and local oscillator, and rectifier.

The frequency-modulated signal from a transmitting station is intercepted by a suitable antenna and fed into L_1, the first r-f transformer, by a transmission line. The secondary of this transformer is tuned by one section of capacitor C_3 and the voltage across its terminals is applied to the grid of the first r-f amplifier stage. After amplification here the signal is fed into a mixer control grid tuned by a circuit consisting of coil L_2 and another section of C_3. In the mixer the r-f signal is heterodyned with the output from the local oscillator whose control circuit comprises a third section of C_3 and coil L_3. The two r-f circuits and the oscillator are aligned by adjustment of their iron cores. The FM components of the output of the mixer are amplified (at 10.7 megacycles as intermediate frequency) and then clipped by the limiter stage. This tube feeds the discriminator circuit, which recovers the audio-frequency modulation from the i-f FM signal. The audio amplifier and power supply are conventional. One section of the dual triode used as the oscillator is an automatic-frequency-control tube which reduces drifting during the warm-up period of the tubes and acts as a fine-tuning adjustment. Details of the operation of the various circuits have been described earlier.

QUESTIONS AND PROBLEMS

1. Point out the main differences between a frequency-modulation and an amplitude-modulation radio system.

2. What are some of the reasons for the noise-reducing properties of frequency-modulation systems?

3. Why is a limiter necessary in a frequency-modulation receiver?

4. What is the frequency-bandwidth for frequency-modulation transmission at present?

5. Compare the variations of output power when modulation level is changed in a frequency-modulation system and then in an amplitude-modulation system.

6. What is the ratio of bandwidth to mid-frequency in the intermediate-frequency amplifiers of a frequency-modulation system as compared with an amplitude-modulation system?

7. Why is it not practical to have frequency-modulation stations in the regular broadcast band?

CHAPTER 15

MONOCHROME TELEVISION

INTRODUCTION

Television enables man to see events which are at a distance or otherwise invisible to him — often while they are actually occurring. The scene being viewed is converted into an electrical signal which varies with respect to time. After suitable modification this signal is broadcast or otherwise conveyed to a receiving set which reconstructs a picture approximating the motion, brightness, and contrast of the original scene. Even the color may be included.

Usually the sound and the visible parts of the scene are transmitted together. In some industrial applications of television principles, such as the remote observation of a manufacturing process, only the picture is needed and the sound is not used. Since sight and sound are independent of each other after being converted into their electrical counterparts, they can be sent by different paths from the point of origin to the receiver. In most cases, however, they are handled at the same time in the equipment. The carriers for sound and picture are spaced 4.5 megacycles apart in frequency as shown in Chapter 11 in the section on television wide-band amplifiers.

A television system utilizes a great many of the techniques and devices already considered in this book. If this fact is kept in mind the complete system will seem less complicated when it is studied in this chapter.

A simplified block diagram of a *monochrome* (black-and-white) television (TV) system is shown in Fig. 15-1. The visible part of the scene to be transmitted is converted into its electrical equivalent (the *video signal*) by a special electron tube in the TV camera while the microphone changes the accompanying sound into the *audio signal*. The electrical output from the TV camera is amplified and sent to the monitoring apparatus. The picture appears here on a picture tube (a cathode-ray tube) and is inspected for quality and content. The audio signal is also monitored at this point. After further amplification both the video and audio signals are used to modulate two separate r-f carriers (spaced 4.5 megacycles apart in the frequency spectrum) and the modulated output from each is fed by a transmission line into the transmitting antenna after passing through a *diplexer*. The diplexer is a circuit to allow the two transmitters to feed the common antenna without interfering with each other. *Synchronizing* (*sync*) signals are generated, used to modulate the picture carrier, and broadcast with the video and audio signals so that the picture at each receiver will stay in step with the picture being televised. Power supplies furnish appropriate voltages and currents to the amplifier tubes in the various parts of the transmitter.

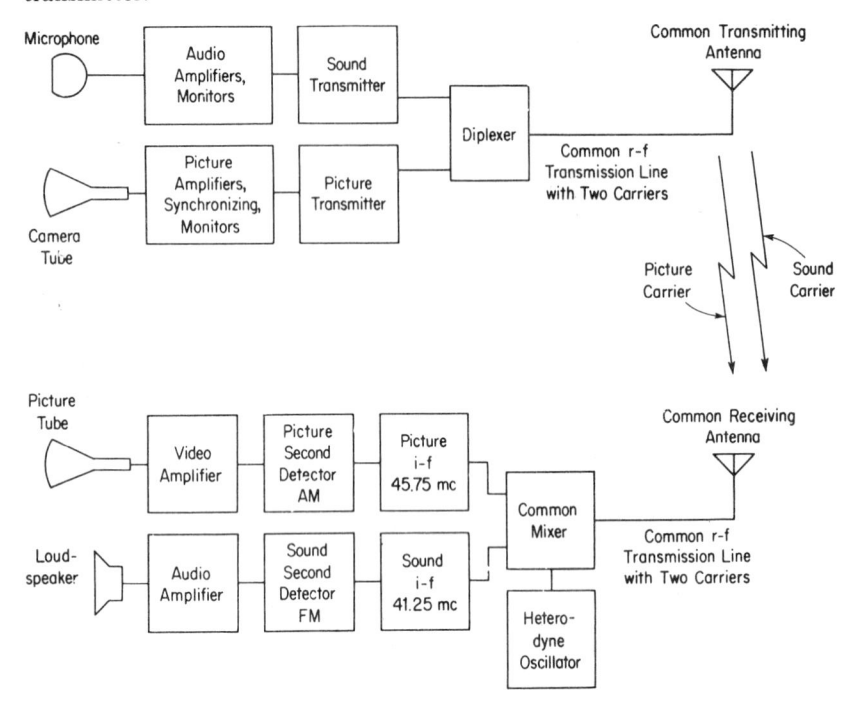

FIG. 15-1. Block diagram of a television system.

At the receiving location the modulated carriers energize the common receiving antenna and are amplified and changed to the i-f waves in a mixer stage. The audio channel may be like that of a conventional FM set although most TV sets use a modification of this system which is described later. The video channel includes the video i-f amplifying stages, video detector, video amplifier, and the picture tube. Sync-signal amplifying and separating circuits drive the deflection circuits of the picture tube through suitably controlled sweep generators and amplifiers. Power-supply units and control circuits also are needed in the complete receiver.

TELEVISION SOUND CHANNEL

The sound system associated with a television channel is like that used in FM broadcasting except for minor differences. Microphones are used to pick up the sound on "location" as in radio broadcasting; however, they must generally be kept out of sight without degrading the sound quality. This is usually effected by mounting a directional microphone on the end of a long boom which can be manipulated by a boom operator with suitable controls.

The sound is amplified and goes through a switching system like the picture (video) switching so that sound and picture are routed by separate but parallel channels. After monitoring the sound goes to an FM transmitter with a maximum frequency deviation of plus and minus 25 kc.

SOUND PRE-EMPHASIS

In a receiver a certain amount of hiss or noise is produced in the input circuits as explained in Chapter 13. If the high audio frequencies are attenuated in the audio amplifier this noise is reduced; however, the audio-frequency quality also is degraded. To avoid this the high audio-frequency response in the transmitter is emphasized in the transmitter by an amount equal to the loss in the receiving audio system. This pre-emphasis is effected by a resistance-capacitance network or filter circuit having a time constant of 75 microseconds. The audio response at the transmitter is increased some 10% at 1,000 cycles and about 600% at 15,000 cycles. The over-all response of the system is thus made substantially uniform

SCANNING

No economical process has yet been devised to transmit an entire picture area by converting it instantly into an electrical signal. Instead it is necessary to *scan* (that is, to explore) the picture area with some device which covers the scene one small area after another and transmits the resulting electrical impulses one after another.

Reading this page is a form of scanning. The eyes are focused on the upper left-hand corner of the reading matter and then move to the right looking along the first line. At the end of this line the eyes look at a point near the beginning of the second line, move to the right, and continue these movements down the page. At the bottom of the page the eyes move so as to start at the upper left-hand corner of the second page and repeat the scanning process there.

HUMAN VISION AND TELEVISION SCANNING

The final result of the video part of the television system is a picture to be seen just as the final result of an ordinary broadcasting radio system is sound to be heard. The eye of the viewer is the final judge of picture quality in the same way that the ear evaluates sound quality. The solutions of many engineering problems in a television system depend upon the properties of the human eye. For example, the number of pictures transmitted per second, the number of lines used for scanning, the colors used in color TV, and many other details of television standards depend upon the eye's behavior.

The eye may be compared to a camera. It has a lens which focuses an image of a scene upon a sensitive surface (the *retina*) inside the eye. The surface of the retina is made up of the endings of many thousands of nerves. When light of sufficient intensity strikes these nerve endings they are stimulated and as a result a small impulse is sent from each through the

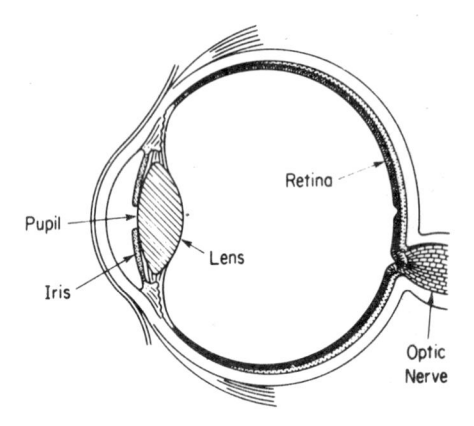

FIG. 15-2. Schematic diagram of the human eye.

optic nerve to the brain. The sensation which registers in the brain is called seeing (*vision*). A million or so of these light-sensitive nerve endings (called *rods* and *cones*) terminate in the retina. The centers of the nerve endings are about 0.0001 inch apart. *Visual acuity*, defined as sharpness of vision with respect to the ability to resolve detail, depends upon this separation between the nerve endings. To *resolve detail* means that images of two separate objects do not appear to merge into one when seen by the eye. For example, the headlights of an automobile appear to be a single spot of light when far away; at closer range they can be seen or resolved into two separate lights. On the assumption that the focal length of the lens in the eye (that is, the distance from the lens to the point where it forms an image of a distant object) is 0.75 in., the angle between lines from

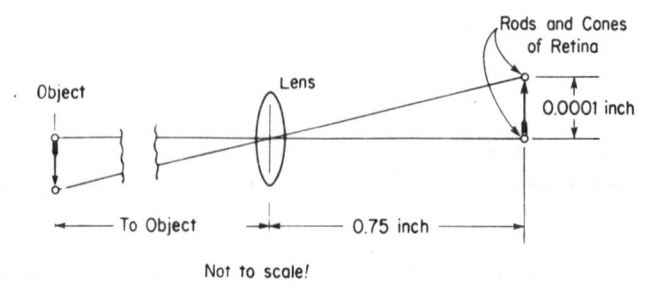

FIG. 15-3. Resolving power of the eye.

two objects that fall on two separate rods or cones can be found from

$$\tan\frac{A}{2} = \frac{a}{2} \times \frac{1}{d} = \frac{0.0001}{2 \times 0.75} = 0.000067, \tag{15-1}$$

approximately, and the corresponding angle A is about a half minute of arc.

Most eyes are not quite as good as this estimate would indicate and the figure of one minute of arc is usually quoted to specify visual acuity. The thickness of the body of a common pin (0.028 in.) held 9 ft away makes an angle of about one minute of arc at the eye. If two items in the detail of a picture make an angle at the eye of less than the resolving power of the eye, then such detail cannot be resolved and it is needlessly expensive to build a television system good enough to transmit it. Such extra refinement is useless if the viewer cannot see it in the picture.

The action of the rods and cones in the eye is not instantaneous. The output of the nerves to the brain requires a small but appreciable time

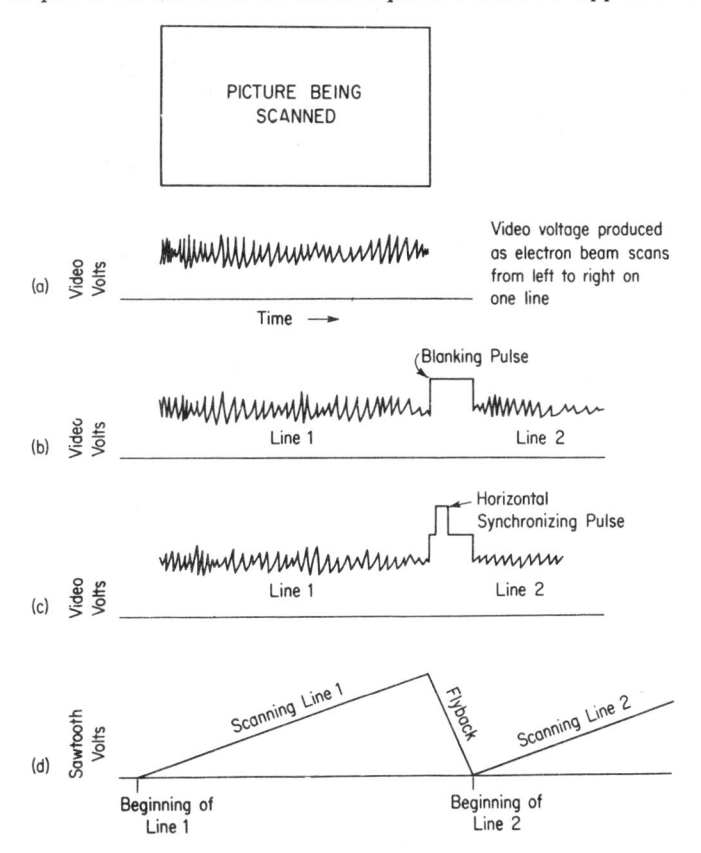

Fig. 15-4. Electronic scanning.

interval to build up and to become ready for the next light stimulus. The result of this time lag in the operation of the eye is called *persistence of vision* or the ability of the eye to retain the sensation (caused by light reflected from an object) for a short period after the stimulus has been removed. If a succession of images (representing stages of motion when a person is walking, for example) is shown to the eye at a rate greater than about 15 pictures per second the eye does not distinguish the separate images but sees what appears to be a smooth change from one to the next. This is the basis for the illusion of motion in moving pictures and also in television.

In the standard monochrome TV system the *repetition rate* (also called *frame frequency*) is 30; that is, 30 complete pictures (*frames*) per second. One thirtieth of a second after a certain image is transmitted a second image is sent and so on. The second image is slightly different from the first because of motion in the scene being viewed.

In a television system an electron beam does the scanning in a way similar to one's eyes in reading. The beam strikes a photosensitive surface in the camera tube and a current flows which is proportional to the brightness of the tiny area upon which the electrons impinge. Starting at the upper left-hand corner of the picture space the beam scans from left to right. As it moves the current changes in proportion to the brightness of the scene and this continues until the beam has scanned the whole line. This is illustrated in Fig. 15-4(a).

HORIZONTAL SCANNING RATE

In the television standards of the United States the horizontal scanning rate for monochrome transmission is 15,750 sweeps or cycles per second. This means that one line is scanned by an electron beam in 1/15,750 second or 63.5 microseconds. Part of this time interval must be used to allow the beam to return to the start of the next line to be scanned without producing any signal. During the return sweep the beam must be *blanked* by a negative bias voltage on a control grid which cuts off the supply of electrons. The bias voltage is supplied by a *blanking pulse* from a suitable circuit. The interval during which the blanking action takes place is called the *flyback*, *return*, or *horizontal retrace* time. This is indicated in Fig. 15-4(b). The time taken by blanking is about 16.5% of the horizontal time or 63.5 microseconds, which leaves about 53 microseconds for active scanning and extracting picture information with 10.5 seconds for the blanking interval.

HORIZONTAL SYNCHRONIZING PULSE

A *synchronizing pulse* is required at the end of each line so that the picture at the receiver will stay in step with the scanning at the transmitter. The sync pulse occurs during part of the blanking interval and since the electron beam is already biased to cutoff, the 25% greater negative voltage causes no visible effect. The horizontal-synchronizing pulse is shown as superimposed on the blanking pulse in Fig. 15-4(c). Synchronizing pulses are the highest voltages produced in a transmitted television signal because it must be possible to separate them from the rest of the signal by their greater amplitude and thus insure that the receiver stays in synchronism with the transmitter at all times. The duration of the horizontal sync pulse is about 8% of the 63.5-microsecond horizontal-line interval; that is, 5 microseconds. This period is shorter than the blanking interval of 10.5 microseconds and so the sync pulse is narrower, timewise, than the blanking pulse, as shown in Fig. 15-4(c). The space on the blanking pulse or *pedestal* preceding the sync pulse is called the *front porch* and the interval following it is the *back porch*, as illustrated in Fig. 15-5.

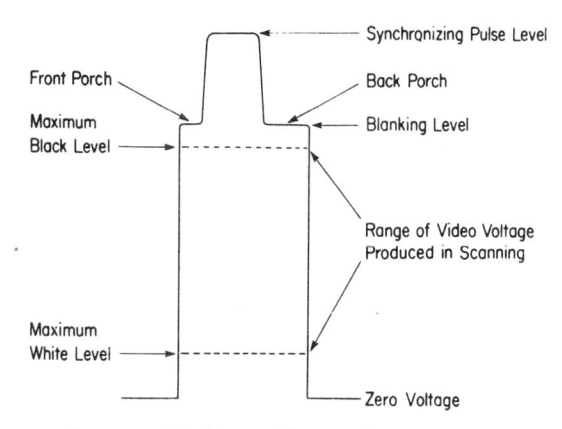

Fig. 15-5. Blanking and horizontal-synchronizing pulse.

The *horizontal-synchronizing pulse* controls the timing of a separate saw-tooth oscillator in the receiver to produce a saw-tooth wave for each pulse, as well as to control the timing of the scanning circuits in the camera at the transmitter. This serrated wave is applied to the *horizontal-deflection* circuit and is shown in Fig. 15-4(d). The beam may be considered to be at rest at the left when there is no deflection (beginning of line 1). The voltage rises during the scan interval and so causes the beam to move to the right. After the saw-tooth voltage reaches its maximum and has moved the beam to the extreme right of the picture it drops rapidly to zero during the flyback time. This returns the beam to the starting point at the left of the picture space. This procedure causes the scanning beam to move at constant speed from left to right; then it is blanked out by the blanking pulse while it returns quickly to the starting side of the picture and starts to scan again in a constantly repeating sequence. However, the beam does not scan the same line but the next in succession. The voltage to cause this latter movement of the beam comes from the *vertical-deflection* system. The timing of the horizontal and vertical scanning saw-tooth waves must be such that they are interlocked and the changes occur at just the correct times with respect to each other. The horizontal and vertical deflection voltages are obtained from a single oscillator in the transmitter (the *synchronizing-signal generator*) which produces all of the control and blanking voltages and times them precisely

VERTICAL BLANKING AND SYNCHRONIZING

When the beam has reached the lower right-hand corner of the picture it has reached the point where a downward excursion has been completed and it must return to the upper left-hand corner to begin to scan another frame. Just as is the case with the horizontal scanning, the beam must be blanked out to prevent its retrace from appearing in the picture. A long vertical pulse is used for this purpose; it is some 15 times the duration of the horizontal-scanning interval. In the vertical blanking and synchronizing sequence the blanking occurs first. After an interval equal to the time of

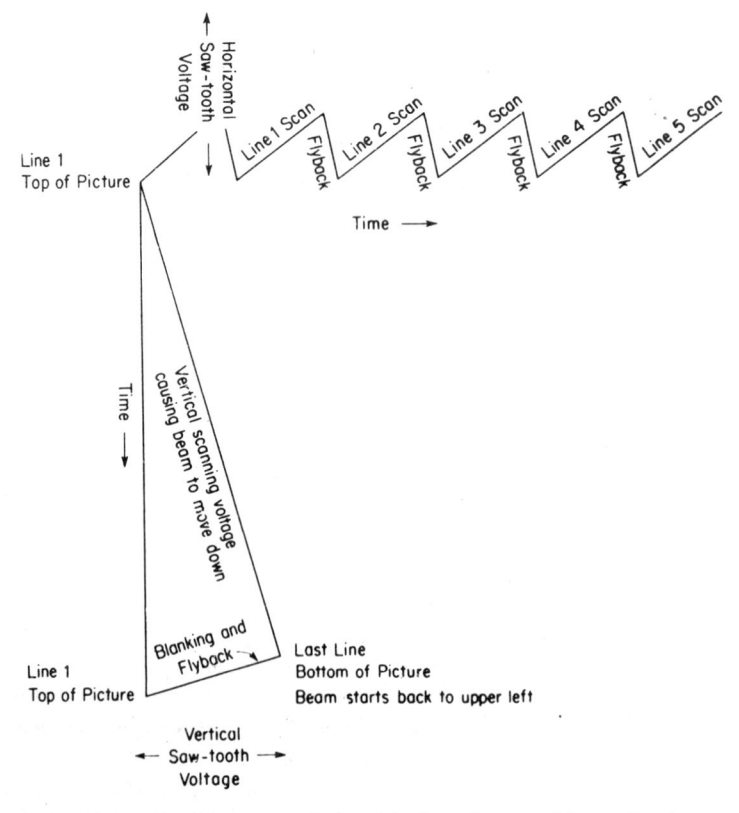

FIG. 15-6. Simultaneous vertical and horizontal sequential scanning (no interlaced scanning—see p. 502).

three horizontal lines the vertical sync pulse is introduced. Like the horizontal pulse this also is about 25% greater in amplitude than the blanking pulse so that it too may be separated by amplitude-sensitive circuits. The vertical-synchronizing pulse, superimposed upon the vertical-blanking pulse also occupies a time interval equal to three horizontal-line scans. This in turn is followed by more blanking equal to about six horizontal scans in time. During this last interval horizontal-synchronizing pulses are again transmitted to be sure that the transmitter and receiver horizontal-saw-tooth oscillators are perfectly in step before blanking is removed and horizontal picture scanning is resumed at the top line.

The vertical and horizontal saw-tooth voltages are shown in Fig.15-6.

SIMULTANEOUS VERTICAL AND HORIZONTAL SYNCHRONIZING

If the horizontal-sync pulses were cut off during the long vertical-retrace interval, the receiver horizontal saw-tooth oscillators might fall out of step with those at the transmitter. When picture scanning resumed the received pictures would not immediately duplicate those at the transmitter. For this reason the horizontal synchronizing signals are transmitted during the entire vertical blanking and synchronizing intervals. They do not interfere with each other because the sync-separating circuits for the horizontal sweep circuits are designed so that they are not affected by the long vertical pulses and vice versa.

EQUALIZING PULSES

Prior to the application of the vertical-synchronizing voltage there is an interval of 190.5 microseconds during which *equalizing pulses* at a frequency of 31,500 cycles are superimposed upon the vertical-blanking pedestal. This is at twice the horizontal scanning rate (15,750 cycles). Alternate pulses of this equalizing group (at 15,750 cycles) actuate the frequency-sensitive *sync separator* circuits which ignore the extra pulses between. This keeps the saw-tooth oscillators at the receiver in step with those at the transmitter all the time.

Following the equalizing pulses the vertical synchronizing pulse is superimposed upon blanking. This is also of 190.5 microseconds duration, three times the horizontal-scanning interval. However, this long vertical synchronizing pulse also has narrow (in time) intervals in it spaced at a frequency of 31,500 cycles (twice line frequency) which produce voltages corresponding to the equalizing pulses. Another interval of 190.5 microseconds with 31,500 cycles equalizing pulses follows. Then there is a series of ordinary horizontal sync pulses. On alternate vertical scans there is an important difference here, however. The horizontal pulses on the odd-numbered scans start a half-line interval sooner than on the even-numbered vertical scans; in the odd-numbered vertical pulses there are six pulses and only five pulses in the even-numbered scans. The details of the composite video wave form are illustrated in Fig. 15-7.

502   *Monochrome Television*   Chapter 15

INTERLACED HORIZONTAL SCANNING

Flicker occurs when the pictures appear to be separate instead of in smooth sequence. The minimum number of pictures per second required depends upon their brightness, since the eye detects flicker most readily at

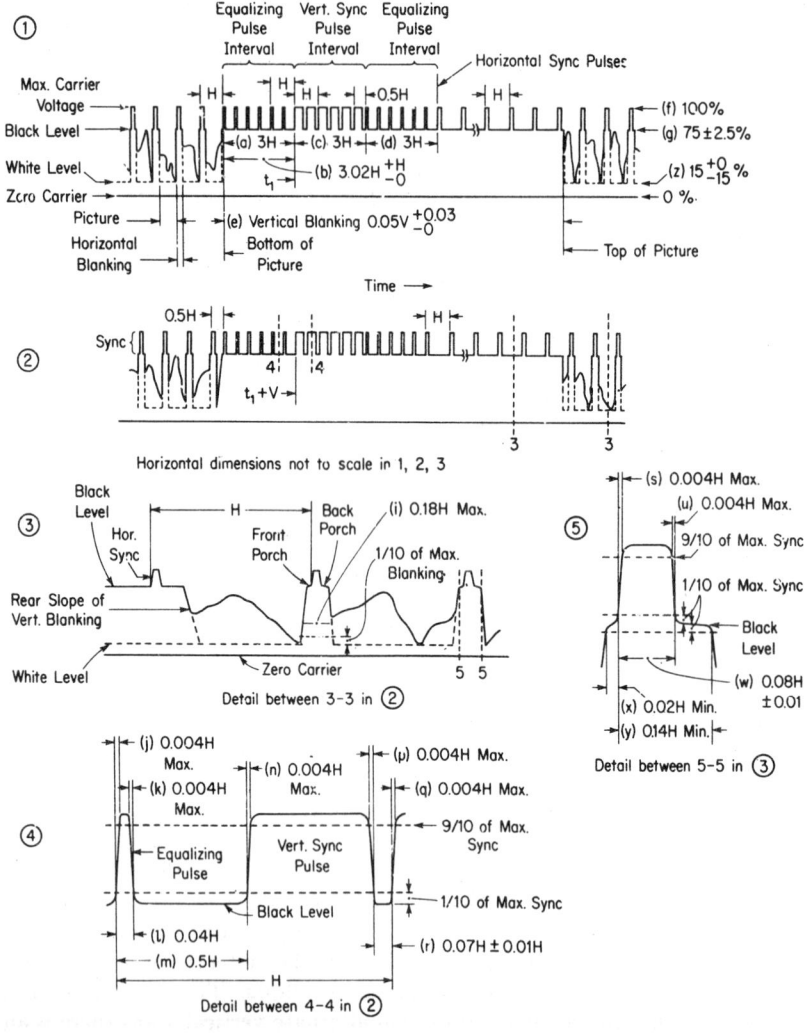

FIG. 15-7. Composite video wave form. *Notes*: (1) $H = $ time from start of one line to start of next line. (2) $V = $ time from start of one field to start of next field. (3) Leading and trailing edges of vertical blanking should be complete in less than $0.1H$. (4) Equalizing pulse area shall be between 0.45 and 0.5 of the area of a horizontal sync pulse.

high brightness levels. To conserve channel space in the frequency spectrum, no more frames per second are displayed than necessary to eliminate this flicker effect. If the picture is scanned line by line consecutively, as many as 50 or 60 frames per second are needed to avoid this annoyance The use of *interlaced scanning* reduces the number of frames required to 30 and still avoids the appearance of flicker.

When the horizontal-scanning beam covers only the odd-numbered lines in one excursion from top to bottom of the picture, a single "flash" is produced. If it then returns to the top and scans the even-numbered lines on the next trip, another flash results. All the picture elements are scanned in two vertical excursions, one-half during each. The eye cannot perceive this operation because the "flashing frequency" is higher than the eye can follow. The number of frames needed is cut in half without degrading the picture quality and at the same time flicker is eliminated. The vertical blanking and synchronizing signal-repetition rates must be doubled compared with those for a 30-frame picture. The horizontal rate remains the same. Each of the 60 excursions of the beam produces one *field*, which is one half of a complete frame.

Since there are 525 horizontal lines in the standard TV picture, there must be 262 complete odd lines and half a line more (262.5 odd lines); the half line means that the scanning ends at the center of the bottom of

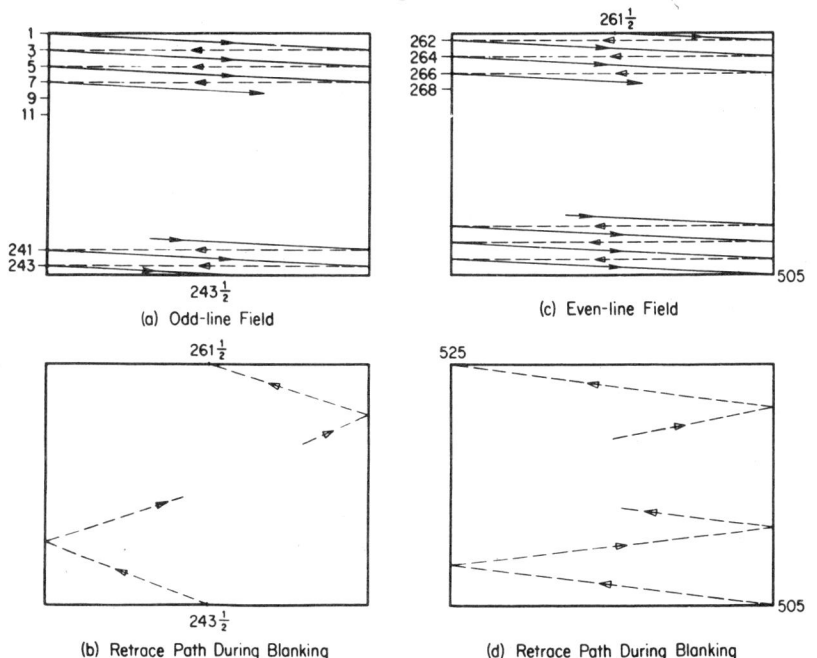

Fig. 15-8. Odd-line interlaced scanning

the picture area. The few lines shown in Fig. 15-8 illustrate the idea. Starting at the upper left-hand corner of the picture (a) the scanning beam moves across the picture area from left to right in about 53 microseconds and returns to the left-hand side of the picture space in some 10.5 microseconds. The scanning beam is blanked out during the return time and the retrace does not show when there is a picture on the screen. This process scans the first line (which is an odd line) and brings the scanning beam into position to scan the second odd line — number 3 of the entire picture. When this line has been scanned the beam goes back to the left-hand edge again and starts to scan line number 5 (the third odd-numbered line) and so on.

Near the bottom of the picture odd line 241, for example, is scanned and on the next trip the scanning beam goes only to the center of the picture — corresponding to line 243.5. The beam is now blanked out and returns to the top of the picture several lines later, arriving at line 261.5, which is the center of the upper edge of the picture space. The beam is turned on again and the scanning continues. The first sweep goes only halfway across the image; upon its return to the left-hand side the scanning of the even-numbered lines begins. The first complete line is the second line of the picture (between lines 1 and 3) and number 262 of the total of lines scanned. The scanning goes on down the picture as shown in Fig. 15-8 and ends with line 505 at the lower right-hand corner. The beam then returns to the top of the picture and is ready to start scanning the next frame. The frame is completely scanned when two successive fields (one odd-line field and one even-line field) have been completed. Since there are 30 frames per second there must be 60 fields per second, 30 odd-line fields and 30 even-line fields. Stated another way, the ratio of the field frequency (60 per second) to the frame frequency (30 per second) is 2 to 1 for interlaced scanning. This gives the effect of smooth motion without perceptible flicker. It is assumed in this description that *vertical-blanking duration* is 6.5% of *vertical-scanning time* (1/60 second); 6.5% of 525 lines are lost during the blanking interval so that 490 lines actually can be seen.

SCANNING RATES

The monochrome standard horizontal-scanning frequency in the United States is 15,750 cycles. The corresponding vertical-scanning rate is 60 per second, with interlacing of lines on successive frames. The corresponding figures for color television differ only slightly from these.

BANDWIDTH REQUIREMENTS OF A TELEVISION SYSTEM

The U. S. standard television system uses 525 scanning lines for each complete frame. The *aspect ratio* (width of picture/height of picture) is 4/3 so that there are $525 \times 4/3 = 700$ *elements* per line. Multiplying the number of *picture elements* by the number of lines gives $700 \times 525 = 367,500$. To repeat this number of picture elements 30 times per second as

Chapter 15   **Monochrome Television**   505

required to send 30 frames per second yields the figure 30 × 367,500 = 11,025,000 elements per second. A certain number of the lines occur when the signal is blanked out during the return sweeps from right to left or from bottom to top of the picture as explained in connection with interlaced scanning. Only about 8 million elements per second contribute to the visible image. The greatest variation in a picture is from black to white or vice versa. If a black picture element happens to lie adjacent to a white picture element both can be transmitted in succession by a single cycle of varying voltage which has a positive maximum for the black element and a negative maximum for the white one with the usual system of polarity. All other variations — the grays or *half tones* — in the picture have smaller changes in amplitude than from black to white. To send 8 or more million elements requires half as many cycles of frequency as there are elements or at least 4 million cycles must be transmitted in one second. Four megacycles per second at least is the highest modulating frequency required under the conditions discussed here.

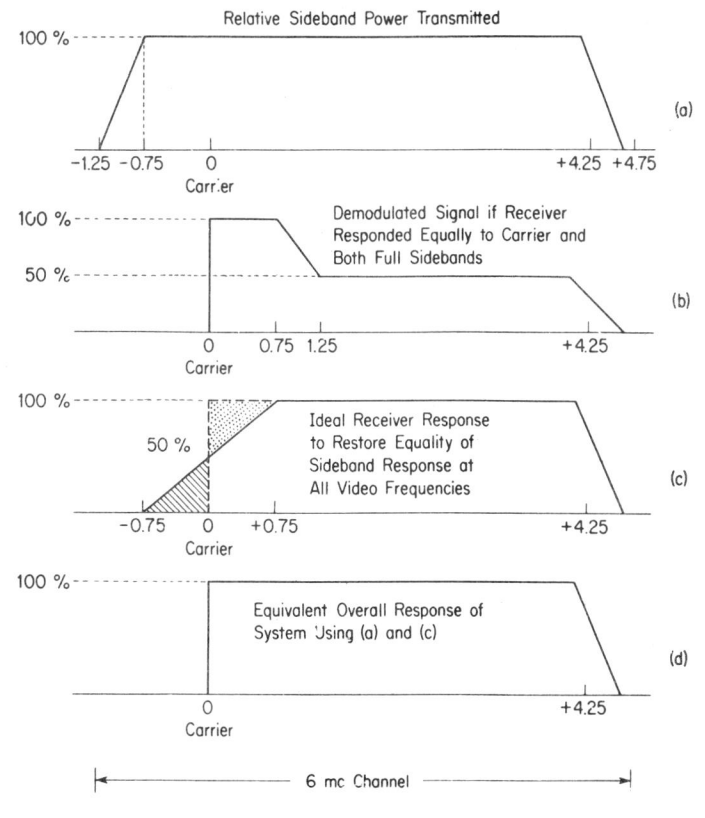

FIG. 15-9. Vestigial sideband transmission

Standard (double-sideband) amplitude modulation requires a channel which contains as many cycles as twice the maximum modulating frequency. To send a 4-megacycle television signal thus requires a radio-frequency channel 8 megacycles in width. It was explained in Chapter 10 that much study was given to methods of minimizing the bandwidth of a television channel. One expedient adopted was the use of vestigial-sideband transmission.

VESTIGIAL-SIDEBAND TRANSMISSION

All of the upper video sidebands are transmitted without alteration. The lower sidebands are transmitted completely only to a frequency of 0.75 megacycle below the carrier in frequency. They are then rapidly attenuated almost to zero at 1.25 megacycles below the carrier. This changes the frequency response at demodulation at the receiver. The response changes because from the carrier frequency to 0.75 megacycle in either direction both sidebands are transmitted and add in demodulation to produce 100% response. Above 0.75 megacycle only one sideband is available so the response from 0.75 megacycle to 4.2 megacycles is reduced to 50%. Equal response at all side frequencies is necessary. This is obtained by modifying the receiver response characteristic so that it is zero at 0.75 megacycle and more below the carrier frequency, 50% at the carrier frequency, and 100% at 0.75 megacycle and higher above the carrier. The sideband power below the carrier (0 to 0.75 megacycle) adds to the sideband power in this same frequency range above the carrier (0 to 0.75 megacycle) to make the sum equal to the sideband power of the single sidebands from 0.75 to 4.2 megacycles. Equality of response is restored at all frequencies at a loss of half the power, which is not important. This may be visualized from Fig. 15-9(c), where the lower sideband power is represented by the hatched triangle. This fills the empty dotted triangle to produce the equivalent response of Fig. 15-9(d) and is called *vestigial sideband transmission.*

VESTIGIAL-SIDEBAND FILTER

Since vestigial-sideband transmission is used, a vestigial-sideband (VSB) filter is often used following the transmitter to eliminate the undesired sideband power at the lower edge of the channel. This filter is constructed of coaxial sections similar to the transmission-line section explained and illustrated at the end of Chapter 12. The response at the lower edge of the channel shown in Fig. 15-9(a) is produced by the VSB filter which is connected directly to the output of the picture transmitter, ahead of the diplexer.

DIPLEXING

At the TV transmitting plant there are two separate transmitters, one for the picture and one for the sound. It is very desirable that the two carriers be transmitted in such a manner that the radio propagation is identical for each or nearly so, to insure that at the receiver both picture and sound signals are received equally well, with the correct ratio of signal intensities. Separate but adjacent transmitting antennas may be used and sometimes are, but ordinarily the two carriers are transmitted from the same antenna. This is accomplished by a filter network called a *diplexer*. This has separate inputs for the picture and sound r-f power but a common output where they appear together. The common output is connected to the r-f transmission lines which convey the signals to the common antenna. From the common transmitting antenna two separate carriers convey the signals to the receiver.

At the receiver a common antenna and transmission line are used also. Separation of the two signals takes place in the receiver after the frequencies simultaneously are heterodyned down to the i-f by means of a heterodyne oscillator and mixer. The sequence is illustrated in Fig. 15-1.

TELEVISION TRANSMITTING ANTENNA

The transmitting antenna must be of a broad-band type to have an essentially constant gain over the 6-megacycle bandwidth. The *horizontal pattern* desired usually is circular and radiation upward is kept to a minimum. Horizontal polarization is standard for both sound and picture radiation; that is, the direction of the electric field set up by the transmitting antenna is parallel with the surface of the earth. Tests have shown that, compared with a vertically polarized signal, horizontal polarization gives better *signal-to-noise ratio*, less *multipath reflections*, and leads to simpler receiving antennas. Mechanically the transmitting antenna must be rigid enough to withstand ice and wind loading and should be self-supporting so that it can be mounted on top of an existing structure, such as a high building or an AM transmitting-antenna tower.

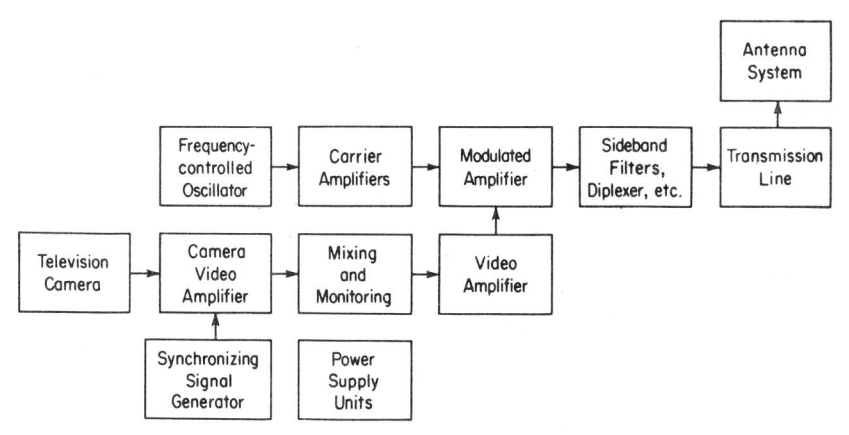

FIG. 15-10. Block diagram of visual trans*x*

TELEVISION TRANSMITTING STATION

The visual transmission chain of a typical television transmitting station is shown in Fig. 15-10. It comprises (a) the television camera; (b) the camera video amplifier; (c) the mixing and monitoring equipment; (d) the transmitter video-amplifying stages; (e) the station frequency-control apparatus; (f) the carrier amplifiers; (g) the modulated amplifier; (h) sideband filters; (i) transmission lines to the antenna system, diplexers, antenna-impedance matching devices, etc.; (j) antenna system; (k) synchronizing-signal generators; (l) operating instruments and test apparatus; (m) power-supply units of various sorts.

TELEVISION CAMERA

The television camera is the eye of the TV system. It has two main duties to perform. The optical system of the camera reduces the scene before it to a small optical image and projects this light image upon a photosensitive surface, where an electrical replica of the light image is formed and scanned electronically. It is thus converted into an electrical (video) signal. The weak output of the photoelectric surface is increased in amplitude by a camera video amplifier and then sent through a cable to the monitoring and control positions as well as to the video amplifiers of the transmitter proper.

TELEVISION CAMERA TUBE

The most widely used type of television-camera tube is the *image orthicon*, which is illustrated in Fig. 15-11. The camera housing normally is equipped with a lens turret carrying several different lenses. These may be used for close-up and long shots; the lenses are mounted on the revolving turret which may be turned conveniently into position to focus an image on the picture tube. A choice is thus available to select the lens of focal length

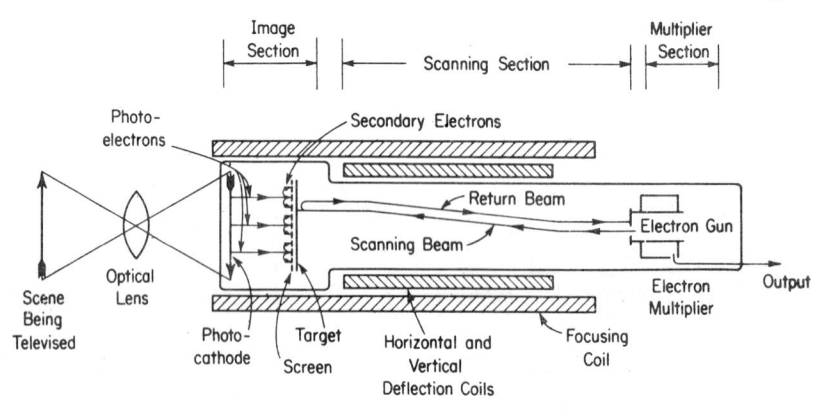

Fig. 15-11. Image orthicon camera tube.

most suitable to meet various picture conditions. The camera contains amplifiers to build up the voltage obtained from the camera tube and also a small kinescope (picture tube) so that the operator may see the picture being produced by his camera. The entire camera mechanism and often the operator, too, ride on a *dolly* equipped with wheels and an elevator device.

In Fig. 15-11 the "scene" is an arrow, head upward. The camera lens (optical) forms an inverted image of the scene through the evacuated glass envelope of the tube upon a photosensitive cathode inside. If the light image is bright at a certain point, the number of electrons released from the sensitive surface at this point is great. At a dark spot almost no electrons are given off. The process is continuous and the released electrons are attracted to the thin conductive-glass target. This is about 300 v positive with respect to the photocathode. The electrons are focused by a longitudinal magnetic field which is established by a long coil outside the glass envelope. When the photoelectrons strike the target they each have enough velocity (kinetic energy) to release several secondary electrons, thus leaving the target with a pattern of positive charges which is an electrical replica of the light image. A very fine mesh screen next to the target collects the released secondary electrons and also acts as the second plate of a capacitor. The back of the target is scanned by the beam from an electron gun (at the right) and just enough electrons are deposited upon the target to neutralize the positive charges. The rest of the electrons in the beam return to the vicinity of the electron gun, where they enter an electron-multiplier amplifier. The target is actually a sheet of glass about 0.001-inch thick which is made so that the charges on opposite sides will neutralize (by conduction through the glass) during the time of scanning but will not spread sidewise toward adjacent parts of the image. The target is at the same potential as the scanning cathode but some 300 v positive with respect to the photosensitive surface. Magnetic deflection is used for scanning, with the coils outside the evacuated glass envelope of the image orthicon. Five stages of *electron multipliers* give a gain of several hundred and reduce the output impedance of the tube to a value convenient for coupling to other circuits. This decreases the effect of thermal agitation and random noise generated in the tube and improves the signal-to-noise ratio. The electron multipliers operate with up to 1,500 v positive potential with respect to the electron-gun cathode.

The output of the camera tube feeds into a video amplifier before being sent over a cable to the mixing and monitoring equipment. The video amplifier must have an output which is substantially constant from 30 cycles to 4 megacycles. Frequency distortion must be kept to a minimum over this great range of frequencies.

In addition it is essential that all the various frequency components of the video signal pass through the video amplifier system in the same elapsed time. This is necessary to preserve the wave shape of the signal. In the ideal case it must take no longer for a 30-cycle component to go through the system than for a 4-megacycle component to do so, nor must it take a shorter time. To illustrate why this must be so, consider a uniform background like the sky, or a plain wall. During a certain part of the program either of these backgrounds may change slowly or not at all. The part of the signal which represents this background lies in the low video-frequency range because of the slow rate of change and uniform character of the background. Fine details, such as the hair and eyes of a subject in a close-up, are carried by the high video frequencies — the light level here varies rapidly from point to point in the picture and the signal output from the image orthicon therefore changes rapidly in the operation of scanning. To keep the fine detail in its proper place relative to the background it is essential that the signal components from both parts pass through the amplifier during equal intervals of time so as to arrive at the output terminals simultaneously. If the high-frequency components are delayed more than a small fraction of a microsecond relative to the low frequencies the fine details in the picture appear to be displaced from their correct location relative to the background upon which they belong.

MONITORING AND MIXING EQUIPMENT

Programs may originate at any one of several points in a studio and the outputs of the various cameras used to cover each scene are fed into the control and monitoring equipment. In addition slide projectors, motion-

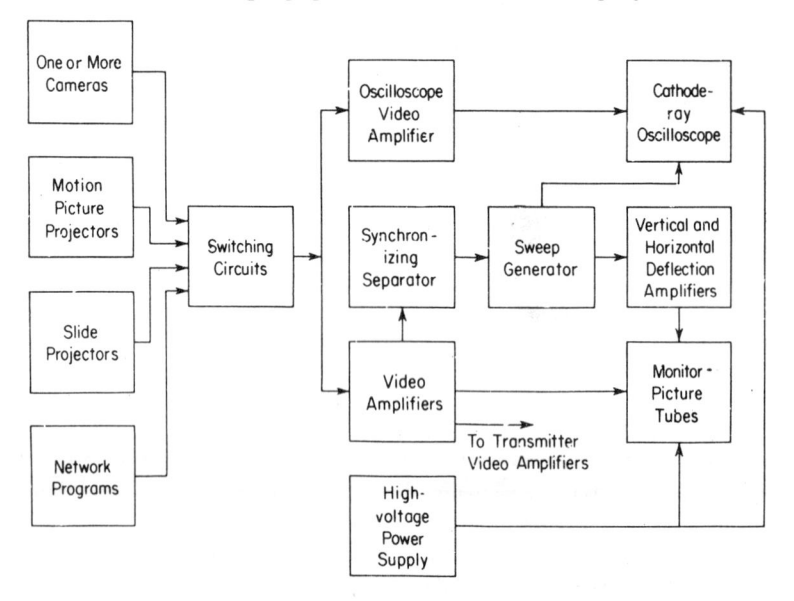

Fig. 15-12. Simplified block diagram of monitoring equipment.

picture projectors, and chain programs from distant locations also feed into this point (Fig 15-12). From these assorted signals the program director and control men select the proper combination of video signals to put on the air, mixing or superimposing there for special effects and cutting from one to another for variety and entertainment value. To aid in this process the control desk has several picture tubes which can be connected into the camera circuit in use or into some other program source. At least one tube shows the picture being transmitted from the antenna. To cut in the different cameras at the correct times the director must watch the action on two or three tubes, switching from long shot to close-up to medium shot according to the script and his judgment, aided by rehearsals. In addition to the picture tube showing the action it is customary to have a standard cathode-ray tube on each channel connected to show the complete (composite) video signal line by line; after some experience the operator can tell a good deal about the general behavior of the system by a look at the lines on the face of this cathode-ray tube.

From the control position the signal is again amplified through a chain of video amplifiers, finally being applied to the carrier through the modulating stage of the transmitter.

NEGATIVE TRANSMISSION

As is shown in Fig. 15-5, negative transmission for video signals is used in the United States. A decrease in the light intensity causes an increase in transmitter power. Stated another way, this means that when the picture is very bright (white) the video amplitude is low; when the picture is dark (or black) the video amplitude is high. An increase in the voltage shown in the diagram (Fig. 15-5) means that the picture is becoming blacker. The black level, at which the scanning beam of the picture no longer strikes the fluorescent screen, is fixed at 75% of the carrier peak voltage (with a tolerance of 2.5% up or down). This value is called the *pedestal level*. A decrease of transmitter power means that the picture is becoming whiter. The whitest part of the transmitted image uses about 12.5% of the full transmitted amplitude.

COMPOSITE VIDEO SIGNAL

The composite television video signal may be built up by the process diagrammed in Fig. 15-13. Line *a* shows the output of the TV camera after being amplified in the camera video amplifier. A set of *blanking pulses b* is added to the camera signal and the combination looks like *c*. The blanking pulses drive the signal to 75% of its maximum voltage after being clipped as shown in line *d* of the figure. Narrower (shorter in time) sync pulses are then added to the clipped wave as in *e* and *f*. These are the sync pulses which trigger the sync generators in the camera and the receivers to start the horizontal sweeps for each line. Since the pedestal level (see diagram

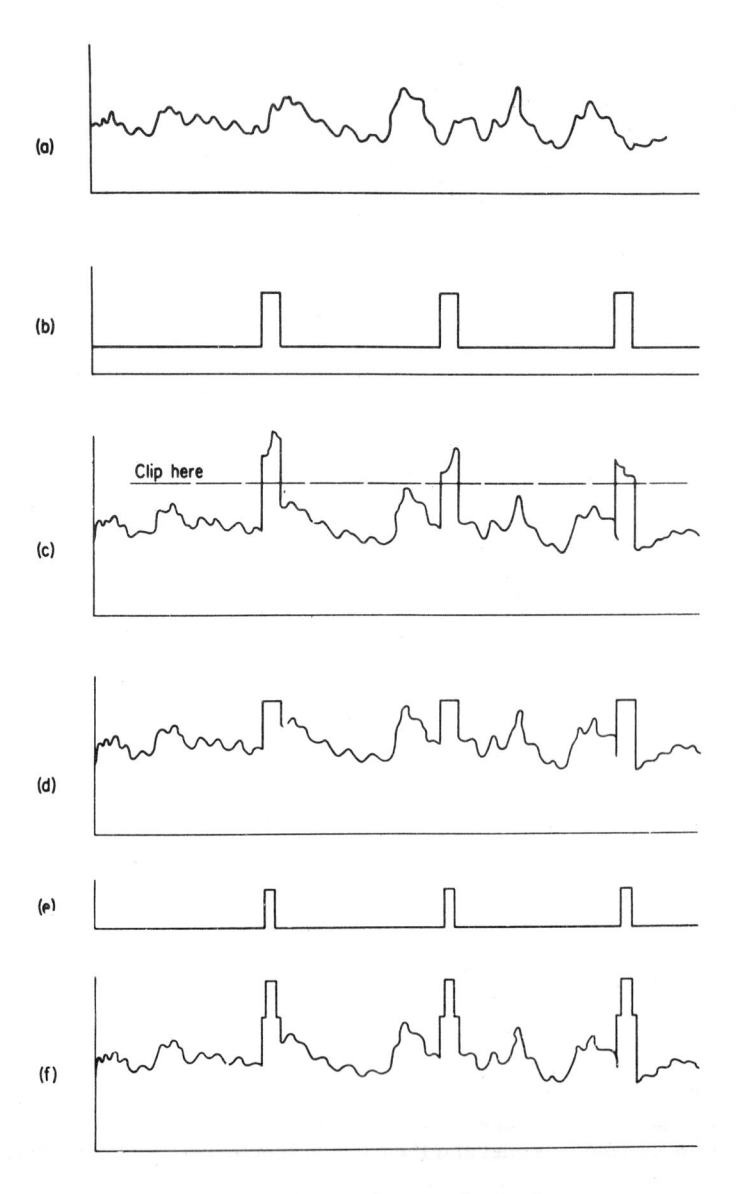

Fig. 15-13. Forming the composite signal.

and Fig. 15-5) represents black in the received picture, the additional voltage (up to 100% of the permissible value of the composite voltage wave at this point in the system) cannot affect the scene as viewed on the receiver screen. The signal is merely driven into the *blacker-than-black* level. The same is true of any noise components which may get into the system during a sync interval. During a line interval noise makes a dark spot in the line. The darkening of the picture is not as easily seen as a white spot would be and this fact is one of the reasons for using negative transmission. The sync-pulse generator must start the pulses at accurate intervals so as to have 15,750 pulses per second. The front porch is included in the signal to make sure that the beam is blanked out before the return sweep begins.

TELEVISION RECEIVERS

Details of the circuits and construction of television receivers vary with different manufacturers but a typical home television receiver includes the items shown in block form in Fig. 15-14. These comprise (a) the television receiving antenna; (b) receiver transmission line; (c) input r-f amplifier; (d) local oscillator; (e) mixer; (f) video i-f amplifier; (g) video detector; (h) video amplifier; (i) picture tube and its control circuits; (j) sync-

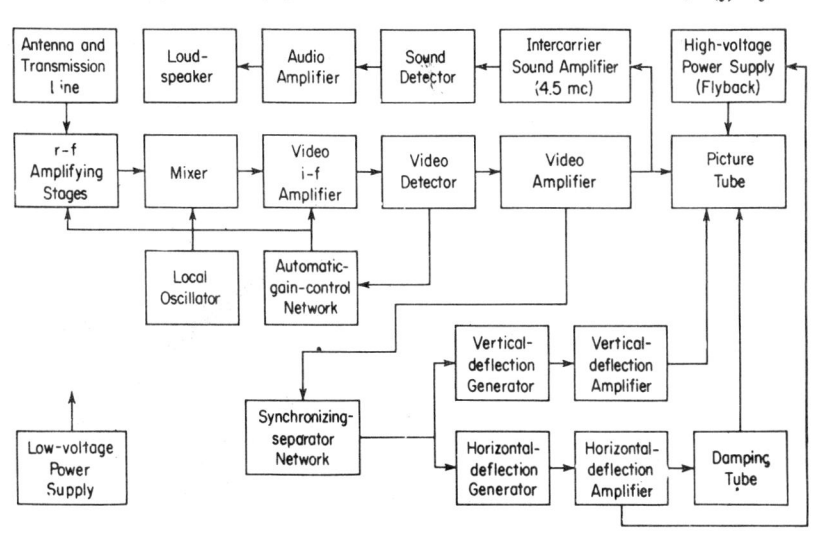

FIG. 15-14. Block diagram of a typical television receiver.

signal separating circuits; (k) vertical scanning (sweep) circuits; (l) horizontal scanning (sweep) circuits; (m) audio i-f amplifier, limiter, and detector or (usually) intercarrier sound amplifier and detector; (n) audio amplifier; (o) automatic-gain-control (a-g-c) circuits; (p) loudspeaker;

(q) high-voltage power supply; (r) medium-voltage power supply; (s) controls — some fixed, some variable, others adjustable at the time a receiver is installed at a given location.

The controls usually provided in a TV receiver are included in five categories: (a) general; (b) sound channel; (c) picture channel; (d) line scanning; (e) field scanning.

Tuning and sound channel controls are similar to those of a conventional broadcast or FM receiver. The picture channel is controlled by the contrast control, brightness control, and focus control. The contrast control varies the gain of the video i-f amplifier and changes the black-and-white contrast between the various parts of the picture. Advancing the control increases the contrast between the light and dark portions of the picture and causes a general increase in the illumination of the picture-tube screen. This may be necessary when the program is being viewed in a room where the general illumination level is high. The background or brightness control regulates the average brightness of the picture-tube screen and is designed primarily to prevent the electron beam from being visible when returning to start another scanning sweep. This control is usually brought out on the rear of the receiver since the user will not need to adjust it in ordinary use of the set when in proper operating condition. The brightness control governs the bias voltage between the grid and cathode of the picture tube. The focus control adjusts the size of the spot which the electron beam makes bright on the screen and is usually set for the smallest spot which can be obtained. This setting should give the finest detail in the picture.

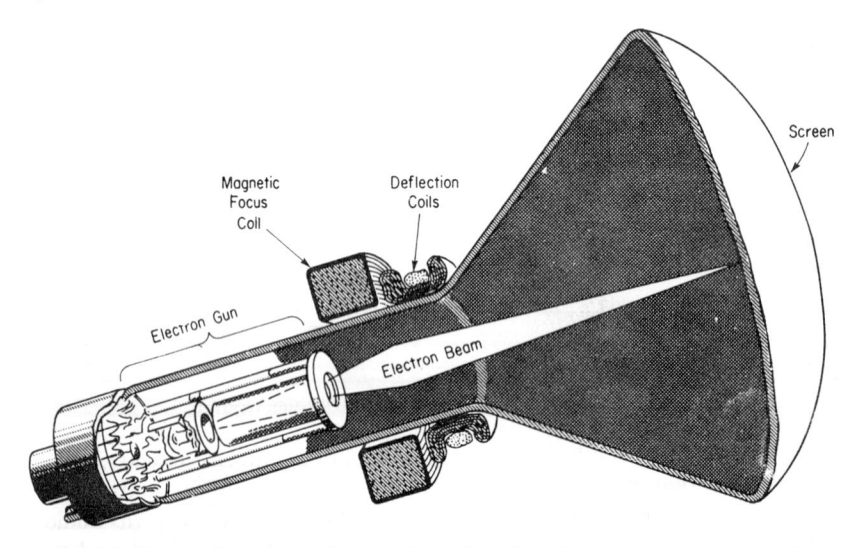

FIG. 15-15. Cross-sectional view of a picture tube.

PICTURE TUBE AND CONTROLS

Basically the picture tube is a cathode-ray tube of specialized design. In size it may reach a maximum screen dimension of two feet or more. Its phosphor screen glows under the impact of the electrons in the scanning beam to give off brilliant white light. The scanning beam can be focused to a very fine spot and retains this sharpness over a wide range of control-grid voltage. Stray light from the cathode heater and from reflections on the inside of the glass envelope are practically eliminated by a thin layer of aluminum, applied after the phosphor is in place. Aluminizing in this way also prevents the formation of the *ion spot* in magnetically focused tubes. This spot is a darkening of the phosphor in line with the axis of the electron gun caused by bombardment from heavy ions among the electrons in the scanning beam. The defect usually occurs near the center of the screen. Since the ions cannot penetrate the aluminum layer, no ion-spot darkening takes place. Permanent-magnet ion traps also are used to remedy this trouble.

SYNC-SEPARATING CIRCUITS

The output of the video detector or that of any of the later amplifying stages is (ideally, at least) an exact reproduction of the composite TV signal at the transmitter. The sync-separating circuits have to perform two principal tasks. The first is the separation of the sync pulses from the part of the composite signal which constitutes the picture itself; that is, the sync pulses must be clipped off the composite signal and the picture components discarded. The second function is to separate the vertical sync pulses from the horizontal ones and to deliver each pulse to its appropriate deflection-signal generator at precisely the correct instant.

FLYBACK HIGH-VOLTAGE SUPPLY

When a steady current flows through an inductor, energy is stored in the magnetic field. The amount of stored energy is $LI^2/2$, where L is the inductance of the coil and I the current flowing at the instant considered. If no current is flowing there is no energy storage in the magnetic field. Now if a current which has been flowing in a coil is suddenly interrupted, the stored energy in the magnetic field must be disposed of somewhere and this often occurs in the spark which forms when the circuit is broken. The voltage which is induced at the break depends upon the suddenness of stopping the current; the smaller the time required to reduce the current to zero the higher the induced voltage. Breaking a circuit which contains an inductance coil, therefore, is one source of a high voltage. A low-voltage source may be used to furnish the steady current which is interrupted.

In the vertical-deflection coils of the picture tube the rate of change of current is slow enough that no very high voltage is likely to be induced. In the horizontal-deflection coils the situation is different because of the rapid sweep and return required for the horizontal deflection of the electron beam. In a typical case the peak-to-peak deflection current is about 1 ampere and this must drop to zero in about 7 microseconds — that is, in less than the minimum horizontal blanking time of 10.5 microseconds. The average rate of change of current is $1/(7 \times 10^{-6})$ or more than 140,000 amperes per second. Since the voltage across the coil is equal to the product of the inductance and the rate of change of current, the resulting voltage across the terminals of a 14-millihenry coil (a typical value) is some 2,000 volts. When stepped up by the matching transformer between the plate of the scanning output amplifier and the deflection coil (assuming a turns ratio of 3 to 1) this voltage becomes 6,000 on the plate of the output tube.

The energy in the magnetic field of the horizontal-deflection coils is often used to provide the high voltage for the anode of the picture tube. A circuit for this purpose is shown in Fig. 15-16. The voltage developed by interrupting the current in the horizontal-deflection coil is applied to the primary

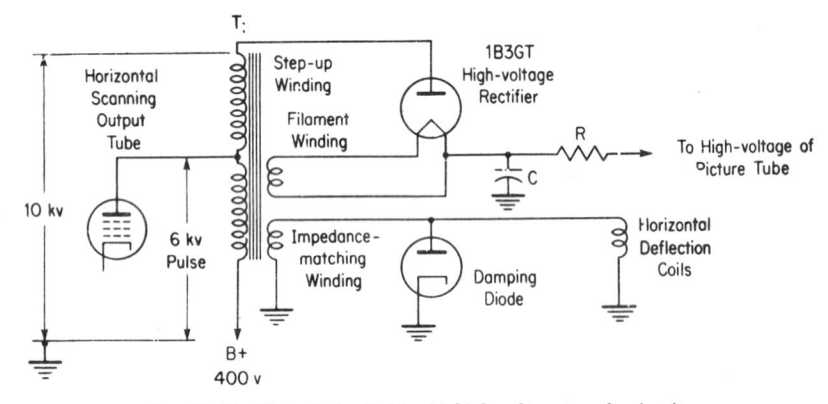

FIG. 15-16. Flyback (or kickback) high-voltage supply circuit.

winding of transformer T_1 through the impedance-matching winding (some 6,000 volts). Another section of the primary winding steps this voltage up to about 10,000 volts by transformer action, and applies it to the plate of a high-voltage rectifier tube (1B3GT/8016, for example). The cathode of this rectifier is heated by connection to another winding on the *flyback* (horizontal-sweep output) transformer. Between the cathode and ground of the system a voltage in the neighborhood of 10,000 volts pulsating d-c appears. This is used on the second anode (*ultor*, or highest-voltage anode) of the picture tube after it has been filtered with a simple R-C network. Since the current required is very small (a few microamperes) the system functions very well and forces the energy stored in the magnetic field to perform a useful service instead of being thrown away.

The flyback high-voltage power supply has very poor regulation but this is an advantage rather than a drawback. A supply with poor regulation is not likely to deliver a lethal shock to a person working on the set. Of course a high voltage should always be disconnected when there is any danger that persons may come in contact with it. All flyback high-voltage power supplies remove the voltage from the highest-voltage anode of the picture tube if the horizontal deflection current fails. This is a safety feature for the picture tube and prevents the undeflected electron beam from burning the fluorescent screen.

Even higher voltages may be obtained from a flyback power-supply unit by using voltage-doubler or voltage-tripler circuits.

DAMPING TUBE

The deflection coils of the picture tube have a certain amount of capacitance between the turns of wire, and there is capacitance with respect to ground as well as inductance and resistance. Unless the resistance is comparatively high (which means a poor design) the coils may form part of an oscillatory circuit that can be shock-excited by the deflection voltage. This results in distortion of the saw-tooth scanning wave if it occurs and makes the *raster* appear uneven. Additional resistance will damp out the oscillation but will also waste energy from the scanning output amplifiers. What is needed is a nearly instantaneous switch to connect a resistance in the circuit only when required and to keep it out of the circuit the rest of the time. A diode (*damping diode* in Fig. 15-16) is almost always used to do this, and is connected so that a large current flows through it only during the time when its anode is positive with respect to its cathode. This additional loading of the circuit with what amounts to resistance of a low value (the plate resistance of the diode) prevents oscillation. This tube is the damping diode.

REINSERTION OF THE D-C COMPONENT

Most video amplifiers use capacitors for coupling between the plate of one stage and the grid of the next; no d-c component of the video signal can pass through such an amplifier. For this reason it is desirable to have a means for *reinsertion of the d-c component* at the grid of the picture tube. Suppose that in Fig. 15-17(a) and (b), two lines of a certain picture are shown together with the corresponding sync pulses. The d-c or average value is the axis through the wave about which the amount of shaded area above and below is the same, as indicated in the diagram. In Fig. 15-17(a) is a part of the scene which is nearly white, and Fig. 15-17(b) represents a line which is nearly black. When these two signals have passed through an amplifier with capacitor coupling, for example, the d-c value becomes the reference axis instead of the peak values as in (a) and (b). Both lines contain approximately the same detail but (a) has a brighter (whiter) background than (b). The latter is darker because the average level of the signal is nearer the black level (the front and back porches of the signal, just where the sync pulses begin; see Fig. 15-5). The camera signal variations, which give detail in the picture, are a-c components; the average value is the d-c component. At the video detector output all the a-c and d-c components appear but the d-c component is lost when the signal passes through the R-C coupled video amplifier stages. The output of this amplifier has the same area between the signal wave both below and above the axis as in

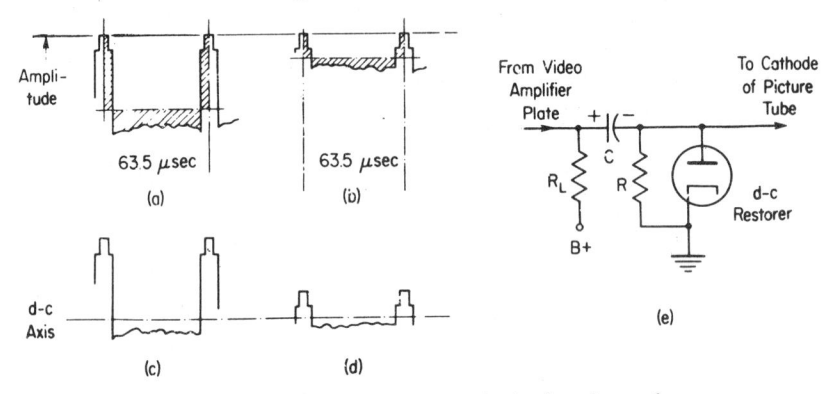

FIG. 15-17. Reinsertion of d-c component; circuit and wave forms.

Fig. 15-17(c) and (d). This means that the average background is darker than it should be. Also the blanking and sync pulses are not aligned to the same level; this is likely to result in poor synchronization. Some of the return traces show on the screen because not enough voltage is developed to drive the grid of the picture tube to cutoff.

One method of d-c reinsertion which is widely used has a diode to charge a capacitor to the peak value of the video signal and then hold it there by allowing the capacitor to discharge only very slowly. Such a circuit is diagrammed in Fig. 15-17(e). At each horizontal sync pulse the capacitor C is charged with the polarity shown substantially up to the peak of the sync pulse. The resistance of the diode when conducting is very low and the load resistor for the last video stage (R_L) is also low. The shunt resistance R is too high to have any effect during the charging operation which takes place during the time of a single sync pulse or about 5 microseconds.

As soon as the amplitude of the video signal drops from its peak value the diode ceases to conduct (its plate is now effectively negative with respect to its cathode) and capacitor C begins to discharge through R. The time constant RC is made large enough (typical value 0.05 second) that C is not discharged to any appreciable extent until another horizontal sync pulse appears and recharges it to the peak of the video wave. Each pulse furnishes just enough charge to replace that which has been lost during the line-scanning time. Of course all this has no effect upon the visible picture because the electron beam is blanked out during the sync pulse. The correct background is restored, however, by the clamping effect of the d-c reinsertion diode, as just outlined. There are no series capacitors between the d-c restoration circuit and the picture so that the average brightness is controlled and restored to the value required by the composite TV signal at the transmitter. This type of d-c reinsertion is sensitive to noise pulses. Since the capacitor charges rapidly and discharges slowly it will charge on a noise peak just as well as upon a sync pulse. The effect of the noise can be reduced somewhat by clipping all signals which exceed the sync pulses in amplitude.

If the grid of the picture tube is driven slightly positive by the sync pulses the grid will draw current just as an amplifier tube does. By using a grid leak for picture-tube control-grid bias the grid acts as a diode in the reinsertion circuit described above. The coupling capacitor charges to the peak of the sync pulse and the grid leak tends to hold the charge during the line-scanning intervals. This causes clamping and reinsertion of the d-c component without the use of an extra tube.

INTERCARRIER-SOUND RECEIVER

Figure 15-1 shows a receiver in which the picture and sound are separated after the first detector. Another widely used type of receiver does not separate them until they have passed through a common i-f system. The sound signals and picture signals may be detected in separate detectors. The type in which all signals pass through the same second detector will be described here.

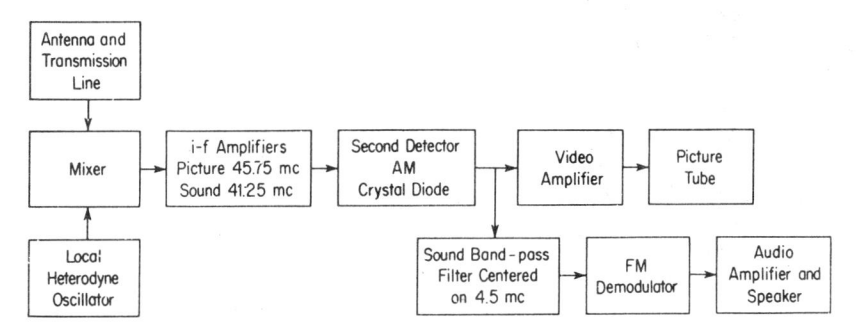

FIG. 15-18. Block diagram of intercarrier-sound receiver.

The picture signal is demodulated conventionally. The frequency-modulated sound signal is removed from the output of the second detector through a narrow band-pass filter centered on a frequency of 4.5 megacycles, the difference frequency between the sound carrier and the picture carrier. The 4.5-megacycle carrier difference is frequency-modulated with the original sound signal. This 4.5-megacycle frequency-modulated carrier is separated from the picture carrier, amplified, and applied to an FM detector in the conventional manner. After recovery in the detector the sound signals are amplified by the usual audio amplifier stages. A circuit to remove the 4.5-megacycle signal is shown in Fig. 15-19. The 4.5-megacycle band-pass filter is a series-tuned circuit connected to ground and tuned to the sound carrier frequency. This provides a short circuit on the

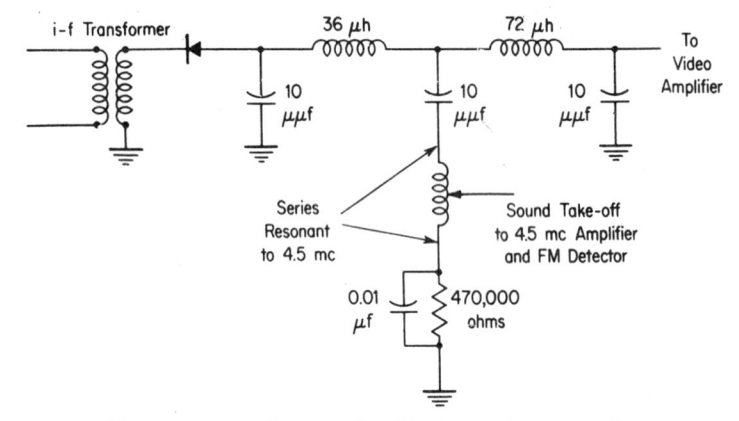

Fig. 15-19. Sound power take-off in intercarrier-type receiver.

video system at and near 4.5 megacycles and thus removes any sound carrier which might cause modulation interference in the picture. The series-tuned circuit is tapped to provide 4.5 megacycles for the sound channel.

TYPICAL TELEVISION RECEIVER CIRCUIT

A circuit diagram of a typical television receiver is shown in Fig. 15-20. The incoming signal from the antenna is applied to the terminals at the left of L_2. The coils L_2 and L_5 may be pretuned and arranged in a turret or otherwise switched to bring in the desired TV channel. The first r-f amplifying tube (6BZ7) is connected in a cascade-type circuit and feeds through L_5 to the mixer stage, which is one half of a 6J6 tube. The other half of this tube functions as the local oscillator and a small variable capacitor (marked *fine tuning*) adjusts the frequency as required for the best reception of the program. The output of the mixer stage, now at the video intermediate frequency, is amplified by the video i-f amplifier, comprising three stages with 6CB6 pentodes and one stage with a 6AU6 pentode. The video detector (a 1N64 diode) feeds into the video amplifier and the intercarrier-sound i-f amplifier. The video amplifier consists of a single 12BY7 pentode which drives the cathode of the picture tube and controls the intensity of the electron beam in it. The *contrast* of the picture is controlled by a variable resistance in the cathode circuit of this tube.

The sound i-f amplifier includes two stages of 6AU6 pentodes and a ratio detector (6AL5). The ratio detector feeds into a conventional a-f amplifier and loudspeaker.

The video output tube also drives the 12AU7 sync clipper and amplifier which excites a 6C4 sync phase inverter. The output of the latter goes to the vertical oscillator which drives the vertical deflection output stage (6S4). The sync phase inverter also drives the horizontal phase detector and the horizontal multivibrator (6SN7). The horizontal multivibrator feeds into the grid of the horizontal output amplifier (6CU6) which is coupled to the horizontal deflection coils through transformer T_3. T_3 also serves as a flyback high-voltage supply. The damper tube, a 6AX4 diode, and the horizontal linearity control are also in this circuit. Both the vertical and horizontal deflection coils are shown in the box marked T_5.

The low-voltage power-supply unit is conventional and a keyed automatic-gain-control circuit tends to keep the signal level constant in spite of changes in radio-wave propagation or other varying factors.

Many other combinations of circuit components may be and in fact have been used to perform the functions just described. The final choice of a suitable circuit often depends upon cost as well as engineering considerations.

LARGE-SCREEN AND THEATER TELEVISION

While picture tubes have been manufactured in increasingly larger sizes as commercial television has developed, it seems unlikely that they will ever be built large enough to compete in size with the motion-picture screen in a theater. To meet this need projection-type picture tubes have been used with a suitable optical system to give full-screen images to be viewed by an entire theater audience. The tube used is similar to a standard cathode-ray tube but the accelerating voltage is about 70 kv for a 7-inch tube and the electron current is correspondingly great. Naturally the image is extremely brilliant. Enlarged by a lens or mirror optical system this image can fill a 9 by 12 foot screen.

The *Schmidt optical system* utilizes a spherical mirror with a correcting lens. The mirror surface may be made from glass in a suitable machine.

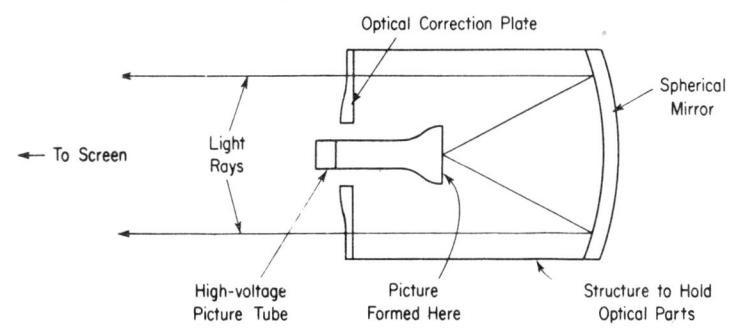

FIG. 15-21. Cross-sectional diagram of the Schmidt optical projection system.

The correcting lens is thin in cross section and is sometimes molded from plastic instead of optical glass. A cross section of a Schmidt optical system as modified for television-picture projection is shown in Fig. 15-21.

INDUSTRIAL AND CLOSED-CIRCUIT TELEVISION

Instead of broadcasting TV signals superimposed upon carriers through the air it is possible to connect the transmitting and receiving apparatus through a suitable cable or other type of wire connection. Such a system has been named *closed-circuit* or *jeep* television. There are many applications of closed-circuit TV in industry and other fields. A few of the uses which may be listed are checking numbers on cars in a railroad yard, carrying on inspection or work with radioactive "hot" material, inspecting a steel furnace during operation, watching prisoners in jail, detecting forest fires, searching under the sea, observing a surgical operation, and many other similar situations where it would be dangerous, difficult, or impossible to view the process in person. Less expensive camera tubes as well as simpler transmitting and receiving equipment than used for broadcasting can be employed — at least in some applications. Television will do an increasing number of similar jobs as time goes on.

TELEVISING MOTION-PICTURE FILM

Motion-picture film is a source of TV program material. A show can be filmed and then stored or shipped to various points for presentation at the convenience of the transmitting station or sponsor.

Standard theater-type motion pictures also can be used as program material. In all cases either 35-mm or 16-mm film must be scanned for transmission.

Motion pictures are projected at the rate of 24 frames per second and a light chopper disk gives the effect of showing 48 frames per second to reduce flicker effect. Television standards in this country call for 30 and 60, respectively, instead of the number of frames used with film. In addition, the film must be pulled down during the blanking interval to show the next frame and it must be held steady during the scanning period.

The storage properties of a camera tube are utilized in scanning motion-picture film; the electrical charges proportional to the light image are allowed to build up on a photosensitive surface and then are scanned and erased by an electron beam. The need to scan 24 frames per second in 60 time intervals per second is met by a sequence of scanning half the frames 2 times and the other half 3 times per frame. Scanning half the frames (12) twice gives 24 images and scanning the other half (12) three times each gives 36 images or a total of 60 per second as required. The operation is carried on by an ingenious mechanism in the motion-picture projector used at the television station.

QUESTIONS AND PROBLEMS

1. What is a picture element? How many picture elements are there in a 441-line picture with a 4/3 aspect ratio? Neglect time required for blanking and sync pulses.

2. What is the effect of a nonlinear sweep voltage upon the picture?

3. What is the purpose of the horizontal sync pulses? What is their frequency? What is the purpose of the vertical sync pulses? What is their frequency?

4. State the primary purpose of interlacing in the scanning process.

5. Calculate the speed in inches per second of the scanning spot on the face of a 21-inch picture tube during the forward sweep of a horizontal line. Recalculate this speed, taking into account the time of scanning lost owing to retrace interval.

6. Draw a diagram of one line as it might be seen on a cathode-ray 'scope and include the blanking and sync pulses for (a) a completely black part of the picture; (b) a white part; (c) a gray picture with a black bar in the center; (d) a gray picture with a white bar in the center.

7. What is meant by negative polarity of transmission? Name one advantage and one drawback of this system.

8. What is the purpose of the front porch in the composite TV signal?

9. Draw a simple block diagram showing the main pieces of apparatus required to transmit (a) a program where news is read by the announcer; (b) a dramatic play with at least four actors; (c) a disaster in a location remote from the studio or transmitter.

10. Describe in general terms the functions of the different parts of a TV receiver.

11. Name the essential parts of a TV picture tube (monochrome). How does it differ from the tube used in the 5-inch laboratory-type oscilloscope?

12. What is the effect of the loss of high video frequencies upon a picture?

13. What is meant by compensation in a video amplifier?

14. What is the highest video frequency in the standard monochrome picture transmission in the U. S.? What is the lowest?

15. What visible effect do the sync pulses have upon the picture?

16. How is the electron beam blanked out during retrace time in the picture tube?

17. What effect on the picture does the loss of horizontal sync pulses have? The loss of vertical sync pulses?

18. What is the time constant of a 1-megohm resistor in series with a 1-μf capacitor?

19. What is the function of an electronic differentiating circuit? An electronic integrating circuit?

20. Why is a saw-tooth wave of applied voltage not suitable for magnetic-deflection picture tubes? What must be done to obtain correct deflection?

21. What is the maximum ultor voltage used in a current-model TV set? (Look this up in a tube manual.)

22. Why is a damping diode used?

23. What are the advantages of a low intermediate frequency? Of a high intermediate frequency?

24. What is the result of radiation from a local oscillator? Why?

25. A ghost on a raster 18 inches wide is displaced 1.8 inches. What time delay does this represent? What additional distance has the ghost signal traveled compared with the direct radiation?

CHAPTER 16

COLOR TELEVISION

INTRODUCTION

The development of color television has followed black-and-white television just as color printing, color photography, and color motion pictures have come after their black-and-white counterparts. Color-TV receivers and transmitters have a good many more tubes and other components than monochrome equipment because the color system is inherently more complicated. However, it can be separated into smaller sub-systems and each one of these can be studied separately. In this way much of the material already learned can be put to use.

SOME PROPERTIES OF LIGHT

When sunlight (that is, white light) passes through a glass prism or through drops of water in the earth's atmosphere we may see a spectrum or rainbow. This consists of a succession of colors of light which go by im-

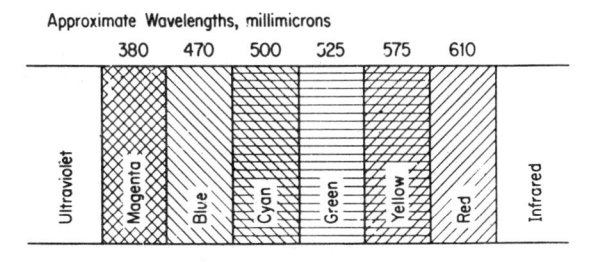

FIG. 16-1. Some colors of the spectrum.

perceptible steps or gradations from violet to red. Some of these colors are indicated in Fig. 16-1 by the hatching on the different sections of the figure. White light can be split into bands of colors arranged in order.

The splitting process can be reversed and by mixing light of three primary colors (such as red, green, and blue) white light can be formed again. This is called *additive color mixing* and must be done with colored light, not paint or pigments. Figure 16-2 shows the combination of light from red, green, and blue sources. In addition to the white in the center yellow is formed by a combination of red and green, magenta (a shade of purple) by combining red and blue, and cyan (a greenish blue) by mixing blue and green. The light source for the red light may be some chemical or electrical source which gives off light of the required primary color or it may be a white-light source with a piece of red glass (an optical filter) in front of it. The ideal optical filter for red transmits only the red portion of the

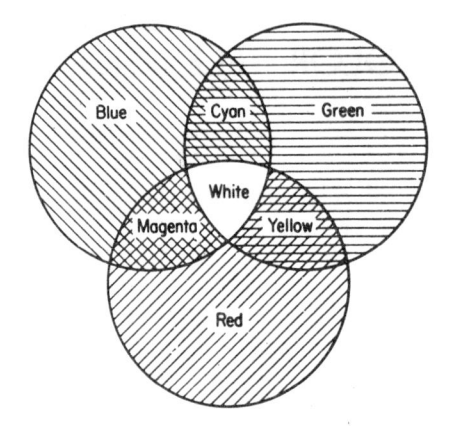

FIG. 16-2. Additive primaries and their combinations.

spectrum and absorbs light of all other colors. Similar green and blue filters can be used over the other sources. The three primary colors used here — red, green, and blue — are called additive primaries because light of these three colors can be added to make white light and, in fact, that of almost all other colors if used in the correct proportions.

Most colors which we see on objects around us are due to reflection of a portion of the white light which strikes the object. If we look at a yellow object through a piece of glass (which is the same sort of yellow in color) the object appears to be yellow. If we look at a blue object through the same yellow filter the blue light reflected from the object is absorbed by the yellow filter and the object appears to be black. Black is no light; that is, the absence of light. If we look at a white object through the yellow glass the object appears to be yellow because all of the reflected light except yellow is absorbed by the yellow optical filter.

PRIMITIVE COLOR-TELEVISION SYSTEM

A primitive color-television system might be made of the above basis as shown in Fig. 16-3. The "scene" is a chart showing some of the spectrum colors and the three cameras with red, green, and blue optical filters, respectively, in front of their lenses scan the entire chart in unison. When the three cameras are focused on the blue segment of the chart the blue camera has a maximum output and the red and green cameras none. When the cyan (greenish-blue) segment is being scanned the blue and green cameras have about equal output (but less than maximum) and the red camera has none. The output for the other colors in the chart is made up in a similar way.

At the receiving end of the system — to the right in the figure — three projectors give light of red, green, and blue color to illuminate the viewing

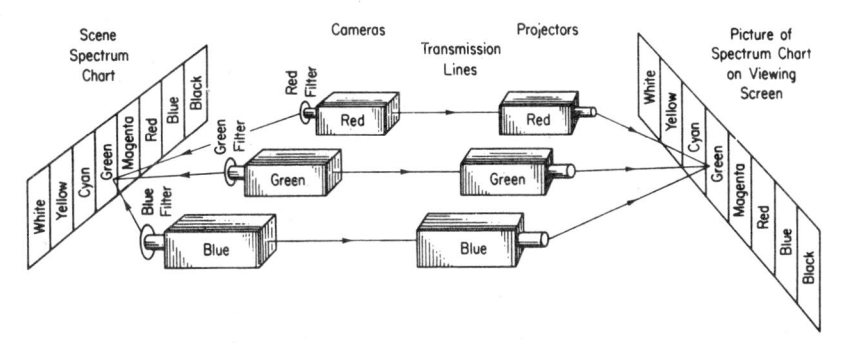

Fig. 16-3. Primitive color-television system.

screen. These sources might be cathode-ray tubes with red, green, and blue phosphors; the images they form are projected through suitable lens systems so as to be superimposed on the screen. The projectors are also arranged to scan the picture area in synchronism with the cameras. When the blue portion of the spectrum chart is being scanned the blue camera sends a signal to the blue projector and the picture on the viewing screen is blue; the other two projectors are shut off. When the cyan segment is being scanned, the blue and green projectors send light to the viewing screen and the red projector is dark. When the cameras scan a white portion of the scene all three projectors operate to give a white picture as indicated in Fig. 16-2. A black segment represents no light in any camera and so no light comes from any projector.

COLOR IN TELEVISION

To understand modern color television it is necessary to study color itself in more detail. The word color has at least three meanings depending upon who uses the term. To the chemist color is a dyestuff or pigment, a chemical compound or mixture like white lead, or vermillion, or one of the coal-tar dyes. To the physicist color means light of a certain wavelength or group of wavelengths; that is, waves occupying a certain portion of the electromagnetic spectrum. The psychologist defines light on a subjective basis — as the sensation which the dyestuff or the spectrum wavelength produces in the brain of a human observer. Television is concerned with color as defined by the physicist and the psychologist to a greater extent than as it is defined by the chemist. All three definitions may be involved in some situations.

Color may be defined as the selective response of the human eye to light. Light is electromagnetic radiation of certain wavelengths of which about one octave (a ratio of 2 to 1 in wavelength) is visible. The wavelengths to which the human eye responds range from about 385 to 760 millimicrons. A millimicron is 10^{-9} meter.

Rather than use the spectrum to indicate the relationship among colors it is convenient to utilize other means. Figure 16-4 shows a *chromaticity* (color) diagram which is used (by international agreement) in the study of color. Along the heavy curved line are the names of a few colors and their

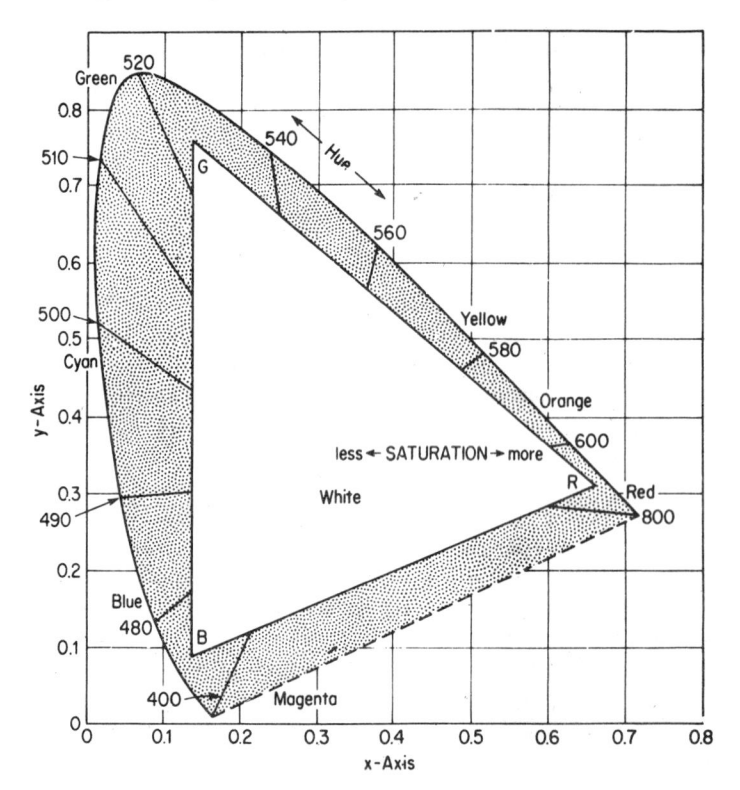

FIG. 16-4. Chromaticity diagram.

approximate locations on the spectrum. The numbers on the curve are the wavelengths of the light in millimicrons. Colors which lie on this curved line are called pure-spectrum or saturated colors. They consist of light vibrations of a single wavelength or a group of wavelengths centered about a dominant wavelength.

White light, which is perceived when the proper combination of two or more (up to all) the colors of light in the spectrum are viewed, is about the center of the area inside the curve, as indicated. The area surrounded by the curve represents all the colors which appear in nature.

The term *hue* is used to refer to the wavelength or combination of wavelengths of which light is composed; hue indicates the redness, yellowness, blueness, etc., of the light. Some hues are indicated around the curve in Fig. 16-4.

The term *saturation* in color-TV language refers to the degree to which the hue is mixed with white light. In Fig. 16-4 a line drawn from the point marked red inward to white describes a series of colors whose hue is red but whose saturation decreases as white is approached. Colors near the white point in the figure are not highly saturated. Colors near the edge of the area inside the curve are more nearly saturated than those near its center.

The term *luminance* is used to describe the brightness or amount of radiated energy in light. The greater the brightness of a light source the greater its energy or luminance. Luminance is not shown on the chromaticity diagram. It may be measured at right angles to the plane of the diagram and a three-dimensional model is required to show chrominance and luminance in the same figure.

Light of a given color can be completely defined by stating its hue, its saturation, and its luminance. Most of the work in the study of color has been experimental and the results are based upon the average of a large number of observations. It can be stated that the total luminance of a colored light is the sum of the luminances of the component lights mixed to produce the given color. Changing the luminance does not appreciably change the appearance of a color over wide limits of luminance.

The eye apparently does not separate the parts of a mixture of light; that is, red light and green light when mixed appear to the eye as yellow light and not as separate colors. In general not more than three differently colored light sources are required to duplicate a given color.

In addition to their position relative to the heavy line marked with the wavelengths of the spectrum in Fig. 16-4, colors may be located in terms of the x- and y-coordinates along the edges of the diagram, running from $x = 0$ to about 0.7 along the bottom and from $y = 0$ to 0.9 on the left side. The chromaticities (colors) of the three primaries specified for color television are

Primary	x	y
Red (R)	0.67	0.33
Green (G)	0.21	0.71
Blue (B)	0.14	0.08

These points are the corners of the inner triangle in Fig. 16-4, marked R, G, and B.

When one unit of red with luminance 0.30 is mixed with one unit of green with luminance 0.59 and also with one unit of blue having luminance of 0.11 the result is white (at $x = 0.310$ and $y = 0.316$ on Fig. 16-4). This white is approximately the color of light from a north sky — slightly bluish white — and it is called *illuminant C*.

A color television system using these three primaries can reproduce any color inside the *RGB* triangle (*color triangle*).

COMPATIBILITY IN COLOR TELEVISION

To avoid making all monochrome sets "blind" to the color pictures, to fit within the existing channel assignments, and to receive monochrome pictures on color receivers the color system is made *compatible* with the monochrome system. This means that:

1. A monochrome receiver can pick up a color transmission and reproduce it as a black-and-white picture.
2. A color receiver can bring in a monochrome signal and reproduce it as a black-and-white picture.
3. A color receiver can reproduce a color broadcast in color.

All of these operations occur without any modification of either type of receiver.

The color information is not divided into three color images (red, green, and blue) as suggested in the primitive system illustrated in Fig 16-3 and then transmitted either simultaneously or in succession. Instead of this a *luminance* (brightness) *signal* and a *chrominance* (color) *signal* are used. The luminance signal carries all the information needed by monochrome

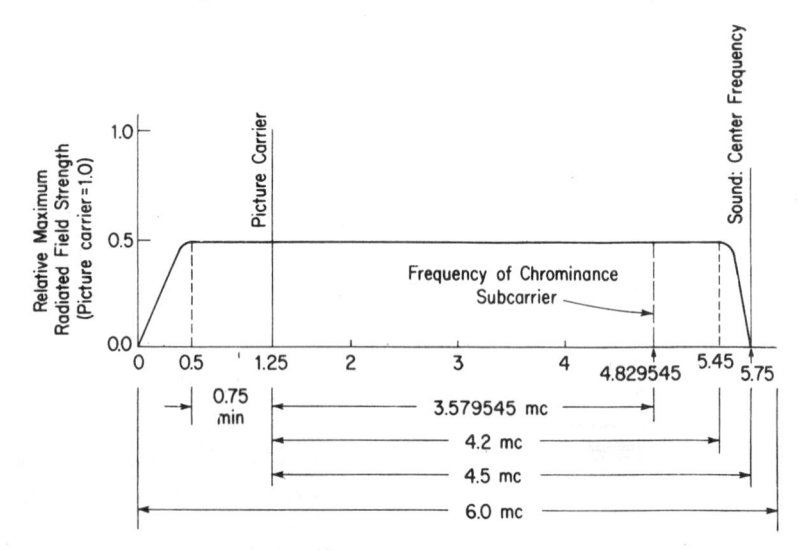

Fig. 16-5. Idealized picture transmission characteristic, showing relative positions of carriers in television channel.

Chapter 16   *Color Television*   531

sets for their operation and it tells the color set how bright a particular spot in the picture must be. The chrominance information is sent in such a way that the monochrome set ignores it entirely. The color-TV receiver extracts the necessary data from the chrominance signal to reproduce the color in the picture correctly.

The additional chrominance information for the color receiver must be transmitted without increasing the channel beyond the 6-megacycle width required by the monochrome system. A process called *frequency interleaving* or *interlacing* permits this.

THE FREQUENCY-INTERLEAVING PRINCIPLE

To transmit a color-TV signal it is necessary to use one carrier for the picture modulation, another for the audio modulation, and then to find a place to put the chrominance modulation. What is called a *subcarrier* is used for the last requirement. The picture carrier is located at a nominal 1.25 megacycles from the lowest frequency in any given TV channel as shown in Fig. 16-5 and amplitude modulation is used to place the video modulation on the carrier. Negative polarity (just as in the monochrome

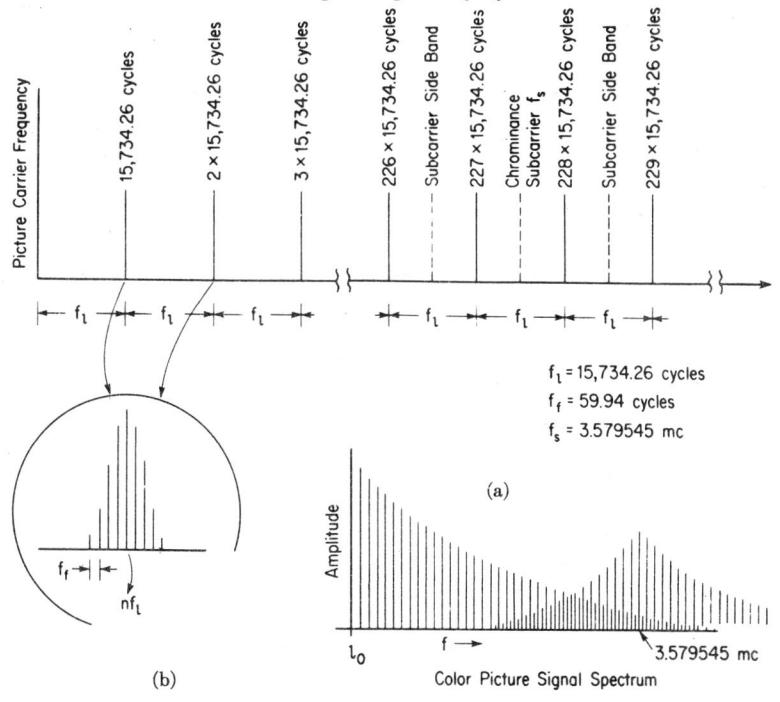

Fig. 16-6. Energy distribution in a TV channel. (a) Spectrum of picture channel showing frequency interleaving; (b) detail of spectrum centered around multiples of line-scanning frequency.

system) is used; an increase in power means that the scene is darker and a decrease in power that it is lighter. The audio carrier is 4.5 megacycles higher in frequency than the picture carrier and frequency modulation is used in the monochrome system.

The energy in the TV channel is not uniformly distributed over every possible frequency in the channel but is bunched around multiples of the horizontal-repetition frequency (approximately 15,750 cycles) with essentially empty spaces between. This is shown in Fig. 16-6, which includes a few of the frequency groups on the high-frequency side of the picture carrier. These groups are shown as separated by multiples of the line-scanning frequency (about $f_l = 15{,}750$ cycles). Each group is made up of a multiple of the field frequency ($f_f = 60$ cycles, approximately), which are quite rapidly attenuated so that most of the frequency spectrum between the groups at the frequency f_l is vacant. This space in the spectrum is used to carry the chrominance information.

Some of the standards of frequency in the black-and-white TV system have been slightly altered for the color standards. Tests have shown that it is desirable to make the line-scanning frequency have an exact harmonic relationship with the sound carrier; the latter is spaced 4.5 megacycles higher in the channel than the picture carrier. To have a harmonic relationship means that the sound carrier spacing relative to the picture carrier (4.5 megacycles) is made an integral multiple of the line frequency. The integral multiple selected is 286; the line frequency then becomes

$$f_l = \frac{4.5 \times 10^6}{286} = 15{,}734.26 \text{ cycles,}$$

This differs from the monochrome standard of 15,750 cycles by only about 0.1 per cent, which is well within the tolerance of a monochrome receiver to changes of scanning frequency. This permits the monochrome receiver to synchronize and receive the luminance component to the color-TV signals in monochrome.

Since there are 525 lines per frame or 262.5 lines per field the field frequency is

$$f_f = \frac{15{,}734.26}{262.5} = 59.94 \text{ cycles,}$$

or just slightly less than the monochrome standard of 60 cycles.

The chrominance subcarrier is modulated with the chrominance signals and its sidebands are used to modulate the picture carrier. It is called a subcarrier because it is actually a modulated modulation component of the composite station signal. The use of a subcarrier locates the color information in the video band at a fixed distance (measured in frequency units) from the picture carrier. This reduces the interference which the chrominance might cause in monochrome pictures and permits the use of demodulating circuits in the receiver which separate the luminance from the chrominance information. Figure 16-6 shows the location of the chrominance subcarrier and some of the sidebands, separated by the line-scanning-frequency intervals.

The subcarrier used to transmit the chrominance information is selected as an odd multiple of half the line frequency; the odd multiple used is 455. This is selected because it is the product of small odd factors ($455 = 5 \times 7 \times 13$); this is desirable in the design of counter circuits used at the transmitter to hold the frequencies in their correct relationships. Half the line frequency is $15{,}734.26/2 = 7867.13$ cycles. This makes the chrominance subcarrier

$$f_s = 455 \times 7867.13 = 3.579545 \text{ megacycles.}$$

The chrominance subcarrier must be held to this frequency within very close tolerance at both the transmitter and the receiver.

Since the frequency of the subcarrier is selected on the basis of one-half the line-scanning frequency, the sidebands of the modulated chrominance subcarrier lie in groups halfway between the spectrum groups of the picture signal (Fig. 16-6). The spacing of these subcarrier sideband groups is the same as that of the picture sidebands, since the scanning rate is the same for both. Thus the chrominance information is "interleaved" with the luminance information; the two signals occupy no more spectrum space than the monochrome signal alone requires.

RESOLUTION OF COLOR BY THE EYE

Psychological tests show that the sensation of color produced depends upon the size of the object seen as well as its actual color. If its area is large, such as the background or a large part of a TV picture, then its hue, saturation, and luminance must be correctly presented for natural appearance. The three primary colors used must be present in the proper proportions for the best results. If the area is small in the picture the eye cannot clearly distinguish the hues involved and it is sufficient to transmit the saturation and luminance information only. Experiments indicate that detail is best if these small areas range in color along a line between greenish-blue (cyan) and a reddish-yellow (approximately orange) on the chromaticity diagram (Fig. 16-4) through white. Colors approximately at right angles to this line (green to magenta) are not resolved as well by the eye and so do not have to be transmitted so accurately.

For very fine detail in the picture it is sufficient to transmit only the luminance information. The finest details in the picture are carried by the high-frequency components in the luminance signal (just as they are in the video signal for monochrome receivers) and these details are in black and white with shades of gray only in both color and monochrome systems.

This is comparable to a water-colored photograph. The fine detail is supplied by the black and white of the print, while the color can be applied in much less detail and yet give the effect of accurate reproduction.

THE COLOR VIDEO SIGNAL

The color video voltage is made up of two parts, a luminance voltage E_Y and a chrominance voltage E_C. These are added to form the color-video signal E_M or

$$E_M = E_Y + E_C.$$

The monochrome signal voltage E_Y is simply a black-and-white video voltage whose amplitude is related to the luminance of the televised scene in essentially the same way as the video in the non-color transmitting system. The chrominance voltage E_C consists of two parts, E_I and E_Q, which are 90° out of phase with each other. The symbols I and Q stand for "in phase" and "quadrature" (90° out of phase), respectively. The three voltages E_Y, E_I, and E_Q are related in magnitude to the camera voltages by a set of so-called matrix equations.

COLOR CAMERA

The color camera may be made up from three monochrome cameras, each responding to a band of light with wavelengths centering about red, green, and blue in the spectrum. A single optical lens may be used with mirrors to split the light into three beams. Each beam is then passed through a red, green, or blue optical filter and shines on the photocathode of one of the three image orthicons in the camera. The electrical output sig-

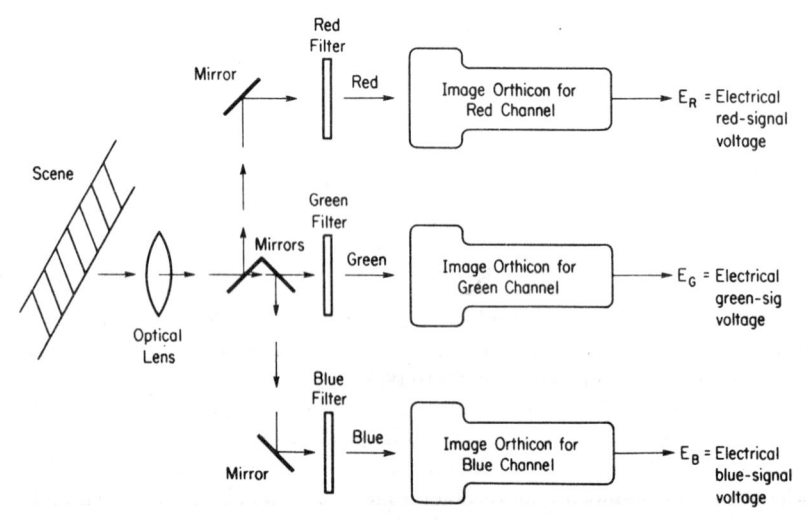

Fig. 16-7. Simplified color-camera diagram.

nals from the three cameras are fed into amplifiers, sync pulses are added to them, and they go through the system to make up the color signal. Three voltages appear at the output of the color camera: E_R is the voltage proportional to the amount of red in the scene, E_G is the voltage proportional to the amount of green, and E_B is proportional to the blue light which strikes the color camera. Any of these voltages may have a value from zero to a maximum depending upon the hue and intensity of the light in a particular portion of the scene.

THE MATRIX EQUATIONS

The *matrix* (or simultaneous) equations are

$$E_Y = 0.30\ E_R + 0.59\ E_G + 0.11\ E_B$$
$$E_I = 0.60\ E_R - 0.28\ E_G - 0.32\ E_B$$
$$E_Q = 0.21\ E_R - 0.52\ E_G + 0.31\ E_B$$

where E_R, E_G, E_B are the three camera voltages and E_Y, E_I, E_Q are the three voltages which make up the color video signal.

These equations mean that, for example, if voltage E_Y is taken as 100% then it is made up of 30% E_R, 59% of the voltage from the green camera section E_G, and 11% E_B voltage. In the same way 60% E_R less 28% E_G with 32% E_B subtracted from the result gives 100% of the E_I voltage. It is not practical to have a negative value of colored light but in the electrical circuits a negative voltage simply means a reversal of polarity. This can be effected by a single-stage vacuum-tube amplifier or by other means. The E_Q voltage is made up in a way similar to that used for the other voltages above.

Now suppose that a test pattern is placed in front of the camera, consisting of a set of color bars as shown in Fig. 16-8. The system is adjusted so that equal camera signals ($E_R = E_G = E_B = 1$) come from the color camera when the white bar is being scanned and at this time the picture appears to be white also. The amplitude of the monochrome (luminance)

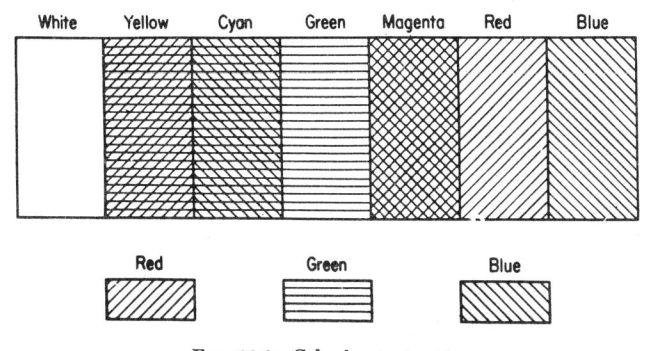

Fig. 16-8. Color-bar test pattern.

and the chrominance signals can be found for the various colors by inserting the correct values of camera signals (E_R, E_G, and E_B) into the matrix equations above and solving. For example, yellow light is made up of equal parts of red and green light with no blue (see Fig. 16-2) or

$$E_R = 1.00, \quad E_G = 1.00, \quad E_B = 0.00.$$

Inserting these values into the matrix equations gives

$$E_Y = (0.30)(1.00) + (0.59)(1.00) + (0.11)(0.00)$$
$$= 0.30 + 0.59 = 0.89$$
$$E_I = (0.60)(1.00) - (0.20)(1.00) - (0.32)(0.00)$$
$$= 0.60 - 0.28 = 0.32$$
$$E_Q = (0.21)(1.00) - (0.52)(1.00) + (0.31)(0.00)$$
$$= 0.21 - 0.52 = -0.31.$$

Since the in-phase (E_I) and quadrature (E_Q) voltages are 90° out of phase with each other they must be added like resistance and reactance to obtain their combined amplitude or

$$E_C = E_I^2 + E_Q^2 = (0.32)^2 + (0.31)^2 = 0.44.$$

These values are calculated for all the colors in the color-bar test pattern and listed in Table 16-1.

Unlike the E_I and E_Q voltages, there is no fixed phase relationship between E_Y and E_C. However, there will be times when they are in phase and other times when they are 180° out of phase. For the yellow bar the maximum occurs when E_C and E_Y are in phase and the minimum when they are 180° out of phase or

$$\text{maximum } E_M = E_Y + E_C = 0.89 + 0.44 = 1.33,$$
$$\text{minimum } E_M = E_Y - E_C = 0.89 - 0.44 = 0.45.$$

TABLE 16-1

Color	E_R	E_G	E_B	E_Y	E_I	E_Q	E_C
White	1	1	1	1.00	0.00	0.00	0.00
Yellow	1	1	0	0.89	0.32	−0.31	0.44
Cyan	0	1	1	0.70	−0.60	−0.21	0.63
Green	0	1	0	0.59	−0.28	−0.52	0.59
Magenta	1	0	1	0.41	0.28	0.52	0.59
Red	1	0	0	0.30	0.60	0.21	0.63
Blue	0	0	1	0.11	−0.32	0.31	0.44
Black	0	0	0	0.00	0.00	0.00	0.00

The envelope of the combined signals ($E_M = E_Y + E_C$) will vary between these limits (1.33 and 0.45). These data for the various colors are tabulated in Table 16-1.

Figure 16-9 shows the monochrome or black-and-white component of the combined signal as viewed on an oscilloscope screen. This is a plot of the

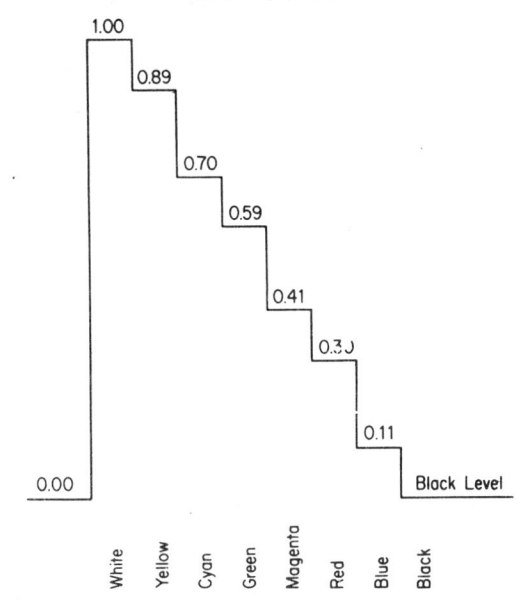

Fig. 16-9. Luminance (monochrome) component of color video signal when scanning color-bar chart (E_Y).

amplitudes shown in the E_Y column of Table 16-1 and goes from a maximum of 1.00 for the white bar through lesser values as the color-bar pattern is scanned to 0.00 for black. This monochrome (luminance) signal is the only portion of the combined signal used by monochrome receivers.

In Fig. 16-10 the chrominance components of the color signal are displayed. These are the same as the values of E_C in Table 16-1. The chrominance component is zero for white but the luminance has a value other than zero. Both the luminance and chrominance are zero for black. For the other colors the magnitude varies according to the color scanned.

The complete color-video signal is a combination of the luminance and the chrominance components and this is worked out in Fig. 16-11.

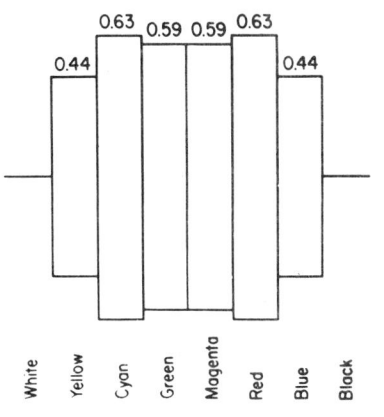

Fig. 16-10. Chrominance component of color video signal (E_C).

TRANSMITTING THE COLOR SIGNAL

A simplified block diagram of a color television transmitting system is given in Fig. 16-12. The audio channel (FM) is the same as for monochrome transmission. The chrominance subcarrier is amplitude modulated with the E_I signal and small-swing frequency modulated with E_Q. In both cases the carriers are discarded and only the sidebands are used to modulate the picture carrier. Two sidebands are used in each case (at least in the range

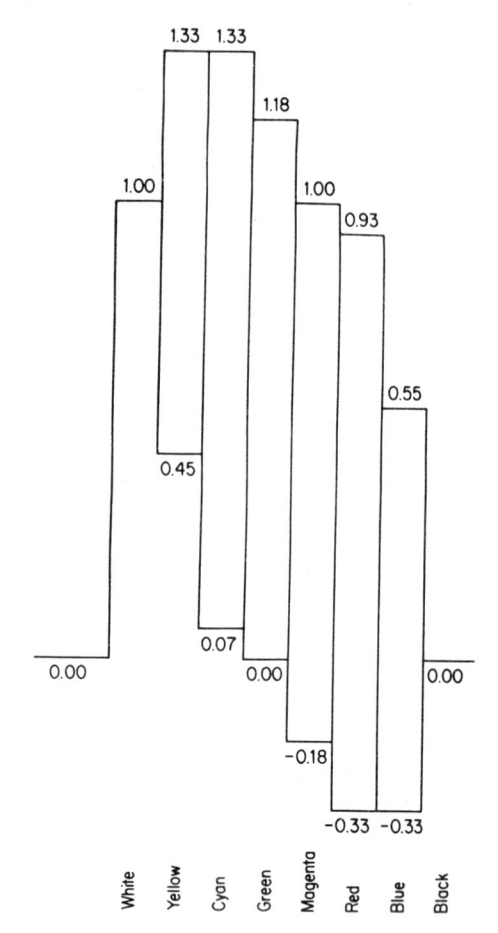

FIG. 16-11. Video modulation voltage from scanning color-bar chart.

where the frequencies overlap) or there will be interference (cross talk) between the I and Q signals. In addition to the subcarrier, interleaved with the wide-band luminance (Y) signal, to which horizontal and vertical sync pulses like those of the monochrome system are added, the carrier is modulated with a reference color-subcarrier sync signal called a *color burst*. All the energy from the transmitter is kept within the assigned 6-megacycle channel.

Chapter 16   **Color Television**   539

The color camera picks up the light from the scene to be televised, divides the light into three component colors by optical filtering, and sends electrical voltages proportional to the amount of light of each color through connecting cables marked E_R, E_G, E_B (Fig. 16-12). Light of a color different from that of the primaries makes its contribution as has been explained.

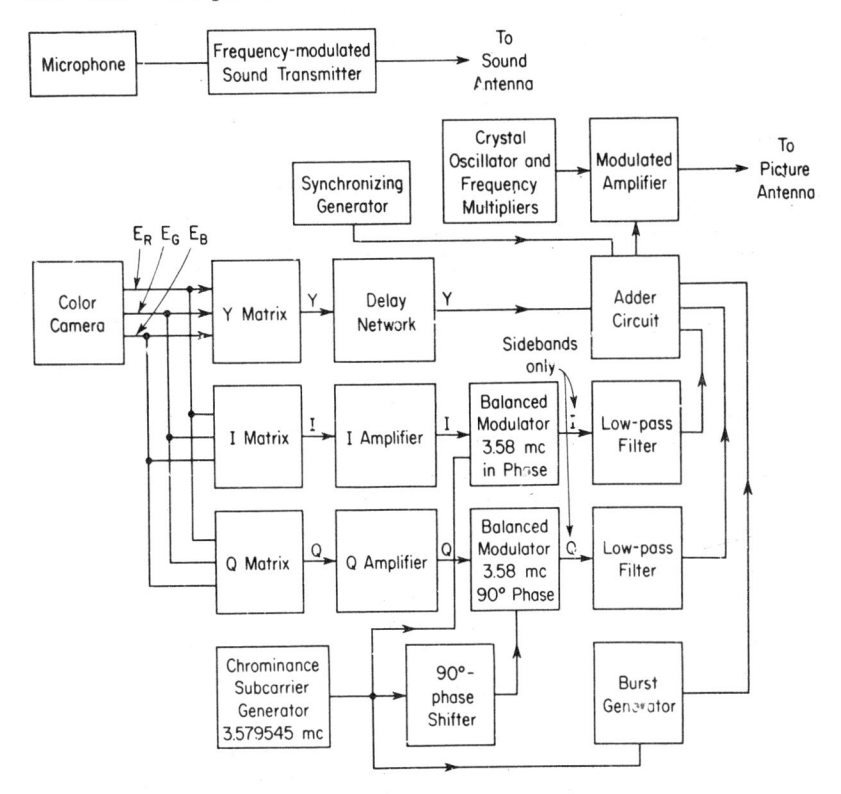

Fig. 16-12. Simplified block diagram of color television picture and sound transmitter.

The voltages from the connecting cables (E_R, E_G, E_B) are applied to a color matrix or color encoder where they are combined electronically in the proportions specified in the matrix equations.

COLOR-MATRIX CIRCUIT

A possible matrix circuit is shown in Fig. 16-13. It consists of three triodes with equal plate and cathode resistors (1,000 ohms). Assume that a signal of one volt is fed into each grid from the red, green, and blue camera circuits. Of course, the voltage actually would depend upon the scene being televised but these values are selected as examples. If the tubes are identical their plate currents will be equal and the voltage drops in the plate and cathode resistors also will be equal. As far as alternating current is concerned the B^+ point (Fig. 16-13) is connected to ground through the large bypass capacitor in the power-supply circuit (C). Current flows from B^+ to the plate, which makes the plate more negative than the B^+ (or ground)

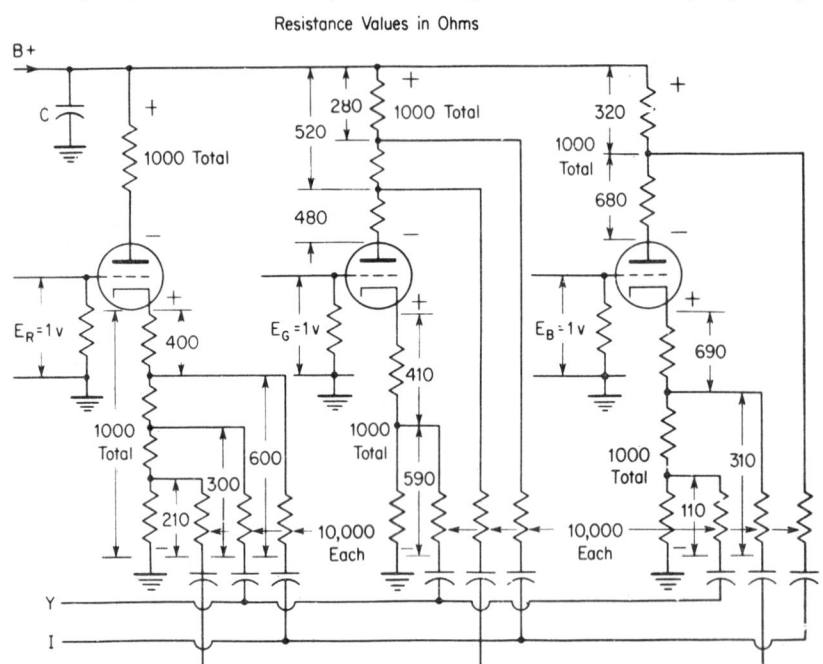

Fig. 16-13. Transmitter-matrix circuit.

point. In the cathode circuit the cathode is more positive than the ground, as indicated by the polarity marks on the diagram (Fig. 16-13). An increase in the plate current makes the plate of each tube more negative with respect to ground than in the quiescent state at the operating point, as explained in the chapter on electron tubes, and the same increase in plate current makes the cathode more positive with respect to ground. A decrease in plate current makes the plate more positive and the cathode more negative than at the quiescent point. Another way of saying this is that the plate and cathode of each tube are just 180° out of phase with each other when varying or alternating voltage is applied to the grid of the tube.

Now suppose that voltages are tapped off the plate and cathode resistors as indicated by taking fractions of the total resistance marked in the diagram (Fig. 16-13). For example, in the first tube to the left a voltage $E_R = 1$ v is fed into the grid from the color camera. This causes a varying plate current of the same wave form as the grid voltage to flow in the plate and cathode resistors. The varying plate current sets up a voltage across the plate and cathode resistors also of the same wave shape as the grid voltage. Near the grounded point of the cathode resistor 300/1,000 of the voltage drop in the cathode resistor is tapped off and fed through a decoupling resistor and coupling capacitor to the Y-signal bus. The voltage between the Y-bus and ground owing to this signal will be 30% of the total voltage developed across the cathode resistor caused by the grid voltage E_R.

In the same way a voltage which is 590/1,000 or 59% of the cathode voltage in the second tube is fed into the Y-signal bus through the decoupling network. In the third tube only 110 ohms are tapped up on the cathode resistor so as to develop 11% of the total cathode voltage to be supplied to the Y-signal bus.

The sum of these three voltages — 30% E_R, 59% E_G, and 11% E_B — is just what is required to satisfy the matrix equation for E_Y. The matrix equations are repeated here:

$$E_Y = 0.30\ E_R + 0.59\ E_G + 0.11\ E_B$$
$$E_I = 0.60\ E_R - 0.28\ E_G - 0.32\ E_B$$
$$E_Q = 0.21\ E_R - 0.52\ E_G + 0.31\ E_B.$$

Negative voltages are tapped off the plate resistors instead of the cathode resistors. The resistances are figured from the common B^+ (or ground) point. Otherwise the process is the same and the output from the Y, I, and Q buses will have the correct magnitudes for the modulation process. The actual values depend upon the scene in front of the TV camera, of course.

MODULATION OF THE CHROMINANCE SUBCARRIER

The video signal from the I output of the matrix circuit is fed into an I amplifier and then into a low-pass filter which removes all components at frequencies higher than about 1.5 megacycles as shown in Fig. 16-14. The filtered output is applied to the subcarrier which has a frequency of 3.579545 megacycles in a balanced modulator circuit which discards the carrier after

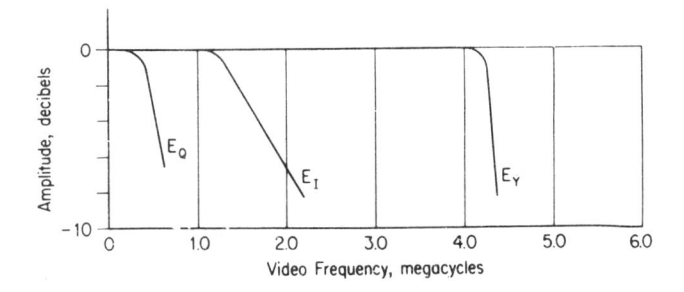

FIG. 16-14. Video frequencies before modulation.

modulation. The sidebands of the I signal go to an adder network where they are combined with the other parts of the composite signal.

The video signal from the Q output of the matrix circuit is fed into a Q amplifier and then through a low-pass amplifier which attenuates all frequencies up to about 0.5 megacycle. The remaining signal modulates a subcarrier which has a frequency of 3.579545 megacycles also but which is 90° out of phase with the subcarrier for the I signal. The subcarrier is again removed and the sidebands are passed on to the adder network to be combined with the rest of the composite signal.

The Y or luminance signal is combined with the rest in the adder network. The full range of frequencies up to 4.2 megacycles is required here because the luminance (Y) signal carries the information for the fine detail of the picture. The I signal conveys information about medium detail and requires frequencies up to only 1.5 megacycles. The Q signal (in quadrature to the I signal) has frequencies only up to 0.5 megacycle because the Q signal is required only for the large areas and coarsest detail of the picture. Since the filter of the I and Q signals introduce delays in their respective channels, an equal delay is placed in the Y channel by a delay network (Fig. 16-12).

SYNCHRONIZING COLOR BURST

The chrominance subcarrier is removed from the composite video signal at the transmitter and must be replaced in the correct frequency and phase at the receiver. Information in the form of a synchronizing color burst of 8 to 11 cycles of sine-wave form at the subcarrier frequency (3.579545 megacycles) is introduced on the back porch of the pedestal just after the horizontal sync pulse. This does not interfere with synchronizing in monochrome receivers because the leading edge of the sync pulse triggers the sweep generators and not the trailing edge. In color receivers the sync burst occurs while the picture is still blanked off the screen. Figure 16-15 shows some details of the synchronizing color burst.

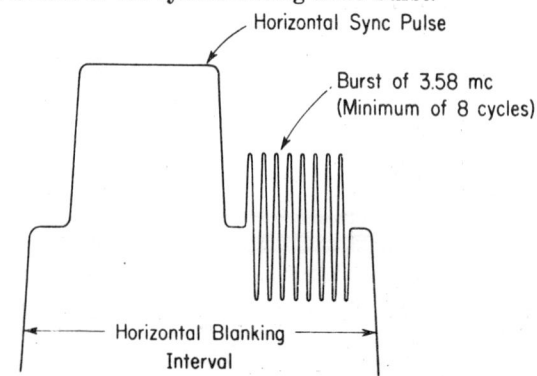

Fig. 16-15. Detail of front porch, sync pulse, back porch, and color burst.

After the Y, I, and Q video plus the chrominance burst along with the horizontal and vertical sync signals are combined in the adder network, the composite signal is used to modulate the picture carrier. The latter is generated by a crystal-controlled oscillator just as in a monochrome transmitter and is stepped up to the correct station frequency through a succession of frequency-multiplying amplifiers.

The resulting distribution of signals in the TV channel is shown in Fig. 16-16. The Y or luminance signal is transmitted with vestigial sidebands as was explained for the monochrome system. The upper sideband is trans-

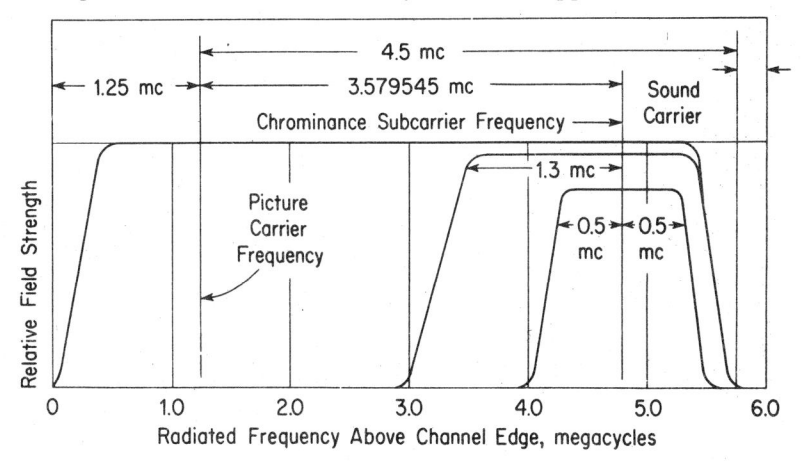

Fig. 16-16. Distribution of signals in a channel. Relative amplitudes of Y, I, and Q signals will vary with picture content.

mitted with practically uniform amplitude up to 4.2 megacycles above the picture-carrier frequency. The lower sideband is attenuated as indicated in the diagram.

The I-chrominance signal is transmitted with unequal sidebands about the chrominance subcarrier (3.579545 megacycles) which extend about 1.3 megacycles above and 0.6 below this frequency. The other set of sidebands — for the Q signal — is transmitted with equal amplitude and extends on either side of the chrominance subcarrier about 0.5 megacycle.

Fine details in the picture corresponding to video frequencies higher than 1.3 megacycles are transmitted only in the luminance part of the signal. In other words, fine detail in the picture is reproduced in black and white, not color. Larger areas corresponding to frequencies between 0.5 and 1.3 megacycles are reproduced in combinations of two colors. On the chromaticity diagram (Fig. 16-4) these lie in the orange to cyan region. Still larger areas of the picture are reproduced with all three primaries (red, green, and blue — their combination gives all the other hues inside the RGB triangle). This compromise method is based upon experiments which demonstrate that the eye does not see fine detail in color but only in shades of gray. The eye is apparently fully sensitive to color only when large areas of it are present in the picture.

COLOR RECEIVER

Figure 16-17 shows a simplified block diagram of a color-television receiver. The blocks with shading are essentially the same as those with the same function in a monochrome receiver and need not be considered here in detail.

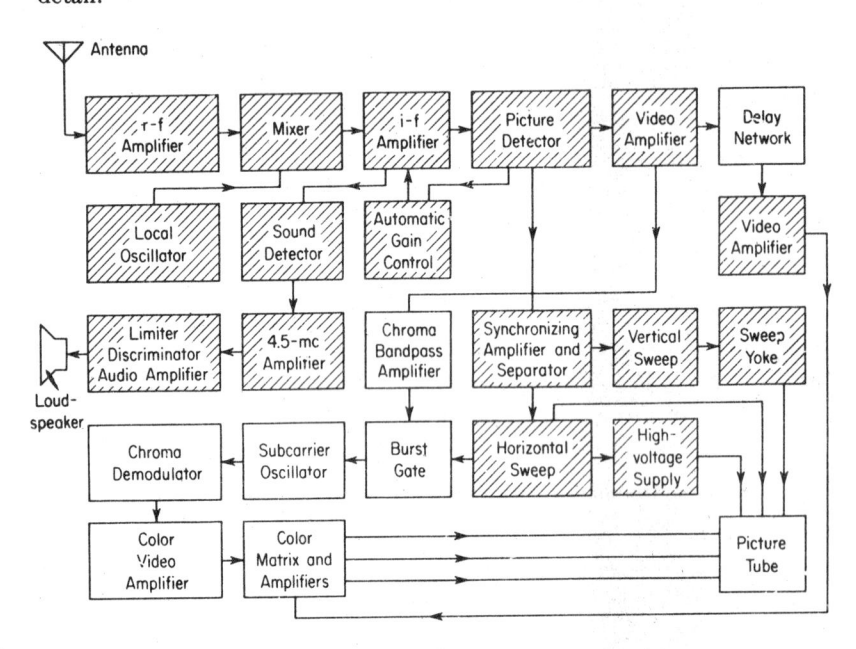

FIG. 16-17. Simplified block diagram of a color-television receiver.

The intermediate-frequency amplifier of the color set must have substantially uniform frequency response up to about 4.2 megacycles. If this is not the case, sideband cutting of the chrominance and luminance signals occurs and picture quality is degraded. The i-f amplifier must offer about 50-db attenuation to the sound i-f to prevent interference which is visible in the picture.

COMPATIBILITY IN RECEIVERS

One requirement of the system which has already been mentioned in connection with transmitters is compatibility between the monochrome and the color television systems. This can be illustrated for receivers by three possible situations.

Chapter 16 — Color Television

When a signal from a monochrome TV station is fed through a color-TV receiver there are no I and Q signals because none are sent from the transmitter. The Y signal passes through the luminance channel and in the matrix circuits of the receiver it is processed so that E_R, E_G, and E_B for the color picture tube have the correct amplitudes to add up to white light when they excite the screen of the viewing tube. A circuit is provided in the receiver to cripple the action of all the I and Q signal stages so that no signals from this source may stray into the picture and cause interference.

A signal from a color transmitter has all components — Y, I, and Q — and when received by a color set all parts of the signal are processed in the receiver matrix to give a picture in color.

When the signal from a color transmitter is received by a monochrome set the entire color video signal is applied to the picture-tube grid. Since the chrominance subcarrier is not transmitted (and the monochrome receiver has no subcarrier oscillator to generate it) only the sidebands are present and these do not appear in the picture.

PICTURE DETECTOR AND COMPOSITE VIDEO AMPLIFIER

The picture i-f amplifier stages feed into the picture detector. From here the signal goes to the first video amplifier, the a-g-c circuit, the sync separators, and the color-burst amplifier. These are shown in Fig. 16-18, a more detailed block diagram of this part of the receiver.

Out of the first video-amplifier stage the video, now the wide-band Y or luminance signal, is delayed by a one-microsecond delay network (required

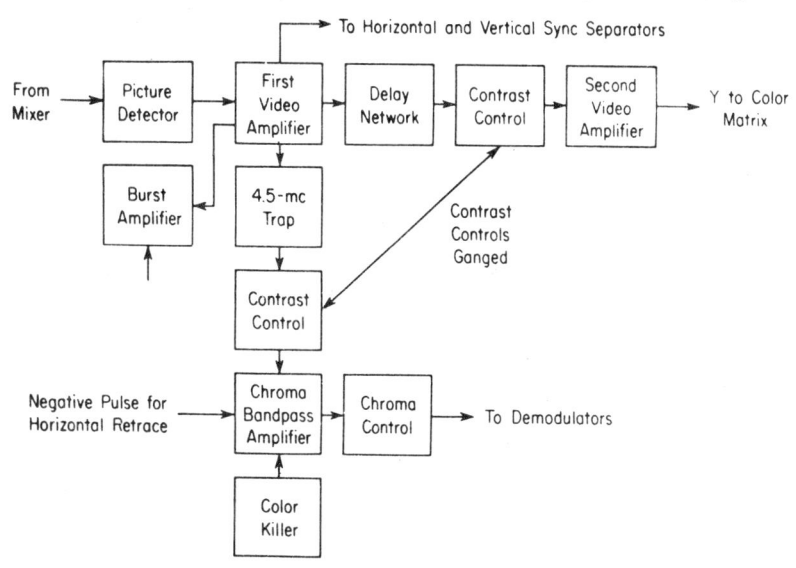

FIG. 16-18. Picture detector and associated circuits.

because of the delay in the *I*-channel and *Q*-channel filters), a contrast (manual gain) control, a second video-amplifier stage, and finally by the color-matrix section of the receiver.

At the plate of the first video-amplifier tube a tuned transformer separates the 3.579545-megacycle color burst from the rest of the signal, and feeds it into the color-burst amplifier; the burst is used to synchronize the chrominance subcarrier at the receiver with that at the transmitter. The sync-separator circuits are fed from the same point and operate in the same way as the corresponding circuits in a monochrome receiver.

The chrominance signal is sent through a 4.5-megacycle trap to keep the intercarrier-sound carrier from the chrominance channel and then through another portion of the contrast control ganged to the first to maintain the correct ratio of luminance and chrominance signals at all settings of the contrast control. The chrominance signal is then amplified by the chroma band-pass (channel width 2.4 to 5.0 megacycles) amplifier.

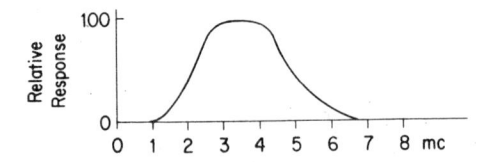

Fig. 16-19. Pass band for chrominance channel.

A color-killer tube is provided which cuts off the chroma amplifier when color-burst information fails to come through. This prevents monochrome signals from going through the chrominance channel during reception of non-color programs. The pass band for the chrominance channel is indicated in Fig. 16-19; luminance information is eliminated but all the color sidebands go through this filter circuit.

The output of the chroma band-pass amplifier is fed through a chroma control (color gain control) which governs the vividness of the colors in the picture as received. From here the signal goes to the chrominance demodulators.

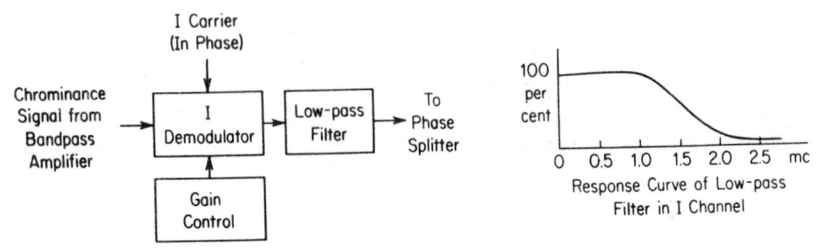

Fig. 16-20. *I* channel demodulator and response of low-pass filter in *I* channel.

Chapter 16   *Color Television*   547

CHROMINANCE DEMODULATORS

The chrominance signal at this point consists of the I and Q sideband voltages. The chrominance demodulator separates these signals after they have been added to I and Q carriers generated in the receiver and synchro-

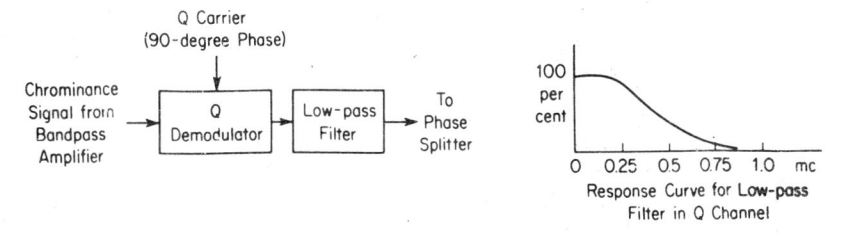

FIG. 16-21. Q demodulator and response of low-pass filter in Q channel.

nized with those at the transmitter by means of the color bursts. The result after passing through low-pass filters is shown in Fig. 16-20 and Fig. 16-21. Four voltages come from the demodulators and phase splitters: $+I$, $-I$, $+Q$, $-Q$. When these are combined with the luminance (Y) signal in the color matrix of the receiver the E_R, E_G, and E_B voltages are recovered. The latter are the voltages which control the brightness of the three color phosphors in the picture tube.

RECEIVER COLOR MATRIX

The color matrix at the receiver — like that at the transmitter — is a miniature electronic computer which solves the simultaneous matrix equations and delivers the answers in terms of voltages. The transmitter matrix yields E_Y, E_I, and E_Q or voltages related to them. The receiver matrix, on the other hand, must take E_I, E_Q, and E_Y, combine them in the correct way and recover E_R, E_G, and E_B. These voltages control the excitation of the phosphors in the picture tube so that light of the proper red, green, and blue colors (and mixtures of these colors of light) appear on the viewing screen.

A receiver color-matrix circuit is shown in Fig. 16-22. The equations given in the transmitter section may be solved for E_B, E_G, and E_R in terms of E_Y, E_I, and E_Q. When this is done the result is

$$E_R = 1.00\ E_Y + 0.95\ E_I + 0.63\ E_Q$$
$$E_G = 1.00\ E_Y - 0.28\ E_I - 0.64\ E_Q$$
$$E_B = 1.00\ E_Y - 1.11\ E_I + 1.72\ E_Q.$$

This means that to get E_R to drive the red phosphor of the picture tube 100% of E_Y is used, 95% of E_I, and 63% of E_Q. For E_G 100% of E_Y is combined with 28% of E_I (with reversed polarity) and 64% of E_Q (with reversed polarity). The voltage E_B can be worked out in similar fashion. If these three voltages are substituted in the matrix equations for E_Y, E_I and

E_Q it will be found that they check within the accuracy of the figures used in this example.

Other circuit arrangements may be used for the color matrix but the principle remains the same. The essential thing is to take three voltages and mix them in the proper proportions to obtain three other voltages, the

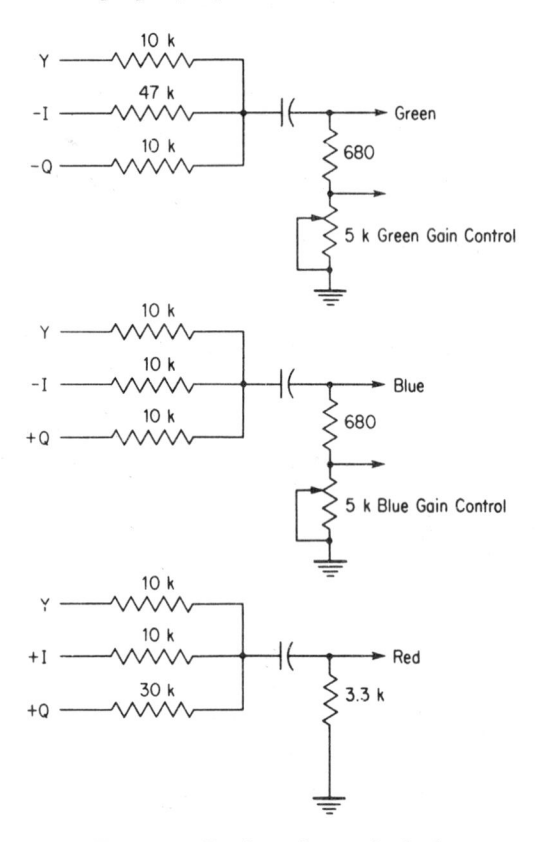

Fig. 16-22. Receiver color-matrix circuit.

latter being desired for use elsewhere in a system. Interaction between the three input voltages must be made small by feeding the network from vacuum tubes or by other suitable circuit arrangements.

It is possible also to transmit color-difference voltages $(E_R - E_Y)$ and $(E_B - E_Y)$ and utilize these in the receiver circuits. These color-difference voltages are related to the E_I and E_Q voltages by the equations

$$E_I = -0.27 (E_B - E_Y) + 0.74 (E_R - E_Y)$$
$$E_Q = 0.41 (E_B - E_Y) + 0.48 (E_R - E_Y)$$

and are displaced 33° (lagging) in phase from E_I and E_Q.

Chapter 16 **Color Television** 549

After the three color voltages E_R, E_G, and E_B have been obtained from the color matrix they are amplified as required to drive the grids of the color picture tube. A gain control is required in the green and blue amplifiers. The red amplifier may operate at maximum gain since it requires more signal voltage to drive the red phosphor than is needed for the others to get the same light output. The gain of the green and blue amplifiers is adjusted as may be necessary.

An adder tube follows each of the resistance networks (one per primary color) of the matrix and each adder is followed by an output stage. A d-c restorer in each of the color circuits operates in the same way as in a monochrome receiver.

CHROMINANCE SYNC CHANNEL

To recover the I and Q signals from their sidebands one precisely controlled a-c voltage must be generated at a frequency of 3.579545 megacycles and a second voltage at exactly the same frequency but 90° out of

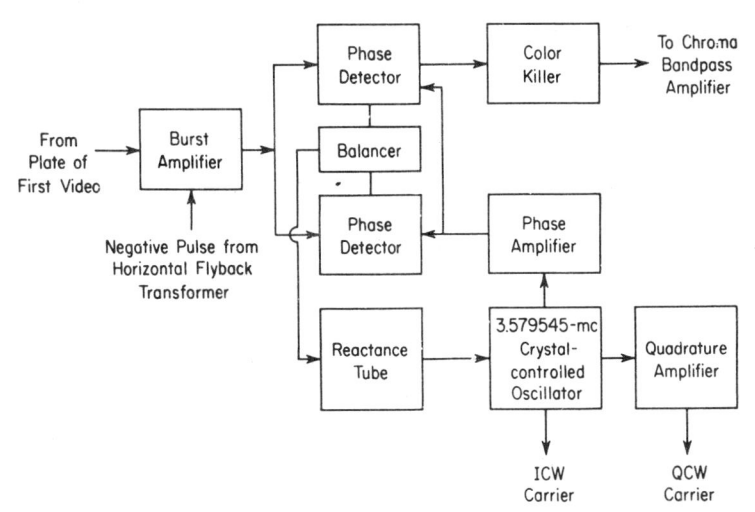

Fig. 16-23. Chrominance sync channel detail block diagram.

phase with the first. The oscillator used for this purpose is crystal controlled as indicated in the block diagram, Fig. 16-23. The phase of these two signals is controlled by comparison with the color-burst signal.

THREE-COLOR PICTURE TUBE

Several methods are available to build picture tubes which display images in color. The essential parts of one model comprise an assembly of three guns as sources of electrons to excite each of three phosphors (red, green, and blue), a control grid in each gun, to regulate the intensity of its electron beam, provision for sweeping all the beams horizontally and vertically, and the necessary accelerating voltages to speed the electrons and shape the beams.

Instead of having a continuous layer of phosphor which gives off white light under impact of the electrons, the three-color tube has red, green, and blue phosphors arranged in groups of three dots (trios) spaced as closely as practicable without overlapping. On some models there are 195,000 trios or

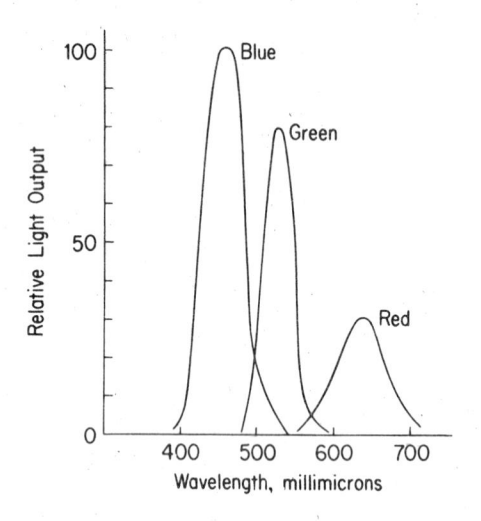

FIG. 16-24. Light output from phosphors.

a total of 585,000 dots on the viewing end of the tube. After the phosphor dots are in place a thin layer of metal (usually aluminum) is deposited to increase the light output, prevent internal reflections of light from striking the viewing screen, and to avoid ion-spot damage. The relative light output of the three phosphors is indicated in Fig. 16-24.

The electron-gun assembly for this picture tube has three closely spaced, parallel-beam guns, built as a unit. The construction provides one gun for the excitation of each of the three phosphors. It is possible to control the brightness of each of the three colors independently of the other two.

Between the gun and the screen of the three-color tube is placed a shadow mask having the same number of round holes as there are dot trios (195,000). This mask is located close to the phosphor dot plate. A diagram (Fig. 16-25) shows how the electron beams pass through the shadow mask and strike the phosphor-dot screen. The beams from the guns are focused in such a way that they converge at the aperture in the shadow mask. As they pass through the hole in the mask the beams diverge from each other so that each strikes a dot of its assigned color and no other. Since the holes

Chapter 16 *Color Television* 551

in the mask as well as the dot trios are arranged in a fixed pattern one beam can strike only red dots over the whole face of the tube. This can be called the red beam and it may be said to come from the red gun. A similar relationship exists for the other guns and beams.

The dot screen and the shadow (or aperture) mask have been built with either flat or curved surfaces. The flat-mask system requires a more complicated circuitry and is not considered here.

Any color within the range of the three phosphors selected for the tube may be reproduced by exciting the proper amount of light from the red, green, and blue phosphors. Since the dot pattern is too small to be seen at normal viewing distance the colors produced appear to blend just as colored lights when mixed as described in the earlier part of this chapter. If the beams from all three guns are cut off, none of the dots gives off light, and the screen appears black at that instant. If all the guns are supplied with the correct bias voltages — which should occur when the camera is pointed at a white surface — the screen appears to be white. If the same proportions of red, green, and blue are maintained but the voltages on the control grids of each gun are changed to reduce the amount of energy in the electron beams (that is, cut off is approached) the screen appears to be some shade of gray. When the screen is scanned by a single beam, say the red one, the screen appears to be red. Other colors are formed in a similar way.

OTHER SYSTEMS OF COLOR TELEVISION

Various other systems of color television have been suggested, tried in the laboratory, or put into limited commercial production. No doubt still other improvements will be devised to make the operation of the present system better. The current standards are the result of a very extensive engineering study of all factors known to be involved and are sanctioned both by industry and by the Federal Communications Commission.

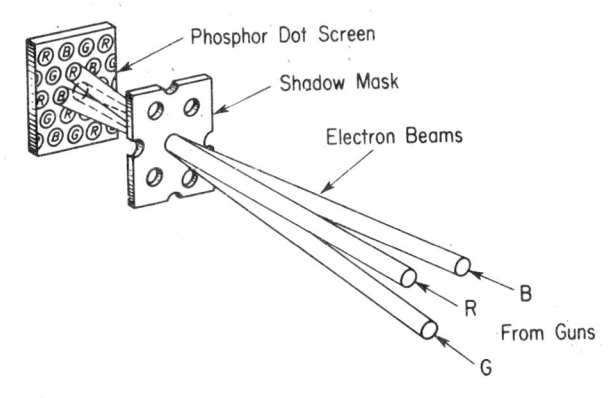

Fig. 16-25. Electron beams converging in shadow mask and striking phosphor dots.

QUESTIONS AND PROBLEMS

1. Name the essential parts of a TV color kinescope.
2. What is meant by "additive color primaries"? How many are required?
3. Define luminance and chrominance.
4. Define hue saturation and brightness.
5. What is the purpose of the color burst on the back porch of the horizontal sync pulses?
6. What is meant by compatibility of color and monochrome TV signals?
7. What additional functions are required in a color TV set compared with a monochrome receiver?
8. What is the purpose of matrix circuits in color-TV receivers and transmitters?

CHAPTER 17

VACUUM-TUBE INSTRUMENTS

Many types of instruments employing electron tubes have been developed for use in the design and maintenance of electrical and electronic equipment throughout the useful range of audio and radio frequencies. Certain of these instruments will be found in every laboratory and among the most common are signal generators, oscillators, vacuum-tube voltmeters, and cathode-ray oscillographs. Other instruments such as tube testers and signal tracers are found useful in electronic maintenance and will also be discussed here.

OSCILLATORS

Vacuum tube oscillators are the usual generating sources for radio- and audio-frequency signals. There are two general classes of oscillators; one group makes use of the property of negative resistance shown by some vacuum tubes. A tube displays negative resistance when its current increases as the voltage applied to the element decreases. The second group of oscillator circuits uses an external network or circuit to couple a portion of the output (plate) voltage back to the input (grid) in proper phase and sufficient amplitude. Most of the oscillators employed in testing equipment are of the second type, and so the negative resistance circuits will not be considered further.

An amplifier which will supply its own input, through feedback of a portion of its output in proper phase and amplitude, will oscillate or generate an a-c output. The circuit will automatically adjust its operation

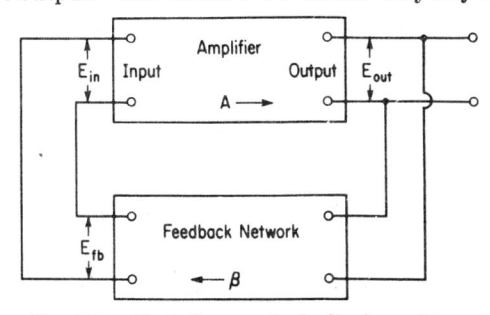

FIG. 17-1. Block diagram of a feedback amplifier.

such that the gain is sufficient to provide the needed input voltage amplitude, and the frequency will adjust to meet the required conditions on the input voltage phase.

A block diagram of a feedback amplifier which may oscillate is shown in Fig. 17-1. It should be noted that no external input is shown, the input being derived from the output voltage through the feedback network. This network transmits only a fraction, β, of the output voltage, and may cause any required amount of phase shift.

The input voltage of the amplifier, E_{in}, is given by

$$E_{in} = E_{fb}$$

where E_{fb} can be found as

$$E_{fb} = \beta E_{out} = \beta(A E_{in}).$$

Combining these equations gives

$$E_{in}(1 - A\beta) = 0. \tag{17-1}$$

If useful output is to be obtained, then E_{in} cannot be equal to zero; therefore for the circuit to operate and produce oscillations

$$1 - A\beta = 0. \tag{17-2}$$

The magnitude of $A\beta$ must then be unity, and any phase shift introduced in A must be compensated by an equal and opposite phase shift in β so that

$$|A\beta| = 1 \tag{17-3}$$

$$\text{(phase shift of } A\text{)} = -\text{(phase shift of } \beta\text{)} \tag{17-4}$$

become the two conditions for oscillations to occur, or for the circuit to produce useful output without an external input.

The first equation forces the circuit to adjust its amplitude of operation or gain A such that the $|A\beta|$ term is equal to unity. In placing the circuit in operation it may be assumed that some disturbance, such as an initial plate current inrush, will occur in the output circuit. A voltage resulting from this disturbance will be fed back into the input circuit, will be amplified greatly, and will appear in the output circuit. This larger output is again transferred back to the input terminals by the feedback network to be further amplified and the process repeated. With the increasing amplitude of the signals which occurs in this process, the amplification A will begin to decrease as it works into the curved portions of the tube characteristics; and when A reaches the value required by Eq. 17-3 the growth of the oscillations ceases, and the amplifier is then in the steady-state oscillating condition.

Equation 17-4 requires that the circuit adjust its frequency of oscillation to a value such that the reactances in the amplifier and feedback network will produce the correct phase angles. In this connection, if an oscillator of good frequency stability is desired, the feedback network chosen should have a rapid variation of phase angle as the frequency is varied. Thus only a small shift of frequency is required at any time to restore the phase angles to the requirements of the second equation.

INDUCTANCE-CAPACITANCE FEEDBACK OSCILLATORS

Many different oscillator circuits employing inductance-capacitance feedback networks have been developed and a few are shown in Fig. 17-2. These circuits as shown employ shunt feed for the d-c power to the plate circuit of the tube and are biased by grid current flowing through a grid leak

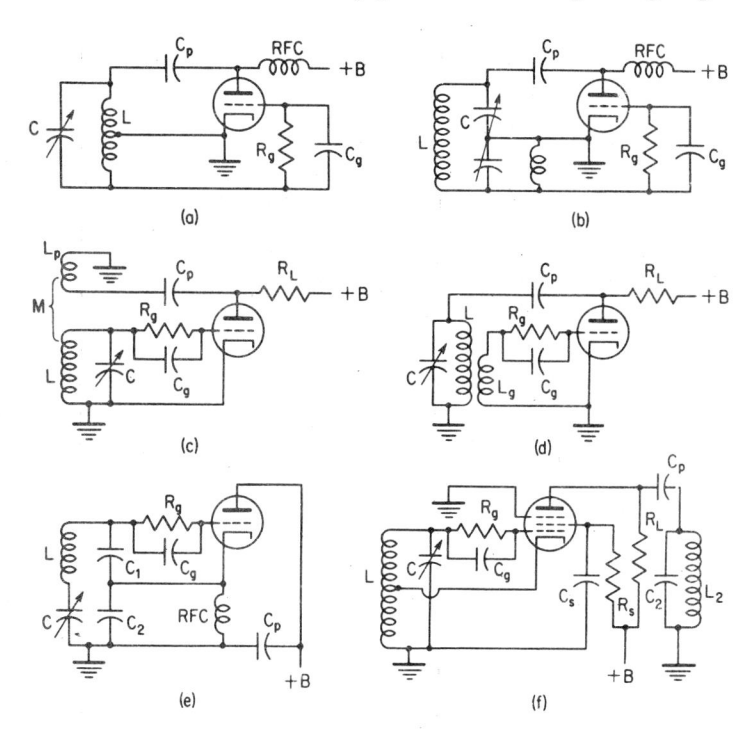

Fig. 17-2. Schematic diagrams of several inductance-capacitance feedback oscillators: (a) Hartley; (b) Colpitts; (c) tuned grid; (d) tuned plate; (e) Clapp; (f) electron-coupled Hartley.

resistor. Series feed of plate power may be used equally well. The control of bias by a grid leak is advantageous since, when the circuit is energized initially, without oscillations present, there is no bias on the tube and the amplification is high. As oscillations build up, the grid current and grid bias increase, thus reducing the gain, and biasing the tube finally to a state in which the amplification is just sufficient to maintain oscillation. Grid leak bias also insures that the circuit may be depended on to start oscillating every time the circuit is energized.

Examination of the Hartley circuit of Fig. 17-2(a), shows that feedback occurs through the mutual inductance M between the portions of the inductance. The frequency of oscillation is found to occur at a value just slightly less than $f = 1/(2\pi\sqrt{LC})$, if M is sufficiently large. Capacitor C_p serves to block the d-c plate supply out of the tuned circuit, and the r-f choke prevents oscillator frequency power from feeding back into the power supply.

The Colpitts oscillator of (b) obtains its feedback through the mutual coupling provided by the circulating currents in the tuned circuit. The tuned-grid and tuned-plate circuits of (c) and (d) have feedback for oscillation provided by the inductive coupling between the coils in both grid and plate circuits.

The Colpitts oscillator has been modified into the series-tuned Colpitts, or Clapp, oscillator of (e).

A common disadvantage of all these circuits discussed is that the frequency of oscillation is somewhat dependent on the tube coefficients, μ and r_p, on the shunted internal tube capacities, and on the external load or power extracted from the oscillator. The electron-coupled oscillator of (f) provides superior isolation of the oscillating L-C circuit from the load by using the cathode, control grid, and screen of a pentode as a triode oscillator with the screen at ground a-c potential. The grounded screen shields the control grid and cathode from the plate and output load circuit. The coupling to the plate and output circuit is through the electron stream, and changes in load applied to the plate circuit have very little effect on the frequency of oscillation.

The effect of internal tube-capacitance change is usually minimized by making the tuning capacitors C quite large so that internal capacitance changes have little effect on the total capacitance determining the frequency of oscillation.

RADIO-FREQUENCY OSCILLATORS OR SIGNAL GENERATORS

Signal generators, as used in laboratory equipment for generating radio frequencies, are usually of the L-C type and may employ any of the circuits discussed. The choice of circuit depends on ease of switching coils, constancy of output over a wide frequency range, and stability of frequency and calibration over long periods of time. A circuit diagram of one oscillator of this type is shown in Fig. 17-3.

In order to make the output frequency of the oscillator more independent of the load connected to the circuit, it is desirable to include a buffer amplifier stage between oscillator and output terminals, to isolate the oscillator from load changes. It is also customary to include some means of amplitude modulation, so that the signal may be traced through the circuits

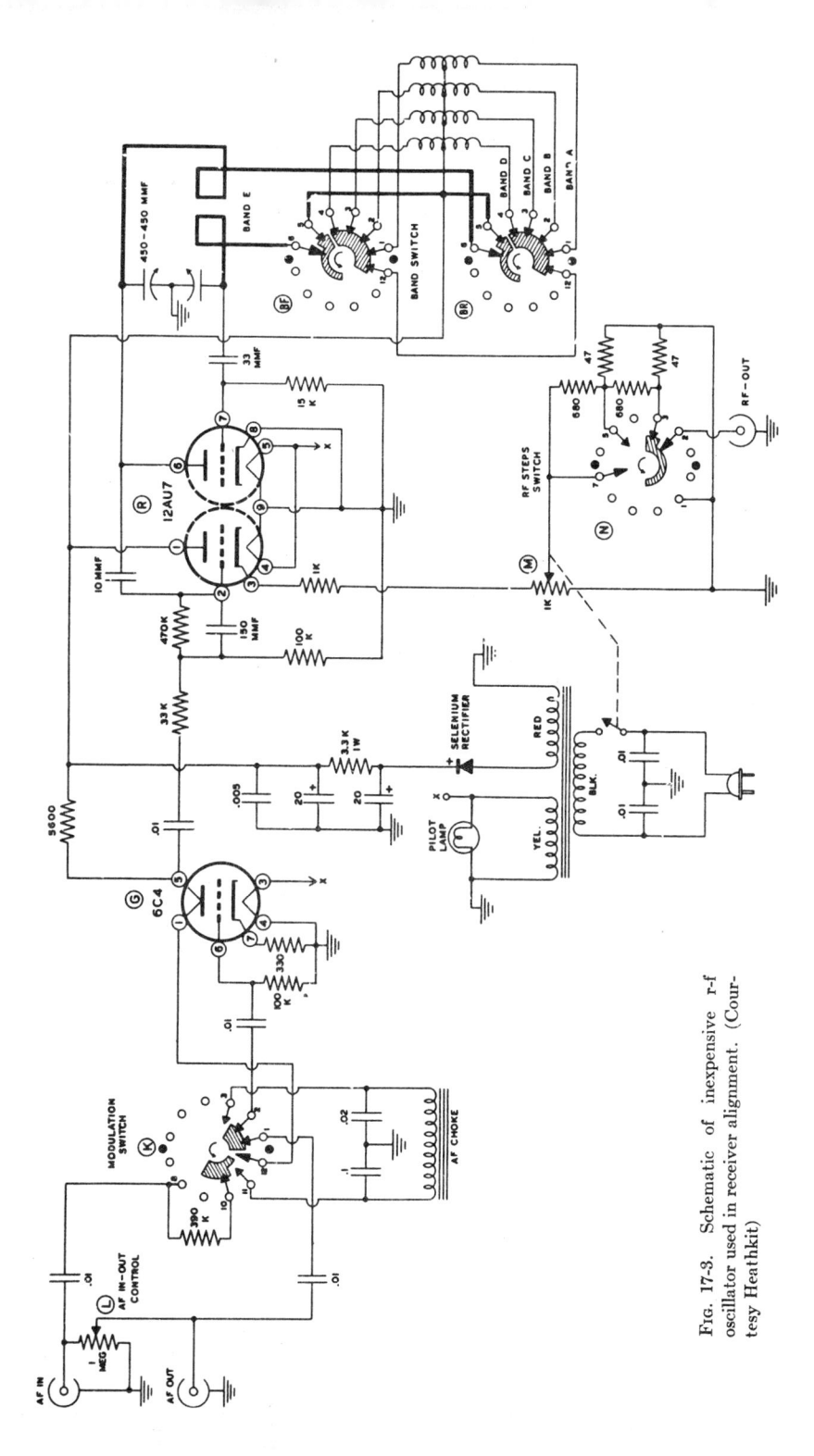

Fig. 17-3. Schematic of inexpensive r-f oscillator used in receiver alignment. (Courtesy Heathkit)

of a radio receiver. In many applications only a small output voltage, in microvolts or millivolts, is required and a low impedance attenuator is included to permit adjustment of output voltage. In more expensive equipment the output voltage is measured indirectly by a vacuum tube voltmeter in conjunction with a calibrated output attenuator.

FREQUENCY-MODULATED OSCILLATORS

Frequency-modulated oscillators are used in testing of frequency-modulation receivers and in visually indicating the performance of bandpass amplifiers and circuits by permitting presentation of the response curve of the circuit under test on a cathode-ray oscilloscope.

The process of frequency modulation requires that the operating frequency of an oscillator be varied about a center value at a rate determined by the frequency of the modulating signal. Several methods of accomplishing frequency modulation in test equipment are available, and the simplest is probably that of direct variation of the L or C of the oscillator tuned circuit at the modulation frequency. A variable capacitor in parallel with C of the tuned circuit may be rotated by a motor, or one plate of a variable capacitor may be connected to and moved by the voice coil of a loudspeaker mechanism. The modulating signal, possibly 60 cycles, is fed to the loudspeaker, and as its voice coil moves in accordance with the modulating signal, the capacitance of the tuned circuit determining the oscillator frequency is changed. The output frequency is thus shifted about at the modulating rate, or covers a band of frequencies which are said to be *swept*. The amplitude of loudspeaker motion determines the magnitude of the

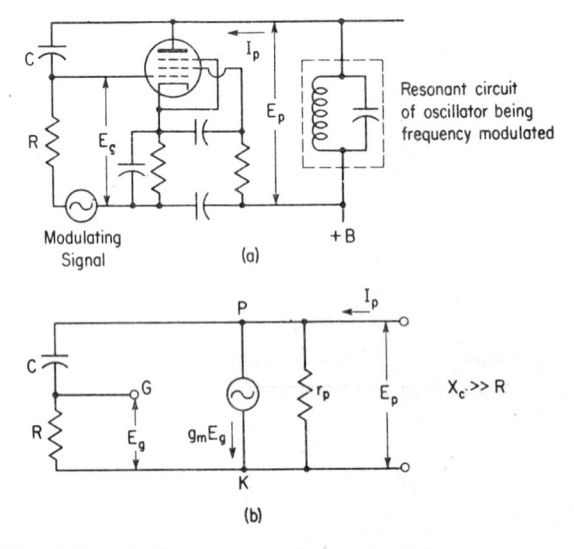

Fig. 17-4. Schematic diagram and equivalent circuit for a reactance tube.

frequency deviation from the center value. This method or that of the motor-driven capacitor is obviously limited to low modulating frequencies, and other means are required for frequencies beyond the audio range.

A less restricted way of obtaining frequency modulation is by use of a *reactance tube,* for which a diagram and equivalent circuit are shown in Fig. 17-4. This circuit makes use of the fact that the plate current of a pentode depends mainly on the control grid voltage and only slightly on the anode voltage (for constant screen voltage). The anode-to-cathode circuit of the pentode is connected directly across the resonant circuit of the oscillator to be frequency modulated, and the alternating plate voltage is the voltage across the resonant circuit.

The grid signal for the pentode is obtained from a series circuit of a resistor and a reactive element (usually a capacitor). If the grid circuit is as shown in (a), Fig. 17-4, the capacitor C is chosen to have a very large reactance at the oscillator frequency. The resistance R is made very small with respect to this reactance (one tenth or less). The current through R and C is determined mainly by the capacitor value and consequently leads the anode voltage by nearly 90 degrees. Since the voltage across a resistor is in phase with the current through it, the grid voltage obtained from R leads the alternating anode voltage by nearly 90 degrees. The plate current, being dependent on the grid voltage, also leads the anode voltage by nearly 90 degrees.

The resonant circuit of the oscillator is supplying a leading current to the reactance tube, just as if the reactance tube were a capacitor. Because the magnitude of current flowing in the reactance tube varies with the transconductance of the tube, the apparent shunting-capacitance effect can be varied by change of grid bias. For the circuit shown, the apparent capacitance of the reactance tube is given by $C_{in} = g_m CR$. If the modulation frequency voltage is inserted in series with resistor R, the tube grid bias can be made to swing, thus varying the apparent capacitance of the reactance tube and modulating the frequency of the oscillator.

Such a frequency-modulation system will show a linear relation between modulation voltage and frequency shift if the value of C_{in} is below 10% of the tuning capacitance C of the oscillator circuit, and if the transconductance of the reactance tube has a linear relation with grid voltage. Using a reactance tube of the type shown, frequency deviations up to about 5 per cent of the center frequency can be obtained with good linearity.

A good frequency-modulated oscillator should be free of any simultaneous amplitude modulation. This requires the grid voltage to be exactly 90° out of phase with the plate voltage. If there is a small in-phase component, the reactance tube will also appear to have a resistive component in its impedance. As this resistance varies, the load on the oscillator will vary, changing its amplitude or voltage output in a cyclic manner. This effect cannot be completely eliminated for the circuit shown but can be for other circuits which are available.

The application of frequency-modulated oscillators to the testing of resonant and band-pass amplifiers will be discussed in connection with applications of the cathode-ray oscilloscope later in this chapter.

RESISTANCE-CAPACITANCE TUNED OSCILLATORS

It is frequently impractical to obtain inductors of sufficiently high Q and small size to permit efficient operation of the L-C oscillator at audio frequencies. A resistance-capacitance oscillator eliminates the inductors, and may be used for frequencies as low as 1 cycle per second.

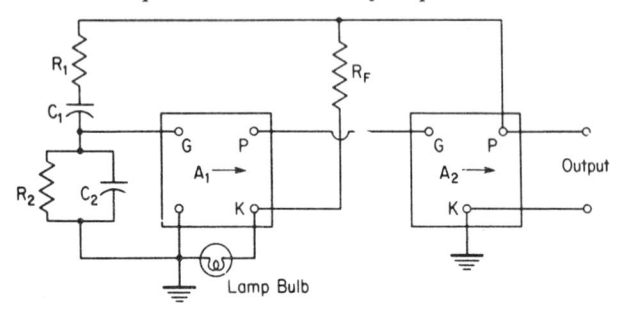

FIG. 17-5. Simplified diagram of a resistance-capacitance tuned oscillator of the Wien bridge type.

A simplified diagram of an R-C oscillator is shown in Fig. 17-5. The oscillator employs two amplifier stages, not necessarily for gain, but to provide the phase shift required by the conditions for oscillation. The frequency of oscillation is given by

$$f = \frac{1}{2\pi\sqrt{C_1 C_2 R_1 R_2}}$$

which is the frequency at which the voltage across the parallel combination of R_2, C_2 is a maximum and at zero phase angle with respect to the a-c plate voltage of the output amplifier. The tungsten lamp bulb in the cathode circuit of the first stage provides a resistor whose resistance increases with increase of current. This lamp bulb and the resistor R_f provide negative feedback which automatically adjusts (by reason of the varying lamp resistance) to reduce the gain so that the required amplification is just obtained, to meet the conditions for oscillation.

By proper selection of circuit parameters and adjustment of the feedback, the amplification can be limited without appreciable wave-form distortion, and an oscillator such as this becomes a good source for sine waves.

If the design is made so that $C_1 = C_2$, $R_1 = R_2$, then the frequency depends inversely on the capacity used for tuning, and not on the square root of capacity as in the L-C oscillators. Since the usual range of variation of a variable capacitor is about 10:1, the R-C oscillator will give a frequency range of 10:1 in one swing of the variable capacitor. By choosing values of $R_1 = R_2$ that are decimally related, decimal ranges of frequency may be obtained with the same dial calibration serving for all ranges.

The R-C circuit is widely used for audio oscillators in laboratory and service testing. The frequency range of a common model is from 20 cycles to 200 kc, in four decimal steps.

GRID-DIP OSCILLATOR

Any of the oscillator circuits of Fig. 17-2 may be used as a grid-dip oscillator by addition of a milliammeter in series with the grid resistor. Because of its convenience in requiring only two terminals for plug-in coils, the Colpitts circuit is most frequently employed. By use of plug-in coils the instrument is made to cover a wide frequency range, and it is constructed so that it is easy to couple the coil to external resonant circuits of signal sources.

The usefulness of the instrument comes from the fact that if the loading on an oscillator is varied, the voltage fed back to the grid current shifts. If the coil of the oscillator is coupled to, or placed near, an external resonant circuit, the grid current will change in the manner shown in Fig. 17-6 as the frequency of the grid-dip oscillator is tuned past the resonant frequency of the external circuit. The tighter the coupling between oscillator and load circuit, the greater the dip in grid current. If the load circuit has a high Q, the grid current dip will be quite sharp.

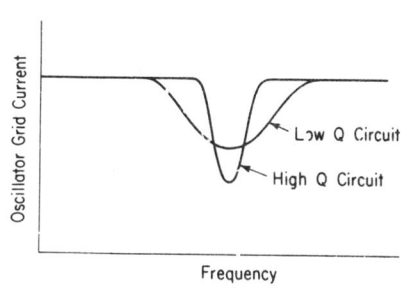

FIG. 17-6. Oscillator grid current as a function of oscillator frequency when the oscillator is coupled to a tuned resonant load.

If the grid-dip oscillator is calibrated for frequency of oscillation it may be used to preset the tuned circuits of a receiver or transmitter to the desired frequency without the necessity of supplying power to the receiver or transmitter. The oscillator may also be used to determine the presence and resonant frequency of unwanted or parasitic tuned circuits which occasionally appear in equipment.

While usually of only fair accuracy, the grid-dip meter can supply information which it is difficult to obtain in any other way.

FREQUENCY-MEASURING EQUIPMENT

Frequency-measuring equipment falls into two general classes. The first type extracts energy from the source whose frequency is being measured, and this energy actuates an indicating device in a frequency-sensitive instrument. These devices are called absorption instruments and are not expected to provide high-accuracy measurements. The grid-dip oscillator of the previous section is such an instrument. The second type matches the frequency of the unknown signal against a standard and known frequency. This is called the *heterodyne* method, and by its use measurements of great precision are possible.

The absorption type is called a wavemeter for historical reasons, and usually consists of a resonant circuit which includes a small lamp bulb or radio-frequency ammeter for indicating the presence of current in the circuit as shown in Fig. 17-7. The circuit is calibrated by comparing its resonance against a standard frequency source, and a frequency scale is applied to the variable capacitor setting. In use, the tuned circuit is coupled loosely to the source being measured, and the wavemeter variable capacitor adjusted until the lamp or meter shows maximum current flowing in the circuit.

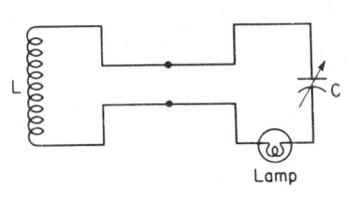

Fig. 17-7. Resonant circuit wavemeter.

The frequency may then be read from the calibrated dial of the capacitor or from charts.

To prevent interaction of the wavemeter resonant circuit with the frequency of the source being measured it is advisable to use the smallest possible coupling which will give an indication on the lamp or meter.

A heterodyne frequency meter is more complicated and expensive but is suitable for much greater accuracy and also for use with signals of very low power, such as those coming into a receiver. A block diagram of a heterodyne frequency meter is shown in Fig. 17-8. The unknown frequency and the signal from the local oscillator are compared in the mixer and the

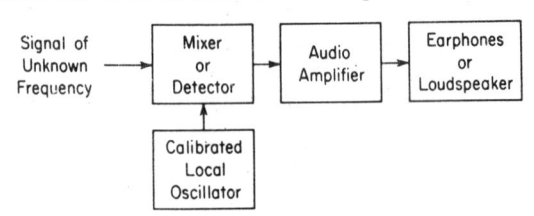

Fig. 17-8. Block diagram of a heterodyne frequency meter.

difference frequency is supplied to the amplifier. When the frequency of the calibrated oscillator is adjusted to be nearly the same as the unknown frequency, an audible beat note will be heard in the earphones. As the calibrated oscillator frequency approaches that of the signal the pitch of the beat note will decrease, and will disappear when the frequencies are exactly the same. The frequency of the unknown signal is then equal to that shown on the calibrated oscillator.

If the local standard oscillator is restricted in frequency range, one of its harmonics may be used to beat against the unknown signal, and measurements can be made over a wide frequency range. The possibility of error is increased because the exact harmonic used may be in doubt. If an absorption wavemeter is used to determine the approximate frequency of the unknown signal then the proper harmonic may be identified and this source of possible error removed. Several crystal oscillators may also be used to secure known marker frequencies.

The standard oscillator must be well built and stable since the measurement can be no more accurate than the calibration of this standard. Its calibration may be periodically checked by comparison with signals from WWV, the station of the National Bureau of Standards, Washington, D. C. This station transmits on 2.5, 5.0, 10.0, 15.0, 20.0, 25.0, 30.0, and 35.0 megacycles with the transmitted frequency accurate to one part in 100 million. With carefully constructed equipment frequency measurements may be made with an accuracy of about 0.01 per cent of the measured frequency. Greater accuracies are possible in work with precise and complex equipment.

CATHODE-RAY OSCILLOSCOPE

A combination of several different types of electronic circuits produce a very useful test instrument in the cathode-ray oscilloscope. A block diagram for such an instrument appears in Fig. 17-9. The two amplifiers are usually resistance-coupled, but are designed for greater bandwidth than used in an ordinary audio amplifier. In some oscilloscopes the amplifiers have frequency ranges extending to zero frequency or d-c, this implying the

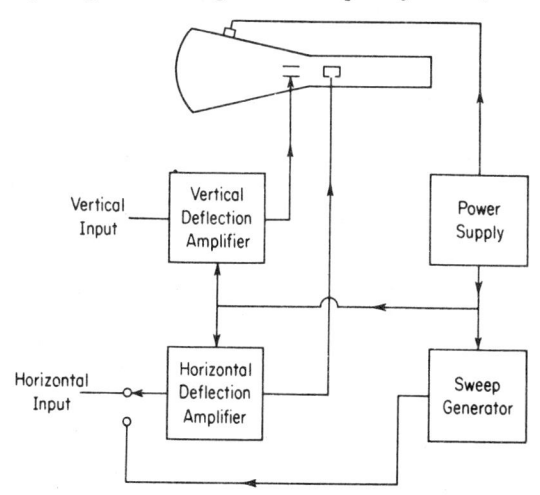

FIG. 17-9. Block diagram of a cathode-ray oscilloscope.

use of no coupling capacitors in the amplifiers. At the high-frequency end of the range, instruments may have amplifiers with good responses to 200 kc, with more expensive equipments reaching 3 or 5 megacycles. The characteristics of the vertical amplifier may be superior to those of the horizontal amplifier in some cases, since the signal to be investigated is usually applied to the vertical amplifier.

The cathode-ray tube is usually of the electric-deflection type, and the most common size for laboratory equipment is the 5-in. diameter. Screens used are the P1 and P5. Several power supplies are usually incorporated, including one for amplifier plate voltages, and a high-voltage low-current source for the cathode-ray accelerating voltage.

The output from the vertical and horizontal amplifiers must be sufficiently large to deflect the beam and spot to any portion of the tube screen without distortion. With high accelerating voltages on the cathode-ray tube to provide high spot brightness, it is necessary to obtain high amplifier output voltage and uniform amplification over a wide frequency band simultaneously. Push-pull amplifiers are usually employed in the final amplifier stages to obtain greater output voltage.

The wiring diagram of a complete oscilloscope for general-purpose use is shown in Fig. 17-10. The major application of the cathode-ray oscilloscope is in the observation of electrical wave forms. Other applications are as a peak voltmeter, frequency comparator, and as an indicator in the alignment of band-pass amplifiers.

For observation of wave forms as functions of time it is necessary to generate internally and apply to the horizontal or x-axis deflection plates sweep voltages or time-varying saw-tooth voltages. Any recurring voltages applied to the vertical or y-axis plates will then appear plotted on the screen as functions of time. Sweep generator circuits used in oscilloscopes may be any of the types discussed in Chapter 8, with less expensive instruments employing a gas-tube sweep circuit, and more expensive ones using vacuum-tube generated sweeps.

In the observation of wave forms the sweep voltage is applied to the x-axis or horizontal deflection plates, and the signal to be studied is connected to the vertical amplifier and vertical or y-axis deflection plates. The result of application of these two voltages to plates causing deflection of the beam in perpendicular directions may be determined graphically, as illustrated in Fig. 17-11 for a sine-wave vertical input signal. If the voltage applied to the horizontal plates is truly linear with time, and if the vertical amplifier does not introduce any distortion, the path traced by the electron beam on the fluorescent screen will be a faithful reproduction of the wave shape introduced to the vertical circuits. By the means shown other wave patterns may be demonstrated graphically.

To use the oscilloscope as a peak voltmeter it is necessary to calibrate the screen. The unknown signal may be first applied and the gain control

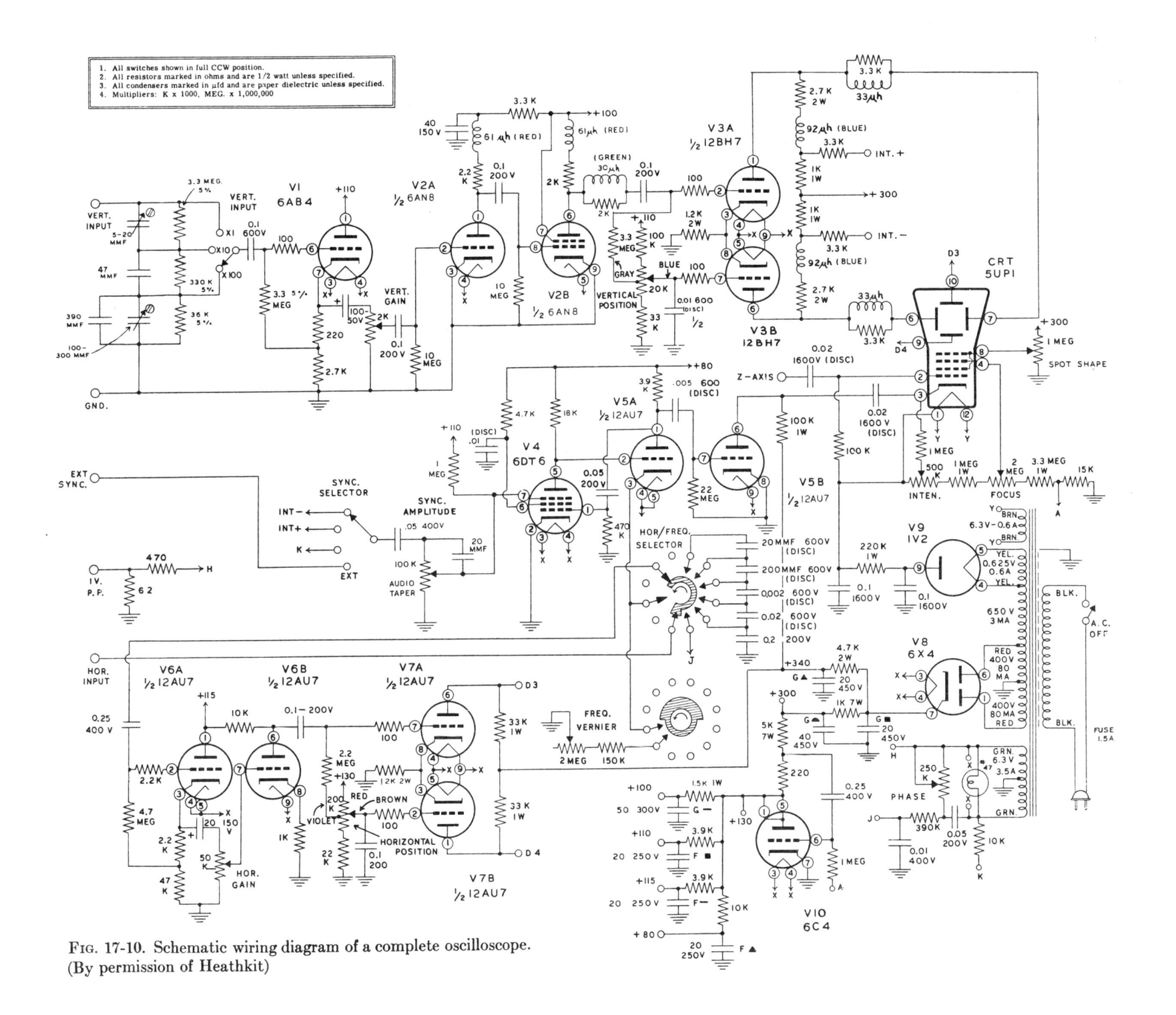

FIG. 17-10. Schematic wiring diagram of a complete oscilloscope. (By permission of Heathkit)

set to provide a suitable screen deflection. The unknown signal is then removed, and a signal of known adjustable magnitude applied. The amplitude of this signal is then adjusted until the same screen deflection is obtained, without touching the oscilloscope gain control. The peak value of

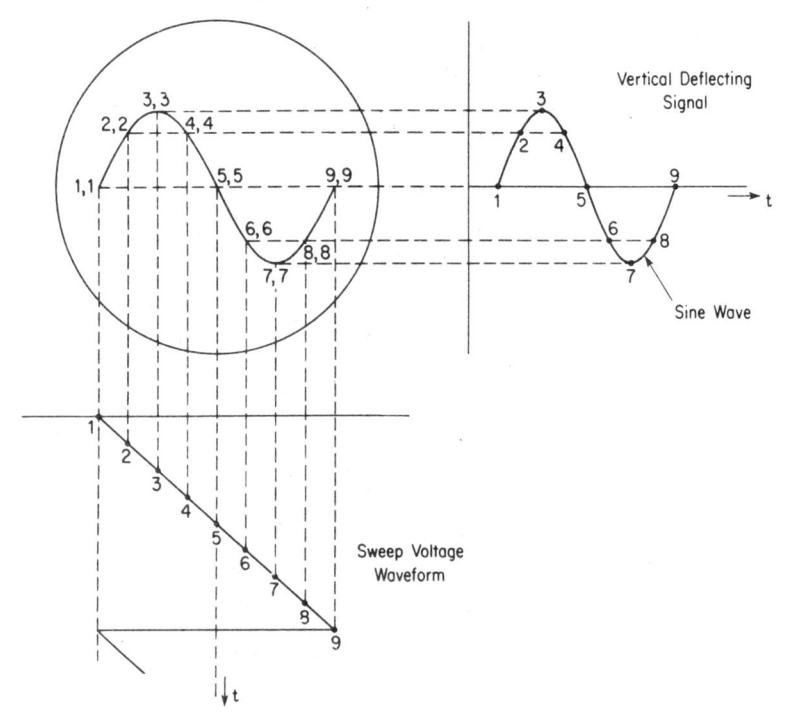

Fig. 17-11. Graphical determination of cathode-ray pattern.

the unknown signal is then equal to the peak value of the known voltage. A square wave is often used as the calibrating signal.

The oscilloscope may also be used for frequency comparisons by applying one signal to the horizontal plates and the second signal to the other pair of plates, either directly or through the amplifiers. The patterns so obtained are called *Lissajous figures* after the man who first obtained them by using a swinging pendulum which left a trace on a sanded surface. If the frequencies of the two signals are not exactly related by a ratio of two integers the pattern will seem to revolve as if on a hollow transparent cylinder, with the rate of revolution reducing as the frequency of one signal is brought closer to a multiple of the other frequency. When the pattern appears to stand still the frequencies are in the ratio of the number of peaks appearing on a horizontal edge of the diagram to the number of peaks appearing on a vertical edge of the diagram. Frequency ratios up to about ten to one can be compared and measured. Above ten to one it is difficult to accurately count the number of cycles of variation of one signal with respect to the other. Typical patterns are shown in Fig. 17-12.

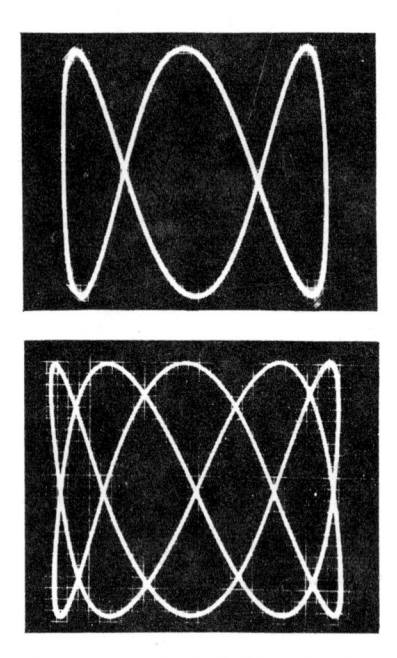

Fig. 17-12. Lissajous patterns obtained with various frequencies related by the ratio of two integers.

The use of an oscilloscope in conjunction with a frequency-modulated or frequency-swept oscillator is shown in Fig. 17-13, where the oscillator, amplifier under test, and oscilloscope are interconnected. The oscillator output

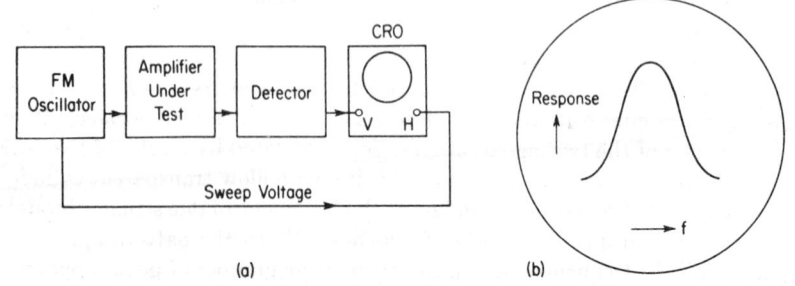

Fig. 17-13. (a) Connections used for FM oscillator and oscilloscope to determine response curve of an amplifier; (b) pattern obtained for radio-frequency amplifier when return trace is blanked out by voltage on CR tube grid.

frequency varies over a range somewhat greater than the response bandwidth of the amplifier being tested. The modulating voltage of the swept oscillator (frequently 60 cycles) is fed to the horizontal or x-axis plates of the oscilloscope, and thus a particular horizontal position of the beam corresponds to a particular instantaneous frequency of oscillator output. The test amplifier output is supplied to the vertical or y-axis circuits. The amplifier output will vary in accordance with its amplification for any particular input frequency, and the vertical deflection of the beam will correspond to the amplification or amplifier response at any particular instantaneous frequency of the oscillator.

Combining these two deflections causes the oscilloscope presentation to represent the amplification as a function of frequency. The circuit behavior while design changes are being made is then immediately apparent, and if the amplifier has doubly-tuned over-coupled stages this method is practically indispensible. The over-all response curve of the radio- and intermediate-frequency amplifiers of a radio or television receiver may be obtained by this procedure.

If a modulated radio-frequency signal is applied to the vertical or y-axis plates, and the voltage causing modulation is connected to the x-axis plates, a figure such as that in Fig. 17-14 is obtained. The degree or percentage of modulation can be measured from the figure and computed as

$$m = \frac{AB - CD}{AB + CD} \times 100\%.$$

The linearity of modulation is indicated by the straightness of the lines AC and BD, while if any phase shift occurs in the modulation process or in the oscilloscope pickup circuits these lines will appear as elliptical curves. It is not necessary for the modulating frequency (usually in the audio range) and the radio-frequency carrier to have any particular ratio in this case because each cycle of the modulating frequency produces a complete figure, and the carrier-frequency lines are not seen except as a shaded area.

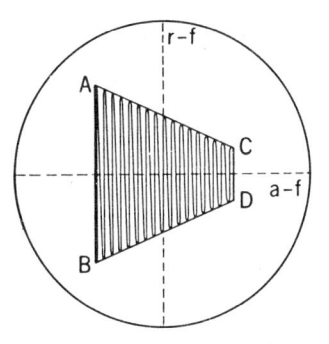

FIG. 17-14. Trapezoid pattern showing modulation.

VACUUM-TUBE VOLTMETERS

High input-impedance voltmeters are required for measurements in most audio- and radio-frequency circuits, if the measuring equipment is not to distort the performance of the circuit and true readings are to be obtained. In addition, the voltmeter indications must not vary with frequency over wide frequency ranges. The vacuum-tube voltmeter meets these requirements more easily than any other type of instrument.

While the earliest vacuum-tube voltmeter employed a triode, the simplest and most common forms of the instrument now use diodes in circuits such as those of Fig. 17-15. The components employed are similar but the characteristics of the circuits differ.

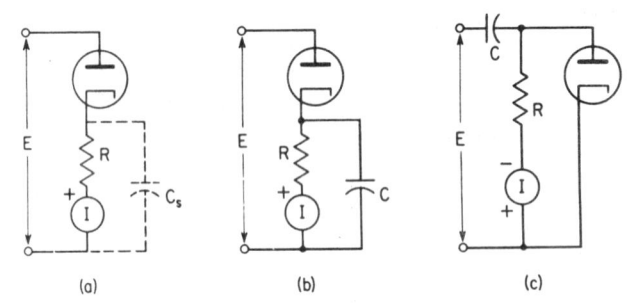

Fig. 17-15. Three basic circuits for diode vacuum-tube voltmeters.

The rectifying properties of the diode are employed in the circuit of Fig. 17-15(a) to give an indication on the direct-current meter in the cathode circuit which is proportional to the average value of the positive half cycle of the voltage being measured. The current through the meter is in the form of half-cycle pulses, as in the half-wave rectifier, and owing to the properties of a direct-current meter the reading of the meter is proportional to the average current flowing. The meter reading is independent of frequency up to some value at which the reactance of stray capacity C_s across R and the meter becomes comparable to R. The upper-frequency limit may be further extended by using a crystal rectifier as the diode, eliminating the shunting capacity between heater and cathode of a thermionic diode. The input impedance of this circuit is almost equal to $2R$, thus R should be large and the meter should have a low current rating.

The performance of the circuit is affected by the presence of capacitance across R and the meter, and in (b) this capacitance is purposely increased to a value such that $\omega RC > 10$. The circuit then behaves as a half-wave rectifier with RC filter, the capacitor C charges to approximately the peak of the applied positive voltage wave form, and diode current flows only in short pulses near the time of the voltage peak. Since the capacitor discharges slowly through R, this discharge current will be nearly proportional to the peak of the applied voltage wave, and the instrument becomes a peak-reading meter. The input impedance of this circuit is reduced to $R/2$, and thus its loading effect on the circuit may be greater.

The circuit at (c) is similar in operation to that at (b) and also gives readings proportional to the positive peak of the applied wave.

Such instruments are frequently calibrated in terms of rms values of applied sine waves. The indications given by the meter are then in error if the wave form of the measured voltage differs from sine form. In that case

Chapter 17   **Vacuum-Tube Instruments**   569

average reading meters will indicate with less error than the peak reading types.

Instruments can be built making use of the fact that a triode may have a grid-plate curve which in a restricted region will show plate current proportional to the square of the applied grid voltage. The applied voltage signal must be restricted to a narrow range, and the calibration is quite dependent on tube voltages, thus such instruments are not often employed

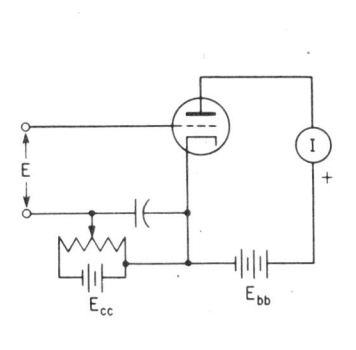

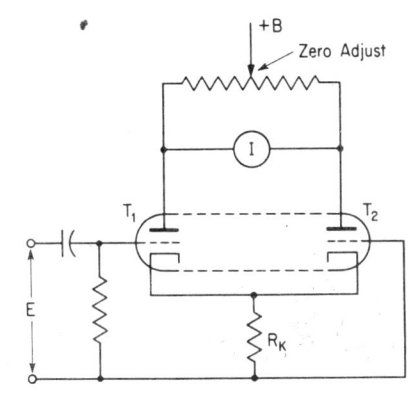

FIG. 17-16. Schematic diagram of a triode square-law vacuum-tube voltmeter.

FIG. 17-17. Balanced vacuum-tube voltmeter using a dual-triode tube.

for accurate work. A typical circuit is shown in Fig. 17-16, and in operation the grid bias E_{cc} is adjusted to bring the tube near cutoff and cause operation in the desired region for squared response. The circuit is then quite sensitive to changes in filament emission, or to values of E_{cc} or E_{bb}.

It is possible to employ triodes as voltmeter tubes, with either linear or square-law response, and avoid many of the instabilities mentioned above. This is shown in Fig. 17-17, wherein tube T_2 acts as a dummy, serving only to maintain the balance of the bridge circuit formed by the two tubes and the two resistors R_L, while the tube T_1 receives the signal voltage. Under no-signal conditions the reading of the meter is adjusted to zero by manipulation of the zero adjustment, thus balancing the bridge. When a signal is impressed on T_1 its plate current (and d-c resistance) changes in proportion to the magnitude of the signal, causing a change in the plate voltage of T_1, and a reading on the meter. If a separate bias for T_1 is chosen close to cutoff, operation may be in the square-law region, while if biased as shown the readings will be linear with input voltage. If the input capacitor is omitted the circuit may be used to measure direct voltages.

The advantage of the circuit lies in the independence of the meter reading with respect to changes in supply voltages. The best performance is obtained when the tubes are identical, but a considerable improvement over single-tube performance can be obtained with somewhat dissimilar tubes.

The circuit diagram of a complete d-c voltmeter of the bridge type is shown in Fig. 17-18. Since the allowable input voltage must be kept below a few volts, an input attenuator or voltage divider is used. To provide for measurement of alternating voltages, a diode or crystal voltmeter may be employed at the input, the d-c bridge voltmeter then measuring the d-c output of the rectifier. To make such a voltmeter useful over a very wide frequency range, the diode or crystal may be mounted in a hand-held probe, only the d-c output leads extending back to the bridge voltmeter. With a germanium crystal in the probe no supply voltages are required, making a very simple device. The voltage measured must be limited to about 20 volts to avoid crystal overload.

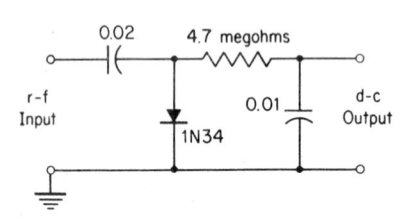

FIG. 17-19. Crystal-probe diagram.

The circuit diagram of a crystal probe is shown in Fig. 17-19. Such a circuit is satisfactory from audio frequencies to several hundred megacycles, with the upper frequency limit dependent on the physical construction of the probe.

TUNABLE VACUUM-TUBE VOLTMETERS

A commonly used instrument modelled after a radio receiver and employing vacuum-tube voltmeter circuits is the Chanalyst. It consists of a tuned radio-frequency amplifier whose output is measured by a vacuum-tube voltmeter. The indication of the voltmeter is given on an electron-ray indicator (magic-eye) tube. The tuned amplifier is calibrated in frequency and tunable over a wide frequency range.

By combining two of these amplifier-magic-eye tube circuits, each tunable over a different range of frequencies, along with an audio voltmeter, a d-c voltmeter, and a wattmeter, it is possible to simultaneously monitor several portions of a radio receiver to determine the location of a failure.

THE Q-METER

A versatile laboratory instrument is the Q-meter. This instrument measures directly the ratio of reactance to resistance or Q of an inductor, as well as measuring inductance and capacity. In simplified form the Q-meter is shown in Fig. 17-20. The Q-meter includes a highly stable variable frequency oscillator E of constant output, and a calibrated variable capacitor C.

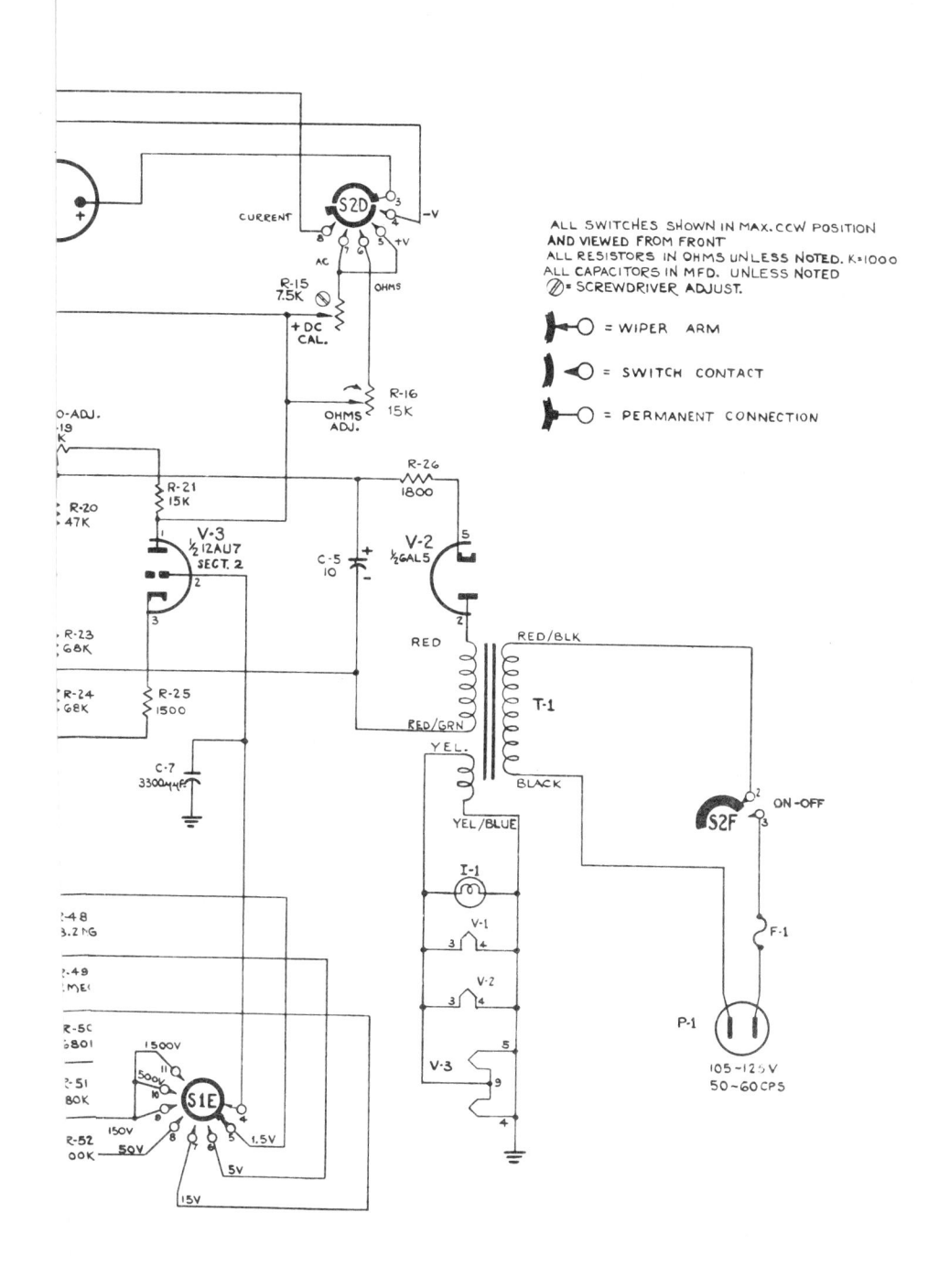

diagram of bridge-type d-c vacuum tube voltmeter.
(of A, copyright proprietor)

Chapter 17 **Vacuum-Tube Instruments** 571

The inductor whose Q is to be measured is placed in series with the variable capacitor C, and the combination resonated at the frequency of the adjustable source E. Resonance is indicated by the rise of voltage across the capacitor, as indicated by a vacuum-tube voltmeter; maximum voltage occurs at resonance. The current in the circuit is then limited only

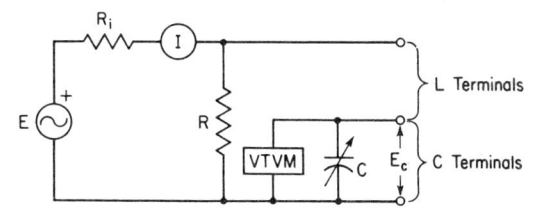

FIG. 17-20. Simplified Q-meter circuit.

by the series resistance of the L-C circuit. This series resistance is assumed to be mainly associated with the unknown inductor under test, since the calibrated capacitor may be made nearly lossless. If the voltage supplied to the resonant combination is E_r, the current I_r, the reactance of the capacitor X_c, and the series resistance of the resonant combination is R, then the current is

$$I_r = \frac{E_r}{\sqrt{R^2 + (X_L - X_c)^2}}.$$

At the resonant frequency $X_L = X_c$, and then

$$I_r = \frac{E_r}{R}.$$

The voltage developed across the capacitor C, at resonance, is the product of I_r and X_c, so that

$$E_c = \frac{E_r X_c}{R} = \frac{E_r X_L}{R} = E_r Q.$$

Thus
$$Q = \frac{E_c}{E_r}.$$

The ratio of E_c to E_r is the Q of the circuit or inductor under test.

The voltage E_r should be developed across a low impedance, so that its value is essentially independent of the condition of tuning of the load. By maintaining E_r at a fixed value it is possible to calibrate the vacuum-tube voltmeter reading E_c, directly in terms of Q. Voltage E_r can then be monitored and maintained constant by measuring the value of the current I.

It is also possible to measure the voltage across C directly by a diode voltmeter, making measurement of I unnecessary.

The value of inductance of a coil may be determined by calculation from the known frequency and the resonating capacitance, as read from the capacitor dial. The measurement of an unknown capacitor can be made by use of a reference inductor of suitable size. Two readings of the capacitance required for resonance and Q are made, one with only the calibrated capacitor, the second with the unknown capacitor in shunt with the calibrated unit. The capacitance of the unknown is then directly given by the difference in readings of the calibrated capacitor for the two tests. Large unknown capacitors may be measured by connecting them in series with the calibrated capacitor. The procedure of this paragraph is known as a *substitution method*.

TUBE-TESTING EQUIPMENT

In the testing of electronic equipment it is necessary to have some means of determining the performance and condition of the tubes, independent of the circuits in which the tubes are to operate. Two basic types of tube testers have been developed, employing tests of electron emission or of transconductance.

One of the most frequent causes of tube failure is low electron emission, and this is a factor easily measured. The emission tester determines the tube condition by measuring the plate current which flows when a standard set of potentials is applied to the tube electrodes. The plate current to be expected from a normal tube is found by testing many good tubes, thus finding the range of variation to be expected from good tubes, and fixing a level of current below which a tube is to be considered as defective.

Unfortunately, a tube may perform well in a given circuit even though the current it passes is below normal for a tube of its type, and it is also possible for a tube with normal emission to perform poorly in a given circuit. Because of this, emission testers are not considered extremely accurate, and a better test is desirable.

Since most tubes are employed as amplifiers, and the amount of amplification obtained is proportional to the tube transconductance this quantity may be measured as a better indicator of tube quality. This may be done by applying a proper set of direct voltages to the tube elements, and then supplying the grid with a fixed alternating voltage, possibly one volt. The alternating grid voltage causes an alternating component of plate current to flow, and the ratio of the alternating plate current to the alternating grid voltage gives the transconductance of the tube. With a fixed alternating grid voltage it is then merely necessary to measure the alternating plate current component to obtain the transconductance, and this may be done with a meter in the plate circuit reading the a-c component only.

Instruments of this type give a good prediction of probable tube performance in a circuit, when calibrated by measurements and averages based on a large number of good tubes.

Chapter 17 Vacuum-Tube Instruments 573

QUESTIONS AND PROBLEMS

1. In each of the circuits of Fig. 17-2, identify the feedback circuit elements required for the circuit to oscillate.

2. In the circuit of Fig. 17-2(b) each section of the split-stator capacitor C has a range of 35 to 365 $\mu\mu$f. Assuming stray capacities are negligible, determine the inductances L needed to provide a frequency range from 50 kc to 30 megacycles, allowing a 10% overlap in frequency from band to band.

3. Show that in the circuit of Fig. 17-5, the gain A_1A_2 must equal 3 if oscillations are to occur.

4. For the circuit of Fig. 17-5, $C_1 = C_2$ and each has a range of 30 to 450 $\mu\mu$f. Determine values of a set of resistors, if $R_1 = R_2$, to provide a frequency range from 40 cycles to 50 kc, allowing 10% overlap in frequency from band to band.

5. In using a reactance tube modulator it is assumed that the equivalent capacity represented by the tube is directly proportional to the g_m of the tube. If this is true, will the frequency of oscillation of a connected oscillator be directly proportional to g_m? Under what restrictions could the g_m-frequency relation be considered linear?

6. Using a graphical construction, show that the Lissajous figure of Fig. 17-12 for 3:2 frequency ratio is obtained when the two deflecting voltages are timed so that both start going positive at $t = 0$. What pattern is obtained if the higher frequency voltage is displaced ¼ cycle later in time? Does this pattern indicate the 3:2 relation clearly?

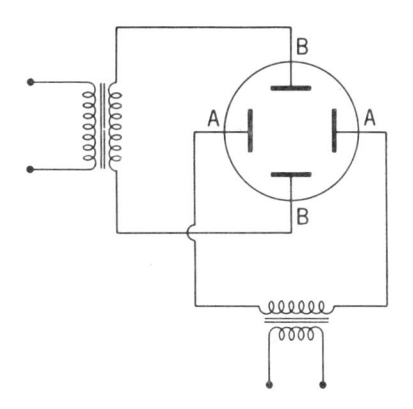

Fig. 17-21.

7. Determine the pattern produced on the cathode-ray screen if a 60-cycle sine wave of voltage is supplied to plates A-A and a 120-cycle sine wave of voltage is supplied to plates B-B of Fig. 17-21. Phase relation is such taht the positive peaks of the two waves are simultaneous.

8. Repeat Problem 7, if the phase relation is such that the zero points of the two voltage waves are simultaneous.

9. A diode vacuum-tube voltmeter similar to Fig. 17-15(b) is to be designed to be peak reading for all frequencies above 80 cycles. The value of R is fixed at 100 k-ohms. What is the minimum value of C which will permit the specified operation?

10. Explain the use of the negative-grid voltage supply in the d-c vacuum-tube voltmeter of Fig. 17-17.

11. The capacitance and equivalent shunt resistance of a small capacitor are to be determined using a Q-meter in a circuit similar to Fig. 17-20. Describe the procedure to be followed and derive the equations for capacitance and shunt resistance, in terms of frequency f, two values of the calibrated capacitor C, and two readings of Q.

CHAPTER 18

ULTRAHIGH-FREQUENCY AND MICROWAVE CIRCUITS

LIMITATIONS OF ORDINARY CIRCUIT ELEMENTS

Ordinary circuit elements, that is resistors, inductors, and capacitors, are used from the lowest frequencies up through the ultrahigh frequencies. However at very high frequencies, and more especially at ultrahigh frequencies certain effects which are negligible at lower frequencies begin to have an important bearing on the operation of the circuit element. For example, a simple coil has an inductance L and an inductive reactance $X_L = 2\pi fL$ which is proportional to frequency. The coil also has a certain "distributed capacitance" between the turns, and this distributed capacitance can be considered as an effective capacitance C_e, shunting the coil as in Fig. 18-1. This effective capacitance presents a capacitive re-

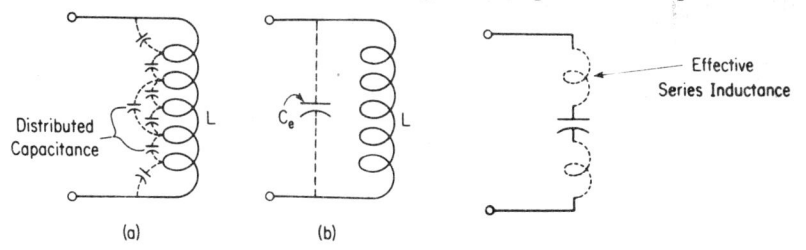

FIG. 18-1. Distributed capacitance between turns considered as an effective shunting capacitance C_e.

FIG. 18-2.

actance $X_c = 1/(2\pi fC_e)$, which is inversely proportional to frequency. At low frequencies this capacitive reactance is ordinarily extremely high and has negligible shunting effect on the inductive reactance of the coil. However, as the frequency is allowed to increase the inductive reactance of the coil increases and the capacitive reactance of the shunting capacitance decreases, so that for a given coil, a frequency can always be reached where these two reactances are equal. At this frequency the coil acts essentially as a parallel-resonant circuit, with an impedance that is resistive. Above this frequency the shunting capacitive reactance of the coil is *less* than the inductive reactance and the input impedance appears as a capacitive reactance. In other words the coil now acts as a capacitor (refer to Problem 1 at end of chapter).

Similarly a capacitor with its leads always has a small series inductance as indicated in Fig. 18-2. This series inductance is made up of the inductance of the leads and the inductance of the plates themselves. At low frequencies the inductive reactance of this series inductance is negligible compared with the capacitive reactance of the capacitor, but as the frequency is increased the inductive reactance increases and the capacitive reactance decreases, so that a point is reached where series resonance occurs. Above this frequency the capacitor may present an inductive reactance, and so act as a coil. (The actual operation at these high frequencies is further complicated by the fact that there is also a shunt capacitance between the leads so that the resulting input reactance may be either inductive or capacitive).

An ordinary resistor in general has some series inductance, some shunt capacitance, and often some dielectric loss, so that its operation at ultrahigh frequencies can be very complicated, and its impedance may have almost any value. However, special uhf resistors have been constructed and these can maintain fairly constant resistance values through the uhf band.

From the above discussion it will be apparent that the operation of simple "ordinary" circuits of L, C, and R can become quite complicated at ultrahigh frequencies. In order to maintain useful values of reactance (say between 100 and 1,000 ohms for parallel tuned circuits) and in order to minimize the effects mentioned above it is necessary to make both coils and capacitors smaller as the frequency is increased. In the limit a tuned circuit may appear as shown in Fig. 18-3 where the coil consists of the shorted capacitor leads.

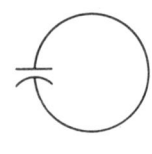

Fig. 18-3. *L-C* circuit consisting of capacitor and shorted leads.

TRANSMISSION LINES AND WAVEGUIDES AS CIRCUIT ELEMENTS

Conveniently, at these ultrahigh frequencies where ordinary circuits become difficult to work with, another type of circuit becomes useful. This second type is the transmission-line or distributed-constants circuit. At still higher frequencies (the microwave or superhigh-frequency region) waveguide types of circuits become of practical size.

Although a primary function of transmission lines or wave guides is the transfer of energy from one point to another, at vhf, uhf and shf it is possible and convenient to use short sections of low-loss transmission lines or waveguides as circuit elements. In order to understand the operation of these distributed-constants circuits it is necessary to reconsider the phase relations of voltage and current along a transmission line under various conditions of the terminating or load impedance.

Chapter 18 *Ultrahigh-Frequency and Microwave Circuits* 577

PHASE RELATION OF VOLTAGE AND CURRENT ON A TRANSMISSION LINE

Figure 18-4(a) shows the relative phase of voltage along a line that is terminated in its characteristic impedance Z_0. The positions of the rotating arrows indicate the phases of the voltages (or currents) at the various points along the line. Figure 18-4(a) is for a given instant of time. As time goes on the arrows all rotate in a counterclockwise direction, as shown, with the frequency f (cycles per second) or the angular frequency $2\pi f$ (radians per second). However their *relative* positions remain unchanged, so Fig. 18-4(a) gives the complete picture of voltage along the line. The instantaneous

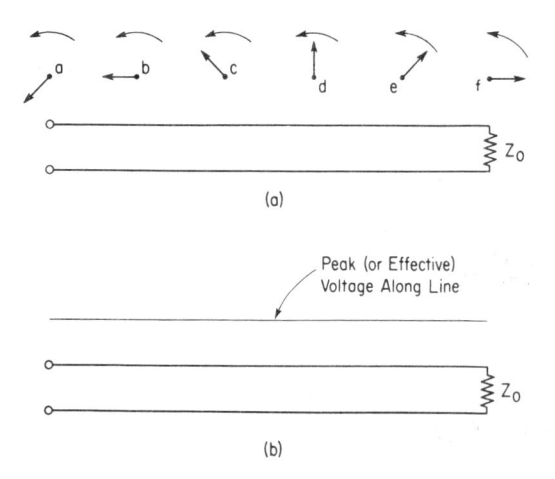

Fig. 18-4. (a) Relative phase of voltage or current along a transmission line terminated in Z_0; (b) relative magnitude of voltage or current along a (low-loss) transmission line terminated in Z_0.

value of voltage at any point is given by the projection of the arrow on the vertical axis. Thus at the instant shown the voltage at d is a positive maximum, at b it is zero, and at f it is zero. At a later instant the voltage at e will be a positive maximum and so on. It is seen that as time goes on the sinusoidal voltage wave along the line moves to the right. Figure 18-4(b) shows the peak (or effective) voltage along the line as would be indicated by a peak reading (or effective value reading) voltmeter. Since the line was assumed to have very low loss and to be terminated in its characteristic impedance which completely absorbs the incident wave, the plot of peak or effective voltage along the transmission line is a straight line as shown.

Figure 18-4 could equally well represent the current distribution along the line. Because the characteristic impedance of a low-loss line is a pure resistance, the voltage and current of the incident wave will be in phase, and so the current arrow would have at every point the same direction as the voltage arrow.

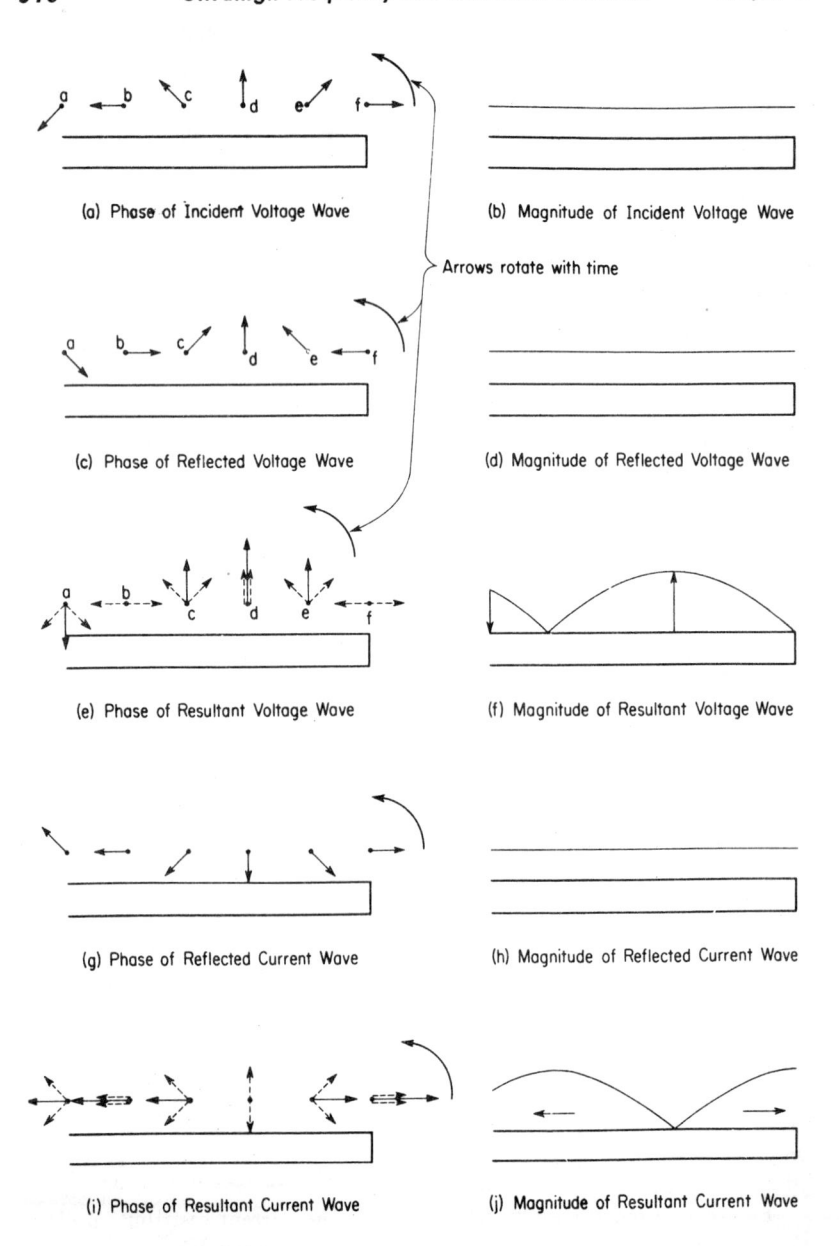

Fig. 18-5. Phase relations of voltage and current on a short-circuited transmission line.

Chapter 18   *Ultrahigh-Frequency and Microwave Circuits*   579

If the transmission line of Fig. 18-4 were terminated in some different value of impedance the *incident* wave voltage and currents along the line would remain unchanged, because the incident wave has no way of knowing what the line termination is. However, there will now be set up at the receiving end of the line a reflected wave which travels back along the line towards the generator. The magnitude and phase of this reflected wave relative to the incident wave depends upon the load impedance. Two special cases are of great importance: a short-circuit termination and an open-circuit termination.

When the transmission line is short-circuited at the receiving end as in Fig. 18-5, a reflected wave is set up at the short-circuit and travels towards the generator end. The magnitude and phase of the reflected wave must be such that the total voltage at the short-circuit is zero, for no voltage can exist across a short-circuit or zero impedance. This condition requires that the reflected wave have the same amplitude but a phase opposite to that of the incident wave at the reflecting point. Figure 18-5(c) shows the phase of the reflected wave along the line. For the reflected wave the phase at point e *lags* that at point f because the wave is traveling from right to left. The total voltage at each point is the vector sum of the voltages owing to incident and reflected waves, and this addition is shown in Fig. 18-5(e). The length of the total-voltage arrows varies sinusoidally along the line. The corresponding peak or effective voltage distribution (as would be indicated by a voltmeter) is shown in Fig. 18-5(f). Voltage nodal points (points where the voltage is *always* zero) occur at points b and f. The relative phase of the *total* voltage *remains constant* along the line except for a phase reversal or change of sign that occurs at the nodal points. This means that the total voltage reaches its maximum value at the same instant of time at all points along the line. As the arrows of Fig. 18-5(e) rotate with time the curve of instantaneous total voltage along the line alternately expands and collapses, but does not progress, and the zero voltage points remain fixed at b and f. This is typical of the *standing wave* distribution that exists on shorted or open lines.

For the waves of current on a short-circuited line, the incident wave will again be represented by the diagram of Fig. 18-5(a). However, at the short-circuit termination the current wave is completely reflected *without change of phase*, so that incident and reflected waves are equal and of the same phase at this point. The reflected wave current traveling toward the generator is then represented by the diagram of Fig. 18-5(g). When the current of the reflected wave is added to the current of the incident wave at each point to obtain the total current the diagrams of Figs. 18-5(i) and 18-6(j) are obtained. Comparing these diagrams with those of (e) and (f) for the total voltage distribution shows 2 differences:

(1) The points of zero voltage occur at the short circuit and at multiples of one-half wavelength from the short (points b and f), whereas the points of voltage maximum occur one quarter (and 3/4, 5/4, etc.) of a wavelength from the shorted end. For the current distribution, points of maximum current occur at the shorted end and at multiples of a half wavelength from the short, whereas zero current points occur at odd multiples of a quarter wavelength from the short. Thus points of maximum current occur at places of zero voltage and vice versa.

(2) Although the incident waves of voltage and current were assumed to be in phase [the same diagram, Fig. 18-5(a), was used for each], it is seen that the arrows representing total voltage [Fig. 18-5(e)] and those representing total current [Fig. 18-5(i)] are displaced 90° from each other. That is, total voltage and total current are 90° out of phase. At the instant when the total voltage is a maximum along the line, total current is everywhere zero, and vice versa. These 90° phase relationships between voltage and current are those that exist with ordinary inductive and capacitive reactances. This fact makes it possible to use sections of open- or short-circuited transmission lines as reactance elements.

TRANSMISSION LINE SECTIONS AS REACTANCE ELEMENTS

Consider conditions at point e on diagrams (e) and (i) of Fig. 18-5. At this point (and all other points between d and f) the resultant voltage leads the resultant current by 90°. This is true regardless of how the line is fed at the input or how long a length of line might exist between the input and point e. In particular, if the line were cut off at the point e, and this point were made the input, then the voltage and current relations at this point would still be as shown. Because the current lags the voltage by 90° at this point, the input impedance of the section of line would appear as on *inductive reactance*. At points to the right of e the total voltage is smaller and the total current is larger, so the value of the inductive reactance given by $X = V/I$ is less. To the left of e the voltage is greater and the current is smaller, so the reactance is greater. At point f the voltage is zero and the current is large, so the reactance is zero, which is correct for a short-circuit. At point d the voltage is large and the current is zero, so the effective input reactance is infinite. Therefore at this point (one-quarter wavelength from the short-circuit) the input impedance of the line looks like an open circuit. To the left of the current nodal point d it is seen that the current *leads* the voltage. Thus at point c, for example, the input impedance of the line would appear as a *capacitive* reactance. This capacitive reactance would vary from an infinitely large value at d to zero value at b.

These results are summarized in Fig. 18-6 which shows the equivalent reactances of shorted lines. Also shown in Fig. 18-6 are the equivalent reactances of open-circuited lines. Open-circuited lines perform as do short-circuited lines, except that at an open circuit the current is reversed in

Chapter 18 *Ultrahigh-Frequency and Microwave Circuits* 581

phase (total current must be zero at an open circuit) and the voltage wave is reflected without phase reversal. Thus in Fig. 18-5, diagrams (e) and (f) can represent the current distribution along an open-circuited line, while diagrams (i) and (j) represent the voltage distribution along an open-circuited line. Within a quarter of a wavelength from the end of the line the current will now lead the voltage so that the input impedance of short open lines is a capacitive reactance as indicated in Fig. 18-6. For open-circuited lines

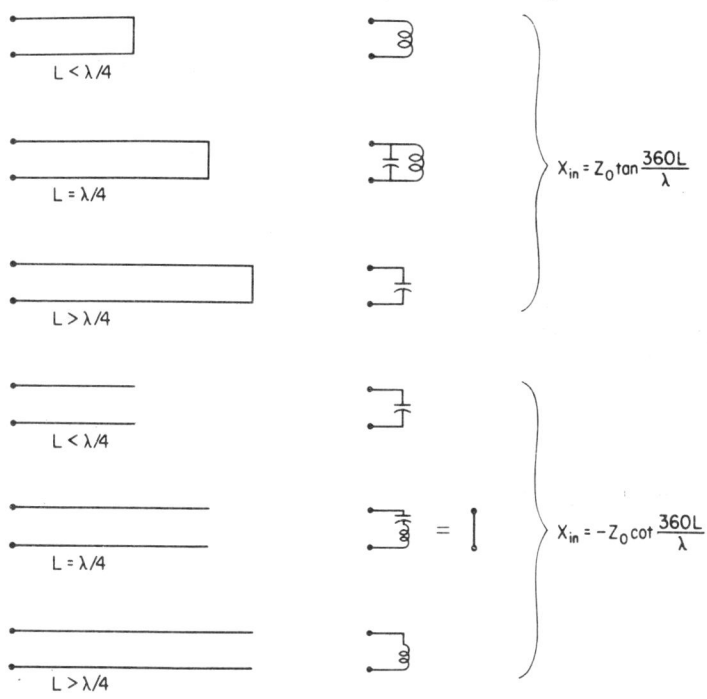

FIG. 18-6. Equivalent input reactances of low-loss shorted and open lines.

longer than a quarter of a wavelength but less than a half-wavelength long the input reactance is inductive. At a quarter of a wavelength from the open circuit, the voltage is zero while the current is a maximum. The input impedance at this point is zero, that is, it appears as a short-circuit.

It is quite easy to calculate the input reactances of shorted and open sections of lines, and the ability to do this increases their usefulness as circuit elements. On a *short-circuited line* the voltage distribution along the line is *sinusoidal* and is given by the formula

$$V = V_{max} \sin 360 \left(\frac{x}{\lambda}\right)$$

where x is the distance along the line from the short-circuit. The quantity

x/λ is this distance measured in wavelengths. The current distribution also has the sine-wave shape, but the maximum current occurs at the short-circuit, that is where $x = 0$. The current at any point x is given by

$$I = I_{max} \cos 360 \left(\frac{x}{\lambda}\right).$$

This distribution is said to be *cosinusoidal*. If the shorted line has a length L, then the voltage and current at the input to the line will have the values given when x is put equal to L, that is

$$V_{in} = V_{max} \sin 360 \left(\frac{L}{\lambda}\right),$$

$$I_{in} = I_{max} \cos 360 \left(\frac{L}{\lambda}\right).$$

The input reactance of the line will be given by the ratio of the input voltage to the input current, that is by

$$X_{in} = \frac{V_{in}}{I_{in}} = \frac{V_{max} \sin 360 \, (L/\lambda)}{I_{max} \cos 360 \, (L/\lambda)} = \frac{V_{max}}{I_{max}} \tan 360 \left(\frac{L}{\lambda}\right).$$

Regardless of the terminating impedance, the ratio of the maximum voltage on the line to the maximum current is *always* equal to the characteristic impedance Z_0 (these points of maximum voltage and maximum current are always one-quarter wavelength apart). That is

$$\frac{V_{max}}{I_{max}} = Z_0.$$

Therefore the input reactance of a short-circuited line of length L is given by

$$X_{in} = Z_0 \tan 360 \left(\frac{L}{\lambda}\right).$$

It will be noted that for values of L/λ less than 0.25 (that is for line lengths less than a quarter wavelength) the tangent of the angle is positive and the reactance will be positive, that is, it will be an inductive reactance. However when L/λ is greater than 0.25 but less than 0.5 (line lengths between a quarter wavelength and a half wavelength) the tangent of the angle is negative and the input reactance of the short-circuit line will be negative or capacitive.

For open-circuited lines the voltage and current distributions are just the reverse of those for shorted lines. The voltage distribution is cosinusoidal with the maximum voltage at $x = 0$, and the current distribution is sinusoidal with the zero current point at the open circuit, $x = 0$. That is for open-circuited lines

$$V = V_{max} \cos 360 \left(\frac{x}{\lambda}\right),$$

Chapter 18   **Ultrahigh-Frequency and Microwave Circuits**   583

$$I = I_{max} \sin 360 \left(\frac{x}{\lambda}\right).$$

The input reactance of an open-circuited line of length L is given by

$$X_{in} = -\frac{V_{max}}{I_{max}} \cot 360 \left(\frac{L}{\lambda}\right) = -Z_0 \cot 360 \left(\frac{L}{\lambda}\right).$$

The minus sign must be used here because for line lengths less than a quarter of a wavelength (L/λ less than 0.25), the input reactance is negative (capacitive). However, for L/λ greater than 0.25 but less than 0.5, the cotangent of the angle is negative and the input reactance therefore becomes positive or inductive.

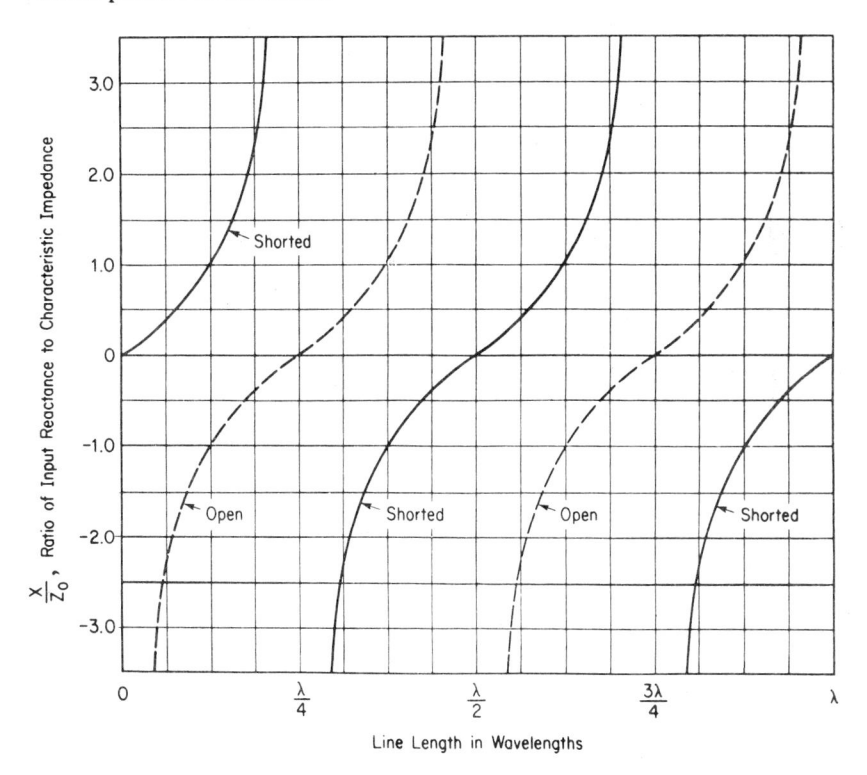

Fig. 18-7. Input reactance of short-circuited and open lines.

In Fig. 18-7 the input reactances of shorted and open lines are shown plotted against the length of the line in wavelengths. These curves are of course just plots of the tangent and cotangent functions. In both cases the reactance always has a value equal to the characteristic impedance Z_0 when the line is an eighth (or $\frac{3}{8}$, $\frac{5}{8}$, etc.) of a wavelength long.

DIRECTION OF ENERGY FLOW

The question of why the voltage wave is reflected with a phase reversal at a short-circuit whereas the current wave is reflected without phase reversal, and vice versa for open-circuited lines, can be answered by considering the direction of energy flow or power flow. The incident wave has associated with it a certain amount of energy (this will be discussed in detail in a later section on the Poynting vector). Because no energy can be absorbed in a short-circuit this incident energy must be reflected back along the line towards the generator. At the short-circuit the phase of the voltage wave must be reversed on reflection because the voltage of the incident wave plus that of the reflected wave must add up to a total voltage of zero at the short-circuit. Since the voltage wave is reversed in phase on reflection, the current wave must be reflected without change of phase. This is so because in order to reverse the direction of power flow the sign of the voltage or the sign of the current must be reversed, *but not both*. This can be shown by the simple example of power flow from a direct-current battery.

In Fig. 18-8(a) a battery delivers power to the load. The conventional symbolism of positive current out of the positive battery terminal has been used. In this case the power flows from the battery to the load on its right. In Fig. 18-8(b) the battery polarity has been reversed so that the top line is

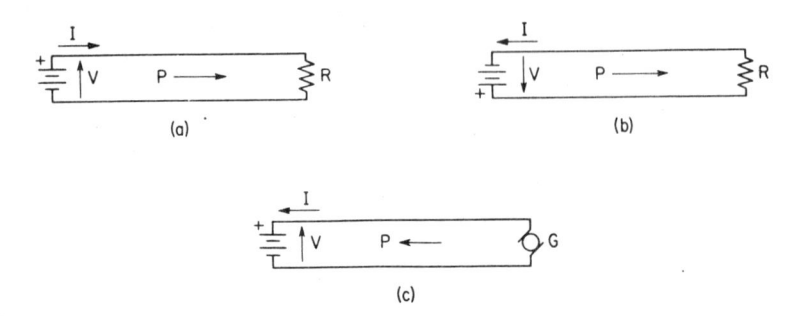

Fig. 18-8. Reversal of direction of power flow P requires a reversal of either V or I, but not both.

now negative with respect to the bottom line, but the current direction has also reversed, and of course the power continues to flow from left to right, that is from battery to load. In Fig. 18-8(c) the original battery polarity of Fig. 18-8(a) has been retained but the current direction has been reversed. This has been accomplished by replacing the load resistance R by a generator G which generates an open-circuit d-c voltage slightly higher than that of the battery. The power now flows from right to left. That is, the generator charges the battery. If now the motor driving the generator is disconnected, the generator will tend to slow down and its generated voltage

Chapter 18   *Ultrahigh-Frequency and Microwave Circuits*   585

will tend to decrease below that of the battery. The current direction will then reverse from that shown in Fig. 18-8(c), the voltage polarity will remain unchanged, and the power flow will be from left to right. The battery will now run the generator as a motor. Thus it is seen that in order to reverse the direction of power flow it is necessary to reverse the sign of the voltage, or the current, but not both.

THE RESONANT QUARTER-WAVE LINE

A quarter-wave section of transmission line, shorted at one end and open at the other, has some interesting and useful properties. It is a *resonant section* with a voltage and current distribution as shown in Fig. 18-9(a). It may be compared with the ordinary lumped-constants resonant

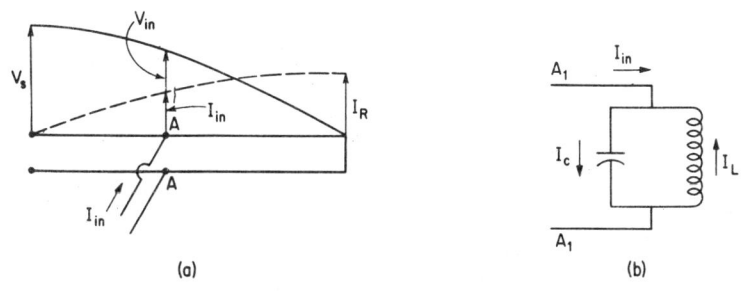

FIG. 18-9. Antiresonant circuits: (a) short-circuited quarter-wave transmission line; (b) lumped-constant equivalent.

circuit of Fig. 18-9(b). In the latter circuit the energy is stored alternately in the coil and capacitor. The voltage and current are 90° out of phase. That is, at the instant the voltage across the capacitor or the coil is a maximum the current through the coil is zero, and a quarter of a cycle later when the current is a maximum the voltage is zero. At resonance the maximum electric energy (stored in the capacitor) and the maximum magnetic energy (stored in the inductor) are equal. During an a-c cycle, energy is alternately transferred from the electric field of the capacitor to the magnetic field of the inductor and vice versa. The only power required from an external source is that required to furnish the ohmic or resistance losses of the circuit.

This means that even for a large circulating current around the circuit, the input current fed in at the terminals A_1-A_1 need be only a quite small amount. This required input current becomes less as the resistance of the circuit decreases and it is zero for the theoretical case of no resistance.

For the resonant line section of 18-9(a) the voltage and current are also 90° out of phase. The current, I_{in}, fed in at a pair of terminals A-A located anywhere along the line will be quite small compared with the current I_R in the line at the shorted end. Again for the theoretical case of no resistance losses the input current (and input power) would be zero.

Both resonant circuits of Fig. 18-9 can be used as step-up or step-down transformers, with connections as indicated in Fig. 18-10. In Fig. 18-10(b) a voltage across the terminals A-A may be stepped down to a smaller voltage across B-B or, conversely, a voltage V_{BB} can be stepped up to a larger voltage across A-A. Similarly in Fig. 18-10(a) an applied voltage V_{AA} can be stepped down to a smaller voltage V_{BB} or, conversely, a voltage V_{BB} can be stepped up to a larger voltage V_{AA}. For proper operation in this manner, the currents into or out of the terminals A-A and B-B should be kept small compared with the circulating current in Fig. 18-10(b) or the line current in

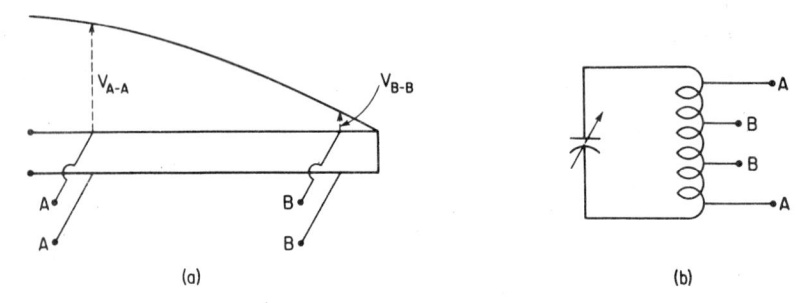

Fig. 18-10. (a) Transmission line and (b) lumped-constant antiresonant circuits used as voltage transformers.

Fig. 18-10(a). This means that the connected loads and generators should have impedances that are large compared with the reactances of the circuit branches across which they are connected. If this is not the case the circuits will be detuned and will have to be readjusted for resonance. In the case of Fig. 18-10(b) this requires a resetting of the capacitor X_c. In Fig. 18-10(a) it requires a readjustment of the line length to something other than a quarter wavelength. In particular if the connected loads are not purely resistive, but have some reactance, the circuits of Fig. 18-10 may be considerably detuned.

In both circuits of Fig. 18-9 the ohmic resistance will ordinarily be quite small and for many purposes may be neglected, that is, considered to be zero. This procedure is permissible as long as the result required does not depend directly on the value of this resistance. Several examples will be considered.

In Fig. 18-9(b) if a voltage V is applied at the terminals A_1-A_1 the current through the capacitor will be $I_c = V/X_c$ and this will be equal in magnitude (very nearly) to the current through the coil $I_L = V/X_L$, and to a first approximation will be independent of the resistance of the circuit. However the *input* current I_{in} will be directly dependent on the resistance in the circuit. This can be seen by considering the power dissipated in the circuit. Assuming the resistance of the circuit to be the inductor resistance, r_L, the power loss will be:

Chapter 18   *Ultrahigh-Frequency and Microwave Circuits*   587

$$I_L^2 r_L = \frac{V^2 r_L}{X_L^2}.$$

The power loss must be supplied by the input power. Since at resonance the input current is in phase with the input voltage, the input power is VI_{in}. From these two relations it is seen that

$$I_{in} = \frac{V r_L}{X_L^2} = \frac{V}{X_L^2/r_L} = \frac{V}{R_{in}}.$$

Thus the effective input resistance has the value

$$R_{in} = \frac{X_L^2}{r_L}.$$

When the inductor resistance r_L is very small the input resistance R_{in} is very large, and R_{in} approaches infinity as r_L approaches zero. Thus the input current approaches zero as r_L approaches zero.

Similarly, in the circuit of Fig. 18-9(a) if the input terminals, A-A, are considered to be at the sending end of the line, the input voltage will be

$$V_{in} = V_S$$

The current I_R through the short bar (a quarter wavelength from V_S) will be

$$I_R = \frac{V_S}{Z_0}$$

and this current is independent of the resistance of the line. However the *input* current depends directly on the line resistance. For the theoretical case of a zero resistance line the input current is zero. For the actual case where the line has some small resistance, power will be dissipated by the line current flowing through this resistance. When this $I^2 r$ power loss is summed up all along the line and equated to the input power, which is $V_S I_{in}$, the input current is found to be

$$I_{in} = \frac{V_S}{2Z_0^2/(rl)}$$

where r is the resistance per meter length of the line (both conductors) and l is the length of the line in meters. Thus the effective input resistance of the shorted quarter-wavelength line is

$$R_{in} = \frac{2Z_0^2}{rl} = \frac{8Z_0^2}{r\lambda}.$$

For the theoretical quarter wave line which has no loss, $r = 0$ and the input resistance $R_{in} = \infty$. For an actual line r has value, but is usually very small, so the input resistance is very large. The resistance r per unit length of line can be computed for a parallel-wire copper transmission line from the formula:

$$r = \frac{8.31 \times 10^{-8}\sqrt{f}}{a} \quad \text{ohms/meter}$$

where f = frequency in cycles per second and a is conductor radius in meters. For concentric lines the corresponding formula is

$$r = 4.16 \times 10^{-8}\sqrt{f}\left(\frac{1}{a} + \frac{1}{b}\right) \quad \text{ohms/meter}$$

where a is the radius of the inner conductor and b is the radius of the outer conductor.

In the circuits of Fig. 18-9 the voltage source was connected in parallel with, or across, the line or circuit elements. This connection produces what is known as a *parallel-resonant*, high impedance circuit. An alternative connection is that shown in Fig. 18-11(a) and (b) where the voltage is connected in series with the line or circuit elements. This connection is known as a

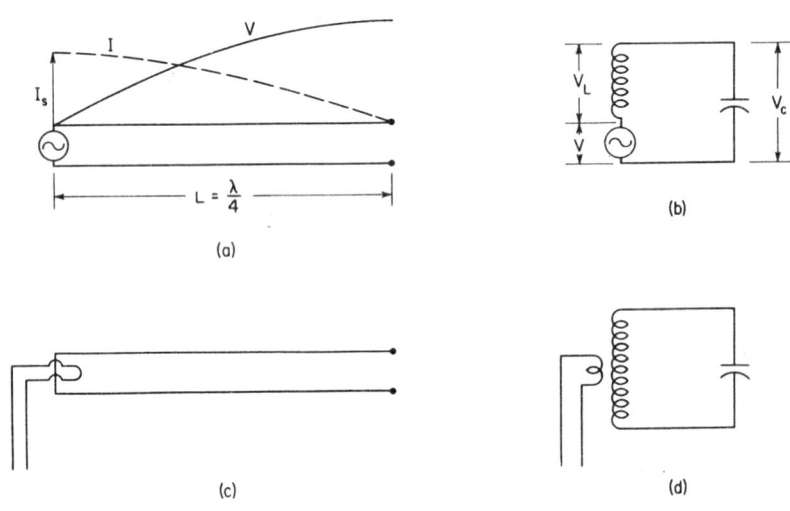

Fig. 18-11. Series-resonant circuits.

series-resonant circuit, and the results obtained with it are considerably different than with the parallel connection. In Fig. 18-11(b) at resonance the reactance of the inductor is equal and opposite to that of the capacitor and the current is limited by the series resistance of the circuit. That is, the current around the circuit will be

$$I = \frac{V}{r_L}$$

where r_L is the resistance of the inductor (assumed to be the only resistance in the circuit). For small values of inductor resistance the current can be very large at resonance. However even for $r_L = 0$ in an actual case the current is

Chapter 18   *Ultrahigh-Frequency and Microwave Circuits*   589

limited to the value of the short-circuit current of the generator, i.e., by the resistance as well as the open-circuit voltage of the generator. The voltage across the coil or capacitor will be

$$V_c = V_L = I\,X_L = V\frac{X_L}{r_L} = QV.$$

The voltage applied is thus seen to be stepped up Q times, where, as usual, $Q = X_L/r_L$ is the quality factor of the inductor, assuming the resistance of the capacitor is negligible.

In Fig. 18-11(a), if the line resistance per unit length were zero the input power required, I^2R_{in}, would be zero, and the input resistance would also be zero. However, when the line is an actual one with a small amount of resistance r per unit length, the input resistance of the line can be found as before by equating input power with power lost. The input resistance in this case is found to be

$$R_{in} = \frac{rl}{2}.$$

It is a small resistance equal to just one-half the total series resistance of the line. The input current I_S will be

$$I_S = \frac{V_S}{R_{in}}$$

and the output voltage V_R will be Z_0 times this. That is,

$$V_R = I_S Z_0 = V_S \frac{Z_0}{R_{in}} = V_S \frac{2Z_0}{rl} = \frac{8Z_0}{r\lambda} V_S.$$

For a low-loss line the series resistance r per unit length is small and the input current I_S and output voltage V_R may become very large. In the limit, for a line with no loss, the input current would be the short-circuit current of the generator, and the output voltage V_R would be Z_0 times this short-circuit current.

Parts (c) and (d) of Fig. 18-11 show commonly used practical circuits that are equivalent respectively to Fig. 18-11(a) and (b). In Fig. 18-11(c) and (d) current in the coupling coil induces a generator voltage effectively in series with the resonant line or resonant circuit.

THE QUARTER-WAVE LINE AS A MATCHING SECTION

The quarter-wave line can be used as a matching transformer to match two impedances one to the other. In Fig. 18-12(a) and (b) are illustrated the voltage and current distributions that exist on a quarter-wave section when it is terminated (a) in a resistance less than Z_0 and (b) in a resistance greater than Z_0. The shape of these curves is independent of the value of the generator impedance Z_g since the shape is dependent only on the relative amplitude of the initial and reflected waves. The impedance-matching properties can be derived as follows. The ratio of output voltage V_R to output current I_R is equal to the terminating impedance Z_R. That is,

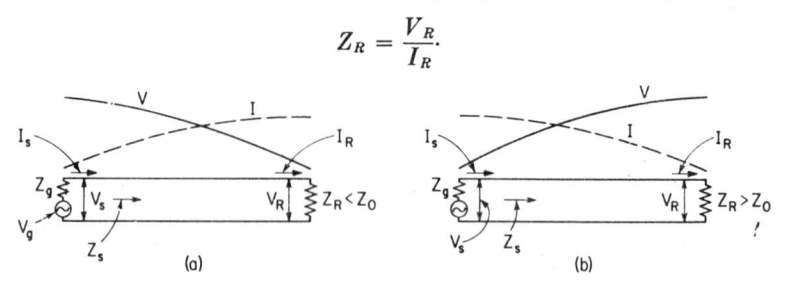

$$Z_R = \frac{V_R}{I_R}.$$

Fig. 18-12. Voltage and current distributions on a quarter-wave line that is terminated (a) in a resistance less than Z_0, and (b) in a resistance greater than Z_0.

Also the sending end impedance, that is, the impedance regarded from the sending end of the line is

$$Z_S = \frac{V_S}{I_S}.$$

The voltages and currents a quarter wavelength apart on a line are always related by the characteristic impedance of the line, that is,

$$V_S = I_R Z_0 \quad \text{and} \quad V_R = I_S Z_0.$$

Combining these relations it is found that

$$Z_S = \frac{V_S}{I_S} = \frac{I_R Z_0}{I_S} = \frac{I_R Z_0^2}{V_R} = \frac{Z_0^2}{Z_R}.$$

Thus the input impedance of a quarter-wave line is equal to the square of the characteristic impedance divided by the terminating impedance. The quarter-wave line is said to be an *impedance inverter* because it inverts the impedance about Z_0. That is, if Z_R is small, Z_S will be large and vice versa. If Z_R is inductive, Z_S will be capacitive and vice versa. In particular if Z_R is equal to zero (a short-circuit), the input impedance of the lossless quarter-wave section will be infinite (an open circuit). Conversely the input impedance of an open-circuited lossless quarter-wave section ($Z_R = \infty$) would be zero (that is, a short-circuit).

EXAMPLES OF MATCHING WITH QUARTER-WAVE LINES

Quarter-wave line sections are often used to match together two lines of different characteristic impedances. This matching problem is illustrated in Fig. 18-13 where a line of characteristic impedance Z_2 is matched to a line of characteristic impedance Z_1. Assuming that line 2 is terminated in its

Chapter 18   **Ultrahigh-Frequency and Microwave Circuits**   591

characteristic impedance Z_2, then its input impedance is also Z_2, and this is the terminating impedance for the quarter-wave matching section. The input impedance of the matching section is then

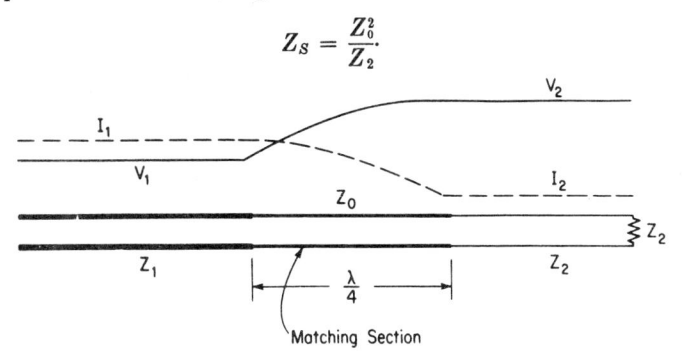

$$Z_S = \frac{Z_0^2}{Z_2}.$$

FIG. 18-13. Quarter-wave section used to match two lines of different characteristic impedances.

The problem is to choose Z_0 to be of such a value that Z_S will be equal to the characteristic impedance Z_1 of line 1. This value is given by

$$Z_1 = \frac{Z_0^2}{Z_2} \quad \text{or} \quad Z_0 = \sqrt{Z_1 Z_2}.$$

It is seen that the quarter-wave line section must have a characteristic impedance that is the *geometric mean* of the two impedances to be matched. For example, if Z_1 were 100 ohms and Z_2 were 400 ohms, the required

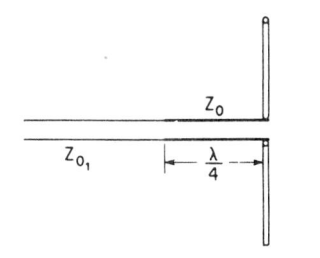

FIG. 18-14. Use of quarter-wave section to match an antenna to its feed line.

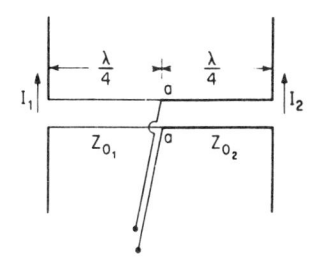

FIG. 18-15.

characteristic impedance of the matching section would be 200 ohms.

A quarter-wave section is often used to match an antenna to its transmission line. Such matching is illustrated in Fig. 18-14. If the antenna has a resistive impedance $Z_a = 70$ ohms, for example, and it is to be connected to a line having a characteristic impedance $Z_{0_1} = 300$ ohms, the required value of characteristic impedance for the quarter-wave section is

$$Z_0 = \sqrt{300 \times 70} = 148 \text{ ohms}.$$

Quarter-wave sections may also be used to cause any desired ratio of currents to flow into several antennas or other loads. This is illustrated in Fig. 18-15 where two antennas are fed from a common junction a-a through quarter-wave sections. Since the current in a line is equal to the voltage a quarter-wave distant divided by the characteristic impedance of the line, the antenna currents will be

$$I_1 = \frac{V_{a-a}}{Z_{0_1}} \qquad I_2 = \frac{V_{a-a}}{Z_{0_2}}.$$

Because the quarter-wave sections are fed with the common voltage V_{a-a} the ratio of currents in the two antennas will be

$$\frac{I_1}{I_2} = \frac{Z_{0_2}}{Z_{0_1}}.$$

The antenna currents are in inverse proportion to the characteristic impedances of the quarter-wave sections which connect them together. This result is independent of the actual antenna impedances, which may or may not be equal to one another.

THE HALF-WAVELENGTH SECTION

The half-wavelength section of transmission line also has some interesting and useful properties. Regardless of the value of the load termination two points one-half wavelength apart on a lossless transmission line are always points of equal voltage. (This statement assumes, of course, that there is no load across the line between the points.) Also the current magnitudes at two points one-half wavelength apart are always equal. Thus a half-wave section of lossless line has the unique properties of a perfect one-to-one transformer, with the input voltage always equal to the output or load voltage, and the input current always equal to the output or load current. The phases of input and output currents also differ by 180°, so the half-wave section is also a phase inverter. Because of the equality of input and output voltages and of input and output currents, the input impedance will always be equal to the output or load impedance as would be expected of a one-to-one transformer.

Figure 18-16 illustrates the use of both quarter-wave and half-wave sections in feeding on antenna array with a given ratio of currents. Because the antennas are fed through quarter-wave sections their currents will depend upon the characteristic impedances of these sections and the voltages at the inputs of the quarter wave sections. That is

$$I_1 = \frac{V_{a-a}}{Z_{0_1}} \quad I_2 = \frac{V_{b-b}}{Z_{0_2}} \quad I_3 = \frac{V_{c-c}}{Z_{0_3}} \quad I_4 = \frac{V_{d-d}}{Z_{0_4}}.$$

However points c-c and d-d are joined by a half-wavelength section and therefore must have the same voltages but opposite phases (similarly for

Chapter 18   **Ultrahigh-Frequency and Microwave Circuits**   593

points a-a and b-b, so that all the quarter-wave sections are fed with voltages of equal magnitudes). Therefore the antenna currents will be inversely proportional to the characteristic impedances of the quarter-wave sections feeding them, and these can be made to have any desired ratios within reasonable limits. If the currents in all antennas should be in phase the alternate connections to the feed line should be reversed as shown in Fig. 18-16.

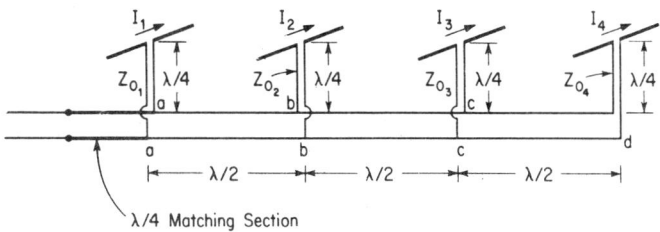

FIG. 18-16. Method of feeding elements of an antenna array with specified current ratios.

As an example, let it be required to feed antennas 1 and 4 with 1 amp each and antennas 2 and 3 with 2 amp each. Then $Z_{0_1} = Z_{0_4} = 2Z_{0_2} = 2Z_{0_3}$. The actual Z_0's chosen should be such as to result in reasonable input impedances. For example, if the antennas are assumed to have resistive impedances of 75 ohms each, suitable values for the Z_0's might be

$$Z_{0_1} = Z_{0_4} = 300 \text{ ohms}, \qquad Z_{0_2} = Z_{0_3} = 150 \text{ ohms}.$$

The input impedances to the quarter-wave sections are then 1,200, 300, 300, and 1,200 ohms respectively. These loads would appear across the main feed line as shown in Fig. 18-17. Because these loads are all multiples of a half-wavelength apart, they have the same voltage across them and are

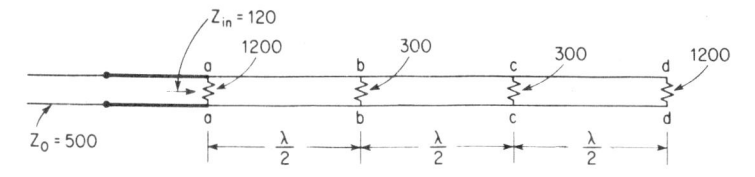

FIG. 18-17. Effective values of circuit of Fig. 18-16 for line values discussed in text.

essentially all in parallel. The total input impedance at a-a is therefore 120 ohms. Note that four times the power would be delivered to each of the central antennas as compared with the end antennas. Assuming the entire array is to be fed with a 500-ohm line, the correct value for a quarter-wave matching section would be

$$Z_0 \text{ (quarter-wave section)} = \sqrt{120 \times 500} = 245 \text{ ohms}.$$

STUB-LINE MATCHING AND DOUBLE-STUB TUNERS

Impedances may also be matched at vhf and uhf by means of stub-line tuners as illustrated in Fig. 18-18. In Fig. 18-18(a) it is desired to match the load impedance R_L to the transmission line having a characteristic resistance Z_0. The voltage curve sketched indicates that the load resistance is less than Z_0 because the voltage is low at the load point. Therefore at a point a quarter-wavelength back from the load the input resistance will be greater than Z_0. It follows that at some point in between, at a distance l from the load, the input resistance will be equal to Z_0. However at this point there will also be some reactance, which may be represented as being in parallel with the resistance. If now this parallel reactance is

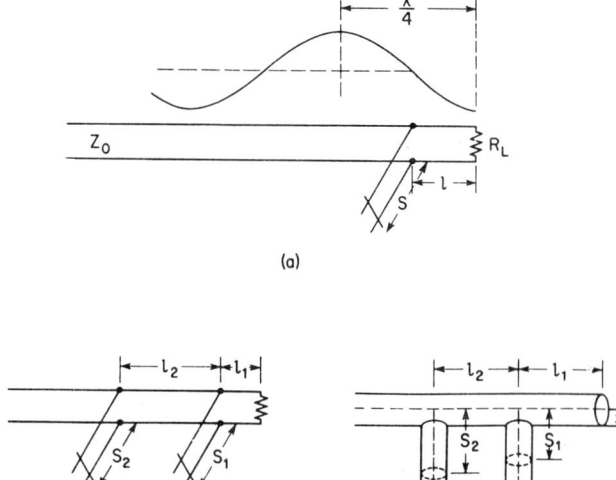

FIG. 18-18. Impedance matching by means of stub-line tuners.

tuned out by means of the parallel stub of variable length S placed at the distance l, the input impedance at l will be a pure resistance of value equal to Z_0. The line therefore will be properly matched at this point and the voltage standing wave curve to the left of the stub will now be as shown by the dotted line. Two adjustments are required to perform this match, i.e., the distance l and the length S. Two adjustments or controls are always required to perform an impedance match because both magnitude and phase, or resistance and reactance, of the impedances must be matched.

Figure 18-18(b) illustrates an alternative method of matching a load to a line. In this case two variable length tuners are placed at fixed positions l_1

and l_2 on the line, and the impedance match is obtained by adjusting S_1 and S_2. This alternative method is especially advantageous in the case of coaxial line arrangments [Fig. 18-18(c)] where adjustments of line lengths l_1 and l_2 would be difficult. The arrangements shown in (b) and (c) are called double-stub tuners. They may be used to match any two impedances within a given range which depends upon the choice of l_2. To match any two impedances whatsoever a triple-stub tuner is sometimes used.

WAVEGUIDE MICROWAVE CIRCUITS

In addition to their primary function of guiding energy from one point to another, sections of coaxial and parallel-wire transmission lines can be used as ordinary circuit elements for impedance matching and other purposes, as has been shown. The range of frequencies where such transmission-line circuits have chief application is from about 30 to 3,000 megacycles, i.e., the vhf and shf bands. Below 30 megacycles the length of a quarter- or half-wavelength section becomes too large for convenience, and above about 3,000 megacycles it becomes difficult to build line sections whose transverse dimensions are small in wavelengths, as they should be for ordinary transmission-line type of propagation. In Chapter 9 waveguides were described which were convenient for the guiding of radio-frequency energy in the frequency range from about 1,000 to 30,000 megacycles. In this same frequency range sections of waveguides can be used conveniently as "circuit elements" in much the same manner as transmission line sections were used at lower frequencies.

Although the operation of these waveguide circuits is similar in many respects to the transmission-line circuits already discussed there are certain differences which must be considered. One of these differences relates to the concepts of voltage and current. On parallel-wire or coaxial transmission lines there are two conductors which carry currents that are equal but oppositely directed. Thus there is no question about what is meant by the magnitude of the current at any point along the line; it is the current that would be read by a suitable r-f ammeter when either line is opened at that point and the ammeter is connected in series. Also the voltage at any point along the line is the voltage that would be measured by a suitable r-f voltmeter connected across the line. Nevertheless practical r-f voltage and current indicators often take the form of small monopole (or dipole) and loop probes connected to a crystal rectifier and microammeter. These indicators or probes actually indicate the strengths of the electric field, which is proportional to voltage, and of the magnetic field, which is proportional to current. For single-conductor waveguides of the form shown in Fig. 9-15(a) and (b) the simple concepts of voltage and current no longer apply. However it is still possible to probe the fields within the guide with a small monopole (or dipole) and loop which indicate respectively the strengths of the electric and magnetic fields. Practical forms of such probes are illustrated in Fig. 18-32 and their use will be discussed later. Hence, with transmission lines it is possible to consider either voltage and current on the one hand, or electric and magnetic fields on the other. With waveguides, attention is focussed on the electric and magnetic fields. The impedance at any point along a transmission line can be defined in terms of the ratio of voltage to current or of E to H. The expression of impedance as the ratio of E to H can still be applied in the case of waveguides, and this ratio is called the *wave impedance* in the guide.

A second difference between transmission line and waveguide circuits is in the measured wavelength corresponding to a given frequency. For an air dielectric, the measured wavelength along a transmission line is always given very closely by the formula

$$\lambda = \frac{v}{f} = \frac{c}{f}$$

where $c = 3 \times 10^8$ m per second is the velocity of light in free space. (For a transmission line with a solid dielectric both the velocity v and the wavelength λ are reduced by a factor $1/\sqrt{\epsilon_r}$, where ϵ_r is the dielectric constant of the solid dielectric relative to that of air.) For waveguides, however, the wavelength and velocity of the wave depend upon the transverse dimensions of the guide. The wavelength and velocity are still related through the formula

$$\lambda_g = \frac{v_g}{f}$$

where λ_g is the wavelength in the guide and v_g is the wave velocity in the guide. Both of these quantities are longer than the corresponding values on a transmission line or for a wave in free space. The *guide wavelength* as it is called can be calculated from the formula

$$\lambda_g = \frac{\lambda_c}{\sqrt{\lambda^2 + \lambda_c^2}}.$$

In this formula λ is the "free-space wavelength" (or wavelength as would be measured on an air-dielectric transmission line) and is given by the usual formula $\lambda = c/f$. The other quantity, λ_c, is the so-called *cut-off wavelength* of the guide. It depends upon the transverse dimensions of the waveguide and upon the mode being propagated. For the rectangular guide of Fig. 9-15(a) carrying the fundamental or lowest order $TE_{1,0}$ mode, the cut-off wavelength is just twice the width of the guide. That is,

$$\lambda_c = 2a.$$

When these two differences are kept in mind the operation of microwave circuits is found to be quite similar to that of transmission line circuits. A simple example of the use of such circuits in combination is shown in Fig. 18-19. Figure 18-19(a) shows a method for matching a coaxial line to a

Chapter 18   **Ultrahigh-Frequency and Microwave Circuits**   597

waveguide. Adjustment of the movable short-circuit plungers on the coaxial line and waveguide makes it possible to obtain an impedance match. An alternative arrangement for matching a coaxial line to a waveguide is shown in Fig. 18-19(b). In this case the length of the projection of the inner

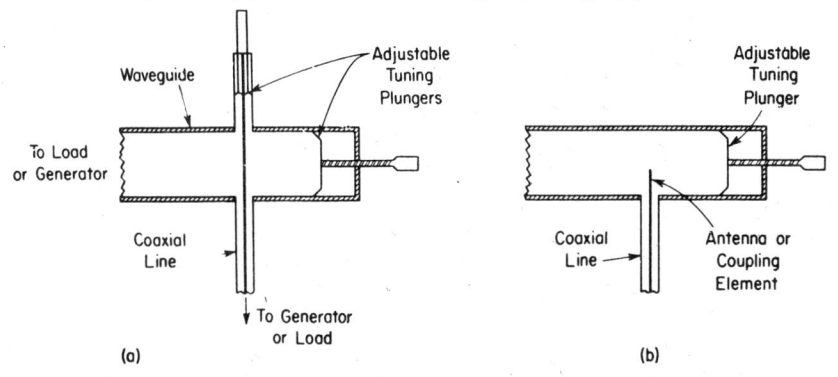

FIG. 18-19. Methods of matching coaxial line and waveguide circuits.

line of the coaxial cable into the waveguide is preset to obtain the correct amount of coupling (the projection acts as an antenna radiating into the guide) and the waveguide shorting plunger is adjusted to tune out the reactance.

PRINTED OR STRIP-LINE MICROWAVE CIRCUITS

Coaxial line and waveguide circuit components are rather expensive to fabricate and do not lend themselves readily to mass production techniques. A type of microwave transmission system which makes possible relatively inexpensive microwave circuit components is the printed or strip-line type of system. Two variations of strip transmission lines are illustrated in

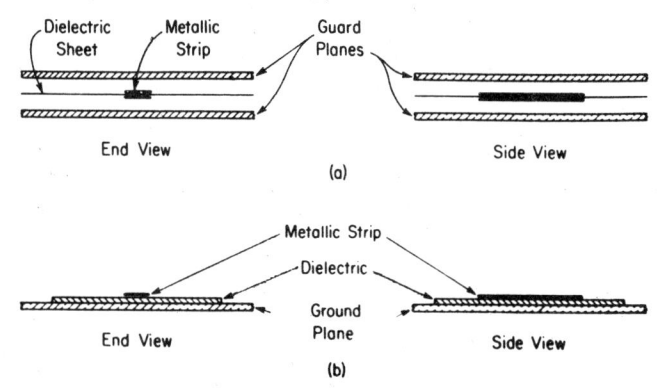

FIG. 18-20. Special microwave transmission lines: (a) strip line; (b) microstrip line.

Fig. 18-20. The first, shown in Fig. 18-20(a), consists of a flat conducting strip supported on a thin sheet of dielectric material which is located midway between two conducting ground planes or guard planes. This variation may be thought of as a coaxial line in which the center conductor is a flat strip and the outer conductor has been opened out to form two flat planes. Because the electric and magnetic fields are concentrated around the flat-strip inner conductor as illustrated, the fact that the outer conductor has been opened up makes practically no difference. Moreover, it is possible to have several such strip lines within the same guard planes with small interaction between them as long as their separation is several times larger than the strip width or guard-plane separation. The flat strip conductor may be stamped to shape by machine or may be printed directly on the dielectric sheet which is usually made of teflon. In practice two flat strips are usually used, one on either side of the dielectric sheet. Under these conditions very little of the electric flux penetrates the dielectric sheet and hence the dielectric losses are very low.

The second variation of strip line, usually called *microstrip*, is shown in Fig. 18-20(b). In this case the strip conductor is separated from a single flat ground plane by a suitable dielectric sheet or strip. Because of the presence of the electrical image of the strip line in the conducting ground plane, this system operates like a parallel wire transmission system consisting of two flat strips separated by a distance $2h$, where h is the height of the strip above the ground plane. If this separation is small the field is concentrated around the strip as before, and interaction between strip lines on the same ground plane will be small. However, just as with parallel wire lines, there exists the possibility of radiation of electromagnetic energy occurring at sharp bends and junctions.

Both types of transmission systems lend themselves admirably to printed-circuit techniques, and it is now possible to print a complete microwave circuit or even a complete piece of equipment such as a microwave receiver in this manner.

UHF AND MICROWAVE GENERATORS

At frequencies below about 100 megacycles the *generators* of radio-frequency energy are almost always conventional vacuum-tube oscillators. Above about 100 megacycles and up to around 1,000, or even 3,000 megacycles in some cases, vacuum tubes continue to be used as generators, although their operation may be rather unconventional. Above about 1,000 to 3,000 megacycles (the dividing line depends on the power requirements) entirely different types of generators are used. In order to understand the need for and method of operation of these unconventional types of generators, the limitations of conventional vacuum-tube generators must be considered.

Chapter 18   *Ultrahigh-Frequency and Microwave Circuits*   599

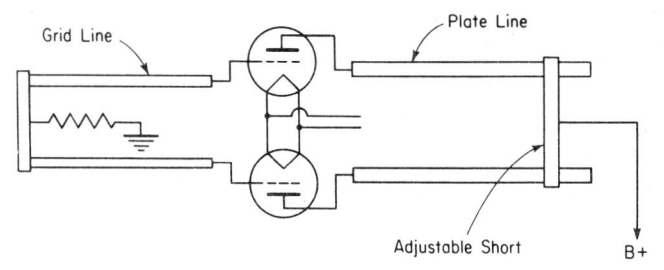

Fig. 18-21. Push-pull oscillator using transmission-line circuits.

(a)

(b)

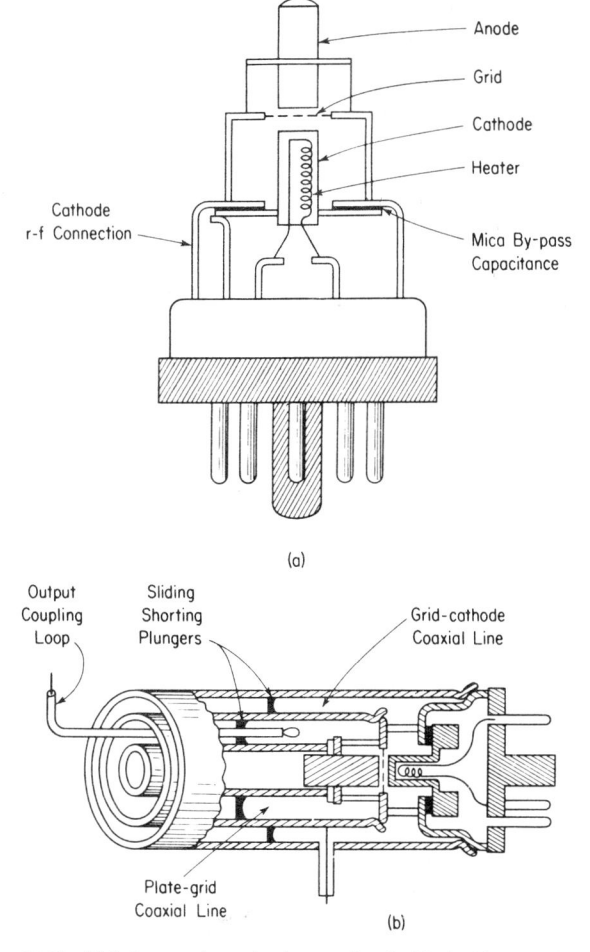

Fig. 18-22. Lighthouse tube and tube associated with double coaxial tuning lines in oscillator.

In the first section of this chapter the limitations of ordinary or conventional lumped-constant circuits were discussed and the possibilities of using transmission-line sections as circuit elements were pointed out. A typical oscillator circuit using line sections is shown in Fig. 18-21. Such transmission-line circuits may be used up into the shf band. However, another limitation which cannot be overcome quite so easily is the effect of electron transit time on the operation of the tube. An electron, traveling from cathode to plate in a vacuum tube, requires a small but finite time to make the transit, and this time is called the *transit time*. Its exact value depends upon the tube geometry and the d-c potentials applied, but it is of the order of 0.001 to 0.01 microseconds for ordinary tubes. This time is so small as to be entirely negligible at audio and low radio frequencies, but at vhf and uhf it can become an appreciable fraction of a period, that is, the time required for one cycle of oscillation. Thus at a high enough frequency, electrons attracted from the cathode would fail to reach the plate before the signal voltages had reversed, and operation in the normal manner would be impossible. (Actually the tube would cease to function at a considerably lower frequency for other more involved reasons.) Difficulties caused by transit-time effects can be reduced by decreasing the electrode spacings and hence reducing the transit time. An example of a vacuum tube designed to operate at relatively high frequencies is the "light-house" tube shown in Fig. 18-22. In this tube the cathode, grid, and plate are plane surfaces with very close spacings. These elements themselves, and their "terminals," which are the metal cap and disk seals, are designed to be integral parts of a double coaxial transmission line as shown in Fig. 18-22. The lines are tuned by means of movable short-circuiting plungers. Typical ratings for a small tube of this type are 1 w at about 1,000 megacycles. However, miniature tubes of this general type have been designed to give small power outputs at frequencies up to the order of 3,000 megacycles.

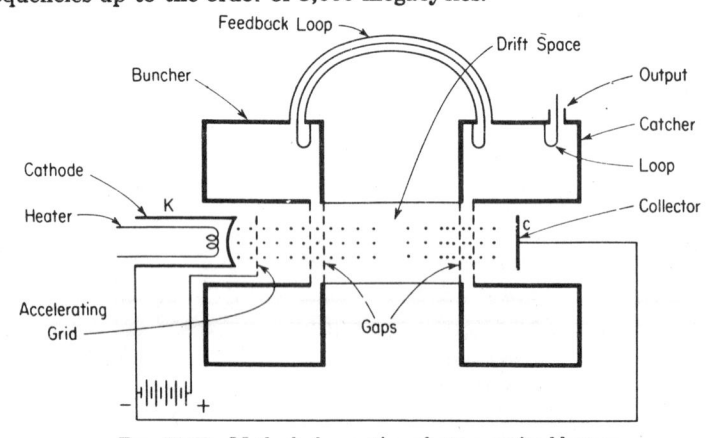

Fig. 18-23. Method of operation of a two-cavity klystron.

When large amounts of output power are required at frequencies above about 1,000 megacycles, other types of generators known as *klystrons* and *magnetrons* are used. The sketch of Fig. 18-23 helps explain the operation of one particular uhf generator, known as a double-cavity klystron. A stream of electrons emitted by cathode K is attracted towards collector c by a high d-c potential. In their passage the electrons traverse the gaps or openings of two cavity resonators, called respectively "buncher" and "catcher." A *cavity resonator* is a metallic cavity with a small opening across which a voltage can be impressed or developed. Such a cavity will resonate at certain frequencies for electromagnetic waves, just as it will for sound waves. For the lowest resonant frequency, the cavity may be thought of, rather roughly, as an inductance which is tuned by the capacitance across its gap. Such a device has the advantage of being a very low-loss, and therefore high Q, circuit, so that for only a small amount of power dissipated large currents will flow (on the inside walls of the cavity) and a large voltage will be developed across the gap.

Continuing the explanation of operation of the klystron circuit of Fig. 18-23, let it be assumed for the moment that a voltage of high radio frequency exists across the gap of the buncher resonator. Then the stream of electrons passing through will be *velocity modulated*. Those electrons that cross the gap during the half cycle when the voltage or electric field is positive or accelerating will be speeded up, whereas those electrons which cross during the negative or decelerating half cycle will be retarded or slowed down. After emerging from the first gap the electrons are allowed to travel or drift in the space between the two resonators, marked *drift space*. Here they will tend to form into bunches, because the electrons that were accelerated will tend to catch up to the slower electrons of the previous cycle, while the decelerated electrons will slow down to join the faster electrons from the following cycle. Of course, if they are allowed to drift still farther the bunches will disperse again, but the electrons will rebunch farther down the tube. The important fact is that by adjusting the d-c potential between cathode and accelerating grid it is possible to arrange to have the electrons crossing the catcher gap in bunches. Under these conditions they will induce across the gap an alternating voltage, the frequency of which will be the same as that of the buncher voltage. If the catcher cavity is tuned to this frequency, the resonant currents in the cavity and voltage across its gap can be quite large. By inserting a small loop into the cavity some of this radio-frequency energy can be fed to a load, e.g., an antenna. The radio-frequency energy which is stored in the cavity and fed to the load must, of course, be abstracted from the electron stream, which in turn acquires its energy from the d-c power supply. The mechanism of abstraction of energy from the electron stream is quite simple. Those electrons which cross the gap when the voltage is accelerating have their velocities increased, and the energy required to do this comes from the cavity or circuit.

Those electrons which cross when the voltage is decelerating are slowed down, and hence give some of their energy to the cavity or circuit. If now it is arranged that the bunches cross the gap during the decelerating half cycle, and the relatively few electrons between bunches cross during the accelerating half cycle, the electron stream will give more energy to the circuit than it will abstract. This excess energy can be radiated or otherwise expended. In an oscillator some of it can be fed back, by means of the feedback loop shown, to excite the buncher. The energy required to excite the buncher is considerably less than that abstracted from the beam by the catcher, since electrons crossing the buncher gap are not bunched. Therefore just as many electrons are decelerated as are accelerated and theoretically no net energy is required to modulate the beam. However the power dissipated by ohmic losses in the cavity must be furnished. Because this required input power is smaller than the output power, the device is an amplifier; when connected as shown to furnish its own input, it is an oscillator.

Generators operating on the above principle are called *velocity-modulation* generators, and the particular type described was a two-cavity klystron. One disadvantage of this type is the necessity for tuning two cavities to the same resonant frequency. This disadvantage does not exist in the *reflex*

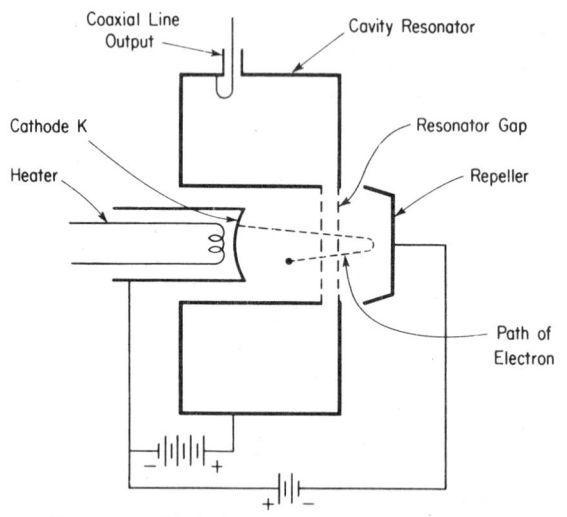

Fig. 18-24. Method of operation of a reflex klystron.

klystron, which uses the same cavity as buncher and catcher. This is accomplished as shown in Fig. 18-24. The electron stream, after emerging from the resonator gap, is turned back upon itself by the repeller, which is at a potential near to but below that of the cathode. The returning beam is again bunched, and these electron bunches give some of their energy to the cavity circuit. A typical commercial reflex klystron is shown in Fig. 18-25.

Chapter 18   *Ultrahigh-Frequency and Microwave Circuits*   603

The term *magnetron* refers to a type of uhf generator which utilizes the motion of an electron stream in crossed electric and magnetic fields. An electron in an *electric* field E experiences a force Ee, where e is the magnitude of the electron charge. Because the electron charge is negative the force is in the direction opposite to the direction of the electric field, and in the absence of other forces the electron will move in a straight-line path in the direction of the force. On the other hand an electron moving with a velocity v in a *magnetic* field of strength B experiences a force Bev which is at right angles to the direction of electron travel. In this case, and in the absence of other forces, the electron will travel a circular path, the radius of which is proportional to v and inversely proportional to B. When an electron moves in crossed (that is, perpendicular to one another) electric and magnetic fields the resulting motion is a combination of straight-line and circular paths as illustrated in Fig. 18-26. These paths, known as *cycloids* or *trochoids*, are the paths traced by a point on the spoke of a wheel which is rolling along a flat surface. The frequency of rotation or oscillation of the electron is proportional to the strength of the magnetic field.

FIG. 18-25. Cut-away view of a typical reflex klystron. (Courtesy Sperry Gyroscope Co.)

Figure 18-27 illustrates in principle how such electron motion might be used to generate oscillations.

Electrons emitted from a space-charge region near a cathode K and traveling the paths of the type shown in the crossed electric and magnetic

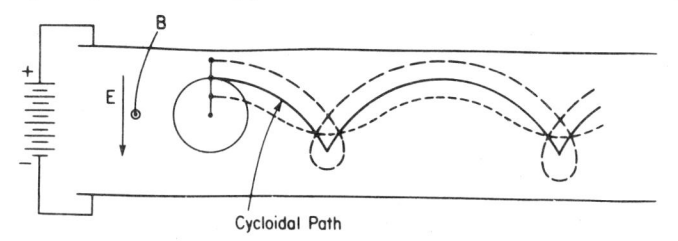

FIG. 18-26. Paths traced by electrons in crossed electric and magnetic fields. (Magnetic field B is perpendicular to plane of paper.)

fields are constrained to move near the gaps in the positive electrode. To these gaps are connected cavity resonators which are resonant at the electron oscillation frequency. The oscillating electron charge traveling across the gap region induces a radio-frequency voltage across the gap and current

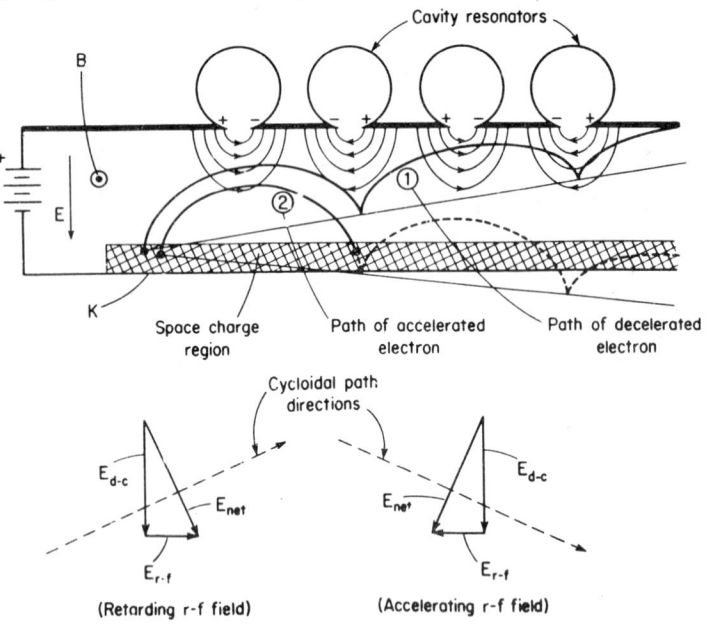

Fig. 18-27. Electron paths in crossed electric and magnetic fields, showing effect of the presence of tuned cavities. (1) Path of an electron which crosses a gap region when the field is decelerating as shown; (2) path of an electron which crosses gap one-half cycle later when field is accelerating.

flows in the walls of the cavity resonator. The induced radio-frequency electric field in the region of the gap would be somewhat as illustrated. If the r-f fields across adjacent gaps are of opposite phase as shown, and if d-c electric and magnetic field strengths are suitably chosen to make the lengths of the cycloidal loops equal to twice the distance between gaps, then it is evident that an electron, crossing a gap region when the field is in a direction to retard or decelerate the electron, will also pass all subsequent gaps during a period of decelerating field, because the r-f electric-field frequency and electron oscillation frequency are the same. Hence such a typical electron, which finds itself in a retarding r-f field on each gap traversal, gives up some of its energy to the external r-f circuit and so aids the production of oscillations. The path followed by this productive electron having this favorable phase is similar to that designated as 1 in Fig. 18-27. Of course, on the average there will be just as many electrons emitted from the space charge region with unfavorable phase, in the sense that they will

Chapter 18   *Ultrahigh-Frequency and Microwave Circuits*   **605**

pass a gap when the r-f field is accelerating, with the result that they abstract energy from the external circuit. However, it can be demonstrated that such unproductive electrons having this unfavorable phase will tend to follow a path of the type indicated as 2 in Fig. 18-27. This means that they will be forced back into the space charge region and effectively be removed from the interaction region. With more productive than unproductive electrons available, there will be a net transfer of energy from the electron cloud to the r-f circuit, and r-f oscillations can result. Explanation of the difference in paths followed by productive and unproductive electrons is as follows: the general progression of the cycloidal path followed by an electron in crossed electric and magnetic fields is always perpendicular to the net electric field. With no r-f field superimposed on the d-c electric field, this direction is parallel to the anode and cathode plates. With an r-f field superimposed upon the d-c electric field, a productive or retarded electron always finds itself in a field that has a net retarding component parallel to the anode, in addition to the d-c field perpendicular to the anode. Similarly, an unproductive electron experiences a net accelerating component parallel to the anode. The directions of the net resultant electric fields are also illustrated in Fig. 18-27, where it is seen that the cycloidal paths for productive electrons will be tilted upwards, whereas for unproductive electrons the cycloidal paths will be tilted downwards.

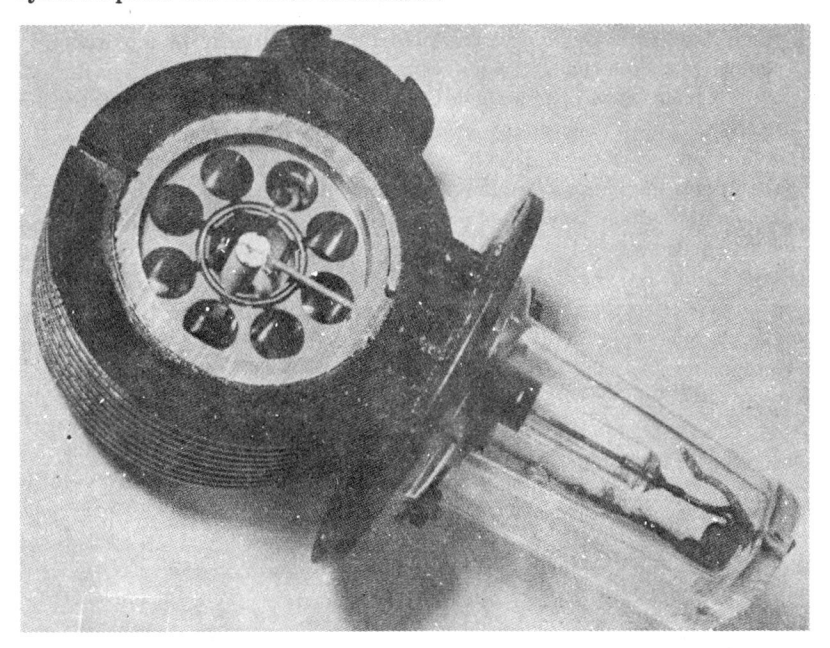

Fig. 18-28. Cut-away view of a 10-cm magnetron, showing cavity resonators. (Courtesy Sylvania Electric Products, Inc.)

A practical embodiment of some of the principles just discussed is the magnetron shown in cut-away view in Fig. 18-28. Here the parallel plane structure of Fig. 18-27 has been warped into a cylindrical form suitable for practical construction. The magnetic field is applied parallel to the axis of the cylinder by a permanent magnet.

The method of operation discussed is only one of many different modes of operation which can be and are utilized in practice. However, all these devices operate by abstracting energy at a high radio frequency from an electron stream.

Klystrons and magnetrons are the most frequently used generators and amplifiers at uhf and shf. Klystrons, usually designed to be tunable over a 10% frequency band, are commonly used as local oscillators and in other applications requiring no more than a few watts. However they can also be designed to deliver kilowatts of power. Magnetrons, with their higher efficiencies (up to about 70%) and with power-handling capabilities from a few watts up to several megawatts of peak power have found extensive application in radar. Both of these devices, utilizing high Q cavities as they do, are quite narrow-band in operation. When larger bandwidths are required, as for example in wide-band amplifiers, *traveling wave tubes* can be used. As with klystrons and magnetrons, the traveling wave tube utilizes the coupling between an electron stream and an electric field. In Fig. 18-27 the electron stream and electric field intensity appear to move along in synchronism with the electron always maintaining approximately the same position from left to right along the tube with respect to the electric field intensity. In this sense the arrangement might be considered a sort of traveling wave tube. Indeed it would be possible to build a traveling wave tube in this manner by "loading" the walls with cavities. The essential requirement is that the wave of electric field intensity travel along the tube with approximately the same velocity as the electron stream. On an ordinary "unloaded" or uniform transmission line the electromagnetic wave or wave of electric field intensity travels approximately with the velocity of light. The velocity of an electron stream depends upon the accelerating voltage and is always less than the velocity of light.

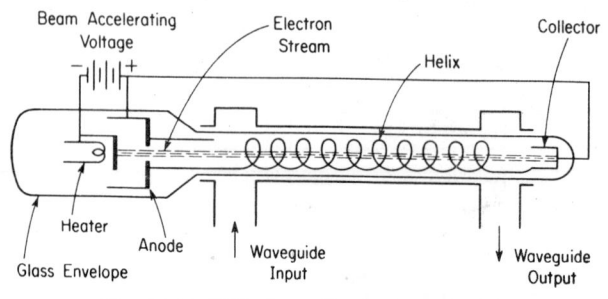

Fig. 18-29. Helical traveling wave amplifier.

Chapter 18 *Ultrahigh-Frequency and Microwave Circuits* 607

For the magnitude of voltages ordinarily used in traveling wave tubes the electron velocity may be only of the order of one-tenth the velocity of light. Therefore it is necessary by some means to slow down the electromagnetic wave along the transmission line or guiding structure. A simple means for doing this in the case of a coaxial line is to make the inner conductor in the form of a helix as shown in Fig. 18-29. Recalling that the velocity of transmission is given by

$$v = \frac{1}{\sqrt{LC}}$$

it will be evident that the velocity can be reduced to low values by making the inductance per unit length of the inner conductor large, which is accomplished by coiling it as shown. An alternative approach is to think of the electromagnetic wave being guided along the helical path with the velocity of light. It then progresses *along the axis* with a smaller velocity which depends upon the diameter and pitch of the helix. In this manner it is quite easy to reduce the velocity of the electromagnetic wave along the axis to be of the same order as that of the electron stream. When this has been accomplished it becomes possible to transfer some of the energy of an electron beam to an electromagnetic wave, and hence increase the amplitude of the electric field intensity.

Figure 18-30 shows the distribution of electric force along a traveling wave tube when a traveling electromagnetic wave is superimposed on the d-c accelerating force. A beam of electrons emitted from the cathode is accelerated by a d-c voltage of about 200-500 volts. If a slowed-down wave of electric intensity is propagated along the helical transmission with the same velocity as that of the electron stream the beam will become bunched as it proceeds along the tube. The explanation for this bunching lies in the

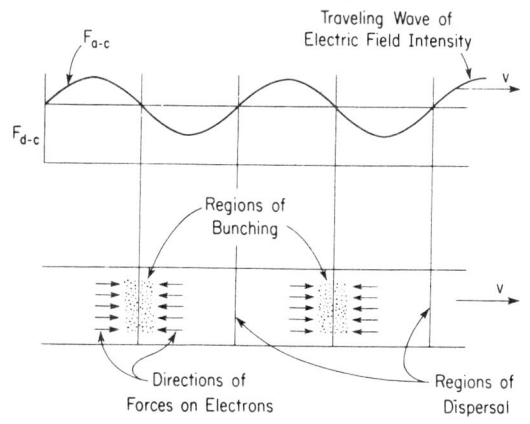

Fig. 18-30. Distribution of electric force along a transmission-line tube, showing regions of bunching.

fact that electrons moving along in a region where the field is accelerating will be speeded up until they are in a region where the a-c field is zero. Similarly, electrons in a region of decelerating electric field will be retarded until they are in a region of zero a-c field intensity. Thus the electrons tend to cluster in bunches which move along with the wave maintaining their positions in alternate regions of zero a-c field intensity. Under these conditions there would be no net transfer of energy from electron stream to wave because, on the average, just as many electrons have been accelerated as have been retarded.

Now let it be assumed that the average velocity of the electron stream is increased slightly, by increasing the d-c beam voltage. Under these circumstances the electron bunches will move into a region of retarding field intensity in the wave, and moreover will *retain* these positions. Correspondingly, regions of accelerating field intensity will now be coincident with the sparsely occupied regions of the electron beam which exist between bunches. Because there are now, on the average, more electrons being decelerated by the electromagnetic wave than are being accelerated by it, there will be a net transfer of energy from the stream to the wave, so that the electromagnetic wave will increase in amplitude as it progresses along the tube. That is, there is an amplification of the signal. If enough of this amplified signal is fed back from the output to furnish the power required at the input, the amplifier will become an oscillator or generator. This feedback of energy can also take place within the tube if the input and output circuits are not properly matched.

It is next necessary to consider how the electron bunches maintain their position in the decelerating regions of the wave, even though the individual electrons are advancing slowly through the wave (under conditions where the average electron beam velocity is made slightly greater than the wave velocity). Because of the varying electric field of the wave, the instantaneous velocity of an electron will vary about its average velocity. It will travel faster in an accelerating region of the wave and will slow down where the field is retarding. Hence there will automatically be a bunching of the electrons in regions of retarding field where the electrons travel more slowly, and a spreading out of electrons in regions of accelerating field where the electrons speed up. An individual electron advancing slowly through the wave is now in a bunch, then in the space ahead of the bunch (which it traverses quickly), then a member of the preceding bunch, and so on. In this manner the bunches retain their positions in the wave and move with the same velocity as the wave, while the individual electrons having an average velocity slightly greater than the wave velocity progress slowly forward through the wave. The fact that the electrons remain longer in a decelerating region than they do in an accelerating region means that effectively there are more electrons being decelerated than accelerated at any time, and hence there is a net transfer of energy from electron stream to

Chapter 18   *Ultrahigh-Frequency and Microwave Circuits*   609

electromagnetic wave as already stated, with a resultant increase in amplitude of the wave. Because the helical transmission line is a broad-band device and there are no tuned circuits involved, the resulting traveling wave amplifier can operate over quite a large frequency band. Figure 18-31

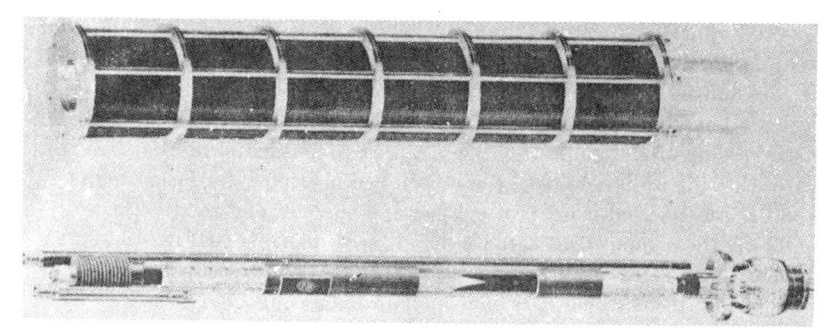

Fig. 18-31. A commercial traveling-wave tube. (Courtesy Sperry Gyroscope Co.)

shows a cut-away view of a commercial traveling tube. The long solenoid shown is mounted external to the tube and is used to produce a longitudinal d-c magnetic field which serves to keep the electron beam from spreading radially. Commercial traveling wave tubes are generally designed for use in the frequency range between 1,000 and 12,000 megacycles, and a typical tube is capable of a gain of 30 to 40 db over a 2 to 1 frequency band. Power outputs range from 1 mw to several kw of peak power.

OTHER UHF AND MICROWAVE DEVICES

The measurement of quantities such as frequency, voltage, current, power, and impedance at microwave frequencies is in some cases quite similar to the corresponding measurement at lower frequencies, but in other cases devices much different from their lower-frequency counterparts are required. Some of the more common microwave-measuring techniques and devices will be mentioned in this section.

The measurement of frequency can be carried out in the usual manner by comparing the unknown frequency with the harmonics of a standard frequency, the only difference being that very high harmonics may be required in the microwave frequency range. An alternative and quite common method of determining frequency at microwaves is through the measurement of wavelength. Using a coaxial line with one end shorted and the other end terminated in a movable shorting plunger, a resonant length can be determined by observing when the amplitude of the field in the coaxial line is a maximum. By moving the movable shorting plunger until the next resonance peak occurs the length corresponding to one-half wavelength is determined. The frequency can then be calculated from the relation $f = v/\lambda$, where v will be very nearly equal to $c = 3 \times 10^8$ meters per second. Alternatively, a high-Q resonant cavity, whose size can be varied by means of a micrometer adjustment, can be calibrated in terms of frequency and makes a very convenient frequency meter.

Measurements of voltage and current are difficult to make and do not have too much significance at microwave frequencies. In a waveguide, for example, the current is distributed unevenly over the walls of the guide and may be flowing in different directions at different places on the same cross section. The voltage between the top and bottom walls of a guide is greater at the middle of the guide (for the $TE_{1,0}$ or fundamental mode) than it is nearer the side walls. As has been mentioned previously, measurements on E and H can be used instead of voltage and current, and relative measurements on these quantities can be made fairly easily. In Fig. 18-32 two common types of probes for measuring E and H are illustrated. The probe of Fig. 18-32(a) consists of a small antenna connected in series with a

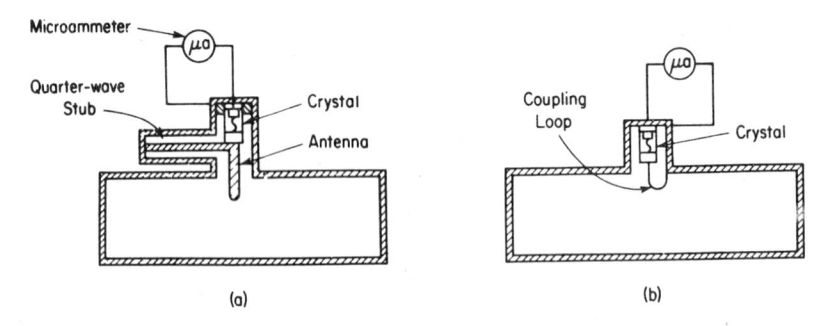

FIG. 18-32. Field strength indicators for measuring (a) electric field strength and (b) magnetic field strength.

crystal rectifier and microammeter. The quarter-wave stub provides a high-impedance support while at the same time allowing a return path for the direct current. This type of antenna probe indicates the relative electric field intensity in the region in which the antenna is inserted. The loop-type probe of Fig. 18-32(b) on the other hand indicates the relative magnetic intensity H in the region. Because E and H are related respectively to voltage and current these probes are often called voltage and current probes. In using these probes it must be remembered that the crystal rectifier is a square-law device (approximately) and that the meter readings will be approximately proportional to the *square* of the electric or magnetic field strength. When higher accuracy is required than is given by the square-law approximation, a calibration curve for the detector must be determined.

Measurement of power at microwave frequencies can be accomplished by terminating the waveguide or transmission line in a *bolometer* or *thermistor*. A bolometer is a resistor, the resistance of which changes with temperature. When microwave power is absorbed by the bolometer its temperature and hence resistance increase, and so a measure of the absorbed power can be obtained by a simultaneous measurement of its d-c resistance, usually by means of a bridge circuit. A thermistor is a semiconductor whose resistance changes rapidly with temperature, making it useful as a bolometer. In order to measure microwave power in this manner it is necessary to match the bolometer to the waveguide or transmission line along which the power is flowing (or make standing-wave-ratio measurements and then perform calculations to determine the percentage power absorbed and the percentage reflected). Sometimes the amount of power involved is too great to be absorbed by the bolometer, and in other cases it may be desired to monitor continuously the power flowing into a load. In these cases a small fraction of the power is tapped off, and measurements are made on this abstracted power. When this is done it is important that the monitoring power measuring device measure only the power flowing in one direction, towards the load, and not be affected by any reflected power flowing in the opposite direction. The device which accomplishes this result is known as a *directional coupler*.

One of many possible types of directional couplers is depicted in Fig. 18-33. This is a two-hole directional coupler, the operation of which is as follows: an auxiliary line or waveguide is loosely coupled to the main line or waveguide at two points a quarter-wavelength apart. In the case of waveguide structures this coupling is achieved by allowing energy to leak into the auxiliary guide through two small holes which have a quarter-wave spacing along the guide. Energy leaking through the first hole induces a weak field

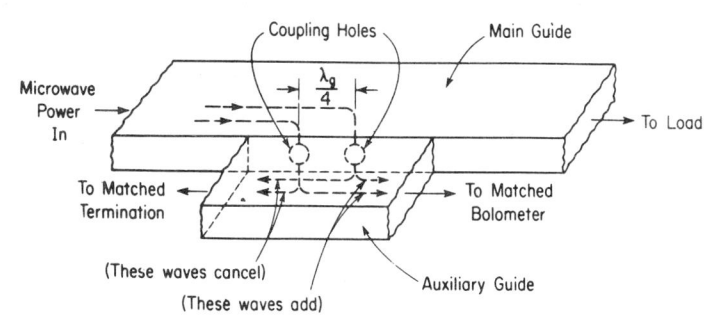

FIG. 18-33. Two-hole directional coupler.

which sets up in the auxiliary guide two waves traveling in opposite directions away from the hole. A similar effect results from leakage through the second hole. For a wave traveling from left to right in the main guide the phases of the fields coupled into the auxiliary guide are such that the waves traveling from left to right in the auxiliary guide add in phase, whereas the waves traveling from right to left in the auxiliary guide are out of phase and hence cancel one another. That this is so can be seen from the fact that for waves traveling to the left in the auxiliary guide the wave through the second hole has one-half wavelength farther to travel than the wave through the first hole. Therefore a wave traveling from left to right in the main guide induces a wave in the auxiliary guide which is also traveling from left to right. Also a reflected wave traveling from right to left in the main guide induces a wave in the auxiliary guide which also travels from right to left. If the auxiliary guide is terminated at the right by a matched bolometer, this bolometer will indicate the amount of power traveling from left to right in the main guide and will be unaffected by a wave traveling in the opposite direction. However, in order to accomplish this result it is necessary to terminate the left end of the auxiliary guide in its characteristic impedance, so that any wave traveling from right to left in this guide will be absorbed and not reflected back towards the bolometer.

Because the operation of the two-hole directional coupler depends on having a quarter-wave spacing between holes, the device is frequency-sensitive and works properly only at the frequency for which it was designed. The *multi-hole* coupler which has many holes spread over about a wavelength is less frequency-sensitive and works satisfactorily over a band

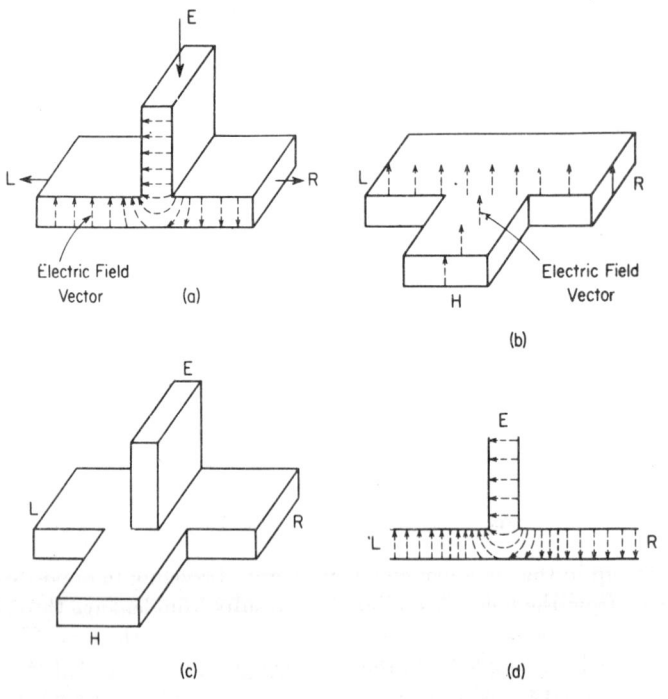

FIG. 18-34. (a) E-plane tee; (b) H-plane tee; (c) magic tee; (d) a signal at E excites the two halves of the H stem in opposite phase.

of frequencies. In the preferred direction the waves through all of the holes add in phase to give a relatively large signal in that direction. In the opposite direction for nearly every wave through any hole there is another canceling wave through a hole which is a quarter-wavelength away, with a resulting very weak signal in that direction.

Another common device useful in microwave circuits is a microwave bridge or magic tee as it is often called. Figure 18-34(a) and (b) shows an E-plane tee (or junction) and an H-plane tee. In the first of these the stem of the tee is in the E-plane or plane containing the electric-field vector in the main guide, whereas in the second the stem of the tee lies in the H-plane. These two types of tees differ in their operation in that a signal fed into the E arm of Fig. 18-34(a) divides equally to produce equal signals in L and R, which however are of opposite phase as depicted by the direction of the arrows. In contrast, a signal fed in at H in Fig. 18-34(b) produces equal signals in L and R having the same phase. Figure 18-34(c) shows a combination E-plane and H-plane tee called a magic tee because of certain special properties that it possesses. As indicated in Fig. 18-34(d) a signal fed in at E will excite the two halves of the H-plane tee stem in opposite phase, with the result that there is zero net excitation. In other words, a signal fed in at E will produce signals at L and R but none at H (and, of course, conversely by reciprocity). This result assumes that the arms L and R are properly terminated in their characteristic impedance so that there are no reflected waves, or at least that L and R are terminated in impedances that are exactly equal. To see what happens when this is not the case, assume that arm R is correctly terminated so that there is no reflected wave but that there is some mismatch at L. Then a signal introduced at E will divide equally into L and R without exciting H. However, because of the mismatch at L, there is now a reflected wave from L which excites the H-plane stem with a field for which the arrows would have the same direction across its mouth, and so power flows into the H-plane stem from this reflected wave only. Hence the signal at H is a direct measure of the degree of mismatch at L. A simple extension of this line of reasoning will demonstrate that there will be no excitation of H by a signal fed in at E as long as the impedances terminating L and R are equal (but not necessarily equal to the characteristic impedance). Therefore any signal at H resulting from a signal at E is a direct measure of the degree of unbalance that exists between impedances at L and R. Therefore this circuit can perform at microwave frequencies all the functions of a common bridge circuit at lower frequencies.

QUESTIONS AND PROBLEMS

1. An inductor has an inductance $L = 10$ microhenrys and an effective distributed capacitance $C_e = 1$ $\mu\mu f$. (a) What is the frequency of self-resonance? (b)

Compare the reactance of this effective parallel circuit with the reactance of the inductor alone (neglecting distributed capacitance) at the following frequencies: 5, 30, and 100 megacycles.

2. A short-circuited quarter-wave parallel-rod line is made of copper conductors having a spacing $b = 5$ cm and conductor radius $a = 0.5$ cm. The operating frequency is 300 megacycles. Determine (a) the characteristic impedance Z_0 of the line; (b) the resistance r per unit length of the conductors; (c) the input impedance R_{in}.

3. A short-circuited section of parallel-rod transmission line having the conductor size and spacing given in Problem 18-2 is to be used as the inductive reactance to tune to antiresonance a capacitor $C = 3$ $\mu\mu$f. The frequency is 300 megacycles. How long should the line section be?

4. A rectangular guide has dimensions $a = 8$ cm, $b = 4$ cm, and carries a $TE_{1,0}$ wave at a frequency of 3,000 megacycles. Determine (a) the guide wavelength λ_g; (b) the cut-off wavelength λ_c and corresponding cut-off frequency f_c of the guide.

CHAPTER 19

RADIO WAVE PROPAGATION

GENERAL NATURE OF PROPAGATION

When a radio wave is radiated from a transmitting antenna it spreads out in all directions, decreasing in amplitude with increasing distance because of the spreading of the electromagnetic energy through larger and larger surface areas. The portion of the energy arriving at a distant receiving antenna may have traveled over any of several possible propagation paths, some of which are indicated in Fig. 19-1.

At frequencies below about 3 megacycles, where vertical antennas erected on the ground are used, the transmission from transmitter T_1 to receiver R_1 may be by means of a *ground wave* or *surface wave* which is guided along the surface of the earth, or it may be by means of a *sky wave* or *space wave* which is radiated up into space and reflected back to earth from a layer of ions and electrons called the *ionosphere*. At higher frequen-

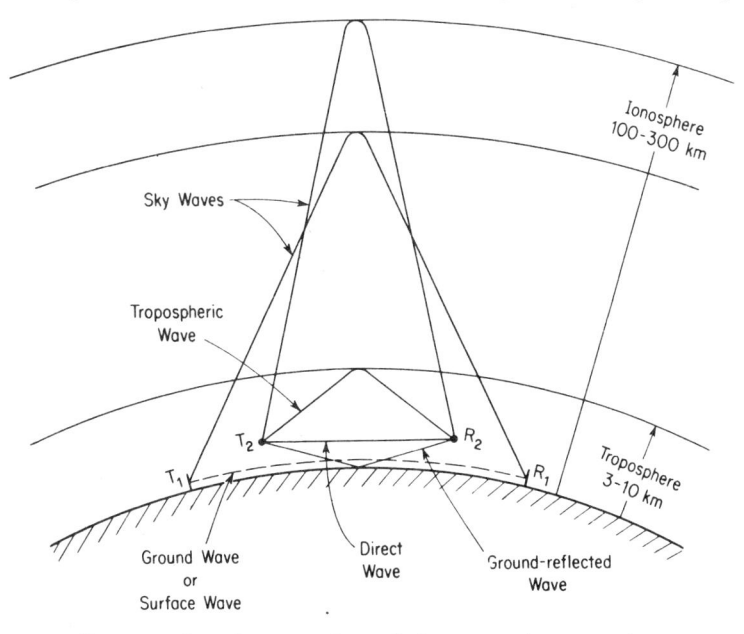

Fig. 19-1. Possible propagation paths from transmitter to receiver.

cies where elevated antennas are used (indicated by T_2 and R_2 in Fig. 19-1) two of the possible paths are a *direct wave* path from T_2 to R_2 and a *ground-reflected wave* which reaches R_2 after reflection from the ground. In addition there is the possibility of the sky-wave signal already mentioned. At ultrahigh frequencies where propagation via all of these paths may be impossible, a *tropospheric wave* which is reflected or refracted in the troposphere (the region of cloud formation) may become of importance.

Which of the several possible paths are operative in any particular instance depends upon the frequency, the separation between transmitter and receiver, and the conditions of the ionosphere and troposphere. While there are cases where several of these paths are operative simultaneously, it will usually be found that propagation is restricted to only one or two paths, or at any rate, that the signal received over a particular path is very much stronger than those received over other paths. In this chapter the factors that affect propagation over each of the several possible paths will be explained in detail.

POLARIZATION

If the signal radiated by a vertical antenna is received a short distance away it will be found that the received signal is a maximum when the receiving antenna is vertical and zero when the receiving antenna is horizontal. This is because the electric field is in the vertical direction at this point. The radiation from the antenna is said to be *vertically polarized*. In a direction at some vertical angle θ to the sending antenna (see Fig. 9-23) the receiving antenna must lie in the direction shown by E for maximum signal pickup. It is then not vertical but it will still be in the *vertical plane* through the antenna and the point P, and there will be zero voltage induced in an antenna perpendicular to this plane. The radiation from the antenna is still said to be vertically polarized, or more correctly, *plane polarized in the vertical plane*. Similarly the radiation from a horizontal antenna is said to be *horizontally polarized* and there will be no direct pickup of such a signal on a vertical antenna.

POWER DENSITY AND POYNTING VECTOR

If one visualizes a radiating antenna located at the center of an imaginary spherical surface, it is possible to consider a certain power density or power per unit area flowing through each element of the surface. If the antenna radiated uniformly in all directions the power per unit area would be $W/(4\pi r^2)$ watts per square meter, where W is the total power radiated and r is the radius of the sphere in meters. Actual antennas are usually directive and do not radiate uniformly in all directions, so that there would ordinarily be more power per unit area through some parts of the surface than others. If the electric field intensity E is known, the power flow per unit area can be calculated (for this simple case where there are no reflected waves) from the formula

$$P = \frac{E^2}{\eta} \text{ watts/square meter}$$

where P is the power per unit area measured in watts per square meter, E is the electric intensity measured in volts per meter, and η is the characteristic or intrinsic impedance of free space which has the value of 377 ohms. The quantity P is a vector quantity having a direction (the direction of power flow) as well as a magnitude and is called the *Poynting vector* after Professor Poynting, who first used this concept. This concept of a certain power flow per unit area is useful in considering the propagation of electromagnetic energy whether it be in free space or along guiding systems such as the surface of the earth or in waveguides.

THE SURFACE WAVE

The part of the radiated energy that travels along near the ground is called the ground wave or surface wave. (The two terms are often used interchangeably, particularly at broadcast frequencies and lower. In general the term ground wave, used in contradistinction to sky wave, is considered to include both surface wave and tropospheric waves). The surface wave is guided along and around the surface of the earth somewhat as a wave is guided along a transmission line. In this guiding process voltages and currents are induced in the ground, and energy is abstracted from the wave. If the ground were a perfect conductor, these ground currents could flow without any losses and the wave would not be affected. However, the ground does have resistance, or a *finite conductivity*, so that energy is required to make these ground currents flow and this energy is absorbed from the wave. The result is that the ground wave is thus *attenuated* or decreased in strength even more than by the distance factor $1/r$ which is due to the spreading out of the wave through larger surfaces as it recedes from the antenna. The amount by which the wave is attenuated due to an imperfectly conducting ground is important in determining how far the wave will travel before the signal becomes too weak to be of any use.

The attenuation caused by the ground depends both upon the conductivity (or resistance) of the ground and upon the frequency being used. A high-frequency wave is attenuated much more than a low-frequency wave over the same ground.

Figure 19-2 illustrates the manner in which the ground-wave signal depends upon frequency and upon the conductivity of the ground. The curves show the field strength in millivolts per meter against distance from the transmitter. The expression millivolts per meter (or microvolts per meter) refers to the voltage that would be induced in a wire 2 m long when placed parallel to the direction of the field at that point. A field strength of 0.1 mv per meter is about the lowest signal strength that will give satisfactory reception, although this depends upon the amount of noise present in the receiving location. A noisy location will require as high as 10 or 100 mv per meter for worthwhile reception, while in a very quiet locality, at a time when static is weak, a signal of 10 μv per meter may be entirely adequate.

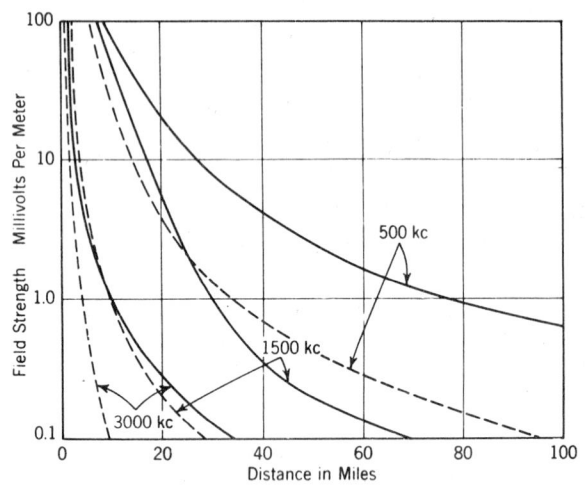

Fig. 19-2. Effect of frequency and ground conductivity upon the strength of the ground-wave signal. Solid lines: fairly good conductivity (10×10^{-3} mho per meter). Dashed lines: poor conductivity (2×10^{-3} mho per meter).

The solid curves in Fig. 19-2 are for a fairly good ground conductivity while the dashed curves are for a poor ground. A conductivity of 10×10^{-3} mho per meter represents a fairly good ground while a conductivity of 2×10^{-3} mho per meter is considered a very poor ground. The ground conductivity seems to depend to a large extent upon the nature of the terrain. Flat prairie country usually shows a high value of conductivity while mountainous or rugged, broken country has a low conductivity.

Because the ground wave is attenuated so much at the higher frequencies its chief usefulness lies in the long-wave and broadcast bands. Daytime reception of broadcast stations is entirely by means of the ground wave.

The ground wave is always vertically polarized because any horizontal component of electric force would be shorted out by the ground. For this reason vertical antennas must be used for ground-wave transmission.

THE SKY WAVE

The energy radiated upwards by an antenna, that is, the sky wave, would be wasted energy as far as radio communication is concerned if it continued on its path and did not return to earth. Fortunately, under certain circumstances it is reflected from the *ionosphere*, or *Kennelly-*

Heaviside layer as it used to be called. The reflected wave may return to earth at distances from the antenna much greater than can be reached by the ground wave, and this reflected wave makes extreme long-distance communication possible.

THE IONOSPHERE

The ionosphere consists of several ionized layers, that is, layers which are electrically conducting. These layers exist at high altitudes, in the upper parts of the earth's atmosphere. Radio waves that strike these conducting layers have their paths changed while passing through the layers. Often the waves penetrate all the layers and are lost, but more often the waves are bent in their paths so much that they return to earth at distant points. The heights of the layers and the *degree of ionization* (that is, the number of ions and electrons in a given volume) determine how far radio waves will go, and what frequencies give the best transmission. The ionized layers are found usually at heights between 50 km (30 miles) and 400 km 250 miles) above the surface of the earth.

In Chapter 4 it was shown that in a gas under very low pressure it is possible to knock one or more electrons out of a molecule of the gas, leaving a positive charge on the molecule. The positively charged molecule is no longer a true molecule, but is an *ion*, and can be attracted or repelled by electric forces. Electrons can be knocked out of a molecule not only by fast moving particles like electrons, but also by certain types of radiation such as ultraviolet rays and cosmic rays. In the high atmosphere, where the pressure is low, conditions are excellent for ionization to take place. The sun constantly gives off ultraviolet rays, and when these reach the upper atmosphere they cause a large proportion of the air particles to become ionized. Cosmic rays are believed to cause some ionization also.

Since the atoms, ions, and electrons in a gas are in constant motion, frequent collisions take place between them. When a positive ion collides with an electron, it may keep the electron to neutralize its charge so that it once more becomes a molecule. This process of *recombination* goes on all the time, so that a molecule, once it has been ionized, does not remain ionized indefinitely. The time that it takes for recombination to occur will depend on several factors, but particularly on the average distance between the particles in the gas. If there are only a few particles present in a given volume, as high in the upper atmosphere, collisions will not occur very frequently, so that the air particles remain ionized for long periods. In the lower parts of the earth's atmosphere, collisions take place so often that the air molecules do not remain ionized for very long. Another reason why there is relatively little ionization in the lower atmosphere is that the ultraviolet rays from the sun are largely absorbed by the upper parts of the atmosphere. As a result there is very little ionization below about 30 miles. Above 250 miles there are so few air particles present to be ionized that the density of ionization is again very low. However, at intermediate heights there is considerable ionization. The region between 30 and 250 miles above the earth is thus the region where the ionosphere exists, and this region therefore has the most influence on the propagation of the sky waves.

Sky waves that return to earth from the ionosphere are found to come from different heights above the earth, depending on the frequency and on the time of reflection. This phenomenon shows that the ionosphere is not one layer, but several layers. The reason why there are several layers in the ionosphere is because the different gases in the earth's atmosphere ionize at different pressures, that is, at different heights above the surface of the earth, and because there are other ionizing agents (cosmic rays, for example), which penetrate to different depths. The number of layers, their heights above the earth, and the amount they bend the sky wave all vary from day to day, from month to month, and from year to year. There are two principal layers, called the E layer and the F layer. The E layer is usually found at a height of 110 km (68 miles), but may vary from 90 to 140 km (55 to 85 miles). The other principal layer is one layer only at night, but splits into two parts during the daytime. The designation F layer is used to refer to the night layer. The designations F_1 and F_2 are given to the two parts of the layer which exist in the daytime, the F_1 layer being the lower one. The F, F_1, and F_2 layers are always above the E layer. Another layer, the D layer, exists only in the daytime at very low heights, but its effects are not so important as those of the other layers, and so will not be considered further. The heights at which the various layers exist are shown in Table 19-1.

TABLE 19-1

Name of layer	Height of layer, miles
E	55 to 85
F (night only)	110 to 250
F_1 (daytime only)	85 to 155
F_2 (summer day)	155 to 220
F_2 (winter day)	90 to 185
D	30 to 55

EFFECT OF THE IONOSPHERE ON THE SKY WAVE

The way in which radio waves are bent by the ionized layers in the ionosphere may be seen with the aid of Fig. 19-3. In this figure the ionosphere is shown as one layer, the dashes representing ions and the number of dashes in any region indicating the density of ionization in that region. Suppose a wave is sent upwards along the path shown, from the transmitter at point A. The path is a straight line until it reaches a region where there are ions. The path of the signal now becomes bent, and bends more and more as it gets into regions of higher and higher ion density. The wave is always bent *away* from regions of high density to regions of low density. If the signal is bent sufficiently in the layer, it finally emerges and returns to earth as shown.

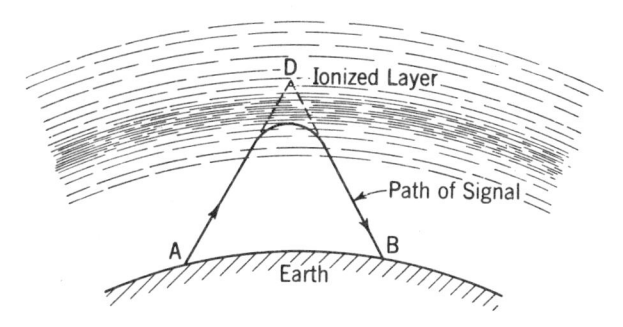

Fig. 19-3. Refraction of the sky wave by an ionized layer.

The actual path of the wave in the ionized layer is a curve, as shown, and it is said to be caused by *refraction* of the wave. This is similar to the refraction of light that takes place in a prism. Since it is usually simpler and more convenient to think of the wave as being *reflected*, rather than refracted, the path can be assumed to be the straight lines AD and DB as indicated in the figure. This assumption is made in measurements of the height of a layer. To measure the height of an ionized layer, a wave is sent out from a transmitter at A and the time taken by the sky wave traveling over the path AD-DB to reach the receiver is compared with the time taken by the ground wave along the direct path AB. From this information, and knowing the distance AB, it is possible to calculate the height of D above the earth. This height is called the *virtual height* of the ionosphere, since it is not the true height. In order to measure the true height of the layer it would be necessary to know the shape of the curved path. In measuring virtual heights the points A and B are usually placed very close together so that the wave is sent nearly vertically upward.

If the frequency of the transmitted wave in Fig. 19-3 is increased sufficiently it will be found that a point is reached beyond which the wave is no longer reflected back to earth. The path taken by the ray in this case is shown in Fig. 19-4. The path is straight until it reaches the ionized region, when it again is bent, but not as much as before. This is because waves of higher frequency are bent less easily in the ionosphere than waves of low frequency. The path is bent *away* from the region of high ion density to that of low density, as shown from points B to C. At C the wave passes through the region of highest ion density and is again bent away from this region to a region of low density. As a result, the path is now curved in the opposite direction and the wave is not bent back towards the earth. The wave emerges from the layer and is lost unless it should be reflected from another higher layer of greater ion density. Whether a wave such as that just considered is reflected from a layer or whether it penetrates the layer depends on the frequency, on the density of the layer, and on the angle at which the wave first strikes the layer. In the example of Fig. 19-4, if the density of ionization in the layer should increase for some

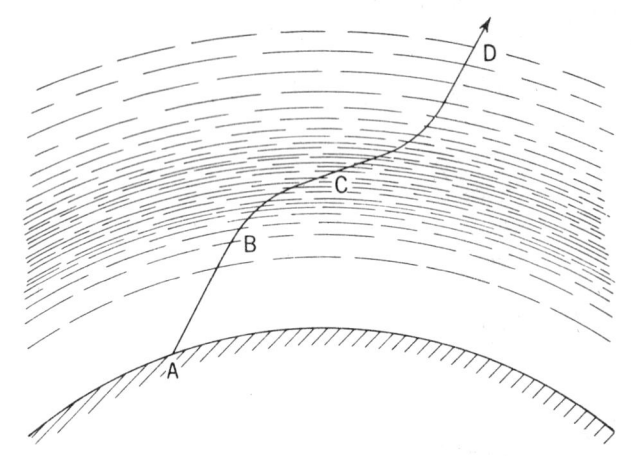

FIG. 19-4. Path of sky wave which penetrates an ionized layer.

reason, it may happen that the wave path will be bent enough to make the ray return to earth, so that higher frequencies would have to be used to make the ray penetrate the layer.

Now suppose that a wave of the same frequency as the signal in Fig. 19-4 is sent out from the transmitter along a path such as AFB (Fig. 19-5) which makes a lower angle with the ground. When the wave strikes the ionized layer, its path begins to bend as before. The amount which the wave is bent is now greater than before, since the wave spends a longer time in the ionized region owing to the low angle at which it is traveling. Thus it may happen that if the angle is low enough, the wave can become bent enough to return the wave to earth again, even though waves of this same frequency would penetrate the layer at steeper angles. Waves sent out at all angles less than that of the wave AFB shown in Fig. 19-5, on the same frequency, will be bent back to earth. Thus there will be an area, say from B to C, in which it is possible to receive the sky wave from the transmitter at A. If a receiver is located at a point nearer to the transmitter, as at D, no sky wave will be received, and unless the ground wave is strong enough, as at E, it will be impossible to receive the signals. There will be therefore an area from E to B in which it is impossible to receive signals for the transmitter, even though points farther away are able to receive. This area of no signal is known as the *skip area*, or *skip zone*, and the distance AB from the transmitter to the point where the sky wave first can be received is known as the *skip distance*.

When a wave is returned to the earth as at *B* or *C* in Fig. 19-5 it can be reflected from the earth, since the earth also is partially conducting. This reflected wave from the earth will then strike the ionosphere and be reflected once more, returning to earth at a great distance from the transmitter. This type of path is known as *two-hop* transmission as compared with the *one-hop* transmission in Figs. 19-3 and 19-5.

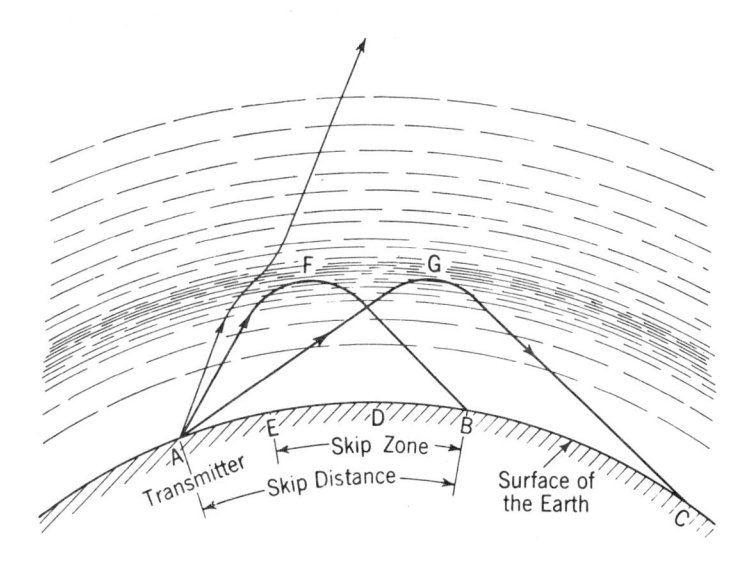

Fig. 19-5. Effects on transmission due to the angle at which the sky wave is transmitted.

CRITICAL FREQUENCIES

It was mentioned in connection with Fig. 19-3 that it is possible to measure the virtual heights of the layers in the ionosphere by sending signals vertically upward and measuring the time taken by the signals to return. Since the speed that radio waves travel is known (186,000 miles per second), it is easy to calculate the distance the wave has traveled. Because the time taken is extremely short, a few thousandths of a second, the signal that is sent up must be a very short pulse in order that the ground wave signal and the reflected signal may be separated. An oscilloscope is used to observe the two signals for measuring the time difference. If such measurements of virtual height of a layer are made at successively increasing frequencies it will be found (as mentioned earlier) that a frequency will be reached for which the waves are no longer reflected back to earth from this layer. The highest frequency for which waves sent vertically upward are returned by a layer is called the *critical frequency* for the layer at the time and place the measurements are made. The critical frequency for a particular layer is *not* the highest frequency that can be used for communication using that layer. As shown in Fig. 19-5, it is only necessary to decrease the angle that the path of the wave makes with the earth in order to have a wave of higher frequency reflected by the layer. For communication between two fixed points the angle the path of the wave makes with the layer depends upon the height of the layer and the distance between the points (see Fig. 19-6). For a given layer height

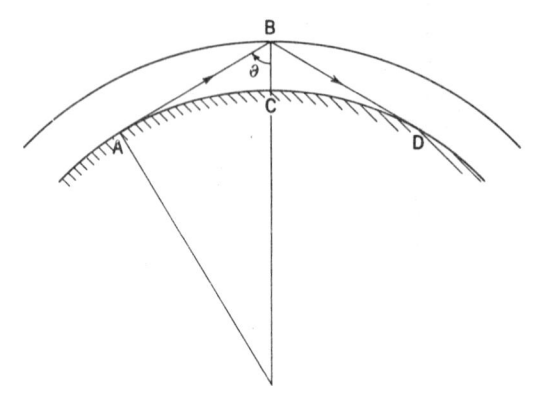

Fig. 19-6. Transmission by means of ionosphere reflection, showing angle θ.

BC there will be a particular angle θ corresponding to each distance AD between transmitting and receiving points. It has been found that the maximum frequency that can be used for sky-wave communication between two such points is given by

$$f_{max} = \frac{f_c}{\cos \theta}$$

where f_c is the critical frequency at the point of reflection and θ is the angle between the ray and the vertical. This maximum frequency that can be used for transmission between two points is called the *maximum usable frequency*.

As the distance between transmitting and receiving points is increased a limit occurs where, owing to the curvature of the earth, the path of the wave is tangent to the surface of the earth at these points. The angle θ corresponding to this limiting distance is about 74 degrees for the F layer. For this case the maximum usable frequency will be

$$\frac{1}{\cos 74°} = 3.6 f_c$$

and this will be the *maximum* frequency that will be reflected back to earth.

If measurements of the virtual height of the ionosphere are made at frequencies higher than the critical frequency for the lowest layer, it is found that the waves are still returned to earth, but from a greater height.

This difference shows that the waves are penetrating the lower layer and are being reflected from a higher layer. As the frequency is increased more, critical frequencies for the higher layers will be found; finally a frequency will be reached for which the waves sent vertically upward are no longer returned to earth, showing that the waves are penetrating all the layers. Exceptions to this behavior occur occasionally when *sporadic E* reflections are present. Owing to reflections at a boundary caused by sharp changes in ionization density, strong reflections from a layer at the height of the E layer sometimes occur at frequencies considerably in excess of the normal critical frequencies for the E layer.

ABSORPTION IN THE IONOSPHERE

In addition to the virtual heights and critical frequencies for each of the layers, the attenuation or absorption of energy from the waves by the ionosphere is an important factor in limiting radio transmission over large distances. When a radio wave passes through an ionized region, it causes the electrons to vibrate. The vibrating electrons collide with neighboring molecules and ions and give up some or all of their energy. This energy is used up in heating the air and is thus wasted. The amount of energy that will be taken from a radio wave and wasted in this way will be greater, the greater the distance the wave travels in the ionized region and the greater the density of the ions and air molecules in the layer. Since ultraviolet rays from the sun cause ionization to be present at lower levels in the daytime than at night, and since there are more air particles at the low altitudes, the absorption of energy will be much greater during the daytime than during nighttime, the absorption occurring mostly in the D and E layers. On account of a sort of resonance condition for the electrons moving in the earth's magnetic field at 1,400 kc, the maximum absorption occurs at this frequency. The further the frequency of a wave is from this resonance frequency, the less the attenuation. For long-distance transmission, frequencies near the maximum usable frequency for the distance are most desirable.

REGULAR VARIATIONS IN THE IONOSPHERE

The characteristics of the ionosphere go through regular variations which affect the propagation of radio waves. These variations can be predicted with fair accuracy. They are of three principal types, called diurnal variations, seasonal variations, and sunspot-cycle (11-year) variations. These changes in the ionosphere are due largely to changes in the radiation from the sun, so that they are mostly changes in ion density in the layers rather than in the virtual heights of the layers. With the exception of the F_2 layer, the heights of the layers show only moderate fluctuations. Therefore the variations in the ionosphere are mostly exhibited as changes in absorption and in critical frequencies.

Changes in the E layer are particularly regular from day to day and from season to season, and depend almost entirely on the position of the sun in the sky. When the sun is directly overhead, the ionization density is highest, as shown by a high critical frequency for this layer. The critical frequency is thus higher during the day than it is at night, and higher in summer than in winter. Variations in the critical frequency for the E layer occur from year to year owing to the 11-year sunspot cycle. The density of ionization in the ionosphere seems to vary with sunspot activity, and is greatest during the most active sunspot periods. Thus the critical frequency and the absorption for the E layer are lowest during a minimum in sunspot activity.

The regular changes in the ionosphere can be predicted fairly accurately. The diurnal and seasonal variations follow regular patterns, so that when allowance is made for the changes brought about by the 11-year sunspot cycle, predictions are readily made. The Bureau of Standards publishes each month ionosphere data including predicted maximum usable fre-

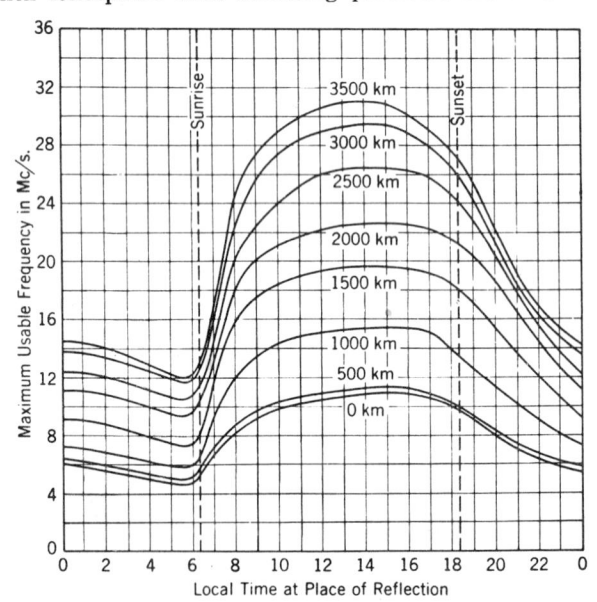

FIG. 19-7. Graph of maximum usable frequency.

quencies for the ensuing month. Figure 19-7 shows a sample graph similar to those published. The graphs show the predicted maximum usable frequency which may be used for transmission over any given distance with one-hop transmission. By making use of these graphs, communication services can lay out ahead of time a schedule of the best frequencies (from among those they have available) to use for communicat-

Chapter 19 — Radio Wave Propagation — 627

ing over particular distances at various times of the day. These predictions of maximum usable frequency are based upon measurements of virtual heights and critical frequencies made regularly by the Bureau of Standards in Washington. These measurements are also published in graph form (see Fig. 19-8) along with predictions of maximum usable frequency.

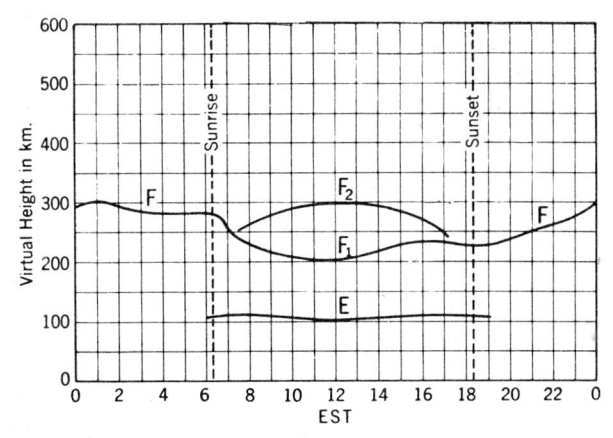

FIG. 19-8. Graph of virtual heights.

FADING

Fading of radio waves is the name given to undesirable changes in the intensity or loudness of the waves at the receiving point, and is caused by variations in the height and density of ionization in the layers of the ionosphere. To see why fading occurs, consider Fig. 19-9. A transmitter at point A is sending out signals at various angles above the surface of the earth. One signal follows the path ABC and is the ground wave from the transmitter. Another wave follows a path such as AEG, being re-

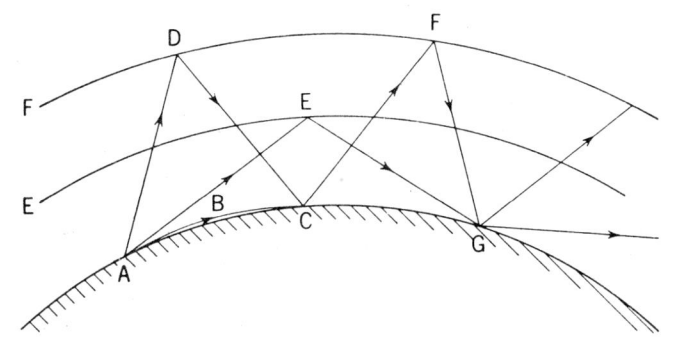

FIG. 19-9. Fading due to interference between waves which have traveled different distances.

flected from the E layer (which is shown as a line for simplicity) and received at G, but not at C. A wave sent out at a higher angle follows the path ADC, since it penetrates the E layer but is reflected by the F layer and is received at points C and G. At point C the received signal is the result of two waves, one a ground wave and the other a sky wave, which have reached that point by traveling different paths. Depending on the differences in the lengths of the paths followed by these two waves, they will add together in phase, giving a loud signal, or they will add out of phase giving a very weak signal. The difference in path will depend greatly on the height of the F layer in this case, so that small changes in its height may change the two signals from the in-phase condition to the out-of-phase condition. This change of phase causes large variations in the received signal strength; that is, the signal *fades*.

In the above illustration it was shown that fading at point C is the result of interference between the ground wave and a sky wave. Fading can also occur as a result of interference between two sky waves, as for example at the point G. Here the received signal is the result of two sky waves which have traveled different paths so that changes in either or both layers can cause fading. One signal arrives at G by one hop and the other by two hops.

As has been explained in earlier chapters, all modulated signals consist of a band of frequencies, not just one frequency, the width of the band depending on the type of signal. Because of the difference between the component frequencies, only part of the signal may fade at any given time, so that, for example, one part of a sideband fades independently of another part, giving a peculiar form of distortion in the audio signal, which is known as *selective fading*.

Fading in a received signal may take place very slowly or it may take place quite rapidly, since the ionosphere varies from a number of causes, some slow and some rapid. It was pointed out above that conditions in the ionosphere go through regular variations from day to day, season to season and year to year. The resulting variations in signals are not usually thought of as fading because they take place so slowly. However, in addition to these regular changes in the ionosphere, other more or less irregular changes take place and cause severe fading.

One of the most startling of the irregular variations in the ionosphere is that known as a *radio fade out*. A radio fade out is the result of a sudden burst of ionizing radiation from the sun, which causes the ionization in the D layer to increase suddenly, which in turn greatly increases the absorption of sky waves of all frequencies. The effect on radio transmission is the sudden fade out of all signals on frequencies above about 1.5 megacycles. The drop in signal strength occurs suddenly and lasts from about ten minutes to an hour or more.

Another and important irregular change in the ionosphere is known as an ionosphere storm. During such a storm the ionosphere becomes quite unstable in its effects on radio waves, causing signals on frequencies above about 1.5 megacycles to drop in level and fade badly. A type of fading known as *flutter fading* takes place, especially at night. The effect of these storms is usually to weaken the sky wave on the broadcast band at night, but sometimes it is increased in strength. Ionosphere storms may last from one or two days on the high frequencies to several weeks on the low frequencies. Radio communication is extremely erratic during these storms.

REDUCTION OF FADING

There are several ways in which fading can be reduced so that usable signals can be received. The most common means employed is *automatic volume control* (avc) in the receiver, described in Chapter 13. In this system, the strength of the carrier of the signal being received is used to control the volume of the receiver so that the output is held reasonably constant in spite of variations in the signal strength. Automatic volume control is not a complete solution to the problem of fading, since the signals often drop so much that they are below the noise and no amount of amplification in the receiver will make the signal usable. Automatic volume control cannot help selective fading since components of the same signal fade out at different times.

One of the best means for reducing fading is known as *diversity reception*, also described in Chapter 13.

Telegraph transmitters sometimes employ *frequency diversity* systems, which are based on the fact that signals spaced even as little as 500 to 1,000 cycles apart fade independently (selective fading). This is done by using 500- or 1,000-cycle modulation on the carrier, and keying this modulated carrier. The modulation on the carrier produces a sideband on each side of the carrier frequency; each sideband may be considered to be another carrier, which conveys the same signal when keyed. Since fading is usually selective as regards frequency, such a telegraph signal is less affected by fading than a single unmodulated keyed carrier.

STATIC AND MAN-MADE NOISE

The output of a receiver which is tuned to a weak signal usually contains some noise in addition to the desired signal. This noise may come from any one of several different sources, natural and man-made. Noise in a receiver is usually the limiting factor in determining the lowest signal strength that can be used for communication or broadcasting purposes.

Static. Noise picked up by the antenna is referred to as *static* or *atmospherics* when it is due to natural causes. Static is caused by natural electrical disturbances, principally thunderstorms, and its energy is found distributed throughout most of the frequency range used in radio. The energy in static decreases as the frequency is increased, so that most static is found at low frequencies. Static is relatively unimportant at ultrahigh frequencies. Since static is really a radio signal produced by nature, it is propagated the same way radio waves are, and is reflected by the ionosphere under suitable conditions. Static impulses may therefore travel great distances under the right conditions, and cause interference in a receiver at a great distance from its origin. A large part of the static heard at any particular place comes from a considerable distance, and the rest is from local thunderstorms and the like.

Since the sky wave on the broadcast band and lower frequencies is greatly attenuated in the daytime, very little daytime static on these bands comes from great distances. Most of the daytime static on these frequencies is due to local thunderstorms. At night, however, the sky wave propagates with less attenuation, so that the static noise level is usually greater than in the daytime. In the short-wave region, the noise level resulting from static is much less than it is at lower frequencies; the higher the frequency, the less the static. Short-wave static can travel great distances in the daytime with very little attenuation, so the static heard at any particular place may have come from some distant point.

In the frequency range from about 9 to 21 megacycles it has been found that even when there is no ordinary static or man-made interference, some noise is still picked up by the antenna. This noise comes from *radio stars* (stars which emit radio waves), and from certain regions of interstellar space. This form of noise is often the limiting noise in this frequency range at a good receiving location.

Above about 30 megacycles no energy is reflected from the ionosphere and so distant static has no effect. Normally there is no static interference in this range except during local thunderstorms.

Man-made noise. Man-made noise is generated by most electrical appliances and electrically-operated devices. Ignition systems, diathermy machines, power-line discharges, sparking brushes on motors and generators all can cause interference with radio reception. In fact, almost any device that produces an electrical spark can interfere with reception. Such noises, once produced by the device, are carried by the power lines connected to the device and are either carried directly into the receiver by the power lines, or radiated in the neighborhood of the antenna and picked up along with the desired signal.

Man-made noise is of two general types, *hiss* types and *impulse* types. Impulse types of noise consist of separate and distinct pulses of very high amplitude and are produced by separated electrical sparks such as occur in ignition systems, a-c power leaks, switch and key clicks, and so forth. In hiss types of noise the pulses occur so closely together that they overlap and sound like a continuous noise. Hiss noise is produced, for example, by commutator sparking in d-c motors and a-c series motors. Static resembles the hiss type of noise in that the separate pulses overlap.

NOISE-REDUCING SYSTEMS

There are several ways in which the noise in a receiver may be reduced in order to improve reception, the system used depending on the type of noise it is desired to reduce. The best noise-reducing system is, of course, the elimination of the noise at its source if possible. This method is often practicable as far as electrical appliances are concerned. An appropriate electrical filter placed in the line at the source of the noise will often eliminate or greatly reduce the interference. A simple line filter consists of a 0.1-μf capacitor placed across the a-c line right at the appliance, or in some cases, of two such capacitors in series with their midpoint connected to ground. When these simple filters do not effect a cure, choke inductors must also be inserted in series with the a-c power line. The sizes of capacitors and inductors for most effective suppression of noise can best be determined by experiment. A power-line filter at the receiver will sometimes reduce noise.

In most cases it is not possible to eliminate the noise at its source, so other means of reducing the noise must be found, the system depending on the type and source of the noise. For static and hiss types of noise, increased selectivity at the receiver will usually reduce the interference. This reduction is possible because the noise energy is spread out over a considerable frequency band, so that the wider the band the receiver accepts, the more noise it picks up. Crystal filters in the receiver are very helpful in reducing hiss types of noise. If the noise is coming from a definite direction a highly directive receiving array will reduce the interference with reception.

Owing to their special characteristics, impulse types of noise require different treatment. Impulse noises are pulses of extremely short duration separated by much longer time intervals. The energy in each pulse, and hence the amount of interference it produces, will depend both on the duration and the amplitude of the pulse. If the noise is sufficient to interfere with a signal, it must therefore have a pulse amplitude very much greater than the signal.

The simplest noise-reducing circuit in a receiver is the *audio limiter*, used only for reception of telegraph signals. Such circuits operate on the principle that a high-amplitude noise pulse causes transient disturbances in the loudspeaker or earphones that last much longer than the pulse itself, so that by limiting the amplitude of the pulse, the transients are reduced. Circuits used for audio limiting include triodes operated with extremely low plate voltage (of the order of 8 to 10 volts), pentodes with low screen voltage (30 to 40), and biased diodes shunted across the output so that any noise or signal exceeding the bias is by-passed by the diodes. Audio-limiting noise-reducing systems are not very effective with low-level signals or with hiss types of noise, and cannot be used for radiotelephone reception on account of the distortion they produce by the limiting process

Superheterodyne receivers often have noise-reducing systems placed at the second detector, which operate on the principle that since the noise pulses in impulse types of noise last for such a short time, no appreciable loss of intelligibility of the signal occurs if the receiver is made inoperative during the pulse. Two types of circuits are used, both employing diodes for their action. In one circuit a low-impedance path for the noise pulse is provided so that the noise is short-circuited to ground. In the other, a diode is connected so that when the noise pulse occurs, the circuit of the audio amplifier is opened and nothing is heard.

A third type of noise-reducing system employs a fast-acting automatic-volume-control system in the intermediate-frequency amplifier to reduce the sensitivity of the receiver during a noise pulse. This type of noise silencer is very effective against impulse types of noise and is of some help with hiss noise. Intermediate-frequency noise silencers are usually used with receivers with crystal filters, so that all types of noise can be reduced.

Mention has been made of crystal filters, which are used to increase the selectivity of receivers. Quartz crystals can be cut so that they act like high-Q resonant circuits and therefore can be used in place of the usual inductors and capacitors in the intermediate-frequency amplifier of a super-heterodyne receiver to increase the selectivity. Special circuits have been devised which allow the selectivity to be varied. Some circuits also allow interfering stations on nearby frequencies to be tuned out.

VHF, UHF AND SHF PROPAGATION

Above about 30 megacycles radio waves traveling upwards are no longer reflected back to earth by the ionosphere and hence there is no skywave propagation in the usual sense. (An exception to this general statement will be considered later.) At these high frequencies attenuation of the surface wave is very high, so that for all practical purposes there is no surface-wave propagation. It follows then that transmission, if it takes place at all, must be by one of the other possible paths discussed below.

DIRECT AND GROUND-REFLECTED WAVES

At the frequencies being considered in this section elevated antennas are used, and transmission is possible by means of a direct wave from transmitter to receiver. For best results when the spacing between transmitter and receiver is large the antenna heights should be made as great as possible. Referring to Fig. 19-10 it will be seen that owing to the curvature of the earth the height of the antennas determine the maximum distance apart which will permit reception of a signal via the direct wave. Figure 19-

Chapter 19   *Radio Wave Propagation*   633

10 shows direct-ray transmission from an antenna at A having a height h_1 above the earth to a receiving antenna of height h_2 located at B. As the receiving antenna is moved further away from A, a point will be reached where the line of sight from A to the receiving antenna will just graze the surface of the earth. This situation is shown by the location C; the distance

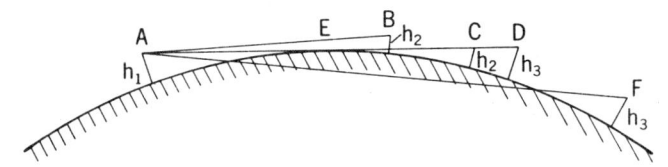

Fig. 19-10. Direct-ray propagation at the ultrahigh frequencies.

from A to C represents the maximum over which this direct line-of-sight transmission can occur for antennas of height h_1 and h_2. However, if the height of either of the antennas is increased, this distance will be extended. In the figure the receiving-antenna height is shown increased to h_3 and the maximum distance of propagation has been increased from AC to AD. When this same height h_3 is used for a receiving antenna located at F no direct-path signal will be received.

The maximum distance for this line-of-sight transmission can be easily calculated from the following formula:

$$d = \sqrt{2h_1} + \sqrt{2h_2}.$$

This formula gives the maximum line-of-sight distance d in *miles* when the transmitting and receiving antenna heights, h_1 and h_2 respectively, are expressed in *feet*. For example, if the antenna heights are 800 ft and 50 ft the maximum line-of-sight distance would be $\sqrt{1{,}600} + \sqrt{100} = 50$ miles.

When the receiving antenna is within sight of the transmitting antenna, as given by the above formula, the received signal is actually the resultant of two waves, one the direct wave shown as A-C in Fig. 19-11, and the other a wave reflected by the surface of the earth and shown as ABC in the figure. These two waves add together at the receiving point C and will *reinforce* or *cancel* each other depending upon whether they arrive *in phase* or *out of*

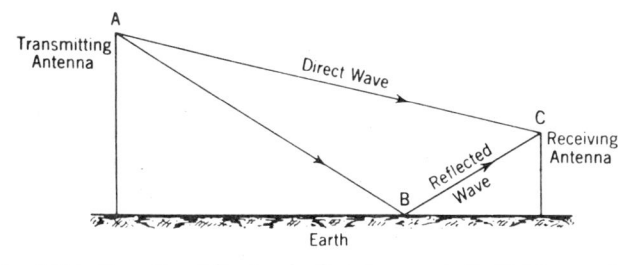

Fig. 19-11. Reception of direct and reflected waves at ultrahigh frequencies.

phase. This addition of waves in phase and out of phase was shown in Fig. 9-3. The reflected wave will be *reversed in phase* upon reflection. This is because the incident wave induces currents in the ground which set up a new wave (the reflected wave) which has the direction of its electric field reversed from what it was in the original wave. That is, the reflected wave is 180° out of phase with the initial wave. This phase reversal *always* occurs when horizontally polarized waves are reflected from the ground but occurs for vertically polarized waves only at the very high frequencies.

Whether the two waves from the transmitting antenna at A arrive at C in phase or out of phase depends upon the relative path lengths of the two waves. If the path lengths are the same (as they very nearly are when the antennas are close to the ground) the waves arrive *out of phase* because of the phase reversal suffered by the reflected ray. However, if the path of the reflected wave is one-half wave length longer than that of the direct ray, it takes one half of a cycle longer to travel from A to C and so arrives in phase. It is evident that it would also arrive in phase if the path difference were three halves, five halves or any odd number of half wave lengths. The signals will arrive out of phase when the path difference is any even number of half wave lengths. Whether cancellation or reinforcement occurs at C, then, depends upon the heights of the antennas and their distance apart.

When the antennas are relatively close compared with their heights (as illustrated in Fig. 19-11) the heights should be picked, if possible, so that the path difference between the two waves is an odd half wavelength and reinforcement will occur. When the distance between antennas is very much larger than their heights (this is the usual case) the path length difference is always much less than one-half wavelength, and the best that can be done is to make the path difference as great as possible by making the antennas as high as possible.

DIFFRACTION AND REFRACTION

As a receiving antenna is moved beyond the line of sight the received signal does not drop abruptly to zero as might be expected, but actually decreases smoothly although rapidly to low values. This reception is due to two factors not yet considered, namely, *diffraction* and *refraction*. It is possible for waves, whether sound waves, light waves or radio waves, to be diffracted or bent around obstacles in their path. The amount of bending or diffraction depends upon the size of the obstacle as compared with the wavelength of the wave. If the obstacle is very large in terms of *wavelengths*, the bending will be small. This case embraces light waves, where the wavelength is so small that nearly all objects are very large compared with it and little diffraction occurs. In the case of sound, however, where the wavelength may be several feet, bending occurs quite readily around most objects. At low frequencies the length of radio waves is sufficiently large compared with ordinary obstacles that the waves bend around them. At very high radio frequencies the wavelength becomes much shorter and only a small amount of bending or diffraction occurs. However, it is sufficient to enable signals to be received at these frequencies several miles beyond the line of sight.

Refraction of radio waves occurs as the result of changes in the density of the air with height and changes in the temperature, pressure, and amount of water vapor in the air. This refraction tends to bend the waves back to earth and has an effect in the lower atmosphere somewhat similar to the effect of the ionosphere in the upper atmosphere. The increase in distance of transmission obtained as the result of normal refraction can be allowed for by considering that the earth is flatter than it really is, that is, by assuming its radius to be increased by about 20 per cent to 35 per cent. Signals received by reason of refraction are not as stable as those for direct-ray transmission, because slight changes in the condition of the atmosphere change the amount of refraction and so produce fading.

TROPOSPHERIC WAVES

If the propagation of radio waves were restricted to the paths or modes which have been considered so far there would be no possibility of reception of vhf, uhf and shf signals at distances greater than say 50 per cent beyond line of sight. Nevertheless, as nearly every owner of a television set knows, it is often possible to receive quite strong vhf and uhf signals at distances much greater than line of sight. It is true such reception may be irregular and as variable as the weather, but there are valid reasons for this behaviour. Reception of this nature is the result of *tropospheric propagation*.

The *troposphere* is that region of the earth's atmosphere where clouds are formed. It extends upward from the surface of the earth to a height of about 10 kilometers or $6\frac{1}{2}$ miles. In this region the temperature decreases with height to a value of about $-50°C$ at its upper boundary. Above the troposphere is the stratosphere where the temperature remains constant at about $-50°C$. Within the troposphere the atmosphere has a dielectric constant (relative to a vacuum) which is slightly greater than unity at the earth's surface where the air is most dense and which decreases to unity at great heights where the air density approaches zero.

The dielectric constant of dry air is slightly greater than unity and the presence of water vapor increases the dielectric constant still further, so the effective dielectric constant depends on air conditions, that is, on the weather. A normal or *standard atmosphere* is one where the dielectric constant is assumed to decrease *uniformly* with height to a value of unity at a height where the air density is essentially zero. Under these conditions waves traveling in the troposphere are refracted slightly towards the earth with the resulting rather small increase of transmission distance previously mentioned. In actuality, so-called *standard* conditions of the atmosphere hardly ever exist. The air is frequently turbulent and at other times there are often layers of air one above the other having different temperatures and water-vapor contents. The former condition leads to a scattering of radio waves from the "blobs" of air and these scattered waves may give usable reception considerably beyond the line of sight. In the case of layer formations, radio waves may be strongly refracted or reflected at the boundary surfaces between layers with a resulting usable signal at distances up to several hundred miles from the transmitter. Still another phenomenon is that of *duct propagation*. Two boundary surfaces between layers of air form a duct or a sort of "leaky waveguide" which guides the electromagnetic wave between its walls. Under these circumstances a receiving antenna within the duct would receive a strong signal even at great distances whereas a receiving antenna outside of the duct would receive a very weak signal. Ducts may be *elevated*, with both walls above the surface of the earth, or in the case of a *surface duct*, the earth itself forms one of the walls.

Layer formations and ducts are common over large flat land masses and occur regularly over certain ocean areas. Hilly or mountainous country on the other hand tends to produce turbulent conditions of the air above it. Because of the direct dependence on atmospheric conditions, this type of tropospheric propagation is very much a function of the weather.

FORWARD-SCATTER PROPAGATION

The possibility of the above-mentioned occasional reception of relatively strong vhf and uhf signals well beyond line of sight has been recognized for many years — indeed ever since 1932, when Marconi successfully transmitted 500-megacycle signals over a distance of 168 miles. However, it is only in quite recent years that experiments have definitely established that it is possible, by using sufficiently large radiated power and high-gain antennas, to achieve *very reliable* vhf and uhf radio communication over distances of 100 to 1,000 miles. This type of reliable beyond-the-horizon propagation has been given the name of *scatter-propagation* or *forward scatter*, because of the mechanism which is believed to be responsible for the phenomenon. Although the physical mechanisms are not yet completely understood, there appear to be two distinctly different modes of transmission involved in forward-scatter propagation. One of these modes is ionospheric, and is believed to result from scattering of the radio waves from the lower E-layer of the ionosphere. The other mode is tropospheric and is thought to be the result of scattering from either blobs or fine layers in the troposphere. It has been suggested that the ionospheric scatter might be from blobs or fine layers at the lower edge of the E-layer, or it could be from the ionized trails of myriads of small meteors which bombard the earth from outer space. It is known that ionospheric scatter permits communication in the frequency range from about 25 to 60 megacycles over distances of 600 to 1,200 miles. At frequencies higher than 60 megacycles the importance of ionospheric scatter decreases, but tropospheric scatter appears to be effective from 100 megacycles up to at least 10,000 megacycles. Because of the large attenuation of the signal along the path, forward-scatter propagation is useful chiefly for point-to-point communication, and radio or television relay links, where high-power transmitters and extremely high-gain transmitting and receiving antennas can be used.

SUMMARY OF RADIO WAVE PROPAGATION

Vlf and lf (15 to 300 kc). At frequencies below 300 kc the surface wave is attenuated very little and it may be used for communication up to distances of 1,000 miles or more. The signals are very stable and show no diurnal or seasonal variations. At greater distances the sky wave becomes of more importance than the surface wave. The sky wave is fairly reliable but there are slight fluctuations owing to variations in the ionosphere. The absorption of the sky wave is less at night than in the daytime, but even in the daytime it is fairly low at the lower frequencies. In the lf band (30-300 kc) sky-wave transmission is good for distances from about 500 miles up to 8,000 miles. At vlf (below 30 kc) the lower surface of the D layer and the surface of the earth appear to act as a waveguide or a concentric spherical cavity resonator within which the electromagnetic energy is constrained. Using a sufficient amount of radiated power, of the order of 1 megawatt, it is possible at 15 kc to lay down a usable signal at any place in the world.

Mf (300 kc to 3 megacycles). This frequency range includes the American broadcast band (550-1,600 kc). In this broadcast band the range of the ground wave varies from 50 miles at the higher frequencies to about 200 miles at the lower frequencies, depending also upon the power of the transmitter. Sky-wave reception is not possible in the daytime owing to high absorption of the sky wave by the D and E layers, but at night the sky wave gives reception at distances from 100 to 3,000 miles. At night there is generally an area where the sky wave and ground wave are of nearly equal magnitudes and in this region severe fading will occur. At greater distances the sky wave alone exists but there will still be some fading caused by variations in the ionosphere. The absorption of the sky wave in the broadcast band increases with frequency up to 1,400 kc. At this frequency the absorption is a maximum. The absorption decreases with decreasing frequency below 1,400 kc and decreases with increasing frequency above 1,400 kc.

Hf (3-30 megacycles). The attenuation of the surface wave at frequencies above 3 megacycles is so great as to render the surface wave of little use for communication except at very short distances of the order of 15 miles or less. The sky wave must be employed, and since it makes use of the ionosphere for its propagation, communication by means of it, although not perfectly reliable, is possible over distances as great as 12,000 miles. The sky-wave absorption in this range decreases with increasing frequency, so that the higher the frequency the more efficient the transmission. It is therefore desirable to use a frequency as near as possible to the maximum usable frequency (defined above). However, if a frequency too near the maximum usable frequency is employed, the irregular changes in the condition of the ionosphere make communication uncertain. Therefore there is a range of frequencies, from about 50 per cent to 85 per cent of the maximum usable frequency for the given distance and time, that may be used satisfactorily. The particular layers utilized in transmitting over a given distance will depend on the distance, the time of day, and so on. For very short-distance communication (say a few hundred miles) using the sky wave, the frequency must be below the critical frequency for the layer used since the sky wave strikes the ionosphere at almost vertical incidence.

Vhf, uhf, and shf (above 30 megacycles). There is no surface wave propagation at these frequencies and except as noted above there is usually no reflection from the ionosphere, so communication must be via the direct and ground-reflected waves for line-of-sight transmission, or by means of tropospheric or scatter propagation for distances much beyond line of sight. Within line of sight for moderate distances the signal will ordinarily be quite stable and free of static.

Beyond line of sight, tropospheric waves can be expected to produce at times comparatively large signal intensities which, however, are likely to be irregular and very much dependent on weather conditions. Finally, forward-scatter, either ionospheric or tropospheric, will give very reliable communication for distances up to about 1,000 miles if high-power transmitters and large transmitting and receiving antennas are employed.

QUESTIONS AND PROBLEMS

1. Why is there a low density of ionization in the atmosphere below about 30 miles and above 250 miles from the earth's surface?

2. What factors determine whether a radio wave is reflected by or penetrates an ionized layer?

3. What is meant by virtual height? Critical frequency? Maximum usable frequency for a given layer?

4. Why is energy absorbed from a wave passing through an ionized layer? How does the amount of attenuation vary between day and night conditions?

5. What causes fading of signals from broadcasting stations at night? Why is fading not present in the daytime on the broadcast band?

Chapter 19 *Radio Wave Propagation* 639

6. What causes selective fading? Flutter fading? Radio fade out?

7. How does the energy in static vary with frequency?

8. What causes fading on ultrahigh frequencies?

9. What factors must be considered by a communication company in deciding which frequency to use for communication between two fixed stations?

CHAPTER 20

RADIO ANTENNAS

FUNCTIONS OF AN ANTENNA SYSTEM

An antenna system usually serves a twofold purpose. The chief function of a transmitting antenna is to radiate efficiently the power furnished by the transmitter. A second function may be to direct this power into directions where it is wanted and to prevent radiation in other directions where it is not wanted. Similarly, the chief function of a receiving antenna is to capture power from a passing radio wave. A second function may be to favor radio waves coming from a particular direction and to discriminate against waves coming from other directions.

For both transmitting and receiving antennas this second directivity function also serves to increase the power radiated in or received from the favored direction. When directional characteristics are relatively unimportant a simple antenna will usually suffice; however, when directivity is a prime consideration a complicated structure may be required.

Practical antennas fall into one of two classes, *elevated antennas* or *grounded antennas*. An elevated antenna is operated some distance above the ground. It may be either horizontal or vertical. A grounded antenna operates with one end grounded through the output of the transmitter or the coupling coil at the end of the feed line. Elevated antennas are used at the higher frequencies, above about 2 megacycles, while grounded antennas are generally used at frequencies below this. At the lower frequencies, a wavelength becomes very long and the necessary size of antenna for efficient radiation becomes quite large. Because of the difficulties of elevating large structures above the ground, grounded antennas are used at these frequencies. Grounded antennas are also used at high frequencies in certain particular applications such as airplane antennas where the airplane itself becomes the ground.

THE ELEVATED HALF-WAVE ANTENNA

When an elevated antenna is used it is generally made a half wavelength long. The antenna is then of a *resonant length*, that is, it is tuned to resonance and its input impedance is a pure resistance equal to its radiation resistance. The current and voltage distributions are similar to those obtained at the end of an open-circuited transmission line. These are illustrated in Fig. 20-1. The current is zero at the ends (necessarily) and the voltage is a maximum at these points. This voltage is the voltage between the two halves of the antenna, one end being positive and the other negative at any given instant. The current distribution is very nearly *sinusoidal*, that is, the current loop shown has the shape of a half sine wave. This distribution occurs because it is obtained by the addition of two traveling sine waves (the outgoing wave and the reflected wave), as illustrated in Fig. 9-5. The voltage wave is also sinusoidal with a node at the center of the antenna, which corresponds to a point one quarter wavelength back from the open end of the transmission line of Fig. 9-5. The voltage loop is generally shown crossing over at the nodal point to indicate the 180° phase difference between the top and bottom halves.

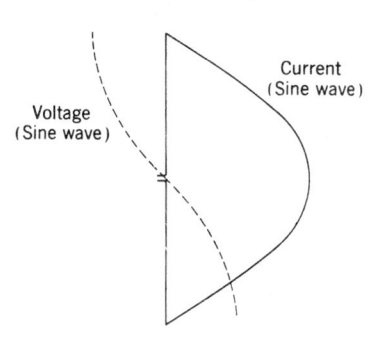

FIG. 20-1. Voltage and current distribution on an elevated half-wave antenna.

Two of many possible methods of feeding an antenna are shown in Fig. 20-2. In part (a) of this figure the antenna is fed by connecting the transmission line in series at the center. The current out of the end of the transmission in this case must equal the antenna current I_a. When the antenna is a resonant length its input impedance is a pure resistance of approximately 73 ohms; so the voltage that must be furnished at the antenna input terminals by the transmission line is 73 I_a. If the characteristic impedance of the transmission line is approximately 73 ohms the antenna will be a matched load for the line and there will be no standing waves on the transmission line. If the characteristic impedance of the transmission line differs greatly from 73 ohms (e.g., if it is 300- or 500-ohm line) the antenna input impedance will not provide a match and there will be standing waves on the transmission line.

An equivalent circuit for the series-fed resonant antenna is shown in Fig. 20-2(b). Because the circuit is *resonant*, the inductive reactance will be cancelled by the capacitive reactance and the impedance of the circuit is simply the radiation resistance of the antenna.

In Fig. 20-2(c) the antenna is fed by inducing a voltage across a small coil connected in series with the antenna terminals. In this case the transmission line current does not have to equal the antenna current. By varying the turns ratio an approximate impedance match can be obtained for any transmission line characteristic impedance. The equivalent circuit for this case is illustrated in part (d) of the figure.

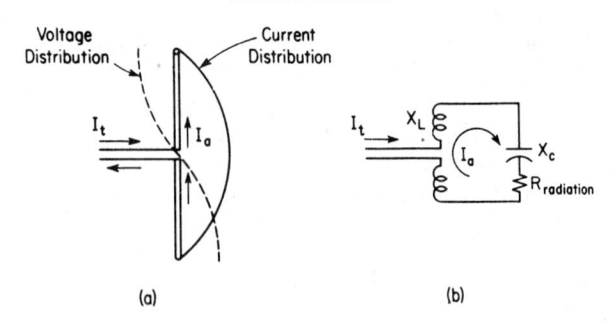

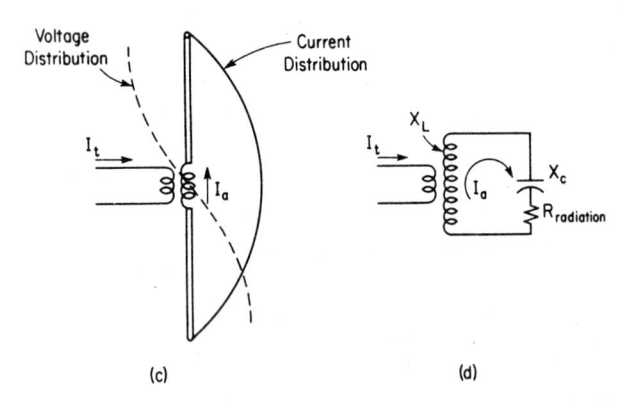

FIG. 20-2. Methods of feeding antennas with their equivalent circuits.

These and other methods of coupling the antenna to its transmission line are described more fully in a later section.

RADIATION CHARACTERISTICS OF A HALF-WAVE ANTENNA

As was mentioned in Chapter 9, an antenna radiates a stronger field at right angles to its axis than in other directions. The *radiation characteristic* of a half-wave antenna far away from the ground is shown in Fig. 20-3. The radiation characteristic is a method of representing graphically the relative field strength in different directions. The distance from the center of the antenna to the curve, along any particular line, represents the relative field strength in that direction. In Fig. 20-3 the line OA is twice the length of OB, so that field strength E in the direction of OA is twice as great as the field strength

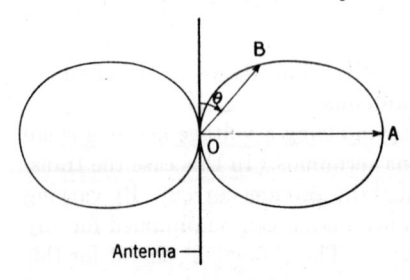

FIG. 20-3. Radiation characteristic of a half-wave antenna.

Chapter 20   Radio Antennas

in the direction of OB. The field is symmetrical about the axis of the antenna and the characteristic shown is merely a cross section of the solid figure, which would represent the three-dimensional characteristic. This solid figure could be obtained by rotating the characteristic of Fig. 20-3 about the axis of the antenna. It is a doughnut shape with practically no hole and with the antenna sticking vertically through the middle.

The characteristic has been shown for a vertical antenna, and for such it is the vertical radiation characteristic; that is, it indicates the relative field strength at various vertical angles. The *horizontal characteristic*, which shows the relative field strength at various horizontal angles, is just a circle of radius OA for this antenna. This is because the antenna radiates uniformly in all directions perpendicular to its axis.

If this same antenna were operated in the horizontal position, the same radiation characteristic would apply, except that it would be turned over with the antenna. The characteristic of Fig. 20-3 would then be the horizontal characteristic. The vertical characteristic looking along the length of the antenna would be a circle.

THE GROUNDED ANTENNA

At the lower frequencies a half wavelength becomes quite long, and an elevated half-wave antenna would be a large and costly structure. This is particularly true because at these frequencies the ground wave is used for transmission and so the antenna must be vertical, or be capable of producing vertically polarized waves.

In the middle of the broadcast band, at a frequency of say 1,000 kc, a half wavelength is 150 m or about 500 ft. The difficulties of constructing a

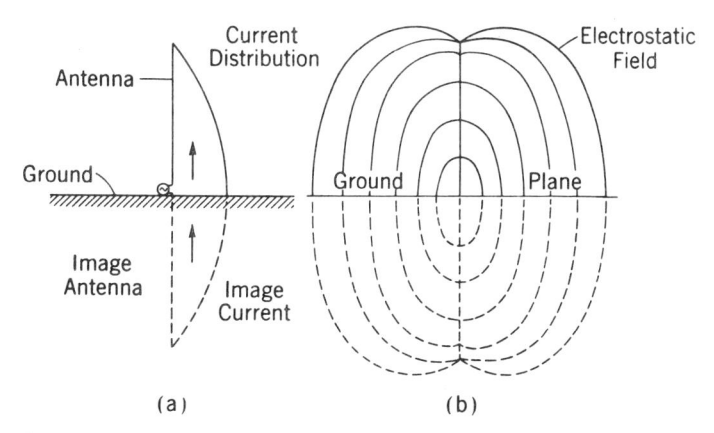

Fig. 20-4. A quarter-wave vertical antenna at the surface of the earth. (a) Antenna and its image, showing current distribution; (b) electrostatic field about the antenna, showing how the image antenna can be used to account for the effect of the ground.

vertical antenna of this length and elevating it above the ground are evident. Fortunately, under these circumstances it is possible to use an antenna only a quarter wave long (or even shorter if inductance loading is used) and operate it with one end grounded. In this case the ground takes the place of the lower half of the antenna. This is shown in Fig. 20-4, where the lower half of the antenna has been replaced by the *image* of the upper half in the ground. It is possible to do this because if the ground were a perfect conductor the distribution of the electric field about the antenna would be as shown by the solid lines of Fig. 20-4(b). This distribution above the ground is exactly the same as that obtained about one half of a half-wave antenna *in free space* (far away from the ground). Therefore, in considering the operation of a grounded antenna it is merely necessary to replace the ground by the image of the antenna and then determine the fields of this *complete* antenna as if it were in free space, that is, remote from the ground.

This behavior can also be understood by considering the current in the antenna. The strength of the electric field at a point in space (not too close to the antenna) *due to any portion of the antenna* is proportional to the current flowing in that portion. The total electric field at the point is the *sum* of all the electric fields produced at that point by the various portions. In Fig. 20-5, to determine the relative electric field strength at the point P,

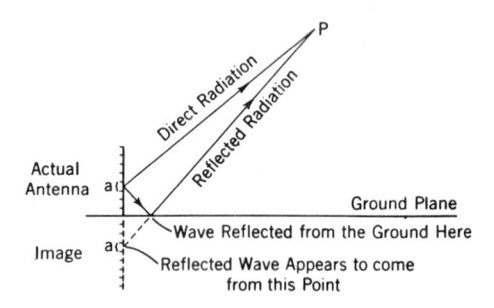

Fig. 20-5. Vertical antenna at the earth's surface, showing how the image can be used to account for the reflected wave.

the contributions of the currents in each of the small lengths a must be summed up. However, besides the energy which reaches P by direct radiation from each of these points, there is the *reflected* energy which is radiated towards the ground and then reflected up to the point P. As far as the electric field strength at the point P is concerned, it is exactly as if this reflected energy were coming from another antenna which is the *mirror image* of the actual antenna. (A *mirror image* is the type of image that is seen in a mirror, with right-hand points on the right, left-hand points on the left, close points close, and distant points far away.)

Chapter 20   Radio Antennas   645

ANTENNAS OF OTHER HEIGHTS

As in the case of the half-wave antenna, a quarter-wave grounded antenna is of a *resonant length* and acts like a circuit tuned to the frequency being used. Essentially this condition means that looking in at the feed point to the antenna the impedance presented is a pure resistance. It also means that with this length of antenna the antenna current for a given applied voltage will be large. If the antenna is lengthened or shortened slightly from this length the current will drop sharply. However, if it is necessary to use an antenna longer or shorter than this resonant length, such an antenna may be *resonated* or tuned by adding capacity or inductance, respectively, in series with it. This tuning is often done with broadcast antennas. In particular, because of cost considerations, it is often desirable to operate with a shorter antenna, say one sixth or one eighth of wavelength. In this case it is necessary to load the antenna with series inductance to tune it to the frequency being used.

DIPOLE AND MONOPOLE ANTENNAS

Elevated antennas with two balanced halves, such as were considered in Figs. 20-1, 20-2, and 20-3, are often called *dipole antennas* or sometimes just *dipoles* (literally two poles). In particular the resonant half-wave elevated antenna is commonly known simply as a *half-wave dipole.* In contrast the grounded (or ground-based) antenna of Fig. 20-4, which is fed against ground and uses the image to replace one half of the antenna, is often called a *monopole antenna* or just *monopole.* In particular the resonant length ground-based vertical antenna is often designated simply as a *quarter-wave monopole.* In later parts of this chapter the various equivalent terms will be used interchangeably as they are used in the field.

LOSSES AND EFFICIENCY

It has already been seen that because an antenna radiates energy it acts like a resistance at the end of the line feeding it. This resistance, because it is a result of the energy or power radiated, is called radiation resistance. However, there are also some other resistances in the circuit. The antenna will be constructed of wires or rods or steel girders, depending upon its size, and these will have resistance. If the antenna is loaded with an inductance coil this coil will also have a resistance, which may amount to several ohms. These resistances are generally called *ohmic* or loss resistances to differentiate them from the radiation resistance of the antenna. The antenna current must flow through these ohmic resistances and so there is a power loss. In addition, there is a dielectric loss in the ground owing to the penetration of the electric field into it. The relative amounts of power lost and power radiated will depend upon the ratio of the loss resistances to the radiation resistance. For this reason it is desirable to have an antenna with a high radiation resistance, since the radiation resistance gives a measure of power radiated or the useful power. The radiation resistance of a half-wave antenna is about 73 ohms. If it is properly constructed its loss resistance may be only 1 or 2 ohms or even less. This means that the losses will be small compared with the useful output power, and the efficiency will be high. For a loss resistance of 2 ohms, the efficiency in this case would be given by

$$\text{Efficiency} = \frac{\text{output}}{\text{input}} \times 100 = \frac{\text{output}}{\text{output} + \text{losses}} \times 100$$

$$= \frac{73}{73 + 2} \times 100 = 97\%.$$

For an elevated antenna the only other losses likely to occur are losses in supporting insulators and *absorption losses* due to currents induced in other conducting structures near the antenna. These can be kept small by placing the insulators at low-voltage points and by keeping the antenna away from nearby conductors.

In the case of grounded antennas the losses *may* be considerably greater than this. The ground that is taking the place of one half of the antenna, and which was assumed to be a perfect conductor for this purpose, actually has resistance that may be quite high. Referring to Fig. 20-4, it will be seen that the current flowing in and out of the bottom of the antenna must also flow out of and into the ground connection. Once in the ground it spreads radially in all directions from the antenna, keeping near the surface or penetrating to considerable depths depending upon whether the wavelength is short or long. These ground currents flowing through the resistance of the ground cause losses which must be supplied from the input power. This loss decreases the efficiency of the antenna system.

As the antenna is made shorter (in terms of a wavelength) the losses become quite large, for three reasons. (1) As the antenna is made shorter it requires more current in it to produce the same amount of radiated power. The losses increase as the square of the current ($P = I^2R$), and so mount rapidly. (2) The antenna must be loaded with an inductance coil to tune it to resonance and the large current flowing through the resistance of this coil may absorb an appreciable portion of the power. (3) The electric field at the base is high, which increases the dielectric losses.

GROUND SYSTEMS

To reduce the losses in the ground which occur with a grounded antenna, a *ground system* is used quite generally, especially with broadcast station antennas. The ground system is often constructed as indicated in Fig. 20-6. Wires of about a half wavelength are stretched radially outward from the ground connection. A common angular separation is 3° so that there are 120 of them. To save wire, alternate radials are often made shorter; this does not affect the operation much. If the wires are soldered or preferably welded together at the joints, the resulting ground system will have very small losses. The dielectric loss at the base can be reduced by placing a screen under the antenna which is also connected to the ground system.

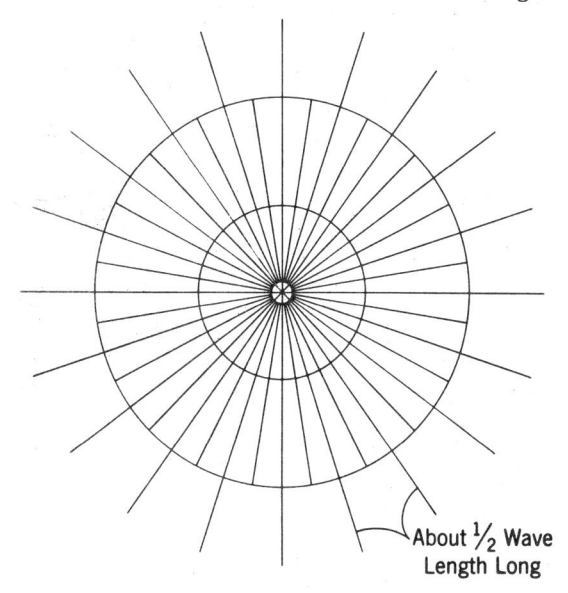

FIG. 20-6. Radial type of ground system used for grounded antennas.

COUPLING NETWORKS

Coupling networks are used to couple or connect the transmitter to the transmission line and the transmission line to the antenna. (If there is no transmission line the coupling network is used to couple the transmitter output directly to the antenna.) Coupling networks serve to isolate the transmission line and antenna from the large d-c potentials at the output of the transmitter. They are used to tune to resonance the circuits they connect, and for this purpose are generally provided with one or more variable elements, such as variable capacitors or variable inductors. Finally, they provide a means for varying the coupling between the circuits, and therefore can be used for impedance matching so that the maximum possible amount of power is transferred from the transmitter to the antenna.

There are many different types of coupling networks, some of which are shown in Figs. 20-7 to 20-12. However, the two functions of tuning and impedance matching can be explained with reference to the simple coupling network shown in Fig. 20-7.

Tuning. In order to obtain large currents in an antenna or any other circuit having reactance, it is necessary to *tune the circuit to resonance*. In

Chapter 3 it was seen that this can be accomplished by adding either inductance or capacitance to the circuit until the inductive reactance is just equal to the capacitive reactance. If the circuits to be coupled together are not already tuned to resonance, the coupling network can be made to tune them.

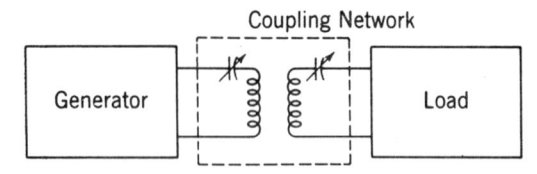

FIG. 20-7. Typical series-tuned coupling network for connecting a low-impedance generator to a low-impedance load.

In the network of Fig. 20-7, the variable capacitors are used to tune the generator and load circuits. Of course it might happen that either or both of these circuits would require added inductance instead of capacitance for correct tuning. However, in this case, since the coupling coils used are now part of these circuits, it is only necessary to make the reactance of these coils large enough so that the circuits (without the capacitors) are inductive, and so can be tuned by means of the capacitors. Variable capacitors are preferred to variable inductances in all low-power applications because they are easier to construct and adjust, but with high-power transmitters the opposite is sometimes true. In that case the capacitors would be fixed and the inductors made variable.

If only one of the circuits being coupled together requires to be tuned (the other already being resonant), then a variable element will be required only in the untuned side. However, if both circuits are detuned and the coupling between them is *loose*, a variable element will be required in both input and output sides of the coupling unit. Coupling is said to be *loose* when it is insufficient. In this case the tuning of one circuit does not have much effect on the tuning of the other, and so the circuits can and must be separately tuned to the resonant frequency. However with *close* coupling, that is, where the coupling is sufficient or greater than critical, the tuning of one circuit affects the tuning of the other so that only one variable element is required for tuning both circuits.

It is necessary nevertheless to adjust the amount of coupling in order to transfer the maximum amount of power from one circuit to the other, and this adjustment affects the tuning so that the correct over-all adjustment is more difficult to obtain. With a tuning element in both circuits the tuning procedure is simpler. In this case the coupling between circuits is made loose and each circuit is then tuned to resonance independently of the other. The coupling is then increased until maximum current flows in the second circuit and the correct adjustment has been obtained. Because of the ease of adjustment that results, two tuning elements are often used even where only one is necessary.

Impedance matching. Besides tuning the circuits they couple together, coupling networks are used to control the amount of power transferred from one circuit to the other. This control is known as *impedance matching*. In Fig. 20-7 the amount of power transferred is controlled by varying the coupling between the two coils. Often, though not always, it is desired to transfer the maximum possible amount of power from one circuit to the other. In this case the coupling is increased until the resistance coupled into the generator circuit by the coupling is just equal to the resistance of the generator circuit alone. It has been previously shown (Chapter 7) that this is the condition for maximum power transfer.

Coupling between two circuits can be varied in ways other than varying the mutual position of two coils. If the coils are fixed or wound on the same form, the coupling may be varied up to the maximum obtainable by providing the coils with taps and tapping on at different points.

The amount of mutual inductance that can be obtained conveniently with air-core coils, such as are used in antenna coupling networks, is quite small. If the resistances of the generator and load are small so that large currents are required for a given amount of power, the large current flowing through the primary will produce a large amount of magnetic flux and sufficient voltage will be induced in the secondary. A *series* circuit, such as that shown in Fig. 20-7, in which the coupling coil and capacitor are in series with the generator or load, is satisfactory in this case. However, when high resistance circuits are to be connected together, only small

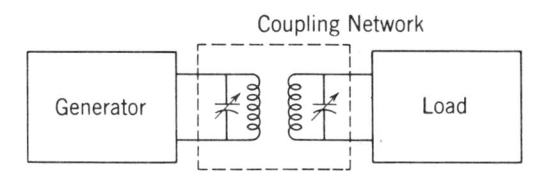

Fig. 20-8. A parallel-tuned network for coupling a high-impedance generator to a high-impedance load.

currents will flow, and it would require a very large number of turns to produce sufficient magnetic flux to induce the required voltage in the secondary. In this case the parallel connection of Fig. 20-8 is used. The inductor and capacitor are in parallel with the generator or load. Here a relatively small current (but correspondingly large voltage) from the high-impedance generator will produce a large circulating current in the primary parallel resonant circuit. This large current produces sufficient magnetic flux to induce the required voltage in the secondary. If both generator and load circuits are high-impedance circuits (that is, high-voltage–low-current circuits) the parallel connection is used on both sides of the coupling network. If one is a high-impedance circuit and the other is a low-impedance circuit, the parallel connection will be used on the high side and the series connection on the low side.

Typical coupling networks. Some typical coupling networks used to couple the transmitter to the transmission line and the transmission line to the antenna are shown in Figs. 20-9 to 20-12.

Figure 20-9 shows methods of coupling from the final stage of the transmitter to a resonant line. The plate circuit of a tube is a high-impedance circuit, and parallel tuning is used on that side. On the secondary

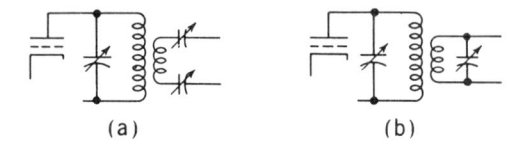

FIG. 20-9. Coupling a transmitter to a resonant line. (a) Series feed for coupling at a low-impedance point; (b) parallel feed for coupling at a high-impedance point.

side series tuning is used, as in Fig. 20-9(a), when the resonant line is being fed at a low-impedance point. If the line is being fed at a high-impedance point, the parallel circuit of Fig. 20-9(b) is used.

When a nonresonant parallel-wire transmission line is being used, the input impedance will be a pure resistance (usually about 500 ohms) and its value will be independent of length. The coupling circuits of Fig. 20-10 are suitable for this. In Fig. 20-10(a) the capacitor is omitted on the line side and the small amount of inductive reactance introduced by the secondary coupling coil is tuned out on the primary side. Close coupling is neces-

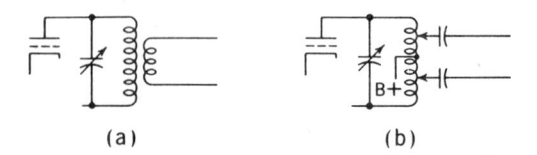

FIG. 20-10. Coupling a transmitter to a nonresonant line. (a) Inductive coupling; (b) direct coupling.

sary to accomplish this tuning. In Fig. 20-10(b) the same result is obtained by direct coupling to the primary coil. The amount of the coupling is varied by moving the taps on the coil. Loose coupling requires only a few turns between taps; close coupling is obtained with a larger number of turns. The fixed capacitors are necessary to isolate the line from the high d-c voltages present in the plate circuit of a transmitting tube.

Figure 20-11 shows a method of coupling a transmitter to a concentric or unbalanced transmission line. Such a line has a low impedance and so series feed is used. A *Faraday shield* is often used in this case. A Faraday shield is a shield which prevents electrostatic coupling between the coils while allowing magnetic coupling.

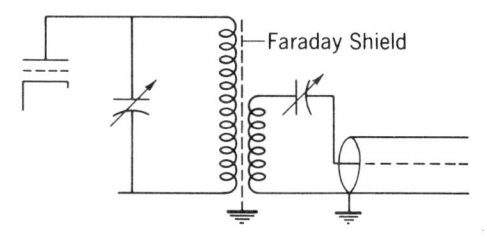

FIG. 20-11. Coupling a transmitter to an unbalanced line.

Coupling the transmission line to the antenna. Several methods of coupling a transmission line to an antenna are shown in Fig. 20-12. Figure 20-12(a) shows a series-feed connection to a vertical grounded antenna. If the antenna is shorter than a quarter wavelength, it will have a capacitive reactance and this can be tuned out by proper adjustment of the loading coil. There is no provision for impedance matching. The radiation resistance of the grounded antenna is about 36 ohms for a quarter wavelength and drops rapidly as the antenna height is decreased, being only about 13 ohms at one sixth of a wavelength and 7 ohms at one eighth of a wavelength. Because of this drop, considerable mismatch may occur and the feed line will be operating as a resonant line with standing waves. Figure 20-12(b) shows a parallel-feed connection to couple a high-impedance (500-ohm) line to a grounded antenna. The coil in series with the antenna and ground is used to tune the antenna circuit to resonance, and the impedance match is obtained by adjusting the coupling between the coils. In Fig. 20-12(c) is shown a shunt-feed arrangement sometimes used with broadcast antennas. The feed line is run up at an angle and connected directly to the antenna tower at a point where the impedance is the same as the line impedance. The impedance to ground along the antenna is zero at the base and increases to a maximum at the top. (This is similar to an elevated half-wave antenna in which the voltage and therefore the impedance between halves is low at the center and increases to a maximum at the ends.)

Connections to elevated antennas are shown in Fig. 20-12(d) to 20-12(g). The center-feed connection (d) makes the feed line a resonant line (with standing waves) when the characteristic impedance of the transmission line differs appreciably from the antenna impedance. However, tuning can be accomplished at the transmitter end of the line, and the on.y effect of the mismatch will be to increase somewhat the losses caused by the standing waves. If the line is to be quite long the direct feed connection of Fig. 20-12(e) is to be preferred, since the impedance of the antenna and line can be matched and standing waves on the line eliminated or at least reduced. The matching comes about in the same way as it does for the shunt-feed arrangement of Fig. 20-12(c). The impedance between two

GROUNDED ANTENNAS

(a) Series feed (Low-impedance line)

(b) Parallel feed (High-impedance line)

(c) Shunt feed

ELEVATED ANTENNAS

(d) Center-feed (Resonant line)

(e) Delta matched (Non-resonant line)

(f) Parallel feed (Non-resonant line)

(g) Direct feed (Single Wire Non-resonant line)

(h) Shorted Stub (Non-resonant line)

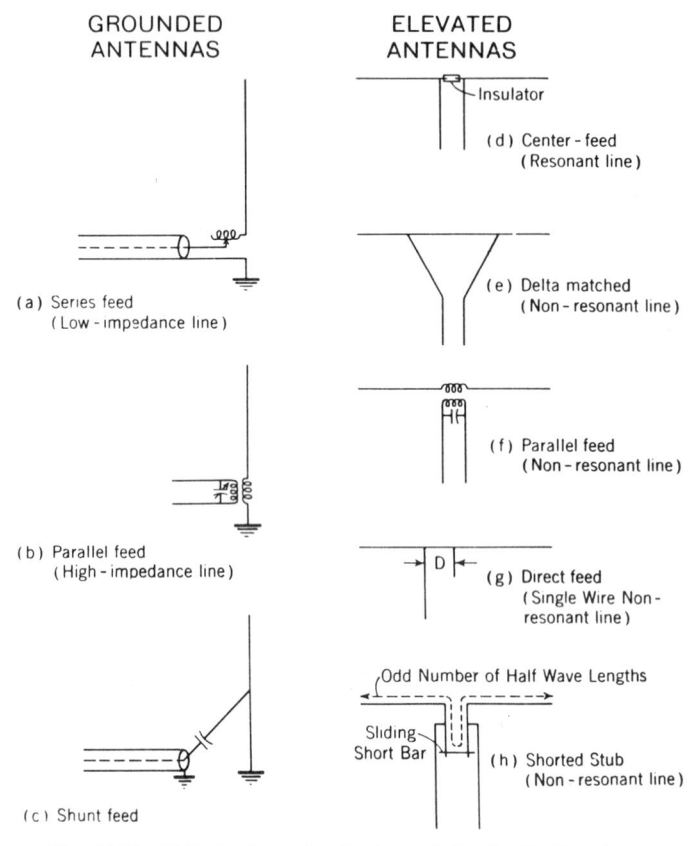

FIG. 20-12. Methods of coupling the transmission line to the antenna.

points on opposite sides of the center of the antenna increases as the distance between the points is increased. By tapping the transmission line to the antenna at points where the impedance is equal to that of the transmission line, an impedance match is obtained. However, because of the loop that then exists at the end of the line, some inductive reactance is present along with the resistance, so that a slight detuning results and there will be some standing wave on the line. Figure 20-12(f) illustrates a parallel type of feed similar to that in Fig. 20-12(b). It is possible to obtain both an impedance match and proper tuning, but the arrangement is not well suited to outdoor work.

Figure 20-12(g) shows a single-wire feed line direct-connected to the antenna. By tapping the line to the antenna at a suitable distance off center, an impedance match can be obtained and standing waves on the feed line eliminated.

Figure 20-12(h) shows a coupling system using a portion of a line as a matching and tuning network. The total length of the antenna from one end down around the shorting bar on the stub and out to the other end should be an odd number of half wavelengths, approximately. This length will make the antenna circuit resonant. Tuning can be accomplished by adjusting the shorting bar. A current loop will appear at the shorting bar and current nodes at the ends of the antenna. A match to the open-wire transmission line can be obtained by tapping the line to the stub at an appropriate distance from the shorting bar.

DIRECTIONAL ANTENNA SYSTEMS

Besides acting as efficient radiators or receivers of electromagnetic energy, antennas may be used to select the directions along which the energy shall be transmitted or received. This selection is usually accomplished by using two or more single antennas properly spaced and suitably fed. Such a system is known as a *directional array*. For transmitting antennas directional arrays are used to direct the energy into certain directions where it is desired and to prevent its radiation in other directions where it would be wasted or where it would create interference. For receiving antennas a directive array can be used to discriminate against signals from directions other than that in which reception is desired.

DIRECTIVITY OF A SINGLE LINEAR ANTENNA

A center-fed linear antenna (that is, a straight wire or rod fed at the center) radiates uniformly in the plane perpendicular to the axes of the antenna but has directivity in a plane parallel to the antenna axis. The directional patterns of linear antennas of various lengths are shown in Fig. 20-13. The antenna is assumed to be vertical so the plane perpendicular to its axis will be the horizontal plane, and in this plane its pattern is a circle since it radiates uniformly in all directions in this plane. For a vertical antenna the vertical pattern will show the relative field strengths produced in various directions in the plane parallel to the axis. For a short dipole this pattern will be a perfect figure eight as indicated in (b). (The field strength is proportional to sin θ, where θ is the angle measured from the axis, and the characteristic is a pair of perfect circles.) For a half-wave dipole the radiation characteristic is a slightly elongated figure eight as shown in (c). This means that the half-wave dipole is slightly more directive than is a short dipole. In (d) is shown the vertical pattern for a full-wave dipole ($L = \lambda$), with the two halves erected in phase, and it is seen to be still more directive with a narrower lobe or beam. When the antenna length is made greater than a wavelength, the pattern "breaks up" with the appearance of new lobes. When the length is two wavelengths the radiation in a direction perpendicular to the axis of the antenna, instead of being a maximum, is actually zero. This behavior can be understood by reference to the current distribution along the antenna which is also shown in the figure. To a first approximation the current distribution

(a) Horizontal Pattern

(b) Vertical Pattern

(c)

(d)

(e)

(f)

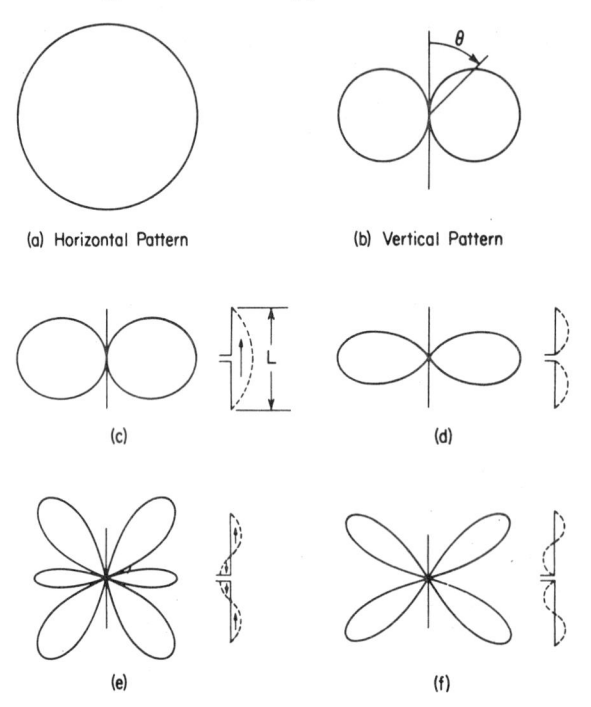

FIG. 20-13. Patterns of vertical antennas of various lengths.

along the antenna is just the same as it would be along an open-circuited transmission line. In this case the transmission line can be considered as having been "opened out" to form the two halves of the antenna. It will be observed that when the antenna length exceeds 1 wavelength (which corresponds to one-half wavelength of opened-out line) there will be standing wave loops of current which have opposite phase (or direction) to the original current loops. The contributions from these reversed phase current loops will tend to cancel the radiation in directions perpendicular to the axis, although in other directions the contributions may be additive, owing to a path-length difference. This pattern break-up that occurs for antenna lengths greater than 1 wavelength is an important factor in the design of television receiving antennas.

Chapter 20   **Radio Antennas**   655

Having considered the radiation pattern of a single antenna or unit, it is now in order to determine what to expect from an array of several such units.

VERTICAL ANTENNAS SPACED ONE-HALF WAVELENGTH

One of the simplest directional arrays consists of two vertical antennas spaced one half wavelength apart and fed in-phase with equal currents. The expression in-phase means that the currents in the two antennas reach their maxima (in the same direction) at the same instant. Figure 20-14(a) shows the horizontal pattern which would be obtained for this array It is the so-called "figure-eight" pattern with the line of zero radiation parallel to the line of the antennas (the line drawn between the antennas). The

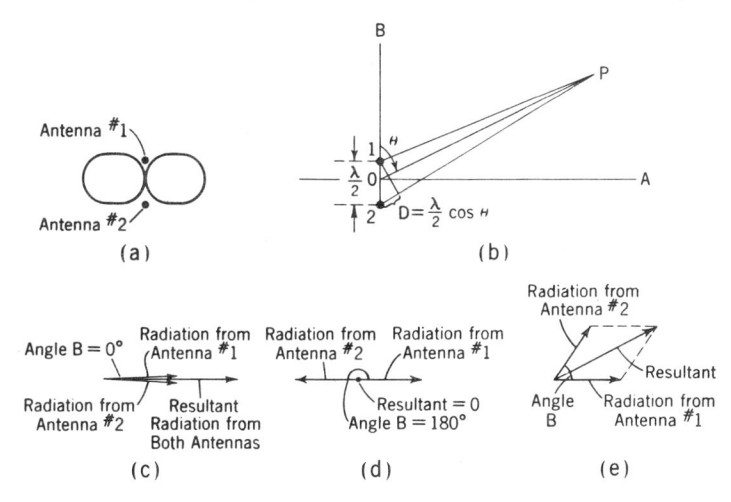

Fig. 20-14. Horizontal radiation pattern for two vertical antennas spaced one-half wavelength and fed in-phase. (a) Radiation pattern; (b), (c), (d), and (e) method of obtaining pattern.

pattern is simply explained by reference to Fig. 20-14(b). For the direction OA the distances from the two antennas to the receiving point are equal so that radiations leaving the two antennas at the same instant will arrive simultaneously at the point A. The currents in the antennas are assumed to be in-phase, so the radiations will leave the antennas in-phase and will arrive at point A in-phase. The addition of sine waves having a phase difference between them was covered in Chapters 3 and 9. The addition is accomplished by drawing vectors with the appropriate phase angle between them. In this case the phase angle between the waves arriving at A is zero (they arrive in-phase) so that the resultant field strength is just twice that from a single antenna [Fig. 20-14(c)].

For radiation in the direction of OB, conditions are different. Antenna No. 2 is one-half wavelength further from the receiving point than antenna

No. 1, so that it requires a longer time for a wave from No. 2 to reach B than it does for a wave from No. 1. A wave leaving antenna No. 2 at a moment of a current maximum will arrive at antenna No. 1 just one half cycle later when the current in No. 1 is a maximum *in the opposite direction*. The two waves which leave No. 1 and travel on together toward point B will then be 180° out of phase and they will cancel each other. Their resultant, as shown by the vector addition in Fig. 20-14(d), is zero.

For waves traveling toward some other point P along a line which makes an angle θ with the line of the antennas, the difference in distance between antenna No. 2 and P and antenna No. 1 and P is something less than a half wavelength. This difference in distance is shown as D in Fig. 20-14(b), and by geometry it will be seen to be approximately equal to $\lambda/2 \cos \theta$ where λ is a wavelength. In this case the phase of the current in antenna No. 1 will change by an angle B which is less than 180° while the radiation from No. 2 is traveling the distance D. The angle B will be given by

$$B = 180 \times \frac{D}{\lambda/2} = 360 \times \frac{D}{\lambda} \text{ degrees.}$$

The waves which travel on together towards point P will differ in phase by B degrees, and so their resultant at P will be as shown in Fig. 20-14(e).

In this manner the radiation at any angle θ can be computed, and if this is done for all angles a directional pattern as shown in Fig. 20-14(a) will be obtained. It is a maximum along OA, zero along OB, and varies from the maximum to zero as the angle θ changes from 90° to 0°.

As an example, the radiation along an angle $\theta = 60°$ will be computed. The distance D for this angle is

$$D = \frac{\lambda}{2} \cos 60° = \frac{\lambda}{4}.$$

The angle B will be

$$B = 360 \times \frac{D}{\lambda} = 90°.$$

For $B = 90°$ the two vectors are at right angles and the resultant vector has a length which is $\sqrt{2}$ times that of a single vector. Therefore the radiation in a direction making an angle $\theta = 60°$ is $\sqrt{2}/2$ times the maximum radiation along OA.

Antennas fed with currents having 180° phase difference. When the same two vertical antennas as above are fed with currents which have 180° phase difference, the resulting radiation pattern is another "figure eight," this time with the zero radiation direction perpendicular to the line of the antennas. This pattern is shown in Fig. 20-15(a). The explanation of this pattern is similar to that of the antennas fed in-phase, except that now radiations which leave the antenna at the same instant are 180° out of phase so that they cancel each other along the line OA. However, along the line OB, the current in antenna No. 1 changes by 180° while the wave from No. 2 travels the distance between the antennas so that the two waves leaving antenna No. 1 and traveling along together towards B are now

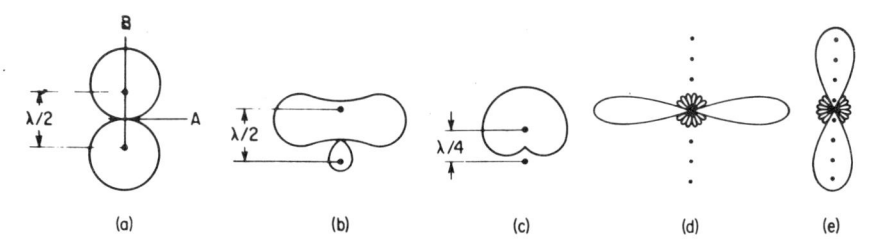

Fig. 20-15. Radiation patterns for commonly used antenna arrays. (a) Two antennas spaced one-half wavelength and with 180° phase difference; (b) same with 90° phase difference; (c) two antennas spaced one-quarter wavelength and fed with 90° phase difference; (d) broadside array (currents in phase); (e) end-fire array (currents 180° out of phase).

in-phase and so produce a large resulting signal in this direction. The radiation in other directions can be computed in a manner similar to that of the in-phase case of Fig. 20-14.

OTHER PHASES AND SPACINGS

When two antennas are fed with equal currents having other phase differences, the directions of zero radiations will be different and different patterns will result. The pattern for antennas spaced one-half wavelength apart and fed with equal currents which have a 90° phase difference is shown in Fig. 20-15(b). A variety of different patterns can also be obtained by changing the spacing between the antennas. An example of particular interest is the *cardioid* or heart-shape pattern obtained with a spacing of one-quarter wavelength and a phase difference of 90°. This pattern, shown in Fig. 20-15(c), is sometimes known as a *unidirectional* pattern because most of the energy is transmitted in one direction. In contrast, the patterns of Fig. 20-15(a), (d) and (e) are *bidirectional*.

When more gain or sharper directivity than can be obtained with two antennas is desired, a line of antennas is often used. A line of antennas spaced one-half wavelength and fed in-phase is known as a broadside array because the energy is radiated broadside to the line of the array. This is evidently so because the radiation from all the antennas would add in phase along the perpendicular to the line. A typical pattern is shown in Fig. 20-15(d). If the same array were used with alternate antennas fed 180° out of phase, the resulting pattern would be as shown in Fig. 20-15(e) and the array would be known as an end-fire array.

PATTERNS IN THE PLANE OF THE ANTENNA AXES

The radiation patterns of Figs. 20-14 and 20-15 are patterns in the plane *perpendicular* to the axis of the antennas. That is, they are the horizontal patterns of vertical antennas *or* the vertical patterns of horizontal antennas. (The effect of the earth as a reflecting plane on these patterns will be considered later.) It is next in order to examine the pattern in the plane in which the antennas lie. This would be the horizontal pattern for horizontal antennas and the vertical pattern for vertical antennas.

In Figs. 20-14 and 20-15 the antennas have been represented by a dot, indicating that the observer is looking along the axis of the antenna in order to see the plane in which the pattern is being shown. When the pattern is examined in the plane *parallel* to the antenna axis the length of the antenna is seen, and is shown as a line as in Fig. 20-13(b) or (c). The radiation pattern of a single half-wave element in this plane is a figure eight, whereas in the plane perpendicular to the axis it is a circle [Fig. 20-13(a)]. Suppose it is now desired to determine the radiation pattern *in the plane of the antennas* of two parallel antennas, spaced one-half wavelength apart and fed in phase. Looking at this plane the antennas would appear as two parallel lines, as in the right-hand figures of Fig. 20-16. If these antennas radiated *uniformly* in this plane the pattern would be known, for it has already been determined as a figure eight with the lobes perpendicular to the line of the array. This pattern is sketched in Fig. 20-16(a) (left-hand side) where it is called the group pattern. However, this pattern

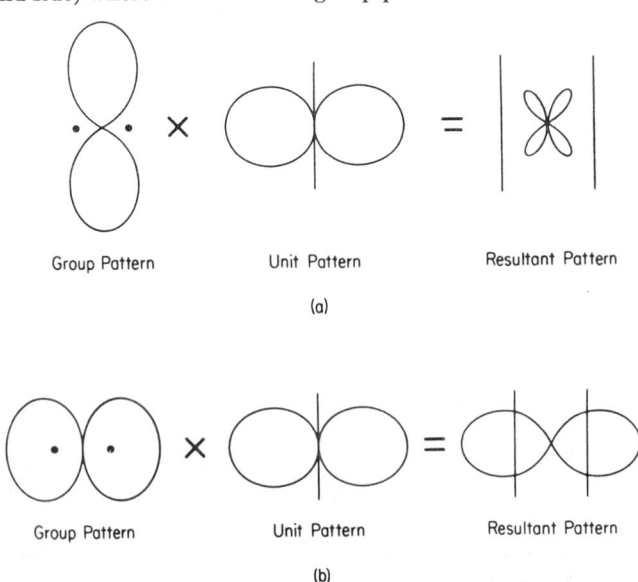

FIG. 20-16. (a) Elements in phase; (b) elements 180° out of phase.

Chapter 20 *Radio Antennas* 659

must now be modified to account for the fact that each of the elements or *units* contributing to field is not radiating uniformly in this plane, but actually has its own directional characteristic indicated by the *unit pattern* of Fig. 20-16(a). This modification is made by multiplying the pattern of two uniform radiators (the group pattern) by the pattern of the individual unit. When this multiplication of patterns is carried out for the in-phase case the pattern indicated as the resultant in Fig. 20-16(a) results. This pattern has four lobes and four nulls (one for each null in the patterns being multiplied) and is not very suitable for most applications. A much more desirable end-fire pattern is obtained for parallel antennas at half-wave spacing with a phasing of 180° between currents. The evolution of this case is shown in Fig. 20-16(b). In this case the lobes of the group pattern have the same direction as the lobes of the unit with a resultant strong signal in these directions for the array.

The necessity for *multiplying* the group and unit patterns rather than *adding* can be understood by noting that the group pattern is obtained as the sum of fields from two uniform radiators. This pattern will have a maximum where the fields add in phase and a null when they add 180° out of phase. However, when each of the fields is zero (as it is along the null of the unit pattern) the resultant must also be zero, regardless of the phase difference between fields in that particular direction. Also, when the individual fields are each reduced to one half along some particular direction the resultant sum must also be reduced to one half. Hence *multiplication* of patterns is called for.

COLINEAR ARRAY

A colinear array consists of linear elements in a line parallel to the axis of the elements as in Fig. 20-17 (right-hand side). The elements are fed in-phase. The spacing between elements is usually made one-half wavelength, center-to-center, and the element length is something less than one-half

Group Pattern × Unit Pattern = Resultant Pattern

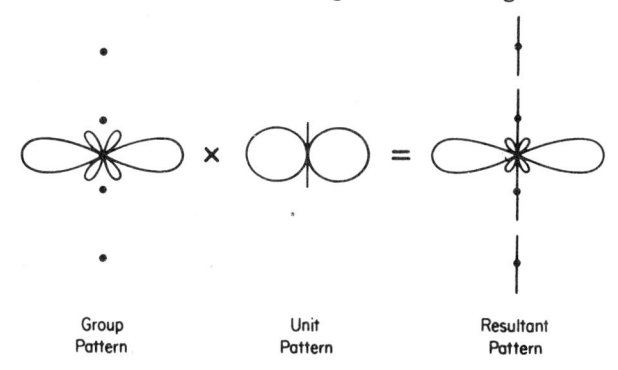

Fig. 20-17. Pattern of a colinear array, obtained by multiplication of the group and unit patterns.

wavelength. A loading coil or short section of shorted transmission line connected in series at the center of each element makes the element a resonant length. The radiation pattern of such a colinear array can be obtained quite easily by again using the principle of multiplication of patterns. The radiation pattern of a line of 4 isotropic or non-directional radiators fed in-phase with equal currents and with one-half wavelength spacing is shown at the left as the group pattern in Fig. 20-17. This group pattern can be calculated by the method used in Fig. 20-14 for two elements. The actual or resultant pattern of the colinear array is obtained by multiplying this group pattern by the unit pattern as illustrated in Fig. 20-17. In this case the resultant pattern is nearly the same as the group pattern, the only difference being a desirable reduction in the size of the secondary or side lobes. The pattern shown is the pattern in the plane of the array. For example, if the colinear array is horizontal, the pattern of Fig. 20-17 is the horizontal pattern. In the plane perpendicular to the line of the array all of the radiations from all elements add in-phase in all directions and the radiation is non-directional. The pattern in this plane is therefore a circle. The three-dimensional radiation pattern of the colinear array is sort of disk-shaped, being the resultant pattern of Fig. 20-17 rotated about the axis.

RECTANGULAR OR MATTRESS ARRAY

For point-to-point communication and many other applications a cone or beam of radiation is more desirable than the disk of radiation produced by the colinear array In other words, vertical directivity as well as horizontal directivity is desired. This result can be obtained from a vertical stack of horizontal colinear arrays as shown in Fig. 20-18(a). Looking down on such an array from above one sees a colinear array of elements, and the horizontal pattern is just exactly the pattern shown as the resultant for the colinear array in Fig. 20-17. Looking along the array from one end a vertical array of non-directional radiators is seen and hence, the vertical pattern in the plane perpendicular to the axis of the elements is just that shown as the group pattern in Fig. 20-17. The three-dimensional pattern of the rectangular array of Fig. 20-18(a) is therefore a bidirectional beam or cone (with a sort of collar representing the side lobes).

When a unidirectional beam rather than a bidirectional beam is desired, this result can be obtained by placing a reflecting screen one-quarter wavelength behind the rectangular array as shown in Fig. 20-18(b). Such an arrangement is sometimes called a "mattress antenna" for obvious reasons. If the screen is on the left side of the array, the radiation from the array which is directed towards the left is reflected by the screen back to the right where it reinforces the radiation to the right from the array. That the field reflected from the screen reinforces the direct radiation to the right, rather than cancelling it, can be determined from a consideration of phase relations. The radiation which is reflected from the screen must travel a path which is one-half wavelength longer than the direct radiation to the right. However, the reflected wave is reversed 180° in phase on reflection [the total electric intensity tangential to a (perfect) reflecting screen must be zero] so the resulting phase of the reflected wave is such as to augment the radiation towards the right.

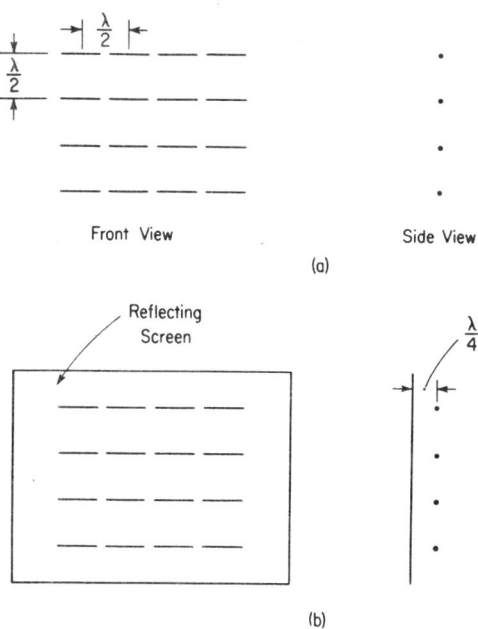

Fig. 20-18. (a) Rectangular array; (b) mattress array.

The effect of the screen could also be determined by making use of the principle of images. This is done in the following section relating to the effect of the reflecting surface of the earth on the radiation patterns of antennas.

EFFECT OF THE GROUND ON VERTICAL RADIATION PATTERNS

The ground is not a perfect conductor and therefore not a perfect reflector of radio waves. However, it can often be considered as such for the purpose of determining its effect on the radiation characteristics of antennas. Figure 20-19(a) shows a vertical antenna elevated one-half wavelength above the ground. As far as the effect at some point P is concerned, the wave reflected from the ground appears to come from an *image antenna* located one-half wavelength beneath the surface. This image antenna is inphase with the actual antenna and the two antennas constitute a directional array. Figure 20-19(b) shows a horizontal antenna elevated one-quarter wavelength (say) above the earth. Again the image appears an equal distance below the surface, but this time its current is 180° out of phase with the current in the actual antenna. This opposite phase occurs because in this case the wave is horizontally polarized and the wave reflected from the ground is *reversed in phase* upon reflection.

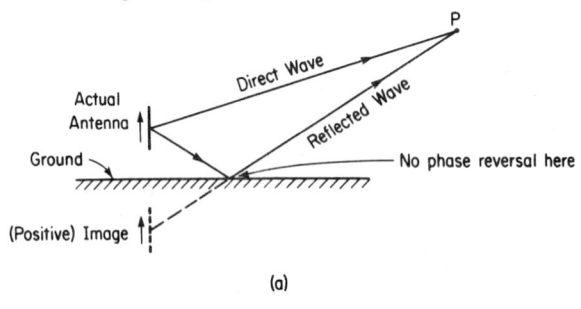

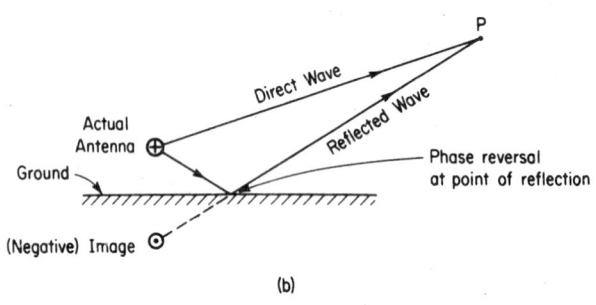

FIG. 20-19. An image antenna produces the same field as a reflecting ground. (a) Positive image for vertical antennas (reflection without phase reversal); (b) negative image for horizontal antennas (180° phase reversal on reflection).

The effect of the presence of the ground may be determined by considering the antenna and its image as a directional array and using the principle of multiplication of patterns to obtain the vertical radiation characteristic. This procedure is illustrated in Fig. 20-20, where the vertical radiation pattern of a horizontal antenna located a quarter wavelength above the ground is determined, as shown. Of course, only the

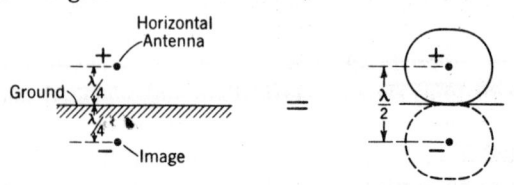

FIG. 20-20 Vertical radiation pattern of a horizontal antenna one-quarter wavelength above the ground, determined by considering the antenna and its image.

upper half of the resulting pattern will apply. The patterns for vertical or horizontal antennas at any other heights above the ground can be determined in a similar manner. The patterns for a horizontal antenna located at heights $\lambda/4$, $\lambda/2$, $3\lambda/4$, and λ above the ground are shown in Fig. 20-21.

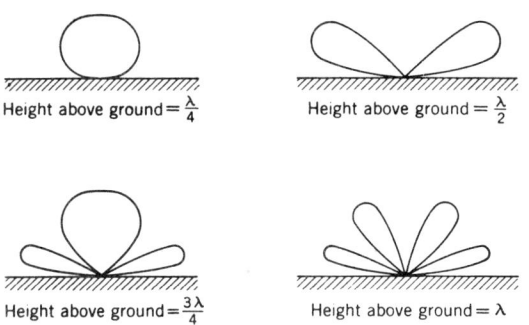

FIG. 20-21. Vertical radiation patterns of a horizontal antenna at various heights above a (perfect) ground.

The patterns of vertical antennas above a perfectly conducting ground can be obtained in a similar fashion. However, in this case there will be two important differences. The first difference is that the image will be in-phase with the actual antenna, resulting in a maximum signal along the surface of the reflecting plane, instead of zero signal as was obtained with horizontal polarization. The second difference is that the group-pattern resulting from the addition of fields from the antenna and its image, assuming them to be uniform radiators, must be multiplied by the pattern of the unit, that is the free-space antenna pattern.

ANTENNA GAIN AND EFFECTIVE AREA

As a result of its directivity a directional transmitting antenna radiates more power in certain directions than would a non-directional antenna radiating the same total amount of power. The directive antenna is said to have a certain amount of *directional gain*. Correspondingly, a directive receiving antenna will receive or collect more power from a wave arriving from certain directions than would a non-directional antenna. The (directional) *gain* of a transmitting antenna *in a given direction* is defined as the ratio of power density (watts per square meter) radiated in that direction to the power density that would have been radiated by an isotropic antenna radiating the same total amount of power. An *isotropic* antenna is an antenna that radiates uniformly in *all* directions, that is, a completely non-directional radiator. Similarly, the gain of a receiving antenna in a given direction is defined as the ratio of the signal power received from a wave arriving from that direction to the signal power that would have been received by an isotropic antenna. Because the radiation pattern of an antenna is the same for receiving as it is for transmitting, it follows that the gain of an antenna is the same for receiving as it is for transmitting. Normally the gain is given for the direction of maximum radiation.

For a short antenna which has the perfect figure-eight or doughnut shape, the gain over an isotropic antenna is 1.5 (or 1.76 db). For a half-wave dipole which has a slightly elongated figure-eight pattern it is approximately 1.6 (or 2 db). (More accurately, the gain of a half-wave dipole is 1.64, or 2.15 db. However, it is convenient to use the rounded-off figure of 2 db, and this is usually done in practice.) In practice also, antenna gain is often referred to the gain of a half-wave dipole as standard. When this is so, it is simply necessary to add 2 db to obtain the gain referred to the isotropic antenna as standard. Transformation from decibels to power ratios or vice versa can be made easily on a slide rule or with the aid of the chart of Chapter 1.

Another term that has significance, particularly in connection with directive receiving antennas, is *effective area*. The effective area of a receiving antenna is defined as the ratio of the power received by a matched load connected to the antenna to the power per unit area of the received wave. When the gain g of an antenna is known, its effective area can be calculated from the relation

$$A_{\text{eff}} = \frac{\lambda^2}{4\pi} g.$$

For large rectangular arrays such as the mattress array described previously, the effective area of the antenna array is approximately equal to its physical area. For other antennas to be described later, which ordinarily are large in square wavelengths, such as horns, parabolas, and lenses, the effective area is usually about 75 to 85% of the actual physical area of the aperture. On the other hand, for wire and rod antennas the effective area bears no relation to the physical area, although the term still has important significance. For example the effective area of a half-wave dipole is given by

$$A_{\text{eff}} = \frac{\lambda^2}{4\pi} \times 1.64 \approx \frac{\lambda^2}{8} = \frac{\lambda}{4} \times \frac{\lambda}{2}$$

which indicates that the half-wave dipole "captures" an amount of power equal to that passing through a $\lambda/2 \times \lambda/4$ rectangle.

MICROWAVE ANTENNAS

In theory, the antennas and arrays considered so far could be used at all frequencies. In practice, the actual form taken by the antenna depends to a large extent upon the frequency band for which it is designed. Although wire or rod antennas can be and are used singly and in arrays at uhf and shf, other types utilizing reflecting or radiating surfaces are often more practical and are used extensively.

Chapter 20   Radio Antennas

In the previous section it was seen that for an antenna of given gain or directivity, for example for a half-wave dipole with a gain of 1.64, or 2.15 db, the effective area is proportional to the square of the wavelength or inversely proportional to the square of the frequency. This means that the actual power received by a matched half-wave dipole for a given transmitted field intensity will be less by 10,000 times (40 db) at 3,000 megacycles than it is at 30 megacycles. In order to offset, at least partially, this reduction in effective area with increasing frequency it is necessary to increase the antenna gain or directivity when the application permits. Fortunately, as the frequency increases (and the wavelength decreases) it becomes easier to construct antenna systems that are large in terms of wavelengths, and which therefore can be made to have greater directivity and gain.

The rectangular or mattress array of dipoles described in an earlier section is one type of high-gain antenna system. Such arrays, having as many as 32 or even more elements, have been used extensively at frequencies from 100 to 400 megacycles for early-warning radar applications. Although this same type of array could be used at 3,000 or 10,000 megacycles, construction would be difficult because of the small size of the elements and the closer tolerances required. At these frequencies the physical size of a reasonably high-gain antenna system becomes small enough to make practical the use of suitably shaped metallic reflectors to produce the desired directivity.

PARABOLIC REFLECTORS AND PARABOLOIDS

The well-known properties of a parabola can be used in the design of directive microwave antennas. Figure 20-22 shows a dipole antenna at the focus F of a parabola. The parabola has the property that the distance from the focus to any point on the parabola and from that point to a plane perpendicular to the axis is a constant. In Fig. 20-22

$$FP_1 + P_1S_1 = FP_2 + P_2S_2 = FP_3 + P_3S_3$$

for all points P_1, P_2, P_3 on the parabola. From this relation it follows that all waves originating at the focal point F and reflected from the parabolic surface will reach the plane SB with the same phase. These waves will therefore add together in phase to produce a relatively strong signal at a distant point along the axis AB. In directions other than along the axis the path lengths for the various reflections will differ and so these waves will tend to cancel one another. Hence, the parabolic reflector produces a narrow beam of radiation along its axis. This same result can be conveniently explained by considering each of the reflections from the surface as originating at an image antenna appropriately located behind the parabolic surface and correctly phased to produce the beam along the axis. Hence, the large reflecting surface with a single feed antenna at the focal point performs like a large multi-element array.

Operation of this system as a directive receiving antenna is also easy to understand. It is evident for a wave arriving along the axis AB, that all parts of the wavefront intercepted by the parabola will be reflected to the focal point and will arrive there in phase, resulting in a very strong field and, hence, large voltage induced in the primary antenna.

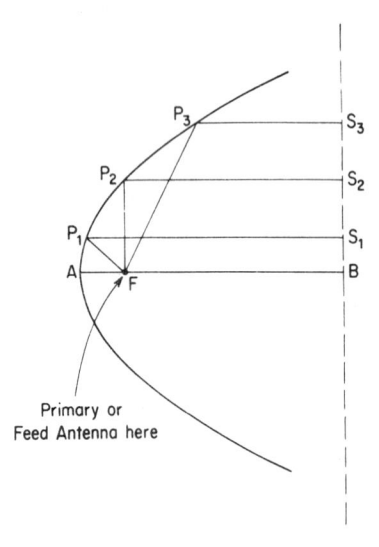

FIG. 20-22. Geometry of the parabola.

When the parabola of Fig. 20-22 is rotated about the axis AB the surface of revolution generated is called a *paraboloid*. Actual parabolic reflectors are usually paraboloids and are so called, although the term microwave "dish" is also used. Important examples of these antennas are shown in Chapter 21.

HORNS

Electromagnetic horns, already mentioned in Chapter 9 [Fig. 9-19(b)] are also much used as microwave radiators. The horn becomes more directive as the size of the mouth of the horn is made larger in wavelengths. However, large horns require careful design if the expected gain is to be realized. Referring to Fig. 20-23 it is seen that the wavefront (on which all points are in phase) is slightly curved at the mouth of the horn. This means that all points on the plane of the mouth (line CAD) are not in phase as they should be to add up to give maximum signal along the axis. In Fig. 20-23 the fields at points C and G are in phase (they are both on the wavefront) but the distance from G to a distant point along the axis is less than the distance from C to the same point by the amount AG approximately. (The observation point on the axis is considered sufficiently distant compared with dimensions of the mouth of the horn that lines to it from C and G are essentially parallel.) If the length AG were as much as one-half wavelength, radiations from fields at G and C would actually cancel one another. Thus it is necessary to design the horn so that AG is less than about a quarter wavelength. For large horns this requires a small flare angle α to

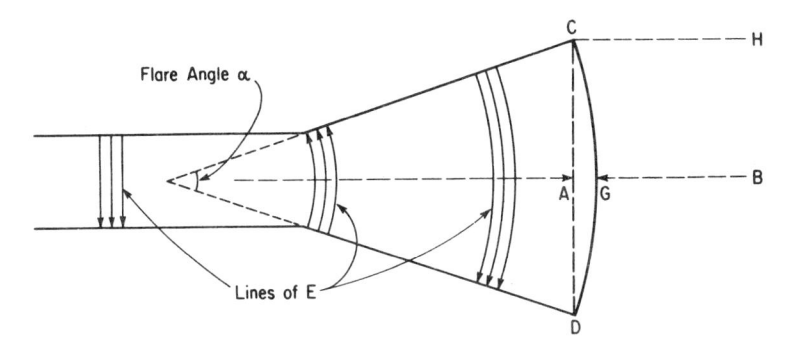

FIG. 20-23. Electromagnetic horn.

keep the curvature of the wavefront small. That is, a horn with a large mouth must also be quite long. When space or other considerations will not permit a long horn, a microwave lens may be used to correct for the curvature of the wavefront.

MICROWAVE LENSES

A microwave lens can be used in conjunction with a horn antenna to correct for the curved wavefront, or it may be used with a non-directional radiator to produce directional radiation. In Fig. 20-23 the difficulties resulting from the curved wavefront are caused by the fact that radiation from point G, say, arrives at a distant point on the axis ahead of radiation from point C. If the radiation from G could be slowed down, or that from C could be speeded up, this situation could be corrected. Figure 20-24

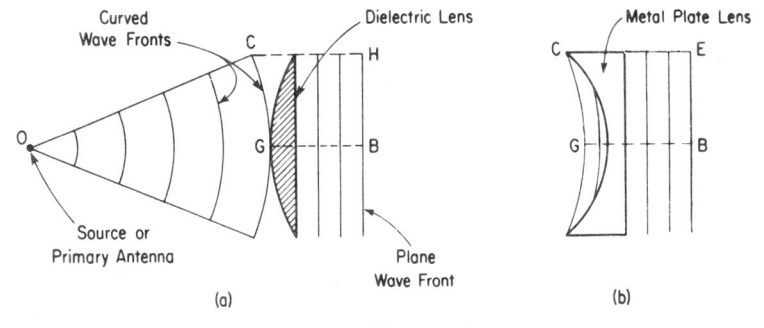

FIG. 20-24. Microwave lenses.

shows two ways of accomplishing this result. In Fig. 20-24(a) a lens of dielectric material, such as polystyrene, has been placed so that the ray GB has a longer path through the dielectric than the ray CH. Because the velocity of an electromagnetic wave, given by $v = 1/\sqrt{\mu\epsilon}$, is less in a dielectric for which ϵ is greater than for air, the wave is correspondingly slowed down in traversing the dielectric. Knowing the velocity of the wave in the dielectric it is an easy matter to so shape the lens that all rays are slowed down the correct amount to make the emerging wavefront plane; that is, on a plane perpendicular to the axis all points are in phase, which is the required condition for having all radiations from this surface arrive in phase at a distant point along the axis.

In Fig. 20-24(b) is shown an alternative way of accomplishing the same result, this time however by speeding up the wave from C rather than slowing down the wave from G. It will be recalled that in a waveguide the wave velocity is greater than the velocity of an electromagnetic wave in air. By causing the wave to pass between parallel conducting sheets which are *parallel* to the direction of the electric field E, this same speeding up of the wave can be realized. It is evident that by shaping the plates somewhat as shown the ray CE is speeded up more than GB, so that the desired plane wave emergent from the lens can be realized.

From Fig. 20-24 it is also evident that if a non-directional radiator is placed at the center, 0, of the curved wavefront, the lens in transforming the curved wavefront into a plane wavefront will transform the non-directional radiator into a directional radiator.

SLOT ANTENNAS

When a metallic reflecting surface is used as in Fig. 20-22 to produce directive radiation, the mechanism of this action may be thought of as follows: the primary or feed antenna adjacent to the reflector excites currents on the metallic surface and it is the radiated fields of these currents which are the reflected waves. In this connection the reflecting surface is often called a *secondary* radiator to distinguish it from the feed antenna which is considered as being the *primary* radiator. An alternative way of

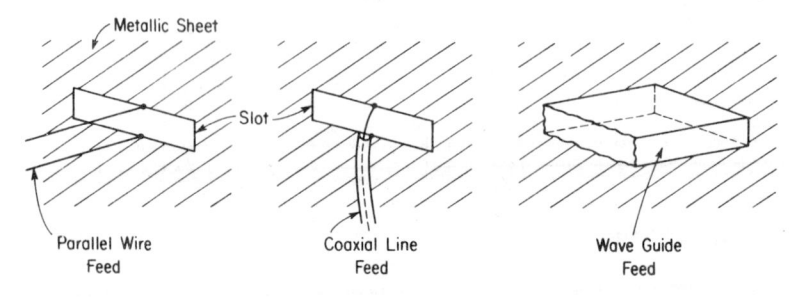

Fig. 20-25. Methods of feeding a slot antenna.

Chapter 20   Radio Antennas   669

exciting currents on a metallic sheet is to cut a slot in the sheet and apply a radio-frequency voltage across the slot possibly by one of the methods indicated in Fig. 20-25. This figure shows a voltage being applied across the slot by (a) a parallel wire transmission line, (b) a coaxial line, and (c) a rectangular waveguide. However the slot is excited, the electric field distribution across the slot and the lines of current flow in the metallic sheet are approximately as sketched in Fig. 20-26(a). Naturally, if the slot is too short the current taking the short path from one terminal to the other is literally a short-circuit current and the slot impedance is low. However, if the slot is of the order of one-half wavelength long the impedance across

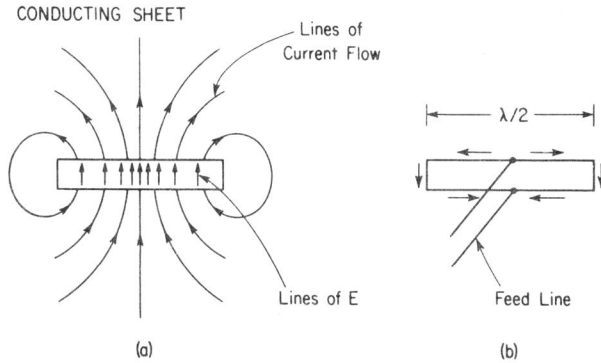

Fig. 20-26. Distribution of electric field intensity and current density for a half-wavelength slot in a conducting sheet.

its terminals is reasonably high. Referring to Fig. 20-26(b), only the metallic edge of the slot has been shown. This metallic circuit would appear to the feed line as two short-circuited transmission lines connected in parallel at the terminals. If these lines are electrically short (that is, short in terms of wavelengths) their input impedances are low. However, if they are each of the order of a quarter-wavelength long, the input impedance will be quite high. The physical arrangement shown in Fig. 20-26(b) would radiate very little power, because equal and opposite currents are flowing close to each other (assuming a narrow slot) with a resulting cancellation of the distant fields. The end currents are flowing in the same direction and these currents will radiate, but because of the short lengths the radiation from them will be quite small. With the physical arrangement of Fig. 20-26(a), the horizontal currents still tend to cancel, but the vertically directed currents are in the same direction on both sides of the slot, and the radiation from them is quite large. Thus a slot of the order of a half wavelength long or longer cut in a metallic surface can result in a very effective radiator. Because there need be no protruding elements this type of radiator is very useful for high-speed aircraft, in that the metallic skin of the aircraft itself can be made to carry the radiating currents. In order to prevent radiation from one side of the slot, in the arrangements of Figs. 20-25(a) and (b), a metallic cavity (of sufficiently large volume) can be used to box in the slot on that side. It will be noted that the radiating currents produced by a horizontal slot are vertical. Consequently a horizontal slot produces a vertically polarized field and of course a vertical slot produces a horizontally polarized field.

WIDE-BAND AND SPECIAL-PURPOSE ANTENNAS

For many purposes it is desirable to have an antenna capable of satisfactory operation over a wide range of frequencies. In communication applications it is often desired to use a single antenna for several channels which are separated in frequency. In television broadcasting the antenna may be designed for one channel only, but this channel is very wide and it is necessary to design the antenna to have an essentially constant impedance over a 6-megacycle band. In television reception the impedance requirements are not so stringent, but here it is usually desired to use a single antenna for more than one channel. For satisfactory operation over a specified frequency range it is necessary that both the radiation pattern characteristics and impedance characteristics of the antenna be suitable over the range.

The impedance characteristics of an antenna can be modified by varying its shape or by using compensating networks or by doing both. A basic principle in the design of antennas for wide bandwidth is that, in general, thicker or fatter antennas are wider band than thin antennas. For example,

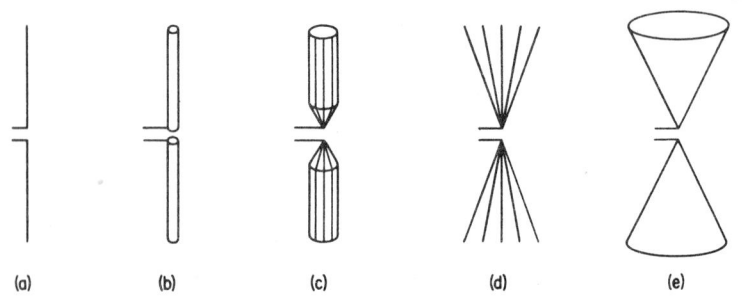

FIG. 20-27. Dipole antennas: (a) thin dipole; (b) thick dipole; (c) cage; (d) fan; (e) biconical.

the bandwidth of a simple dipole antenna is considerably greater when the antenna is constructed using a thick rod or tube or even a cage of wires than when made with a thin wire. Two main effects result from making the antenna fat. The anti-resonant impedance which occurs when the total length of the antenna is approximately one wavelength is reduced from the value of several thousand ohms which it has for thin antennas to only a few hundred ohms. Also the reactance variations on either side of resonance are much less for the fatter antennas. Figure 20-27 shows some thin and fat dipole antennas with the wider bandwidth antennas on the right side of the figure.

A simple compensating network for increasing the bandwidth of an antenna is that shown in Fig. 20-28(a). The parallel circuit of L and C is made anti-resonant at the half-wavelength resonant frequency of the dipole, and at this frequency its impedance is a pure resistance of rather high value. Below this resonant frequency the antenna reactance becomes capacitive but the reactance of the coil and capacitor combination becomes inductive. Similarly, above the resonant frequency where the antenna

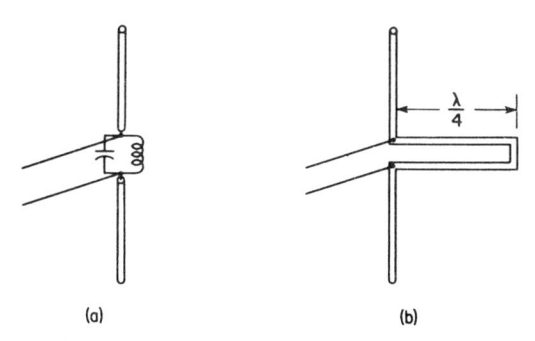

Fig. 20-28. Compensating networks for increasing antenna bandwidth.

reactance is inductive, the reactance of the coil and capacitor combination is capacitive. Thus there is a tendency to compensate for the reactance variations of the antenna over a small frequency band about the resonant frequency. If a small resistance is included in series with the L-C circuit, the slope of the reactance curve of this circuit can be made of such value that it will produce almost perfect compensation for the antenna reactance variations over a small frequency band. In Fig. 20-28(b) is shown the transmission-line equivalent of the compensating network of Fig. 20-28(a). The short-circuited quarter-wave line connected in parallel with the antenna as shown acts in a similar manner to the parallel L-C circuit in compensating for reactance variations. The folded dipole antenna, to be described later, provides this type of compensation. For compensation of both resistance and reactance variations over a wider band of frequencies, more complicated lumped-constant or transmission-line type circuits are often used.

FOLDED DIPOLE ANTENNA

An antenna that has some interesting and useful properties is the *folded dipole* shown in Fig. 20-29. It consists of two elements which are a half wavelength long at the center frequency and which are joined together at their ends. They therefore have the same voltages between their ends and as far as the radiation fields are concerned they are essentially two half-wave elements in parallel. If the elements have the same diameter they will carry equal radiating currents which are in the same direction as shown. For 1 amp in each element the told effective radiating current will be 2 amp and the power radiated will be

$$W = I^2 R_r = 4 \times 73 = 292 \text{ w}$$

where 73 ohms has been used as the radiation resistance for a half-wave dipole. However, although there are effectively 2 amp of radiating current

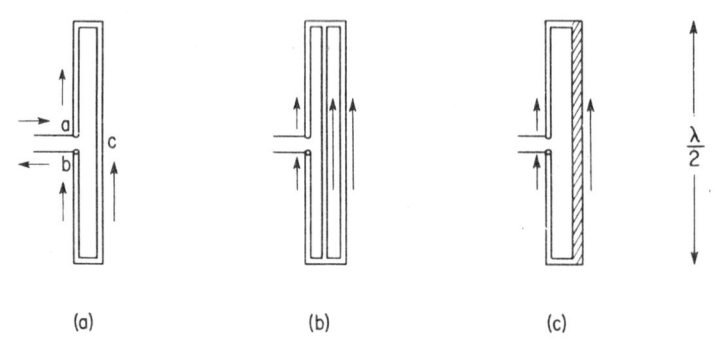

(a)   (b)   (c)

FIG. 20-29. Folded-dipole antennas.

the current furnished at the terminals a-b is only 1 amp. Because all the power which is radiated must be supplied at these terminals, it is evident that the effective *input* resistance must be given by

$$I_1^2 R_{in} = 292$$

or $R_{in} = 292$ ohms, just 4 times the radiation resistance of a simple half-wave dipole. If three elements had been used as in Fig. 20-29(b) only one third of the total radiating current would be supplied at the input and the input resistance would be 9 times that of the simple half-wave dipole, or approximately 657 ohms. It is evident that this antenna has built-in impedance transforming properties which make it easy to match to the characteristic impedance of the transmission line feeding it. As an alternative to adjusting the number of elements, the input impedance can be altered by using only 2 elements of different diameters as indicated in Fig. 20-29(c). The currents in the 2 elements are then unequal, with the larger current flowing in the thicker element. By this device it is possible to obtain almost any input resistance that might be desired.

The folded dipole has another useful property, in that it also has a built-in reactance compensation network of the type shown in Fig. 20-28(b). Referring to Fig. 20-30 it is seen that the configuration of Fig. 20-29(a) may also be viewed as two shorted quarter-wave transmission lines which are connected together at point C and fed in series. The expected transmission line currents would be distributed and have directions indicated by I_t, whereas the "antenna currents" already considered have the distribution and directions indicated by I_a. In fact, *both* sets of currents flow simultaneously in response to a voltage applied between the terminals *a-b*. The transmission line currents, flowing in opposite directions in adjacent rods, contribute nothing to the radiated power. However, they do affect the input impedance, which can now be considered as the "antenna impedance" of the folded dipole in parallel with the reactance of the series-connected shorted transmission lines. As explained in the previous sections, the reactance variations of the transmission line sections tend to compensate for the reactance variations of the antenna, resulting in a considerably more constant impedance around the center frequency of the antenna. It should be noted, however, that at twice this center frequency the shorted transmission-line sections are one-half wavelength long, which places a short across the antenna terminals, rendering the antenna useless at this frequency. This fact is of importance in connection with television receiving antennas, where frequency ranges of more than two-to-one are encountered.

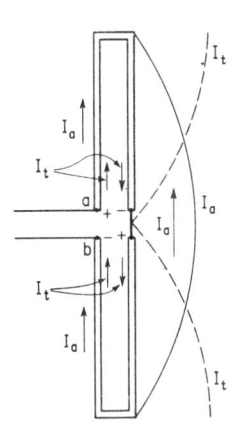

Fig. 20-30. Folded dipole showing antenna currents I_a and transmission-line currents I_t.

TELEVISION TRANSMITTING ANTENNAS

Antennas used for broadcasting television programs must fulfill certain special requirements. They must radiate horizontally polarized signals. (This polarization is standard for television in the United States. In England and certain other countries, vertical polarization is standard.) Usually the antenna is required to be "omnidirectional"; that is, it should radiate uniformly in all directions *in the horizontal plane*. Finally, the antenna impedance should be essentially constant and equal to the characteristic impedance of the transmission line feeding it over the full channel width of 6 megacycles. If this last requirement is not satisfied there is a strong possibility that that portion of the signal which is reflected back along the transmission line because of the impedance mismatch at the antenna will be again reflected at the transmitter terminals to make a second journey down the transmission line. If the transmission line has considerable length the radiation of this re-reflected signal will appear as second image (or ghost) because of the time delay between it and the original signal. Therefore it is extremely important to have a nearly perfect impedance match for the television transmitting antenna.

The polarization and directional requirements, when taken together, rule out the use of a single dipole, either vertical or horizontal. However, a simple loop with vertical axis meets these requirements and various forms of vhf and uhf loops have been designed. A vertical slot in a vertical cylinder also satisfies the first two requirements. It will be recalled that a vertical slot produces horizontal radiating currents, and if the metal sheet in which the slot is cut is wrapped into a cylinder the resulting antenna will radiate nearly uniformly in all horizontal directions as long as the diameter of the cylinder is not greater than about one fifth of a wavelength.

An antenna of rather different design that has proven very satisfactory for television broadcasting is the "superturnstile" or "batwing" antenna, one pair of arms of which is shown in Fig. 20-31(a).

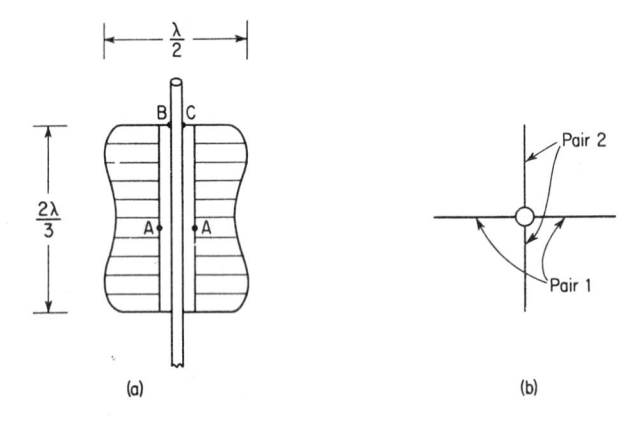

Fig. 20-31. "Superturnstile" or "batwing" antenna.

The configuration of Fig. 20-31(a) may be thought of as the arms of a flat horizontal dipole which is very fat or wide. The length of the dipole (at its longest point) is approximately one-half wavelength and its width is of the order of two thirds of a wavelength. It is effectively a sheet radiator but in practice the flat sheet is replaced by a number of parallel tubes to reduce wind resistance. It is fed at terminals A-A, and the short-circuit connections at B and C cause the input impedance of the antenna to be shunted by the input impedances of short-circuited transmission lines, which provides a certain amount of built-in compensation. An alternative and perhaps more valid visualization is to view the arrangement as vertical slot radiator cut in a flat sheet. The shape of the sheet has been selected to provide a good distribution of current over the sheet and optimum impedance characteristics.

Whatever method of visualization is used for the antenna of Fig. 20-31(a), it is evident that although it can be expected to radiate a horizontally polarized signal, the radiation will not be omnidirectional or uniform in the horizontal plane. In fact the horizontal pattern is very closely just that of a horizontal dipole or vertical slot, i.e., a slightly elongated figure-eight pattern. The desired omnidirectional radiation is obtained by using two such units mounted at right angles to each other as shown by the top view in Fig. 20-31(b) and fed with equal currents which are 90° out of phase. It

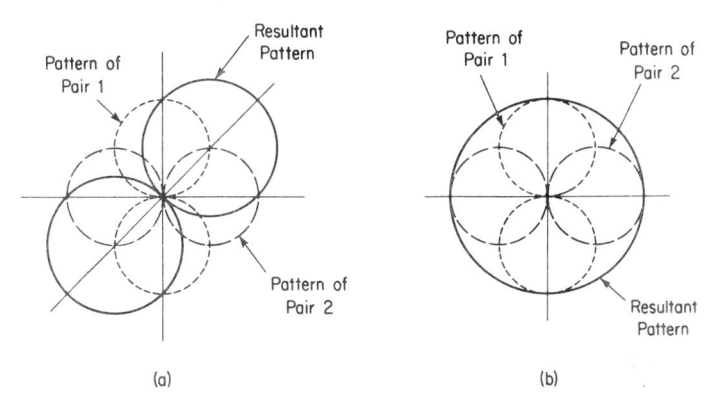

FIG. 20-32. Radiation patterns produced by crossed dipoles when dipoles are fed (a) in phase, (b) with 90° phase difference.

is left as a problem for the reader to verify that two crossed figure-eight patterns fed in phase add to produce another figure-eight pattern along the 45° line [Fig. 20-32(a)], but two crossed figure-eight patterns fed with 90° phase difference result in a figure eight which rotates with time, hence producing a circular or uniform radiation pattern as in Fig. 20-32(b).

TELEVISION RECEIVING ANTENNAS

The general considerations involved in selecting an antenna type for television reception are the same as they are for other applications, that is, impedance characteristics and directional characteristics. It is desirable that the antenna impedance be of the same order of magnitude as that of the transmission line and receiver input circuit because this condition ensures that an appreciable fraction of the signal power intercepted by the antenna gets delivered to the receiver (instead of just being reradiated into space). However it is not essential that a close impedance match be obtained (as in the case of television transmitting antennas) because there will be no reflections on the transmission line, with the consequent possibility of ghosts, as long as the receiver input circuit is properly matched to characteristic impedance of the transmission line. Directivity may be desirable for either or both of two reasons. Increased directivity gives increased gain with a corresponding improvement in signal-to-noise ratio. This factor is important in fringe-area reception. Closer to the transmitter, increased gain may be of no consequence but directivity may still be desirable to help eliminate reflected waves coming from different directions which may cause ghosts.

The special problems encountered in the design of antennas for television reception are associated with the wide frequency range that must be covered. It is a relatively simple matter to design a television receiving antenna for single channel coverage. A fat dipole or folded dipole is adequate in many cases. If more directivity is desired a Yagi array (described

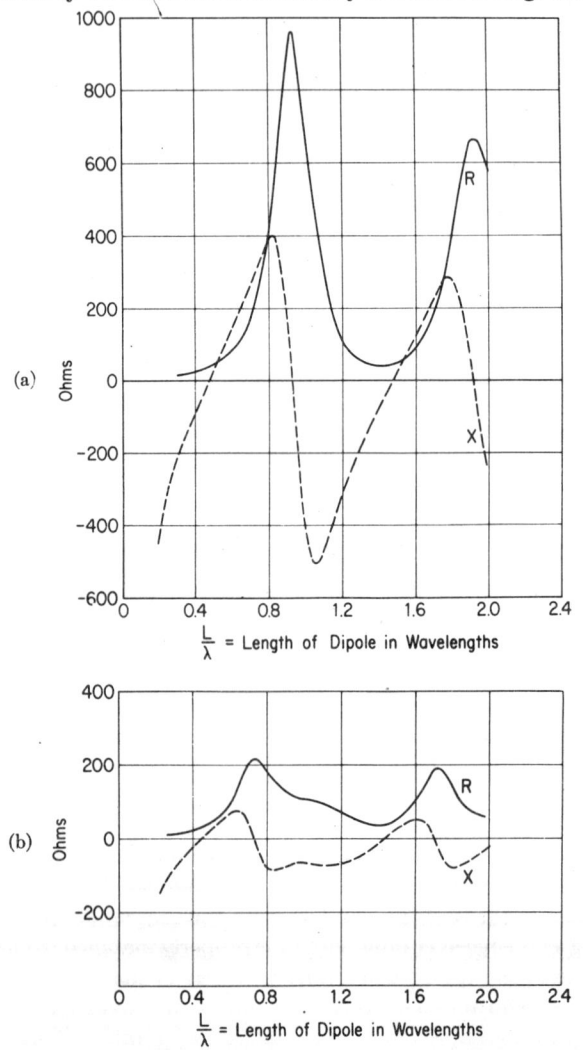

Fig. 20-33. (a) Resistance and reactance of a typical thin dipole as a function of length of the dipole in wavelengths; (b) resistance and reactance of a typical fat dipole as a function of length of the dipole in wavelengths.

below) or a standard multi-element array can be used. However when it is desired to receive several different channels with the same antenna the design problem becomes complicated. As an example of the sort of problems encountered, suppose an attempt is made to use a single dipole to receive all channels in the so-called vhf television band, which at present extends from 54 to 216 megacycles, a four-to-one frequency range. (The "low" band extends from 54 to 88 megacycles and the "high" band extends from 174-216 megacycles). It is evident that if the antenna is cut to be nearly one-half wavelength at the lowest frequency in order to have reasonable radiation resistance and signal pick up, it will be two wavelengths long at the highest frequency. Referring to Fig. 20-33(a) and (b) the antenna impedance is observed to go through extremely large variations, particularly for the thin dipole. For a fat dipole the variations are less extreme, and it is possible that the impedance characteristics of a fat dipole might be adequate in areas of good signal strength. However the directivity characteris-

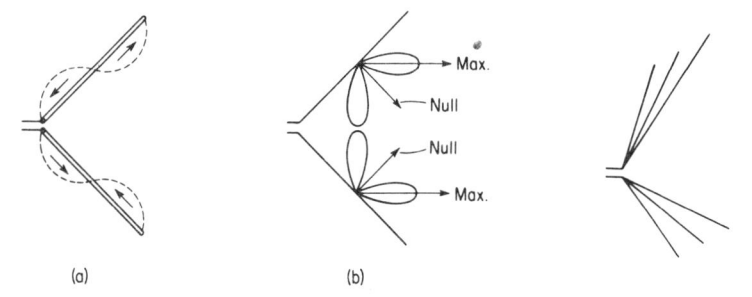

FIG. 20-34. Vee antenna: (a) current distribution and (b) radiation pattern of each arm, when length of each arm is 1 wavelength.

FIG. 20-35. Vee-fan antenna.

tics must also be considered. Referring to Fig. 20-13 it is seen that for antenna lengths appreciably greater than one wavelength, the direction of maximum signal pick-up is no longer perpendicular to the antenna, and when the antenna length is two wavelengths there is a null in this direction. Hence the antenna would have to be rotated to receive the higher channels. This defect can be remedied by the simple expedient of tilting the arms of the antenna forward to form a horizontal *Vee* antenna [Fig. 20-34(a)]. This modification makes only a slight difference at the low frequency end of the range but it does remove the deep broadside null which previously occurred at the upper frequencies. The reason for this difference can be understood by reference to Fig. 20-34(b) which shows that the directions of the nulls which occur for each element (when the element is one wavelength long) are no longer coincident and the contributions from the two halves of the antenna add to give a strong response in the direction towards which the ends are tilted. This Vee antenna can now be improved in its impedance characteristics by making each half of it fan-shaped (using two or more elements) as discussed previously. This Vee-fan antenna shown in Fig. 20-35 is the prototype of a large number of antenna types used for television reception.

An alternative method of covering the four-to-one frequency range is to use two antennas, one for the high band and one for the low band. If these are on the same mast and fed with the same transmission line it is necessary to ensure that the presence of one does not affect the proper operation of the other. One method of achieving this result is illustrated in Fig. 20-36 where a fan-shaped high-frequency antenna is connected in parallel with a simple dipole low-frequency antenna. If the dipole elements are made approximately four times as long as the fan elements, and if the dipole is cut for half-wave resonance at the lowest frequency, the fan will be in half-wave resonance at the highest frequency. At the lowest frequency the fan elements (or "whiskers") will each be only 1/16 wavelength long and so will have a high reactance and hence negligible shunting effect on the relatively low input resistance of the half-wave dipole. Moreover, at the highest frequency, where the fan is in half-wave resonance, the dipole will have an over-all length of 2 wavelengths. Consequently the input impedance of the dipole will be very high [see Fig. 20-33(a)] and it will have negligible shunting effect on the relatively low input resistance of the half-wave fan. Thus, in effect, the dipole operates at low frequencies and the fan operates at high frequencies. This *dual-frequency* arrangement is also the prototype of many television receiving antennas.

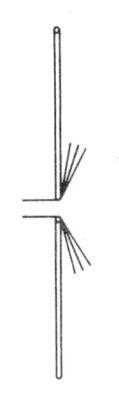

FIG. 20-36. Dual frequency dipole fan.

When only single-channel reception is required but more directivity is desired than is given by a simple dipole, a *Yagi* antenna (named after its inventor, Prof. Yagi) is sometimes used. A Yagi antenna consists of one driven element and one or more parasitically excited "reflector" or "director" elements as indicated in Fig. 20-37. The parasitic elements are not connected to the transmission line, but obtain their excitation from voltages induced in them by the current in the *driven* or *fed* element. The phase and magnitude of the current that flows as a result of this induced voltage depends on the spacing between elements and upon the reactance of the elements. This latter quantity can be varied by adjusting the length of the parasitic antenna. In general, for the small spacings of 0.1 to 0.15 wavelengths that are usually used, a parasitic element longer than one-half wavelength acts as a *reflector* causing the main beam to lie in the opposite direction, whereas an element shorter than half wavelength acts as a director, causing the main beam to lie on the same side of the driven element as does the parasitic element. A common arrangement is to use one reflector and one or more director elements. A typical radiation pattern for a properly adjusted 3-element Yagi is shown in Fig. 20-37(b). The presence of the parasitic elements reduces the input impedance of the driven element so a folded dipole is often used for the driver because of its impedance matching capabilities.

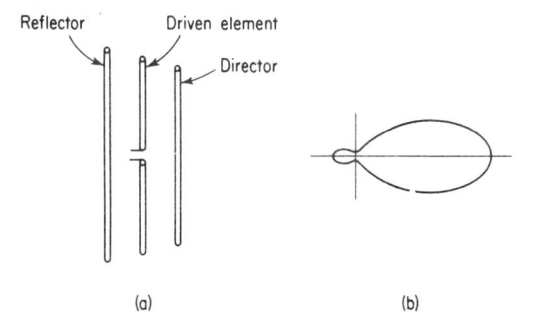

Fig. 20-37. (a) Yagi antenna; (b) typical radiation pattern.

Because the pattern of the Yagi array depends upon the length of the parasitic elements *in wavelengths*, the pattern of a given array changes markedly with frequency. For this reason good results should not be expected on channels other than the one for which the array was designed. However adequate operation is sometimes achieved on adjacent channels.

In the uhf television band (presently 470-890 megacycles) the receiving antenna problem is somewhat simpler owing to the fact that, although very wide in megacycles per second, the frequency range covered is slightly less than 2 to 1. Also at these frequencies where the wavelengths are reasonably short (the wavelength of 500 megacycles is only 60 cm or 2 ft) it becomes easier to construct directive antennas. Most of the antenna types used in

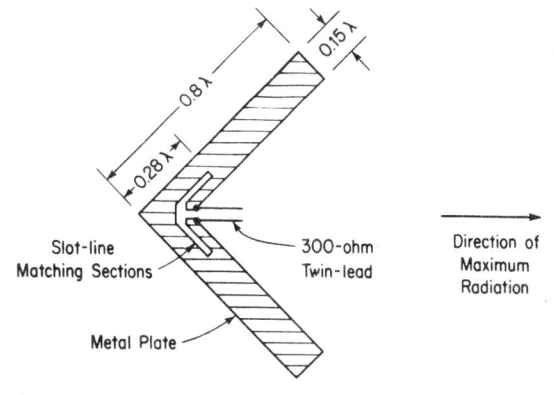

Fig. 20-38. V-plate antenna for uhf TV reception. (Dimensions shown are for mid-band wave length.) (Courtesy D. E. Royal)

the vhf television band can also be used in the uhf band, with an appropriate change in dimensions, and in addition, others which would be impractically large at the lower frequencies become feasible at uhf. A simple and effective antenna for use at uhf is the V-plate antenna shown in Fig. 20-38. Because of its shape and the large width of conductors used, this antenna has excellent pattern and impedance characteristics over the entire uhf band. The slots cut in the metal act as shorted transmission line sections and are designed to provide impedance compensation. The antenna delivers a single beam forward and provides a gain which ranges from 10 db at the low-frequency end to 12 db at the high-frequency end, with a good impedance match to a 300-ohm line throughout the band.

NAVIGATIONAL ANTENNAS

An antenna that has useful directional characteristics for navigational purposes is the loop antenna shown in Fig. 20-39. The dimensions of this antenna are usually small compared with a wavelength, so that the currents are in phase all around the loop (as shown). The two vertical sides of such a loop are then equivalent to two vertical antennas having a spacing that is a fraction of a wavelength and having currents in *opposite phase* (one current is up when the other is down). The resulting horizontal pattern is a figure eight with the zero line perpendicular to the plane of the loop. The top and bottom parts of the loop are equivalent to two horizontal antennas, also 180° out of phase, so that their pattern is also a figure eight (this time vertical) having its zero line perpendicular to the plane of the loop. This sharp null or zero-signal line makes the loop very useful in the direction-finding applications.

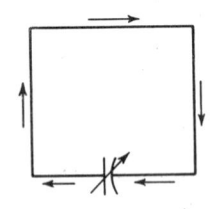

Fig. 20-39. Loop antenna.

DIRECTION FINDERS AND RADIO COMPASSES

Radio direction finders and radio compasses which depend upon the simple properties of a loop are among the simplest and most useful navigational aids. They are used in both air and marine navigation. By using a directional receiving antenna which can be rotated, the direction from which a signal is coming can be determined. When used for receiving, the pattern of a loop antenna is a figure eight. When the plane of the loop is at right angles to the direction from which vertically polarized waves are coming, the same voltage will be induced in each of the vertical sides of the loop; but since these voltages send currents around the loop in opposite directions, they will cancel and no signal will be heard. The same is true for waves that come from the opposite direction. Now, suppose the loop is rotated 90° so that one side of the loop is nearer to the transmitting station than the other. The voltages induced in each side of the loop are now not quite in phase opposition because of the time the radio wave takes to go from one side to the other. Therefore the two voltages will not quite cancel each other and a signal will be heard. Signals will be heard for waves that come from the opposite direction also. In order to increase the signal output from a loop antenna, it is usual to use a capacitor at the terminals to tune the loop to resonance.

If a loop antenna is mounted so that it can be rotated about a vertical axis and tuned to a station, the loudness of the signal received will depend on the direction of the loop with respect to the station. If the plane of the loop is at right angles to the line to the station, no signals will be heard, but if it is turned 90° from this position, maximum signals will be heard. Thus, either the position of zero signal (the *null* position) or the position of loudest signal could be used for direction finding. The null is usually used since it gives a sharper indication.

When a loop is turned to either the null position or the maximum-signal position, there is still an uncertainty as to whether the signal is coming from the front or the back directions of the loop. This uncertainty exists because there are two positions of the loop for zero signal. Often it is known in which general direction the station lies, especially in marine navigation, so that the uncertainty is removed. However, in cases where this is not known, other means must be used. To determine the *sense* of the reading of the loop (that is, which of the two possible directions is the correct one), *sense antennas* are sometimes used. A sense antenna is a small vertical antenna that picks up a signal which is then fed into the loop in such a way as to unbalance the loop. The unbalance of the loop makes one of the directions of maximum signal louder than the other. To use the sense antenna, a bearing using the null position is first taken with the sense antenna disconnected. Then, to determine the sense, a switch is closed and the loop is turned until the direction of loudest signal is found. A pointer on the loop then shows the true direction of the station.

Loop antennas are subject to certain errors, most of which can be avoided by proper construction and calibration. If a loop antenna is not balanced with respect to the ground, unequal stray currents will flow in the sides of the loop, changing the directions in which zero signal occurs. These two directions of zero signal are no longer exactly 180° apart, so that incorrect bearings will be obtained, depending on which null direction is used. Errors from this cause are eliminated by using a loop that is shielded electrostatically and carefully balanced with respect to ground. Sometimes small compensating capacitors are used to eliminate any residual unbalance. The presence of wires, large metal objects, and conductors in the neighborhood of a loop can cause errors in the bearing indications.

There will be some horizontally polarized waves present in signals that have been reflected from the ionosphere, as the plane of polarization may be rotated on this path. These waves are coming downward instead of traveling horizontally as do the ground waves. The horizontal component of the wave induces voltages in the horizontal arms of the loop, thus affecting the null position and giving incorrect bearings at night. This *night error* can be largely eliminated by using an *Adcock antenna*. An Adcock antenna is simply a pair of very short vertical antennas, spaced a small distance apart and crossed over at their centers to form an H. This antenna is shown in Fig. 20-40. The action of the antenna is just the same as that of a loop as far as vertically polarized waves are concerned. Horizontally polarized waves, such as those present in downcoming sky waves from the ionosphere, induce voltages in the horizontal parts of the antenna in such a way that they cancel and do not produce any signal. Small Adcock antennas give very low signals since they are equivalent to a loop antenna with only one turn. Adcock antennas give accurate bearing indications under conditions that render loop antennas completely useless.

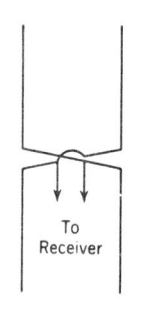

Fig. 20-40. Adcock antenna.

It is possible to attach a meter or other indicator to a loop antenna in such a way as to give indications on a dial telling the pilot whether the plane is headed directly towards the transmitting station or whether it is flying to the right or left of this direction. An arrangement of this type is called a *radio compass*. One type of compass uses a fixed loop at right angles to the line of flight. Another vertical antenna is used with it to give a maximum signal from the front of the plane. When the plane is flying directly towards the station, the needle on the meter rests in the center of the dial. If the plane deviates from this course the needle moves from the center position, showing which way the plane has deviated from its course. This arrangement is often called a *homing device*, since it guides the plane to its home base. Homing devices are very useful in guiding planes back to an aircraft carrier.

Another form of radio compass operates on the same principle as that described above, but with the difference that the loop can be rotated. This rotation allows the pilot to take bearings on stations off his course without changing the direction of flight. An important advantage of this type of compass is that it allows corrections to be made for drift, which cannot be done with fixed-loop homing devices. In the *automatic radio compass* a servomechanism arrangement keeps the loop oriented so that its null direction is always turned towards the incoming wave (direction of zero pickup). It is only necessary then for the pilot to tune to the desired transmitter and the loop with its associated pointer will automatically indicate the direction of the transmitter.

Chapter 20 — Radio Antennas

QUESTIONS AND PROBLEMS

1. (a) Sketch the voltage and current distributions on an elevated half-wave antenna and on a quarter-wave grounded antenna. (b) Sketch the vertical and horizontal radiation characteristics of these antennas (in the vertical position).

2. Why should an antenna be *tuned?* How can this be accomplished? What is meant by *loading* an antenna?

3. Why can the effect of the ground be simulated by an *image* antenna?

4. What are the power losses which can occur in an antenna system? Why is a short antenna likely to be less efficient than one which is a quarter or half wave long?

5. Explain fully why long resonant lines are less efficient than nonresonant lines of the same length.

6. Define *characteristic resistance* of a transmission line.

7. Explain the use of quarter-wave matching sections.

8. Number 12 wire has a diameter of 0.081 inch. Compute the characteristic impedance of line having two such wires spaced (a) 6 inches apart; (b) 3 inches apart. *Answer:* (a) 600 ohms, (b) 520 ohms.

9. The plate circuit of a vacuum tube is to be connected to a 500-ohm transmission line. Show a coupling network suitable for this connection. Why would a parallel resonant circuit rather than a series resonant circuit be used on the tube side of the network?

10. Explain how airplanes can be kept *on course* by means of radio beacons.

11. How do direction finders operate? What is a radio compass? A homing device?

CHAPTER 21

RADAR, RADIO RELAYS, RADIO AIDS TO NAVIGATION, PULSE COMMUNICATION

The radio systems explained in this chapter have many features in common. A number of them utilize "microwaves" in the thousands of megacycles, highly developed directional antennas of unique design, pulse transmission, and unique designs of tubes for generating carrier power. They constitute a group of services born of new knowledge of apparatus, theory, and techniques largely gained since the beginning of World War II. Although they are the most recently developed services, they follow the basic principles of radio communications as expounded in this book. One of the most important and useful of the new services, radar, will be described first.

RADAR

Consider a flashlight. Its small batteries and tiny bulb do not actually produce much light but the reflector and lens so concentrate and focus it in one direction that it produces an intense light in a small area. A person searching in darkness for an object moves the light beam about until the light strikes the object sought, illuminates it, and it reflects some of the light back to the person's eyes; he then sees it, sees its direction and even estimates its distance. A radar system operates by using a sharp beam of radio waves, instead of light waves, to search the darkness and fog for the unknown.

The term *radar* signifies *radio detection and ranging*. It is one of the great radio developments of the World War II period. Conceived and developed in its early phases prior to the war, tremendous advances took place under wartime pressure because of its great value in military operations. Its applications in peacetime air and marine transportation and national defense are many and growing.

As explained briefly in Chapter 10, in a pulse radar system extremely powerful but very short pulses are transmitted at intervals that are very long compared with the duration of each pulse. During the waiting time of the transmitter, between successive pulses, the receiver operates. Echoes from nearby objects arrive first, followed by others from greater distances, and so on. When adequate time has passed for all echoes to be received, another pulse is transmitted and the sequence of events is repeated. Such reflected wave energy makes it possible to establish the distance, altitude, speed of motion, and under some conditions, the size and shape of the reflecting object.

ESSENTIAL COMPONENTS OF A SIMPLE RADAR SYSTEM

The essential components of a simple radar system are:

1. A radio transmitter which produces extremely short but powerful pulses of carrier power.

2. A moveable beam antenna for both transmitting and receiving which concentrates the transmitted radio waves in a narrow confined beam and which receives reflected radio waves back with the same directional selectivity.

3. A radio receiver which demodulates the received reflected carrier waves delivered by the antenna.

4. An indicating device calibrated in yards or miles to show directly the distance of the object which caused the waves to be reflected back.

5. A send-receive switch which connects the antenna to the transmitter when transmitting pulses and connects the antenna to the receiver at all other times.

Figure 21-1 shows in block diagram form the essential components as connected for the simplest radar system. In the upper left corner is the pulse generator which produces the extremely short pulses at a predeter-

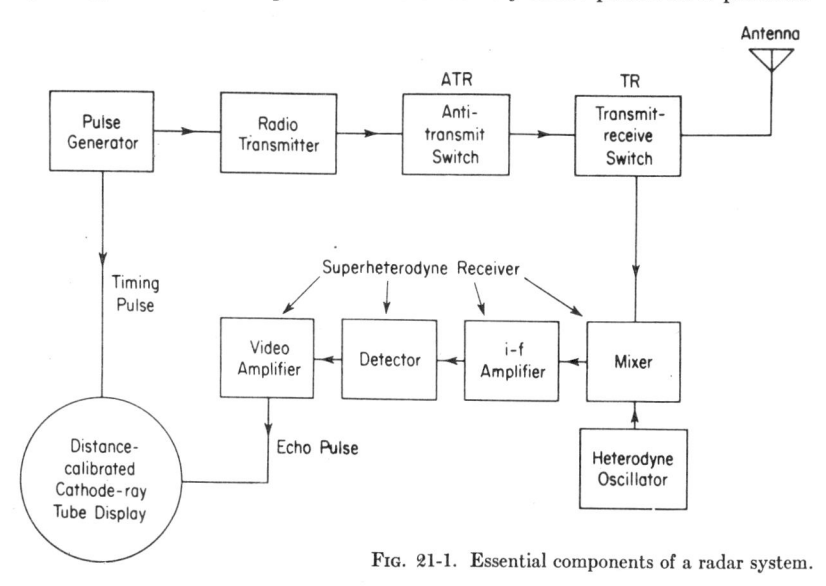

FIG. 21-1. Essential components of a radar system.

mined repetition rate. The sequence of events which takes place when a single pulse is transmitted will be explained. It is repeated for all succeeding pulses.

The pulse is produced. It travels to the radio transmitter which is inoperative between pulses. When actuated by the pulse the radio transmitter produces a burst of very powerful carrier waves at carrier frequency. The burst of powerful carrier waves passes through the gaseous one-way switch (ATR), the function of which will be explained below. The carrier waves then go to the TR switch (transmit-receive) which passes them through to the antenna. During this time the receiver is disconnected by the transmit-receive switch. The short burst of carrier cycles has now been transmitted by the antenna and the radio waves are on their way in space. When they strike an object some of the wave energy will be reflected back to the radar station antenna, so the system must be placed quickly in operating condition to receive the reflected energy.

Immediately after the pulse is transmitted the TR switch connects the antenna to the receiver. The heterodyne oscillator, mixer, intermediate-frequency amplifier, detector, and video amplifier are nothing more than a superheterodyne receiver. The heterodyne oscillator injects a signal into the mixer which combines with the reflected incoming pulse to produce the desired intermediate frequency which is amplified in the i-f amplifier, detected in the detector, and amplified in the video amplifier. The reflected wave energy then goes to the calibrated device which indicates the distance of the reflected object. The range indicator is a cathode-ray tube in which time is the abscissa and amplitude the ordinate. One thing is still missing. To determine the time between pulse transmission and reception the exact time of transmission has to be established. This is accomplished by bleeding a small amount of power from the pulse generator and inserting it in the cathode-ray scanning circuit with the received pulse. The display consists of an electron beam moving from left to right at a known speed. When the pulse generator sends down a marker timing pulse it will appear on the screen as a "pip," a vertical line. When the reflected pulse comes back, time will have elapsed and the electron beam will have moved to the right. The pip from the reflected wave therefore will be at the right of the timing pip by a distance proportional to the time out and back of the transmitted-reflected-received pip. The excursions of the electron beam from left to right are synchronized with the pulse generator so the pips will be stationary on the screen. The time scale is calibrated in yards, miles, etc., as desired for the particular application of radar being used. This type of display is commonly used in cathode-ray tube oscillography and is called Type A scanning. It is the simplest form of presentation. Coming back to the ATR (anti-transmit) switch, its function is to pass transmitting pulses with negligible loss during transmission intervals but automatically to disconnect the radio transmitter during receiving intervals to avoid losing any of the feeble and precious received signal.

Both TR and ATR are gaseous devices specially developed for the purpose. Mechanized switches could not operate fast enough. Gaseous discharge is triggered off by the carrier burst from the transmitter and continues until the burst ends. TR, during burst transmission, passes power from the transmitter to antenna, automatically disconnecting the receiver.

VIDEO AMPLIFIER

In referring to the receiver the demodulated signal was said to be amplified in the video amplifier. Why video instead of audio for radar? A radar modulating pulse has a rectangular shape and is very short. The accuracy of the system depends upon short pulses. But to create and reproduce a short pulse a wide band of component frequencies is required. The pulse is made up of sine waves of fundamental and harmonic frequencies all added together. If a sufficient number of harmonic components is not included, the pulse shape is widened and reduced in amplitude and loses its character and usefulness for radar. Hence, for retaining the pulse shape the amplifier must pass a very large number of high-order harmonics. The bandwidth must be beyond audio-frequency limits and get into the area of television video frequencies, in the megacycle range. Therefore, the amplifier is called a video amplifier.

PULSE REPETITION RATE

What is the proper pulse repetition rate? For simple cases a considerable latitude is available. As soon as a complete sequence of events for a single pulse transmission and reception is completed, another pulse may be transmitted. The practical limit is linked with the time for one sequence.

It is known that radio waves travel about 186,300 miles per second. If a radar pip were received from a distance of 300 miles, the round trip distance would be 300 miles out and 300 back, or 600 miles. Knowing the speed of travel of radio waves through space, it is obvious that

$$\text{time interval} = \frac{600 \text{ miles out and back}}{186,300 \text{ miles per second}} = 0.0032 \text{ sec.}$$

Since the time per round trip is 0.0032 sec, simple arithmetic shows that the round trip journey is completed in 1/310 sec.

It would not be feasible to send a pip every 1/310 sec because it would start at the instant when an echo was arriving and smother it. Furthermore, the receiver would be deactivated when the echo arrived. But if a two to one margin of time were allowed, then 165 pulses per second could be transmitted and there would be no conflict between one sequence and any other. All echoes would have been received, the cathode-ray tube beam would have been returned from the end of its sweep to the starting point and the system again would be ready. For shorter range radars the repetition rate could be greater because the time required for the weakest detectable pulse echo to return is shorter, and the system may be restored and ready sooner.

The cathode-ray tube display of a simple distance measuring radar would appear as shown on Fig. 21-2. Two pips are shown as received, one from an object 43 miles distant, the other from an object 70 miles distant.

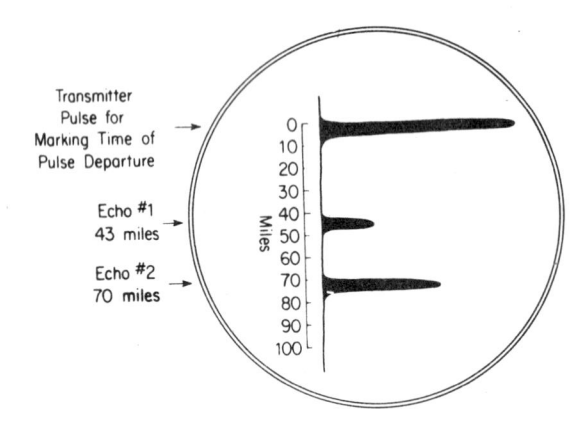

Fig. 21-2. Type A radar display. Cathode-ray tube display for simple distance-direction measuring radar.

Judging by the relative sizes, the more distant object would be the larger because of the greater amount of energy received even from the longer distance.

TIME BASE

Most radar installations are equipped to adjust the time base for accurate measurement. For example, an object at two miles could not be read accurately on Fig. 21-2, nor could one at 150 miles. By changing the speed at which the electron beam moved from left to right, and correspondingly changing the calibration, either the 2-mile or 150-mile distances could be seen more accurately.

PULSE DURATION — SEPARATION OF OBJECTS

The duration of the pulse is important, particularly for measurement of short distances. This is due to the fact that the system must convert from the "transmit" position to the "receive" position before any echo pip is admitted to the receiver. This determines the minimum range at which an object may be accurately located. For example, a pulse two microseconds in duration occupies a radial distance in space of about 2,000 feet. If the system were an airport radar set, it could be desirable to be able to measure through fog or darkness the distance to objects as close as 500 ft. But by the time the two microsecond pulse transmission ended a substantial part of the leading edge would have been reflected back and found the set still transmitting and unable to receive. A portion of the echo might be received but the accuracy would have been lost. If an object is less than half as distant as one pulse wave train, obviously part of the echo must be lost. Five hundred feet is about the shortest distance measurable with reasonable accuracy with a 1-microsecond pulse. Correspondingly, for shorter distances, such as 100 ft, a pulse as short as a fifth of a microsecond would be necessary.

The pulse duration also determines the ability of a radar system to differentiate between objects almost equally distant. This is called "range discrimination." As an example, if two distant airplanes were only a short distance from each other, a short pulse would be required to provide two separate reflections which could be identified readily. Too long a pulse would make the two aircraft appear as one because the reflection from the first object would overlap that from the second object and the two reflections would appear as only one. In a properly designed system a 1-microsecond pulse makes it possible to identify two aircraft in a given direction if the distance between them is more than about 500 feet. With a 5-microsecond pulse the separation must be about 2,500 feet. On the other hand, longer pulses have certain advantages. More energy is transmitted in a longer pulse which may be more easily detected. A long pulse contains fewer sideband components and therefore the receiver bandwidth can be made narrower. This reduces the amount of noise which is proportional to the receiver bandwidth. Hence the signal-to-noise ratio is improved. For this reason long-range radar systems employ longer pulses than intermediate or short-range systems.

A receiver must have a bandwidth which depends upon the duration of the pulse, or the amplitude of the pulse will be reduced by elimination of essential sideband components. Since very short pulses contain wider sidebands than long pulses, a wider pass band is required in the receiver. The pass band required for good reproduction of the pulses is inversely proportional to the pulse length and should be from one to two times the reciprocal of the pulse length in seconds.

From the foregoing statements it follows that the free space range of a radar set depends upon the product of the pulse *power* and the pulse *length*, or the *energy* in the pulse. It also follows that a compromise is often necessary between the needs for *range discrimination* and *maximum range*.

The minimum detectable signal power varies directly with receiver bandwidth, when the latter is correct for the pulse length, and it varies approximately inversely with the square root of the number of pulses which reach the target on each scan.

RADIO FREQUENCIES USED FOR CARRIER

The frequencies most applicable to radar cover a very broad part of the radio spectrum but there are practical limits. For example, very sharp directional beam patterns are desirable for the antenna. The sharper the beam is made, the larger must be the antenna in terms of wavelength. Physical size is an important consideration. The same degree of beam sharpness can be attained at 5,000 megacycles with an antenna only one tenth as large as a beam antenna at 500 megacycles, and one one hundredth as large as a beam antenna at 50 megacycles. An ideal radar antenna at 50 megacycles becomes prohibitively large, heavy, expensive, and cumbersome for rapid movement of scanning.

For maximum range the preferable frequencies are those below about 7,500 megacycles because above that frequency atmospheric moisture and rain may cause heavy losses of wave energy in space.

The lower limit of desirable frequencies would not include those at which propagation occurs over extreme distances owing to ionospheric reflection. It is desirable to have the waves completely dissipated at distances not greatly exceeding the maximum useful range of the system so that interference is not caused or received and so that a degree of military secrecy may be maintained at distances beyond those of useful significance. Radar frequencies are used considerably above 7,500 megacycles for short-range services where the atmospheric moisture attenuation is not a serious limitation. In general radar utilizes frequencies between about 300 megacycles and 12,000 megacycles, with diminishing activity in the lower frequency range.

BEAM ANTENNAS

The performance of directional beam antennas was discussed in Chapter 20. Paraboloids are widely used. A parabolic reflector is shown in Fig. 21-3(a). All wave energy departing to the left of the source of energy is reflected to the right in parallel lines and the efficiency in that direction is sharply improved. But much energy radiated from the source but not

(a) Parabolic Reflector   (b) Power Lost   (c) Double Parabolic Reflector

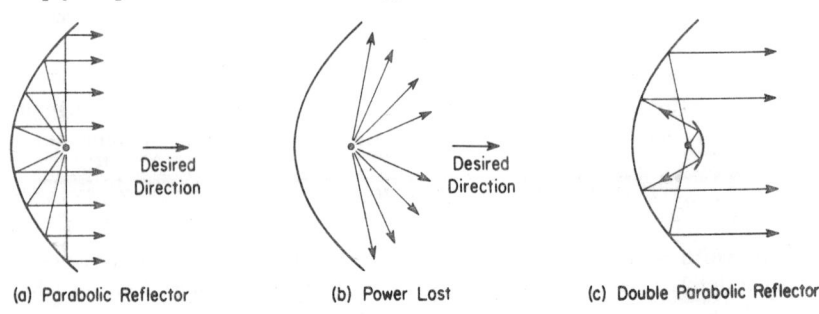

FIG. 21-3. Improvement in a simple parabolic antenna.

going to the left into the parabolic reflector is still lost as shown in Fig. 21-3(b). A further useful concentration is possible by the addition of another smaller reflector shown in Fig. 21-3(c) to concentrate this wave energy into the large parabolic reflector where it will be combined usefully with the energy shown in Fig. 21-3(a). In most radar equipments other methods are employed to prevent loss of power and will be discussed. The example used here applies to antennas at the lower range of frequencies and was used to illustrate the importance of conserving and making use of all of the power generated in the transmitter.

ANTENNA POWER GAIN

A general discussion of antenna power gain was covered in Chapter 20.

In practice the maximum power gain that may ordinarily be achieved with radio antennas is approximately 50,000, for two reasons. First, the wave energy does not go out of the antenna in a geometrically perfect straight-line beam pattern. Some power is lost at the edges of the pattern in the form of minor lobes of power which have been referred to as "halo." Second, the surface is never geometrically perfect.

Figure 21-4(a) shows the beam pattern of a parabolic antenna 10 ft in diameter operating at 4,000 megacycles. The field intensity is plotted in terms of per cent-of-maximum at various angles from the true focal point.

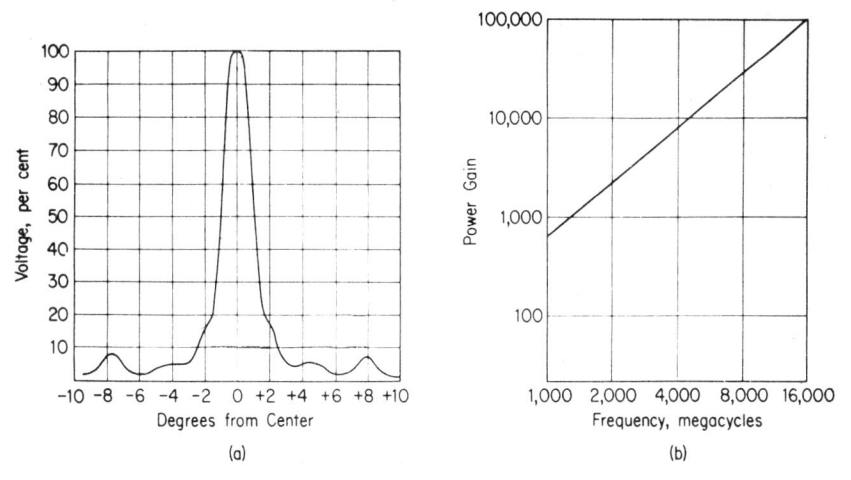

Fig. 21-4. Characteristics of a ten-foot parabolic antenna. (a) Beam pattern; (b) power gain.

Figure 21-4(b) shows the power gain of this antenna at various frequencies. The power gain of such an antenna compared with a point source isotropic antenna is directly proportional to its area expressed in wavelengths, so if the frequency is doubled the power gain ideally is also doubled. This applies to either transmission or reception with such an antenna. Figure 21-5 shows one of the largest parabolic receiving antennas ever built. It is a Bell Laboratories unit 60 ft in diameter and may be compared with the Bell System microwave radio relay type of horn at the left. Proper performance

FIG. 21-5. A 60-ft diameter parabolic microwave antenna at right, with radio relay type antenna at left. (Courtesy Bell Telephone Laboratories)

of such huge antennas depends upon precision in design and manufacture because small errors in mechanical dimensions are large in terms of wavelengths. Such errors may cause a change in path length in part of the optical path and cancellation instead of addition could result.

High-gain antennas vary in type and form to meet various requirements. They consist of large horns, parabolic reflectors, and variations of each to produce a variety of beam shapes for special purposes. An interesting example of a variation of the circular parabolic wave shape is shown in Fig. 21-6. This consists of a horn into which r-f power is fed at the throat. The curved reflecting surface of the upper part of the horn is parabolic in shape and may be considered a section of a conventional parabolic reflector as shown in Fig. 21-7. However, in this case the power is not dispersed over a complete 360° parabolic reflector. The same energy concentration and power gain is achieved by focusing all of the power into a sector of a complete parabolic reflector, reducing the size and cost. This horn antenna is used by the Bell Telephone System in its "microwave" radio relay system. The horn weighs 1,700 lb, is $20\frac{1}{2}$ ft high, 11 ft wide, and 9 ft deep. It has a

Chapter 21 *Radar* 693

power gain of 10,000 at 4000 megacycles, 20,000 at 6,000 megacycles, and 63,000 at 11,000 megacycles. The last types are unusually high and are achieved only because the contour of the reflecting surface is accurately

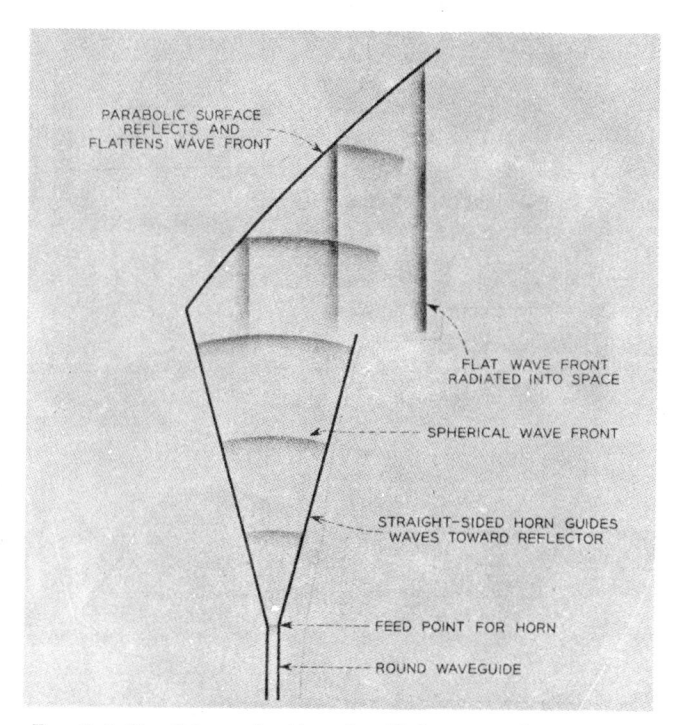

Fig. 21-6. How flat wave front is produced in horn type microwave antennas used in radio relays. (Courtesy Bell Telephone Laboratories)

shaped within about 1/16 of an inch. It should be noted that in a horn of this kind the power is prevented from escaping in useless directions and is efficiently guided by the metal throat to the reflecting surface.

ANTENNA BEAM WIDTH

In sections to follow special beam antennas will be described. Some of the reasons for different designs are explained here. Obviously, an object anywhere within the beam of a radar set will cause a reflection and a pip. But the location up or down in the beam or to the right or left in the beam will not be known. For the most exact pin-pointing of an object the sharpest possible beam would be needed. But then the area scanned is more limited, and a target could be missed by being outside the beam area. A long and growing list of special radar uses has led to unique designs. A search radar system may scan a vertical spiral. As it makes complete revo-

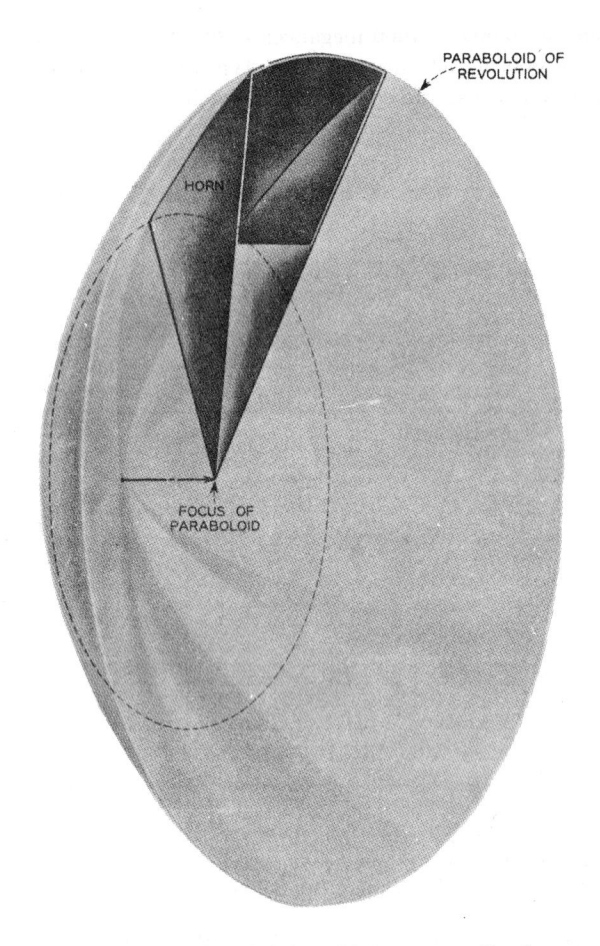

FIG. 21-7. How section of parabola is used in horn type radio relay antenna. (Courtesy Bell Telephone Laboratories)

lutions in the horizontal direction, the beam may be continually and slowly raised, revolution by revolution, until all space around the station has been searched. It then depresses the beam and repeats the cycle. Another method is to use separate antennas and beams for the vertical and horizontal scanning.

In height-finding radar it is particularly important that the beam be sharp and narrow in the vertical direction. ·On the other hand, for accurate location in the horizontal directions it is particularly important that the beam be narrow and sharp in the horizontal plane. In the plane of secondary interest it may be an advantage to have the beam wider so that the target may more easily be located. Figure 21-8 shows the principle applied to a height-finding radar system. In Fig. 21-8(a) movement of the beam up or down will produce reflection and a pip so long as the target aircraft remains anywhere within the beam. The height may be measured only with poor accuracy. In Fig. 21-8(b) the beam angle is much narrower and the

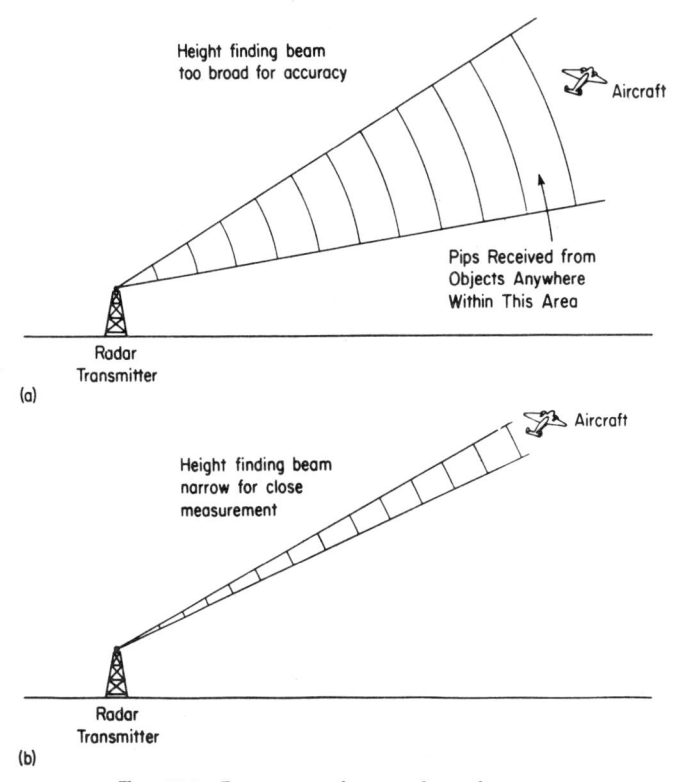

FIG. 21-8. Importance of narrow beam for accuracy.

accuracy is correspondingly much greater. The same principles apply to object location in the horizontal direction except that the beam width must be sharp and narrow in the horizontal plane.

A beam sharp in one plane and broad in the other will have an oval outline. In the sharp plane, for example the horizontal, the antenna will have a narrow width to provide for sharp focusing and a minimum of lost power, but in the vertical plane will have considerable height. Figure 21-9 shows a radar system using separate antennas for vertical and horizontal scanning. These antennas have the oval shape referred to above. Each has a beam which is sharp in one plane and relatively wide in the other. The vertical oval is for altitude measurement and the horizontal oval is for azimuthal measurement.

Figure 21-10 shows an airborne type of radar antenna in which the beam may be made either fan-shaped or pencil-shaped, as desired, by rotating the lower section. In the photograph the fan-shaping section is in place. By rotating the assembly 180° the pencil-shaping section moves into place.

Fig. 21-9. Portable precision approach radar (GCA) using separate vertical and horizontal antennas and displays. (Courtesy Laboratories for Electronics, Inc.)

The latter, as may be seen, is the lower section of a parabola. For fan shaping, the lower section combines with the upper section to approximate an oval-shaped reflector with a narrow horizontal beam and a wider vertical beam. This flexibility is of value in scanning earth terrain, clouds, and for other aircraft. With the desired lower section locked in place the whole assembly may be rotated at fast or slow speed through 360° or a selected sector of less than 360°. The antenna may be tilted up, down or straight ahead. The power is fed to the reflecting surface through the metal waveguide which projects in front of it.

REFLECTION FROM OBJECTS

Radio waves are reflected by all objects but most effectively by those of metal. Naturally, the larger the object the more energy it will reflect.

Objects which reflect include airplanes, ships, birds, trees, land, buildings, bridges, thick clouds, water, etc. The reflection efficiency depends upon the surface texture and contour, the angle of incidence of the waves, the reflection coefficient of the material, etc. Ordinarily, objects of radar search, such as aircraft, have irregular surfaces. One of the most effective means of confusing radar performance in military operations has been the dropping

FIG. 21-10. Compact "turtle shell" antenna of airborne Radar scans earth's terrain, clouds, and other aircraft. This two-in-one unit can be changed at will to fan beam or pencil beam. Versatile motion includes fast or slow full 360-degree and sector scans; selective tilts looking up, down, or straight ahead; and beam switching. Scale shows miniature size. (Courtesy Sperry Gyroscope Co.)

of strips of thin metal foil from aircraft. It produces pips at the receiver which are difficult to differentiate from the targets being sought because the surface is metal and large, it falls slowly with the wind and may be carried and dropped in large quantities. Under almost any conditions of position and shape and size of an object, there is some reflection of wave energy back in the direction of the receiver. The effective range of detection depends largely upon these factors.

In detecting ships the waves of a rough sea produce a background of overlapping relatively weak pips out of which strong individual pips from the ship must be identified.

RADAR DISTANCE RANGE

Various factors limit the effective maximum distance of radar operation. The transmitted beam is sharp and powerful but, being wedge-shaped, its width increases with distance. This causes the intensity of the waves to be dispersed over a larger area. So the transmitted energy arriving at the aircraft becomes weaker with increasing distance coincident with a lengthening distance for the reflected wave to travel to return to the receiver.

In Chapter 13 the ultimate exploitation of receiver sensitivity was shown to be limited by thermal agitation and shot noise produced in the receiver itself. This limitation is no less final in radar reception.

The maximum range of a radar system depends not only upon its electrical design but also upon the distance to which the transmitter can "see" the distant object without its being much below the horizon. The distance to the horizon from an observer was given in Chapter 18.

It is not difficult to calculate that a radar on the ground may "see" an aircraft at an altitude of 20,000 ft at a distance of about 175 miles. If the radar antenna were elevated on a 50-ft tower, the optical range would be about 183 miles. If the radar were in an aircraft at 20,000 ft altitude, it could "see" another aircraft at the same altitude at a distance of about 350 miles. Maximum ranges of very high powered radar sets at practical altitudes of 50 or 100 ft may be of these orders with reasonable reliability for aircraft detection. The U. S. Army in 1946, with specially prepared radar equipment, transmitted approximately $\frac{1}{2}$-second pulses to the moon and received return pips at the proper interval of about $2\frac{1}{2}$ seconds required by the waves to make the round trip.

At the higher frequencies used in radar, operation is usually conducted under free space conditions where the beamed signals are confined to vertical angles where they do not strike the earth. At the lower frequencies where antenna size is limited and sharp beams are unattainable because of large mechanical dimensions this is not always feasible. In some applications where a target is on the sea or the ground it is sometimes impossible at any frequency.

When part of the beam wave energy strikes the earth it is reflected. In the process it is shifted in phase and some of it may be lost or scattered. The reflected energy in addition traverses a longer path than the energy which travels directly to the target. As a result the portions of the waves which traverse these separate paths combine in space with phase differences which depend upon the differences in path lengths. Where the phases are alike, the signal intensity is increased. Where the phases are in opposition, the intensity is reduced. The phase of the reflected signal reverses 180° for horizontal polarization. For vertical polarization the shift diminishes with angle of incidence.

As a result of this effect and the constantly changing difference in path length with distance and altitude, a complex field intensity pattern is created in space. In general a radar beam operating under these conditions would appear in space as a stack of uptilted alternately high-intensity and low-intensity "beavertails" which repeat, within the confines of the beam, whenever the path length difference is a multiple of a wavelength. The effect of this wave interference pattern exists both in transmitting to the target and in receiving from it. Detection of distant targets then is critically dependent upon both altitude and range. As an aircraft target flies through such a changing field pattern, the received signal will alternately vanish and reappear as the high and low intensity "beavertails" are penetrated by the target. In the high-intensity "beavertails" the range of the system will be somewhat increased at times over free space conditions because of the addition of the direct and reflected wave components.

Because of the geometry of the paths, the lower frequencies (longer wavelengths) are at a disadvantage compared with the higher frequencies in searching areas close to the earth. At the lower frequencies the lowest useful "beavertail" is at a greater altitude. These effects are not taken into consideration in the *radar equation* presented in the following section, because only free space conditions are included in the derivation.

THE RADAR EQUATION

The range of a radar system may be estimated by computation. The effective radiated power in the beam would be the product of the antenna power gain times the transmitter power in kilowatts. The signal power intensity at the target area S_T may be shown to be

$$S_T = \frac{\text{effective radiated power}}{4\pi \times \text{distance squared}}.$$

The signal power intensity at the receiver area is proportional to

$$S_A = \frac{\text{radar cross-section area of target}}{4\pi \times \text{distance squared}}.$$

The signal power delivered by the receiving antenna will be proportional to its effective cross section:

$$\frac{\text{power gain of antenna} \times \text{wavelength squared}}{4\pi}.$$

The complete equation for received signal power then is

$$S_{\substack{\text{power}\\ \text{received}}} = \frac{ERP}{4\pi D^2} \times \frac{\text{target radar cross section}}{4\pi D^2} \times \frac{\text{receiving antenna power gain} \times \text{wavelength squared}}{4\pi}.$$

An inspection of the last equation shows that the signal power received is inversely proportional to the fourth power of the distance. The equation may be rewritten as

$$S_{\substack{\text{power}\\\text{received}}} = \frac{PG^2\lambda^2 C}{(4\pi)^3 D^4},$$

where P = transmitted power, G = power gain of antenna, λ = wavelength, C = radar cross section of target, D = distance.

The power of a large antenna is distributed over a defined area as shown in Fig. 21-4(a). The radiating system may be considered an aperture of large area which is excited by a plane wave. The maximum power gain (G_{max}), the area of the aperture A, and the wavelength are related as follows:

$$G_{max} = \frac{4\pi A K}{\lambda^2},$$

where K is a constant usually from 0.5 to 0.6.

If account is now taken of the dimensional relationships of the antenna, as represented by the last equation, and substitution is made for G in the previous equation, it becomes

$$S_{\substack{\text{power}\\\text{received}}} = \frac{PA^2 C K^2}{4\pi D^4 \lambda^2}.$$

The maximum range of a radar system may be computed by rearranging the last equation to show D, assuming the *minimum* usable received signal power is known.

$$D_{max} = \left(\frac{PA^2 C K^2}{4\pi \lambda^2 S_{\substack{\text{min usable}\\\text{signal power}}}}\right)^{1/4}$$

MINIMUM USABLE RECEIVED SIGNAL POWER

In a radar set, as in any other radio communications system, the fundamental limit to the distance range is reached when the received signal is too weak to be detected through the noise produced in the input circuits of the receiver. In a radar system employing Type A scanning, noise, when present, causes the electron beam to be deflected upward in the same manner as it is deflected by received pips, as it progresses from left to right. Noise consists of fluctuating voltage. Therefore, on the face of the CRT it has the appearance of an indistinct field of grass under constant agitation. Radar pips, when received, are of longer duration than any noise pulse, are fixed in position, and are readily visible if the amplitude is at least comparable with or higher than the noise pattern. In following sections a different form of scanning will be described, called the PPI system, in which the received signal voltage causes only a change in *brightness* of the CRT trace. In this type of set the noise appears as flashes of light, or scintillation, along the line of the sweep.

The minimum value of usable received signal power in a radar set is difficult to specify because it depends upon pulse length, receiver noise factor, pulse repetition rate, type of scan, speed of sweep, persistence of luminance of the CRT phospher, the skill of the person operating the receiver, and other factors. As a rough indication it can be stated that under the control of a skilled operator a 3,000-megacycle system with a 50-ohm receiver input impedance may give identifiable pips with only a few microvolts at the input to the receiver, equivalent to 1×10^{-13} w.

Chapter 13 explained and discussed receiver noise factor, the degree to which the noise in a receiver exceeds that of a theoretically perfect receiver. Broadcast-type receivers have available a continuous uninterrupted carrier when a station is tuned in. However, in a radar receiver a carrier is present only when r-f pulses are being received. Since noise produced in the input circuits of receivers is not of uniform amplitude, but fluctuates at random over wide limits from instant to instant, the reception of a very short r-f pulse may occur at an instant when the noise voltage is either high or low compared with other instants when other pulses are being received. Therefore, the instantaneous minimum usable signal power will vary from instant to instant depending upon random fluctuations of the noise voltage.

If a large number of pulses are received for each scan of the target the over-all effect is to smooth out, or average, the noise voltage. Also, a large number of pulses provide more certain identification when external interfering noise or pulses from other radar sets are present. There are also other factors which bear upon the value of minimum usable signal power, one of which is the ability of an observer to integrate successive repetitive pulse images on the CRT screen. It has been established that the minimum usable signal power in a radar receiver varies approximately inversely with the square root of the number of pulses transmitted and received per scan of a target.

How may one derive an equation that would give at least an approximate figure of minimum usable signal power?

It has been shown previously in Chapter 13 that the noise voltage produced in a circuit is given by

$$E = \sqrt{4KTRB},$$

where T = degrees Kelvin temperature, B = bandwidth in cycles, K = the constant 1.37×10^{-23}, R = resistance in ohms.

If both values are squared this becomes

$$E^2 = 4KTRB.$$

If R is transposed this becomes

$$\frac{E^2}{R} = 4KTB.$$

Since E^2/R is the expression for power, the noise power is
$$P_N = 4KTB.$$
The noise factor of the receiver should be taken into account because it represents the increase in noise over the antenna thermal noise represented in this equation, so it is included as a factor:
$$P_N = 4KTBN.$$
If allowance is now made for various factors unique to a radar system, by making some arbitrary assumptions, this expression could be rewritten
$$P_N = KTBN.$$
If it is assumed that $T = 290°$, $B = 1.5$ megacycles, $N = 6$, then
$$P_N = 3.6 \times 10^{-14} \text{ w.}$$
If radar pips could be identified by a skilled operator when the voltages equalled the noise voltage, the signal power shown above would be the minimum usable received power.

DISTANCE RANGING AND DIRECTION

The radar system described in preceding sections is the simplest type used for locating individual objects in space. The display is a cathode-ray tube face on which the transmitted timing pulse appears at the left and the reflected echo pips appear in positions at the right. Calibration in distance may be accomplished by securing by adhesive a calibration scale to the external surface of the tube face. However, a better and commonly used method is to insert into the cathode-ray scanning circuit carefully timed pulses which cause distance calibrating pips to appear at selected intervals as the electron beam travels to the right. This method of distance calibration is more accurate since changes in power line or plate supply voltages or linearity of sweep, for example, affect the calibrating pips and the echo pips equally.

In a simple radar, using Type A scan, the direction of objects is established by a circuit from the antenna mechanical tracking system to the receiver display which indicates at all times the exact direction in space to which the antenna beam is pointed. Thus, the measurements are those of distance and direction for individual objects. The area scanned may be a full circle and from earth to zenith, a narrow sector with a restricted vertical range, or any intermediate degree of horizontal or vertical area. But this is not the most convenient method of presenting the information received. A much more useful system, plan-position radar, is in common use.

PLAN-POSITION RADAR

Plan-position radar systems may produce on a cathode-ray tube a complete monochrome plot of all fixed and moving objects surrounding a radar station or may do so for any selected sector of the area with a variable range and scale. The plan-position indicator, called the PPI, is somewhat more complex than the system previously described but still is relatively simple in its operation.

Assume that the antenna has a very narrow horizontal beam width and a very broad vertical beam width covering a vertical range from the earth to high altitudes. A pulse transmitted in a fixed direction will bring reflected echo signals from every object in its path, including the earth, water, buildings, ships, aircraft, etc. The coefficient of reflection, angle of incidence, contour and texture of the objects will differ, causing variations of the intensity of the echo. The closest object reflections will be received first, followed in turn by the reflections from the increasingly more distant objects.

The simple Type A scan electron beam is caused to move only from left to right across the CRT tube face, and the signal is used to cause vertical deflection. On the other hand, if we cause the electron beam to start in the center of the face and move toward the edge in a direction corresponding to that in which the antenna is pointing, the sweep will be along a radial or spoke. The part of the spoke near the tube-face center, the starting point, will correspond to the close echoes. Progress along the spoke corresponds to increasingly distant objects. If, now, the beam brightness is controlled by the echoes so that only when an echo is received is there any illumination of the face, the presence of the echoes will cause the spoke to have bright points at ranges corresponding to the distances of the reflecting objects. This is accomplished by impressing the received signal on the grid of the CRT so that it controls the brightness.

If the antenna is swung clockwise to cover an adjoining sector, and the electron beam is turned off and returned to the center of the tube face, and the electron-beam deflection coil is swung clockwise by the same angle as was the antenna, the next transmitted pulse will bring reflections from the adjoining sector now being scanned. If the antenna and electron beam repeat this sequence, sector by sector, a complete circle will be traversed and on the face of the tube there will appear a complete map of the area surrounding the station. The sequence then for the PPI radar is:

1. A pulse is transmitted and simultaneously the electron beam starts to move from the center of the tube face toward the edge.

2. The intensity of the beam is modulated, while it is moving, by the echo signals received as they arrive from increasingly distant points, delayed in time of arrival. The most distant echoes arrive as the electron beam comes to the edge of the face.

3. The radial spoke is now completed.

4. The electron beam is momentarily cut off and moved back to the center while the antenna and electron-beam deflection directions are shifted slightly to a different direction.

5. A second pulse is transmitted and the cycle repeats until all echoes are back and this cycle ends.

6. These cycles are continued until the full circle, or any desired portion of it, has been filled in.

The complete "pie" may have been cut into 360 pieces, each one degree wide, each of which was plotted individually in turn. If the antenna and electron-beam directions are synchronized and calibrated for absolute compass direction, the map will show the true bearing for every one of the objects "painted" on the tube face. In practice, the upper vertical spoke is often made to correspond with north and the tube face is permanently marked with angular bearings.

Fig. 21-11. Artist's conception of the way airborne radar reproduces what the "turtle shell" antenna sees on the surface. Area depicted is the familiar outline of the region near Cape Cod, Mass. The dark plastic wedge at top swings to any compass heading for exact measurement of drift. (Courtesy Sperry Gyroscope Co.)

Such a cathode-ray tube face is shown in Fig. 21-11. It is combined with an artist's illustration of the antenna of Fig. 21-10 scanning the Cape Cod area outlined on the map, and the corresponding radar display. Faint distance range calibration circles may be seen on the display.

SPEED OF ROTATION OF BEAM

The rotation of the beam is very slow compared with the pulse-rate times. Hence, the amount of rotation is negligible between the time a pulse leaves and the echoes return. Actually the antenna rotates at constant speed. For long-range sets the antenna may rotate once in each ten seconds. For shorter-range applications it may rotate as fast as once per second. In some special radar applications the sector to be covered is quite narrow, and it has been possible to swing the beam direction of a mechanically fixed antenna by varying the electrical beam-producing characteristics of the antenna circuits.

IMAGE-PERSISTENCE TIME

In PPI radar cathode-ray tubes, it is essential that the image not die out quickly but persist and retain its glow long enough for the complete map to be seen clearly. Therefore, the phosphors used are selected to remain fluorescent for rather long periods.

A SHIPBOARD RADAR SYSTEM

Figure 21-12 shows the components of a typical PPI shipboard radar installation. The antenna is located on a mast or other high point and the remainder of the components are in the pilot house or nearby. The unit has the following general characteristics.

Range scales	1-2-4-8-20-40 miles
Range accuracy	2% average for all ranges
Minimum range	55 yards or less
Bearing resolution	1.5°
Bearing accuracy	One degree
Pulse lengths	0.25 to 0.65 microseconds
Repetition rates/second	800 for 8-20-40 miles
	2,000 for 1-2-4 miles
Carrier frequency	3,070 megacycles
Transmitter power output	20,000 w peak
Beamwidth	Horizontal 1.9°, vertical 15°
Antenna rotation speed	Once in 10 sec
Variable range marker	0.5 to 20 miles
Receiver gain	Down to noise level
Receiver noise factor	14 db or less
Receiver bandwidth	2.5 and 8 megacycles
CRT face diameter	16 in.

An almost identical system differs in only a few respects, namely operation on 9,375 megacycles and the generation of 40 kw peak transmitter output power.

Both of these sets are designed primarily for ocean-going and lake ships. The large vertical beam width is provided to assure continuity of scanning when the ship pitches and rolls in heavy seas and to assure adequate downward coverage of nearby objects.

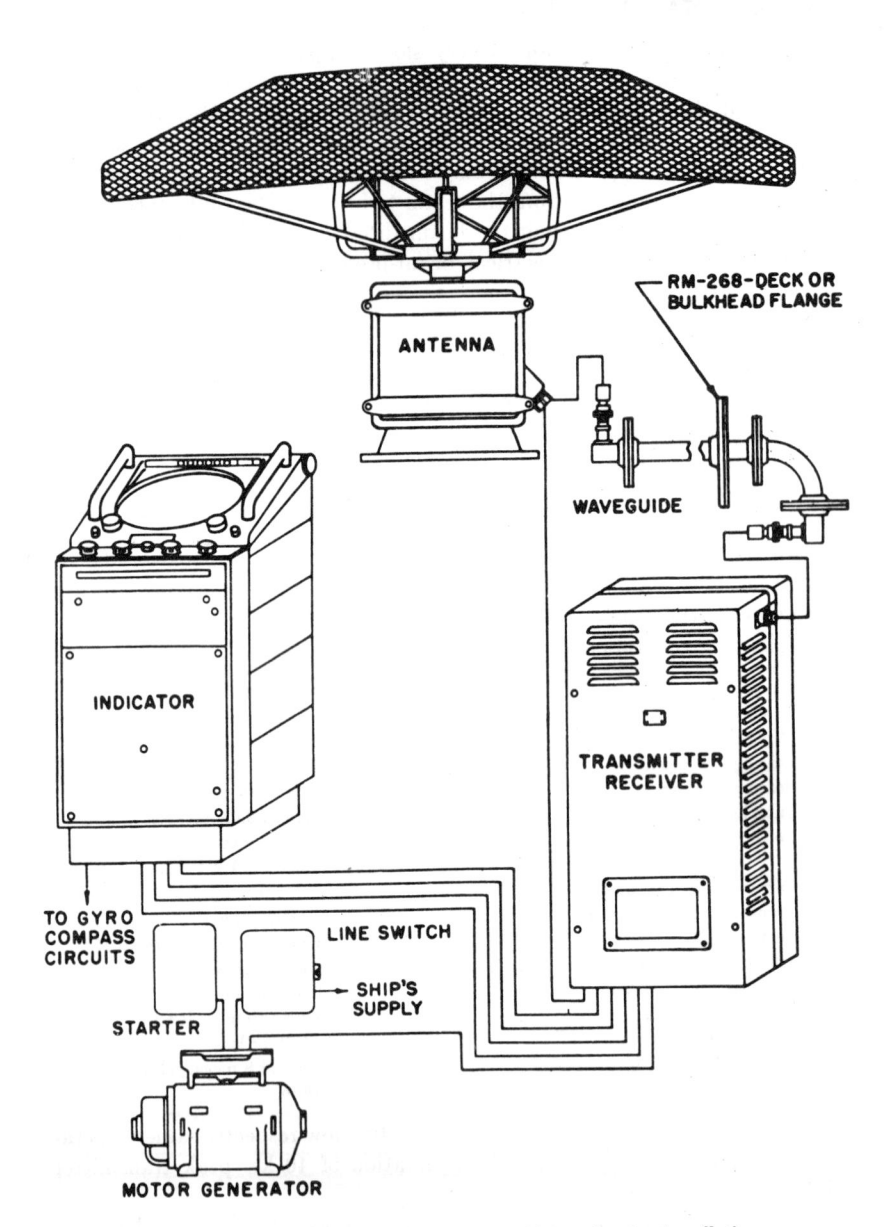

FIG. 21-12. Components of a typical PPI shipboard radar installation. (Courtesy Radiomarine Corp. of America)

Chapter 21 *Radar* 707

WEATHER RADAR

PPI radar systems are also used extensively for observing cloud formations, such as thunderstorms, hurricanes, line squalls, etc., following their courses and intensities and, for aircraft and ships, detecting non-turbulent paths through or around them. Frequencies around 5,500 megacycles have been found to be about the best for penetrating rainfall and observing targets beyond, such as clouds, land and water masses, mountains and other obstructions, and for identifying landmarks.

Aircraft radar for this purpose includes a beam antenna in the nose of the aircraft, protected by a plastic, radio-transparent cover, transmitting and receiving apparatus in the fuselage section nearby, and the cathode-ray tube display and control unit at the pilot's elbow. Figure 21-13 shows such an aircraft radar equipment. It will "see" through 15 miles of heavy rain

FIG. 21-13. Airborne radar set of type used by airlines. (Courtesy RCA)

and produce satisfactory warning of hail shafts. This unit complete weighs less than 125 lb, including the 75 kw peak power transmitter-receiver which weighs only 44 lb, the accessory unit which weighs 33 lb and the antenna which weighs only 25 lb with its motion mechanisms. Figure 21-14 shows how the antenna is mounted in the nose of an airplane so that full antenna rotation of 360° is obtainable.

The display produces a 360° continuously rotating PPI plot. In the nose of an aircraft the effective coverage approximates 270°, depending upon the type aircraft. The indicator controls consist of cursur, range-marks, beam intensity, and lights. The apparatus control unit in the cockpit, in addition to the display, includes range, antenna up or down beam tilt, contour gain and stabilization, etc. The antenna is line-of-sight stabilized to within 2 degrees when the aircraft has a roll plus pitch of as much as 20 degrees plus or minus, at a rate up to 20 degrees per second. The antenna beam may be tilted up 10° or down 15°. The range adjustments are 20, 50, and 150 miles. The transmitter delivers pulses of

Fig. 21-14. Manner in which radar antenna is mounted in nose of airliner. (Courtesy RCA)

75 kw peak power. The pulse width is 2 microseconds and the pulse repetition rate is 400 per second. The ranges are accurately calibrated within 2 per cent. The antenna scans at the rate of one rotation in 4 seconds. Antenna dishes of 22, 30, or 34 inches diameter may be used, depending upon the room available. The beam width of the small dish is 7°. The larger dishes have correspondingly greater power gain and narrower beam width.

Figure 21-15 shows a CRO tube and its controls mounted in the upper center of the instrument panel of an aircraft. Figure 21-16 shows the radar transmitter-receiver unit with the cover removed to illustrate the method of construction and assembly.

ISOECHO CONTOUR

On a cathode-ray tube display of a storm or rain area, the field intensity of the echo signals will vary depending upon the concentration of moisture

FIG. 21-15. Manner in which radar display and controls may be mounted in aircraft cockpit.

FIG. 21-16. Airborne radar transmitter-receiver showing internal construction. (Courtesy RCA)

which produced the reflection. While the intensity of the echo signal may vary over a very wide range, the normal PPI display of the cloud formations may appear almost uniform. This is due to the inability of the cathode-ray tube phosphorescent material to cover a wide range of brightness. This condition may be circumvented by the use of an isoecho contour circuit, which, when energized, alters the PPI display to show heavy rainfall as black holes, the reverse of the brightness spots which would be desirable but which are often unobtainable.

The action of the isoecho contour circuit is simple. Radar echo field intensities above a certain preselected level cause the CRO electron beam to be biased off, automatically producing, instead of a limited brightness, a black spot. Areas of greatest turbulence in thunder storm cloud formations are those with the greatest rainfall rate and it is essential to distinguish the areas of greatest turbulence. Thus, when this circuit is switched on, areas of light rainfall, producing relatively weak echo signals, are shown by relatively weak light on the CRO display and areas of heavy rainfall in the center become black. Only when a very strong echo signal intensity is received, above the preselected value, will the screen be caused to go black because lower signal intensities will not trigger the beam cutoff circuit.

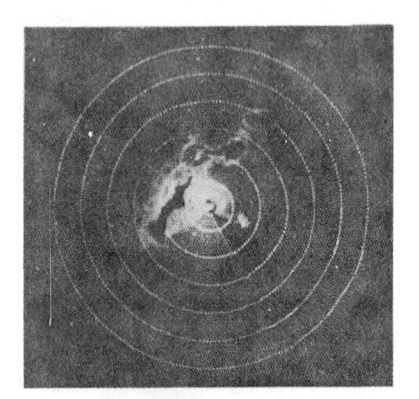

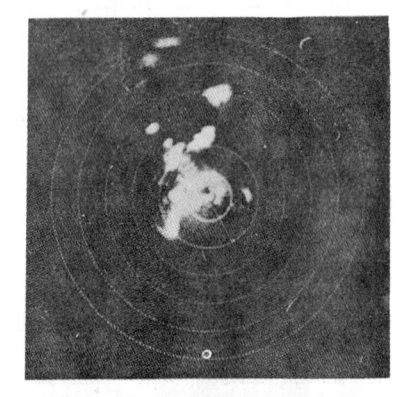

FIG. 21-17. Line squall shown on radar with 5-mile range circles. Figure at left shows isoecho display switched on to show areas of greatest turbulence. (Courtesy RCA)

Figure 21-17 shows examples of cloud formations on a weather radar screen. The aircraft is at the center of the pattern. The sector in which the radar beam is cut off during the time the antenna is facing backwards toward the airframe bulkhead may be identified by lightly defined radial lines between which no cloud pattern exists. These figures show a line squall on the 30-mile range adjustment. The right-hand figure shows clouds appearing to produce uniform echo signal intensity. Actually, the CRO brightness saturates and fails to show the area of maximum turbulence. The left-hand figure shows the same storm with the isoecho circuit in operation. The areas of greatest turbulence are now black and identifiable. The range marks shown as dotted circles consist of calibrating pips introduced during each pulse sequence and show 5-mile intervals.

PRECISION APPROACH RADAR

Radar is used for ground-controlled approaches to airport runways. It is abbreviated to GCA. Such precision approach systems provide accurate information to the ground observer concerning the distance, altitude, and direction of an aircraft desirous of landing under conditions of low visibility. The ground observer may "talk down" the pilot by informing him over a voice radio circuit of his location, the courses and distances to follow, the point to start letting down, deviations from the correct approach, and other information.

In such a system the altitude is measured on one antenna especially designed with a beam pattern resembling somewhat a flat "beavertail," and the azimuthal direction and distance on another antenna specially designed with a pattern resembling somewhat a vertical "beavertail" to obtain accurate observations in both vertical and horizontal planes. The transmitter-receiver picture tube is alternately switched back and forth between these antennas so that on one tube face both the elevation sector and the azimuth sector are observed simultaneously, by virtue of the persistence of the phosphor. Figure 21-9 whose antenna was discussed previously shows such a radar system. The antennas scan their respective planes at a rate of 2 cycles with a phase difference of 90° between the motions of the reflectors. The beam widths are 0.8° in the plane of scan. For an on-course target the deviation error for either elevation display does not exceed $\frac{1}{2}\%$ of range from touchdown plus or minus 20 ft. The power gain of the antennas is about 2,000, the pulse repetition rate is 2,000 per second, the peak transmitter power is 50 kw, the carrier frequency is 9,080 megacycles, the range is 10 miles for any type aircraft, the range marks are for one-mile spacing at 0.2% accuracy, and the operation of the system may be performed remotely up to a distance of 10,000 ft using one coaxial video cable to convey the pulse information and 15 wires for control and indication.

Equipment such as this may be mounted in a trailer for mobility. This form of radar is installed at most large airports with heavy air traffic. Auxiliary units to which the transmitter-receiver-antenna group are connected but which are not shown on the figure are the PPI indicator, the power supply system with regulated voltages, and the transmitter pulse-forming unit.

OTHER RADAR APPLICATIONS

Many other applications of radar are in use and a few are described briefly in the following paragraphs.

Radar altimeter. Radar is employed to show accurately the height of aircraft above the terrain over which they are flying, or the terrain ahead. It does not differ markedly from the radars previously described except that the power may be relatively low and the radiated power is directed toward the earth. The short range and large coefficient of reflection from the earth or water do not require high power nor high gain directive antennas.

Surveillance radar. This form of radar is employed to indicate continuously the presence and location of all aircraft within a range of about fifty miles around airports. The specifications do not differ greatly from the weather radar set described except that the peak power may be 600 or 700 kw, the antenna will be much larger for greater gain and beam sharpness, perhaps 12 ft wide and 8 ft high, etc.

Early warning radar. This form of radar achieves maximum range with great power output and very large antennas of optimum design. Its function is military.

Airborne surface detection. This form of radar has been used for detection of submarines. It usually operates at frequencies around 10,000 megacycles with a range of about fifty miles.

Night interception radar. This form of radar is employed on fighter aircraft to locate enemy aircraft at night or in poor visibility. The search plane often was guided to the general area of the enemy by long-range radar after which the shorter range intercepter radar located the target and closed in on it.

Fire control radar. Radar is used in connection with automatic electronic computers for automatically locating, tracking, and predicting the positions of enemy aircraft and ships, and aiming and firing weapons to destroy them. In some of the World War-II naval actions, radar such as this enabled ship targets to be located, ranged and sunk, the entire action being visible on the radar display with neither ship ever having sighted the other visually.

RADAR BEACONS

A radar beacon is used to provide information about distance, direction, and bearing to its known fixed location. A ship or aircraft directs radar pulses toward a radar beacon. The beacon is equipped with a receiver which in turn triggers the beacon transmitter into responding with a codified sequence of pulses unique to itself which identify the beacon when they are observed on the receiver display. The responsive pulses from each

Chapter 21 *Radio Relay Systems and Radio Aids to Navigation* 713

station on a PPI display are in a close group. The beacon may operate on a frequency different than that of the questioning stations, in which case a separate antenna and receiver may be required for receiving the responsive pulses. In co-channel operation a PPI display may show simultaneously several radar beacon responsive pulse groups, each in its correct position as though they were conventional echoes.

Radar beacons may be portable and have been used by paratroops and other combat units to reveal their locations. In military operations, aircraft and ships were equipped with "identify friend or foe," abbreviated IFF, apparatus. When an interrogating signal was directed toward a distant friendly aircraft so equipped, its receiver triggered a small transmitter which responded with a return signal of identification. Lack of a proper responsive signal identified an enemy.

TUBES USED IN RADAR

Tubes used to generate large amounts of power for radar transmitters differ from conventional triodes, tetrodes, or pentodes because the frequencies employed are far above the range at which conventional tubes can function. Magnetrons and klystrons are used. These are described elsewhere in this book. Transmission of carrier frequency power is accomplished from unit-to-unit in the transmitter and to the antenna in waveguides, which are hollow rectangular metal pipes.

RADIO RELAY SYSTEMS

Radio relay systems are in extensive and increasing use for conveying information between distant terminals. The carrier frequencies most adaptable to this service for long distances are those between about 1,500 and 11,000 megacycles, but for some applications at shorter distances frequencies between 100 and 200 megacycles are used. These relays conventionally consist of unattended stations separated by 20 to 50 miles along the relay path. At the transmitting terminal the signal is transmitted by a very sharp radio beam to the first relay point where it is received on an antenna with high power gain, amplified perhaps 500,000 times, and then retransmitted to the next relay station over another sharp beam. The transcontinental radio relay network of the Bell System employs nearly 200,000 channel-miles of radio relays. Between New York and Los Angeles there are used simultaneously in cascade 107 geographically separated stations located on mountain tops, towers, and tall buildings. At times, transmission of television over the Bell relay system involves the use of over 300 stations. Scores of other radio relay systems are in use for television relaying, telephone communication, aviation, communication and telemetering along power lines and pipe lines, for telegraph and facsimile transmission by companies such as the Western Union Company, for police communications along vehicular turnpikes, thruways, etc.

714   Radio Relay Systems and Radio Aids to Navigation   Chapter 21

Portable radio relay systems are used extensively for military operations and by private companies. The systems range from the simple two-station type used for television program remote pickups to the vast Bell System chain.

RADIO RELAY ANTENNAS

In the section on radar the principles underlying the design of high-power-gain "microwave" antennas were explained and the Bell System horn-type relay station antenna was used as an illustration. In radio relaying the purpose of the transmitting antenna is to deliver signal to a single point, the distant receiving antenna. Therefore, "pencil" type beams are most effective. Radar-type antennas producing "beavertail" shaped beams are not advantageous. The parabolic reflector having a beam pattern and power gain similar to that shown in Fig. 21-4 is used. The antenna illustrated in Fig. 21-6 is a form of parabolic reflector.

Antennas which are advantageous for transmitting are equally advantageous for receiving, and in computing the progressive power levels of a radio relay system the receiving antenna power gain is equally as important as the transmitting antenna gain in the equations. Normally the transmitting and receiving antennas are identical in design and construction and are mounted in close proximity facing in opposite directions. Since relay systems usually are made to transmit a wide frequency pass band, the antennas are required to be quite uniform in gain over a wide range of frequencies. But since the carrier frequencies used in radio relaying are in the thousands of megacycles, the pass band is but a small fraction of the carrier

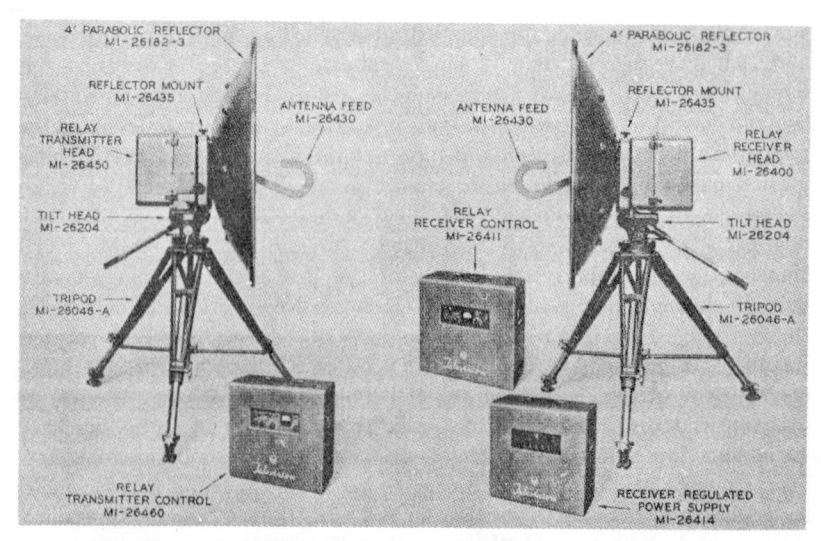

FIG. 21-18. Seven thousand megacycle television relay equipment for mobile or fixed operation. Range about 25 miles. (Courtesy RCA)

Chapter 21 *Radio Relay Systems and Radio Aids to Navigation* 715

frequency and the wide-band transmission may be achieved without unusual difficulty.

EXAMPLE OF A SIMPLE TELEVISION RADIO RELAY SYSTEM

Mobile radio relays are employed extensively in television pickups of sporting events, etc., which take place outside of the studio plants. At a scene of action cameras, amplifiers, and other auxiliaries are brought in by truck or boat and produce the composite television signals which are to be distributed to the networks and the transmitting stations. These signals

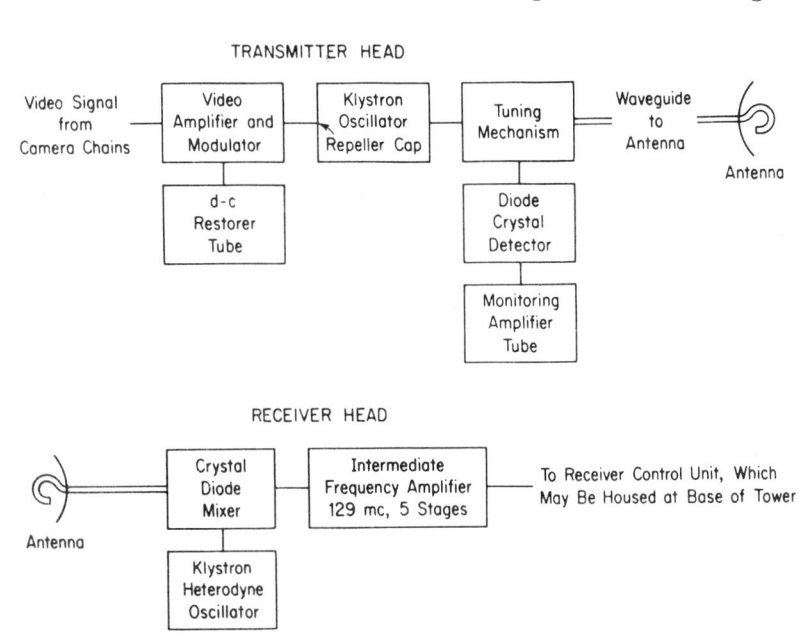

TRANSMITTER HEAD

Video Signal from Camera Chains → Video Amplifier and Modulator → Klystron Oscillator Repeller Cap → Tuning Mechanism → Waveguide to Antenna → Antenna

d-c Restorer Tube

Diode Crystal Detector

Monitoring Amplifier Tube

RECEIVER HEAD

Antenna → Crystal Diode Mixer → Intermediate Frequency Amplifier 129 mc, 5 Stages → To Receiver Control Unit, Which May Be Housed at Base of Tower

Klystron Heterodyne Oscillator

FIG. 21-19. Seven thousand megacyle mobile or fixed television radio relay system.

occupy a frequency band of about 5 megacycles and cannot be transmitted over conventional telephone lines, even when available, as can the sound signals. Figure 21-18 shows a specially developed radio relay system. The carrier frequency is about 7,000 megacycles, the carrier power is one watt, and the antenna power gain per unit is about 5,000 for a 4-ft reflector, as shown, or 11,500 for a 6-ft reflector. Frequency modulation of the carrier is employed with a deviation of 5 to 6 megacycles each side of the carrier center frequency. The transmitter input terminals require 2 v peak-to-peak for full modulation, and the receiver delivers an equal voltage without appreciable distortion. The superheterodyne receiver has an i-f of 129 megacycles. The range of this relay is about 25 miles with the 4-ft reflectors when line-of-sight conditions exist between the transmitting and receiving antennas. The transmitter employs a klystron oscillator for producing carrier power and, in combination with the power gain of the antenna, an effective radiated carrier power of 5,000 to 11,500 w is achieved in the beam. The over-all frequency response is uniform within 5% between 60 cycles and 7 megacycles. Monitoring voltage is obtained from a crystal diode detector coupled to the resonant cavity. This relay system may also be used for fixed service. The transmitter and receiver heads are weatherproof. The i-f from the receiver head may be fed from its elevated position to the control unit in enclosed space at the base of a tower. The receiver employs a klystron as a heterodyne oscillator. Figure 21-19 shows a block diagram of the transmitter and receiver heads which constitute the items of immediate interest.

MULTIPLEXING

A radio relay may be viewed as a super-highway for conveying all forms of messages or signal information be it speech, television, pictures, teleprinter, music, etc. It has the advantage that it can handle a very wide band of frequencies and many channels of message traffic simultaneously. Private systems, for example, may be designed to meet a specified need for 24 voice channels simultaneously. How may all of these signals be transmitted simultaneously on one carrier without becoming hopelessly and

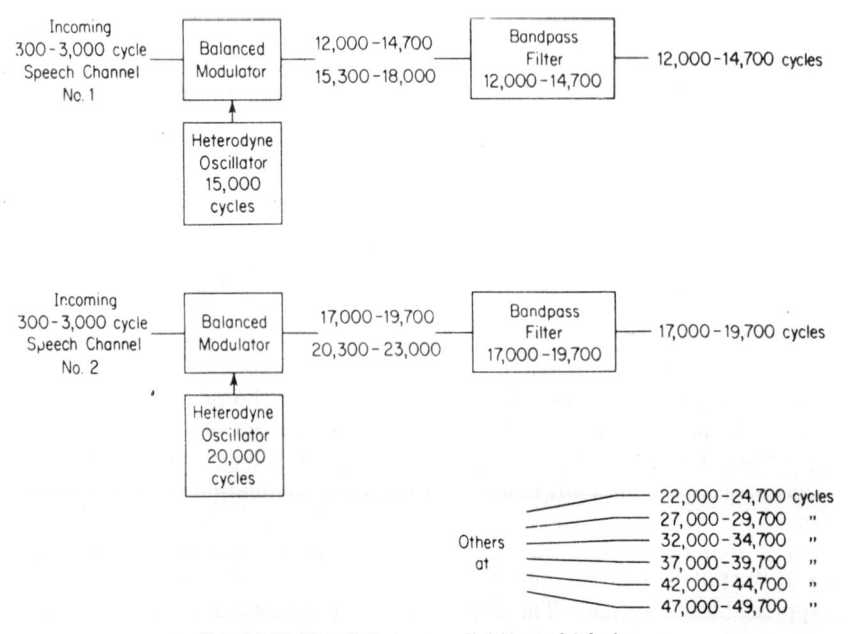

Fig. 21-20. Simple frequency division multiplexing.

Chapter 21 Radio Relay Systems and Radio Aids to Navigation

irretrievably scrambled together? A commercial radio relay system for doing so will be described. It is the type shown in Chapter 10 for carrying message traffic between New York City and Riverhead and Rocky Point.

The carrier frequency is approximately 1,800 megacycles, frequency modulation of the carrier is used with a peak deviation of plus and minus 1.5 megacycles, the transmitter carrier power is 3 w, the receiver bandwidth is 6 megacycles, the modulation frequency range is from 3 kc to 135 kc, and, of most immediate interest, the system handles simultaneously 24 individual speech channels, each of 300 to 3,000 cycles pass band. These 24 speech channels are handled simultaneously without interference or cross talk within the overall radio relay system pass band of 3 to 135 kc. It is accomplished by *multiplexing* and with the aid of the heterodyne frequency changing principle, the importance of which was stressed in Chapter 10 and elsewhere in this book.

The multiplexing is accomplished in a single sideband, suppressed carrier, frequency division multiplexing system designed primarily for multi-channel radio relay use on one radio relay carrier.

Multiplexing is accomplished by heterodyning each individual voice frequency band upward to a new frequency band different from any other voice channels also heterodyned up. In the multiplexing arrangement a carrier is provided at 15 kc and at each higher 5 kc interval. Incoming speech channel number 1 is used to modulate the 15-kc carrier in a balanced modulator which produces upper and lower sidebands but suppresses the carrier as shown at the upper section of Fig. 21-20. These sidebands are applied to a band-pass filter which passes the lower sidebands but not the upper sidebands. We now have speech in speech channel number 1 converted to the frequency band 12,000 to 14,700 cycles.

Speech channel number 2 is used to modulate the next higher carrier of 20 kc in another balanced modulator, the upper sideband is suppressed in a 17,000-19,700 cycle band-pass filter, and we have left speech in speech channel number 2 converted to the frequency band 17,000-19,700 cycles. The third speech channel modulates a 25-kc carrier and is similarly filtered to reappear as 22,000 to 24,700 cycles. Speech channels 1 through 8 inclusive are heterodyned and "stacked up" in frequency in this manner between 10 and 50 kc as Group 1, as shown in Fig. 21-20. Two additional groups of 8 each are "stacked up" at higher frequencies to make a total of 24, occupying adjoining frequency segments of the range of 12,000 to 135,000 cycles.

GROUPING

Separation into groups of 8 channels each permits flexibility. Each group may be viewed as a block which may be stacked up in a desired band of frequencies, as described. In the system under discussion the three 8-channel groups may be stacked up separately in the same frequency band as shown in Fig. 21-21. Then it is possible in one operation to heterodyne a whole group to a different frequency band. Group 1 may be left between 10 and 50 kc. Group 2 may then, as a whole unit, be heterodyned up to the block 55 to 95 kc, and Group 3 to 95 to 135 kc. The process may be reversed at the receiving end. Grouping substantially reduces the number

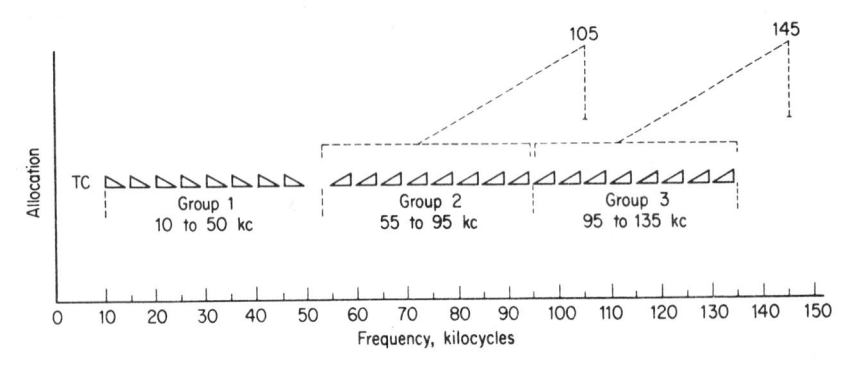

FIG. 21-21. Frequency allocation chart. Typical group arrangement shown provides three 8 channel groups between 10 and 135 kilocycles.

and variety of apparatus units, and less stringent filtering requirements are necessary. There is the added advantage that a whole group of 8 channels *en bloc* may be routed through the system individually, tied in with other systems, taken off the relay chain at intermediate relay stations, etc.

These channels are all combined to modulate the 1,800-megacycle relay carrier which goes to the receiving terminal through the intermediate relay stations. There the reverse process takes place. The 1,800-megacycle signal is demodulated to produce the band 12,000-135,000 kc, the individual groups and channels pass through band-pass filters to separate them from all other channels, the missing carriers are added, the resulting signals are detected, and the original speech channels are restored.

SUBDIVISIONS OF MESSAGE FREQUENCY CHANNELS

Each of the 3-kc speech channels may in turn be divided and contain multiplexed supervisory control or telemetering functions numbering up to 16 each. The Riverhead-Rocky Point-New York relay has many of its 3-kc channels divided into subchannels for teleprinter and code circuits.

RELAY FREQUENCY STAGGERING

When an amplifying system feeds power from its output to its input, self-oscillation may take place. In radio relay stations some power from the transmitting antenna may reach the receiving antenna.

In addition, a signal from one relay station may overreach the following station and produce interfering signals to the third station. To avoid

Chapter 21   *Radio Relay Systems and Radio Aids to Navigation*   719

difficulties from these sources it is standard practice to change the carrier and sideband frequencies a few per cent by heterodyne action as they pass through each relay. In the multiplexed system last described, the frequencies are shifted 40 megacycles at each relay. The shift is usually made sufficient to stagger the incoming and outgoing pass bands. When two-way systems are used, as is usually the case, four frequencies may be employed at each station. Frequency staggering has made it possible to carry several channels through a common antenna at each station and in duplex operation to use a common antenna for transmitting and receiving.

RELAY DESIGN

It is not required that demodulation to signal frequencies and remodulation take place in a radio relay station and it is not done in normal practice. It introduces areas where distortion may be compounded, is expensive, and complicates the installation. By heterodyne action the carrier frequency and sidebands are reduced to the area of perhaps 30 to 70 megacycles, amplified more easily at these frequencies, heterodyned back up to a carrier frequency slightly different than that originally received, and further amplified.

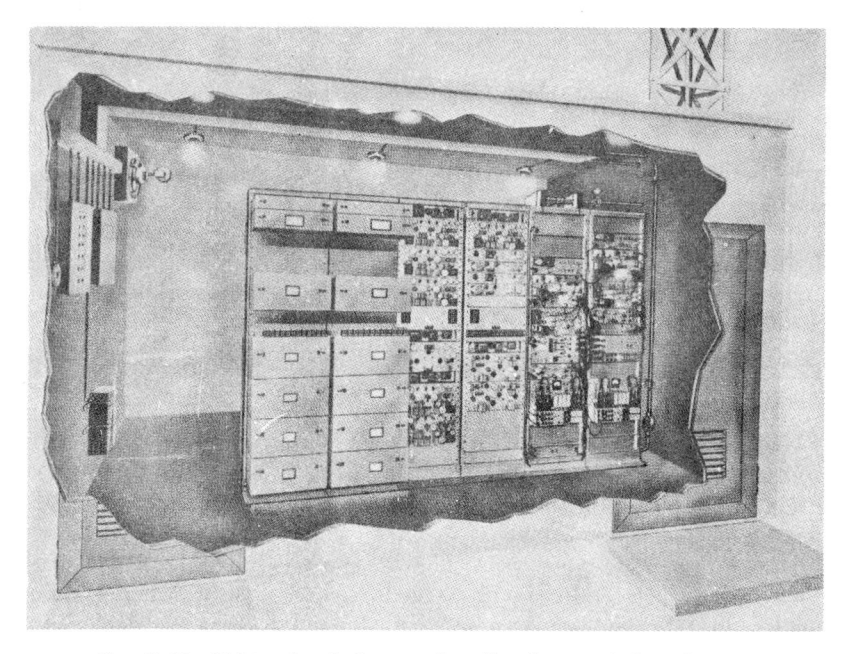

FIG. 21-22. Eighteen-hundred megacyle radio relay terminal equipment including multiplexing equipment at left and transmitting-receiving equipment at right. (Courtesy RCA)

Figure 21-22 shows a photograph of a relay terminal using equipment such as was last described. The racks at the left contain multiplexing apparatus and at the right transmitting and receiving apparatus. For simple intermediate relay stations only the two racks at the right are required since message frequencies as such are not processed or used. Figure 21-23 shows a complete mobile radio relay terminal station which may be

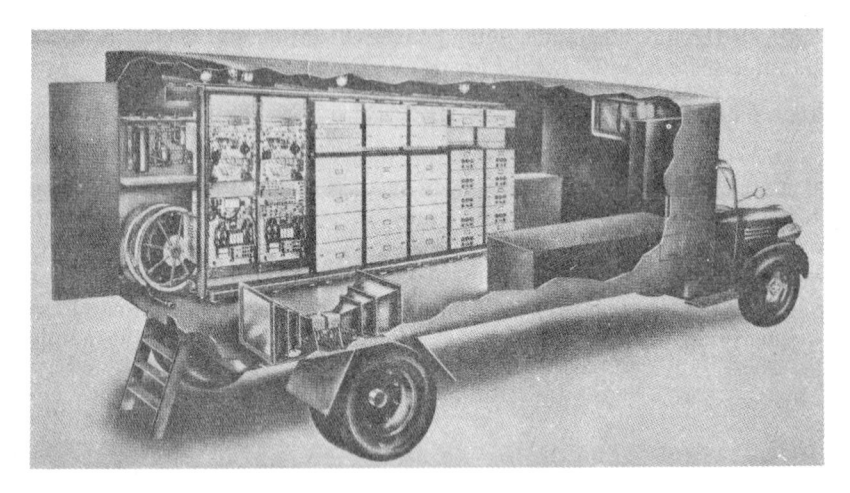

FIG. 21-23. Eighteen-hundred megacycle molbie radio relay system. (Courtesy RCA)

used to tie in to a regular wire line system or into a fixed microwave system. The two racks at the right provide for 16 teletype channels, the 3 in the center for 24 voice channels, and the two at the right for regular and standby duplicate transmitter-receiver assemblies. Two horn antennas, flexible transmission lines, and many other auxiliaries including a workbench make a complete plant.

BELL SYSTEM RADIO RELAY SYSTEM

The radio relay system of the Bell System is capable of providing six television channels in each direction, each as much as 8 megacycles in width. The stations operate on carrier frequencies in the 3,700-4,200 megacycle range utilizing frequency modulation. The available 500-megacycle band is divided into 12 one-way channels separated by 40 megacycles between mid-frequencies. These are employed to create six two-way channels in which the frequency-staggering pattern shown in Fig. 21-24 is used. This relay employs horn antennas, the most recent model of which has been described. They are connected to the radio equipment by rectangular brass waveguides having dimensions of about $1\frac{1}{4}$ inches by $2\frac{1}{2}$ inches and losses of about 15 per cent in power per hundred feet of length. In adjacent repeater stations the same frequencies are used but with the

Chapter 21 *Radio Relay Systems and Radio Aids to Navigation* 721

directions reversed to conserve spectrum space and minimize cross talk.

The terminal equipment for each message channel permits it to be used for a single television signal or it may be used for the simultaneous transmission of 720 one-way telephone conversations multiplexed, or "stacked up," in the manner described in the section on multiplexing. The transmitting terminal utilizes an initial 70-megacycle carrier frequency-modulated by the television signal or the multiplexed voice signals. It is hetero-

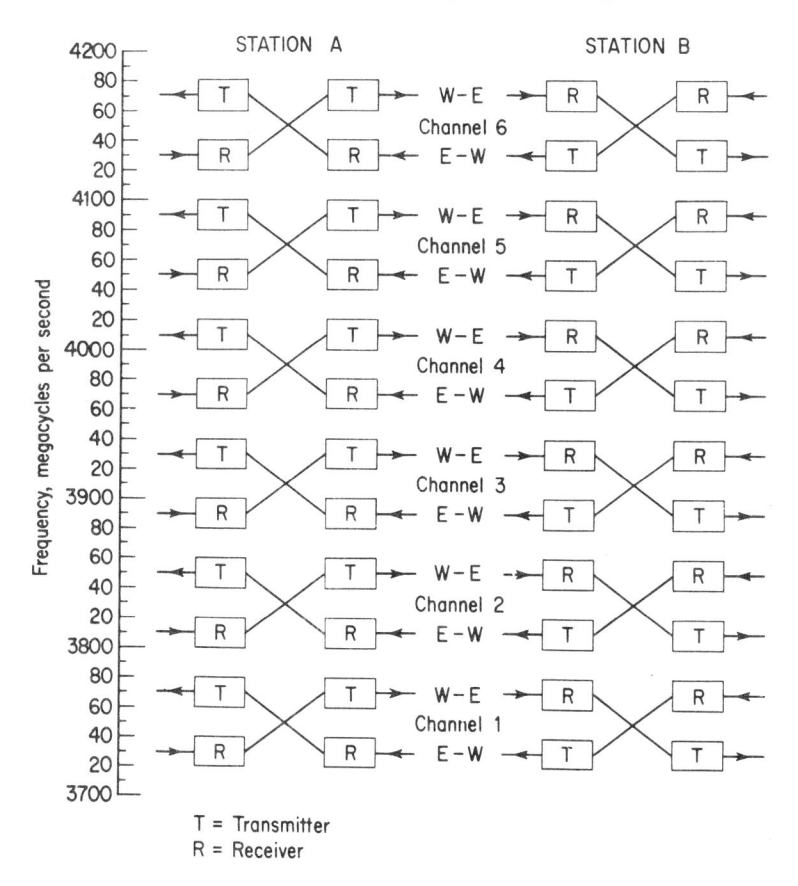

Fig. 21-24. Manner in which Bell System employs frequency staggering in radio relay stations.

dyned to the 4,000-megacycle carrier and amplified to a power level of about one-half watt. This modulated radio carrier is then fed to a channel branching filter. It is then combined with the outputs of other carrier channels and all are fed to a single transmitting antenna.

At each repeater the individual incoming carrier channels are segregated by means of filters, heterodyned down to an i-f of 70 megacycles, amplified,

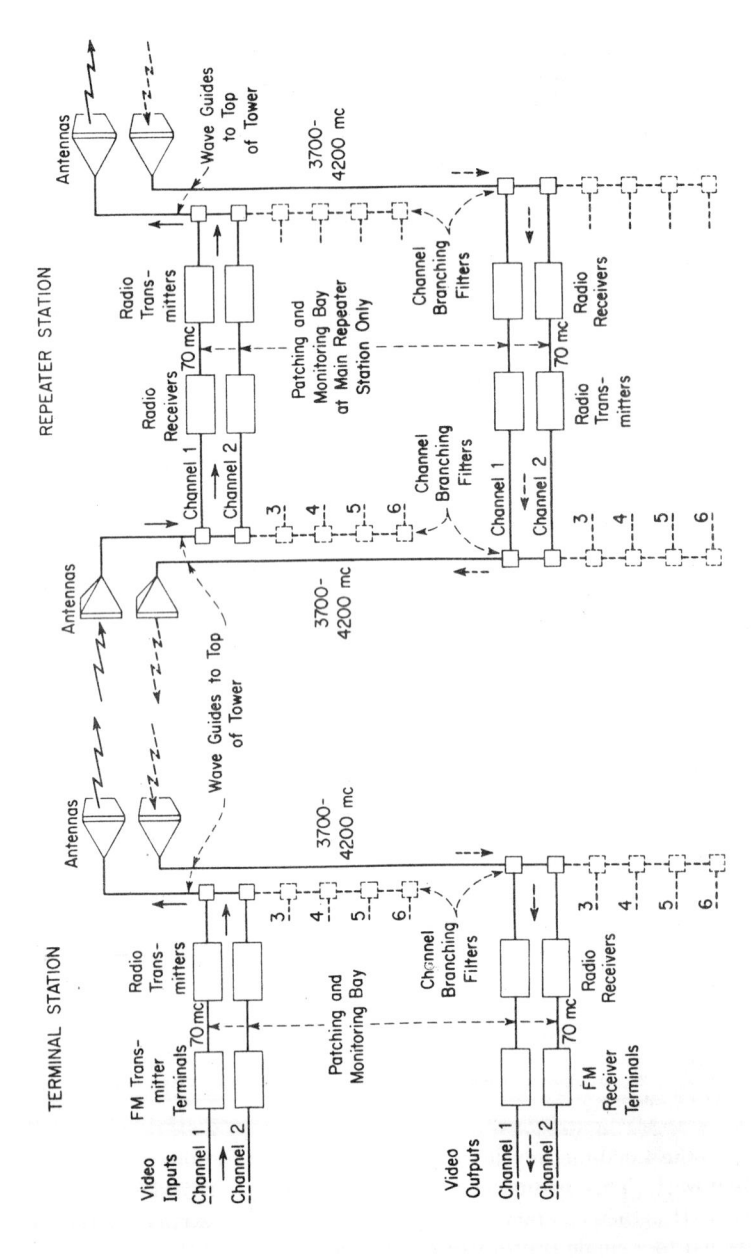

Fig. 21-25. Schematic block diagram of a terminal and repeater station in Bell System relay.

Chapter 21 *Radio Relay Systems and Radio Aids to Navigation*

heterodyned up to the new outgoing carrier frequency, and amplified. The schematic block diagram of a terminal and repeater station is shown in Fig. 21-25.

The normal radio path loss over an average 28-mile hop is about 138 db to which there is added waveguide and filter losses of about 5 db. This total 142-db loss is made up partly by the power gain of the receiving and transmitting antennas totalling about 78 db, leaving 65 db to be made up by repeater amplification. Automatic gain control with a range of about 30 db maintains constant power output during nominal fading and tube aging. Frequency modulation "noise quieting" is maintained up to a limit of about 40 db fading.

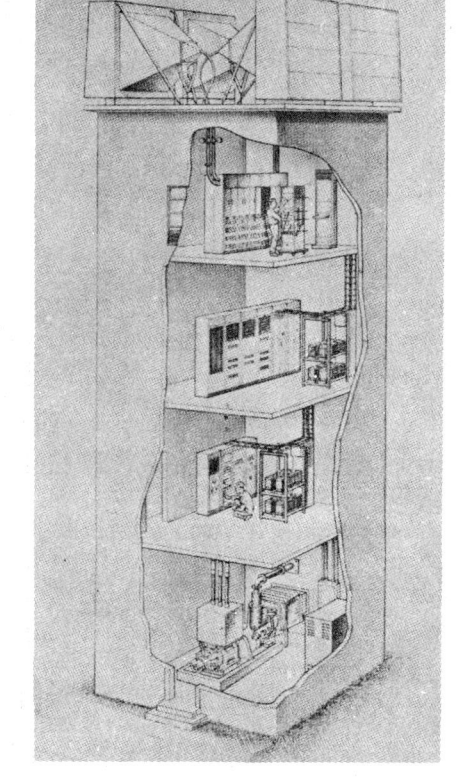

FIG. 21-26. Typical Bell System radio relay station of which there are many hundreds in the United States: 3700-4200 megacycles. (Courtesy Bell Telephone Laboratories)

Figure 21-26 shows a typical Bell System radio relay station. An auxiliary emergency power generator, automatically turned on when local public utility power fails, is shown on the lower floor. At the top floor the radio receiving-transmitting apparatus is shown. The 4 horn antennas of early design are shown on the roof. Below the radio room there is shown the FM terminal apparatus. On the second floor low frequency control apparatus is shown. Figure 21-27 is a photograph of the Cisco-Butte, California, relay station on a remote high Sierra-Nevada mountain location.

The relay system described includes elaborate automatic remote indicating and control apparatus. Forty-two possible alarm indications may actuate a remote alarm center. From that point inquiry signals may

FIG. 21-27. Cisco-Butte, Cal., station of the transcontinental radio-relay system. This station is located in a valley high in the Sierra Nevada Mountains. (Courtesy American Telephone & Telegraph Co.)

be sent to any relay and replies are automatically returned which identify the location and nature of the abnormality. An interesting exercise for the student is the computation of the total power gain for a chain of relays, such as 250 which are used between Los Angeles and Miami. Assume an average loss per hop of 128 db.

RADIO RELAY NOISE

On a long radio relay it is inevitable that a minute amount of noise will be added at each relay station. To maintain over-all circuit noise within tolerable levels each station in the chain must be designed and operated to minimize its contribution. At the Bell System relay stations the signal-to-noise ratio must be maintained at about 66 db, which is equivalent to 2,000 in voltage or 4,000,000 in power. The system was engineered for frequency modulation to make this possible.

The noise power added at a frequency-modulated repeater station may be computed as follows:

$$P_n = \frac{NF \times K \times T \times \Delta f}{3 \ (F_s/F_m)^2} = \text{watts},$$

where NF = noise factor of receiver

$KT\Delta f = 0.8 \times 10^{-14}$ = thermal agitation noise in watts per megacycle

Δf = effective bandwidth in megacycles

F_s = maximum frequency deviation

F_m = maximum modulation frequency.

For free space propagation the power received at the receiver input is reduced from the transmitter power by

$$P_r = \left(\frac{4\pi D}{G\lambda}\right)^2,$$

where D = distance from transmitter in feet

λ = wavelength in feet

G = power gain of antenna over isotropic radiator

P_r = power gain required at receiver to transmit the original power.

PULSE COMMUNICATION

Frequency division multiplexing was described and explained in connection with radio relays. Another form of multiplexing makes use of *time division*. In time division, multiplexing pulse code transmission may be employed to advantage.

Intelligence may be transmitted by making changes in an otherwise unvarying succession of short, rectangular pulses identical to each other in form and spacing. Assume that speech is to be transmitted. Figure 21-28(a) shows an audio envelope. Constantly spaced pulses are caused to change amplitude in accordance with the audio envelope voltage. In other words, the envelope voltage is sampled frequently and the samples are transmitted as pulses of varying amplitude. This is called pulse amplitude modulation, and is commonly abbreviated to PAM.

Figure 21-28(b) shows how pulse width may be employed. The pulse amplitude remains constant but the pulse width varies with audio voltage. This is called pulse width modulation, or PWM.

Figure 21-28(c) shows how the pulse repetition rate may vary. This is called pulse frequency modulation, or PFM.

Another form of pulse transmission causes the pulse to change position from a normal axis, called pulse position modulation, or PPM. Another causes a change in pulse shape, called pulse shape modulation, or PSM.

Another causes the number of pulses to change and is called pulse number modulation or PNM, or pulse code modulation, PCM. Another causes a change in pulse intervals and is called pulse interval modulation, or PIM.

The object of a transmission system is to reproduce at the output any function of time which appears at the input. It has been established that a frequency f can be reproduced if the sampling is accomplished at a fre-

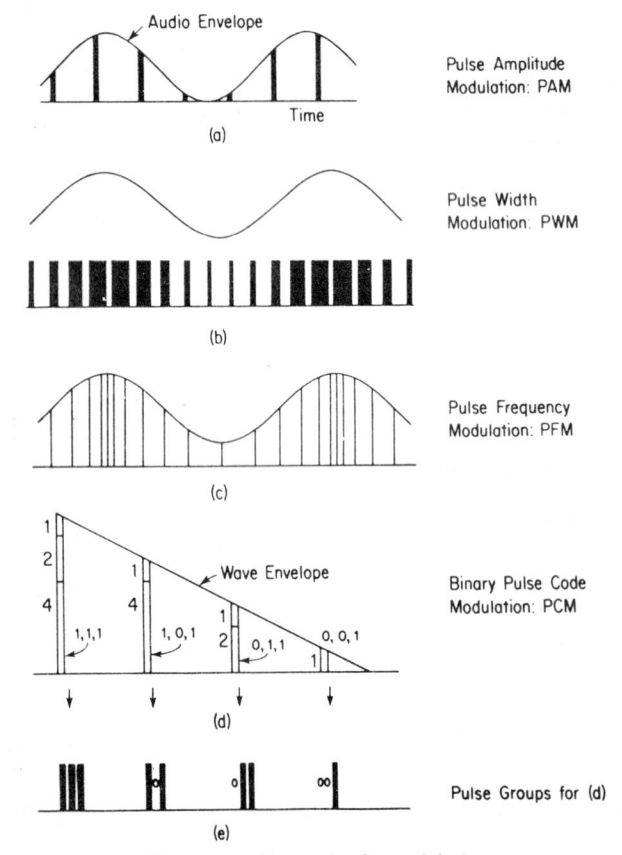

FIG. 21-28. Forms of pulse modulation.

quency of $2f$. To reconstruct the sampled signal, each sample is caused to generate a proportional impulse, and the regularly spaced series of impulses is passed through a low pass filter of cutoff frequency f. The pulses thus are smoothed out to reproduce the original signal. When the sampling frequency is at least twice the highest frequency present in the message signal, the distortion which may be created by this sampling falls outside of the band 0-f and is removed by the low pass 0-f filter. Were it necessary to transmit all amplitudes of a sampled wave it would be necessary to sample at such a high rate that the advantages of pulse code modulation would be diminished or lost.

PULSE CODE MODULATION

In considering how it is possible to recreate a signal using a sampling rate as low as $2f$, it should be borne in mind that a complex signal, while having an irregular-shaped envelope, is composed of sine waves of different frequencies. Frequency f is the highest frequency sine wave present. Sampling of any lower frequency component naturally would be more frequent per cycle.

CODE GROUPING

A sample may consist of a single pulse varying in amplitude, width, or position. It is possible also to identify a single value of amplitude by transmitting a varying group of pulses which remain in fixed positions and each of which represents a different amplitude value. For example, three pulses may be used to represent eight different amplitude values, including zero, as shown below. A pulse may be transmitted (1) or omitted (0) depending upon the need for it in the code.

Amplitudes Represented	Code	Interpretation
0	000	$0+0+0 = 0$
1	001	$0+0+1 = 1$
2	010	$0+2+0 = 2$
3	011	$0+2+1 = 3$
4	100	$4+0+0 = 4$
5	101	$4+0+1 = 5$
6	110	$4+2+0 = 6$
7	111	$4+2+1 = 7$

In the digits in the code column, the number at the right represents an amplitude of 1. The center number represents an amplitude of 2. The number at the left represents an amplitude of 4. When each column contains the figure 1, the amplitude represented is the sum of the three, or $4 + 2 + 1$, making 7 amplitude units.

For 011 the amplitude is the sum of $0 + 2 + 1$, or 3. For 101 it is the sum of 4, 0, and 1, or 5. For convenience the amplitudes represented by the digits are totaled under "interpretation."

To decode such a group there must be generated a pulse which is the linear sum of all pulses transmitted, each multiplied by its place value (1, 2, or 4). Figure 21-28(d) shows how a triangular wave would be sampled using this binary counting. At the decoder the pulses may be caused to charge resistor-capacitor combinations which recreate the original amplitudes. These are then sampled. The time constant is such that the charge is dissipated between sampling times. Figure 21-28(e) shows how succeeding pulse groups would appear on a time scale.

PULSE POWER

A pulse-modulated carrier is transmitted only when sampling pulses are present. The power is zero at other times. The intervals between pulses are relatively long. Maximum circuit reliability and efficiency may be achieved when each pulse is transmitted at maximum amplitude because weaker pulses are more vulnerable to noise and interference. Therefore, there is a preference for width, coding, or spacing control of pulses rather than the use of variable amplitude.

Pulse widths are dimensioned in microseconds duration. The duty cycle of a pulse transmitter is the ratio of the average power during a single repetition cycle to the average power of a pulse. Because the duty cycle is low compared with continuous carrier transmission, the heating produced is lower and the tubes normally can be worked harder. The pulse widths utilized may be approximately 0.5 microsecond in duration. In pulse-position modulation a pulse is caused to move horizontally from its normal axis over an interval of perhaps 5 microseconds on either side for maximum modulation. With allowance made for additional space between adjacent message channels, the duty cycle for PPM may be less than 10%.

TIME DIVISION MULTIPLEXING

In a pulse-type transmission system the input signal passes through a band limiting filter to exclude frequencies higher than those required. The signal is then sampled at a rate at least twice as frequent as the highest frequency to be transmitted. These samples are then encoded. This type of transmission is most useful in multiplexing arrangements where many message channels are transmitted simultaneously on one carrier, one transmitter, and one receiver. And synchronous time division multiplexing is normally most advantageous. How does it function?

Consider an 8-channel system utilizing pulse-position modulation. There are available 8 separate channels with sampling pulses for each, representing 8 speech circuits of 0-3,000 cycles each. Each speech signal is sampled 10,000 times each second. The pulses of 0.4-microsecond duration are caused during maximum modulation to shift horizontally in either direction over a distance representing 5 microseconds time, as shown in Fig. 21-29(b).

In time division multiplexing the eight signal samples are transmitted one after another in succession. When signal number one is to be transmitted, a gate is opened at the transmitter, the sample modulates the transmitter, the signal goes to the receiver where a corresponding signal gate has opened, and the sample goes to its demodulating system. After the sample of signal number one has been transmitted, the transmitting and receiving gates close, those for the number 2 signal sample open, and the sample then goes through the system to its separate demodulator. This sequence is repeated for each channel sample after which the original condition is restored and the 8-channel sequence is repeated. It occurs 10,000 times per second.

Chapter 21 **Pulse Communication** 729

For this system to function, it is apparent that the eight gates must open and close synchronously at the correct times, and the samples must go through while they are open and maintain their average axis positions. This requires a synchronizing pulse to be present to control the sequence synchronously throughout the system. The synchronizing pulse is trans-

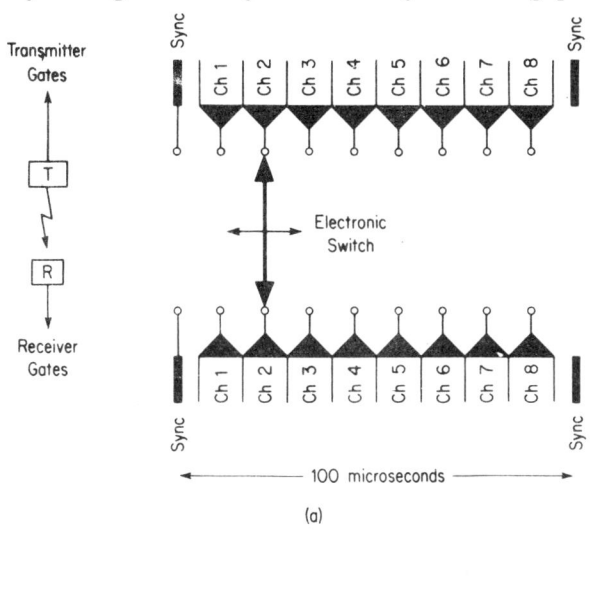

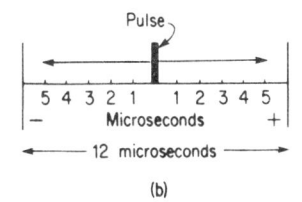

FIG. 21-29. (a) Time interval for one transmitting sequence in 8-channel PPM multiplex system; (b) time intervals for one channel.

mitted prior to the beginning of each 8-channel sequence to lock the system. These pulses are of 2 microseconds duration, and recur 10,000 times per second. The interval is obviously 100 microseconds.

Figure 21-29(a) shows the time block for each sequence. The gating system may be viewed as an electronic switch which successively starts at the left of Fig. 21-29(a) and connects corresponding channel numbers for 12 microseconds each before going on to the next position. The significant time intervals for an individual channel are shown in Fig. 21-29(b). The pulse changes horizontal position during modulation. The 10,000 sequences per second for which the system was designed provide for 10,000 samples per second to be transmitted for each of the eight channels.

PERFORMANCE

In pulse code modulated systems more bandwidth and less power is required than for conventional AM systems. There are substantial advantages in signal-to-noise ratio compared with other wide-band systems such as FM. The binary on-off PCM described is capable of functioning under conditions of noise of such large amplitude that the pulses may be almost indistinguishable. This form of transmission is particularly attractive for multiplex circuits of high quality and reliability.

Pulse transmission has the advantage over some other forms that, in transmission over long relays in which the pulse shapes are degraded, they may be used in regenerative repeaters to act as triggers to create new properly shaped pulses. Signal-to-noise ratios which have been degraded may thus be *improved* at a relay point, an advantage seldom encountered. The improved protection against noise achievable in well-designed pulse communications systems is the most important reason for their use.

LORAN NAVIGATIONAL AID

The *loran* system is a modern method of determining positions accurately and quickly by radio. The word *loran* is derived from the initial letters of the words *long range navigation*. The principle is very simple. It is based upon the difference in travel time (in millionths of a second) to the observer of pulsed radio signals from two transmitting stations spaced several hundred miles apart. Since radio waves travel at a constant speed of 983.24 ft per microsecond, or require 6.1838 microseconds to travel one nautical mile, a direct and unvarying relationship exists between travel time and distance. In essence, measurement of radio wave travel time *is* measurement of distance.

Consider Fig. 21-30. Q, P, and R are radio staions. If P and Q were each to transmit a pulse at exactly the same instant the pulses would travel toward a distant point at the same speed. But if one of the stations is closer to that distant point, its pulse will not have so far to go and will arrive first, followed by the pulse from the more distant station. If the difference in arrival time is measured, the true distance is still not known, but the difference in path length is accurately established. If one pulse were 6.1838 microseconds ahead of the other one, the difference in path length would be one nautical mile.

A moment's reflection will show that a difference in path length of one nautical mile could exist at very many distant points. But it will also show that these points would fall on a smooth curve, actually a hyperbola, which is known as a loran line of position. Hence, when a navigator has obtained a time difference measurement from a pair of loran transmitters, he will know that his absolute position will lie on some point on a particular loran line of position. Such a line is shown in Fig. 21-30 as 2L5-3000. Now, how may the absolute distance and location be learned? If another pair of loran transmitters located at different points were to give the navigator another loran line of position, the lines would intersect and the point of intersection would have to be the navigator's position. A second loran line of position may be established if a third station were added, at R for example, and P were common to Q and R. This is satisfactory so long as PQ paired pulses may be clearly distinguished from PR paired pulses. Now the second line of position from PR is identified as 2L6-2500. A position, instead of only a direction, has now been established as X.

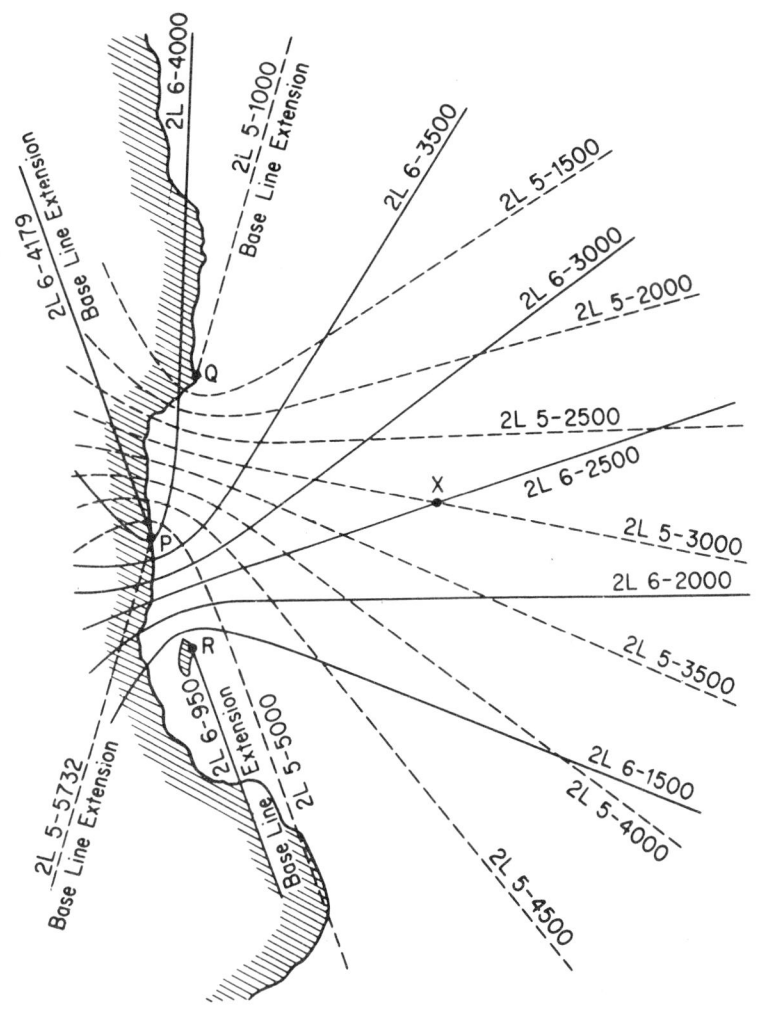

Fig. 21-30. Simplified loran chart. Two lines-of-position from two pairs of loran stations provide a loran fix. P is a double-pulsed master station common to both pairs. Q and R are slaves. Lines with prefix 2L5 are formed by P-Q. Lines with prefix 2L6 are formed by P-R. The meaning of the prefixes is explained in the text.

The loran system establishes accurate positions at sea to distances of about 900 miles by day and 1,500 miles at night.

IDENTIFICATION OF STATIONS AND LINES OF POSITION

The location of loran transmitting stations is shown in loran tables and charts. The available stations in any given area may be easily located on charts or index chartlets in the tables. Refer to Fig. 21-30 which is a simplified loran chart. Note that the station at P is a double pulsed *master* station and Q and R are called *slave* stations. Loran lines of position are shown full for the pair P-R and dotted for the pair P-Q. The lines are numbered for identification in the following fashion.

1. The first number represents the receiving frequency channel of the loran pair.

Channel	Frequency
1	1,950 kc
2	1,850 kc
3	1,900 kc
4	1,750 kc

2. The letter following the first number represents the basic pulse recurrence rate.

S (slow).................20 per second
L (low).................25 per second
H (high).................33⅓ per second

3. The single number following the letter denotes the specific station pulse rate assigned for individual station identification.

0, 1, 2, 3, 4, 5, 6, 7 (following S, L, or H)

4. The series of numbers following the dash indicates the microseconds time difference as read on the loran receiver-indicator for the particular loran line of position.

For example, in Fig. 21-30 the point x lies on a loran line of position marked "2L6-2500." A ship located at x would show the following:

2 means Channel 2 — 1,850 kc

L means basic rate "L"

6 means station rate "6"

2,500 means time difference reading — 2,500 microseconds

FUNCTION OF THE RECEIVER INDICATOR

The loran indicator is the instrument on which the navigator measures the difference in times of arrival of the pulsed loran signals. The equipment includes a radio receiver especially adapted for loran use which feeds the pulse signals into the "display tube" or cathode-ray oscilloscope of the indicator. Pulses appear on the screen as vertical lines set on top of "pedestals" on two horizontal traces. Controls on the front panel are set to receive the proper frequency channel, basic recurrent rate, and specific station rate of the desired loran station pair. The pulses from this pair will then appear practically stationary on the scope. Pulses from other stations having different recurrence rates will drift across the screen and be neglected.

Figure 21-31 shows the loran receiving apparatus. The power supply is shown above the receiver, the junction unit is at the lower right, and the antenna coupling unit is at the upper right. The receiving antenna is a

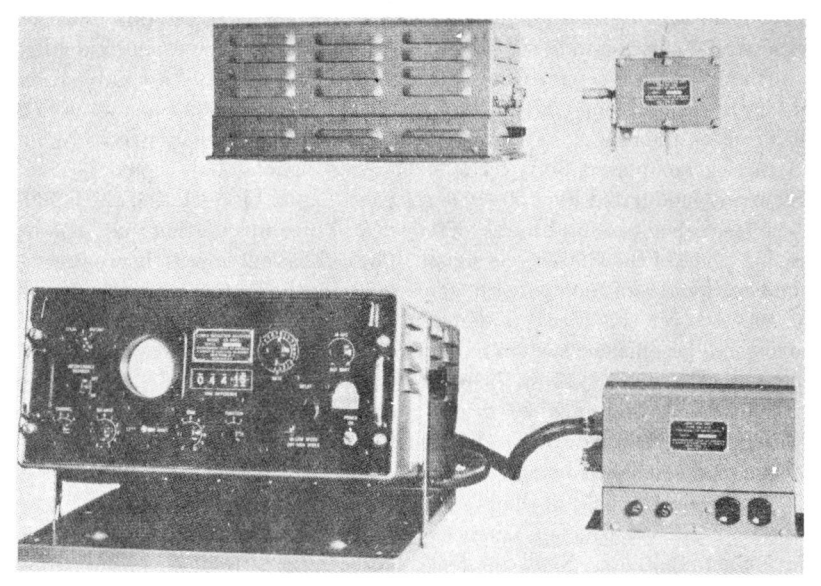

FIG. 21-31. Direct-reading loran indicator receiver. (Courtesy RCA Marine Corp.)

vertical wire up to 125 ft long rigged from a yardarm and connected to the upper terminal of the antenna coupling unit.

Loran stations operate at frequencies on assigned channels around 1,850 kc. The sensitivity of the receiver shown is $\frac{1}{2}$ microvolt. The receiver i-f is 1,100 kc and the heterodyne oscillators are crystal controlled, a very desirable feature made convenient by the limited number of fixed loran channels.

In the loran system the master and slave pulses actually are not transmitted simultaneously. Each slave pulse is delayed by a carefully controlled amount so that the corresponding master pulse is always received first. This eliminates any ambiguity in identifying the pulses and gives time differences which increase continually from a minimum value at the slave station to a maximum at the master station.

A number of loran stations operate on the same radio frequency, but the number of pulses transmitted each second (the recurrence rate) differs for each pair. By setting the equipment for the proper frequency and recurrence rate, any desired pair of signals can be observed

VHF OMNIDIRECTIONAL RANGE SYSTEM (VOR)

The *very* high frequency *omnidirectional* radio *range* operates on channels between 112 and 118 megacycles as a navigational aid to aircraft. It is identified by the abbreviation VOR, representing the italicized words. The basic principle of operation is quite simple. It compares the phase of two separate 30-cycle signals which are transmitted and received individually.

It employs two separate azimuthal radio field patterns. A combination of four horizontal loop transmitting antennas are located on a square with a fifth unit in the center. The center element is independently driven by an r-f carrier modulated 30% by a 9,960-cycle signal. This signal is itself frequency modulated by a 30-cycle reference signal by deviating the 9,960-cycle frequency plus and minus 480 cycles. Voice modulation may also be used in place of the 9,960-cycle signal. The central antenna thus produces a nondirectional radiation pattern which provides the system reference signal. At the receiver this signal is demodulated, the 30-cycle component is recovered to become the system reference, and it is applied to one terminal of the indicator meter system.

The corner antennas are caused to produce a variable phase pattern corresponding to a rotating radio beacon, the carrier being suppressed and 30-cycle r-f sidebands being radiated.

As the electric field of the beacon rotates, two events take place at a receiver. First, the rotating beam sweeps across the receiving antenna producing a time index. Next, the fixed reference field produces an indication of when the beam crosses north. The time of one full beam revolution is known precisely. The difference in the time intervals then is used automatically to compute and indicate the angle from north at which the receiver happens to be. By adjusting the phase of the reference signal, the receiver can be made to show "on course" indication. Under such conditions the indicator pointer will remain vertical when the aircraft is on course, or deviate to the right or left to show the direction and degree in which the craft is off course to or from the range station.

In flying an airway the pilot selects one range station after another by tuning to the proper r-f channel. The bearing indications are accurate to within about 1.5 degrees. The powers used are 200 w which provide a reliable distance range of from about 75 miles to 250 miles, depending upon altitude.

It will be recalled that the only r-f carrier transmitted is the reference carrier, which has its 9,960-cycle subcarrier, which in turn was frequency modulated by 30 cycles.

The rotating beam contains no carrier but only the 30-cycle sidebands which remain from the suppressed carrier. These 30-cycle r-f sidebands combine with the reference carrier and are demodulated to 30 cycles. This constitutes the variable signal which, when applied to the metering system as the second signal, produces an indication calibrated in degrees from north or, if desired, from the desired course.

DME: DISTANCE-MEASURING EQUIPMENT

In the section on radar a radar beacon was described. An interrogation pulse signal is transmitted from a craft in motion, which triggers a responsive pulse from the ground-based radar beacon transmitter. The interval between transmission and reception indicates distance. If the moving ship employs a rotating beam as does a radar set, the direction of the beacon is also shown. The exact identification of one beacon, among several, may be established by a pulse code response signal. This system is a navigational aid known as *distance measuring equipment* or DME.

OTHER NAVIGATIONAL AIDS: ILS

A large number and variety of radio aids are used in navigation. In addition to systems herein described, others make use of the directional properties of loop antennas which have sharp nulls, or utilize two or more CW signals for comparisons of phase or amplitude or modulation; others compare pulse signals. One additional system from the large group will be described briefly, the *instrument landing system* for aircraft, abbreviated ILS.

At the end of an airport runway, or close to it, two directional signals confined to narrow sectors by directional antennas are transmitted for guiding aircraft to the runway. One directional signal is directed at a slight angle above the horizontal, passing upward toward the area from which aircraft would approach. This beam exactly coincides with the correct vertical descent path and gives an indication to the pilot when he is following the proper vertical course. The other beam provides the same function for the horizontal course. In the aircraft a meter with movable vertical and horizontal pointers appears before the pilot. When the crossed vertical and horizontal pointers cross at the center the approach is correct. By correcting the course when the vertical and horizontal indicator lines are not correctly related and aligned, the aircraft will arrive at the runway sought. In order to inform the pilot when to start his descent, fan marker beacons fixed at appropriate distances from the runway and in the approach path give signals which correspond to known ground points as he flies over them. The fan markers operate around 75 megacycles and provide a short-range signal directed upward in a fan-shaped beam for aircraft reception, with the fan across the course. In addition to approach markers, others identify the point to start descending. The ILS system operates around 335 megacycles. Manual control of the aircraft is conventional in making ILS approaches but automatic control systems have been developed.

QUESTIONS AND PROBLEMS

1. Explain the operation of a simple radar system.
2. How does a PPI radar system operate?
3. What determines the radar pulse repetition rate? Why?
4. What is effective radiated power? How is it computed?
5. Why are short radar pulses used?
6. How many different radar applications can you describe?
7. Describe multiplexing. How is it accomplished?
8. What is time division multiplexing? Frequency division?
9. How many forms of pulse transmission can you describe?
10. What are the inherent advantages of pulse transmission? The disadvantages?
11. How is heterodyning employed in radio relays?
12. Describe the loran system.
13. What frequencies are preferable for radar? Why?
14. What frequencies are preferable for radio relays? Why?
15. What is frequency staggering? Why is it used? Where?

CHAPTER 22

INDUSTRIAL APPLICATIONS

INTRODUCTION

Although the techniques of electronics were first developed in connection with radio and communication, they have been so extensively applied to industry that industrial electronics is now an important part of the electronics industry.

An important use of high-frequency electric currents is for heating. Since the high-frequency sources are usually electronic oscillators, the manufacture and servicing of these devices is a significant phase of the electronic industry. By the use of high-frequency currents important advances have been made in placing heat "where it is wanted when it is wanted," without undesirable distribution of heat in the wrong places, or long time delays in getting the heat to the right places.

A second major use of electronics is for purposes of measurement. Here the technique is to generate or control currents proportional to some physical quantity, then amplify these currents to amounts which can operate an indicating instrument to measure the values of or the variations in temperature, speed, force, pressure, or other industrial quantities.

A third use is in the control of industrial operations. This application is at the heart of the modern trend toward "automation." It is closely allied to measurement because measurement is normally the first step in control. After the measurement is made and has been compared with a desired standard, action must result to control the operation of a valve, the change of position of a roller, or the feed of a lathe. The technique of converting the weak signals of the measuring amplifier to powerful motor currents usually involves the use of thyratron or ignitor tubes, amplidynes, or magnetic amplifiers. These techniques not only reduce the manual labor required in industrial operations, but they also make possible the more precise measurement and control in chemical processes and other manufacturing operations.

A fourth development is in the field of new electronic instruments for research. The electron microscope is helping solve many problems in medicine and biology. The mass spectrometer and electron diffraction instruments are opening new areas of development in chemistry and allied sciences.

It is impossible to cover completely the wide field of industrial electronics, but illustrative examples in the above four major branches will serve to introduce this most interesting side of electronics.

HIGH-FREQUENCY INDUCTION HEATERS

High-frequency currents are extensively used to heat the surfaces of gears and rollers for heat treating and also for heating localized areas for brazing and soldering. The high frequency is supplied by induction from a coil of water-cooled copper tubing to the part to be heated. Thus, if a small gear is placed in the center of a coil as shown in Fig. 22-1 and a high-

Fig. 22-1. Inductively heating a 1-in. diameter gear for hardening, using a 15-kw electronic heater. Part of the surface-hardened gear is polished and etched to show the depth of the hardened zone.

frequency current flows in the coil, the coil acts as the primary of a transformer and the gear itself acts as the secondary. A current will tend to flow in the secondary (or gear) which will oppose the magnetomotive force of the primary current. In this type of transformer the current in the primary is approximately constant, and so the secondary current tends to neutralize most of the magnetomotive force of the primary and thus hold that flux to a minimum. In order to do this the secondary current will flow just as close to the coil as is possible, and in a gear or cylindrical object placed in a solenoid the current will therefore flow on the surface. This is particularly true of iron and steel, which are good magnetic conductors.

With high frequency and high current flow it is possible to heat the surface of gears to hardening temperature without permitting the heat to penetrate more than a few hundredths of an inch into the part. This is shown in Fig. 22-1(b), where a section of the gear has been cut and polished to show the depth of the hardening. The hardening of the tooth root is prevented and the teeth thus retain their proper strength. This tendency of the secondary current to flow on the outside surface is sometimes called *skin effect*.

Where high frequency is used to provide localized heat for soldering and brazing joints, the high-frequency coils should follow the joint as closely as possible because, even with non-magnetic materials, the current in the secondary or part being heated will flow as close to the coils as possible. In this way, the currents in the secondary may be controlled so as to produce the heat where it is most effective. This is shown in Fig. 22-2, where the primary coils are so placed as to concentrate the heating along the joint to be soldered or brazed.

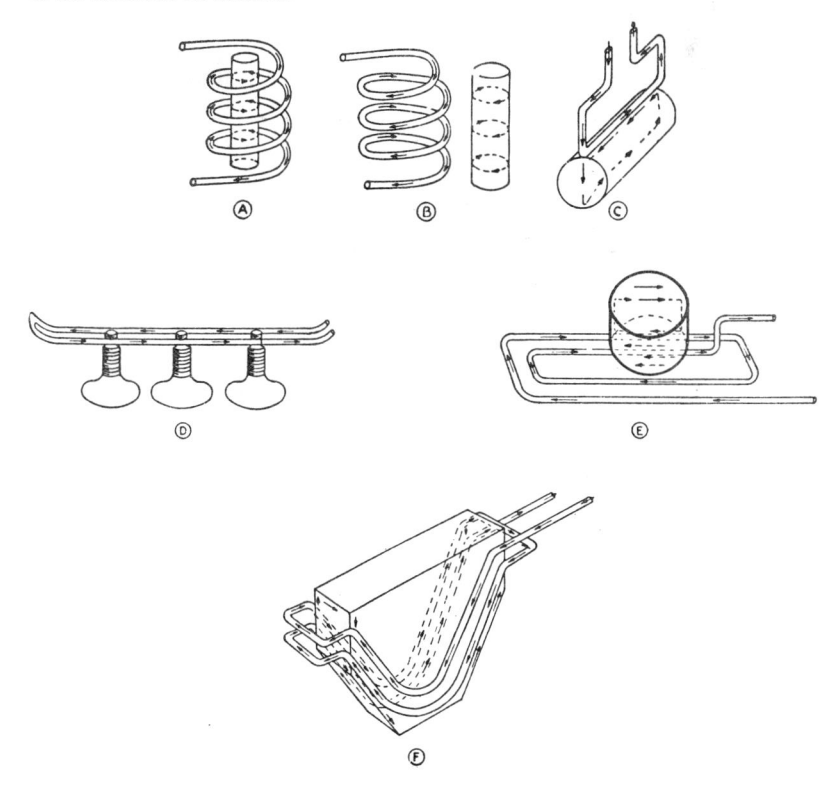

FIG. 22-2. Sketches showing direction of current flow in electronic heater coils and in parts being heated.

740　　　　　　　　　*Industrial Applications*　　　　　　　Chapter 22

Although the final solution to the design of high-frequency coils for heating usually involves an experimental procedure, the following rules may be helpful as a guide.

1. The coil should take the approximate shape of the part to be heated.
2. The part should be centered in the coil if possible.
3. Sharp corners will tend to heat first, because usually these corners are nearest the coil and, in addition, they have a minimum of mass. Therefore, the shape of the coil should be adjusted to clear the corners as far as possible.
4. Where dissimilar metals are to be brazed together, the current must be concentrated on the one with the lower resistance and permeability. Thus, copper and silver because of their lower resistance heat most slowly,

Fig. 22-3(a). A 15-kw heater: exterior view with work table and quenching fixture adapted to surface-hardening a wide variety of small parts.

Chapter 22　　　　　*Industrial Applications*　　　　　741

while brass heats more rapidly and steel heats most rapidly. Therefore, the coil should be kept close to the silver, copper, and brass, and away from the steel.

5. The material to be brazed should come to temperature first and draw the brazing alloy into the joint. Therefore, the brazing alloy should be kept as far away from the coil as possible.

6. To obtain uniform heating around the circumference of a part, it may be necessary to rotate it while it is being heated.

In order to limit the heating to a very thin surface layer for case hardening, or to heat a joint for soldering or brazing, it is necessary to supply high-frequency power in the frequency range of 200,000 to 500,000 cycles, and so vacuum-tube oscillators are required. Figure 22-3 shows both the exterior and interior views of such a power oscillator. The rectifier tubes

Fig. 22-3(b). A 15-kw heater: interior view of heater showing transformer, rectifier tubes, oscillator tubes, and wiring.

742 *Industrial Applications* Chapter 22

are shown in the upper right portion of the interior view, and the high-frequency oscillator tubes are in the central left portion of the interior.

A diagram of the oscillator circuit is shown in Fig. 22-4. This includes the rectifier to supply unidirectional plate voltage to the oscillator tube, the oscillator tube and its circuit, and the tank circuit, which is the oscillatory circuit stimulated by pulses every cycle from the oscillator tube. In this

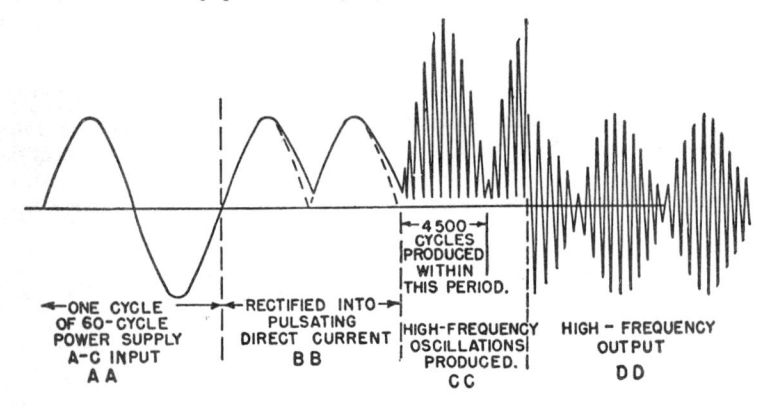

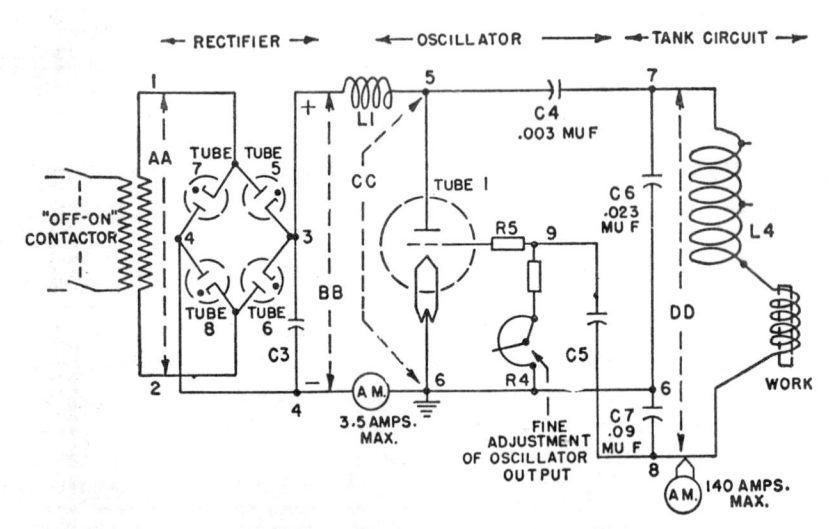

Fig. 22-4. Elementary diagram of an electronic heater.

circuit the capacitive portion of the tank circuit is split so that a portion of the voltage may be fed back to the grid to maintain the oscillations. A variation of the grid-leak resistance is used to vary the self-biasing effect of the oscillator tube and thus to control the output. This variation is indicated in the fine adjustment of the oscillator output. The wave diagram at the top of the figure indicates the type of current flow in the various portions of equipment.

Similar equipment is used for the heating of dielectric materials. The internal loss involved in stressing the dielectric first in one direction and then in the other is very small. It is thus necessary to have a very large number of reversals per second if adequate heating is to be obtained. The frequency range for dielectric heating is much higher than that for metal heating, being normally from 1 to 100 megacycles.

An example of the advantageous use of dielectric heating is in the bonding of plywood. The glue may be made quite lossy while the dry laminated wood produces small loss. Thus the glue can be heated quickly without burning the wood. The earlier method was to put the laminations between thermally heated presses. The time required to heat the glue without burning the wood is very much greater with this non-electronic method.

The electronic design of the heater is similar to that of Fig. 22-4 except that the work is placed between the plates of the tank circuit capacitor C_7. The work coil for the inductive heater then becomes a small inductive coil in the tank circuit with such value as will give the proper operating frequency.

High-frequency heating equipment is expensive, and the power efficiency is generally low. Nevertheless, its ability to concentrate the heat in a specific place makes it possible to obtain results that can be achieved in no other way. This makes possible production economies that pay big dividends on the money invested in the equipment.

DIATHERMY

The use of high-frequency electric generators to produce heat in living tissue is an important tool for physicians. There is no fundamental physical difference in the heat generated in body tissues and in other materials that have high resistivity. Body organs have some difference in their electrical characteristics, but these are not sufficient to make possible selective heating within the body. Furthermore, the blood circulation coupled with the normal system of body temperature regulation makes it difficult, if not impossible, to raise the local temperature appreciably above that of the body as a whole. It does appear, however, that the physiological effects of local internal heat generation by high-frequency electric and magnetic fields are very beneficial in treating certain infections and diseases.

These electromagnetic fields may be applied either by capacitance-type electrodes, as described above for the heating of lossy dielectrics, or by the use of coils to produce eddy currents in the body tissue.

When the electric field is used, it is customary to space the electrodes one or two inches from the tissue. This helps to produce a more favorable ratio between deep tissue heating and superficial or skin heating. This type of diathermy requires a frequency of the order of 40 to 50 megacycles, which corresponds to a wavelength of 7.5 down to 6 meters. The necessity of careful adjustment of the electrodes in order to obtain the desired deep-tissue heating makes this equipment somewhat less suitable for handling by technicians.

The use of a coil, which is either wrapped around the tissue to be heated or formed into a pancake type of coil that is laid on the surface where the heat generation is desired, is less critical of adjustment. The heating is accomplished by eddy currents resulting from the magnetic field; and since the body and air have the same permeability they form an essentially

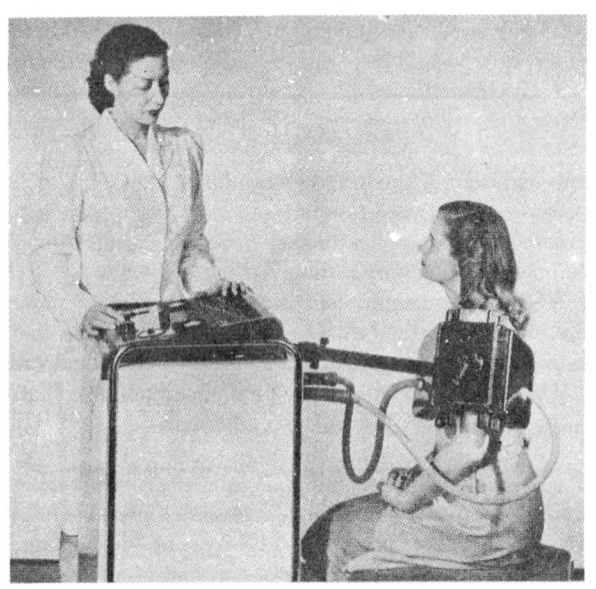

Fig. 22-5. Inductive diathermy being applied to shoulder of a patient. (Courtesy The Burdick Corp.)

homogeneous magnetic material and so the heat penetrates quite effectively. The frequency in this case should be reduced to about 10 or 15 megacycles in order that the inductive effects of the coil predominate.

The high-frequency generator is similar to a crystal-controlled transmitter but without provision for modulation. The case and method of operation are shown in Fig. 22-5. The coil is supported by an adjustable arm so that the patient does not bear the weight. An alternate coil that may be wound around the tissue is normally provided in addition to the adjustable pancake coil shown.

ELECTRON TIMERS

One of the important problems in industrial production is the measurement of very small periods of time. Mechanical timers operate quite satisfactorily down to a few seconds, but below that value the errors become a large percentage of the total time. For industrial applications requiring very short periods, the use of electronic timers has become almost universal. These are available in a wide variety of forms from different manufacturers. It will be impossible to cover all of these, so the equipment of one of the larger companies will be described as illustrative of the general form.

First, a very elementary time-delay circuit will be analyzed. In Fig. 22-6(a) a simple triode is shown connected so that no current flows in the plate circuit as long as the switch S is closed, because of the high negative potential of the grid. When the switch is opened, the grid immediately assumes

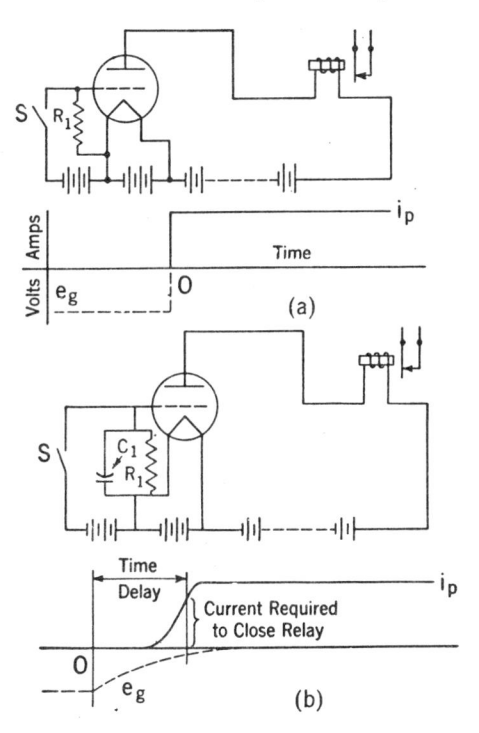

FIG. 22-6. Elementary time delay circuit.

the potential of the cathode and permits full plate current to flow. This is shown in the time diagram just below the circuit. Here the grid voltage is negative, and the plate current is zero until time 0 when the switch S is opened. At this time the grid voltage jumps to zero and the plate current assumes full value. In Fig. 22-6(b) a capacitor C_1 is placed in parallel with the high resistor R_1. In this case, when the switch S is opened, the grid does not immediately assume the potential of the cathode because it is necessary for the capacitor to discharge through the resistor before the grid voltage assumes the same potential as the cathode. This is shown in the time diagram just below the circuit. When the switch is opened, the capacitor immediately starts to discharge through R_1, and its voltage decreases as explained in Chapter 3. No plate current flows until the grid voltage rises to a less negative value than the cutoff voltage. The plate current will then rise rapidly so that full plate current is reached by the time all of the capacitor charge has been dissipated.

The time delay is determined by the size of the capacitor C_1, the time being greater when the capacitance is larger. It is also determined by the value of the resistor. When the circuit resistance is larger, the time delay is also greater. This circuit is the basis for many electronic timing circuits. As shown above, it requires several batteries and is not adapted to a-c operation.

The diagram of Fig. 22-7 is a slightly simplified form of the commercial timer. Several features must be understood, and these will be taken up in sequence. The first consideration is the functioning of grid bias dependent

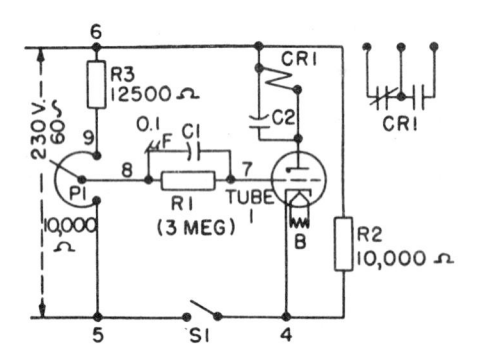

FIG. 22-7. Electronic timer circuit.

upon electron flow from the cathode to the grid in the portion of the cycle during which (with the switch S_1 open) the grid is more positive than the cathode.

In Fig. 22-8 the voltage of point 8 of Fig. 22-7 is assumed as the reference potential, and the voltage of point 6 and point 5 are plotted with respect to point 8 when the potentiometer is connected to point 9. The 12,500-ohm resistance R_3 in series with the 10,000-ohm potentiometer divides the voltage so that approximately 130 v are across R_3 and 100 v across the potentiometer. These voltages are rms values, so the maximum values would be 185 v and 140 v, as shown in the diagram. When the switch S_1 is open, the cathode assumes the voltage of point 6. As soon as the cathode becomes more negative than the grid, the grid attracts the electrons and a current flows. The current i_g charges the capacitor C_1. This charge tends to leak off during most of the cycle, but the 3-megohm resistance is so large that it leaks off very slowly. In this way a small impulse of grid current occurring once each cycle, as shown in Fig. 22-8, is sufficient to hold the grid at a negative potential of about 165 v with respect to point 8 of the circuit. If the potentiometer setting should be changed so

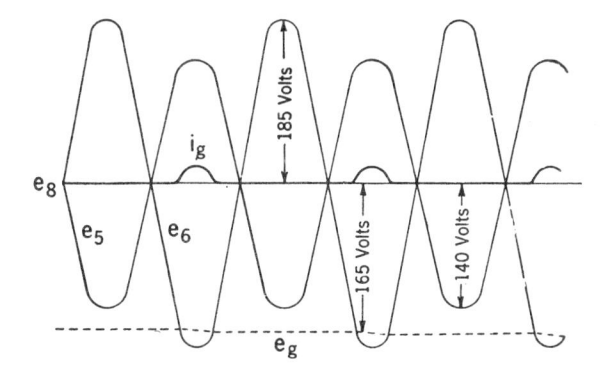

Fig. 22-8. Diagram of grid-control currents and voltages of an electronic timer.

that the maximum voltage between point 8 and point 6 would be about 245, the grid could be expected to assume a negative voltage of about 235 with respect to point 8. The voltage of 5 with respect to point 8 under these conditions has a maximum value of about 80 v.

The second feature of the operation is the decrease of the grid-bias voltage when the source of that bias voltage is removed. An elementary form of this was shown in the circuit of Fig. 22-6(b). It has been assumed in the above analysis of the circuit of Fig. 22-7 that switch S_1 is open, which is the normal condition of the circuit. When the operation to be timed is started, this switch is closed. If the condition studied in Fig. 22-8 exists, and if the cathode voltage is now assumed as the reference voltage, then the grid voltage will be the sum of the 100 v (rms) from point 5 to point 8 and the bias voltage across the resistance R_1. This gives the grid voltage shown in Fig. 22-9. When the switch is first closed, the minimum negative grid voltage is approximately 25 v. The recharging of capacitor C_1 does not occur each cycle, as in Fig. 22-8, since the switch S_1 is closed. The

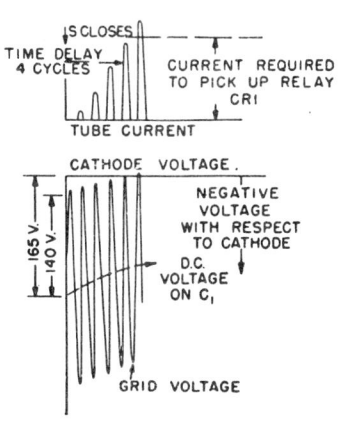

Fig. 22-9. Voltage and current variations in an electronic timer (short time setting).

negative grid bias, therefore, continues to decrease. As the bias decreases, the minimum negative voltage becomes less and plate current begins to flow in pulses as indicated in the top portion of the diagram of Fig. 22-9. When the current pulses reach the value necessary to operate the relay coil, the contactor closes. In this case the relay will operate after 4 cycles, which is the minimum time for the combination of circuit parameters herein described.

If the potentiometer is adjusted so that point 8 divides the voltage in proportion of 173 and 57 v, the initial or starting grid bias is 235, as shown in Fig. 22-10. The a-c voltage superposed on this grid bias will be only 80 v maximum, and so the plate current will not operate until 14 cycles have

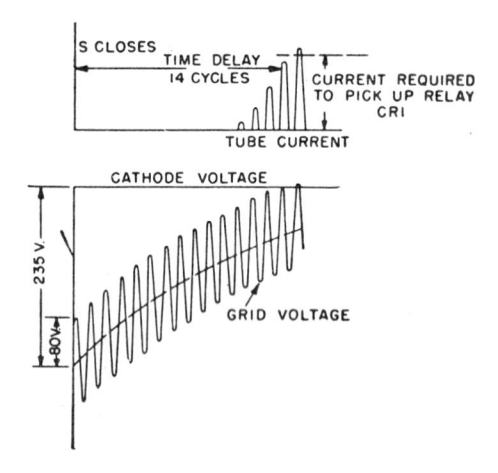

FIG. 22-10. Voltage and current variations in an electronic timer (long time setting).

passed. The potentiometer is therefore used to control or adjust the length of time delay. With the constants indicated in the circuit, the timer may be adjusted to operate from 4 to 65 cycles.

The above description explains the major portion of the circuit operation. There are in addition some minor items in the circuit which are of interest. The capacitor C_2, which is connected across the energizing coil of the relay, is placed there to make it possible to maintain a reasonably constant current in the relay coil in spite of the pulsating character of the plate current, and thus prevent chattering of the contacts. The three terminals in the upper right-hand corner of Fig. 22-7 are connected to the relay contacts. When the relay operates, one of these contacts opens and the other closes.

If a larger capacitor is placed in the circuit at C_1, then a larger band of time delays will be obtained.

Chapter 22   *Industrial Applications*   749

CYCLE TIMERS

In industrial operation it is frequently necessary to perform a series of operations which follow in a specific order and each of which is accurately timed. The combination of several electron timers arranged in chain or sequence will often control these operations and taken as a group will act as a timing unit. Such process timers are often called *cycle timers* because they control a cycle of operations.

Cycle timers are not necessarily composed of electronic timers. They may be in the form of a group of cam-operated switches which initiate the various process operations. They may in fact take any of a large number of forms, but the electronic timer group is easily changed and adjusted and so provides an especially flexible unit for experimental work.

ELECTRONIC METHODS OF INDUSTRIAL MEASUREMENT

The advantage of electronic methods when applied to industrial measurement is in the extreme sensitivity of instruments resulting from the use of amplifiers. This increased sensitivity gives improved accuracy over using the older and more orthodox measuring techniques. More important, however, is the ability to use new types of units to convert variations of industrial quantities to variations of electrical quantities.

Electronic tubes have an important limitation when used in measuring circuits. Their characteristics, being dependent upon the position and physical condition of the tube elements, and the degree of vacuum, are not always the same in different tubes of the same type, nor will these characteristics remain constant throughout the life of the tube. This lack of uniformity and constancy with time makes it undesirable to include electronic tubes directly in the measuring circuit, if precise results are desired.

Sensitive cells or *transducers* convert a variation in some quantity that is being measured into a variation of voltage or of electric circuit characteristics. A variation of circuit characteristics can, in turn, be converted into a variation of voltage by circuit manipulation. The electronic techniques are then often used to detect these voltage variations and to control devices that produce a known and calibrated standard voltage to balance or neutralize the unknown voltage.

The industrial quantity may then be measured in terms of the calibrated voltage required to neutralize the voltage variations in the sensitive cell or transducer.

TEMPERATURE MEASUREMENT USING ELECTRONIC METHODS

One of the most common transducers of the voltage type is the *thermocouple*, which is used for temperature measurement. It will be used as typical for voltage-type transducers, and two electronic systems will be discussed as illustrative of this type of measurement.

A thermocouple is made by welding two different metals together. The connections to these metals are usually then made to copper wires at room temperature. When the junction of the two dissimilar metals is heated, a small voltage is developed between the two metals. Since the magnitude of this voltage is approximately proportional to the temperature (over the recommended operating range), it may be used as a measure of temperature. (Stated more exactly, the magnitude of the voltage is approximately proportional to the difference in temperature between the *hot junction* and the *cold junctions* of the thermocouple wires.) The voltages generated by six commonly used thermocouple materials are given in Table 22-1 for various hot junction temperatures when the cold junctions are kept at the temperature of melting ice.

TABLE 21-1

VOLTAGES GENERATED BY COMMON THERMOCOUPLES
(Temperature of Cold Junction—0° C)

EMF, mv	Degrees C			EMF, mv	Degrees C		
	Platinum to platinum- (10% rhodium)	Platinum to platinum- (13% rhodium)	Copper to constantan		Chromel to alumel	Iron to constantan	Chromel to constantan
0	0	0	0	0	0	0	0
2	265	259	49	5	122	93	80
4	478	457	94	10	246	182	153
6	678	638	136	15	367	272	221
8	861	806	176	20	485	362	286
10	1,037	964	213	25	602	453	350
12	1,206	1,114	250	30	720	543	413
14	1,374	1,259	285	40	966	711	537
16	1,543	1,404	319	50	1,232	865	661
18	....	1,550	353	60	....	....	786
				70	....	....	915

Since thermocouples have a tendency to develop resistance at their junction, the current flow must be very small if accurate measurements are to be obtained.

MILLIVOLTMETER WITH ELECTRONIC FOLLOWER

One way of overcoming this effect of a variation in thermocouple resistance is by use of a high-resistance millivoltmeter, and this method is used by one representative manufacturer. With these restrictions, however, it is impossible to obtain sufficient torque on the meter to drive a recording pen or to operate a temperature controller. In order to do this, a

Chapter 22 — *Industrial Applications* — 751

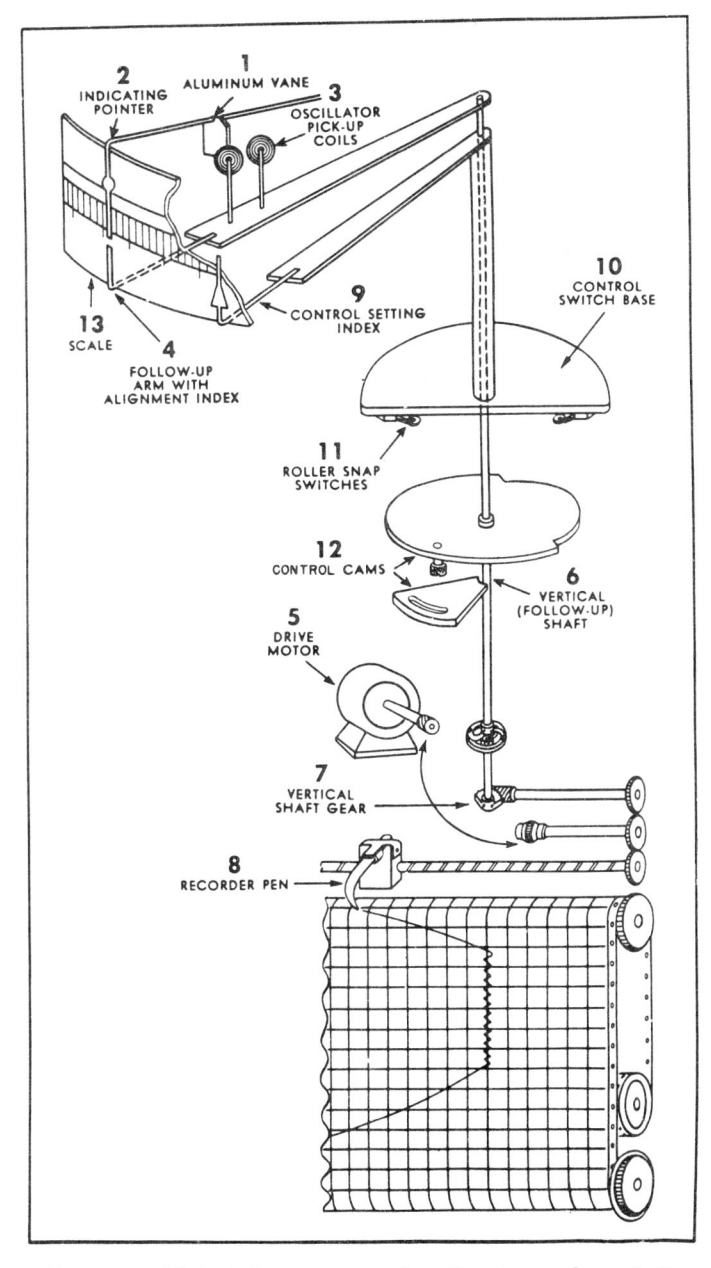

FIG. 22-11. Mechanical arrangement of a Wheelco recorder controller.

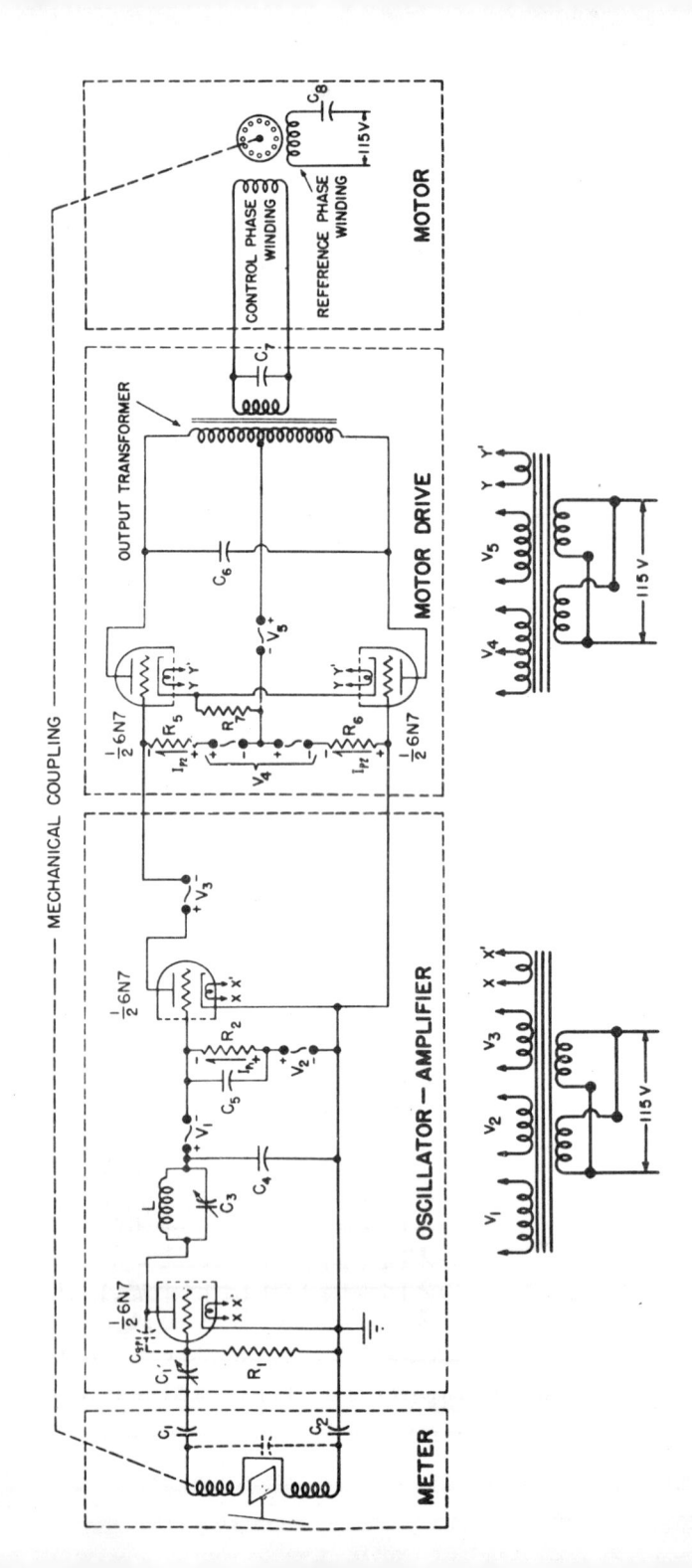

FIG. 22-12. The electronic control circuit for a Wheelco recorder controller.

combination of mechanical and electronic control is used. The mechanical arrangement is shown in Fig. 22-11. The high-resistance millivoltmeter pointer is shown at 2 with an aluminum flag or vane on it. A follow-up arm has two small coils mounted on it that provide the inductance in the grid of a tuned-plate tuned-grid oscillator. This follow-up arm is mounted on the same axis as the millivoltmeter movement, and the motor drive is controlled by the electronic circuit so that it will follow the indicating pointer until the center of the coils reaches the near edge of the aluminum vane. In this way the motor keeps this follow-up arm in exact alignment with the millivoltmeter pointer, but does not place a mechanical load on the pointer. The motor drive also controls the position of the pen arm and the cams and switches that turn the heat on and off in the furnace.

It is the electronic control that is of particular interest, and the circuit for it is shown in Fig. 22-12. Two double diode tubes are used. One half of the first tube is a tuned-plate tuned-grid oscillator that operates at about 15 megacycles. The movement of the aluminum flag or vane into the space between the coils acts in two ways. In the first place, the eddy currents in the vane add losses to the circuit and reduce the effective Q of the coils. In the second place, the vane increases the stray capacitance and changes the resonant frequency. If the grid circuit is tuned with the flag remote from the coils, the movement of the flag into the coils will produce an increased impedance in the series resonant circuit across the grid resistor R_1. The effect of this detuning is transmitted through the amplifier as a variation in the average plate current. In this circuit 60 cycles are used as the plate supply throughout, so that the resultant plate currents are at that frequency.

By balancing the push-pull circuit driving the second tube at the midpoint of plate current in the amplifier tube (the range of current is from 2 to 50 ma), zero current in the output transformer will be obtained for this condition. When the flag is inserted between the coils to a lesser degree, transformer current is obtained which is used to drive a small capacitor-type induction motor. The motor then drives the follow-up arm until the balance point is again reached. When the flag is too far into the coils the phase of the output current is reversed, and this reverses the direction of motor rotation. In this way the follow-up arm is maintained in correct alignment with the pointer of the millivoltmeter.

POTENTIOMETER VOLTAGE MEASUREMENTS

An alternate method of measuring thermocouple voltage with very little current flow is to balance this voltage with an equal and opposite voltage. This opposite voltage can then be used as a measure of the thermocouple voltage and thus of the temperature. The most common method of making this balance is by means of a potentiometer, using a galvanometer or very sensitive millivoltmeter as the detector. Such balance procedures may or may not use electronic circuits.

CAPACITOR POTENTIOMETER METHODS

An alternate method is to charge a capacitor with the thermocouple voltage, and measure the charge on the capacitor by comparing it with the charge on another capacitor supplied with a known standard or reference voltage. To understand this method, a quick review of fundamental theory will be made. In Chapter 3 it was learned that the charge on a capacitor is proportional to the product of the voltage and capacitance. The charge on the capacitor C_1 produced by the thermocouple voltage is

$$Q = EC_1.$$

The same charge can be obtained by different combinations of voltage and capacitance. Thus

$$Q = E_s C_2,$$

where C_2 is the capacitance of a variable capacitor, and E_s is a standard voltage. Substituting this value of Q in the equation for an unknown voltage with the known capacitance C_1, it is evident that the unknown voltage can be measured in terms of the standard voltage and the ratio of the capacitances.

$$E_x = \frac{Q}{C_1} = \frac{E_s C_2}{C_1} = E_s \frac{C_2}{C_1}$$

A comparison of the charges on two different capacitors can be made by connecting them in series; and if the charges are the same, they will neutralize each other and no residual charge will be left to produce a voltage. If the charges are not the same, a voltage will result and the polarity of this voltage will indicate which charge is the larger.

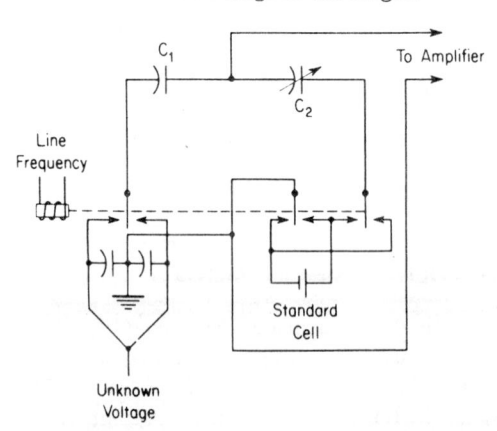

Fig. 22-13. The voltage measuring circuit of Foxboro Dynalog instruments.

Chapter 22   *Industrial Applications*   755

The manufacturer employing this technique uses a multiple pole vibrator switch to alternately charge and then compare these charges in a circuit such as shown in Fig. 22-13. The electronic amplifier and its as-

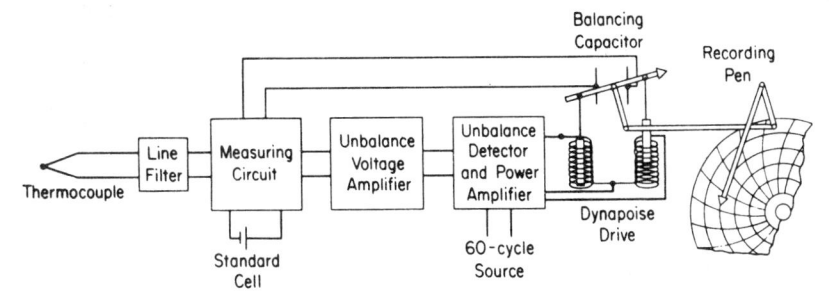

FIG. 22-14. Block diagram of electronic detector and solenoid balancing motor of a Foxboro Dynalog instrument.

sociated electromagnetic motor adjusts the variable capacitor C_2 (which is charged by the standard cell) until a balance is reached. The block diagram of the amplifier and detector unit is shown in Fig. 22-14.

TYPES OF VOLTAGE-SENSITIVE ELEMENTS

Other voltage-producing sensitive elements include electromagnetic generators to measure the speed of shafts, piezoelectric crystals to measure the variation in pressure, and calibrated electrodes for pH measurement. The above and many other types of electronic equipment are used to measure, record, and control industrial quantities using these voltage-sensitive transducers.

IMPEDANCE-VARYING TRANSDUCERS

The variation of impedance in a transducer or sensitive element may be used for the measurement of many industrial quantities. The use of resistance-varying transducers is probably more extensive than any other type. The variation of a resistance bulb with temperature is used almost exclusively for temperature measurements below 400°F. A typical resistance coil in its protective well is shown in Fig. 22-15. Many other forms are also

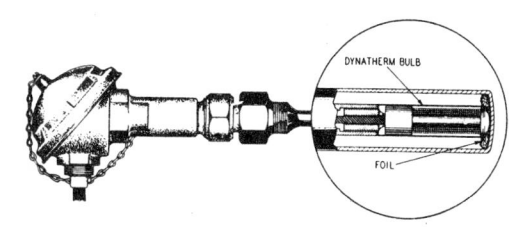

FIG. 22-15. Typical resistance bulb for temperature measurements.

available for special purposes. One of these, shown in Fig. 22-16, is composed of very fine wire mounted on a thin paper backing that can be cemented to the surface whose temperature is to be measured.

To measure strain (on machine parts), force, or pressure, a resistance strain gage may be used. This transducer element is very similar in appearance to that shown in Fig. 22-16 and is also cemented to the surface. The unit is placed on the surface so that when the machine part is stretched (or strained) the wires are also stretched and this increases the resistance of the transducer.

Conductivity cells are extensively used as transducers. They measure the resistance (or conductivity) between two carefully insulated and calibrated electrodes immersed in a liquid. In this way the concentration of acid, salt, and alkaline solutions can be measured.

Movement may be measured by the difference in capacitance of two insulated surfaces as they move with respect to each other, or by the variation in inductance as the air gap in a magnetic circuit is increased or decreased.

Fig. 22-16. Surface-temperature resistance element.

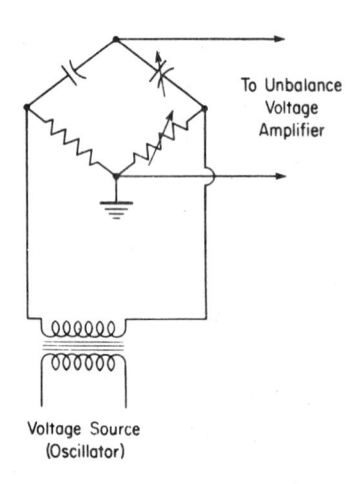

Fig. 22-17. Capacitor bridge for resistance measurement.

BRIDGE MEASUREMENT OF IMPEDANCE

The measurement of impedances is usually accomplished by means of a bridge circuit. The same type of instruments that are used for null measurements on potentiometer circuits may also be used for null measurements on bridge circuits. These instruments may or may not use electronic circuits to obtain increased sensitivity. One bridge circuit using electronic balancing is shown in Fig. 22-17. In this circuit the variation in resistance is balanced by a corresponding percentage change in the capacitor arm of the bridge.

With a-c applied to the bridge the phase of the residual voltage is reversed when the variable capacitor changes from too small to too large. It is thus possible to feed this residual voltage into a conventional voltage amplifier, and then into a phase-sensitive detector circuit that is used to control the variable capacitor, as was illustrated in Fig. 22-14.

STRAIN GAGE MEASUREMENTS

Where vibration is being studied with the use of a strain gage, a simple voltage divider circuit similar to that in Fig. 22-18 may be used. As the movement of the surface stretches the wires of the strain gage, its resistance is changed and the voltage across it varies. This variation in voltage is transmitted across the capacitor C to a conventional stabilized amplifier and measured by an oscilloscope.

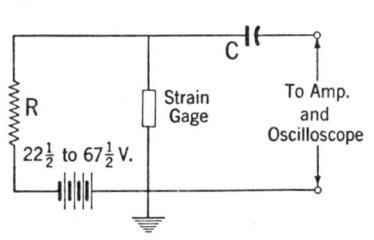

FIG. 22-18. Voltage-divider circuit for dynamic-strain measurement.

The cathode-ray oscilloscope discussed in Chapter 16 is a most useful and convenient instrument for the study of rapidly varying quantities that have a regular recurrence rate. This instrument is used with either piezoelectric or resistance strain gage elements to measure the cylinder pressure in gasoline engines, and with displacement elements to study engine vibration. Such measurements are typical of the extensive use of the cathode-ray oscilloscope in solving production and research problems in industry.

THE TOOLS OF CONTROL

As indicated at the beginning of this chapter, control usually requires large power and this often involves thyratrons or ignitron tubes. In order to understand their operation in a control circuit a brief discussion of the theory and operating characteristics will be given. Magnetic amplifiers are an alternate source of large power currents and although they are not electronic, they are so extensively used with electronic controls that a brief introduction to the theory of these devices will be given.

Many simple control devices are used in industry and two of these will be illustrated. The closed loop system of control or servomechanism is the basis for many of the automatic controls in industrial and military use. A paragraph on the meaning of servomechanisms is provided so that the place of electronics in the system can be described.

GAS TRIODES OR THYRATRONS

In Chapter 4 it was found that if gas molecules were left in a vacuum tube, they would produce ionization. This ionization would permit the current to increase almost indefinitely, once the ionization potential of the gas had been reached. This characteristic is used to advantage in gas-filled rectifier tubes, since the low voltage drop across the tube tends to reduce the power loss in the rectifier. When a grid is included in a tube filled with a small amount of gas of low ionization potential such as mercury, operating characteristics are obtained that are very useful for control purposes.

When the grid is highly negative with respect to the cathode, all of the electrons that are evaporated from the cathode are repelled, and none of them attain sufficient velocity to ionize the gas atoms. There is, therefore, no appreciable current flow in the tube, regardless of the plate voltage. As the negative grid potential is reduced, eventually a few electrons will escape past the grid, and these are accelerated to the ionizing velocities.

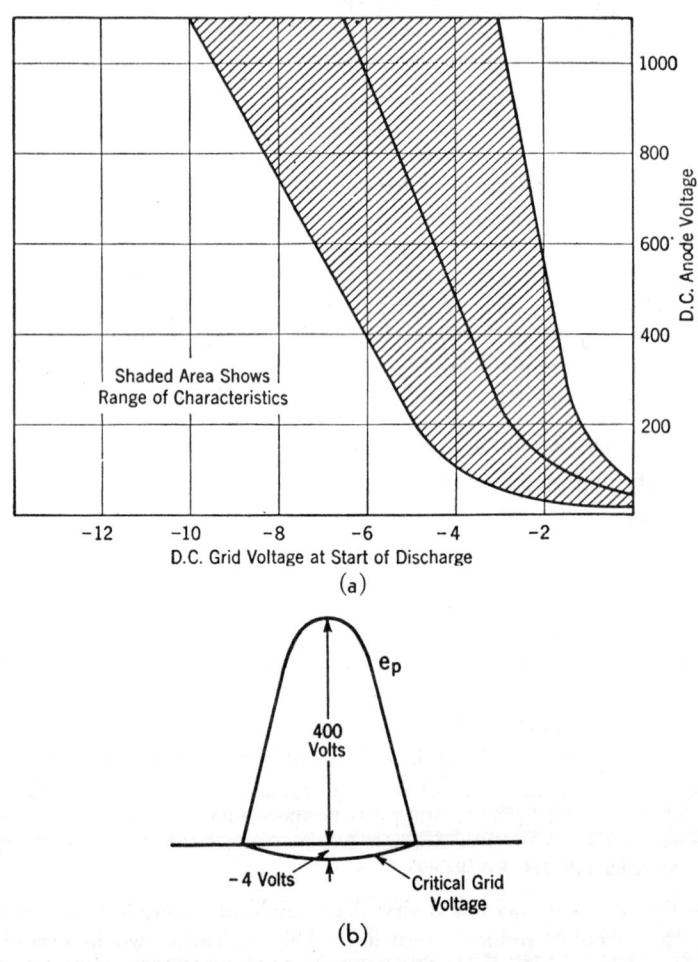

Fig. 22-19. (a) Typical thyratron control characteristics; (b) critical grid-voltage characteristics of a thyratron tube with an a-c plate voltage.

As soon as ionization occurs, the tube becomes highly conducting and the voltage difference between the cathode and anode drops to about 20 volts, the minimum voltage necessary to produce ionization. When ionization occurs, the tube is said to *fire*. It acts as an open or closed switch with a constant voltage drop of about 20 volts. The grid voltage at which ionization will occur is somewhat dependent upon the magnitude of the anode voltage, and this relationship is shown in Fig. 22-19(a). Here the graph shows the negative values of grid voltage at which the tube will fire with different anode voltages. These values are called the *critical grid voltages* because with grid voltages more negative, no current flows, and with grid voltages less negative, complete ionization is obtained. In other words, the tube characteristic is discontinuous at this point.

When the tube is once ionized, it is impossible to block the flow of current with the grid because the positive ions form a neutralizing cloud around the grid when it becomes negative and ionization continues. In order to *turn the switch off* or to stop the current flow in a thyratron, it is necessary to remove the positive voltage on the anode long enough for the tube to de-ionize. Since this involves the combination of a large number of electrons with positive gas ions, the time required is in the order of one tenthousandth of a second. Because it is necessary to remove the anode voltage in order for the grid to regain control of the tube, it is common practice to use an a-c voltage on the anode. With 60 cycles, it is then possible for the grid to gain control 60 times per second.

It is often convenient to plot the critical negative grid voltage against time with an a-c voltage on the anode. This is done in Fig. 22-19(b) and indicates clearly the grid voltage necessary for the tube to fire at any time during the half cycle that the anode is positive. In a typical thyratron, a negative grid potential of 4 v will prevent firing with an anode voltage of 400 v, while if the anode voltage is raised to 500 v, the negative grid voltage must be increased to 5 in order to continue to prevent ionization.

The gas triode makes an excellent relay since it operates at a fixed predetermined grid voltage. When it fires, full-load current is immediately obtained. It has high current-carrying ability with low voltage drop between plate and cathode. The grid may be controlled by a photoelectric tube, a vacuum-tube triode, a sensitive contacting device, or any other method that will give a grid voltage less negative than the critical when it is desired to have the tube operate.

The thyratron may also be used to supply a unidirectional current of variable average magnitude for various control purposes. This is usually done by controlling the phase of the grid potential so that the tube will fire at different portions of the half cycle during which the plate is positive. This is shown diagrammatically in Fig. 22-20. The method of obtaining the phase shift is not indicated, but the effect of such a phase shift on the average current flow is shown for several phase positions of the grid voltage. In

Fig. 22-20(b) the grid voltage is in phase with the plate voltage and the maximum current flow is obtained.

The average current is proportional to the area under the current curve. In part (c) of the diagram, the grid potential has been given a lag of 90°, so that the critical negative grid voltage is not reached until the anode voltage

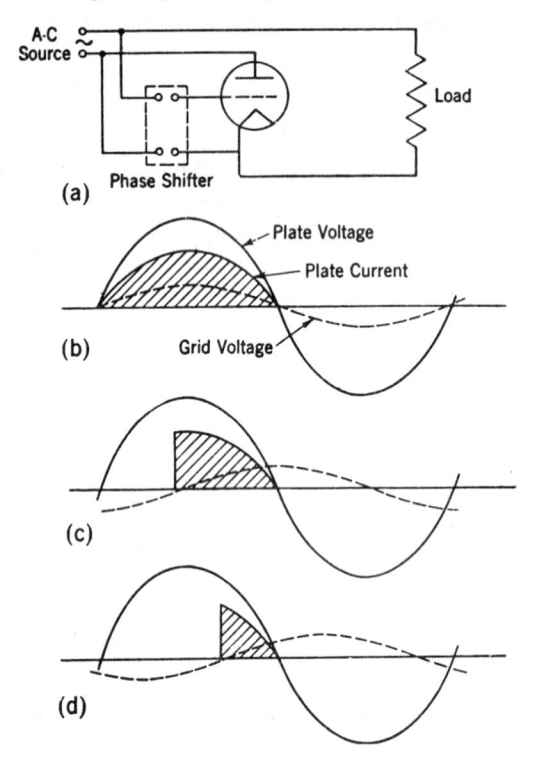

FIG. 22-20. Phase shift grid control of a thyratron tube.

has reached almost maximum value. The area under the current curve is thus reduced to almost half of the previous value. When the phase of the grid voltage lags still further, the time of firing of the tube is delayed still more and the area under the curve (the average value of current) is reduced to a small value indeed. It is thus possible to vary the average value of current to a motor or other device over a wide range by the variation of the phase of the grid voltage of a thyratron. This can be done with small loss and often with comparatively inexpensive equipment.

A PHASE-SHIFTING CIRCUIT

The use of phase-shifting circuits is so general that one type of such a circuit will be discussed. In Fig. 22-21 a voltage is supplied by a transformer which is in phase with the main voltage source. This is shown as V_{ab} in (b) and (c) of the diagram. The output of this transformer is impressed on a circuit composed of a constant inductive reactance and a variable high resistance. When the resistance is large as compared to the inductive reactance, the current flow will be small and lagging by only a small angle, as indicated by the diagram at (b).

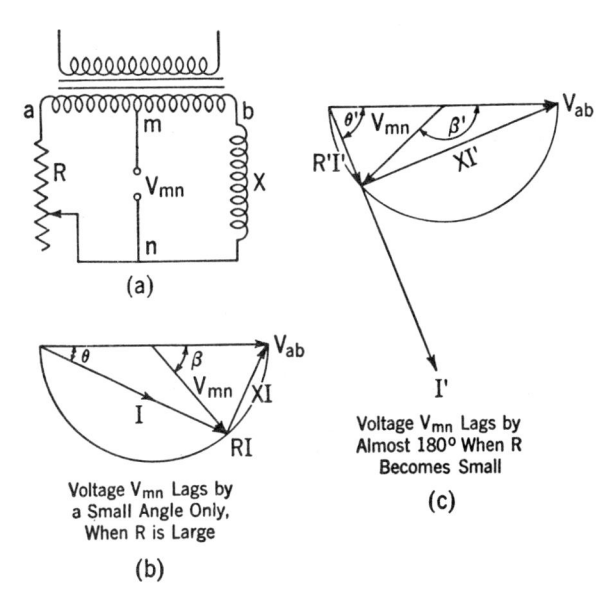

FIG. 22-21. A phase control circuit.

The voltage V_{mn}, measured from the midpoint of the transformer secondary to the connection between X_s and R, will lag by an angle β which is twice as large as the angle of current lag θ. When the resistance has been greatly reduced, the current flow in the circuit will be much larger and will lag by a large angle θ'. The voltage V_{mn} will continue to have the same magnitude but will lag by an angle β' which is still equal to twice θ'. Thus the voltage V_{mn} is constant in magnitude but varies in phase position as the resistance value is changed. This voltage can be used to control the grid of the thyratron tube. Similar phase-shifting may be obtained by using a fixed capacitor in place of the inductor. A fixed resistor and variable inductor or capacitor will also function satisfactorily.

THE IGNITRON TUBE

The ignitron tube is a three-element tube having characteristics similar to the thyratron except that it can be manufactured to carry currents up to several thousand amperes.

The diagram of Fig. 22-22 shows the construction of the ignitron tube. The cathode is a mercury pool located in the bottom of the tube. The anode is usually a graphite cylinder supported by an insulator in the top of the

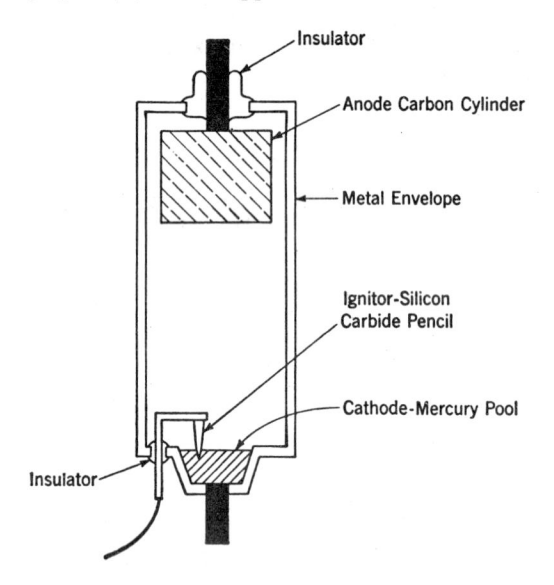

FIG. 22-22. The construction of an ignitron.

tube The main current connections are shown by the heavy black conductors at the top and bottom of the tube. The ignitor is the silicon-carbide (carborundum) pencil which dips into the mercury.

In operation a current pulse is applied from the ignitor to the mercury when it is desired to fire the tube, or to cause it to conduct. The action of this pulse is such that the current flow causes one or more minute arcs at the contact surface of the carborundum pencil and the mercury, thus providing an initial supply of electrons which starts the ionization process. This pulse is usually supplied by a thyratron, as it is necessary to fire the tube each cycle when it operates on alternating current. Various methods are used for timing the pulses, depending upon the application of the tube.

The ignitron has two main industrial uses. The first is in the precision control of current in spot welding and seam welding many different types of metal. The second is in the rectification of alternating current for industrial use. An important advantage of the ignitron as an electrical relay is that there are no moving parts, and therefore low maintenance results. Also, special foundations are not required and fire hazards are reduced.

THE MAGNETIC AMPLIFIER

A result which is quite similar to the time-controlled currents from the thyratron can be obtained, without the use of vacuum tubes, with a device

Chapter 22   *Industrial Applications*   763

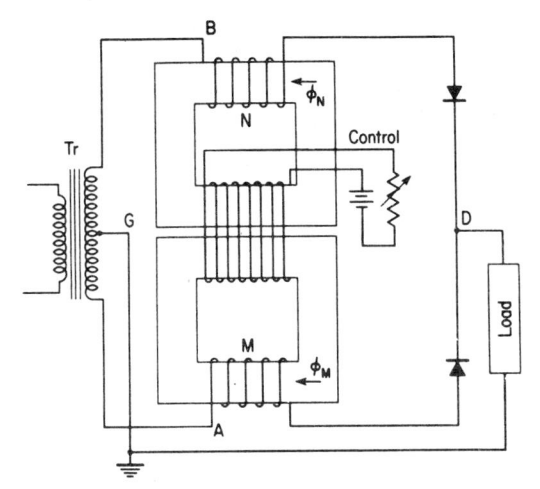

FIG. 22-23. A simple magnetic amplifier circuit.

known as a magnetic amplifier. It consists of two or more blocking inductors wound on separate iron cores with rectifier elements in the lines so that current can flow in one direction only. The iron cores are of special steel that has high permeability until saturation is reached and then very little additional flux is possible.

A simple single-phase circuit will be analyzed to show how this device operates. A diagram of this equipment is shown in Fig. 22-23 and consists of a power supply transformer Tr, the two cores with blocking coils M and N, a common control winding, two rectifier units, and the load resistor. The cores are assumed to have a hysteresis loop as shown in Fig. 22-24.*

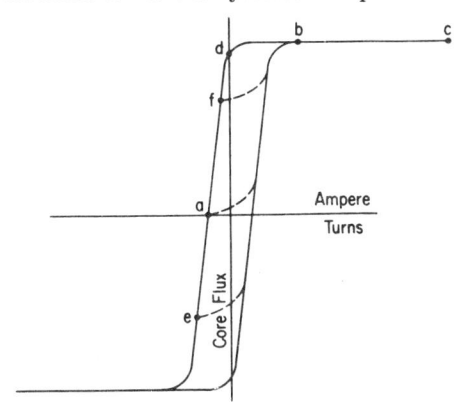

FIG. 22-24. The hysteresis loop for the core of a magnetic amplifier

* See Chapter 2 for a discussion of hysteresis loops.

This loop is slightly idealized in that the sides of the loop are assumed to be straight and the saturation is complete, so only a fixed magnitude of flux can be reached.

Let it be assumed that there is a sufficient current flowing in the control winding to bring the flux in the core back to zero as indicated at a. In Fig. 22-25(a) the time variation of voltages, fluxes, and currents in core M are plotted. The analysis is started at time t_1 when e_{AG} is at zero value and increasing. This voltage starts a current from A to D and back through the

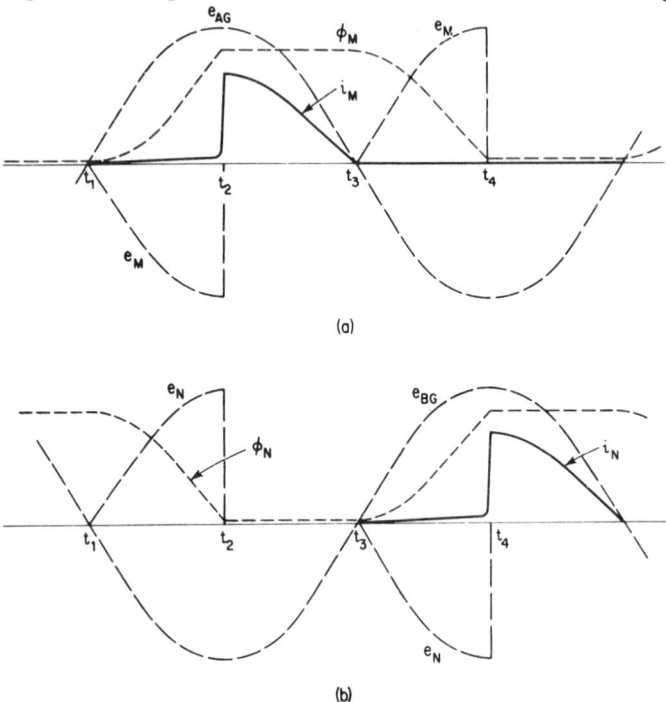

Fig. 22-25. Time phase variations of voltage, flux, and current in a magnetic amplifier with load current at half value.

load. The current, however, gives a magnetic force or ampere turns that causes the flux to build up along the dotted line from a to b in Fig. 22-24. This increase of flux causes a counter-voltage to be established in the coil M, which limits the current to a very small value.

The rate of change of flux will be sinusoidal as shown by the dotted line. If properly designed the limit of core flux will be reached at t_2 when the voltage wave has reached its maximum value. No further increase in flux is possible, so the counter-voltage in the coil collapses and full voltage is impressed on the load through the rectifier unit. The current jumps to a maximum value giving ampere turns indicated at c, Fig 22-24. The current then reduces to zero along with the voltage, and the ampere turns also reduce to zero. The flux does not, however, reduce to zero, but remains at the value indicated at d, Fig. 22-24. When the voltage reverses, the current is prevented from flowing by the rectifier unit. It remains to be shown how the flux returns to zero at a, and this will be done in connection with the analysis of the next half cycle for coil N.

The time variation of voltages in coil N is shown in Fig. 22-25(b), which has the same time scale as part (a) but is displaced downward in order to prevent confusion with a large number of waves on one diagram. At time t_3 the voltage e_{BG} is just starting to become positive and the sequence that occurred at t_1 in coil M now takes place in coil N. The load current is, therefore, a series of pulses coming each half cycle.

Attention may now be turned to the reaction of the control coil which is normally composed of a large number of turns of fine wire. The rate of change of the flux in core N immediately after t_2 might be expected to produce a high voltage and a secondary current. A very small current in the control winding, however, produces large ampere turns, and so this

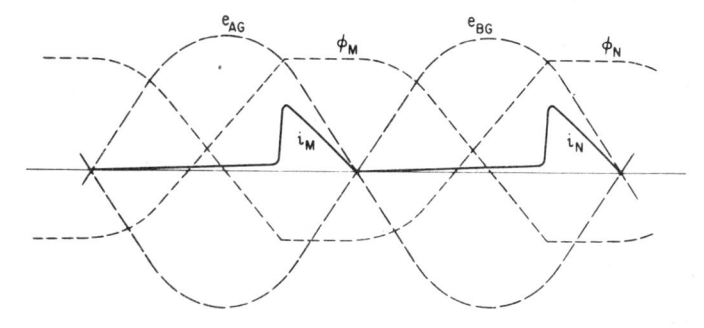

Fig. 22-26. Time phase variations in a magnetic amplifier with small load current.

provides the magnetic force to cause the flux in core M to reduce to zero. If this flux reduction just neutralizes the increase of flux in core N, then no voltage is induced in the control winding and this is the condition that exists. The counter-voltage in coil M set up by this flux variation essentially neutralizes the negative voltage e_{AG} so that there is very little voltage across the rectifier until time t_4 when the flux change in core M ceases. The reduction of flux in core N occurs when the flux in core M is again increasing.

If the control current is increased, the flux in the core is forced negative. Let it be assumed that the starting flux is at point e of Fig. 22-24. The increase in flux now follows from e to b, and this change of flux, being much greater, permits the coil M to block the current for much longer. This is shown in Fig. 22-26; obviously the load current is greatly decreased.

If the control current is decreased until the starting flux is at point f, then only a small increase in flux is possible. This permits blocking by the coil for only a small part of the cycle, and a large average load current results, as shown in Fig. 22-27.

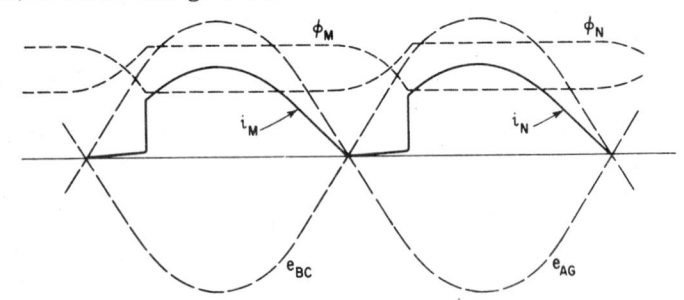

Fig. 22-27. Time phase variations in a magnetic amplifier with large load current.

The load current is thus controlled by small variations of current in the control coil and acts as a power amplifier. Several control coils may be wound on one amplifier unit. So far as the operation of the unit is concerned these several coils produce one equivalent magnitude of ampere turns. The magnetic amplifier is not usually used for audio and radio frequency amplification, but for industrial controls it is exceedingly satisfactory.

PHOTOELECTRIC MEASUREMENT AND CONTROL

A typical measuring and control device is the phototube with its associated amplifier and relay circuit. The phototube is a form of resistance varying transducer that responds to the intensity of illumination. It may be used to measure the temperature of the blast from a Bessemer converter in a manner similar to the measurement of temperature with a resistance bulb.

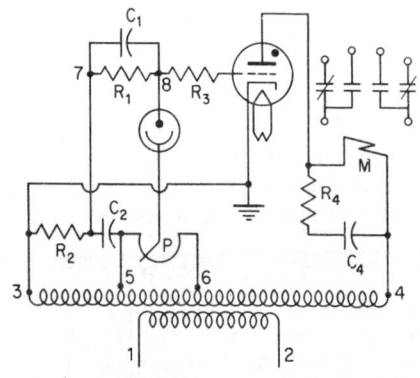

Fig. 22-28. The electronic circuit for a photoelectric relay.

As a control device a photoelectric cell may be used for opening doors when a light beam is broken by a pedestrian. It can count automobiles entering a tunnel and sort beans by knocking the discolored ones out of line. A typical circuit for a photoelectric relay is shown in Fig. 22-28. Power to operate the relay is supplied from a 115-v, 60-cycle line to the primary of the transformer numbered 1 and 2. The phototube controls a thyratron tube that in turn operates a magnetic relay. The voltages at the various points of the circuit are shown in Fig. 22-29. Point 3 on the diagram is at ground potential and is used as the reference. The anode voltage is the full

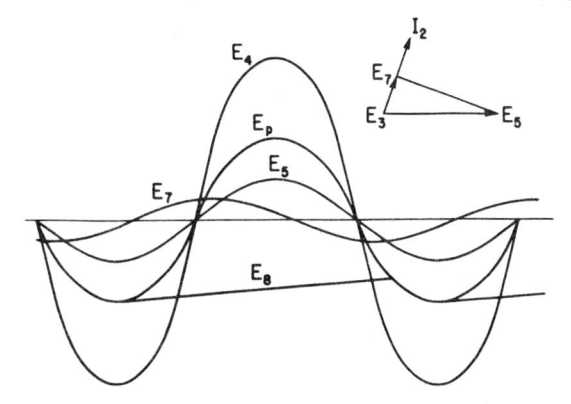

FIG. 22-29. Voltage distribution in an electronic photoelectric relay.

potential of the secondary winding of the transformer. The voltage from point 3 to point 5 is impressed upon C_2 and R_2 in series and is shown in Fig. 22-29 as the instantaneous voltage curve for point 5. The phasor diagram for this portion of the circuit is shown in the upper right-hand corner of Fig. 22-29. Here it is seen that the current I_2 leads the voltage 3-5, and the resistance drop in R_2 will also lead this voltage but will be of considerably smaller magnitude. This instantaneous voltage is shown as the voltage curve for point 7 and is observed to lead the anode voltage and so is positive when the anode voltage is changing from negative to positive. The thyratron tube thus fires on the first portion of the positive cycle and so supplies maximum current to the actuating coil of the relay M in Fig. 22-28. This analysis has assumed that the grid of the tube was at the same potential as point 7. Such would be the case when no light falls on the phototube. Under these conditions the phototube acts as an open circuit, the thyratron tube fires, and the actuating coil of the relay is therefore energized.

When light strikes the cathode of the phototube, it releases electrons and the tube becomes conducting. The rate of flow will depend upon the number of electrons released and, therefore, upon the intensity of the light that strikes the phototube. When a large amount of light strikes the tube, it becomes a better conductor. Current will flow, however, only when the anode (which is connected to point 8) is more positive than the cathode, which is connected to the potentiometer. The capacitor C_1 will tend to hold the point 8 at the potential that is obtained when the potentiometer is at its maximum negative value, since R_1 and R_3 both have very high resistances. The charge leaks off gradually, as is indicated by the curve E_8 in Fig. 22-29, which assumes zero drop in the phototube as a result of the intense light which strikes it. The negative grid voltage on the thyratron prevents it from firing, and the relay coil is therefore inactive.

When less light reaches the tube, the negative potential of point 8 cannot be so great, since there is an insufficient current flow to charge the capacitor C_1. Thus the grid voltage will be somewhere between the curves E_8 and E_7, depending upon the light intensity. As the light is decreased in intensity, a point will be reached where the current flow in the phototube is insufficient to charge the capacitor and maintain the required negative grid value to block the thyratron tube. When this point is reached, the tube will fire and the relay will operate. The magnitude of the current flow in the phototube will depend (among other factors) upon the voltage of the potentiometer. By adjusting this voltage it is therefore possible to control the critical intensity of light at which the relay will operate. The relay marked M is provided with a double set of contacts, one set of which is open when the coil is energized and closed when it drops out. The other set closes when the coil is energized and opens when the light blocks the tube and drops the relay armature. The capacitor C_4 and resistor R_4, which are in parallel with the relay coil, are used to prevent chattering of the relay as explained in the electronic timer circuit.

ELECTRONIC CONTROL OF MOTORS

The electronic control of motors has become a most important contribution of electronics in industry. These motors are usually d-c motors supplied from a-c lines by electronically controlled thyratron rectifiers. A simplified diagram of this type of control is shown in Fig. 22-30. At the right of the diagram the field circuit of the motor is supplied by a small rectifier using two gas diodes. (In many units this rectifier would be electronically controlled.) In the center portion of the diagram the motor armature is supplied by a thyratron rectifier of high current capacity. This rectifier has phase control on the grids of the thyratrons so that the time of firing of the thyratrons can be controlled, and in this way the average armature current flow is subject to the control element of the unit shown at the left of the diagram. Phase control is accomplished by means of a phase shifting circuit using a saturable reactor as the control element. The control current to the saturable reactor is adjusted electronically, but in the diagram this control is personalized for simplicity. The electronic unit would thus control speed, or torque, or some combination of these quantities in accord with a predetermined plan.

Chapter 22 — *Industrial Applications* — 769

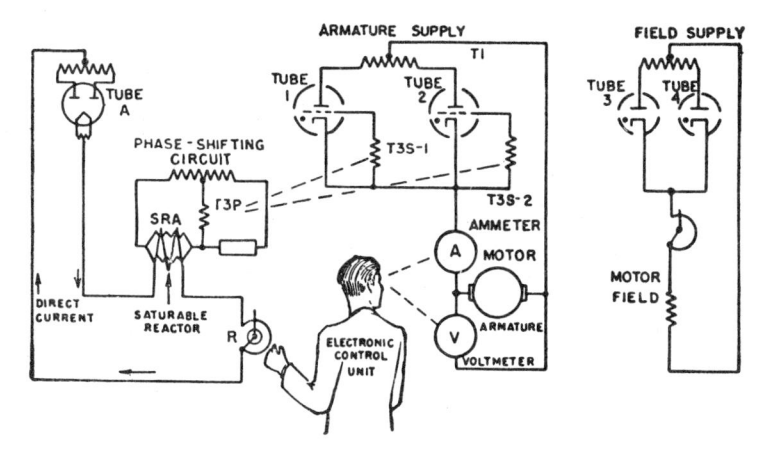

FIG. 22-30. Simplified diagram of thyratron motor control.

SERVOMECHANISMS

In the preceding paragraph it was stated that the direction or regulation of the motor was often controlled by an electronic device rather than by a person. Such an electronic controller can maintain the tension on a paper roll constant or vary the speed of the motor in accord with some predetermined pattern. The function of the control would be to compare the output or performance of the motor with the desired standard, measure the error, and then act to correct the error. Such a device is often called a closed loop control system. A servomechanism is a power-amplifying closed loop system and is usually associated with the control of movement or position. The control of antiaircraft gun positions from a radar station and the control of the rudder position on a ship from a small, easily operated wheel on the bridge are typical illustrations.

A block diagram of the functions of the various parts of a servo-system is given in Fig. 22-31. In the first place, there must be some standard or desired position or quantity. This may be fixed or varying according to some pattern that is externally determined as, for instance, the position of the radar antenna in the gun control, or the position of the control wheel on

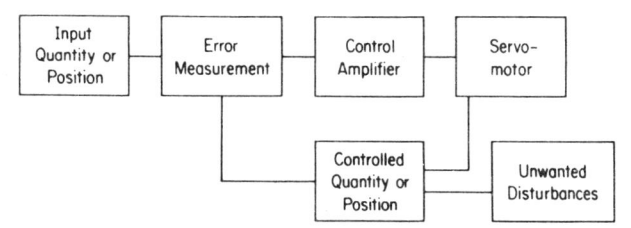

FIG. 22-31. A block diagram of the functions of a closed-loop control system or servomechanism.

the bridge of the ship. Then there is the position of the controlled quantity which is determined by past history, external disturbances, and the movement of the servomotor. Exact correspondence between the input and controlled positions is seldom achieved, so the error-measuring device compares these two and determines the error. The small signal from the error-measuring unit is then amplified and evaluated in the control amplifier. The evaluation often consists not only in measuring the amount of the error, but also in determining whether the error is increasing or decreasing, and sometimes also considers the movement of the input or standard. The resultant signal goes to the servomotor that changes the position or quantity of the controlled unit to agree with that of the input.

The theory of the stability of such a system is beyond the scope of this book. Most servo-systems use the same electronic circuits that have been described in previous chapters. In many cases the time constants of the circuits will be longer so that time delays in the electronic circuits can compensate for large mechanical inertias.

A servo-system of this type has already been discussed in this chapter. The millivoltmeter with electronic follower that is used to measure the thermocouple voltage in the Wheelco recorder-controller is such a device. The indicating pointer is the standard or input. The position of the aluminum vane in the tuning coils acts as the error-measuring element. This is designated as the meter section in Fig. 22-12. The oscillator-amplifier unit acts as the amplifier controller, and the motor drive unit and the associated motor represent the servomotor unit.

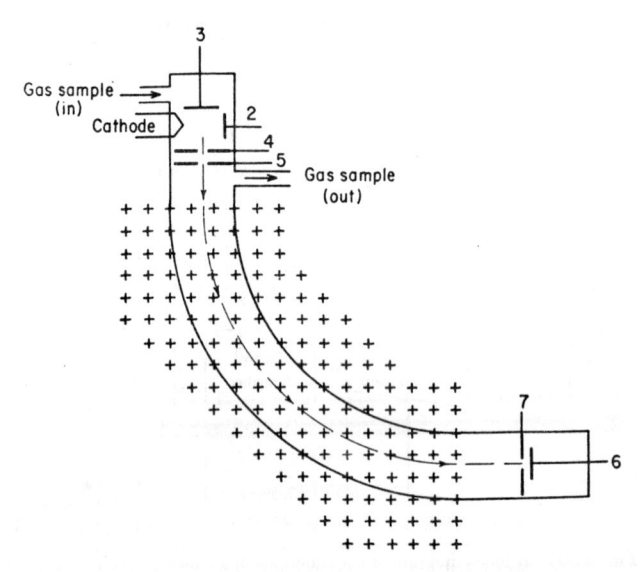

Fig. 22-32. Simplified diagram of a mass spectrometer.

Most servo-systems may be broken up into their functional groupings as this has been. Such a procedure often assists greatly in analyzing the performance of such equipment and in locating the cause of improper performance.

ELECTRONIC RESEARCH INSTRUMENTS

Several truly electronic instruments are also of great importance to industry. The first is the *mass spectrometer*, which measures the mass of the ions by deflection of a stream of ions in a magnetic field. A simplified diagram of this instrument is shown in Fig. 22-32. The sample to be studied, whicn is in the gaseous state at very low pressure, is fed through the evacuated tube that is in the heart of the instrument. The low-pressure gas is ionized by electrons moving from the cathode to the positively charged electrode 2, as shown in the figure. The positively charged ions are accelerated downward by the difference in potential of electrodes 3 and 4. Most of them will strike the negatively charged electrode 4 and will regain their missing electron. Part of them, however, will pass through a slit in the

Fig. 22-33. Control and recording panel on a mass spectrometer. (Courtesy General Electric Co.)

772 *Industrial Applications* Chapter 22

electrode and will be further accelerated by the negatively charged electrode 5. These positive ions are thus projected into the curved portion of the vacuum tube. This curved portion of the tube is located in a uniform magnetic field, as shown by the crosses on the diagram. A force will act on the ions that is perpendicular to both the movement and the magnetic field. This will cause them to travel in a circular path, the radius of which is dependent upon the mass of the ion. Only those ions having the proper mass will pass through the exit slit in plate 7 to reach collecting plate 6.

By a careful balance of accelerating potentials and the magnetic field, the various ions can be determined from the controls of the instrument, and the relative abundance of ions of different mass in the gas sample can be determined from the current to the collecting electrode.

The actual instrument contains many refinements as indicated by the

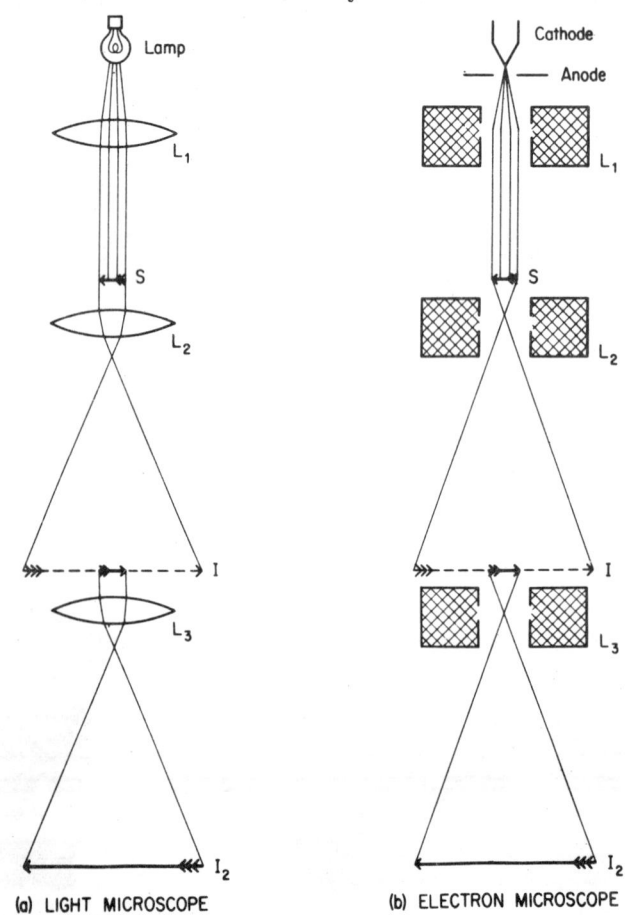

FIG. 22-34. Focusing of the electron microscope is analogous to that of the light microscope.

Chapter 22 — *Industrial Applications* — 773

FIG. 22-35. Large electron microscope for research.

photograph of Fig. 22-33. This type of instrument is extensively used for quick chemical analysis in petroleum refining, in the synthetic rubber industry, and in other similar process industries.

A second electronic instrument that is of considerable importance is the *electron microscope*, shown in Fig. 2-35. It is possible to produce a narrow beam of electrons by the focusing action of electric and magnetic fields.

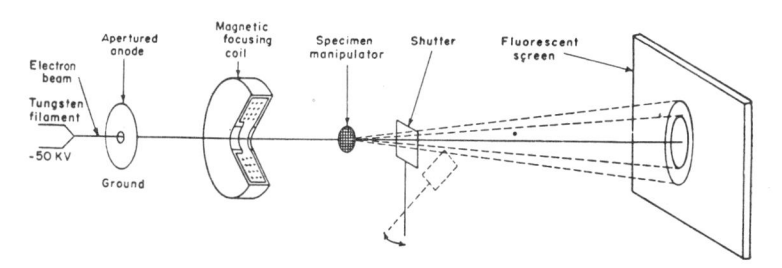

FIG. 22-36. The component parts of an electron diffraction instrument.

This beam of electrons can then be used as the basis of a shadowgraph microscope. The similarity to the light microscope is shown in Fig. 22-34. The definition with the electron stream is much greater than that of light, and so, much smaller objects can be viewed than with light. It has been possible, in fact, to obtain pictures of some of the more complex synthetic-rubber molecules.

The *electron diffraction instrument* is a third instrument using electronic principles directly. Its component parts are shown in Fig. 22-36. In this instrument a small beam of electrons is directed at a very thin specimen. Most substances are crystalline in their nature, and the crystalline structure of the specimen diffracts the electron beam so that the pattern on the fluorescent screen or photographic plate becomes a series of concentric rings, as shown in Fig. 22-38(a). The chemical composition of very minute samples can be determined from a study of such a pattern and a knowledge of the velocity of the electrons in the beam. A commercial form of this instrument is shown in Fig. 22-37.

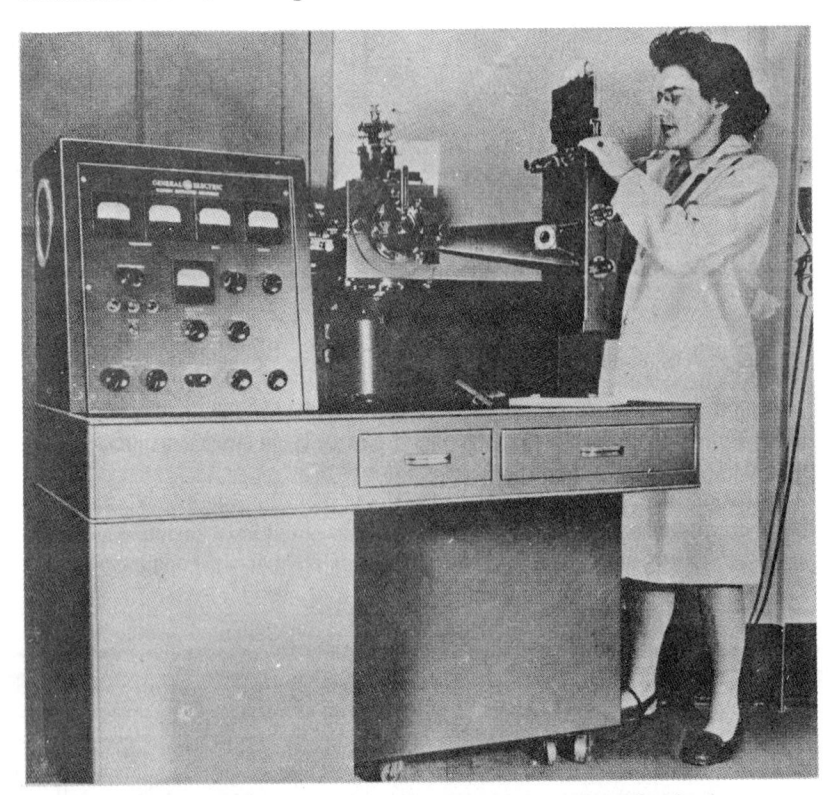

Fig. 22-37. Loading the camera of a commercial electron diffraction instrument. (Courtesy General Electric Co.)

Chapter 22 *Industrial Applications* 775

(a)

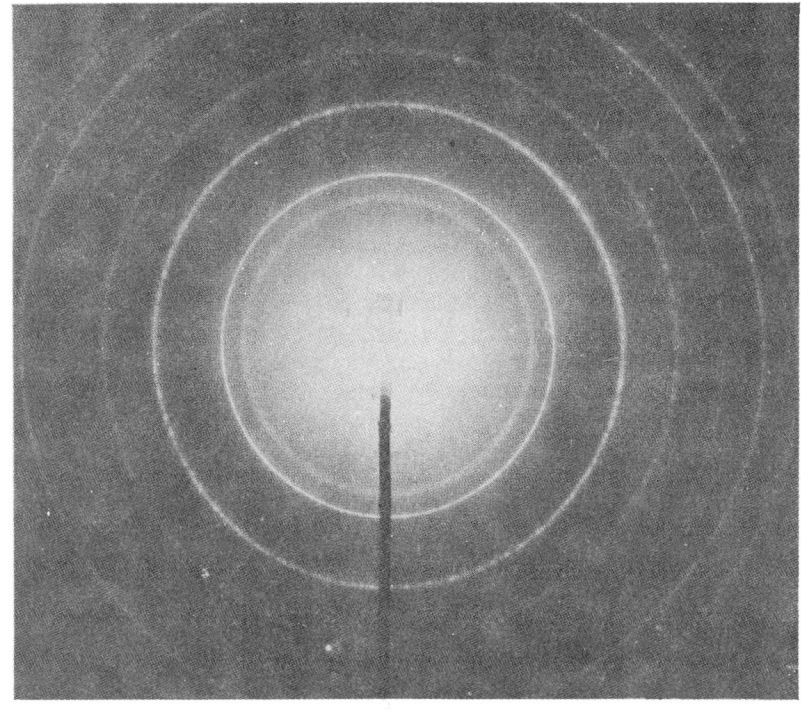

(b)

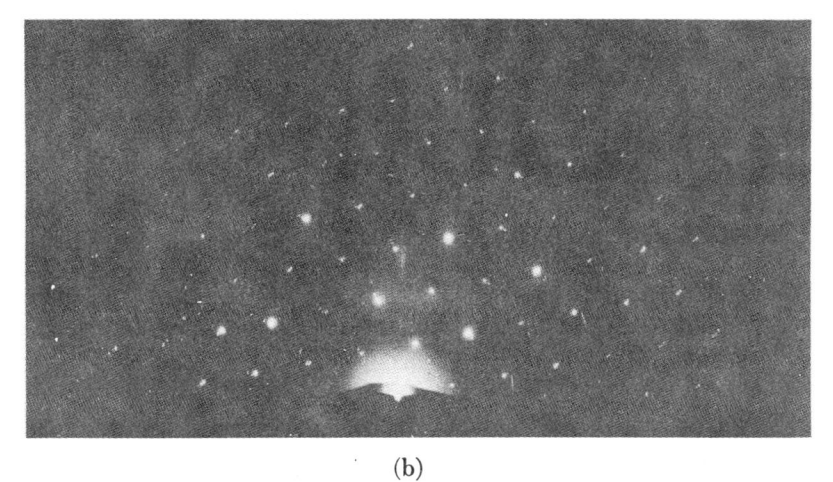

Fig. 22-38. Typical electron diffraction patterns. (a) The pattern for magnesium oxide, obtained by transmission of the beam; (b) the pattern for mica, obtained by reflection of the electron beam. (Courtesy of RCA)

Surface contamination may be studied by obtaining a reflected pattern from the electron beams, as shown in Fig. 22-38(b). This instrument is extensively used in the oil, steel, and other chemical industries.

The use of the mass spectrometer, the electron microscope, and the electron diffraction instrument requires highly skilled technicians guided by qualified scientists. Research scientists and production control engineers find in them, however, measurement tools that will make new manufacturing methods possible.

APPENDIX

SAFETY AND SPECIAL RADIO SERVICES

The present Safety and Special Radio Services may, for convenience, be grouped into four general categories as follows:

Safety services: marine, aviation, police, fire, forestry-conservation, highway maintenance, special emergency, State Guard, and point-to-point public service stations in Alaska.

Industrial services: power, petroleum, forest products, special industrial, low-power industrial, relay press, motion picture, agriculture, and radiolocation-land.

Land transportation services: railroad, motor carrier, taxicab, and automobile emergency.

Amateur, disaster communications, and citizens services.

These major classifications are in turn broken down into subgroups which form the little-known "networks" devoted to the protection of life and property and to the business, professional, and personal interests of so many people.

The following pages give a brief description of the some 50 radio services which comprise the Safety and Special Radio Services.

AVIATION SERVICES

Aeronautical radio is vital to the protection of life and property in the air, and to the maintenance of an adequate system of navigational aids on the ground and aloft. The necessity of radio in connection with aircraft operation is shown not only by the fact that it is legally required for airlines, but also by the large growth of voluntary installations by private aircraft.

The Aviation Services are concerned with the licensing and regulation of non-governmental aircraft radio stations, aeronautical enroute and aeronautical fixed stations, airdrome control stations, aeronautical utility mobile stations, radionavigation stations, aeronautical advisory stations, flying school stations, flight test stations, aeronautical public service stations, and Civil Air Patrol stations. A brief description of these various classes of stations follows:

Aircraft radio stations are essentially any type of radio transmitter installed aboard an aircraft. Except the public service type, such stations are used for operational and safety purposes. Because of the nature of their services, aircraft radio stations are divided into three categories: namely, air carrier aircraft, private aircraft, and public service aircraft.

Air carrier aircraft stations are used aboard commercial aircraft engaged in transporting passengers or cargo for hire. This class of station includes all scheduled, non-scheduled, and cargo carriers. To insure safe and efficient operation, air carrier aircraft are required to use communication and navigation equipment manufactured to meet high standards.

Private aircraft stations provide radio communication on aircraft used for pleasure and business other than the carrying of passengers and cargo for hire. This is the largest class of aircraft station.

Aeronautical public service stations furnish communication between aircraft in flight and the ground. They enable persons in aircraft to connect with the land line telephone system through public coastal stations. Public service aircraft stations on transport planes engaged in intercontinental service operate on the frequencies available to ship-telephone and ship-telegraph stations in the same manner as vessels offering public service.

Aeronautical enroute and aeronautical fixed stations provide the radio communication service necessary for the safe, expeditious, and economical operation of aircraft. Aeronautical enroute stations communicate between the ground and aircraft, whereas aeronautical fixed stations furnish point-to-point communication. In the continental United States, aeronautical fixed radio stations are used primarily as "back-up" circuits for land line facilities; however, in international operations, and operations in areas where land line facilities are not adequate, radio provides the primary service. Domestic air carriers are required to maintain two-way ground-to-air radiotelephone communication at terminals and other points to insure satisfactory communication over the entire route. Such a system is independent of safety radio facilities provided by governmental agencies.

Airdrome control stations make communication possible between an airdrome control tower and aircraft or aeronautical utility mobile stations; and further, exercise control over aircraft within the control zone of an airport in addition to controlling traffic, both aircraft and vehicles, at the airport. Such control consists of directing arriving and departing planes so as to avoid collisions and maintain an efficient flow of traffic into and out of the airport. Airdrome control stations at principal airports are normally operated by the Civil Aeronautics Administration.

Aeronautical utility mobile stations are installed aboard crash, maintenance, fire and other vehicles that operate on an airdrome, and are usually under control-tower direction. These stations provide two services: (1) communication by routine maintenance vehicles necessary to the operation of an airport; and (2) communication by emergency vehicles in the case of an accident on the field. This service is useful to both municipal and private airports.

Radio Navigation stations establish, by radio means, the traffic lanes of the air and provide information so that aircraft may determine their position, course, heading, distance from a station, etc. They are, for the most part, operated by the government; however, the type of navigational stations licensed by the FCC includes stations which furnish navigation, instrument landing, direction, distance, and altitude information.

Flying school stations are employed on the ground or on board aircraft for communicating instructions to students or pilots operating aircraft.

Flight test stations, ground or aircraft, are used for communication in connection with the testing of aircraft and aircraft components. Newly designed equipment can be tested under flight conditions. Communication with the ground is essential to log pertinent data and instructions pertaining to these flight tests.

Aeronautical advisory stations provide advisory communication between an airport operator and private aircraft so that airmen may ascertain the condition of the runways, fuel available, wind conditions, weather, or other needed information. Aeronautical advisory stations are not used for the control of aircraft in flight. Authorizations are issued only to the owner or operator of a landing area not served by an airdrome control station.

Civil Air Patrol stations — The Civil Air Patrol is a civilian auxiliary of the United States Air Force, but its radio stations are licensed by the Federal Communications Commission. It utilizes Air Force frequencies for communicating with land or mobile stations while carrying out search, rescue, training, or other activities for which this organization is responsible.

MARINE RADIO SERVICES

The use of radio on ships is the oldest of the safety radio services and the one with which the public perhaps is most familiar, mainly through publicity which has attended its performance under emergency conditions at sea.

The *Maritime Mobile Service* employs radiotelephone and radiotelegraph communication, and is broken down into ship-to-ship and ship-to-shore communication, including navigational aid communication.

Broadly speaking, maritime radio uses may be divided into those which are required by law for safety purposes and those which are voluntary on the part of ship owners (combining safety with other purposes, such as navigation and commerce), and those which are available for public correspondence.

All radio stations on ships of United States registry (other than government stations not on board vessels of the Maritime Administration) are required to be licensed by the Federal Communications Commission. The latter regulates ship radio equipment and operation and is responsible for administration and enforcement of the law, including standards for compulsorily installed equipment for safety purposes. Periodic ship inspections are required to enforce this mandate, which extends to the inspection of radio equipment on foreign ships touching U. S. shores.

Important from the compulsory safety standpoint (in addition to the Communications Act) are the International Safety at Sea Convention (London, 1948), which applies to ships making international voyages, and the Agreement for the Promotion of Safety on the Great Lakes by Means of Radio (effective November 13, 1954), which applies to ships on the Great Lakes only.

Radio stations on vessels utilizing telegraphy, telephony, or both, communicate with other ships, with aircraft, and with coast stations to transmit and receive signals and messages relating to safety of life and property and to assist navigation. In turn, coast stations transmit reports on weather and hazards to navigation. In addition, most shipboard stations handle messages for passengers and crew.

Under the Communications Act, all cargo vessels of 500 or more gross tons and all passenger vessels navigated in the open sea are required to carry radiotelegraph installations (radiotelephone may be substituted on cargo vessels under 1,600 gross tons) unless exempted by the FCC under certain conditions. The Safety Convention, among other things, requires cargo vessels between 500 and 1,600 gross tons on international voyages to be equipped with either radiotelegraph or radiotelephone. A large number of vessels on the Great Lakes are required to be equipped with radiotelephone installations under the new Great Lakes Agreement. Other ships, not coming within the compulsory classifications, are free to select either radiotelegraph or radiotelephone if radio communication is desired.

The use of frequencies and the operating procedures employed in the Maritime Mobile Service requires international coordination, inasmuch as ships traveling over the world must enjoy universality of communication. The first international convention for this purpose was held in Berlin in 1906, and the most recent one at Atlantic City in 1947. Many provisions of the latter came into force on January 1, 1949, and others on subsequent dates.

One of the important actions at Atlantic City was designating 2,182 kilocycles as the world-wide distress and calling frequency of radiotelephony for ships, and a long distance emergency frequency 8,364 kilocycles for radio-equipped lifeboats and other survival craft.

In addition, the very high (short-distance) frequency 156.80 megacycles was designated for calling, safety, intership, and harbor control purposes. Specific frequency bands were established for shipborne radar.

As a result of World War II, improvements were made in ship radiotelegraph and radiotelephone equipment looking toward conserving spectrum space and providing more efficient service. Perhaps more significant for the future was the wartime development of very high frequency radiotelephone equipment for short-distance communication and radar, loran, and other electronic devices and systems used for navigational purposes.

It is demonstrated, for example, that vessels equipped with *radar* may, if such radar is properly utilized, enter and leave ports during periods of reduced visibility with less danger of collision and running aground. Radar can fix the position of a ship independently of any other ship or shore stations when the ship is within approximately 50 miles of an identified shore line

The *loran* system is a longer distance radio navigational aid for determining the geographic position of a ship or aircraft at any point in its journey within the loran service area. Loran is especially valuable when weather conditions make celestial observation impossible. This system, as further distinguished from radar, is dependent upon the continuous transmissions from other stations, usually on land at known locations. The range of loran transmission over seawater is from 500 to 700 nautical miles during the day and up to 1,400 nautical miles at night. (See Chapter 21.)

Ramark is a navigational aid used in conjunction with radar but is dependent upon transmissions from a known fixed location. It enables the position of a ship or aircraft to be determined by means of bearings visibly presented on the radar indicator scope, thus helping to lead the vessel or aircraft to its specific destination.

Coast radio stations, normally established at fixed points on land, are used primarily for furnishing public communication service with ships at sea and on inland waters These stations are further classified according to the type of communication used (telegraph or telephone) and according to communication range.

The services of public coast stations are open to the general public. The services of limited coast stations are not. A portion of the time of a public coast station is devoted to transmitting weather reports to ships, expediting distress, urgency, and safety traffic, relaying messages, and such other services to ships as may be necessary. The rest of its time is spent in handling public correspondence.

Coast stations in the Continental United States and some coast stations in U. S. Territories which provide telephone communication to ships connect directly with public land line telephone systems. *Alaskan coast radio stations* are low-power medium range stations transmitting messages mainly for fishing, logging, fur trading, and other enterprises which utilize radio for safety and business communication, in localities where there are no public service, land line, or ship-to-shore facilities. These radio stations in general are open to public correspondence. *Fixed public stations* in Alaska provide safety and public service communication between Alaskan communities and with the Alaska Communication System. The latter (ACS), under the Department of National Defense, operates the main intra-Alaska communication system and routes Alaska message traffic to and from all parts of the world. The Commission maintains liaison with the ACS.

PUBLIC SAFETY RADIO SERVICES

These services comprise the police, fire, forestry-conservation, highway maintenance, special emergency, and State Guard radio services. They are available, primarily, to governmental agencies directly concerned with the public welfare.

The *Police Radio Service* is the oldest of these services, and probably is the best known. It serves municipal, county, and state police departments. Among other things, it provides communication between police land stations and mobile units, including police aircraft and police ships. Even the foot patrolman is part of this network when provided with a portable transmitter-receiver combination.

Police radio stations have been established in every state, in nearly every county, and most cities over 5,000 in population now have some police radio protection. Many of these, in turn, are integrated into regional and nation-wide police communication circuits.

Most police radio stations utilize radiotelephone, but radiotelegraphy is also employed by zone and interzone stations. The radiotelephone normally provides three-way communication — from a fixed land station to mobile units, from mobile units to land stations, and from one mobile unit to others. For each land station there may be a dozen to several hundred mobile units.

Radiotelegraphy, on the other hand, provides communication between cities and between states for the exchange of important information among police departments. Teletype has come into general use in areas not served by radiotelegraph stations.

Police frequencies are in the following frequency ranges: 1610-7,935 kc; 37-47, 72-76, 154-162, 450-460 megacycles, and in the microwave region over 952 megacycles.

The *Fire Radio Service* plays an important public role in the prevention and control of fires. It is used to maintain contact between fire headquarters and fire fighters.

The importance of an independent fire radio service has been recognized by the country's larger municipalities. Many of them have established their own systems to replace service heretofore provided by police radio stations. Volunteer fire departments also use radio.

Eligibility to communicate over fire radio frequencies now extends to "governmental subdivisions" (states, territories, counties, cities, etc.) and persons or organizations charged with specific fire protection activities. Applications from persons other than governmental must be accompanied by an endorsement from the public body holding local jurisdiction.

Users of fire radio frequencies require two distinct types of communication: namely, that employed between headquarters and the fire apparatus, and secondly, between the fire chief and individual fireman at the scene of the blaze. The first permits headquarters to maintain contact with all fire apparatus out on call. The second enables a squad chief at a fire to direct his men within or around a building. The former consists of a station located at a central point — such as a fire house which transmits essential messages to radio-equipped emergency apparatus. "On-the-scene" communication is provided by light-weight and low-powered packsets carried by individual firemen.

Fire-radio frequencies are in the following ranges: 1,630 kc; 33-34, 46-47, 72-76, 153-155, 159-162, 166-171, 450-460 megacycles, and the microwave region above 952 megacycles.

The *Forestry-Conservation Radio Service* provides communication networks essential to the prevention, detection, and suppression of forest fires and the conservation of wild life and natural resources. Forestry-conservation radio facilities are operated by the same kind of governmental authority which may utilize fire radio frequencies. Individuals or private organizations responsible for protecting large tracts of wood also are eligible. Its facilities are now used in game law enforcement, protection of forests from insects and disease, reforestation, flood and erosion control, as well as protection of timber from fire.

Basically, forestry-conservation radio communication systems are similar to police and fire radio networks. They consist of a land station at a fixed location, mobile units attached to trucks, and pack units carried by foresters or game wardens. In addition, stations can be set up in lookout towers during fire hazard seasons.

Forestry-Conservation Radio Service frequencies are in the following ranges: 2,212-2,244 kc; 30-47, 72-76, 156-162, 170-173, 450-460 megacycles, and in the microwave region above 952 megacycles. Because it is generally recognized that forest-fire control should take precedence over any other type of conservation work, a large number of the frequencies allocated for forestry-conservation use are primarily for forest-fire-fighting.

The *Highway Maintenance Radio Service* provides communication primarily between base stations and mobile units, and between the latter. Base stations also communicate with each other but usually only on a secondary basis, i.e., non-interference to mobile communications.

Highway maintenance communication is employed to coordinate activities at, or speed units to, the scene of snow-laden roads, landslides, road blocks, and similar emergency situations. It also permits highway departments to maintain instantaneous contact with road crews during floods and other natural disasters.

The Highway Maintenance Radio Service is assigned frequencies within 33-48, 72-76, 156-162, 450-460 megacycles, and in the microwave region above 952 megacycles.

Special Emergency Radio Service authorizations are issued only to the following: (1) establishments located at remote distances from other communication facilities; (2) emergency relief organizations (such as the Red Cross) which have disaster communication plans drawn up and ready for operation; (3) physicians normally practicing in remote areas; (4) ambulance services; (5) beach patrols engaged in life-saving operations; (6) school bus operators of regular routes in rural areas; and (7) *communications common carriers* and other operators of safety and public communications facilities.

The telephone and telegraph companies use this service in emergencies involving breaks in wire lines. Such situations are met by rushing trailers equipped with portable radio units to the scene and maintaining radio communication between each end of the break until repairs are made.

Special Emergency Radio Service frequency assignments are in the bands 2,726-3,201 kc; 33-48, 72-76, 157-162, 450-460 megacycles, and in the microwave region over 952 megacycles.

The *State Guard Radio Service* was operative during World War II for emergency purposes. State Guard stations handle, primarily, emergency communications relating to public safety and the protection of life and property and, secondarily, communications essential for training and organization maintenance. Low-power portable or mobile units together with higher powered base stations are used for these purposes.

The State Guard is not to be confused with the National Guard, which uses federal military radio facilities. When the National Guard is called into federal service, the State Guard constitutes the only military security force left to the particular state.

State Guard operation is on the frequency 2,726 kc.

DISASTER COMMUNICATIONS SERVICE

World events made necessary the establishment of a Disaster Communications Service by the Commission in 1951. The purpose of this service is to provide essential communication "incident to or in connection with disasters or other incidents which involve loss of communication facilities normally available or which require the temporary establishment of communication facilities beyond those normally available." This covers occurrences which involve the health or safety of a community or a larger area. Examples are floods, earthquakes, hurricanes, and even armed attack.

Both government and non-government stations are eligible. Thus, any fixed, land or mobile station can qualify; also amateurs and commercial operators. Authorization, however, is on the basis of participation in a recognized local, regional, or national disaster communications plan.

Its operation is assigned to the frequency band of 1,750-1,800 kc.

AMATEUR RADIO SERVICE

The Amateur Radio Service is one of the largest radio services in point of number of licensees and, at the same time, it is one of the oldest and most active radio groups.

An amateur station may not be used to transmit or receive messages for hire, or be used in connection with any commercial enterprise. However, this service provides interested and qualified citizens with a means of obtaining purely voluntary technical training and experience in the field of radio.

The amateur service has no age limits. Though the average age of self-styled "hams" is about 34, teen-agers are numerous, and even seven-year-olds have qualified for the novice license.

Although nominally a personal hobby, the amateur service has a high degree of public value. It constitutes a pool of self-trained radio technicians and operators upon which the country can draw in time of war. Amateurs furnish emergency communication during hurricanes, floods, wide-spread fires and other disasters.

There are separate licenses for amateur stations and operators. They are however, covered in a single combination license card. Applications for both are made on the same simple form.

To qualify for an amateur operator license, the applicant must pass a prescribed code test and technical examination. There are six classes of amateur operators according to degree of qualification. Most examinations are given in the field by Commission representatives.

RADIO AMATEUR CIVIL EMERGENCY SERVICE

An important part of the amateur's public service is his participation in the Radio Amateur Civil Emergency Service (RACES). This is a service which makes use of the amateur, his equipment, and portions of his normal frequency bands in time of war or other national emergency. At such times, safety radio services will be burdened with such a load of emergency communications concerning their own special field of activity that additional functions must depend almost entirely upon the RACES for their radio communications needs.

Each civil defense organization desiring to participate in the RACES must submit a communications plan describing the proposed operation. When such a plan has been approved, station authorizations may be issued to current holders of amateur station licenses who are members of the civil defense organization.

Although only an amateur may be a station licensee in RACES, certain grades of commercial radio operators as well as amateur radio operators may operate such stations or their units provided that they have received certification as to their loyalty and reliability from their civil defense organization. Each communications plan submitted to the Commission must include the certification of a Civil Defense Radio officer, who must be a holder of a commercial or amateur operator license above a certain grade and who will be responsible for the radio communication facilities for civil defense use.

LAND TRANSPORTATION RADIO SERVICES

The Land Transportation Radio Services consist of railroad, taxicab, motor carrier, and automobile emergency radio services.

The Railroad Radio Service: Of the 47 frequencies in the 153-162 megacycle region assigned to the Land Transportation Radio Services, 40 are allocated to the Railroad Radio Service for use in the Chicago railroad terminal area by 32 railroads operating in and out of that city.

Thirty-nine of these 40 frequencies are available for use in areas outside of Chicago. These 39 frequencies also may be used by the Public Safety Radio Services on a secondary sharing basis in areas where they will cause no interference to railroad radio operations.

Railroad communication needs fall roughly into two general categories — safety and operational. As a safety device, the primary use of radio is on main-line operations for end-to-end (from caboose to engine cab, for example) and wayside point-to-train communication (from a wayside station to the train en route).

As distinguished from this purely safety aspect, radio is further used to increase efficiency and economy of yard and terminal operations. Yard operation is essentially a local service requiring a communication range generally less than five miles. Terminal operation is also local in nature, although requiring a wider coverage — as much as 35 miles in some instances.

Frequencies available to the Railroad Radio Service are within 72-76, 159-162, 450-460 megacycles, and in the microwave region above 890 megacycles.

The *Taxicab Radio Service* is one of four services in the Land Transportation Radio category.

Persons licensed in any one of the Land Transportation Radio Services may render dispatching service on a cost-sharing non-profit basis to any other person engaged in the same type of transportation activity. This permits several companies to make joint use of the same base station, thereby reducing original investment and operating cost. Taxicab companies were among the first to adopt this method of operation.

Frequencies available to the Taxicab Radio Service are in the following ranges: 152-158, and 452-453 megacycles, and between 2,450 and 12,200 megacycles in the microwave region. The microwave frequencies are available only for developmental operations.

| Appendix | Safety and Special Radio Services | 787 |

The Automobile Emergency Radio Service: Masses of automobiles on crowded highways require prompt maintenance and repair service if the roads are to be kept clear. Since one disabled car can produce a serious and dangerous traffic jam, its speedy removal is essential. In this task, radio is of material assistance because it permits the rapid dispatch of service vehicles and tow trucks.

Two frequencies in the 30-40 megacycle band are shared by public garages which provide emergency road service, and two frequencies are provided in the 450-460 megacycle band for use by automobile associations, exclusively.

The *Motor Carrier Radio Service*, as its name implies, is intended for operators of land motor vehicles engaged in providing a common- or contract-carrier service for the transportation of passengers or the goods of others for compensation as a regular occupation or business. Its primary purpose is to furnish communication between terminals and vehicles operating on the streets or highways, either for the carriage of passengers or freight, or for certain supervisory or service activities in that connection. Local package delivery services are specifically excluded from this service, as are the operators of taxicabs, livery vehicles, school buses, or vehicles utilized only for sightseeing or special charter purposes.

Frequencies in the 72-76 and 450-460 megacycle bands, together with frequencies in the microwave region (above 890 megacycles), are shared by those operating in this service, and with other services as well. In addition, common and contract carriers of passengers operating within a single urban area have available frequencies between 30.66 megacycles and 31.14 megacycles; common and contract carriers of passengers operating between urban areas have available frequencies between 43.70 megacycles and 44.06 megacycles; and common and contract carriers of property operating between urban areas have available (for interurban operations only) frequencies between 43.98 megacycles and 44.42 megacycles.

INDUSTRIAL RADIO SERVICES

The industrial radio group embraces power, petroleum, forest products, motion picture, relay press, radiolocation, low-power industrial, and special industrial radio operations.

The privately operated radio communication system represents a new industrial tool in the national economy. Savings are made in time spent by employees on certain projects and in the ability to summon assistance to trouble spots before they become serious. But not to be overlooked are the radical changes in operating procedure which are brought about by the utilization of radio communication.

For example, a radio communication system eliminates the need for holding a fleet of repair trucks at base for emergency use. With radio, trucks can be dispatched on routine assignments to all parts of a city, and in the event of an emergency they can be rerouted to the scene more rapidly than otherwise possible.

Industries are recognized individually for purposes of allocating frequencies and determining eligibility. All, however, are required to follow similar licensing and operating procedures and to use transmitting equipment which conforms, in general, to common minimum technical standards of performance.

The following paragraphs describe in brief how the individual Industrial Radio Services operate:

The *Power Radio Service* provides facilities for electric, gas, water, and steam public utilities. Formerly, this type of service was within what was known as the "Utility Radio Service" and before that — although on a more restricted basis — by the "Special Emergency" class of station. At that time, utilities could use frequencies only on a showing of actual emergency. Under the 1949 revision of the rules, those limitations were liberalized.

Today, the most important application of radio by public utilities generally is in connection with restoring service interrupted by fire, storm, flood, and accident, although the principal volume of messages concerns routing maintenance activities not necessarily of an emergency nature.

Routine messages are transmitted to coordinate such construction activities as cable-pulling, wire-stringing, and pipe-laying. Special messages also are exchanged between load dispatchers and supply sources, such as generating stations, gas storage areas, and pumping stations.

Power radio frequencies are in the bands 2,292, 2,398, and 4,637.5 kc; 35-50, 72-76, 153-159, 450-460 megacycles, and in the microwave region above 890 megacycles.

Petroleum Radio Service frequencies are available to the petroleum and natural gas industries, with the exception of retail distributors. (The latter can, to a limited extent, use frequencies allocated to common-carrier general mobile services.)

Petroleum radio is helpful to persons employing geophysical methods to probe beneath the earth's land and water surface for accumulations of gas and oil. It is used also in drilling for, producing, collecting, or refining oil and gas, and in transporting these fuels and their by-products through pipelines from supply sources to distribution points. Efficient operation of pipeline systems depends upon continuous and reliable information about pressures, rates-of-flow, etc.

Petroleum is usually found in remote areas where the construction of wire lines would be impractical. During drilling operations it is desirable that continuous communication be maintained between well site, field headquarters, and mobile units.

Fire, explosion, well blowouts, accidents, equipment failures, and other emergencies require immediate coordinated action by medical, fire-fighting, mud-conditioning, and well-cementing services.

In addition to these emergency requirements, supervision of drilling operations calls for adequate communication facilities for transmitting information essential to the successful completion of a well. Radio communication is likewise important to roving pipeline repair crews and to aircraft patrolling the lines.

Microwave, often referred to as the "last frontier" of unused spectrum space, is being used for voice communication from one end of a pipeline to the other. The magnitude of such an operation is readily apparent in lines running for 1,000 miles or more.

Frequencies for the Petroleum Radio Service are in the following bands: 1,602-1,700, 2,292, 2,398 kc. and 4,637.5 kc; 25-50, 72-76, 153-159, 450-460 megacycles, and in the microwave region above 890 megacycles.

The *Forest Products Radio Service* is among the services established by the Commission in 1949. It places in the hands of privately owned timber and logging companies radio communication facilities similar to those employed by federal and state governments to detect, prevent, and suppress fires. Forest products radio facilities also are used in the interest of safer, more efficient and economical logging operations.

Radio is used in connection with actual logging operations — the trucking of logs from the area, the fluming of logs by water, their transportation over logging railroads, and the maintenance and use of specialized heavy machinery in remote areas.

The Forest Products Radio Service uses frequencies in the bands 1,676, 1,700, and 2,398 kc; 29-50, 72-76, 153-159, 450-460 megacycles, and in the microwave region above 890 megacycles.

The *Relay Press Radio Service* is employed by newspapers through operation of a central transmitter and mobile radiotelephone equipment in automobiles carrying reporters and photographers. One of the advantages of this type of radio system lies in the ability to send press representatives on routine assignments with the knowledge that they can be contacted instantaneously for further instructions or change in assignments.

Relay press frequency assignments are in the ranges of 72-76, 173-174, 450-460 megacycles, and in the microwave region above 890 megacycles.

The *Motion Picture Radio Service* is available only to the motion picture industry. In the main, it is used on location to tie parties to the nearest wire line for purposes of safety of life and property, to expedite shipment of supplies, also to coordinate action taking place on outdoor sets. The low-power radio equipment used for the latter purpose serves a useful function in coordinating the "shooting" of scenes, frequently to the point where time-consuming retakes are unnecessary.

Motion Picture Radio Service frequencies are in the following ranges: 1,628, 1,652, 2,292, 2,398, and 4,637.5 kc; 49-50, 72-76, 152-153, 173-174, 450-460 megacycles, and in the microwave region above 890 megacycles.

The *Low-Power Industrial Radio Service* is open to all industrial and commercial concerns having a need for short-range (localized) radio communication is the conduct of business. It provides for operation of any desired number of portable transmitter-receivers which are restricted to very low power in order to reduce interference between units and thereby permit many to operate on a few frequencies.

The range of the units employed by low-power industrial radio users varies from a few hundred feet to about a mile, depending on the positions of the communicating parties with respect to each other, surrounding terrain, buildings, and other factors.

Frequencies available to the Low-Power Industrial Radio Service are 27.255 and 27.51 kc; 33.14, 35.02, 42.98, and 154.57 megacycles.

The *Special Industrial Radio Service* was also activated in 1949. It provides longer distance communication than is possible in the Low-Power Industrial Radio Service, but may only be employed under certain conditions and circumstances.

The service is open to businesses engaged in production, fabrication, construction, agriculture, manufacturing, and mining, also some specialized activities such as those incident to petroleum operations, delivery of fuel oil, butane gas, and ice; servicing and repairing of heavy machinery, etc. Operation is confined to a plant area, construction project, or locations outside of cities of 500,000 or more population.

Special industrial radio is distinct from low-power industrial radio in that the former has a range up to about 30 miles, depending on terrain, the shielding effect of buildings, and other technical factors. Radio communication in urban plant area operations coming under the special industrial radio rules can be used only when the prospective user satisfies the Commission that low power is unsuitable for his purpose.

Frequency assignments are in the following ranges: 2,292, 2,398, and 4,637.5 kc; 27-50, 72-76, 152-155, 456-458 megacycles, and in the microwave region above 890 megacycles.

The *Industrial Radiolocation Service* provides for the use of radio in conjunction with geographical, geological, and geophysical activities engaged in by persons having need to establish a position distance or direction for purposes other than navigational. Oil exploration parties and persons engaged in rain-making activities on a commercial basis are included in those licensed in this service.

CITIZENS' RADIO SERVICE

The Citizens' Radio Service is designed primarily to afford private two-way short-range communication service.

Citizens' radio may be employed for communication on farms, such as between house and workers in the field; on ranches, such as between range shacks and fence riders; between work parties in remote areas; and with automobiles or other moving vehicles over limited ranges. In addition, it offers opportunity for radio-control devices, such as opening gates and garage doors; control of model airplanes, boats and other models; display signs, and, in fact, almost anything not expressly prohibited by the rules.

The Commission is allowing the widest possible latitude in developing the Citizens' Radio Service commensurate with provisions of treaty, law, and regulation. However, citizens' radio stations are not permitted to charge for messages, carry broadcast material, transmit directly to the public, or communicate with stations in other services.

The Citizens' Radio Service operates primarily in the band 460-470 megacycles, but also has the frequency 27.255 megacycles for low-powered operation limited to the remote control of objects or devices by radio.

REGULATION OF THESE STATIONS

Stations in the Safety and Special Radio Services are, as in the case of other radio services, basically regulated by the Federal Communications Commission according to the provisions of the Communications Act of 1934, as amended and, technically, by their respective covering rules and regulations.

Persons interested in the operating details of a certain service are urged to obtain a copy of the rules covering that particular service. These rules are not distributed by the Commission but are sold by the Superintendent of Documents, Government Printing Office, Washington 25, D. C. A list of Commission rules and other printed publications available from that source will be furnished on request to the Commissions' Washington office or any of its field offices.

After Safety and Special Radio Services stations are licensed in accordance with the rules, they are inspected regularly by engineers attached to the Commission's field offices located throughout the country. The stations must comply with the terms of their authorization regarding frequency tolerance, power limitations, permissible communications, call signals, etc. Wilful or repeated violations may result in revocation of license or criminal prosecution, or both.

In general, call signals are assigned in an order determined by blocks of calls made available for that purpose. Thus a station may have a call ranging from three letters and one digit (aeronautical land) to two letters and four digits (ship telephone). Because of their number, amateur calls are more complicated. The combinations identify the kind of station (for example, experimental stations are distinguished by an "X") and, in some cases, further indicate the regional location of the station.

The Commission's field staff investigates interference complaints, monitors transmissions for adherence to the assigned frequencies and quality of emission, and examines operators for the various classes of licenses. Complaints of illegal operation, unlawful interception of calls, etc., are received from time to time, and these matters are then further investigated by the field force. Unauthorized radio transmission is prohibited.

Any person who is legally, technically, and financially qualified and who can show that the public interest, convenience, or necessity will be served by a grant of his application, is eligible for a license if his proposed operation is of a type permitted under the rules of the Commission.

HOW TO APPLY FOR A LICENSE

The first step in seeking to operate a station in the Safety and Special Radio Services is to file an application with the Commission for authority to render a particular service which is provided for in the rules.

This authority, when granted, authorizes the station to operate for a stated period. In most Safety and Special Radio Services the normal license period is four years. Exceptions are coast stations and private and public service aircraft, where licenses run for two years; and five years for RACES, amateur, disaster, and citizens' station licenses.

Each application must be specific and complete. It should contain, among other things, information about the station location, proposed equipment, power, antenna height, and operating frequency.

Further application procedure must be followed in order to obtain a *renewal of license*. There are also forms for requesting additional time to construct a station, to modify a permit or license, and for permission to assign or transfer control of a permit or station license.

Application forms are available from any of the Commission's field offices, or by addressing the "Secretary, Federal Communications Commission, Washington 25, D. C." The Commission maintains engineering field offices in Boston, New York, Philadelphia, Baltimore, Norfolk, Atlanta, Savannah, Miami, Tampa, New Orleans, Mobile, Houston, Beaumont (Texas), Dallas, Los Angeles, San Diego, San Pedro, San Francisco, Portland (Oregon), Seattle, Denver, St. Paul, Kansas City (Missouri), Chicago, Detroit, Buffalo, Honolulu (Hawaii), San Juan (Puerto Rico), Anchorage and Juneau (Alaska), and Washington, D. C.

Applications, and all related correspondence, should be addressed to the Secretary of the Commission. Exceptions are applications for the Citizens' Radio Service and for amateur and commercial radio operators, which can be made to the area field office, and applications for service in Alaska, which should be made to the district office at Seattle.

The license privilege is extended by the Communications Act only to citizens of the United States. It is denied to corporations in which any officer or director is an alien or of which more than one fifth of the capital stock is owned or recorded or voted by aliens or foreign interests.

INDEX

A

Abbreviations, use of, 64
Abscissa, 39
Absorption losses, 646
Acoustic feedback, 239
Adcock antenna, 682
Additive color mixing, 525
Additive primaries, 526
Admittance, 116
Air-core transformers, 121
Algebra, 24
Alnico magnet, 74
Alternating current:
  adding, 98f.
  circuits carrying direct and, 192
  defined, 83
Alternating-current bridges, 127
Alternating-current circuits, 97-98
Alternating-current meters, 124-126
Alternating-current power sources, 104
Alternating-current waves, values of, 99-101
Alternation, frequency of, 97-98
Ammeter:
  defined, 51
  use of, 79
Ampere, defined, 49
Ampere-turn, as unit of magnetomotive force, defined, 71
Amplidyne, 738
Amplification factor, 139
Amplifier(s):
  audio, 225
    defined, 194
    and radio, differences between, 350
  cathode-coupled, 397
  cathode-follower, 219
  Chiriex, 391
  chroma band pass, 546
  Class A, 359, 360, 368
  Class B, 359, 376, 379, 386
  Class C, 359, 381, 386
  classes defined, 195
  compensated video, 227
  design of, 350, 351, 354, 355, 356, 360, 363, 366, 370, 372, 373, 374, 376, 381
  direct coupled, 230
  direct current, 230
  Doherty, 389
  frequency response, 205

Amplifier(s) (cont.):
  fundamentals of, 194
  grounded-grid, 395
  over-all gain, 205
  parallel, 373
  pentode, 201
  power, 215
  public address, 237
  push-pull, 374
  radio, 349ff. (*see also* Radio amplifiers)
    and audio, differences between, 350
  resistance-capacitance coupled, 196, 199
  shielding of, 370, 435
  transformer coupled, 210
  vacuum-tube, 194
  video, 206, 225
  voltage, 194
Amplifier application notes, 360
Amplitude, definition of, 283
Angle, 33
Antenna(s):
  Adcock, 682
  dimensions of, 303
  dipole, 645
  dual frequency, 678
  electromagnetic, 303
  elevated, 640
  elevated half-wave, 640
  folded dipole, 671
  grounded, 640, 643
  half-wave, radiation characteristics of, 642
  loop, 680
  microwave, 664
  microwave beam, 690ff.
    beam width, 693
    examples, 693, 694, 708, 724
    power gain, 691, 705
    relay, 714
    television relay, 715, 719
  monopole, 645
  navigational, 680
  radar, 690, 691, 693
  receiving, 307
  reflector, 678
  sense, 681
  slot, 668
  special-purpose, 670
  television receiving, 675
  television transmitting, 507, 673

Index

Antenna (*cont.*):
  uhf, 679
  V-plate, 680
  wide-band, 670
Antenna axes, patterns in the plane of, 658
Antenna gain and effective area, 663
Antenna system(s):
  directional, 653
  functions of, 640
Antenna tower lighting, 449
Antenna tuning and matching, 449
Antilogarithm, 11
Antinodes, defined, 287
Area covered by FM signals, 483
Armature core, drum type, 78
Armstrong system, transmitter, FM, 485
Array:
  colinear, 659
  mattress, 660
  rectangular, 660
Aspect radio, 504
Atmospherics, 629
Attenuator, 233
Audio amplifier:
  gain of an, 200
  nonlinear distortion, 206
  transistor, 240
Audio limiter, 631
Audio transformers, 121
Automatic radio compass, 682
Automatic volume control, 149
Automation, 737
Average values of a-c waves, 100

B

Back porch, TV, 499
Bandwidth for TV, 504
Bar magnet, magnetic field of, 69
Base:
  binary number system, 20
  common logarithms, 10
  decimal number system, 7
  Naperian logarithms, 45
Base region, 165
Batteries, types of, 64–65
Beam power tube, 149
Bias circuits:
  fixed, 252
  self, 252
  for transistors, 251
Bidirectional pattern, 657
Binary number system, 20
Binomial squares, 29
Blacker-than-black level, 513
Blanking interval, TV, 498
Blanking pulse, TV, 498
Blocking oscillator, 273
Body organs, electrical characteristics of, 743
Bolometer, definition of, 610

Bonding of plywood, 743
Boundary conditions, 288
Bridges, alternating current, 127
Brushes, 76

C

Calculus, integral, 44
Camera, TV, 508
Camera tube, 508
Capacitance:
  measurement of, 127f.
  resistance, inductance, and, in parallel, 116f.
  resistance, inductance, and, in series, 107f.
  resistance and, in series, 107, 109
Capacitator, coupling, 198
Capacitive circuit response, 95–97
Capacitive reactance, 105ff.
Capacitor:
  capacitance of, 91
  defined, 91
  determining factors in capacitance, 92
  electrostatic characteristics of, 91
  relation between voltage and current in, 105–106
Capture effect, FM, 481
Carrier wave concept, 309
Cascade, 202
Cathode:
  equipotential, 134
  physical construction, 133
Cathode coupled amplifiers, 397
Cathode-follower amplifier, 219
Cathode-ray, screens for, table, 155
Cathode-ray tube, 152
Cavity resonator, definition of, 601
Channel multiplexing, 716, 717, 718
Characteristic impedance, 239, 293
Chiriex amplifier, 391
Chroma band-pass amplifier, 546
Chromaticity, 528
Chrominance demodulator, 547
Chrominance signal, 530
Chrominance sync channel, 549
Circuits:
  alternating current, 97–99
  clamping, 266
  clipping, 268
  closed TV, 522
  color matrix, 539
  common-base, 241
  common-battery, 189
  common-cathode, 222
  common-collector, 241, 249
  common-emitter, 241, 248
  common-grid, 222
  common-plate, 222
  direct current, 48–82
  grounded-cathode, 222, 241

Index

795

Circuits (*cont.*):
  grounded-grid, 222, 241
  grounded-plate, 222, 241
  horizontal deflection, TV, 499
  inverse feedback, 218
  parallel, 53, 115–116
  pulse, 263
  push-pull, 214
  push-pull transistor, 257
  resonant, 112ff.
  scaling, 275
  self-bias, 252
  series, 52, 104f., 107f., 112ff.
  series-coupling, 649
  series-parallel, 54–56
  series-resonant, 112ff.
  sync-separating, 501
    TV receiver, 515
  transistor, 241
  vacuum tube, 207
Circuit elements, limitations of ordinary, 575
Circuit mixing, 233
Clapp oscillator, 556
Close coupling, 648
Closed-circuit TV, 522
Coaxial transmission lines, 297
Coefficient, numerical, 25
Coercive force defined, 74
Colinear array, 659
Collector, 165
Color, resolution of, by the eye, 533
Color-bar test pattern, 535
Color burst, 538, 542
  synchronizing, 542
Color camera, TV, 534
Color-killer tube, 546
Color matrix, receiver, 547
Color-matrix circuit, 539
Color picture tube, 549
Color receiver, TV, 544
Color television, 525ff.
Color triangle, RGB, 530
Color video signal, 534
Colpitts oscillator, 556
Commutator, explanation of, 76
Compatible color TV, 530
Complementary symmetry, 257, 258
Computer:
  digital, 20
  to "program" a, 20
Conductance, defined, 53
Conductor:
  defined, 48
  iron as a, of magnetism, 71ff.
Constant, dielectric, defined, 91ff.
Contactor switch, electric, 71
Contrast, picture, 520
Coordinates, rectangular, 39
Cosine of angle, 34

Coulomb, defined, 49
Countervoltage (*see* Self-induction)
Coupling:
  close, 648
  coefficient of, of coils, 122f.
  loose, 648
  selectivity and, 123f.
Coupling networks, 647
  typical, 650
Crosby system, FM, 485, 486
Crystal filters, 632
Crystal probe, 570
Current:
  components of, in tubes, 129
  rate of change, in sine wave, 101f.
  relation between voltage and, in capacitor, 105f.
Current amplification, 166
Current feedback, 252
Cycle timers, 749
Cycloids, definition of, 603

D

Damping diode, 517
Damping tube, 517
D'Arsonval, Arsene, 78
Decibel, 15, 191
Decimal number system, 1
Decoupling filter, 257
De-emphasis, FM, 488
De Forest, 131
Degree of ionization, 619
Delay distortion, 182
Demodulation, 338
  diode detectors, 340
    crystals, 340
    tubes, 341
  filtering, 339
  grid, 346
  heterodyne, 347
  ICW, 349
  message restoration, 340
  plate, linear, 343
  plate, square law, 344
  rectification, 338
Derivative, 40ff.
Detectors, 338–340
  crystal, 340
  grid circuit, 346
  heterodyne, 349
  plate, linear, 343
  plate, square law, 344
  tube diode, 341
Deviation ratio, 479
Diathermy, 743–744
Dielectric constant, 91ff.
Dielectric heating, 743
Diffraction, 634
Digit, 1

796 Index

Diode, 134
  damping, TV, 517
  junction, 162
  semiconductor, 163
Diplexer, TV, 493
Diplexing, 506
Dipole antennas, 645
Direct current, circuits carrying alternating and, 129
Direct-current circuits, 48ff.
Direct-current generators, 74, 77
Direct-current meters, theory and construction of, 78ff.
Direct-current motor, principle of, 77
Direct-current restorer, 266
Direct waves, 632
Direction finders, 680
Directional antenna systems, 653
Directional array, 653
Directional coupler, definition of, 611
Directional gain, 663
Directivity of a single linear antenna, 653
Discriminator, FM, 488, 489
Distortion, 121, 181
Distortion in radio amplifiers, 363
Diversity receivers, 473
Diversity reception, 629
Doherty amplifier, 389
Dolly, 509
Donor, 161
Double-stub tuners, 594
Driven element, 678
Dry-cell battery, construction of, 65f.
Dual-frequency antennas, 678
Dual-purpose tubes, 150
Duct:
  elevated, 636
  surface, 636
Duct propagation, 636
Dynamometer type of meter, 124f.
Dynamotors, principle of, 77f.

E

Eccles-Jordan trigger circuit, 274
Effect of ionosphere on sky wave, 620
Effect of the ground on vertical radiation patterns, 661
Effective area, of antenna, 664
Effective values (*see* Root mean square)
Efficiency (of antennas), 645
Electric current, magnetic effect of, 70f.
Electric field, 288
  direction of, 306
  strength of, 289
Electric force, lines of, 289
Electrical quantities, 49
Electrical sheets, 74
Electromagnetic antenna, 303
Electromagnetic horn, 302, 667

Electromagnetic radiator, 303
Electromagnetic waves, 282, 289
  on wires, 288
Electromagnetism, 69ff.
Electromotive force, defined, 74f.
Electron diffraction instrument, 774–776
Electron emission, 131
Electron-hole pair, 160
Electron microscope, 737, 773–774
Electron multipliers, 509
Electron theory of electricity, 48f.
Electron tubes, limitations of, 749
Electronic methods of industrial measurement, 749–757
  impedance variation method, 755–757
  pH measurement, 755
  pressure measurement by piezoelectric crystals, 755
  temperature measurements, 749–755
Electronic mixer, 239
Electronic research instruments:
  electron microscope, 773–774
  electronic diffraction instrument, 774–776
  mass spectrometer, 771–773
Electronic timers, 744–749
  cycle timers, 749
  time-delay circuits in, 745–748
Electrostatic coupling, 651
Elevated antennas, 640
Elevated duct, 636
Elevated half-wave antenna, 640
Emission, 132
  constants for, 132
Emitter, 165
Energy flow, direction of, 584
Equalizing pulses, 501
Equation, 29
  quadratic, 31
Error:
  maximum, 5
  percentage of, 5
  relative, 5
Exciting current, 120
Exponent, 3
Exponential e^t, 44

F

Factor, 25
Factorial, 45
Factoring, 27
Fading, 627
  flutter, 629
  reduction of, 629
  selective, 628
Farad, defined, 91
Faraday shield, 651
Feedback:
  degenerative, 203
  inverse, 216

Index

Feedback (cont.):
  in oscillator, 553
  regenerative, 203
Field, TV, 503
Figures, significant, 5
Figure-eight pattern, 655
Filter, decoupling, 257
Filter, vestigial sideband, 506
Filter circuits, 170
Fixed-bias circuits, 252
Flash, TV, 503
Flashing frequency, 503
Flicker, TV, 502
Fluid theory of electricity, 48
Flutter fading, 629
Flux, symbol for, 122
Flyback, TV, 498
Flyback high-voltage, 515, 516
FM-FM system, telemetering, 484
FM receivers, 488
FM transmitters, 485
Folded dipole antenna, 671
Forming, 163
Forward-scatter propagation, 636
Forward transfer resistance, 242
Fraction:
  common, 8
  decimal, 8
Frames, TV, 498
Frame frequency, TV, 498
Frequency (-ies), 283, 290
  of alternation, 99–101
  critical, 623
  in diathermy, 743
  effects of, on inductive reactance, 102f.
  in induction heating, 741, 743
  lower half-power, definition of, 201
  maximum usable, 624
  in thyratrons, 759
  upper half-power, definition of, 201
Frequency deviation, 498
Frequency distortion, 181
Frequency diversity, 629
Frequency measurement:
  equipment for, 561
  Lissajous figures, 565
Frequency modulation (FM), 478ff.
Front porch, TV, 499
Functions, trigonometric, 33
Functional relationship, 39
Fuses, uses of, 61

G

Gas triode (*see* Thyratrons)
Gas tubes, 158
Generator(s):
  direct current, principle of, 74ff.
  internal resistance, 238
  maximum power output, 238

Graphic characteristics, 247
Grid, floating, 197
Grid, grounded, amplifiers, 395
Grid-dip oscillator, 561
Grid-leak resistor, 198
Ground-reflected waves, 632
Ground systems, 646
Grounded antennas, 640, 643
Group pattern, 659

H

Half tone in TV pictures, 505
Half-wavelength section, 592
Harmonics, 84f.
Hartley oscillator, 556
Heat, generation of, 100
Heat treating, 738–739
Heating of dielectric materials, 743
Height, virtual, definition, 621
Henry, defined, 87
Heterodyne, 562
Heterodyning, 326, 330, 347
High-fidelity, 183
  FM, 483
High-field emission, 132
High-frequency coils, design of, 740–741
High-frequency heating equipment:
  applications, 738–739, 743
  power efficiency, 743
High-frequency induction heaters, 738–743
  applications, 738–739, 743
  design of coils for, 740–741
  power efficiency of, 743
Hiss noise, 630
Hole, 159
Homing device, 682
Horizontal deflection circuit, TV, 499
Horizontal deflection coils, TV, 515
Horizontal polarization for TV, 507
Horizontal retrace time, 498
Horizontal scanning rate, 498
Horizontal sync pulse, 498, 499, 501
Horns, 666
Hue, 529
Hum and tube noise, 204
Hybrid coils, 192
Hysteresis, defined, 73
Hysteresis loops, 73f.

I

I-channel demodulator, 546
I-signal, color TV, 545
Iconoscope, 155
I-f amplifier, FM, 482, 488, 492
Ignitron tube:
  characteristics, 761–762
  uses, 762
Illuminant C, 530
Image antenna, 661

798   Index

Image orthicon, 508
Impedance, 105ff.
    bridge measurement of, 756–757
    characteristic, 293
    transistor input, 242
    transistor output, 242
Impedance inverter, 590
Impedance matching, 117f., 237, 649
Impedance-varying transducers, 755–757
    bridge-measurement type, 756–757
    conductivity cells, 756
    resistance strain gage, 756, 757
Impulse noise, 630
Increment, 42
Induced voltages, magnitude of, 86
Inductance:
    defined, 87
    magnitude of, 89f.
    measurement of, 127f.
    mutual, 118ff.
    resistance, capacitance, and, in parallel, 116f.
    resistance, capacitance, and, in series, 107f.
    resistance and, in series, 104f.
    unit, defined, 87
Inductance coil, power in, 104
Inductive reactance, 102ff.
Industrial TV, 522
Insulator, defined, 49
Integer, 6
Integral, 43
Integration, 44
Intercarrier-sound receiver, 519
Interlaced horizontal scanning, 502
Interleaving, frequency, 531
Internal resistance:
    of battery, 69
    of generators, 238
Interpolate, 11, 35
Intrinsic conduction, 160
Inverse feedback, 216
Inverse-feedback circuits, 218
Ion, 619
Ion spot, 515
Ionization, 158
Ionization potential, table, 159
Ionosphere, 618, 619
    absorption in the, 625
    definition of, 615
    regular variations in the, 625
Iron:
    as conductor of magnetism, 71ff.
    magnetic characteristics of, 72ff.
Iron-core transformers, 121
Iron-vane type of meter, 125
Isotropic antenna, 663

J

Jeep TV, 522
Joule, defined, 58

K

Kennelly-Heaviside layer, 618–619
Kickback high-voltage supply, 515, 516
Kilowatt-hour, defined, 58
Kinescope, 152
Kirchhoff's laws, 61
Klystron, 157
    definition of, 601

L

Large-screen TV, 521
Lattice, 160
Lenses, microwave, 667
Lenz's law:
    application of, 103, 118, 119
    statement of, 86
License, radio station, how to apply for, 792
Limiter, FM, 488
Line-of-sight transmission, 633
Lissajous figures, 565
Load line, 141
Logarithm, 10
    characteristic of, 10
    common (base 10), 10
    mantissa of, 10
    Naperian, 45
Longitudinal waves, 284
Loops or antinodes, defined, 287
Loop antenna, 680
Loose coupling, 648
Losses and efficiency (of antennas), 645
Loud-speakers, 188
Luminance, 529
Luminance signal, 530

M

Magic-eye tube, 155
Magnet:
    alnico, 74
    bar, 69
    horseshoe, 69
    permanent, 73, 74
Magnetic amplifier, 738, 762–766
Magnetic coupling, 651
Magnetic field:
    described, 69f.
    direction of, 306
    energy stored in, 89
    strength of, 71, 289
Magnetic flux:
    direction of, rule for finding, 71
    lines of, 69ff.
Magnetic structure, laminated type of, 74
Magnetism, 69ff.

Index

Magnetism (cont.):
  residual, 72
Magnetization curve, 74
Magnetomotive force, ampere-turn as unit of, 71
Magnetron, 157
  definition of, 601, 603
Magnitude of induced voltages, 75
Majority carriers, 163
Man-made noise, 629
Mass spectrometer, 771–773
Matrix circuit, 539, 540
Matrix equations, color TV, 535
Maxwell, definition of, 70
Meters:
  alternating current, 124ff.
  d'Arsonval type, 78
  dynamometer type, 124
  iron-vane type, 125
  rectifier type, 125f.
  thermocouple type, 126
  wave, 118
Mho, defined, 53
Microphones, 183
Microphonic noise, 205
Microwave antennas, 664
Microwave circuits:
  printed or strip-line, 597
  waveguide, 595
Microwave devices, 609
Microwave generators, 598
Microwave lenses, 667
Mirror image, 644
Mixer, 233
Mixing equipment, TV, 510
Mobility, 160
Modes, 300
Modulation:
  of chrominance subcarrier, 541
  velocity, 157
Modulation index, 478
Modulators:
  Class B, 379, 407, 421
  forms of:
    amplitude, 309
    frequency, 311
    modulation, 309
    phase, 311
    pulse, 725, 727, 728, 730
    single sideband, 319
    suppressed carrier, 320, 439
    vestigial sideband, 319
  types of, 325, 376
    balanced, 385
    Chiriex, 391
    Class B, 379
    Doherty, 389
    grid, 399
    high level, 376

Modulators, types of (cont.):
    low level, 376
    phase, 311
    plate, 376, 401
    pulse, 311, 725, 727, 728
Monitoring equipment, TV, 510
Monopole antennas, 645
Motors:
  direct current, principle of, 77
  electronic control of, 768
Multi-hole coupler, 612
Multipath reflections, TV, 507
Multistage audio amplifier, 202
Multistage transistor amplifiers, 253
Multivibrator, 276
  synchronization of, 279
Music waves, 128
Muting of receiver noise, 472
Mutual inductance, 118ff.

N

Navigational aids, 730
  distance measuring (DME), 735
  identification, 732
  indicator function, 732
  lines of position, 732
  loran, 730
  other forms, 735
  vhf range (VOR), 734
Navigational antennas, 680
Negative resistance, 553
Negative transmission, 511
Networks, coupling, 647, 650
Night error, 682
Nodes, 287, 288
Noise, man-made, 630
Noise-reducing systems, 631
Noise reduction in FM system, 482
Nonlinear distortion, 182, 210
Notation:
  place, 2
  scientific, 2
Null position, 681
Number:
  negative, positive, 1
  rounding off a, 6
Number system:
  binary, 20
  decimal, 1

O

Odd-line interlaced scanning, 503
Ohm, defined, 49f.
Ohmmeter, principle of, 79
Ohm's law, applications of, 51ff., 58, 99
Operating point, 142
  selection of, 209
Operation, symbols of, 7
Ordinary circuit elements, limitations of, 575

Ordinates, 39
Orthicon, 156
Oscillations, parasitic, 393, 401
Oscillators, 553
  blocking, 273
  feedback, 553
  frequency-modulated, 558
  grid-dip, 561
  inductance-capacitance, 555
  negative resistance, 553
  radio frequency, 556
  resistance-capacitance, 560
  swept, 558
Oscillograph, sweep voltage for, 269
Oscilloscope:
  cathode-ray, 563
  complete, 564
Oxide-coated emitters, 133

P

Parabolic reflectors, 665
Paraboloid, 666
Parallel circuits, 53, 116f.
  resistance and inductance in, 116f.
  resistance, inductance, and capacitance in, 116f.
Parallel resonance, 116f.
Parallel-resonant, definition of, 588
Parallel-tuned network, 649
Parasitic oscillations, 393, 401
Patterns in plane of antenna axes, 658
Peak inverse voltage, 151
Peak values, a-c waves, 99f.
Pedestal, TV, 499
Pedestal level, TV, 511
Pentode, 145
  equivalent circuit, 146
  variable-mu, 148
  voltage gain, 146
Per cent modulation, measurement, 567
Permalloy, magnetic characteristics of, 74
Permanent magnet, 73, 74
Persistence of vision, 498
Phase:
  of voltage and current, 294
  in wave motion, 285
Phase angle, 102
Phase difference, 285
Phase distortion, 182
Phase inverter, 223
Phase modulation, 479
Phase relation of voltage and current on a transmission line, 577
Phase-shifting circuits, 760–761
Phasor, 39
  definition of, 97ff.
Phasor circuit techniques, 108–111ff.
Phonograph pickups, 184

Photoelectric cell, 767
Photoelectric emission, 132
Photoelectric measurement and control, 767–768
Picture detector, 545
Picture element, 504
Picture tube, 515
  color, 549
Plate characteristic, 139
Plate currents, 214
Plate dissipation, 151
Plate resistance, 136
  definition, 140
Plate resistor, 197
Polarization, 69, 616
Polynomial, 26
Potentiometer, defined, 52 (*see also* Voltage divider)
Power:
  and energy, derivation of, 58
  electric, generation of, 75ff., 100
Power amplifiers, 194
  definition of, 195
Power density, 616
Power ratings:
  tube, 409
  on voltage dividers, importance of, 60
Power sources, alternating current, 100
Power transformers, 121
Poynting vector, 616
Preamplifier, 232
Pre-emphasis, FM, 479, 488
Primary battery, use of, 64
Primary colors, 526
Primary radiator, 668
Printed microwave circuits, 597
Progressive wave, 288
Propagation:
  forward-scatter, 636
  general nature of, 615
  radio wave, 637
Public-address amplifier circuit, 237
Public-address system, 232
  operation of, 239
Pulse circuits, 263
Pulse communication, 725ff.
  code grouping, 727
  performance, 730
  pulse amplitude (PAM), 725
  pulse code (PCM), 727
  pulse frequency (PFM), 725
  pulse position (PPM), 725
  pulse shape (PSM), 725
  pulse width (PWM), 725
  time division multiplexing, 728
Pulse resolution, 275
Push-pull circuits, 214
Push-pull transistor circuits, 257

Index

Q

Q, definition of, 105
Q-meter, 570
Q-signal, color TV, 545
Quarter-wave lines, 589, 590
Quarter-wave monopole, 645

R

Radar, 684ff.
  antennas for, 690, 691, 693
  carrier frequencies used in, 690
  components, essential, 685
  distance range, 698
  distance ranging, direction, 702
  equations, 699–700
  image persistence time, 705
  minimum usable signal, 700
  plan position, 702
  pulse duration, 688
  pulse repetition rate, 687
  reflection from objects, 696
  time base, 688
  tubes used, 713
  uses, 707, 708, 711, 712
Radian, 38
Radiation, mechanism of, 305
Radiation characteristics of a half-wave antenna, 642
Radiation patterns:
  bidirectional, 657
  effect of the ground on, 661
  unidirectional, 657
Radiation resistance, 304, 305
Radiator, electromagnetic, 303
Radical, 32
Radio amplifiers, 349ff.
  amplitude distortion, 363
  applications, 360
  bias voltage supplies, 357
  cathode-coupled, 397
  Chiriex, 391
  circuits, 354
  class A pentode, 368
  Doherty, 389
  filtering supply circuits, 371
  grid-modulated, 399
  grounded-grid, 395
  high-frequency problems, 393
  high-level modulation, 376
  low-level modulation, 376
  neutralization, 366
  parasitic oscillations, 401
  push-pull, 374
  resonant-coupling, 351
  shielding, 370
  single-ended, 356

Radio amplifiers (cont.):
  television wide-band, 386
  tetrode, 371
  tube complements, 372
Radio compasses, 680, 682
Radio communications systems, 309ff.
  channel width, 321
  components, 312
  examples, 330, 334, 715, 720
  government regulation, 323
  radio channel, 321
  radio spectrum, 322
  shared use, 323
  scope, 317
Radio fade out, 628
Radio frequencies, 298
Radio frequency spectrum, 322
  carrier modulation, 309, 311
  communication channel, 321
    channel width, 321
  message carrier concept, 309
  regulation of use, 323
  shared use, 323
  single sideband, 319
  static, 452
  suppressed carrier, 320
  vestigial sideband, 319
Radio receivers, 325ff., 451ff., 488ff.
  AM, 451ff.
    automatic volume control, 457
    diversity, 473
    figure of merit of AVC, 458
    image frequency response, 462
    multi-band, 467
    multiple fraction, 465
    muting of noise, 472
    noise, 452
      hum, 456
      static, 452
      thermal, 452
      tube, 453
    noise factor, 454
      measurement, 455
    sensitivity, 451
      useful, 453
    superheterodyne, 459
      examples, 462
    transistorized, 467
      examples, 470
    transoceanic, 475
    tuned r-f, 459
    useful sensitivity, 453
Radio relay system, 713
  antennas, 714
  carrier frequency staggering, 718
  design, 719
  examples, 715, 720
  multiplexing, 716

Radio relay system (*cont.*):
  noise, 724
  performance, 730
  power, 728
  pulse code, 727
    code grouping, 727
  pulse communication, 725
  subdivision of channels, 718
  time division multiplexing, 728
Radio station license, how to apply for, 792
Radio wave propagation, 637–638
Radius vector, defined, 98
Raster, 517
Rate of change, 40
Ratio detector, FM, 488, 490
Reactance:
  capacitive, 107
  characteristics of, 102
  inductive, 102f.
Reactance elements, transmission line sections as, 580
Reactance tube, 559
Reactance-tube system, FM, 486
Reactance voltage, 103
Receiver:
  radio (*see also* Radio receivers)
    AM, 451ff.
    FM, 488ff.
    simple, 325f.
    superheterodyne, 326ff., 459
    television, 513, 520, 544
      color, 544
      color matrix at, 547
Receiver color matrix, 547
Receiver selectivity, 328
Receiving antenna, 307
Recombination, 619
Rectangular or mattress array, 660
Rectifier, 137
  bridge, 176
  full-wave, 170
  half-wave, 169
  voltage-doubler, 174
  voltage-multiplier, 175
Rectifier tubes, 173
Rectifier-type meters, 125f.
Reflected waves, 286
Reflection, microwaves from objects, 696
Reflectors, parabolic, 665
Reflector antenna, 678
Reflex klystron, 602
Refraction, 634
Regulated power supplies, 177
Reinsertion of d-c component, 517
Repeaters, 192
Repeating coil, 189
Repetition rate, TV, 498
Reproducers, 187
Residual magnetism, 72

Resistance:
  alternating-current circuit with, 99
  capacitance and, in series, 107
  defined, 49
  determination of, 56ff.
  effect of temperature on, 58
  inductance and, in parallel, 116f.
  inductance and, in series, 104f.
  inductance, capacitance, and, in parallel, 116f.
  inductance, capacitance, and, in series, 107f.
  radiation, 305
Resistance data, 57
Resistance mixer, 233
Resistor, size and rating of, 59ff.
Resolve detail, TV, 496
Resonance:
  general concepts of, 112ff.
  parallel, 116f.
Resonance testing methods, 127
Resonant circuits:
  parallel, 116f.
  series, 112f.
Resonant frequency, 114
Resonant-length antennas, 640
Resonant quarter-wave line, 585
Return time, TV, 498
Reverse transfer resistance, 243
Richardson's law, 132
Right-hand rule, 71
Ripple, 171
Rms (*see* Root mean square)
Root mean square, values, a-c waves, 100

S

Saturation, color, 529
Scaling circuit, 275
Scanning, electronic, 497
Scanning rates, TV, 504
Scatter propagation, 636
Schmidt optical system, 521
Screen-grid tube, 144
Screen voltage, 202
Secondary battery (*see* Storage battery)
Secondary coil, effect of, mutual inductance, 119
Secondary emission, 132
Secondary radiator, 668
Selective fading, 628
Selectivity, 123f.
Self-bias circuits, 252
Self-induction, voltage of, 87
Semiconductors, 159
Sense antenna, 681
Sensitive cells (*see* Transducers)
Series circuits, 52, 104f., 107f., 112ff.
  resistance and capacitance in, 107

Index

Series circuits (*cont.*):
  resistance and inductance in, 104f.
  resistance, inductance, and capacitance in, 107f.
Series coupling circuit, 649
Series-parallel circuits, 54ff.
Series resonant circuits, 112ff.
Series-tube regulators, 178
Serrated wave, 499
Services, types, 314
  aids to navigation, 317
  amateur emergency, 785
  aviation, 777
  broadcasting, 314
  citizens, 790
  control, 317
  disaster, 784
  emergency, 785
  facsimile, 316
  industrial, 787
  land transportation, 787
  marine, 779
  public service, 782
  radar, 316, 684
  relaying, 719, 720
  telegraphy, 314
  telemetering, 317
  telephony, 315
  teleprinter, 315
  television, 316
Servomechanisms, 769–771
Shf propagation, 632
Shielding of amplifiers, 370, 435
Shunt-feed, for antennas, 651
Sideband suppression, 318, 319
  advantages, 320
  examples, 386, 439
  single, 319, 439
  suppressed carrier, 320, 439
  vestigial, 319, 386
Side frequencies, FM, 480
Signal generator, 556
Signal-to-noise ratio, TV, 507
Significant figures, 5
Sine of angle, 33
Sine curve, construction for, 37
Sine waves:
  in nature, 97
  radius-vector representation of, 98
  rate of change of current in, 101f.
  representation of, 83ff.
Single-wire feed, 653
Sinusoidal waves (*see* Sine waves)
Skin effect, 739
Skip area, 622
Skip distance, 622
Skip zone, 622
Slide rule, 17
Slope of curve, 41, 43

Slot antennas, 668
Solid jumpers, 61
Sound, 180
Sound system, TV, 494
Sound waves, 289
  and electromagnetic waves, compared, 289
Source impedance, effect of, 117f.
Space charge, 135
Spectrum:
  FM, 483
  radio frequency (*see* Radio frequency spectrum)
Standard atmosphere, 635
Standard WWV frequencies, 563
Standing waves, 286, 292
Standing wave ratio, 295
Static, 452, 629
Stationary wave, 288
Steels:
  dynamo and transformer, magnetic characteristics of, 74
  hardened carbon, magnetic characteristics of, 74
  permanent magnet, characteristics of, 74
Storage battery, construction of, 66ff.
Strain gage measurements, 755
Strip-line microwave circuit, 597
Stub-line matching, 594
Subcarrier, 531
  chrominance, 532, 541
Substitution method, 572
Superposition:
  method of, 62f., 128f.
  principle of, 62
Superscript, 3
Suppressor, 145
Surface duct, 636
Sweep voltage, 269
  synchronization of, 271
Symbols and abbreviations, circuit elements, 63ff.
Sync-separating circuits, 501
Sync-separating circuits, TV receiver, 515
Sync-signal generator, TV, 500
Synchronizing color burst, 542
Synchronizing pulse, 498
Synchronizing signals, TV, 493, 498, 500, 501

T

Tangent of angle, 34
Tape recording, 187
Telemetering, FM-FM system, 484
Telephone circuits, 189
Telephone line, 190
  losses in, 191
Telephone receivers, 187
Televising motion-picture film, 522

804   Index

Television:
  bandwidth required for, 504
  color, 525ff.
  monochrome, 493ff.
Television receivers, 513
Television receiving antennas, 675
Television transmitting antennas, 507, 673
TV transmitting station, 508
Temperature measurements, 749–753
Tetrode, 143
  beam-power, 149
  plate characteristic, 144
Theater TV, 521
Thermal agitation, 205
Thermionic emitters, 133
Thermistor, definition of, 610
Thermocouple:
  capacity-potentiometer type, 754–755
  cold junction of, 750
  hot junction of, 750
  millivoltmeter with electronic follower, 750–753
  potentiometer voltage measurement type, 753–754
Thermocouple-type meters, 126
Theta (θ), 37
Thoriated tungsten, 133
Thyratrons, 270, 738, 757–768
  in control of motor, 768
  operation of, 757–760
  phase-shifting circuit, 760–761
Time constant, 88, 96, 198, 265
  of a coil, 88–89
Time-delay circuit, principle of operation, 745–748
Time-phase angle, 102
Timers, electronic, 744–749
  cycle timers, 749
  time-delay circuits in, 745–748
Transconductance, 140
  testing of, 572
Transducers, 749–757
  defined, 749
  impedance-varying type, 755–757
  pH-measuring type, 755
  piezoelectric-crystal type, 755
  voltage-type, 749–755
Transfer characteristic, 138
Transfer characteristic, dynamic, 142
Transformer, use and characteristics of, 120f.
Transistor, 131, 165
  common-base equivalent circuit for, 241
  current amplification, 166
  junction, 162
  trigger circuit, 279
Transistor amplifier:
  multistage, 253
  R-C coupled, 255

Transistor amplifier (*cont.*):
  transformer coupled, 256
Transistor audio amplifiers, 240
Transistor circuits, 241
  complementary symmetry, 257
  push-pull, 257
Transistor input impedance, 242
Transistor output impedance, 242
Transistorized receivers, 467, 470
Transit time, definition of, 600
Transmission lines, as circuit elements, 576
Transmitters, 405ff., 485, 508
  audio amplifiers, 428, 429
  carrier frequency stability, 407
  designing for purpose, 405
    examples 428, 443
  FM, 485
  frequency multipliers, 437
  inverse feedback, 426
  keying methods, 436
  modulator, 429
  neutralization, 424
  performance specifications, 405
  piezoelectric effect, 413
  power rating, 409
  power supplies, 429
  power supply regulation, 425
  quartz crystals, 413
  quartz crystal oscillators, 410
  radio amplifiers, 417, 419, 420
  r-f transmission lines, 446
    coaxial, 449
    open wire, 446
  safety interlocks, 435
  shielding, 435
  single-sideband system, 439
  telegraph, 436
  television, 508
Transmitter matrix circuits, 540
Transoceanic receivers, 475
Traveling wave, 288
  definition of, 294
Triangle, 33
Trigger circuit, 274
  transistor, 279
Trigonometry, 33
Triode, 137
  equivalent circuit, 142
  gas, 270
Trochoids, definition, 603
Troposphere, 635
Tropospheric waves, 635
Tube:
  beam power, 149
  cathode-ray, 152
  dual-purpose, 150
  electron, limitations of, 749
  gas, 158
  internal heat dissipation in, 366

Index   805

Tube (cont.):
  rectifier, 173
  vacuum, principles, 131ff.
Tube power ratings, 409
Tube-testing equipment, 572
Tungsten, 133
Tuning, 647

U

Uhf, antenna, 679
Uhf devices, 609
Uhf generators, 598
Uhf propagation, 632
Unidirectional pattern, 657
Unit inductance, defined, 87
Unit pattern, 659

V

V-plate antenna, 680
Vacuum-tube circuit, 207
Vacuum-tube oscillators, frequency range in heating applications, 741–743
Vacuum-tube testing equipment, 572
Vacuum-tube voltmeter, 567
  balanced, 569
  diode, 568
  square-law, 569
  tunable, 570
Valence electrons, 160
Value, absolute, place, 2
Variable, independent, 43
Variable-mu pentode, 148
Vector, 39
Velocity of propagation, 284
Velocity modulation, 157
Velocity-modulation generators, 602
Velocity of the wave, 282
Vertical blanking, 500
  duration of, 504
Vertical-deflection coils, 515
Vertical-deflection system, 499, 500
Vertical-scanning time, 504
Vertical synchronizing, 500
Vestigial-sideband filter, 506
Vestigial-sideband transmission, 506
Vhf propagation, 632
Video signal:
  composite, 511
  TV, 493
Video wave form, 502
Visual acuity, 496
Voice waves, 128
Volt, defined, 50
Volt-ohm-milliampere meters, 81f.
Voltage:
  alternating, 83
  and current on a transmission line, phase relation of, 577
  generation of, 74ff.
  relation between current and, in capacitor, 105f.

Voltage divider:
  defined, 52
  power ratings on, 60
  use of, 60
Voltage-doubler rectifier, 174
Voltage-feedback, 252
Voltage gain, 143
Voltage-sensitive elements:
  calibrated electrodes for pH measurements, 755
  piezoelectric crystals, 755
Voltage-type transducers, 749–755
Voltmeter:
  caution on use of, 80f.
  defined, 51
  principle and use of, 79
  resistance requirements, 80f.
Volume control, 233
  constant impedance, 234
  master, 233
Volume unit, 191

W

Watt, defined, 58
Wattmeters, 124f.
Watt-second, defined, 58
Wave(s):
  direct, 616
  electromagnetic, 282, 290
  ground-reflected, 616
  longitudinal, 284
  music and voice, 128
  nature of, 282
  reflected, 286
  sky, 618
  standing, 286
  surface, 617
  in three dimensions, 301
  transverse, 284
  tropospheric, 616
  velocity of, 282
Waveguides, 298
  as circuit elements, 576
Waveguide microwave circuits, 595
Wave impedance, definition of, 596
Wavelength, defined, 283
Wave meters, 118
Wave motion, phase of, 285
Weber, definition of, 70, 86
White light, color TV, 526
Wide-band antennas, 670
Wires, electromagnetic waves on, 288

Y

Y-signal, color TV, 545

Z

Zener limit, 165